Lecture Notes in Computer Science 6136

Commenced Publication in 1973
Founding and Former Series Editors:
Gerhard Goos, Juris Hartmanis, and Jan van Leeuwen

Joan Martí Arnau Oliver
Jordi Freixenet Robert Martí (Eds.)

Digital Mammography

10th International Workshop, IWDM 2010
Girona, Catalonia, Spain, June 16-18, 2010
Proceedings

Springer

Volume Editors

Joan Martí
Arnau Oliver
Jordi Freixenet
Robert Martí

Institute of Informatics and Applications
University of Girona
Girona, Catalonia, Spain

E-mail: {joan.marti, arnau.oliver, jordi.freixenet, robert.marti}@udg.edu

Library of Congress Control Number: 2010928388

CR Subject Classification (1998): I.4, I.5-6, H.3, J.3, I.2, H.4

LNCS Sublibrary: SL 6 – Image Processing, Computer Vision, Pattern Recognition, and Graphics

ISSN 0302-9743
ISBN-10 3-642-13665-6 Springer Berlin Heidelberg New York
ISBN-13 978-3-642-13665-8 Springer Berlin Heidelberg New York

springer.com

Printed in Germany

Typesetting: Camera-ready by author, data conversion by Scientific Publishing Services, Chennai, India
Printed on acid-free paper 06/3180

Preface

This volume of Springer's *Lecture Notes in Computer Science* series comprises the scientific proceedings of the 10th International Workshop on Digital Mammography (IWDM), which was held June 16–18, 2010 in Girona, Catalonia. The IWDM meetings traditionally bring together a diverse set of researchers (physicists, mathematicians, computer scientists, engineers), clinicians (radiologists, surgeons) and representatives of industry, who are jointly committed to developing technology, not just for its own sake, but to support clinicians in the early detection and subsequent patient management of breast cancer. The IWDM conference series was initiated at a 1993 meeting of the SPIE Medical Imaging Symposium in San Jose, CA, with subsequent meetings hosted every two years by researchers around the world. Former workshops were held in York, England (1994), Chicago, IL USA (1996), Nijmegen, The Netherlands (1998), Toronto, Canada (2000), Bremen, Germany (2002), Durham, NC, USA (2004), Manchester, UK (2006) and Tucson, AZ USA (2008). Each of these scientific events was combined with very successful and focused industrial and research exhibits, which demonstrated the milestones of digital mammography over the years.

A total number of 141 paper submissions from 21 countries were received. Each of these four-page abstract submissions was reviewed in a blind process by at least two members of the Scientific Committee, which led to a final selection of 46 oral presentations and 57 posters during the two and one-half days of scientific sessions. At this point I would like to acknowledge the excellent work of the Scientific Committee in guaranteeing scientific significance by means of providing feedback to the authors for the final papers. Thus, the 103 final papers included in this proceedings volume (*LNCS* 6136) constitute a comprehensive state of the art in breast imaging today.

As in the previous meetings, the 2010 IWDM program reflects not only the major challenges over the past that still remain active (CAD, lesion detection, image segmentation, image registration, risk assessment), but also the current trends and efforts being made to improve digital mammography for the early detection of breast cancer, paying special attention to volumetric imaging and digital breast tomosynthesis. We were also very honored to have as keynote speakers such internationally recognized researchers as Melcior Sentís, from UDIAT, Spain, who discussed the "Transition to Digital Mammography: Challenges and Future Trends," Ingvar Andersson from Malmö University Hospital, who raised the question "Breast Tomosynthesis: Mammography of the Future?," and Sir Michael Brady from the University of Oxford, UK, who spoke on "Information from Breast Images." The three keynote speakers provided a window to give us a glance at the future in breast imaging by means of understanding the past and current achievements.

The 10th IWDM also took advantage of the participation of industrial partners who both exhibited at the meeting and provided sponsorship of various conference events, thus adding considerable quality and visibility to the meeting.

Organizing an event like the IWDM meeting is not feasible without a bunch of active people working behind the scenes. Among them, Laura Batlle, who worked many hours to recruit the industrial partners and helped during the registration process; Àric Monterde, responsible of all the 2010 IWDM designs, including the logos and poster; Xavier Lladó, who took care of all the details related to the conference center; Jordi Freixenet, in front of the social events; Arnau Oliver, who timely produced the final version of the proceedings; and Joseta Roca, always in search of financial support. The contribution of Robert Martí as scientific and technical advisor became essential when planning and executing the phases of the meeting. Many thanks also to Reyer Zwiggelaar, who conveniently pushed me to apply for the organization of the conference during my research stay at Aberystwyth University (Wales) in 2008.

June 2010 Joan Martí

Organization

The 10th International Workshop on Digital Mammography (IWDM) was organized by the *Computer Vision and Robotics Group* at the University of Girona (UdG).

Workshop Chair

Joan Martí	University of Girona, Spain

Scientific Committee

Susan Astley	University of Manchester, UK
Ulrich Bick	Charité University, Germany
Michael Brady	University of Oxford, UK
Hiroshi Fujita	Gifu University, Japan
Maryellen Giger	University of Chicago, USA
Nico Karssemeijer	University of Nijmegen, The Netherlands
Elizabeth Krupinski	University of Arizona, USA
Andrew Maidment	University of Pennsylvania, USA
Robert Martí	University of Girona, Spain
Heinz-Otto Peitgen	University of Bremen, Germany
Etta Pisano	University of North Carolina, USA
Martin Yaffe	University of Toronto, Canada
Reyer Zwiggelaar	Aberystwyth University, UK

Local Organizing Committee

Joan Batlle
Marc Carreras
Xavier Cufí
Jordi Freixenet
Rafael García
Xavier Lladó
Arnau Oliver
Marta Peracaula
Pere Ridao
Joaquim Salvi
Laura Batlle
Àric Monterde
Joseta Roca

Table of Contents

Oral Session 1: CAD

Oral Session 2: Image Processing and Analysis I

Oral Session 3: Breast Imaging Physics

Poster Session 1

Oral Session 4: Physics Models

Oral Session 5: Clinical Experiences

Oral Session 6: Breast Density

Oral Session 7: Digital Breast Tomosynthesis

Poster Session 2

Oral Session 8: Image Processing and Analysis II

Oral Session 9: Lesion Detection

Oral Session 10: Registration

Impact of CAD with Full Field Digital Mammography on Workflow and Cost

Paul Taylor[1], Henry Potts[1], Louise Wilkinson[2], and Rosalind Given-Wilson[2]

[1] Centre for Health Informatics and Multiprofessional Education, University College London, Archway Campus, Highgate Hill, London, N19 5LW, United Kingdom
{p.taylor,h.potts}@chime.ucl.ac.uk
[2] St. Georges Healthcare NHS Trust, Tooting, London, SW17 0Q, United Kingdom

Abstract. The cost-effectiveness of single reading with CAD as an alternative to double reading was assessed in a national screening programme using CAD with full field digital mammography. The impact of CAD on the time taken to read screening films (n=5710) and on the proportion of films referred for arbitration and for assessment was measured (n=3064). No evidence was found of a change in the time taken to read films and no evidence of a change in rates of referral or recall. Estimates of the cost implications were made under three different scenarios for screening units. We conclude that single reading with CAD is likely to be a cost-effective alternative to double reading in terms of radiologist time. Published data however shows increased recall rates using CAD and no significant increase in sensitivity for CAD use over single reading. Any decision to introduce CAD instead of double reading should take into account the impact of sensitivity and specificity on women attending for screening.

Keywords: cost analysis, workflow, CAD, screening.

1 Introduction

Computer aided detection (CAD) systems for mammography are widely used in the United States to improve cancer detection by a single reader. In Europe, however, where double reading is more common, the argument commonly made for CAD is that it could allow a single reader to achieve a sensitivity equivalent to that of double reading. This would be desirable either if some screening centres are, in practice, unable to perform double reading or if replacing double reading by single reading with CAD would save money.

Existing research has concentrated on the sensitivity of CAD and relatively little is known about its cost-effectiveness or its impact on workflow. We have previously conducted an economic assessment of CAD as part of an early study of its impact on sensitivity and specificity.[1] That study failed to show an improvement in decision-making with CAD, so inevitably the economic assessment was that CAD was cost-increasing. Lindfors et al. use a Markov model simulation to show that adding CAD to a mammographic screening programme based on single reading resulted in a mean cost per years of life saved of $19,058, which we would regard as an encouraging estimate for CAD.[2] Their model, however, includes an assumption that the impact

J. Martí et al. (Eds.): IWDM 2010, LNCS 6136, pp. 1–8, 2010.

of CAD is to reduce false negatives by 77%, a figure taken from a study by Warren Burhenne et al. which showed that CAD prompted 77% of previously missed cancers.[3] However prompted cancers may still be missed by radiologists and this figure should be interpreted as the maximum potential impact of that CAD implementation. The Centre for Evidence-based Purchasing recently published an 'Economic Report' on CAD which concluded that double reading was cheaper than single reading with CAD.[4] Their study relied on estimates of reading time that were obtained using analogue film. Mammography is in the process of adopting digital technology, which dramatically simplifies the provision of CAD. The purpose of our evaluation was therefore to assess the impact on workflow and cost of using CAD with Full Field Digital Mammography (FFDM). The evaluation was performed in a UK screening unit that uses a protocol of double reading with arbitration in which any film that either reader considers suspicious is referred for arbitration by two film readers. The unit employs a mix of film readers, all of whom have specialist training in screening mammography. Some are consultant radiologists, others are radiographers (technicians) who have elected to specialize in mammography and train as screening film readers.

2 Impact on Workload

2.1 Method

For the purposes of the study a Hologic Selenium FFDM with CAD was installed in the South West London Breast Screening Unit in August 2007. Initially the system was used without CAD to allow readers to become familiar with the technology. Problems with the reliability of the system led to the machine being removed. A newer version of the Selenium was installed in November 2008. Data was initially collected again without using CAD. CAD was used from December 2008 to March 2009. Prior to CAD use readers received the following training:

1. Manufacturer's standard training (one session with an applications specialist and an e package of instructions was available)
2. A training roller consisting of a set of cancer cases with CAD prompts and feedback of radiological findings at screen reading, assessment and definitive pathological findings at surgical treatment.

The South West London Screening Unit uses an automated data entry system, integrated with a national information system, for analogue screening. Since this system was not integrated with the FFDM system, readers used paper forms to record the outcomes of digital films, rather than attempt to use two different computerized systems for a single task.

Data were collected to assess the impact of CAD on (a) reading times and (b) the percentage of cases sent to arbitration and assessment. Data were collected in two time periods. Seventy-three sessions were observed between 22nd August 2007 and 4th Jan 2008. Each session was double-read giving a total of 146 observed sessions. There was missing data on timing in 30 of these, leaving a total of 116 measurements. During this period film readers used the FFDM machine without CAD. Seventy seven sessions were observed between 21st October 2008 and 14th March 2009, making 154 observations. Sixteen film readers participated, including 11 consultant radiologists and five film-reading radiographers (technicians).

Twenty-one sessions had missing data on reading times, 14 were not considered because fewer than 5 sets of films were read (the time per case is unlikely to be measured accurately in these cases) and four were considered outliers. (274 of the sessions in this study were read at under 1 minute 45 seconds per film, four were read at between 2 minutes 20 seconds and 3 minutes 55 seconds per film. It was assumed that either that the timing data were recorded incorrectly in these four sessions or that some additional activity had taken place during the sessions.) Of the four outliers, one was read with CAD, three without. Of the included observations, 25 did not use CAD, 91 did.

In addition to comparing timing data for readers with and without CAD, we compared timing data for radiologists and film-reading radiographers, for first and second readers and for readers in the earlier and later phases of the study.

2.2 Results

The times taken by radiographers and by radiologists to read films were compared in the pre-CAD phase. The mean time per case for a radiologist (n=11) was 67 seconds (CI: 55-80), compared to 84 seconds (CI: 55-113) for a radiographer (n=5), suggesting that there was no significant difference between time taken by a radiologist or by a radiographer. However, the mean time per film for a radiographer-read session (n=23) was 74 seconds (CI: 61-88), compared to 56 seconds (CI: 52-61) for a radiologist-led session (n= 96). Plotting the number of sessions read against the time taken per case for each reader (Figure 1) suggested that this is because a small number of readers who read large numbers of films on FFDM were much faster, and that more of these readers were radiologists.

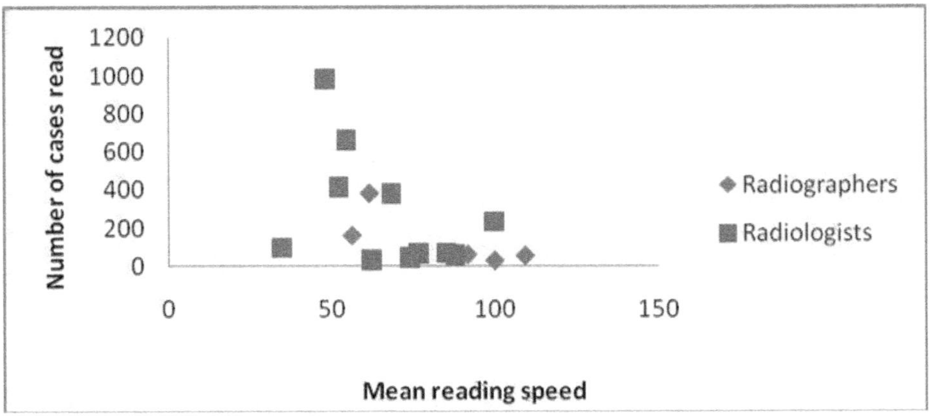

Fig. 1. Plot of reading speed against volumes read with FFDM for film-reading radiographers and radiologists

Timing data for the two periods is summarised in Table 1. None of the observable comparisons (with CAD vs without CAD, first reader vs second reader, 2007-8 vs 2008-9) are statistically significant, and none of the interactions are significant.

Table 1. Timing data for cases read with and without CAD in the two periods, 2007-8 and 2008-9. Data are presented for first and second readers separately.

Reader	Period	Protocol	No. of Cases	Mean time per case (secs)
First	2007-8	Without CAD	1711	62
	2008-9	Without CAD	185	63
	2008-9	With CAD	849	60
Second	2007-8	Without CAD	1871	57
	2008-9	Without CAD	216	53
	2008-9	With CAD	878	54

Outcomes for these cases are summarised in Table 2. Analysis of the data on screening outcomes for cases handled with FFDM during the study period was carried out excluding some clinics where CAD was used by one reader and not the other. No statistical differences were detected on comparison of clinics read with and without CAD for any of the outcomes measured (referral to arbitration, referral to assessment or cancers detected). It should be noted that this was not a trial of the sensitivity of CAD: the numbers assessed for recall from screening and detected cancer are small and the proportions are not a reliable assessment of overall performance.

Table 2. Outcomes for women handled using FFDM during the study period

Period	Protocol	No. of cases	Proportion sent to arbitration (per 1000 cases)	Proportion sent to assessment (per 1000 cases)	Cancers detected (per 1000 cases)
2007-8	Without CAD	2087	156.7	49.3	6.2
2008-9	Without CAD	174	137.9	74.7	11.5
2008-9	With CAD	803	130.7	43.6	10

3 Impact on Cost-effectiveness

3.1 Methods

Calculations of cost effectiveness are based on the balance of cost of radiologist time saved or spent in screening and post-screening assessment by using CAD with single reading instead of double reading and the cost of CAD workstations. To estimate the cost implications of using CAD, we considered the possible impact on three hypothetical screening centres, chosen as the lower quartile, median and top quartile in terms of numbers of women screened in 2007/08.[5] We used the median national value for initial recall rate, 4.5%, in that year for all three centres. We calculated the amount of time that would be saved in moving from double reading to single reading with CAD, assuming CAD has no impact on time taken to read films (59 seconds) and no impact on the number of cases sent to arbitration.

Estimates of recall rate vary, we consider three estimates from a recent multi-centre randomized controlled trial in the UK comparing double reading to single reading with CAD: the highest and the lowest figures from the centres studied, and the mean.[6] Since estimates of the cost implication of additional recalls also vary, we calculated impact under three different assumptions: assuming each additional recall for assessment required 20, 40 or 60 minutes of radiologist time.

All costs are calculated assuming a time horizon of 7 years, assuming this to be a reasonable lifetime for this kind of system, taking the Treasury target rate of 2% as an estimate for inflation and using the recommended discount rate for future expenditure of 3.5%[1]. [7] Figures are presented as the average annual cost over the seven years. Figures for radiologist costs (£163 per hour in 2007/08) were obtained from a standard reference source. We assume all films are read by consultant radiologists and that there are no other costs associated with double reading or with additional recalls. Cost quotes for R2 licenses and maintenance were provided by Medical Imaging Systems Ltd.[9] The costs include three elements: cost of a licence for a CAD workstation, cost of an additional licence allowing CAD to be read on an additional softcopy workstation and cost of maintenance. The costs of the two licences vary according to the number of licences purchased. We present costs for a range of configurations, factoring in the cost of a all-inclusive maintenance contract. Again we assume no additional costs implied by CAD other than the initial purchase and a maintenance contract covering upgrades.

3.2 Results

Potential cost reductions are shown in Table 3 for a small, medium and large screening unit, on the assumption of a 3%, 15% and 37% increase in recall to assessment and assuming 20, 40 and 60 minute appointments for assessment. A minus sign indicates a negative saving, i.e. a cost increase. CAD may be cost-effective as an alternative to double reading if the mean annual cost is less than the amount saved through the reduced use of radiologist time. Table 4 shows the mean annual cost for a variety of possible configurations. A comparison of the Tables 3 and 4 shows under what circumstances a CAD installation will be cost-effective. No installation will be effective if the impact on recall rate is high and recall appointments are long. If, for example, 8 workstations are required to handle throughput in a large screening unit, then the average annual cost will be £41,374. For there to be a saving, either the appointments will have to average less than 40 minutes, or the impact on recall rate will have to be no more than 15%. There is scope for a cost-effective use of CAD.

---

[1] Discounting is a technique used in economic assessments to take account of the fact that, all other things being equal, one would prefer to defer payments. Therefore future costs are 'discounted' at a rate intended to reflect the strength of that preference.

Table 3. Mean annual reduction in radiologist salary costs achieved by moving from double reading to single reading

Size of unit	Relative increase in recall rate	Time taken per patient at assessment		
		20 mins	40 mins	60 min
Small (12,000 films p.a.)	3%	31,516	30,701	29,910
	15%	28,002	23,567	19,262
	37%	21,522	10,410	-375
Medium (19,000 films p.a.)	3%	47,350	46,126	44,937
	15%	42,071	35,407	28,940
	37%	32,335	15,641	-563
Large (27,000 films p.a.)	3%	68,937	67,154	65,423
	15%	61,251	51,549	42,133
	37%	47,077	22,771	-820

Table 4. Mean annual costs of purchase plus maintenance for possible CAD installations. Note each CAD machine supports one acquisition system by default but can be configured, on the purchase of additional licences, to support up to three additional acquisition systems.

		Number of base CAD systems purchased			
		1	2	3	4
Licenses for additional workstations per base system	0	7,201	12,740	18,279	23,818
	1	10,780	19,898	29,017	38,135
	2	14,359	27,057	39,755	52,452
	3	21,518	41,374	61,230	81,085

4 Discussion and Conclusions

The evaluation reported here found that CAD slows readers down or encourages them to recall more women for arbitration or for referral. All cases in this study were double read and the knowledge that another reader would also look at the case may have influenced readers' response to CAD. It is possible that a reader single reading with CAD would take longer than the readers in this study.

It is clear that using CAD with FFDM is a much more straightforward proposition than using CAD with analogue films. This is because the digitization step is not required and no modification of the usual display technology is required. There is no discernable effect on reading time. In an earlier study of CAD by this group, using analogue films, reading time was 25 seconds without CAD and 45 seconds with CAD so there was very little time saved by using CAD rather than double reading. In fact extra time was consumed since there were extra recalls. Although the timings obtained from that study cannot be compared to the timings in this study, since the recording processes were different, we can conclude that the negative impact of CAD on reading times observed in that study, was not observed now.

In addition, in this study, there was no evidence of an increase in recall rates. If this finding were robust, this would suggest that overall film reader time would be saved by replacing double reading by single reading with CAD. However the evidence on recall rates from this study should be considered in conjunction with that from other studies - such as the Gilbert et al. study and the Taylor and Potts systematic review of published studies up to 2008 - both of which suggest that CAD significantly increases recall rates.[6,10] That said, there is sufficient heterogeneity in the observed results in both those studies to suggest that some centres are able to introduce CAD while controlling the impact on recall rates.

The impact of CAD on cancer detection also remains uncertain with conflicting evidence from the Taylor and Potts review, which showed a non significant increase (odds ratio 1.04, CI 0.96-1.13) and Gilbert et al. showing equivalence to double reading on a single multicentre study. The study reported here lacked the statistical power to show an impact on cancer detection rates and the cost-effectiveness calculation has been based on the assumption that single reading with CAD and double reading would detect equal numbers of cancers.

CAD seems to save radiologist time and costs, unless the impact on recall rates is at the higher end of that identified in the studies seen so far, and if the additional recalls are especially expensive in terms of radiologist time. Our calculations - and our conclusions - differ from those of the Centre for Evidence-based Purchasing.[4] They use a figure of £72 an hour for radiologist time, based on a published figure for annual salary and an assumption of a 40 hour week. We took a figure of £163 an hour from thea standard reference source.[8] This figure includes not just salary but on-costs (employer's health insurance) and other overheads including the investment in pre-registration training, annuitized over the consultant's expected working life. The figure is also calculated so that the cost of the time spent on direct patient-related activity reflects the normal ratio of such activity to other tasks. We feel that this figure provides a more realistic assessment of the true costs of radiologists' time. However, we accept that this takes a broader perspective on the issue than some decision-makers might want. The Centre for Evidence-based Purchasing also base their assessment of the time taken to read films on the data from Khoo et al. study which used digitized analogue films, an ergonomically different task from that of reading FFDM films with or without CAD.[11]

Our calculation assumes that the only extra cost incurred if additional women are recalled for assessment is the cost of the radiologist's time. The NHS Purchasing and Supply Agency, however, include elements for the cost of consumables and the time of other staff, including administrators and radiographers. This is a sounder basis on which to cost recall rates. The final estimate of the cost of an assessment visit is £125. Note that they estimate the cost of double reading to be £1 per case, and the cost of single reading with CAD to be £0.92, based on very different timing figures to those used here (25 seconds per case without CAD, 45 seconds with CAD). Given these figures, even a moderate increase in rate of recall to assessment would obliterate any possible savings from replacing double reading with single reading with CAD.

For the purposes of the cost-effectiveness calculation, we have assumed that all films are read by radiologists. In fact, 21% of the films in our study were read by film-reading radiographers. The Personal Social Services Research Unit's *Unit Costs of Health and Social Care 2008* estimates that the cost per hour of patient related activity - allowing for on-costs, overheads and the annuitized cost of the investment in training - is £48 for a radiographer, 29% of the cost of a consultant radiologist. The evidence of this study is that the time taken to read films was a function of experience with FFDM rather than professional qualification. Other studies report that radiographers are able to read screening mammograms at least as well as radiologists.[12] The use of radiographers in double reading clearly reduces the savings that might be obtained by moving to single reading with CAD and may well be a more cost-effective alternative.

The decision to move from double reading to single reading with CAD should be made on an analysis not just of the costs in radiologist time, but of the impact on the women attending screening on sensitivity and specificity for cancer detection.

Acknowledgments. The assistance of staff at the South West London Screening Centre is gratefully acknowledged. Medical Imaging Systems provided cost data at the request of Dr Maureen Gillan from the University of Aberdeen. The work was carried out as part of an evaluation supported by NHS Breast Screening Programme.

References

1. Taylor, P., Champness, J., Given-Wilson, R., et al.: Impact of computer-aided detection prompts on the sensitivity and specificity of screening mammography. Health Technol. Assess. 9(6), iii (2005)
2. Lindfors, K.K., McGahan, M.C., Rosenquist, C.J., Hurlock, G.S.: Computer-aided detection of breast cancer: a cost-effectiveness study. Radiol. 239(3), 710–717 (2006)
3. Warren Burhenne, L., Wood, S., D'Orsi, C., et al.: Potential contribution of computer-aided detection to the sensitivity of screening mammography. Radiol. 215, 554–562 (2000)
4. Centre for Evidence-based Purchasing Economic report: Computer-aided detection in mammography CEPP09040 (February 2009), NHS Purchasing and Supply Agency, `http://www.pasa.nhs.uk/pasa/Doc.aspx?Path=%5BMN%5D%5BSP%5D/NHSprocurement/CEP/CEP09040.pdf`
5. NHSBSP Breast Screening Programme, England 2007-08 The Health and Social Care Information Centre, `http://www.cancerscreening.nhs.uk`
6. Gilbert, F.J., Astley, S.M., Gillan, M.G., et al.: Single reading with computer-aided detection for screening mammography. N. Engl. J. Med. 359(16), 1675–1684
7. HM Treasury The Green Book: Appraisal and Evaluation in Central Government, `http://www.hm-treasury.gov.uk/data_greenbook_index.htm`
8. Personal Social Services Research Unit Unit Costs of Health and Social Care (2008), `http://www.pssru.ac.uk/pdf/uc/uc2008/uc2008.pdf`
9. Medical Imaging Systems personal communication
10. Taylor, P., Potts, H.W.: Computer aids and human second reading as interventions in screening mammography: two systematic reviews to compare effects on cancer detection and recall rate. Eur. J. Cancer. 44(6), 798–807 (2008)
11. Khoo, L., Taylor, P., Given-Wilson, R.: Computer-aided detection in the UK National Breast Screening Programme: prospective study. Radiol. 237, 444–449 (2005)
12. Wivell, G., Denton, E.R., Eve, C.B., et al.: Can radiographers read screening mammograms? Clin. Radiol. 58(1), 63–67 (2003)

Performance of Triple-Modality CADx on Breast Cancer Diagnostic Classification

Neha Bhooshan, Maryellen L. Giger, Karen Drukker, Yading Yuan, Hui Li, Stephanie McCann, Gillian Newstead, and Charlene Sennett

Department of Radiology, The University of Chicago
5841 South Maryland Avenue, Chicago, Illinois 60637
bhooshan@uchicago.edu

Abstract. The purpose of this study is to evaluate the potential of computer-aided diagnosis (CADx) methods utilizing three breast imaging modalities: full-field digital mammography (FFDM), sonography, and dynamic contrast-enhanced magnetic resonance imaging (DCE-MRI) for breast lesion classification. Three separate databases for each modality were retrospectively organized: FFDM (255 malignant lesions, 177 benign lesions), ultrasound (968 malignant lesions, 158 benign lesions), and DCE-MRI (347 malignant lesions, 129 benign lesions). From these single-modality databases, three dual-modality databases were constructed as well as a triple-modality database (31 malignant lesions, 17 benign lesions). Our computerized analysis methods consisted of several steps: (1) automatic lesion segmentation; (2) automatic feature extraction; (3) automatic feature selection; (4) merging of selected features into a probability of malignancy. Stepwise linear discriminant analysis using a Wilks lambda cost function in a leave-one-lesion-out method was used for feature selection. The selected features were merged using a Bayesian artificial neural network (BANN) with a leave-one-lesion-out method. The classification performance was assessed using receiver-operating characteristics (ROC) analysis. Results showed that the computerized analysis of breast lesions using image information from all three modalities yielded an AUC of 0.95±0.03. The observed trend of increasing performance as information from more modalities is included in the classifier indicates that the use of all three modalities can potentially improve the diagnostic classification of CADx.

Keywords: computer-aided diagnosis, mammography, ultrasound, DCE-MRI, multi-modality.

1 Introduction

Breast cancer is the most common cancer and the second leading cause of cancer death in women in Western countries[1]. Imaging has become a key part of a patient's workup for diagnosis, treatment decisions, and therapy monitoring. Although mammography is the primary breast imaging modality, sonography and DCE-MRI have emerged as complementary modalities. Different modalities have different strengths and weaknesses, and radiologists have found that using all three modalities helps in

J. Martí et al. (Eds.): IWDM 2010, LNCS 6136, pp. 9–14, 2010.

their interpretation task. In addition, computer-aided diagnosis (CADx) schemes have been developed to further aid radiologists in analyzing images [2].

We have previously developed and reported on single-modality [3-9] and dual-modality [10,11] CADx for breast lesions. The purpose of this current study is to combine computer-extracted features from all three modalities (mammograms, sonograms, and MRI images) to generate a single computer-estimated probability of malignancy and evaluate its performance and potential role in distinguishing malignant from benign lesions.

2 Materials and Methods

2.1 Database

All images (mammography, sonography, MRI) were acquired at the University of Chicago Medical Center (UCMC) between 2002 and 2006 and retrospectively collected under institutional review board (IRB) approved protocols with HIPAA compliance.

The full-field digital mammogram (FFDM) database consisted of 255 malignant lesions and 177 benign lesions. The number of images per lesion ranged from 1 to 3 including standard and special views. All images were acquired on a GE Senographe 2000D system (GE Medical System, Milwaukee, WI) at 12-bit quantization with pixel size of 100μm. The ultrasound database contained 968 malignant and 158 benign lesions, and all images were obtained from a Philips HDI 3000 scanner (Philips, Andover, MA). The DCE-MRI database consisted of 347 malignant and 129 benign lesions. The images were acquired with a T1-weighted spoiled gradient echo sequence using Gd-DTPA on a 1.5T GE Signa MRI scanner (GE Medical System, Milwaukee, WI). Each case had one precontrast and three to five postcontrast series at intervals of 68 seconds, and each series contained 60 coronal slices.

In previous studies, we created three dual-modality databases (mammography-ultrasound, ultrasound-MRI, mammography-MRI) from the single-modality databases. The mammography-ultrasound database contained 40 malignant and 60 benign lesions while the ultrasound-MRI database had 56 malignant and 33 benign cases. The mammography-MRI database contained 168 malignant and 45 benign lesions. The triple-modality database had 31 malignant and 17 benign lesions. It should be noted that the number of cases decrease due to the criteria that the lesion be imaged by either both or all three modalities.

2.2 Computerized Analysis and Performance Evaluation

Our computerized analysis method consisted of several steps: (1) automatic lesion segmentation; (2) automatic feature extraction in terms of mathematical descriptors; (3) automatic feature selection; (4) merging of selected features into a probability of malignancy [3-9]. For each type of image, a different segmentation algorithm was used and different features were calculated. For the mammograms, a dual-segmentation based on the active contour model was used to segment the lesion from the surrounding parenchyma, and then fifteen features are extracted. Similarly, the lesion on each ultrasound was segmented and forty features describing shape, margin, texture, and posterior acoustic behavior were calculated. Fuzzy c-means (FCM) was used to automatically segment the lesion in the DCE-MR image and then to extract the lesion's characteristic kinetic curve. Thirty-one features including textural, morphological, kinetic, and spatial enhancement variance features were generated.

Stepwise linear discriminant analysis using a Wilks lambda cost function [12] in a leave-one-lesion-out (LOLO) method was used for feature selection. Feature selection was initially performed on each of the large single-modality databases as well as on the dual-modality databases. Due to the small size of the triple-modality database, only the features selected from the single-modality databases were input to the multi-modality feature selection process. The selected features were merged using a Bayesian artificial neural network (BANN) [13] with a leave-one-lesion-out method, to generate a probability of malignancy. The classification performance was assessed using receiver-operating characteristics (ROC) analysis [14].

3 Results

We have already completed previous studies on single-modality and dual-modality CADx using mammography, sonography, and DCE-MRI. To summarize, we achieved AUC values of 0.75±0.04, 0.88±0.01, and 0.79±0.04 for mammography, ultrasound, and MRI, respectively. For the dual-modality databases, we achieved higher AUC values of 0.87±0.03, 0.92±0.03, and 0.93±0.04 for mammography-MRI, mammography-ultrasound, and ultrasound-MRI, respectively. These performances as well as the associated selected features are shown in Figure 1.

For the triple-modality database, feature selection yielded features from all three modalities: two ultrasound features (texture and ratio in average gray in/rim), one mammography feature (spiculation based on region of interest (ROI)), and one MRI feature (time to peak). The LOLO cross-validation using a BANN to merge the selected features yielded an AUC of 0.95±0.03. These results are also tabulated in Figure 1. An example lesion from the triple-modality database with corresponding mammogram, ultrasound, and DCE-MRI images is shown in Figure 2.

In order to perform LOLO for each combination of modalities within the same database, we repeated the LOLO on the single-modality and dual-modality using only the cases in the triple-modality cases (31 malignant, 17 benign). As shown in Table 1, we achieved AUC values of 0.76 ± 0.07, 0.81 ± 0.06, and 0.73 ± 0.07 for mammography, ultrasound, and MRI, respectively. For the dual-modality classification, we achieved AUC values of 0.89 ± 0.05, 0.85 ± 0.06, and 0.87 ± 0.05 for mammography-MRI, mammography-ultrasound, and ultrasound-MRI, respectively.

It is important to note that due to the limited size of the triple-modality database at this time, one can only assess trends in performance as more modalities are added to the classification, and not perform statistical analyses.

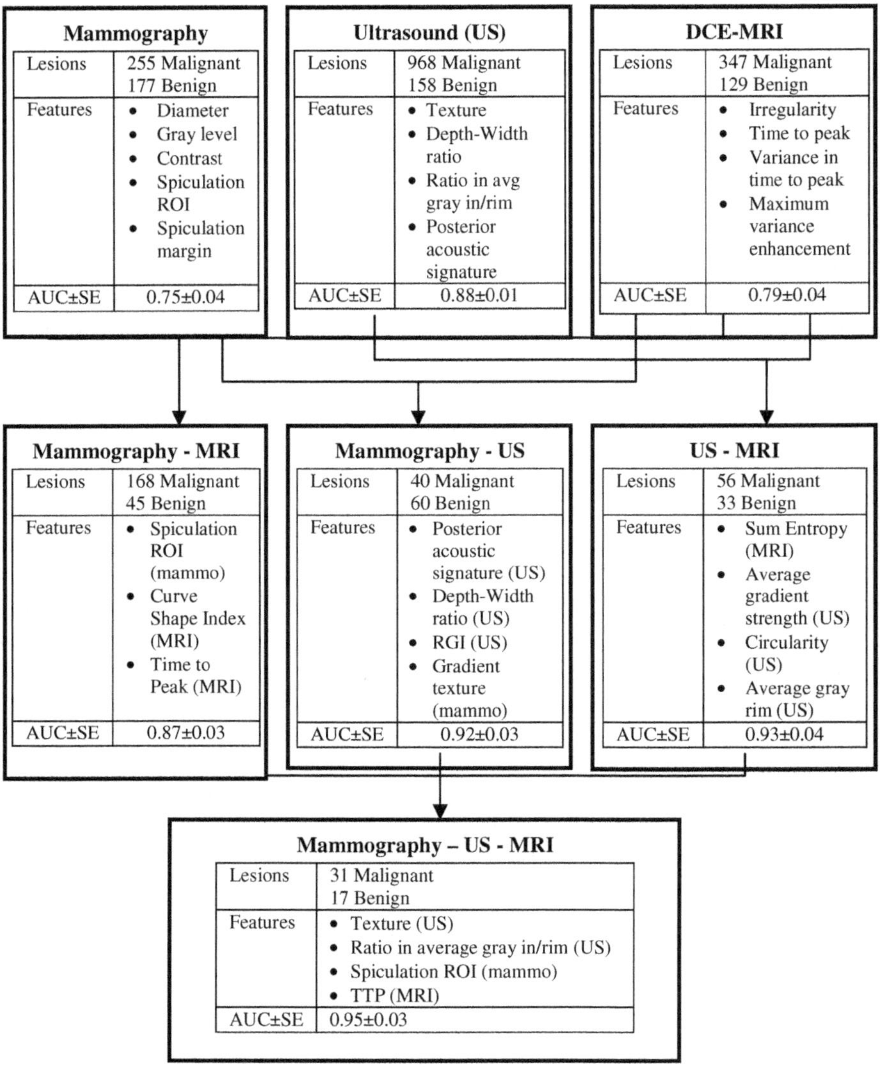

Fig. 1. Database, selected features, and preliminary performance for single-modality, dual-modality, and triple-modality CADx for breast lesion diagnostic classification in the task of distinguishing between malignant and benign breast lesions. Due to limited database size, feature selection for the triple-modality CADx included only features previously selected by the single-modality analyses. AUC=area under the ROC curve, SE=standard error.

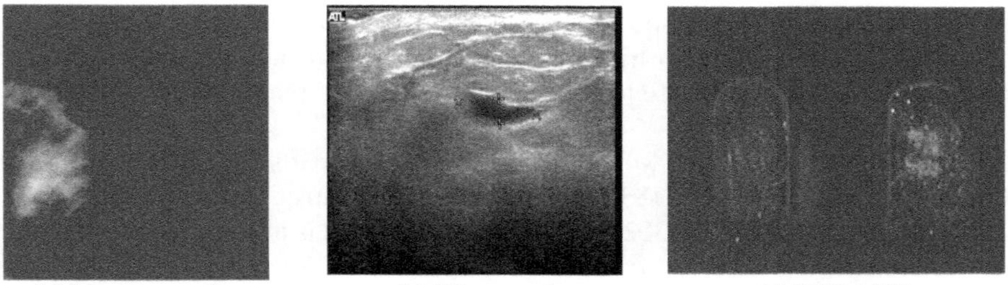

(a) Mammography (b) Ultrasound (c) DCE-MRI

Fig. 2. Example lesion in triple-modality database: (a) mammography, (b) ultrasound, and (c) DCE-MRI images of the same lesion

Table 1. Performance of single-modality, dual-modality, and triple-modality classification only within the triple-modality database (31 malignant, 17 benign lesions). AUC=area under the ROC curve, SE=standard error.

Database	**Selected Features**	**AUC ± SE**
Single Mammography	Contrast (mammo) Average Gray Level (mammo) Spiculation Margin (mammo)	0.76 ± 0.07
Single US	Ratio in average gray in/rim (US) Posterior Acoustic Signature (US)	0.81 ± 0.06
Single DCE-MRI	Curve Shape Index (MRI) Variance in Margin Sharpness (MRI)	0.73 ± 0.07
Dual Mammography-MRI	Variance in Margin Sharpness (MRI) Spiculation ROI (Mammo)	0.89 ± 0.05
Dual Mammography-US	Ratio in average gray in/rim (US) Texture (US) Spiculation ROI (Mammo)	0.87 ± 0.05
Dual US-MRI	Posterior Acoustic Signature (US) Texture (US) Curve Shape Index (MRI)	0.85 ± 0.06
Triple Mammography-US-MRI	Texture (US) Ratio in average gray in/rim (US) Spiculation ROI (mammo) TTP (MRI)	0.95 ± 0.03

4 Summary

In this study, we evaluated our automatic computerized methods in the task of breast lesion classification using computer-extracted features from three modalities: full-field digital mammography, ultrasound, and DCE-MRI. Features from all three modalities were selected as the most effective features for distinguishing malignant from benign lesions in the triple-modality database, and the resulting performance of our CADx method yielded an AUC of 0.95±0.03.

A limitation of the study is the small sizes of the dual-modality and triple-modality databases. However, we believe this preliminary study shows a promising trend in the improvement in diagnostic classification of breast lesions from single-modality and dual-modality CADx to triple-modality CADx. We are currently collecting a larger triple-modality database to validate these observed trends.

Acknowledgements. This work was supported in parts by NIH grants R33-CA113800 and P50-CA125183, a NIH Medical Scientist Training Grant (MSTP) grant, and DOE grant DE-FG02-08ER6478. M. L. Giger is a stockholder in R2 Technology/Hologic and receives royalties from Hologic, GE Medical Systems, MEDIAN Technologies, Riverain Medical, Mitsubishi and Toshiba. It is the University of Chicago Conflict of Interest Policy that investigators disclose publicly actual or potential significant financial interest with would reasonably appear to be directly and significantly affected by the research activities.

References

1. Jemal, A., Siegel, R., Ward, E., Hao, Y., et al.: Cancer statistics. Cancer J. Clin. 58, 71–96 (2008)
2. Giger, M.L., Chan, H.-P., Boone, J.: Anniversary paper: History and status of CAD and quantitative image analysis: The role of Medical Physics and AAPM. Med. Phys. 35, 5799–5820 (2008)
3. Drukker, K., Giger, M.L., Vyborny, C.J., Mendelson, E.: Computerized detection and classification of cancer on breast ultrasound. Acad. Radiol. 11, 526–535 (2004)
4. Horsch, K., Giger, M.L., Venta, L.A., Vyborny, C.: Computerized diagnosis of breast lesions on ultrasound. Med. Phys. 29, 157–164 (2002)
5. Huo, Z., Giger, M.L., Vyborny, C.J., et al.: Automated computerized classification of malignant and benign masses on digitized mammograms. Acad. Radiol. 5, 155–168 (1998)
6. Li, H., Giger, M.L., Yuan, Y., Chen, W., Horsch, K., Lan, L., Jamieson, A., Sennett, C.A., Jansen, S.A.: Evaluation of computer-aided diagnosis on a large clinical full-field digital mammographic dataset. Acad. Radiol. 15, 1437–1445 (2008)
7. Chen, W., Giger, M.L., Bick, U., Newstead, G.M.: Automatic identification and classification of characteristic kinetic curves of breast lesions on DCE-MRI. Med. Phys. 33, 2878–2887 (2006)
8. Chen, W., Giger, M.L., Lan, L., Bick, U.: Computerized interpretation of breast MRI: Investigation of enhancement-variance dynamics. Med. Phys. 31, 1076–1082 (2004)
9. Chen, W., Giger, M.L., Li, H., Bick, U., Newstead, G.M.: Volumetric texture analysis of breast lesions on contrast-enhanced magnetic resonance images. Mag. Reson. in Med. 58, 562–571 (2007)
10. Drukker, K., Horsch, K., Giger, M.L.: Multimodality Computerized Diagnosis of Breast Lesions Using Mammography and Sonography. Acad. Radiol. 12, 970–979 (2005)
11. Yuan, Y., Giger, M.L., Li, H., Bhooshan, N., Sennett, C.: Multi-modality computer-aided breast cancer diagnosis with FFDM and DCE-MRI. Acad. Radiol. (in press)
12. Johnson, R.A., Wichern, D.W.: Applied Multivariate Statistical Analysis, 3rd edn. Prentice-Hall, Englewood Cliffs (1992)
13. Kupinski, M.A., Edwards, D.C., Giger, M.L., Metz, C.E.: Ideal observer approximation using Bayesian classification neural networks. IEEE Trans. Med. Imag. 20, 886–899 (2001)
14. Metz, C.E.: ROCKIT (computer program), version 0.9b (1998), `http://www-radiology.uchicago.edu/krl/roc_soft.htm`

Needle Path Planning Method for Digital Breast Tomosynthesis Biopsy Based on Probabilistic Techniques

Laurence Vancamberg[1,2], Anis Sahbani[2], Serge Muller[1], and Guillaume Morel[2]

[1] GE Healthcare, Buc, F-78530
laurence.vancamberg@ge.com, serge.muller@ge.com
[2] ISIR-CNRS, Pierre & Marie Curie University, Paris 6, UMR 7222, Paris, F-75005
anis.sahbani@upmc.fr, morel@robot.jussieu.fr

Abstract. Needle insertion planning for breast biopsy has the potential to improve patient comfort and intervention safety. However, this is a challenging task because of the infinite possibilities of insertion points and the breast tissue deformations during the procedure. In this paper, we present a novel approach that couples probabilistic motion planning methods with Finite Element Simulation in order to find an optimal path taking breast deformation into account. This method reduces the error (*i.e.* the distance between the needle tip and the lesion) meanly by 80% and the proposed planner divides the planning time by 5 in comparison to a classic Rapidly-Exploring Random Tree planning method.

1 Background

In stereotactic biopsy procedures, the radiologist chooses the needle insertion point according to the lesion position into a breast quadrant and tries to minimize the path length in order to limit risks of infections and haematomas. Deurloo *et al.* [1] showed that displacements of breast tissue and needle deviation during this type of interventions cause an error of 2.4 mm, limiting the achievable diagnostic accuracy.

The introduction of digital breast tomosynthesis (DBT) modality can offer new possibilities for screening and diagnosis [2] but also for needle biopsy procedures. DBT gives access to spatial coordinates of any voxel in the reconstructed volume of the breast instead of having only the 3D coordinates of the target with the classic stereotaxy procedure. Using the reconstructed volume, a 3D model of the breast can be established and some specific zones to be preserved, such as vessels, can be segmented.

Therefore, DBT biopsy can be considered as a searching problem for needle paths avoiding forbidden zones. It can be then formulated as a motion-planning problem for a needle progressing into deformable tissue where vessels are the obstacles. Breast needle path planning is intrinsically a challenging problem because of the infinity of possible insertion points and because of the mutual dependence between the final position of the lesion and the needle insertion point (respectively final and initial configuration in the motion planning formalism).

Combining motion-planning techniques with the simulation of soft tissue deformations to determine a needle trajectory has already been proposed in the literature,

J. Martí et al. (Eds.): IWDM 2010, LNCS 6136, pp. 15–22, 2010.

often applied to brachytherapy [3]. Nevertheless the authors are restricted to 2D environments [3][4] or to biopsy guidance devices with 2 degrees of freedom [5]: a translation and a rotation along the needle axis, which is not the case for breast biopsy guidance devices at the moment (only one translation). Dehghan *et al.* have presented in [6] a 3D planning method in the case of brachytherapy taking soft tissue deformations into account. Nevertheless, even if there is an infinity of possible needle insertion points as in our case, their optimization method is only well-suited to aim at 3 (or more) quasi-aligned targets.

2 Method

The aim of the proposed approach is to find an optimal insertion point and an optimal needle path constrained regarding the application specificities: the needle orientation is fixed during its insertion, there is no real-time image feedback that could be used in visual servoing architecture and breast deformation should be taken into account.

Traditionally, motion-planning algorithms were based either on exact and complete methods (the planner always produces a feasible path, when one exists), or on heuristic methods. The complexity of the first techniques, in particular regarding high dimension systems, has limited their use. The second ones are not complete. Recently, non-deterministic approaches were proposed in the literature, some of them are based on the exploration of the admissible configurations space (*CS*); the others try to catch the connectivity of *CS*. These techniques, called probabilistic methods, overcome the exponential complexity of the problem (in terms of degree of freedom) and they also verify probabilistic completeness propriety (the probability that they will produce a solution approaches 1 as more time is spent). These probabilistic methods are less sensible to the dimension of the research space and turns out to be very efficient [7].

In addition, two types of physical deformation models are used to analyze deformations of anatomical parts: mass-spring model and Finite Element (FE) method. The FE method produces realistic deformations for a continuum compared to the discrete mass-spring model, whose stiffness coefficients are often difficult to determine [8].

Therefore, in this paper, we present a hybrid method combining probabilistic motion planning techniques with Finite Element (FE) simulation to solve our problem.

First, the radiologist selects in a slice of the DBT reconstructed volume, the lesion (s)he wants to biopsy. The coordinates of the target in the planning system, noted $L=(X_L, Y_L, Z_L)$, are calculated from the selected pixel and the reconstruction scale factors. To perform an accurate puncture, the slice should be selected so that the target is in focus. To guaranty a precision of 1 mm, a 0.5 mm sampling in depth dimension is done (Shannon criterion).

From the DBT volume, a triangular mesh of the patient breast surface *BS* and a patient-specific tetrahedric mesh *M* of the breast are generated using VTK [9], as shown in Figure 1. This mesh *M* is remeshed in order to have one node corresponding to the target. Knowing the geometry of the biopsy device, a subset of *BS*, *BS'*, is defined corresponding to the surface of the breast where the needle can be inserted. In Figure 1, *BS'* is represented in pink. Synthetic vessels (obstacles) are added to the environment in red, as well as the detector and compression paddle.

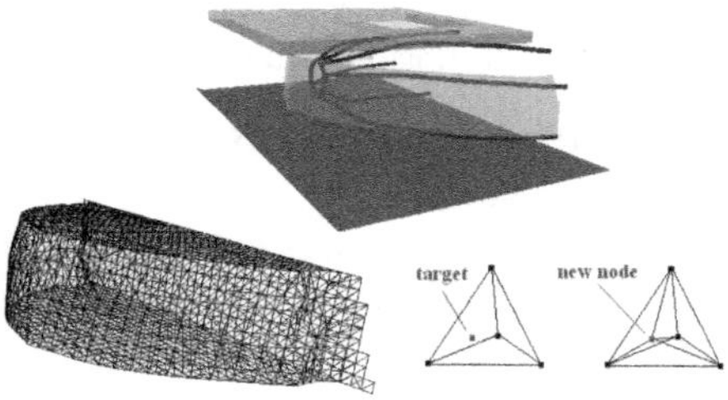

Fig. 1. Up: Simulator graphic environment. Down: Tetrahedric mesh of the breast M and the remeshing process.

The proposed planning approach is articulated in four steps:

- Generation of a low-cost path P' with relaxed constraints
- Needle insertion simulation along P' and $Z_{P'}$ definition
- Family $F_{P'}$ of optimal candidate paths generation
- Optimal path P^* selection

Each step will be detailed next.

- Step 1: Generation of a low-cost path P' with relaxed constraints

Without taking breast deformations into account, N free needle paths (meaning paths presenting no collision with any forbidden zones) are computed using probabilistic planning techniques. The path generation algorithms used are described below.

Then, a cost function C is defined. C depends on the radiologist preferences: length of the path, distance to the chest wall, … For each path, P, the cost c is computed:

$$C(P) = \alpha \times - \min_{ni \in P} (\mathrm{d}(ni, \mathrm{Obst})) + (1 - \alpha) \times \mathrm{L}(P)$$

where ni are the nodes of the path P, d is a metric measuring the distance between the nodes and the obstacles, L is the length of the path and α, a weighting factor.

The free needle path with the lower cost, noted P', is selected.

Figure 2 illustrates the generation of the free needle paths and the selection of the lower cost path P'.

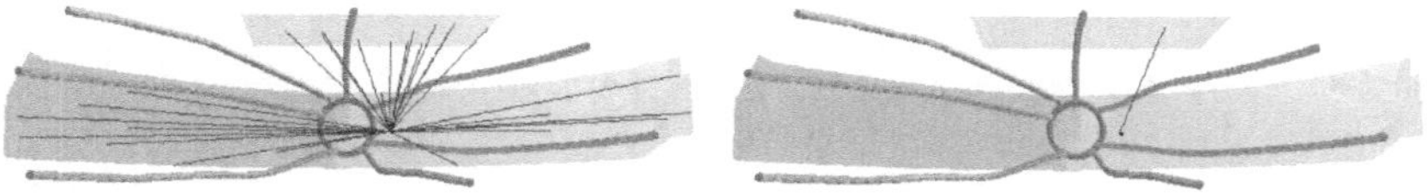

Fig. 2. Left: N free paths are generated. Right: the lower cost path P' is selected. The breast is seen from the back.

In this paper, we focus on a new path generation algorithms used for this first step. In our previous work [10], we first extended to our problem a technique developed for prostate treatment: a Rapidly Exploring Random Tree (RRT) method [11] with back-chaining [12] has been adapted. The contribution of this paper is a novel approach to find a free path going from the lesion *L* to *BS'*. Instead of sampling randomly from the configuration space, we proposed to bias the sampling toward the initial surface *BS'*.

Initial Surface Biased Algorithm: a point $S \in BS'$ is randomly selected. As the planning problem is holonomic, the optimal value for the input *u* applied to the tree root *L*, can be easily chosen by the calculation of a parametric equation of the trajectory going through *L* and *S*.

In our application, two types of needle trajectories are generated. A symmetric-tip needle exerts forces on the tissue equally in all directions so it follows a straight line when it is inserted into homogeneous tissue. A bevel-tip needle exerts forces asymmetrically and bends in the direction of the bevel. It follows an arc of circle [13] whose curvature depends on the needle itself and tissue characteristics. Figure 3 illustrates the different types of needle trajectories.

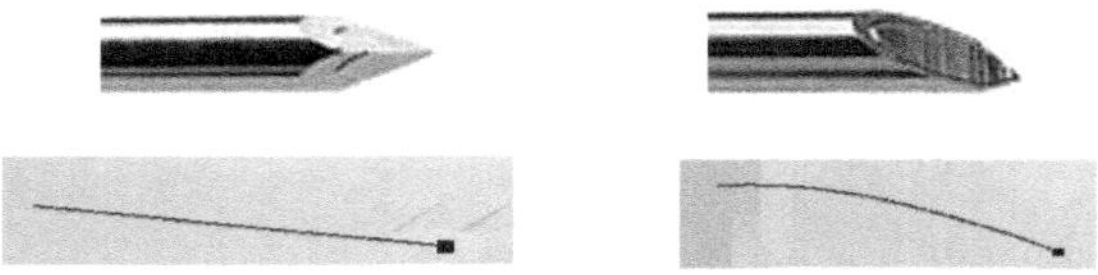

Fig. 3. Trajectories generated by non-steerable needles (orientation fixed). Left: a symmetric rigid needle follows a straight line. Right: a bevel-tip flexible needle follows an arc of circle.

Therefore, two types of parametric equation have to be calculated. If L = (X_L, Y_L, Z_L) and S = (X_S, Y_S, Z_S), for symmetric stiff needles, the parametric equation of the path is:

$$\begin{cases} X(t) = X_L(1-t) + t \times X_S \\ Y(t) = Y_L(1-t) + t \times Y_S \qquad t \in [0;1] \\ Z(t) = Z_L(1-t) + t \times Z_S \end{cases}$$

For bevel-tip needles, the parametric equation of the path is:

$$\begin{cases} X(\theta) = R\cos(\theta) \times \frac{(X_M - X_C)}{\|CM\|} + R\sin(\theta) \times \frac{(X_S - X_M)}{\|SM\|} + X_C \\ Y(\theta) = R\cos(\theta) \times \frac{(Y_M - Y_C)}{\|CM\|} + R\sin(\theta) \times \frac{(Y_S - Y_M)}{\|SM\|} + Y_C \\ Z(\theta) = R\cos(\theta) \times \frac{(Z_M - Z_C)}{\|CM\|} + R\sin(\theta) \times \frac{(Z_S - Z_M)}{\|SM\|} + Z_C \end{cases}$$

where $\theta \in [-\arcsin(\frac{\|SL\|}{2R}); \arcsin(\frac{\|SL\|}{2R})]$, C = ($X_C$,$Y_C$,$Z_C$) is the center of arc of circle and M the middle of [SL]. The coordinates of C can be calculated from the equations:

$$\begin{cases} X_C = \|CM\|\cos(\alpha)\times\dfrac{(Y_S - Y_L)}{N} + \|CM\|\sin(\alpha)\times\dfrac{(X_S - X_L)\times(Z_S - Z_L)}{(N\times\|SL\|)} + X_M \\ Y_C = \|CM\|\cos(\alpha)\times - \dfrac{(X_S - X_L)}{N} + \|CM\|\sin(\alpha)\times\dfrac{(Y_S - Y_L)\times(Z_S - Z_L)}{(N\times\|SL\|)} + Y_M \\ Z_C = \|CM\|\sin(\alpha)\times\dfrac{-(X_S - X_L)^2 - (Y_S - Y_L)^2}{(N\times\|SL\|)} + Z_M \end{cases}$$

where $N = \sqrt{(X_S - X_L)^2 + (Y_S - Y_L)^2}$ and α is randomly selected in $[0;2\pi]$.

If $N = 0$, the coordinates of C can be calculated from the following equations:

$$\begin{cases} X_C = \|CM\|\sin(\alpha)\times\dfrac{-(Y_S - Y_L)^2 - (Z_S - Z_L)^2}{(N2\times\|SL\|)} + X_M \\ Y_C = \|CM\|\cos(\alpha)\times\dfrac{(Z_S - Z_L)}{N2} + \|CM\|\sin(\alpha)\times\dfrac{(Y_S - Y_L)\times(X_S - X_L)}{(N2\times\|SL\|)} + Y_M \\ Z_C = \|CM\|\cos(\alpha)\times\dfrac{-(Y_S - Y_L)}{N2} + \|CM\|\sin(\alpha)\times\dfrac{(X_S - X_L)\times(Z_S - Z_L)}{(N2\times\|SL\|)} + Z_M \end{cases}$$

where $N2 = \sqrt{(Z_S - Z_L)^2 + (Y_S - Y_L)^2}$ and α is randomly selected in $[0;2\pi]$.

When the path is generated, the collision detector checks that it is at a certain distance *D* from the obstacles. *D* is defined by the user and can be adjusted to take into account the needle diameter.

This new initial surface based algorithm generates free paths going from the lesion *L* to a breast surface point *S*.

- Step 2: Needle insertion simulation along *P'* and $Z_{P'}$ definition

The needle insertion along *P'* is simulated using FE analysis and a zone $Z_{P'}$ around the final target position is defined. The technique used for needle insertion simulations is described in step 4. Figure 4 illustrates the displacement of the target. The needle is represented in red, the initial position of the target is in dark red (needle tip) and the final position of the target taking into account breast deformations is brighter and is the center of the illustrated cubic zone $Z_{P'}$.

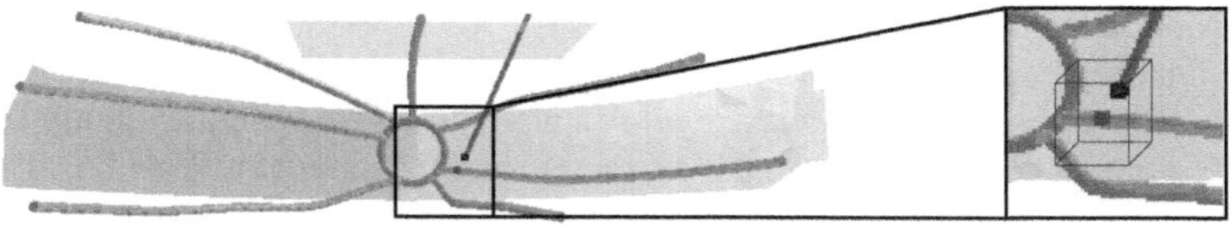

Fig. 4. Left: Displacement of the target. Right: zoom on the target final position and definition of $Z_{P'}$.

- Step 3: Family of optimal candidate paths generation

A family of optimal candidate paths $F_{P'}$ is generated from the lower-cost path *P'* and the previous defined zone $Z_{P'}$. Therefore, an orientation constraint is added to the previous path generation algorithm in order to obtain a family of paths similar to *P'* arriving in $Z_{P'}$. Figure 5 illustrates the candidate path generation.

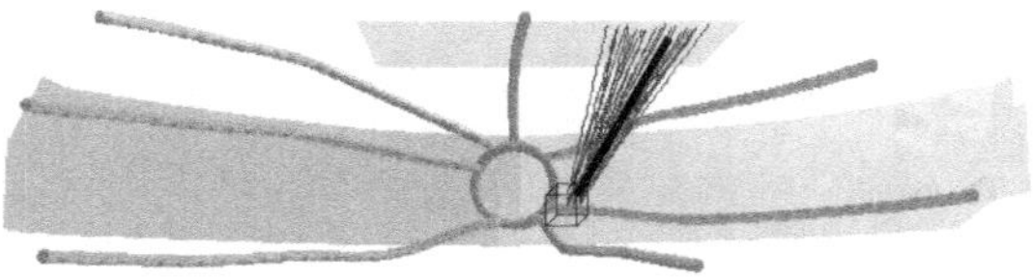

Fig. 5. Optimal path candidates generated from *P'* (in black)

- Step 4: Optimal path P^* selection

Finally, the needle insertions along the paths of the family $F_{P'}$ are simulated using the breast deformation model in order to select P^*, the one that minimizes the positioning error, *i.e.* the Euclidean distance between the target and the needle tip final positions.

$$P^* = \min_{P \in F_{P'}} (e)$$

Figure 6 illustrates the mesh deformation. The error, e, is represented in the right zoom image. The initial target position is shown in white.

Fig. 6. Left: Needle insertion simulation along a path *P*. Right: zoom on the target.

For step 2 and 4 of our planning method, we used Finite Element (FE) method to simulate breast tissue deformations during the needle insertion. We modified *Artisjokke* simulator [14] to simulate needle insertion and to have a patient specific application. Forces resulting from the needle insertion deform the mesh *M* and make the lesion move. We constraint the nodes linked to the detector and the compression paddle to be motionless. We considered a homogeneous, linear and isotropic elasticity model. Young's modulus describing the biomechanical properties of the tissue has been found in the literature [15].

We finally proposed a complete approach providing a solution to our problem: we first find a low cost path with relaxed constraints P', then we generate a family $F_{P'}$ of multiple optimal candidate paths and test them using the simulator. Finally we select the best path P^* that minimizes the final distance between the needle tip and the target. This optimal path has an orientation that is not a priori known. Therefore, we can imagine that the future needle guidance device should be poly-articulated.

3 Results

The proposed planning method has been validated on 12 targets in a breast biopsy phantom (*CIRS-Stereotactic Needle Biopsy Training Phantom*).

We first made a DBT acquisition of the phantom and located the position of the targets in the volume. The breast mesh M has been constructed and our planning algorithm has been run to find an optimal insertion point and an optimal path.

The efficiency of the planning has been evaluated by comparing:

- the error computed for *P'*, the lower cost path found without taking breast deformation into account and noted error'
- and the error obtained for the optimal path, P^*, finally found by the planning algorithm and noted $error^*$.

The simulation results show that the algorithm reduces the error, *i.e.* the Euclidean distance between the needle tip and the target, by 80% with a standard deviation of 13% for both trajectory types. Table 1 presents the results calculated for bevel-tip needles.

Table 1. General Algorithm Evaluation for Bevel-tip Needles

Lesion number	1	2	3	4	5	6	7	8	9	10	11	12
error' (mm)	2.32	1.28	0.41	1.13	1.76	2.56	3.16	2.44	0.98	1.44	2.40	1.66
error* (mm)	0.09	0.06	0.12	0.10	0.23	0.72	0.17	0.22	0.11	0.21	0.15	0.13

The planning computation time has been also compared between our first path generation algorithm based on RRT and the novel approach. 1000 paths in various cluttered environments were generated using both methods. Our initial surface biased approach divides the planning time by about 5, as shown in Table 2.

Table 2. Path Generation Algorithm Evaluation (Environment n°24)

	Symetric-tip needle				Bevel-tip needle			
	T_{min}	T_{max}	T_{mean}	T_{stdev}	T_{min}	T_{max}	T_{mean}	T_{stdev}
RRT method (ms)	0.0	590.0	35.6	40.0	0.0	250.0	29.2	28.8
BS' biased approach (ms)	0.0	30.0	4.9	5.3	0.0	20.0	5.2	5.2

4 Discussion

We have proposed a novel method combining motion-planning algorithm with Finite Element (FE) analysis to find an optimal insertion point and an optimal needle path in a 3D environment. The proposed approach takes soft tissue deformations into account. In addition, the DBT acquisition is used to create a patient-specific application.

We have shown that this method reduces the error meanly by 80%. In addition, the novel paths generation algorithm accelerates path research step by a factor of 5.

Part of our perspective work is to evaluate the simulation accuracy. We plan to compare the simulation results with the deformations obtained on an anthropomorphic biopsy phantom. We also plan to evaluate our method with a breast deformation model taking into account breast inhomogeneity.

Up to now, we only used 3D synthetic vessels but work is going to automatically segment the vessels from the DBT slices in order to get realistic and patient-specific obstacles.

Acknowlegments. This work has been partially funded by ANRT under CIFRE grant n°1035/2008.

References

1. Deurloo, E., Gilhuijs, K., Kool, L., Müller, S.: Displacement of breast tissue and needle deviation during stereotactic procedures. Investigative Radiology 36 (2001)
2. Genaro, G., Baldan, E., Bezzon, E., La Grassa, M., et al.: Clinical Performance of Digital Breast Tomosynthesis Versus Full-Field Digital Mammography: Preliminary Results. In: Krupinski, E.A. (ed.) IWDM 2008. LNCS, vol. 5116, pp. 477–482. Springer, Heidelberg (2008)
3. Alterovitz, R., Goldberg, K., Okamura, A.: Planning for Steerable Bevel-tip Needle Insertion Through 2D Soft Tissue with Obstacles. In: IEEE International Conference on Robotics and Automation (ICRA), pp. 1652–1657 (2005)
4. DiMaio, S.P., Salcudean, S.E.: Needle Steering and Motion Planning in Soft Tissues. IEEE Transactions on Biomedical Engineering (2004)
5. Hauser, K., Alterovitz, R., Chentanez, N., Okamura, A., Goldberg, K.: Feedback Control for Steering Needles Through 3D Deformable Tissue Using Helical Paths. Robotics: Science and Systems V (2009)
6. Dehghan, E., Salcudean, S.E.: Needle Insertion Parameter Optimization for Brachytherapy. IEEE Trans. on Robotics 25, 303–315 (2009)
7. Latombe, J.-C.: Robot Motion Planning. Kluwer Academic Publishers, Boston (1991)
8. Gibson, S.F.F., Mirtich, B.: A survey of Deformable Modeling in Computer Graphics., TR-97-19, Mitsubishi Electric Research Laboratory (1997)
9. Schroeder, W.J., Avila, L.S., Hoffman, W.: Visualizing with VTK: a tutorial. IEEE Computer Graphics and Applications 20, 20–27 (2000)
10. Vancamberg, L., Sahbani, A., Muller, S., Morel, G.: Needle Path Planning for Digital Breast Tomosynthesis Biopsy. In: IEEE International Conference on Robotics and Automation, ICRA (2010)
11. LaValle, S.M.: Rapidly-Exploring Random Trees: A New Tool for Path Planning, TR 98-11, Department of Computer Science, Iowa State University (1998)
12. Xu, J., Duindam, V., Alterovitz, R., Goldberg, K.: Motion Planning for Steerable Needles in 3D Environments with Obstacles Using Rapidly-Exploring Random Trees and Backchaining. In: IEEE International Conference on Automation Science and Engineering (2008)
13. Webster III, R.J., Cowan, N.J., Chirikjian, G., Okamura, A.M.: Nonholonomic Modeling of Needle Steering. In: 9th International Symposium on Experimental Robotics (2004)
14. Nienhuys, H.W., van der Stappen, A.F.: A computational technique for interactive needle insertions in 3D nonlinear material. In: IEEE International Conference on Robotics and Automation, ICRA (2004)
15. Krouskop, T.A., Wheeler, T.M., Kallel, F., Garra, B.S., Hall, T.: Elastic Moduli of Breast and Prostate Tissues Under Compression. Ultrasonic Imaging 20, 260–278 (1998)

X-ray Mammography – MRI Registration Using a Volume-Preserving Affine Transformation and an EM-MRF for Breast Tissue Classification

Thomy Mertzanidou[1], John H. Hipwell[1], M. Jorge Cardoso[1], Christine Tanner[2,1], Sebastien Ourselin[1], and David J. Hawkes[1]

[1] Centre for Medical Image Computing, University College London, Gower Street, WC1E 6BT, London, UK
[2] Computer Vision Laboratory, ETH Zürich, Sternwartstrasse 7, 8092 Zürich, Switzerland
t.mertzanidou@cs.ucl.ac.uk

Abstract. Registration of MR volumes to X-ray mammograms is a clinically valuable task, as each modality provides complementary information on normal and abnormal breast tissue structure and function. We propose an intensity-based technique with a 3D volume-preserving affine transformation. An important part of our framework is the use of an Expectation-Maximization (EM) algorithm, with a Markov Random Field (MRF) regularization, that is used for breast tissue classification and subsequently the mapping of the MR intensities to X-ray attenuation. Initially, the proposed framework was tested on simulated X-ray data, where the goal was to register the original undeformed MRI to a simulated X-ray that was produced using a real compression image, acquired from volunteers in the MR scanner (8 cases). Since the ground truth in this case can be estimated from individually defined landmarks, we have evaluated the mean reprojection error, which was $3.83mm$. The algorithm was then applied and evaluated visually on 5 cases that had both X-ray mammograms and MRIs.

Keywords: multimodal registration, 2D – 3D registration, breast tissue classification.

1 Introduction

X-ray – MRI registration has the potential to aid radiologists in the diagnosis, staging and surgical planning of breast cancer. However, little work has been reported on this task ([1], [2], [3]). The authors in [1] and [2] follow a feature-based registration approach. These techniques, although they do not require long computational time, are usually less robust, as they fail in cases where the selected features are not visible in both images and also when one or more features are mismatched. The third technique [3] uses a subject-specific Finite Element Method (FEM) for modeling, to simulate the deformation of the breast during mammography. The large variability in meshing techniques and the material

J. Martí et al. (Eds.): IWDM 2010, LNCS 6136, pp. 23–30, 2010.

properties used, makes breast FEM modeling a user-dependent, time-consuming task that is not practical in clinics. In this paper we describe an extension of our previous work [5] that aims to overcome the above problems, while providing clinically useful accuracy (less than $10mm$). Intensity-based techniques were widely used for rigid 2D-3D registration tasks in the past, mainly confined to rigid body orthopaedic or vascular applications, achieving good results ([7], [8]).

A key contribution of this paper is the use of realistic X-ray simulations from the MR volume, due to improved breast tissue classification. The techniques introduced in the literature for this task aim mainly to segment the fibroglandular tissue to provide breast density estimation from the MRI. To our knowledge these classify the MR voxels based on the intensity information only, using either manual thresholding [9] or a fuzzy c-means technique [10]. The EM-MRF approach combines both intensity and spatial information and was initially introduced for classification of brain tissue types. The implementation that we use is similar to the one proposed by Van Leemput et al. [4], with a modification that incorporates anatomical information.

Another difference to our previous work is the use of a volume-preservation constraint, to avoid the breast volume expansion previously observed.

The next section describes in detail our methodology and more specifically the tissue classification approach (section 2.1) and the registration framework (section 2.2). Experiments and results are discussed in section 3. Finally, section 4 contains the conclusions and future work.

2 Methodology

Our registration technique requires the simulation of an X-ray image using the MR volume, in order to directly compare it with the real mammogram in 2D and drive the registration. The volume is projected iteratively in the registration framework using the updated transformation parameters. These simulated images use the projection of the X-ray attenuation values instead of the MR intensities to provide a realistic X-ray image.

The mapping of the MRI to an X-ray attenuation volume is done off-line, before registration and requires first the classification of the voxels into breast tissue categories (glandular and fat). We can then calculate the new volume intensities by weighting these classes with different factors in order to simulate the difference in X-ray attenuation. The intensity of voxel i in the X-ray attenuation volume is given by: $(w_G \cdot P_G^i + w_F \cdot P_F^i)$, where w_{class} are the weights of each tissue type and P_{class}^i is the classification result for voxel i, for each one of the classes $(0 \leq P_{class}^i \leq 1)$. The choice of the most appropriate weights was performed empirically. The goal was to produce simulations with similar contrast to digitized film mammograms. The classification method is described below.

2.1 Breast Tissue Classification Using an EM-MRF Approach

As in [4], our method integrates an intensity model, a spatial regularization scheme and bias field inhomogeneity correction in the same framework.

The incorporation of spatial information has been shown to improve classification results in the past as it provides robustness to noise and it allows the use of anatomical information. Specifically for the breast tissue classification, the MRF regularization is considered an appropriate choice due to the anatomy of the fibroglandular tissue. Since this is connected in a tree-like structure inside the breast, our hypothesis is that the voxels containing glandular tissue are more likely to appear connected to other glandular voxels rather than isolated inside the fat (and similarly for fat voxels).

The intensity model assumes three classes (for glandular, fat tissue and background) and the bias field is modeled using a third order polynomial basis function. Instead of considering Gaussian distributed intensities corrupted by a multiplicative bias field, log-transformed intensities are used to make the bias field additive. For K classes let $\theta_k = \{\mu_k, \sigma_k\}$ denote the normal probability distribution with mean μ_k and variance σ_k^2 of a voxel belonging to class k and let $z_i = e_k$ be the tissue type of voxel i, where e_k is the unit vector with the k-th component equal to 1 and the others equal to zero. For J basis functions $\phi_j(x)$, $C = \{c_1 ... c_J\}$ denotes the bias field parameters. The probability density for voxel i, with intensity y_i, given it belongs to class k is:

$$f(y_i|z_i = e_k, \Phi_y) = G_{\sigma_k}(y_i - \mu_k - \sum_j c_j \phi_j(x_i)), \quad (1)$$

where $\Phi_y = \{\theta_1, ..., \theta_k, C\}$ is the intensity model parameters and $G_\sigma()$ is a normal distribution with mean zero and standard deviation σ. The model parameters are optimized using an EM algorithm under a Maximum Likelihood formulation. Due to the large variation of glandular structures in the breast across the population, there are no anatomical priors available. If m is the iteration number, then the ML estimation gives:

$$\mu_k^{(m+1)} = \frac{\sum_{i=1}^n p_{ik}^{(m+1)} (y_i - \sum_{j=1}^J c_j^{(m)} \phi(x_i))}{\sum_{i=1}^n p_{ik}^{(m+1)}} \quad (2)$$

$$(\sigma_k^{(m+1)})^2 = \frac{\sum_{i=1}^n p_{ik}^{(m+1)} (y_i - \mu_k^{(m+1)} - \sum_{j=1}^J c_j^{(m)} \phi(x_i))^2}{\sum_{i=1}^n p_{ik}^{(m+1)}}, \quad (3)$$

where

$$p_{ik}^{(m+1)} = \frac{f(y_i|z_i = e_k, \Phi_y^{(m)}) f(z_i = e_k)}{\sum_{j=1}^K f(y_i|z_i = e_j, \Phi_y^{(m)}) f(z_i = e_j)}. \quad (4)$$

The intensity model alone can only give accurate results when the different distributions are well separated. This is not the case for the glandular and fat tissue due to many voxels containing both tissue types (partial volume effect). The use of an MRF regularization scheme improves the overall robustness of the model parameter estimation and provides spatial consistency. Voxels are thus classified based also on the current classification of the neighboring voxels. In this case, equations 2 and 3 remain the same, while 4 is now given by:

$$p_{ik}^{(m+1)} = \frac{f(y_i|z_i = e_k, \Phi_y^{(m)}) f(z_i = e_k|p_{N_i}^{(m)}, \Phi_z^{(m)})}{\sum_{j=1}^{K} f(y_i|z_i = e_j, \Phi_y^{(m)}) f(z_i = e_j|p_{N_i}^{(m)}, \Phi_z^{(m)})}, \tag{5}$$

where

$$f(z_i = e_k|p_{N_i}^{(m)}, \Phi_z^{(m)}) = \frac{e^{-\beta U_{mrf}(e_k|p_{N_i}^{(m)} \Phi_z^{(m)})}}{\sum_{j=1}^{K} e^{-\beta U_{mrf}(e_j|p_{N_i}^{(m)} \Phi_z^{(m)})}}, \tag{6}$$

with $\Phi_z = \{G, H\}$ the MRF model parameters and $U_{mrf}(z|\Phi_z)$ the energy function that depends on Φ_z. G and H are $K \times K$ matrices that define the transition energy between classes. Further details of the bias field parameter estimation and detailed explanations of the other equations can be found in [4].

The above regularization makes the classification more robust to noise and to isolated misclassified voxels (e.g. isolated voxels classified as fat and surrounded by glandular tissue). Instead of estimating the MRF parameters from the image as in [4], we use a two-level MRF with its parameters derived from the anatomical properties of the breast. In the first level, the interclass MRF energy is the same for all classes, thus the MRF only adds global spatial consistency and robustness in the parameter estimation. In the second stage, after the EM converges, the MRF energy matrices (G and H) are altered in order to include more anatomical knowledge (e.g. the cost of having glandular tissue next to the background is higher than having fat next to the background) and the classification is restarted again until convergence. This modification allows an unbiased and robust parameter estimation in the first step followed by a second step that enforces more anatomical knowledge and topological constraints. The values of the MRF energy matrix are chosen empirically, in order to produce realistic X-ray mammogram simulations.

2.2 2D – 3D Registration Framework

For registration, we use a 3D volume-preserving affine transformation to approximate the deformation of the breast during mammography. At each iteration of the algorithm, the similarity measure is calculated in 2D between the real X-ray mammogram (*target*) and the perspective projection of the X-ray attenuation volume (*source*). The value is then passed to the optimizer which updates the parameters of a 3D affine transformation. In all the experiments we have used the Normalized Cross Correlation (NCC) as similarity measure. This has been tested before for suitability in 2D-3D registration tasks and was shown to work best [5].

An advantage of the affine transformation is that it can be implemented directly in the ray-cast projection function, avoiding the additional computation of transforming and interpolating the 3D volume at each iteration. To avoid non-physical expansion of the volume (for CC views in the superior-inferior and also in the posterior-anterior direction) we have included a volume-preservation constraint, by ensuring that the product of all scaling factors across the 3 dimensions is unity ($s_x \cdot s_y \cdot s_z = 1$). This is done by constraining the scaling on the direction of the projection (superior-inferior for a CC view) to be $s_z = 1/(s_x \cdot s_y)$. This removes one degree of freedom from the optimization process, reducing the size of the search space and potentially enhancing the robustness of the registration. Our previous results have shown that without this constraint, the volume increases in a physically unrealistic way [5].

3 Experiments

3.1 EM-MRF Classification Results

Figure 1 shows the results of the EM-MRF algorithm, as opposed to a histogram-based classification that uses manual thresholding ([5],[6]). We can see that the proposed method contains more details of the glandular tissue. It is also fully automated and gives reproducible results. The only requirement is that the pectoral muscle is segmented from the volume. In our semi-automated method the user defines landmarks on the boundary between the pectoral muscle and the breast through which a parametric surface is fitted. Intensity-based techniques are more prone to errors for this task, as this boundary is not well-defined in many cases, especially when the glandular tissue is very close to the chest wall, or when organs with intensities similar to fat (such as the liver) are attached to the rib cage.

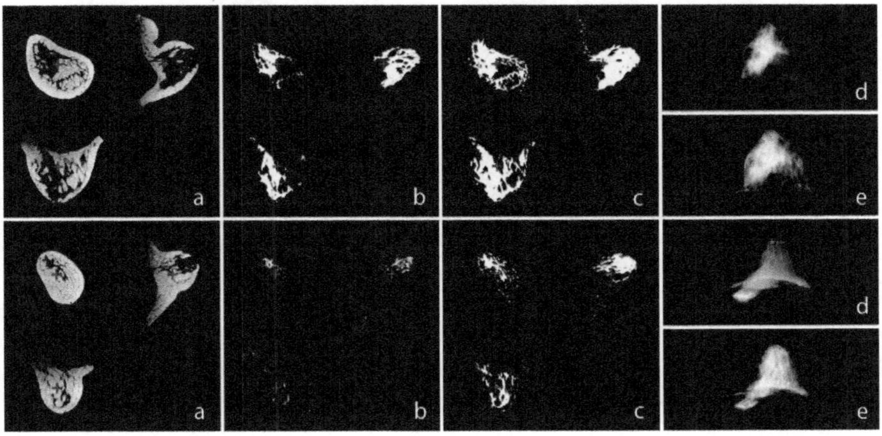

Fig. 1. (a) Original MRI, (b) X-ray attenuation volume using manual thresholding and histogram-based classification, (c) using the EM-MRF algorithm, (d) Simulated X-ray mammogram from the undeformed volume using manual thresholding, (e) using EM-MRF. The two rows correspond to two patients. The red cross indicates the position of a corresponding coordinate in each image.

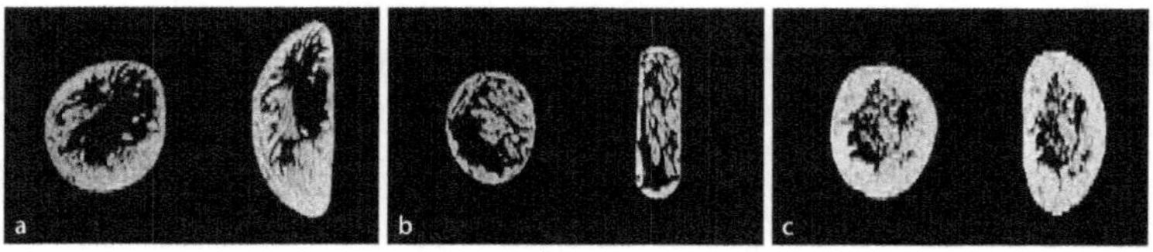

Fig. 2. Coronal slices of the data used for evaluation for 3 volunteers (a)-(c). From left to right in each image: volume before and after compression.

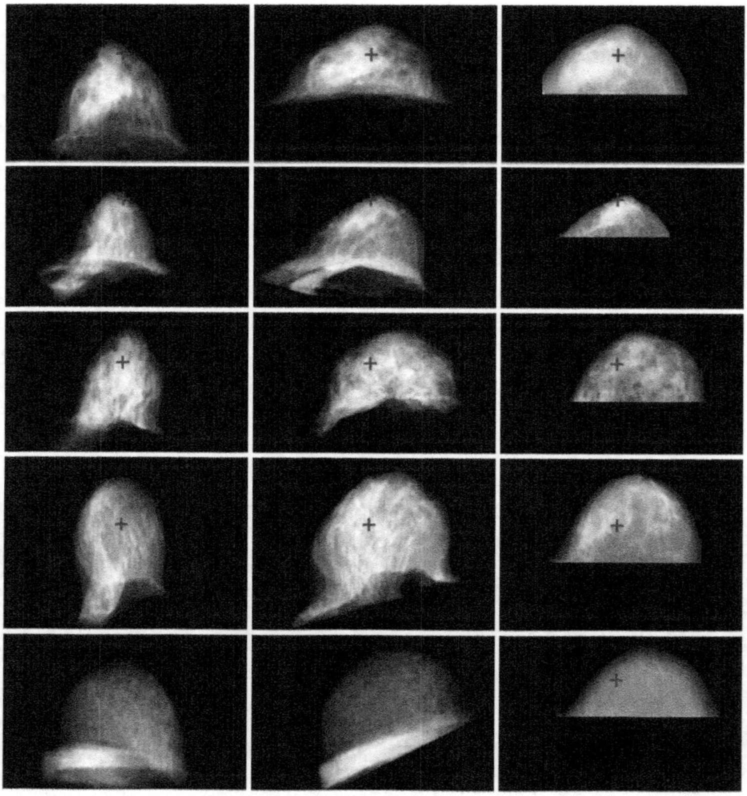

Fig. 3. Registration results on real data (5 cases/rows). From left to right: projection of the source volume before registration, after registration and real X-ray mammogram. The red cross indicates the position of a corresponding coordinate in each image.

3.2 Registration Results

In all registration experiments we have used as source image the 3D X-ray attenuation volume that was calculated using the pre-contrast MRI. The proposed algorithm was tested on two sets of data. The first set used simulated mammograms for evaluation, while the second set, containing real MR and X-ray data, provided a visual assessment of the algorithm's performance.

Evaluation. For evaluation, we used only simulated X-rays and a series of real MR compressions of the breast from 8 volunteers [11] in the lateral to medial direction. Figure 2 shows some examples of these compression images (volume sizes: $1mm \times 1mm \times 2.5mm$). The goal was to register the uncompressed volume to the simulated X-ray that was produced from the compressed MRI. The ground truth of the correspondences in this case was estimated, by manually picking 3D landmarks between the undeformed and the compressed MRI. The mean reprojection error [8] for these experiments was reduced to $3.83mm$ (with a standard deviation of $1.59mm$) after registration, from an initial $11.58mm$ (std $6.65mm$) misalignment.

Experiments on Real X-ray Data. After validating our algorithm on simulated mammograms, we tested it on real X-ray (CC views) and MRIs from a different population. We have used 5 cases and the results were evaluated visually as estimating 2D-3D ground truth correspondences is not a straight-forward task. Figure 3 shows the registration results. We can see that the similarity between the real mammogram and the projection of the source volume after registration is greatly improved, with the breast volume expanding to the medial and lateral direction.

4 Conclusions

We have presented a new framework for X-ray mammography – MRI registration, using a volume preserving affine transformation. We have also shown an improvement in X-ray simulations, which is a prerequisite for visual correlation of MR with X-ray structures and may improve registration performance. The results on real compression data show that this is a promising technique that can give clinically valuable accuracy, while the low number of degrees of freedom and the lack of FEM or feature extraction makes this approach potentially robust, accurate and reproducible.

An essential part of future work is the use of a radiologist's annotations to evaluate the results on real X-ray mammograms. Establishing ground truth X-ray/MRI correspondence on normal cases is not always possible, but can be done for datasets that include lesions (benign or malignant) or cysts. Another part of our framework that we plan to evaluate is the breast tissue classification. It is particularly difficult to provide quantitative validation for this task, as there is no ground truth available. Furthermore, we will investigate and assess the benefit of incorporating spatial information in the classification technique. Finally, a natural future step is the use of a non-rigid transformation model, with more degrees of freedom than the affine.

Acknowledgements. This work was funded by the European 7th Framework Program, HAMAM, ICT-2007.5.3 and EPSRC grant EP/E031579/1. The authors would like to thank the Cambridge University Hospitals NHS Foundation Trust, the UK MR Breast Screening Study (MARIBS) [12] and UCLH for providing the data for this study.

References

1. Behrenbruch, C., Marias, K., Armitage, P., Moore, N., English, R., Clarke, J., Brady, M.: Fusion of contrast-enhanced breast MR and mammographic imaging data. Medical Image Analysis 7, 311–340 (2003)
2. Marti, R., Zwiggelaar, R., Rubin, C., Denton, E.: 2D-3D correspondence in mammography. Cybernetics and Systems 35, 85–105 (2004)
3. Ruiter, N., Stotzka, R., Muller, T., Gemmeke, H., Reichenbach, J., Kaiser, W.: Model-Based registration of X-ray Mammograms and MR images of the female breast. IEEE Transactions on Nuclear Science 53, 204–211 (2006)
4. Van Leemput, K., Maes, F., Vandermeulen, D., Suetens, P.: Automated Model-Based Tissue Classification of MR Images of the Brain. IEEE Transactions on Medical Imaging 18, 897–908 (1999)
5. Mertzanidou, T., Hipwell, J., Tanner, C., Hawkes, D.: An intensity-based approach to X-ray mammography – MRI registration. SPIE Medical Imaging: Image Processing, 7623–106 (2010)
6. Hipwell, J., Tanner, C., Crum, W., Schnabel, J., Hawkes, D.: A New Validation Method for X-ray Mammogram Registration Algorithms Using a Projection Model of Breast X-ray Compression. IEEE Transactions on Medical Imaging 26, 1190–1200 (2007)
7. Penney, G., Weese, J., Little, J., Desmedt, P., Hill, D., Hawkes, D.: A Comparison of Similarity Measures for Use in 2-D–3-D Medical Image Registration. IEEE Transactions on Medical Imaging 17, 586–595 (1998)
8. Hipwell, J., Penney, G., McLaughlin, R., Rhode, K., Summers, P., Cox, T., Byrne, J., Noble, A., Hawkes, D.: Intensity-based 2D–3D registration of celcelebral angiograms. IEEE Transactions on Medical Imaging 22, 1417–1426 (2003)
9. Wei, J., Chan, H., Helvie, M., Roubidoux, M., Sahiner, B., Hadjiiski, L., Zhou, C., Paquerault, S., Chenevert, T., Goodsitt, M.: Correlation between mammographic density and volumetric fibroglandular tissue estimated on breast MR images. Medical Physics 31, 933–942 (2004)
10. Nie, K., Chen, J., Chan, S., Chau, M., Yu, H., Bahri, S., Tseng, T., Nalcioglu, O., Su, M.: Development of a quantitative method for analysis of breast density based on three-dimensional breast MRI. Medical Physics 35, 5253–5262 (2008)
11. Tanner, C., White, M., Guarino, S., Hall-Craggs, M., Douek, M., Hawkes, D.: Anisotropic behaviour of breast tissue for large compressions. In: Proceedings of the IEEE International Symposium on Biomedical Imaging, pp. 1223–1226 (2009)
12. Leach, M., Boggis, C., Dixon, A., Easton, D., Eeles, R., Evans, D., Gilbert, F., Griebsch, I., Hoff, R., Kessar, P., Lakhani, S., Moss, S., Nerurkar, A., Padhani, A., Pointon, L., Thompson, D., Warren, R.: Screening with magnetic resonance imaging and mammography of a UK population at high familial risk of breast cancer: a prospective multicentre cohort study (MARIBS). The Lancet 365, 1769–1778 (2005)

Interactive Multi-scale Contrast Enhancement of Previously Processed Digital Mammograms

Fabian Zöhrer[1], Markus T. Harz[1], Anke Bödicker[2], Heike Seyffarth[3], Kathy J. Schilling[4], László Tabár[5], and Horst K. Hahn[1]

[1] Fraunhofer MEVIS, Universitätsallee 29, 28359 Bremen, Germany
fabian.zoehrer@mevis.fraunhofer.de
[2] MeVis BreastCare GmbH & Co. KG, Universitätsallee 29, 28359 Bremen, Germany
[3] MeVis Medical Solutions AG, Universitätsallee 29, 28359 Bremen, Germany
[4] Boca Raton Community Hospital, 800 Meadows Road Boca Raton, FL 33486, United States
[5] Department of Mammography, Falun Central Hospital, 79182 FALUN, Sweden

Abstract. Manufacturers of digital mammography systems include proprietary and unique post-processing algorithms to enhance the image contrast for diagnostic presentation. We use a wavelet-based enhancement method to both automatically and interactively adjust the contrast of these previously processed mammograms. Apart from enhancing small but relevant structures in dense breast tissue, this method allows for the automatic homogenization of cross-vendor contrast appearance, supporting current-prior comparison tasks in the screening and diagnostic setting when different systems were used for acquisition. An optional smooth crossfading between originally processed and enhanced images can be used interactively to adapt the presentation view to individual situations, for example to facilitate the detection or interpretation of mammographic features in dense breast tissue. A user preference study with experienced readers revealed that our work might positively impact prior-current comparisons and the diagnosis of mammographic images.

1 Background

Mammography is considered the most important modality in breast cancer screening and diagnosis. In dense breasts, however, the process of detecting subtle signs of cancer such as architectural distortions, masses and asymmetries is hampered by their reduced contrast in dense breast tissue. Additionally, it has been observed that an increased density of the breast is linked to a higher risk of developing breast cancer [1]. There has been significant work on the field of mammographic image enhancement [2,3,4,5] and it has been shown that these techniques can improve the detectability of important features in mammographic screening.

Nowadays, manufacturers of digital mammography systems include their proprietary post-processing algorithms to enhance digital mammograms for diagnostic presentation, which gives these processed mammograms an unique appearance and contrast. In [6] Chen et al. compared the diagnostic abilities of two

J. Martí et al. (Eds.): IWDM 2010, LNCS 6136, pp. 31–38, 2010.

post-processing methods provided by the GE Senographe DS System, premium view (PV) and tissue equalization (TE). Their study showed that PV provided better diagnostic information compared to TE, particularly for patients with malignancy in dense breast.

During screening or therapy, patients frequently undergo examinations with mammography systems of different manufacturers. In the process of screening, a patient's current mammograms are compared to the prior mammograms in order to aid detecting changes in breast morphology, which can be an indication of a growing lesion. Snoeren and Karssemeijer [7] presented a gray-scale and geometric registration of full-field digital "for processing" mammograms to film-screen mammograms based on a parametric model of the acquisition aspects. However, in a clinical setting the availability of "for processing" images is not always granted for a number of reasons including system restrictions and external image acquisition. Our work presents methods and an initial assessment for the homogenization of "for presentation" mammograms acquired with different machines and treated with different post-processing methods. We aim to ease diagnostic assessment of those prior-current mammogram pairs.

2 Method

Wavelet based multi-scale analysis has formerly been applied to our task [3], but we will attempt a simpler and faster multi-scale approach here.

2.1 Interactive Multi-scale Analysis

A schematic view of the method utilized is shown in Fig. 1. Each decomposition step consists of constructing an approximation/low-pass image LPn and a detail/high-pass image HPn from the input image. The first approximation image LP1 is obtained by convolving the input image two times with a Gaussian 3×3 filter kernel and subtracting it from the input image yields the detail image HP1. The detail image contains the high-frequency spatial information suppressed by the blur filter. This difference of Gaussians approximates the mexican hat wavelet. This method is numerically equivalent to a wavelet transform for the scales and datatypes we used in the implementation. Note that the approximation image is not subsampled but retains its size. The decomposition is repeated in step 2 by constructing LP2 and HP2 from LP1. The filter kernel size is increased in every decomposition step n to 2^n+1 to cover a different image frequency. By repeating the decomposition N times we obtain a set of N detail images at different scales and a final approximation image LPN. Following, each of the N detail images HPn are multiplied with a weight factor ϵ_n to adjust their strength, while a offset value δ_N is added to the approximation image LPN to adjust the mean gray-value of the whole image. After the enhancement step, the final image is reconstructed by adding all weighted detail images HPn^* to the adjusted approximation image LPN^*. A mask covering only foreground pixel corresponding to the breast is derived from the original image and applied to

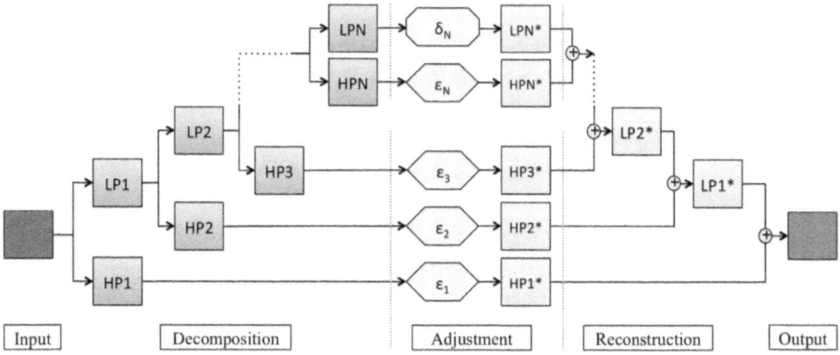

Fig. 1. The input image is divided through N steps into a final approximation/low-pass image LPN and N detail/high-pass images HPn. Before reconstruction, each detail image can be weighted separately with a multiplicative factor ϵ_n to adjust its influence on the overall image contrast characteristics. Additionally, the approximation image LPN can be adjusted with the additive value δ_N to lower or raise the mean gray value of the output image.

the output image such that background values corresponding to air remain unchanged. This method allows for the interactive adjustment of the values ϵ_n and δ_N and the reconstruction of the enhanced image in real time and thus allows to quickly adapt the image contrast to individual tasks like the enhancement of dense tissue contrast. Additionally, we included a tool to smoothly change from the original to the enhanced image via mouse interaction. This crossfading between images should give the user the ability to the control the overall adjustment by one scalar value mapped to one dimensional mouse movements.

2.2 Automatic Homogenization of Current-Prior Mammograms

We implemented an automatic homogenization method, which adjusts the contrast characteristics of a prior mammogram (source) to a current mammogram (target). Above decomposition is applied to both mammograms to compare their detail and approximation images. For each detail image HPn of target and source mammograms the mean of absolute values excluding background is calculated as well as the mean gray value for both approximation images LPN. Following, the respective ratio between a detail image pair of the same level n is used as weight factor ϵ_n, while the difference between both approximation images' mean gray value is used as additive offset value δ_N. By this simple adjustment the frequency strengths of the source mammogram should be adjusted to those of the target mammogram.

2.3 Pre-processing of the Digital Mammograms

The set of digital mammograms was acquired at Boca Raton Community Hospital with a Senographe DS system from GE Medical Systems and a Hologic LORAD Selenia system. The available post-processing methods for the Senographe DS were "proc 1"(P1) and "premium view"(PV), while there was only one post-processing (HP) available for the Hologic images. The set featured 50 mammograms of eight patients, who underwent annual screenings. Five patients were scanned with a GE system with P1 and PV post-processing, while three patients additionally had scans on a Hologic system.

As the detectors of mammography systems differ in size and number of detector elements the digital mammograms will be different in their pixel sizes (Hologic: $0,07^2\,\mathrm{mm}^2$ and GE: $0,094^2\,\mathrm{mm}^2$) and image size (Hologic: 3310×2728 pixel and GE: 1914×2294 pixel). As a result the breast and its morphology is displayed in a higher resolution and thus by more pixel in the Hologic images. As the above automatic method compares both images frequency wise, it is important for the morphology in both images to be at comparable pixel sizes. For the automatic calculation of the adjustment values ϵ_n and δ_N, the current mammogram was resampled by using Lanczos interpolation to fit the pixel size of the prior mammogram. Note that for viewing, the current mammogram was not resampled.

Often, the appearance of the mammograms' gray values is changed for the presentation on the screen by applying a sigmoid LUT on the graphic card. The DICOM tags of GE (P1 and PV) mammograms feature sigmoid representation with different options (softer, normal, harder), while the Hologic pixel values seem to be post-processed already in this fashion and are displayed in a linear LUT. We applied the LUT information of the "normal"sigmoid found in the DICOM tags to the gray values of both GE image types to take into account the way the images should be displayed based on the DICOM informations.

2.4 Design of the User Preference Study

We conducted a user preference study to evaluate the automatic adjustment of current-prior mammograms on the one hand and the smooth transition between the originally processed and the enhanced mammograms on the other hand. We started displaying both current and prior mammograms in original "for presentation"view, i.e. as they would be seen in any digital mammography reading station. The prior could be switched via mouse click from original to the adjusted image and back again. After viewing a current-prior pair, the reader had to decide whether to view the prior mammogram of the next pair either in its original or the adjusted representation. Crossfading between original and enhanced images was achieved via replacing the original mammogram on the graphic card between 0 and 100% with the enhanced image. This crossfading could be handled manually via mouse interaction. Note that this study was not held in a clinical setting, the room was not darkened and we used a standard 23"TFT color monitor for viewing. However, it was possible to view the mammograms in full resolution mode and to zoom into and out of the images.

3 Results

The automatic adjustment was evaluated by four experienced readers (two US american, two European) for a series of current-prior mammogram pairs of different post-processing types or systems. After viewing the first current-prior pair in both original and adjusted setting, all readers preferred to start viewing the next pair with the adjusted prior. Two readers stated that it would be enough to see the priors only in the adjusted mode, while the two others wished for the option to change back to original view. The Fig. 2 and 3 show the results of the automatic adjustment of prior to current mammograms. Window and level settings for the original and adjusted images are accordant to the DICOM tags.

The crossfading between original and enhanced image was also evaluated by the four readers. While the enhancement of dense areas was appreciated, the main aim was to evaluate the value of an interactive tool to continuously change the degree of enhancement. The reduction to one parameter was especially appreciated as a possible alternative to the two dimensional window/level interaction. Figure 4 shows some steps in the smooth transition from an original GE PV mammogram to the fully enhanced image.

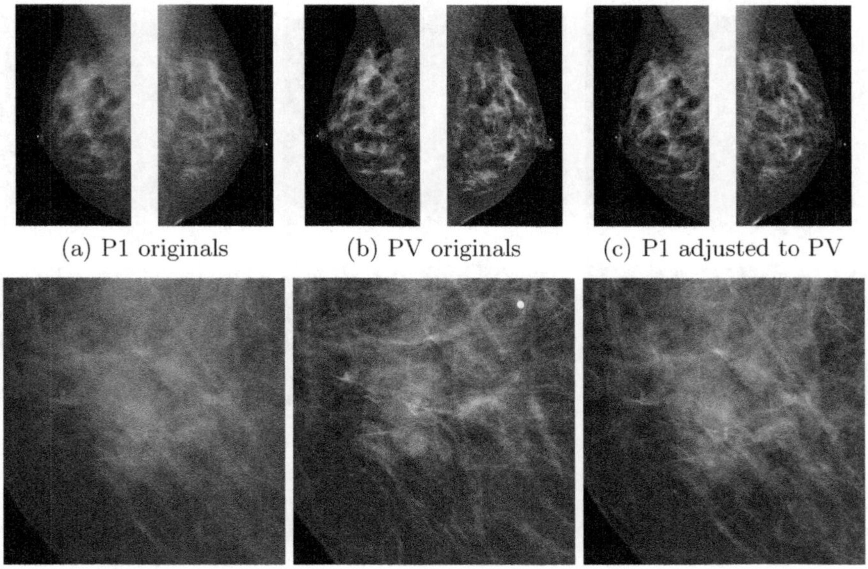

(a) P1 originals (b) PV originals (c) P1 adjusted to PV

(d) P1 original (e) PV original (f) P1 adjusted to PV

Fig. 2. The upper row shows the automatic homogenization between original GE P1 prior (a) and GE PV current (b) MLO pairs. Image (c) shows the results of the automatic adjustment of the contrast characteristics of the prior P1 mammograms to the current PV mammograms. The lower row features close ups of an original GE P1 prior (d) and a GE PV current (e) and again the result (f) of the automatic adjustment of the contrast characteristics is shown.

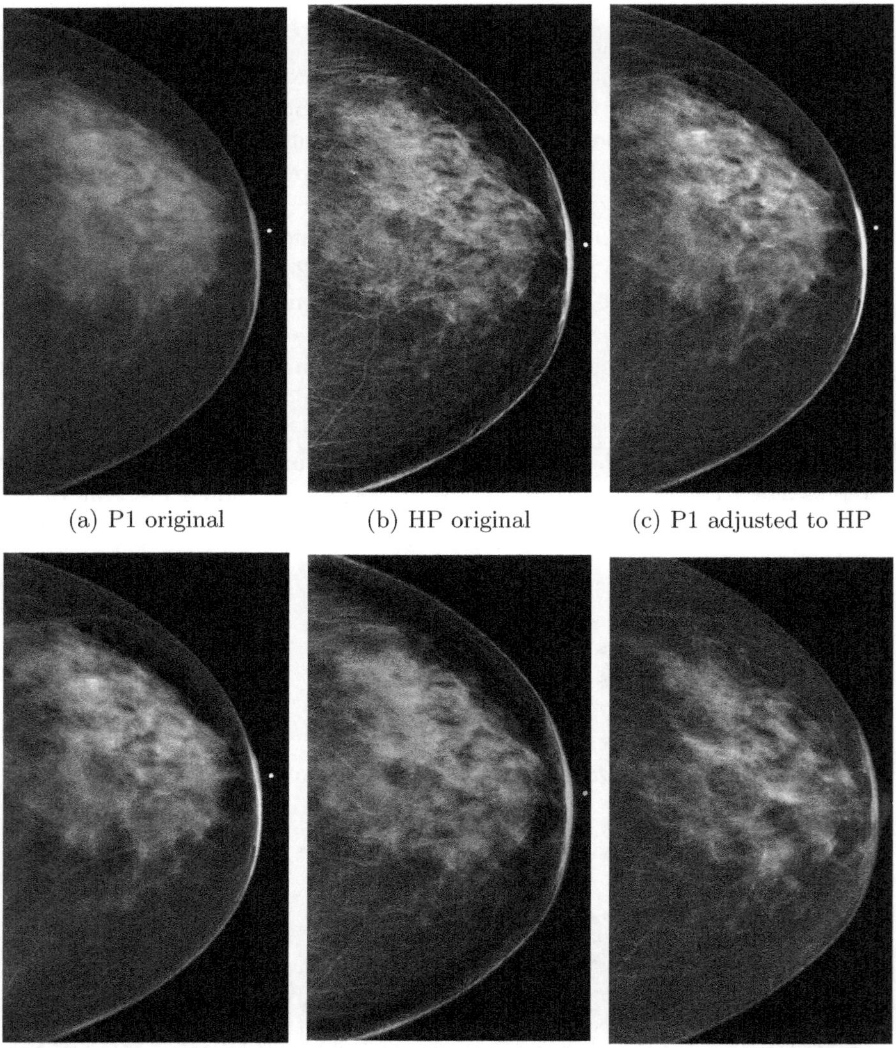

(a) P1 original (b) HP original (c) P1 adjusted to HP

(d) P1 adjusted to PV (e) HP adjusted to PV (f) PV original

Fig. 3. These six images display three original mammograms of the same breast and the results of three different homogenizations of the contrast appearances between these current-prior mammograms. All three original mammograms feature a different original post-processing and were acquired during annual screening starting with the GE "proc 1"(P1) and followed by the Hologic (HP) and the GE "premium view"(PV) mammograms. Image (c) shows the result of the automatic adjustment of the contrast characteristics of the P1 prior (a) to the HP follow up (b), while images (d) and (e) show the automatic adjustment of the original P1 (a) and HP (b) mammograms to the PV (f) current mammogram. Note that the window/level settings for all originals are accordant to their DICOM tags.

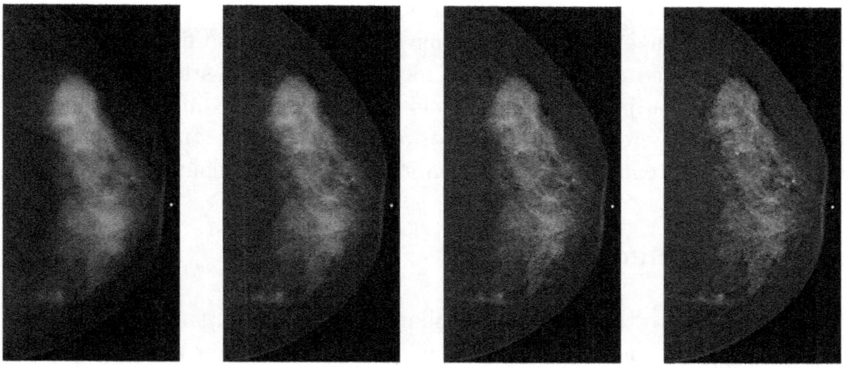

(a) PV original (b) Enhanced 33% (c) Enhanced 66% (d) Enhanced 100%

Fig. 4. Four different steps in the smooth transition from original GE PV mammogram (a) to the fully enhanced result (d). Images (b) and (c) show the enhancement strength of 33% and 66%. The transition is handled interactively via mouse interaction.

4 Discussion

We have described a fast multi-scale enhancement method which allows for the adjustment of the contrast appearance of mammograms to those post-processed with different algorithms or acquired on different systems. One advantage of such homogenization is a possible facilitation of the current-prior comparison of differently processed mammograms. Besides this homogenization, the adjustment can also be used interactively to enhance the contrast of certain image features, which might be especially beneficial in dense breast tissue. Another aspect of our application is the interactive smooth transition between the original input image and the enhanced image. During the process of crossfading between these contrast settings the change of luminance is strongest in those image structures, which are changed the most by the adjustment. This possibly can assist the detection of features and due to the smooth transition there is no abrupt change of the image. The reduction to one dimensional interaction can be considered as an advantage above the common window/level handling with well-defined lower and upper limits for the parameter range.

It is noteworthy that this method does not yet include any acquisition parameters or knowledge of the manufacturers post-processing and that the automatic adjustment relies on simple statistical values. In some cases, we experienced saturation effect induced by the automatic adjustment, as we reduce the gray value range to the original 0 and 4095. This cutting of higher values has to be addressed in future improvements, as it can lead to some small flat image parts which destroys information. However, as the adjustment is only done for priors which are used for comparison with the currents, this problem was rated as bearable by the readers compared to the advantages of the adjustment. Furthermore this method does not include the different skin enhancement processing applied by the manufactures and the positioning of the breast during scan. Differences between the amount of pectoralis imaged in current and prior for example will lead to different image statistics. Further research and modifications will include a possible geometric registration and a more sophisticated statistical comparison for the automatic adjustment as well as an inclusion of different enhancement functions. Currently we are planning for another observer study with cases from a larger set of different manufacturers and an evaluation using phantom images.

Acknowledgements

The authors want to thank Roland Holland, Ulrich Bick, André-Robert Grivegnee, Robert Schmidt and Marco Rosselli del Turco for their participation in the user preference study and the invaluable input they gave us in many discussions. Part of this work was funded by the European FP7.

References

1. Ursin, G., Hovanessian-Larsen, L., Parisky, Y., Pike, M., Wu, A.: Greatly increased occurrence of breast cancers in areas of mammographically dense tissue. Breast Cancer Res. 7(5), R605–R608 (2005)
2. Chan, H., Vyborny, C., MacMahon, H., Metz, C., Doi, K., Sickles, E.: Digital mammography: ROC studies of the effects of pixel size and unsharp-mask filtering on the detection of subtle microcalcifications. Investigative Radiology 22(7), 581–589 (1987)
3. Laine, A., Schuler, S., Fan, J., Huda, W.: Mammographic feature enhancement by multiscale analysis. IEEE Transactions on Medical Imaging 13(4), 725–739 (1994)
4. Morrow, W., Paranjape, R., Rangayyan, R., Desautels, J.: Region-based contrast enhancement of mammograms. IEEE Transactions on Medical Imaging 11(3), 392–406 (1992)
5. Pisano, E., Cole, E., Hemminger, B., Yaffe, M., Aylward, S., Maidment, A., Johnston, R., Williams, M., Niklason, L., Conant, E., et al.: Image Processing Algorithms for Digital Mammography: A Pictorial Essay1. Radiographics 20(5), 1479 (2000)
6. Chen, B., Wang, W., Huang, J., Zhao, M., Cui, G., Xu, J., Guo, W., Du, P., Li, P., Yu, J.: Comparison of tissue equalization, and premium view post-processing methods in full field digital mammography. European Journal of Radiology (2009)
7. Snoeren, P., Karssemeijer, N.: Gray-scale and geometric registration of full-field digital and film-screen mammograms. Medical Image Analysis 11(2), 146–156 (2007)

Lesion Segmentation in Breast Sonography

Joan Massich[1,⋆], Fabrice Meriaudeau[2], Elsa Pérez[3], Robert Martí[1], Arnau Oliver[1], and Joan Martí[1]

[1] Computer Vision and Robotics Group, University of Girona, Spain
jmassich@atc.udg.edu
[2] University of Burgundy, Le2i Laboratory CNRS UMR 5158, Esplanade Erasme, 21000, Dijon, France
[3] Dept of Radiology, Hospital Josep Trueta of Girona, Spain

Abstract. Sonography is gaining popularity as an adjunct screening technique for assessing abnormalities in the breast. This is particularly true in cases where the subject has dense breast tissue, wherein widespread techniques like Digital Mammography (DM) fail to produce reliable outcomes. This article proposes a novel and fully automatic methodology for breast lesion segmentation in B-mode Ultra-Sound (US) images by utilizing region, boundary and shape information to cope up with the inherent artifacts present in US images. The proposed approach has been evaluated using a set of sonographic images with accompanying expert-provided ground truth.

Keywords: ultrasound, breast cancer, segmentation, segmentation evaluation.

1 Introduction

Breast cancer is one of the leading causes of death for women in developed countries and is most effectively treated when detected at an early stage [5].

Considering this, DM is, and remains the major screening tool for breast cancer [3]. However, in the recent past, studies [6,7] have shown that US images of the breast can help supplement mammography by detecting breast cancers that may not be visible in a traditional mammogram. This is particularly true in the cases where the subjects have dense glandular breast tissue, which tends to shield the presence of a tumor in a mammogram. In addition, US images are non-invasive with no side effects, rendering sonography as an attractive adjunct to digital mammography and heading a re-emergence of interest in understanding how to do image segmentation applied to ultrasound data [4].

This reserach proposes a novel and fully automatic technique to segment breast lesions for conventional B-mode US images, by utilizing region, boundary and shape information to cope with the inherent artifacts present in these images.

⋆ We gratefully acknowledge the help of Dr. Gururajan and Dr. Sari-Sarraf from Texas Tech University. This research was partially supported by the Spanish Government MEC grant nb. TIN2007-60553 and the University of Girona BR grant nb. 09/22.

J. Martí et al. (Eds.): IWDM 2010, LNCS 6136, pp. 39–45, 2010.

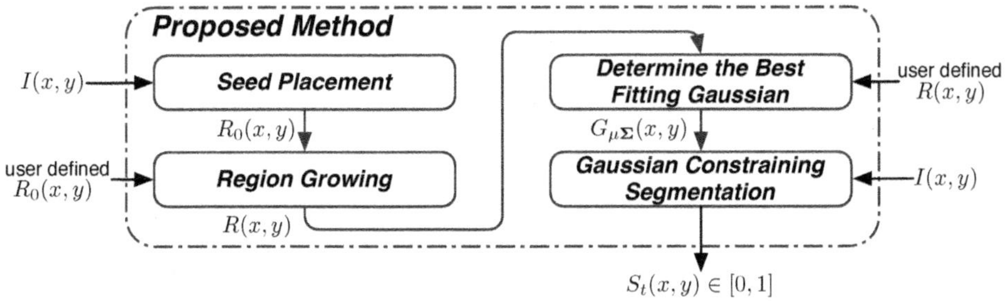

Fig. 1. Block diagram of the proposed methodology. When user interaction is used (only for semi-automatic segmentation), it overwrites the previous input.

The performance evaluation of the presented segmentation approach is conducted using a set of sonographic breast images that have accompanying ground truth provided by multiple experts. In this regard, this paper also presents a modified version of the Simultaneous Truth and Performance Level Estimation (STAPLE) [8] algorithm that is employed to extract the gold-standard for an image that has been delineated by multiple experts.

2 Proposed Approach

In this paper, a new segmentation procedure for breast lesions US images is proposed, allowing different levels of user-interaction, and ultimately leading automation of the entire approach. The block diagram of the proposed approach is shown in Fig. 1. The primary step involved in this process is the Gaussian Constraining Segmentation (GCS), which is a boundary-based technique whose key input is a multivariate Gaussian function that describes the shape, position and orientation of the lesion, derived from the one proposed by Horsch et al. [1] This Gaussian function is determined by fitting an ellipse to the $R(x, y)$ blob. This $R(x, y)$ preliminar segmentation of the lesion can be an interactive segmentation input or it can be determined by a fuzzy region growing from a seed region $R_0(x, y)$. Such region initialization $R_0(x, y)$, can also be a given input or it can be automatically determined by the seed placement procedure (a simplification to the seed placement proposed by Madabhushi and Metaxas [2]) leading to a fully automatic segmentation procedure.

The GCS step (see Fig. 2), outputs a binary mask in accordance with Eq. (1) and this mask corresponds to the final segmentation result. The GCS consists of two steps: (a) finding $\Psi(x, y)$ which is an intensity dependent function ($f(I(x, y))$) constrained by a Gaussian ($G_{\boldsymbol{\mu\Sigma}}(x, y)$) that best represents the lesion, and (b) selecting the proper threshold for $\Psi(x, y)$ to segment the lesion.

$$S_t(x, y) = threshold(\Psi(x, y), t) \quad \text{where} \quad \Psi(x, y) = f(I(x, y)) \cdot G_{\boldsymbol{\mu\Sigma}}(x, y) \quad (1)$$

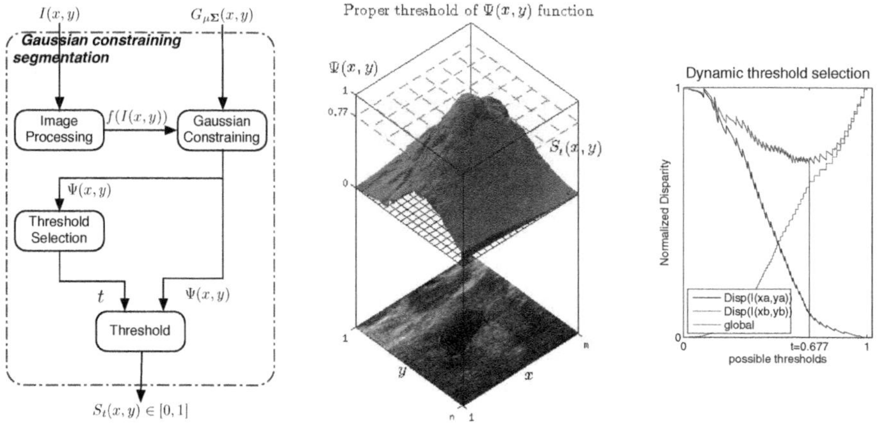

Fig. 2. Gaussian Constraining Segmentation (GCS) outline: (a) block diagram (b) graphical representation for the GCS. The figure shows a 3D representation of the lesion shaped potential funcion $\Psi(x,y)$, and its projection on the sonogram obtained by thresholding $\Psi(x,y)$. (c) threshold selection upon the minimization of the global disparity measure.

For a given dataset, threshold t can either be manually tuned up for the whole set, or dynamically determined for each image based on its characteristics. When a dynamic thresholding is used, the threshold t is selected from a set $\{t_1, .., t_n\}$ as the best candidate accordingly to equation 3.

$$t = \min_t (\ \text{Disparity}(I(x_u, y_u)) \ + \ \text{Disparity}(I(x_d, y_d)) \ ) \tag{2}$$
$$\textit{where} \quad (x_u, y_u) \in \{(X,Y) \mid \Psi(x,y) \geq t\}$$
$$(x_d, y_d) \in \{(X,Y) \mid \Psi(x,y) < t\}$$

Figure 2(c) illustrates how the disparity measure varies along the different thresholds. The optimal threshold for each image given the lesion shaped potential function $\Psi(x,y)$ corresponds to the minimum global disparity.

The main advantage of the proposed method over that presented by Horsch et al.[1] is: (a) fully automatic capability without any information of the lesion, (b) better lesion description and delineation. Lastly the advantage over the reserarch done by Madabhushi and Metaxas [2], is the incorporation of a faster lesion refinement step after the region growing.

3 Experiment Description

In order to evaluate the proposed approach, a set of 25 sonograms were acquired in the *Hospital Dr. Josep Trueta* of Girona. Each image has seven ground truth delineations provided by different radiology experts. In the first step, the STAPLE algorithm [8] is used to obtain the underlying ground truth from the multiple expert delineations. STAPLE states that the ground truth and performance levels of the experts can be estimated by formulating the scenario as a missing-data problem , which can be subsequently solved using an Expectation Maximization (EM) algorithm. The EM algorithm, after convergence, provides the Hidden Ground Truth (HGT) estimation that has been inferred from the segmentations provided by the experts.

The proposed segmentation approach is evaluated as follows. For a given image, the segmentations produced by the evaluated methods and STAPLE on the ith pixel can be denoted as S_i and G_i respectively. Within this framework, we propose a performance coefficient based on the True-Positive Ratio (TPR) or Jaccard coefficient weighting its numerator by the HGT. Such coefficient is calculated accordingly to Eq. (3), where p_i is the HGT probability of each pixel.

$$\text{coeff} = \frac{\sum (S_i \cap G_i) \cdot p_i}{\sum (S_i \cup G_i)} \tag{3}$$

For evaluation purposes, the results obtained by applying the proposed approach (on its three user interaction levels: no user input, given a seed region $R_0(x, y)$, and, given a blob $R(x, y)$) have been compared to the results obtained by applying other techniques: Horsch et al.[1] (on its partially automatic version which needs the lesion center and the four corners), Madabhushi and Metaxas [2]. The proposed approach has been tested both with dynamic and manually tuned-up threshold selection.

4 Results

The results and dicussion will focus on the performance of the proposed segmentation technique versus that presented in [1,2] over the entire dataset.

Figure 3 shows the segmentation results on four sonograms from the dataset. In the figure, the white dashed line indicates the delineation of the lesion after the segmentation, while the different $\Psi(x, y)$ isolines are represented by means of colored curves.

Previous to the segmentation procedure, a rank value filtering method is applied by means of a 3×3 median filter, in order to reduce the impact of speckle noise.

To quantitatively assess the segmentation approach, the metric proposed in equation 3 is used to compare the agreement between the gold-standard and the obtained segmentation. This is done for all the user interaction levels available in our proposal and for the whole dataset. The distribution of the dataset segmentation performance or agreement reward, is illustrated as a box-plot in figure 4. The first six entries represent our method results, differentiating manual tune up thresholding from dynamic selection thresholding for each level of user interaction level. The last two entries, plot the performance of the bibliographic methods [1,2].

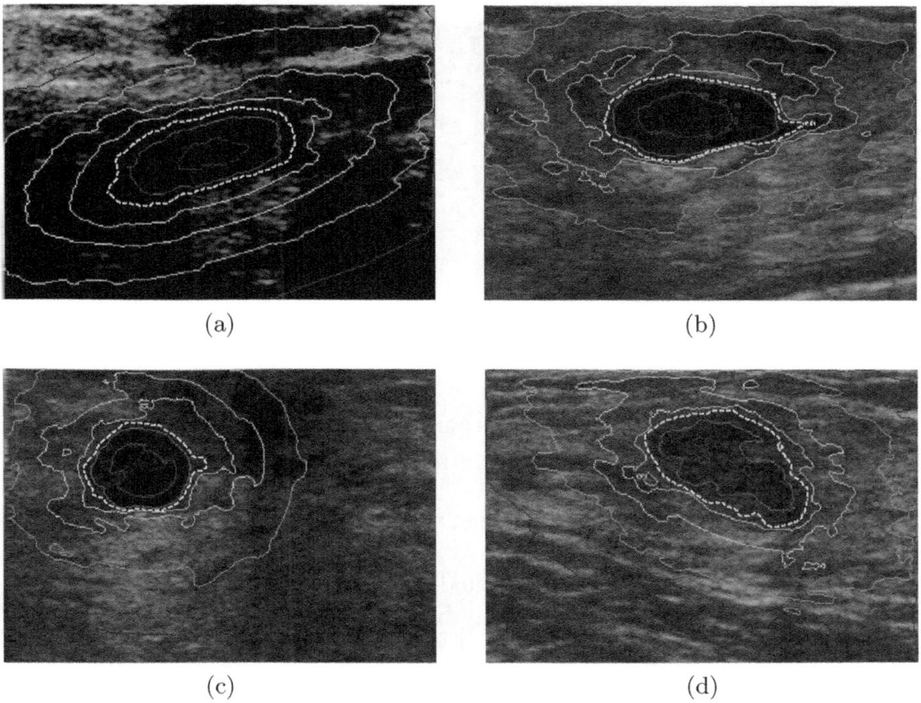

Fig. 3. Lesion segmentation results for 3 different images from the used dataset. Different level slices of the $\Psi(x,y)$ function are represented by different colour. The white dashed line indicates the lesion segmentation.

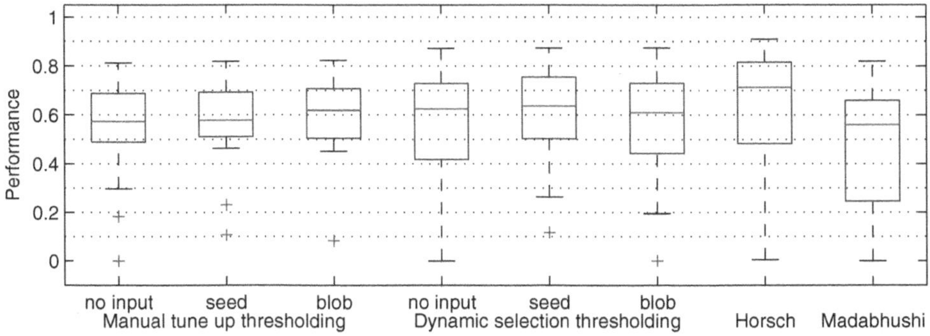

Fig. 4. Box-plot representation of the tested methods performance for the current dataset

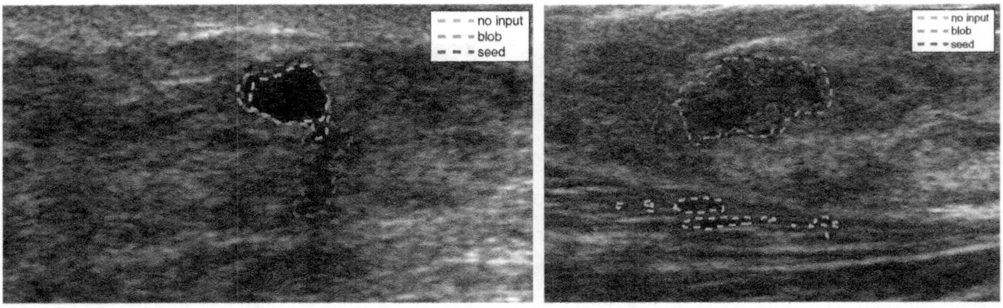

(a) (b)

Fig. 5. Segmentation difficulties: (a) shadowing problem, (b) non-lesion structures segmentation

The results show that the segmentations obtained by the proposed method fit the gold-standard up to 85% in some cases, while the average values are around 60%. Figure 4 shows that there is less variability when a manually tuned-up threshold is selected. However, the mean values for the segmentation results are higher when thresholding is dynamic.

While comparing the proposed method with the methods from the bibliography [1,2], our method performs slightly better than the one proposed by Madabhushi and Metaxas [2]. On the other hand, although the method proposed by Horsch et al.[1] shows better performance when considering the mean value, our method behaves better in terms of variability.

The quantitative comparison has been performed using the metric proposed in equation 3. Notice that the obtained results are lower than the expected ones when using a TPR metric, because the metric used penalizes or rewards depending on which part of the segmentation S overlaps the gold-standard G. The metric used only behaves as TPR when all the elements of the overlapping area have probability 1 of being lesion (see eq. 4).

$$\frac{\sum (S_i \cap G_i) \cdot p_i}{\sum (S_i \cup G_i)} = \text{TPR} \qquad \textit{if and only if} \qquad p_i = 1 \ \forall i \tag{4}$$

5 Discussion

The shadowing problem, as shadow being part of the lesion, is not particularly addressed or treated here. However, as long as the region growing output does not contain the shadow, the proposed method does not consider that shadow as a part of the lesion by means of the multivariate attenuation of the gaussian. Figure 5(a) illustrates how the shadow gets included as a part of the lesion for a seed guided segmentation.

This misdelineation is explained by the region growing initialization $R_0(x, y)$ (supplied by the user) since it has lead to grown up region $R(x, y)$ containing the shadow. Subsequently this region $R(x, y)$ has lead to a wrongly fitted gaussian ($G_{\boldsymbol{\mu}\boldsymbol{\Sigma}}(x, y)$) that misdescribes the lesion.

In the same way, the proposed method does not explicitly address the problem of segmenting other non-lesion structures present in the sonogram (like glandular or fatty tissue) as a lesion. Our does not address a situation where the seed region is misplaced on a structure that has some similarities to the lesions. For the present experiment, this only has happened in one image of the dataset. Figure 5(b) shows the sonogram where the automatic seed placement misplace the initial region for the region growing. Hence, the obtained segmentation by using the fully automatic version of the method results in a muscular region being extracted as a lesion. However, in most of the cases the seed region is correctly placed, so the method does not segment structures other than the lesions.

This research has presented and tested a breast lesion segmentation framework or methodology applied to ultrasound images. Although the results are satisfactory, further research of all the variable elements of the proposed method should be done to improve the results. For the testing, the same experiment can be done with a larger dataset to achieve more reliable conclusions.

References

1. Horsch, K., Giger, M.L., Venta, L.A., Vyborny, C.J.: Automatic segmentation of breast lesions on ultrasound. Medical Physics 28, 1652–1659 (2001)
2. Madabhushi, A., Metaxas, D.: Automatic boundary extraction of ultrasonic breast lesions. In: Proceedings of 2000 IEEE International Symposium on Biomedical Imaging, pp. 601–604 (2002)
3. Moore, S.K.: Better breast cancer detection. IEEE Spectrum 38(5), 50–54 (2001)
4. Noble, J.A., Boukerroui, D.: Ultrasound image segmentation: A survey. IEEE Transactions on medical imaging 25(8), 987–1010 (2006)
5. Sivaramakrishna, R., Powell, K.A., Lieber, M.L., Chilcote, W.A., Shekhar, R.: Texture analysis of lesions in breast ultrasound images. Computerized medical imaging and graphics 26(5), 303–307 (2002)
6. Stavros, A.T., Rapp, C.L., Parker, S.H.: Breast ultrasound. Lippincott Williams & Wilkins (2004)
7. Taylor, K.J.W., Merritt, C., Piccoli, C., Schmidt, R., Rouse, G., Fornage, B., Rubin, E., Georgian-Smith, D., Winsberg, F., Goldberg, B., et al.: Ultrasound as a complement to mammography and breast examination to characterize breast masses. Ultrasound in Medicine & Biology 28(1), 19–26 (2002)
8. Warfield, S.K., Zou, K.H., Wells, W.M.: Simultaneous truth and performance level estimation (STAPLE): an algorithm for the validation of image segmentation. IEEE Transactions on Medical Imaging 23(7), 903–921 (2004)

Mammogram Compression Using Super-Resolution

Jun Zheng[1], Olac Fuentes[1], Ming-Ying Leung[1], and Elais Jackson[2,⋆]

[1] The University of Texas at El Paso, El Paso, Texas, 79968
jzheng@miners.utep.edu, ofuentes@utep.edu, mleung@utep.edu
[2] Tennessee State University, Nashville, TN 37209

Abstract. As mammography moves towards completely digital and produces prohibitive amounts of data, compression plays an increasingly important role. Although current lossless compression methods provide very high-quality images, their compression ratios are very low. On the other hand, several lossy compression methods provide very high compression ratios but come with considerable loss of quality. In this work, we describe a novel compression method that consists of downsampling the mammograms before applying the encoding procedure, and applying super-resolution techniques after the decoding procedure to recover the original resolution image. In our experiments, we examine the trade-offs between compression ratio and image quality using this scheme, and show it provides significant improvements over conventional methods.

1 Background

As mammography moves towards completely digital, technological advances in data storage and transmission have not kept up with the tremendous growth of digital data. This creates serious challenges for long-term storage and efficient transmission of mammograms. For example, a typical mammogram can be 4500×4500 pixels. If stored in uncompressed 16-bit per pixel (bpp) format, it would take about 40 MB for storage and approximately half an hour for transmission using a high-speed modem [12]. Thus compression will play an increasingly important role in Picture Archiving and Communication Systems (PACS) to reduce file sizes while maintaining relevant diagnostic information.

In recent years, there has been discussion about which type of compression techniques, lossy or lossless, is better for mammogram compression. Although current lossless compression methods provide very high quality images, the compression ratios are very low, typically from 1.5:1 to 3:1. On the other hand, several lossy compression methods provide acceptable compression ratios but come with considerable loss of image quality and diagnostic information [3,4,5,6,7,8,12].

In this work we describe a novel lossy compression method that consists of downsampling the mammograms before applying the encoding procedure, and

⋆ This work was supported in part by NSF grants DMS0800272 and IIS0852066, and NIH grants 2506GM008012-39 and 2G12RR008124-16A1.

J. Martí et al. (Eds.): IWDM 2010, LNCS 6136, pp. 46–53, 2010.

applying super-resolution techniques after the decoding procedure to recover the original resolution image.

2 Methods

Incorporating super-resolution into compression aims to maximize compression ratio while maintaining relatively high-quality images. The flowchart of the proposed method is shown in figure 1. The method first downsamples a high-resolution image $HR(x)$ to a low-resolution one $LR(x)$, then it uses an algorithm similar to JPEG to encode it. The decompression process first decodes the stored low resolution image to obtain $LR'(x)$ and then applies a super-resolution algorithm to produce $HR'(x)$.

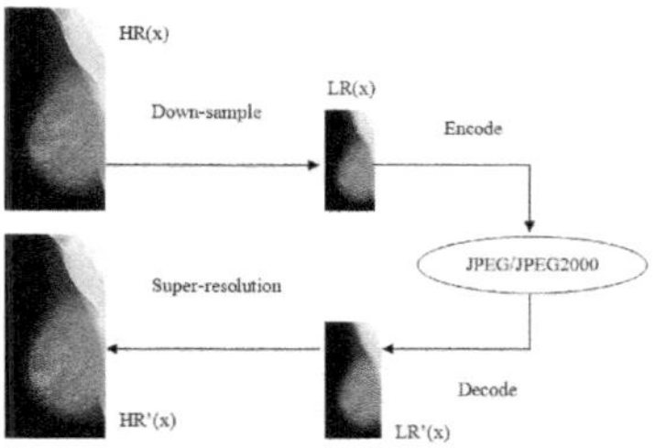

Fig. 1. Flowchart of the method

In the pre-processing stage, we downsample the mammograms using bilinear downsampling. In the post-processing stage, the super-resolution algorithm generates high-resolution mammograms from low-resolution decoded mammograms with no manual registration. The super-resolution algorithm consists of four main steps. The first step consists of automatically aligning the breasts to a standardized position. The second step uses a process called eigentransformation to infer a global model representing the low-frequency information in the image. In eigentransformation, Principal Component Analysis (PCA) is used to fit the input images as a linear combination of the low resolution images in the training set. The HR images are then inferred by replacing the LR training images with HR ones, while retaining the same combination coefficients. In the third step, a patch-based one-pass algorithm generates the high-frequency contents of the HR images. The fourth step remaps the breasts back to their original position.

2.1 Automatic Alignment

Image alignment is the key to the success of our automatic mammogram super-resolution algorithm. In practice, we cannot assume that any low-resolution mammogram has been accurately aligned, even though the approximate positions of the breasts in mammograms are given by mammography sensors.

Therefore, in preprocessing, we automatically align the mammograms to make sure each breast is exactly in the same position, and then perform the super-resolution. The automatic alignment process consists of two parts, segmentation-based initialization and 2-pass mesh warping [11].

The 2-pass mesh warping algorithm accepts a source image and two 2-D arrays of coordinates. The first array, S, specifies the coordinates of control points in the source image. The second array, D, specifies their corresponding positions in the destination image. The first pass is responsible for resampling each row independently. It maps all (u, v) points to their (x, v) coordinates in the intermediate image I. For each pixel P in intermediate image I, the value of P is evaluated as a weighted sum from the left most boundaries of P in S, x_0, and the rightmost boundaries of P in S, x_1.

$$P = \frac{\sum_{x=x_0}^{x_1} k_x S_x}{x_1 - x_0}$$

where k_x is the scale factor of source pixel S_x, and the subscript x denotes the index that lies between $floor(x_0)$ and $ceil(x_1)$. The scale factor k_x is defined as

$$k_x = \begin{cases} ceil(x) - x_0 & \text{if } floor(x) < x_0 \\ 1 & \text{if } x_0 \leq x < x_1 \\ x_1 - floor(x) & \text{if } ceil(x) > x_1 \end{cases}$$

The second pass then resamples each column in I, mapping every (x, v) point to its final (x, y) position. This process is virtually identical to the first pass; we just need to substitute (x, v) for (u, v), and substitute (x, y) for (x, v) [11].

The key to apply 2-pass mesh warping is to build the 2-D arrays of coordinates. The segmentation-based initialization builds the 2-D arrays of coordinates automatically. The initialization for 2-pass mesh warping consists of 5 steps (Figure 2):

1. Convert the input image to binary image.
2. Use erosion to remove the labels on the binary image.
3. Convert the binary image to a gradient image to find the edge.
4. Use skeletonization to reduce the edge to a single line.
5. Sample points on the line and make the 2D- array of coordinates for 2-pass mesh warping automatically.

2.2 Global Modeling

In global modeling we use an algorithm called eigentransformation, which was originally introduced by Wang [9]. The eigentransformation is a simple and powerful technique for image enhancement based on principal component analysis (PCA). It assumes that we have a training set of pairs of images $\langle(L_1, H_1), \ldots, (L_n, H_n)\rangle$, where each pair (L_i, H_i) contains a low resolution image L_i and its corresponding high-resolution counterpart H_i. The eigentransformation allows

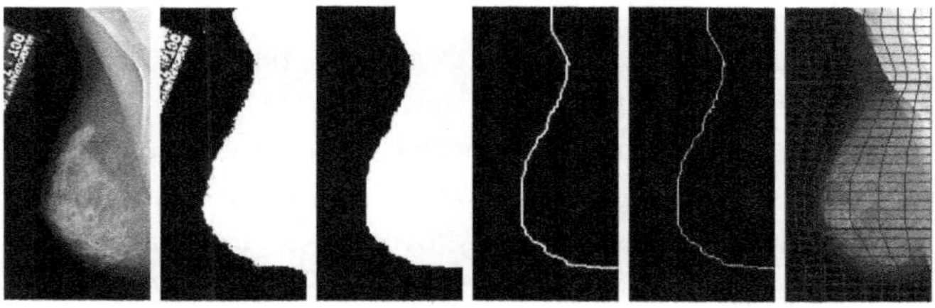

Fig. 2. Automatic registration: (a) Original image. (b) Converting the original image to a binary image. (c) Using erosion to remove the labels in the binary image. (d) Converting the binary image to gradient image to find the edge. (e) Using skeletonization to reduce the edge to a single line. (f) Sampling points on the line and making the mesh.

any image to be represented as a linear combination of images in the training set. When given a low resolution image L, it finds the vector of coefficients $[c_1, \ldots, c_n]$ so that

$$L = \Sigma_{i=1}^{n} c_i L_i + \mu_L$$

where μ_L is the mean low-resolution images.

Given the vector $[c_1, \ldots, c_n]$, the approximate high resolution image H can be computed by

$$H = \Sigma_{i=1}^{n} c_i H_i + \mu_H$$

where μ_H is the mean high-resolution image.

Because the coefficients are not computed from HR training data, some noise-like distortion may be introduced. To reduce distortion, we apply constraints by bounding the projection onto each eigenvector by its corresponding eigenvalue, then the synthesized image is reconstructed from these constrained coefficients.

2.3 Local Modeling

Given a global model, to construct the corresponding local model, we first filter the global model with a Gaussian high-pass filter, and then subdivide the filtered global model into patches, which we call the low-frequency patches of the high-resolution (HR) images, by scanning a window across the image in raster-scan order. Similarly, we also filter and subdivide the HR images in the training set into patches which we call high-frequency patches of the training HR images.

To construct a local model, for each low-frequency patch, a high-frequency patch of the training HR image is selected by a nearest neighbor search from the training set based on local low-frequency details and adjacent HR patches previously determined. The selected high-frequency patch should not only come from a location in the training images that has a similar corresponding low-frequency appearance, but also agrees with the overlapping pixels, which we call high-frequency overlap, at the edges of its previously determined high-frequency neighbors. This ensures that the high-frequency patches are compatible with those of the neighboring high-frequency patches.

In this work we compute the local model with an algorithm similar to Freeman's [1][2] one-pass algorithm. We first concatenate the pixels in the low-frequency patch and the high-frequency overlap to form a search vector. The training set also contains a set of such vectors. Then we search for a match by finding the nearest neighbor in the training set. When we find a match we extract the corresponding high-frequency patch from the training data set and add it to the initial global model to obtain the output image.

Mathematically, this process can be described as follows. Suppose we have a training data set

$$\{(x^{(i,j,k)}, y^{(i,j,k)}, z^{(i,j,k)}),$$

$$i = 1, 2, \ldots, l; j = 1, 2, \ldots, m; k = 1, 2, \ldots, n\}$$

where $x^{(i,j,k)}$ is the low-frequency patch at the ith row and j^{th} column of the kth training HR image, $y^{(i,j,k)}$ is the corresponding high-frequency overlap and $z^{(i,j,k)}$ is the corresponding high-frequency patch of the training HR image, l is the number of rows of patches in a training image, m is the number of columns of patches in a training image and n is the number of training images.

Given an input LR patch $\overline{x}$, we need to find an HR patch $z^{(i',j',k')}$ such that

$$z^{(i',j',k')} = \min_{i,j,k}(d(\overline{x}, x^{(i,j,k)}) + \alpha * (d(y^{(i,j,k)}, y_N^{(i,j,k)})))$$

where $d(x, y)$ is the Euclidean distance between x and y, $y_N^{(i,j,k)}$ is the overlap of $z^{(i,j,k)}$ with the adjacent, previously determined high-frequency patches, which are the patches above and to the left of the current high-frequency patch in the local model, α is a user-controlled weighting factor, and $z^{(i',j',k')}$ is the selected high-frequency patch.

3 Experimental Results

We use DDSM (Digital Database for Screening Mammography) for our experiments. DDSM is a standard dataset used by mammography image analysis research community. The database has about 2,500 cases. Each case includes two images of each breast, along with some associated patient and image information. In this work, we use 100 normal left Mediolateral Oblique (MLO) images from DDSM for training and 10 normal left MLO images images for testing.

In this paper, the quality of a super-resolution image is defined as the similarity of the decompressed image with the original high-resolution image. We use Peak Signal-to-Noise Ratio ($PSNR$) and Mean Structural Similarity ($MSSIM$) index to measure the quality of results.

Let X and Y be two images to be compared. The $PSNR$, which is most commonly used as a measure of quality of reconstruction, is defined as

$$PSNR(X,Y) = 20 \times log_{10} \frac{255}{RMSE(X,Y)}$$

where $RMSE$ is the root mean square error between the two images.

The structural similarity (SSIM) index [10] is an implementation of the idea of structural similarity, from an image formation point of view, which takes into account contrast, luminance, and structure to determine similarity between two images. $SSIM$ is defined as

$$SSIM(x,y) = \frac{(2\mu_x\mu_y + C1)(2\sigma_{xy} + C2)}{(\mu_x^2 + \mu_y^2 + C1)(\sigma_x^2 + \sigma_y^2 + C2)},$$

$$\sigma_{xy} = \frac{1}{T-1}\sum_{i=1}^{T}(x_i - \mu_x)(y_i - \mu_y)$$

where x and y are subimages of X and Y, T is the total number of pixels in each subimage, μ_x is the average of x, μ_y is the average of y, σ_x is the standard deviation of x, σ_y is the standard deviation of y. $C1 = (k_1L)^2$ and $C2 = (k_2L)^2$ are two variables to stabilize the division with small denominators, L is the dynamic range of the pixel values (typically this is 255), $k_1 = 0.01$ and $k_2 = 0.03$ by default. The mean SSIM (MSSIM) is then simply the mean of the SSIMs for all subimages. A value of MSSIM of 1 indicates perfect similarity [10].

Figures 3 and 4 show the image quality measures for our algorithm. The downsample factors in this experiment are 4, 8, 16, and the number of Discrete Cosine Transform (DCT) coefficients used in an 8×8 block for JPEG encoding are 1, 2, 3, 5, 10, and 15. Table 1 compares the results of our method (downsample factor = 16, 15 DCT coefficients) with the results of JPEG2000 and lossless JPEG. Sample results of different resolutions are reported in Figure 5.

Table 1. Average compression results for 10 mammograms

Compression methods	PSNR	Compression ratio
JPEG2000	41.95	80:1
Lossless JPEG	lossless	3:1
Our method	35.9023	20480:1

4 Discussion

From figures 3 and 4, we can see that when we just use 1 or 2 DCT coefficients, which results in very blocky images, we get a PSNR of 33.458 after super-resolution of this kind of blocky images. These results indicate that super-resolution can also attenuate the JPEG compression artifacts when the

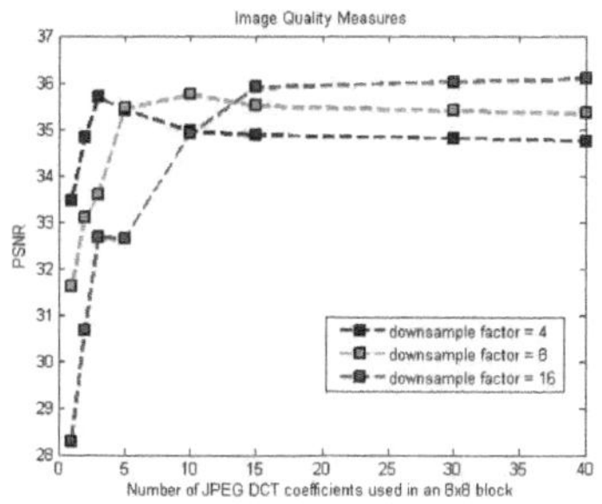

Fig. 3. PSNR of the decompressed images after super-resolution

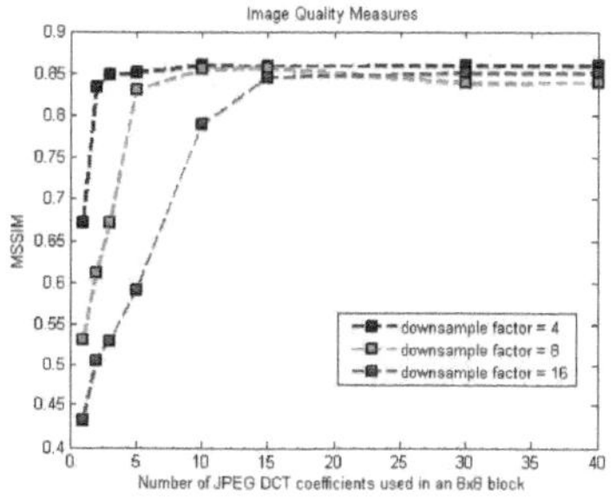

Fig. 4. MSSIM of the decompressed images after super-resolution

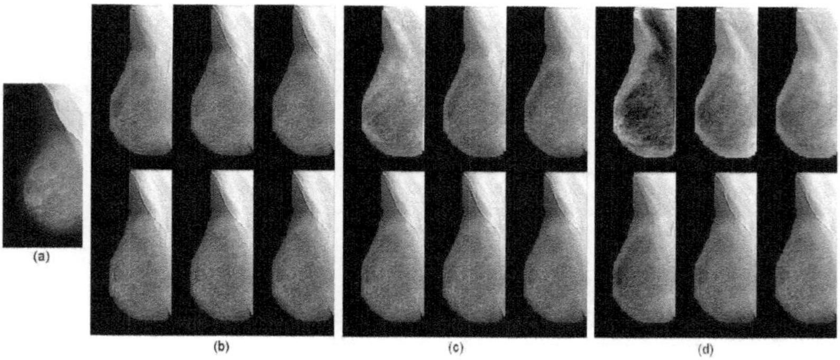

Fig. 5. (a) Original image. (b) The decompressed images after super-resolution (Coeffs=1,2,3,5,10,15; downsample factor=4).(c) The decompressed images after super-resolution (Coeffs=1,2,3,5,10,15; downsample factor=8).(d) The decompressed images after super-resolution (Coeffs=1,2,3,5,10,15; downsample factor=16).

compression ratio is high. From Table 1, we can see that compared with the results of JPEG2000 and lossless JPEG, the compression ratio of our method is thousands of times higher than lossless JPEG, and hundreds of times higher than JPEG2000 with a slightly lower PSNR. However, some details of the decompressed images are different from the original images. Current work by our group consists of determining whether super-resolution affects clinical diagnostic performance, by both humans and computers.

For future work, we will study a hybrid scheme that is appropriate for accurate compression of mammograms. We will use lossless compression in the region of interest, and lossy compression with super-resolution in the other regions. This strategy has the potential to achieve high compression ratios without reducing the diagnostic quality of the mammograms.

References

1. Freeman, W.T., Jones, T.R., Pasztor, E.C.: Example-based super-resolution. IEEE Computer Graphics and Applications 22(2), 56–65 (2002)
2. Freeman, W.T., Pasztor, E.C., Carmichael, O.T.: Learning low-level vision. International Journal on Compter Vision 40(1), 25–47 (2000)
3. Grinstead, B., Sari-Sarraf, H., Mitra, S., Gleason, S.: Content-based compression of mammograms for telecommunication and archiving. In: 13th IEEE Symposium on Computer-Based Medical Systems (CBMS 2000), p. 37 (2000)
4. Li, X., Shen, Y., Ma, J.: An efficient medical image compression scheme. In: Proceedings of the 2005 IEEE Engineering in Medicine and Biology 27th Annual Conference, Shanghai, China, pp. 3437–3439 (January 2006)
5. Ravi Kumar, M.S., Koliwad, S., Dwarakish, G.S.: Lossless compression of digital mammography using fixed block segmentation and pixel grouping. In: Proceedings of indian conference on computer vision, graphics and image processing, pp. 201–206 (2008)
6. Perlmutter, S.M., Cosman, P.C., Gray, R.M., Olshen, R.A., Ikeda, D., Adams, C.N., Betts, B.J., Williams, M.B., Perlmutter, K.O., Aiyer, J.L., Fajardo, L., Birdwell, R., Daniel, B.L.: Image quality in lossy compressed digital mammograms. Signal processing 59(2), 189–210 (1997)
7. Persons, K., Palisson, P., Manduca, A., Erickson, B.J., Savcenko, V.: An analytical look at the effects of compression on medical images. Journal of Digital Imaging 10, 60–66 (1997)
8. Scharcanski, J.: Lossless and near-lossless compression for mammographic digital images. IEEE International Conference on Image Processing, 2253–2256 (2006)
9. Wang, X., Tang, X.: Hallucinating face by eigentransformation. IEEE Transactions on Systems, Man and Cybernetics, Part C: Applications and Reviews (2005)
10. Wang, Z., Bovik, A.C., Sheikh, H.R., Simoncelli, E.P.: Image quality assessment: From error visibility to structural similarity. IEEE Transactions on Image Processing 13(4), 600–612 (2004)
11 Wolberg, G.: 2-Pass Mesh Warping. In: Digital Image Warping, pp. 222–240. IEEE Computer Society Press, Los Alamitos (1990)
12. Zukoski, M.J., Boult, T., Iyriboz, T.: A novel approach to medical image compression. International Journal of Bioinformatics Research and Applications (2006)

Development of an Ensemble of Digital Breast Object Models

J. Michael O'Connor[1,2], Mini Das[2], Clay Didier[1], Mufeed Mah'd[1], and Stephen J. Glick[2]

[1] Biomedical Engineering & Biotechnology Program, University of Massachusetts Lowell, One University Avenue, Lowell, Massachusetts, 01854, USA
[2] Department of Radiology, University of Massachusetts Medical School, 55 Lake Avenue North, Worcester, Massachusetts, 01655, USA
michael.oconnor@umassmed.edu

Abstract. In the investigation of emerging tomographic breast imaging methods using flat-panel detectors (FPDs), digital breast object models are useful tools. These models are also commonly referred to as digital phantoms. We have created an ensemble of digital breast object phantoms based on CT scans of surgical mastectomy specimens. Early versions of the phantoms have been used in our published research. Recently we have improved some of our processing tools such as the use of 3-D anisotropic diffusion filtering (ADF) prior to segmentation, and we have evaluated breast object models generated with different methods including power spectral analysis, ROI statistics and an observation study.

Keywords: Flat-panel detector (FPD), anisotropic diffusion filter (ADF), Digital Breast Object Model.

1 Introduction

A recent report on breast screening from the U.S. Preventative Service Task Force gained wide attention [1]. The recommendation against universal mammographic screening for women aged 40-49 without known risk was based on evidence that this screening method for this age range has a high rate of false-positives [2]. This recommendation emphasized that despite the past success of mammographic screening, improved breast imaging techniques would be beneficial. With this goal in mind, a number of institutions are actively investigating breast tomosynthesis (BT) and dedicated breast computed tomography (BCT) [3-8].

Although these new breast imaging modalities are promising, further studies need to be performed to assure that they are operating with optimal performance. In pursuit of this goal, one topic of interest is the development of breast object models. As suggested by Barrett and Myers [9], any meaningful approach to optimizing an imaging system must include a definition of; 1) the specific task to be performed, 2) the observer, 3) an object model representing the objects to be imaged, and 4) a measure of performance used to evaluate task performance. This paper focuses on the third item,

J. Martí et al. (Eds.): IWDM 2010, LNCS 6136, pp. 54–61, 2010.

or more specifically the development of a methodology for generating a breast object model. A number of 3D breast object phantoms have been previously proposed [10-15]. All of these proposed breast object models have useful properties; yet, their realism compared to actual patient tomographic breast reconstructions is somewhat lacking.

In order to produce realistic 3D breast objects, we began work to use high resolution, low noise CT images of surgical mastectomy specimens acquired on a bench-top CT breast imaging prototype. The first report of this work was made in early 2008 [16]. In that same conference session another research group reported work with a similar goal to produce a 3-D breast phantom based on breast CT data acquired from patients during a clinical trial [17]. These same researchers in a subsequent conference announced a future goal to incorporate geometric shapes into their imaged-based model [18]. Their method is more fully described in a recent publication [19]. We have also further described our work [20]. Subsequently, we have resolved methodology issues and have continued to produce an ensemble of digital breast object phantoms.

2 Methods

Under an approved institutional review board (IRB) protocol [21] and with informed patient consent, fresh mastectomy specimens were obtained immediately following surgery and prior to tissue gross pathology. Using a previously described prototype system [22], each specimen was imaged by placing it in an appropriate sized holder suitable for either of two basic positions; 1) uncompressed, modeling the pendant position proposed for dedicated CT breast imaging (CTBI) systems or 2) upright compressed, modeling the breast position used for tomosynthesis and/or mammography. While various techniques have been used, typical acquisition parameters were 300 projections over a single 360° trajectory with five projections taken at each angular step with an x-ray technique of 40kVp, .5mAs. The projection set was software flat field corrected and then averaged at each angle with pixels binned two-by-two. The projection set was then reconstructed using previously described simulation software [23] which implements the filtered back-projection (FPB) algorithm by Feldkamp, Davis and Kress (FDK) [24]. Post-reconstruction processing steps typically performed were: 1) remove the specimen holder from reconstructed volume; 2) apply corrections to reconstruction voxel values (values are in units of un-calibrated linear attenuation coefficients (μ)); 3) apply a 3D anisotropic diffusion filter to the reconstructed volume and 4) segment the filtered reconstruction volume into a breast object. Removal of the specimen holder was threshold-based and, as needed, utilized a prior holder-only reconstruction. Correction for intra-slice voxel variation was followed by inter-slice correction using the method of Altunbas et al. [25] (Note: variations are caused by scatter, beam-hardening, and insufficient angular sampling with circular orbit cone-beam reconstruction). The classic Perona & Malik anisotropic diffusion filter (ADF) [26] was implemented for a 3-D volume using the National Library of Medicine's Insight Segmentation and Registration Toolkit (ITK) [27]. Using prior knowledge of the tissue histogram for each slice, the filtered reconstructed volume was segmented into a tissue density phantom. In some instances, after segmentation, the breast object phantom was multiplied by a shape template in order to produce a more realistic shape.

Two phantom model types were considered. The first model generates phantoms with only two discrete voxel values, one for adipose tissue density and the second for glandular tissue density. The second model allows for a range of tissue densities between peak adipose and peak glandular densities. The first phantom type will be called the binary model and the second phantom model type will be referred to as the multi-value model. While both types of models appear to give good results, the multi-value phantom has been selected as the standard. This selection was made after evaluation of three factors: 1) comparison of Regions of Interest (ROI) statistics, 2) evaluating the slope of a power law spectrum, similar to that described by Metheany et al. [28], and 3) an observation similarity study. The ROI methodology was described in the previous IWDM conference [20]. Using a previously described computer simulation of breast CT [23], 12 reconstructed volumes were generated from both phantom models. ROI's of 25 x 25 pixels were randomly positioned within approximately uniform adipose or fibroglandular regions of simulated reconstructions from both phantom models, as well as within the original specimen reconstruction. Measurements of ROI contrast and fractional standard deviation in both homogeneous adipose ROIs and homogeneous glandular ROIs were computed. The second evaluation was based on analysis of CT slice frequency spectrum as recently described in Metheany's research [28]. Her technique is an extension of classical work by Burgess et al. [29] which showed that mammogram projections follow a power-law spectrum over a certain scale and fractal dimension. The power law describes isotropic, Gaussian noise with a power spectrum for a frequency (f) spectrum defined by the equation $P(f)=\frac{\alpha}{f^{\beta}}$ where α is signal magnitude and β is the power-law exponent. Metheany et al. developed the framework for applying this same power-law analysis to CT breast slices. Power spectrum analysis was used to evaluate and compare the two phantom models.

The final evaluation was a subjective four observer preference study to evaluate similarity between simulated reconstructed slices and the original specimen CT slice. The simulation parameters were chosen to be as close as possible to the measurement technique used in the original specimen acquisition. Observers were given three guideline criteria. Those criteria were 1) edge evaluation, 2) tissue texture and 3) preservation of fine structure in the simulated reconstruction.

3 Results

The ultimate goal of this work is to produce useful phantoms for observer studies that use breast simulations therefore evaluation of the two model types was made with respect to original reconstructed specimens. As a general question we wanted to know how similar is a simulation to the original specimen upon which the phantom was based. The specimen was corrected for intra-slice and inter-slice variation. The simulations were made with parameters similar to the original specimen acquisition. No additional filtering was performed on either specimen or simulations. Several objective metrics were determined and one subjective study was performed.

The first objective examination was fractal or β evaluation. This was an assessment of the image power spectrum of the specimen in comparison with the image power spectrum of simulated reconstructions based on the two phantom types (binary and multi-value). The aim was to see if one could predict better similarity by an observer by using fractal analysis as a figure of merit. In this work β is simply a proxy for fractal dimension and is used as that figure of merit, where β is the absolute value of the slope of the power law spectra or simply the power-law exponent. This value is determined by the fit to the log-log plot of power over the frequency range .1 to .45 cycles/mm. β values for twenty uncompressed specimens were compared with both the corresponding reconstructions based on the binary phantom and the reconstructions based on the multi-value phantom. For those twenty cases, β values for the specimen reconstruction were close to β values of the reconstructions based on multi-value phantoms as well as to the β values for the simulations based on the binary phantoms. The average β for all specimen reconstructions was 2.32 (standard deviation .42). The average absolute difference between the reconstructed specimens and the reconstructions based on the multi-value phantoms was 0.42 while the average absolute difference between the specimen reconstructions and the simulations based on the binary phantoms was .25.

The second objective examination was an evaluation of ROIs. In this analysis a comparison was made between homogeneous regions in the original specimen with corresponding homogeneous regions in reconstructions based on binary models and regions in reconstructions based on multi-value models. To determine homogeneous ROIs twevle multi-value phantoms were automatically scanned for 25^2 pixel regions that were either glandular homogeneous or adipose homogeneous. The ROI area was $5mm^2$. In all twevle phantoms over 30,000 ROIs were determined (11,910 glandular ROIs and 21,218 adipose ROIs). In the selection, care was taken not to select contiguous regions (i.e. there was a minimum gap of 2mm between ROIs). While the selection is based on the multi-value phantom, when comparing ROIs in the simulated reconstructions, the corresponding ROI in the specimen was checked to ensure that the region did not have values that would indicate that the ROI contains non-tissue values (i.e. air or metal or micro-calcifications). As stated, one cannot see a significant difference between similar homogeneous ROIs in simulated reconstructions based on either phantom type. For comparison the figures of merit were Contrast and Fractional Standard Deviation. Contrast here is defined as (Glandular Mean – Adipose Mean)/Adipose Mean. Fractional Standard Deviation is the Standard Deviation/Mean. The mean contrast for all ROIs in all twelve specimens is .409 with standard deviation of 0.061. The mean contrast for simulations based on the multi-value phantoms was 0.347 with standard deviation of 0.025. The mean contrast for simulations based on the binary phantoms was 0.351 with standard deviation of 0.024. The contrast of the specimen does not differ significantly from simulations based on either of the two phantom types and there is little difference between the contrast in the two simulations. For the Fractional Standard Deviations the average for all regions in all twelve specimens and reconstructions is provided in Table 1. There are small differences between the specimen regions and either of the two simulation cases. Fractional Standard Deviations for both simulation types are close.

Table 1. Fractional Standard Deviation: MEAN is the average fractional standard deviation for all ROIs in all twelve cases. STDev is the average standard deviation of the fractional standard deviation for all ROIs in all cases.

Tissue Type	SPEC MEAN	SPEC STDev	MULTI. MEAN	MULTI. STDev	BIN. MEAN	BIN STDev
ADIPOSE	0.063	0.034	0.059	0.025	0.060	0.026
GLANDULAR	0.052	0.022	0.045	0.021	0.045	0.021

A subjective image preference study was also performed to compare different phantom models. In this study twenty slices from each of twenty specimens were selected for a total of four hundred CT slices. The corresponding slices from both the simulated reconstruction based on the binary phantom and the simulated reconstruction based on multi-value phantom were placed in a matching set. Four observers were randomly presented sets of simulations and asked to indicate which one of the set was most similar to the specimen reconstruction based on similarity of edges and apparent tissue texture as well as which one best preserved the fine structures from the original specimen CT slice. In one hundred responses the observers selected the simulation based on the multi-value model in eighty-five cases. In thirteen cases observers found no difference. In only two cases was the reconstruction based on the binary model selected. In consideration of the selection criteria, in fifty of the eighty-five cases in which the multi-value model was selected, the observer indicated that the multi-value was superior in all criteria. In none of the cases was simulation based on the binary phantom deemed to be best for all three criteria. In eight of one hundred instances the binary model was marked with better texture. In seven cases the binary model was judged as having better edges. In only one case of the hundred was the simulated reconstruction based on the binary model thought to be superior for the preservation of fine structure. This study does decidedly favor the multi-value phantom over the binary phantom. It also suggests the major shortcoming of the binary phantom, which is that it does not preserve fine structure as well as the multi-value phantom.

Since 2006 over 70 patients have consented to having their mastectomies imaged on our prototype CT breast imaging system. While the majority of studies were multiple images made using one mastectomy specimen, some were bi-lateral specimens from one patient. Most mastectomies were suspected of having cancer; some were removed for prophylactic treatment and thought to be disease free. The current ensemble of breast phantoms includes twenty uncompressed phantoms and numerous compressed phantoms (more phantoms will have been developed by the time of the conference). Within the ensemble are a variety of sizes and tissue densities. Figure 1 shows sagittal and coronal CT slices from; A&D) a reconstruction of the mastectomy specimen, B&E) the resultant object model generated using this specimen, and C&F) simulated reconstructed images made by inputting the breast object model to the computer simulation software to make simulated projections, and then subsequently to reconstruct these projections using Feldkamp filtered backprojection (FBP). Figure 2 shows sagittal, coronal and transverse slices of a compressed breast phantom, as well as a simulated mammogram using this breast phantom.

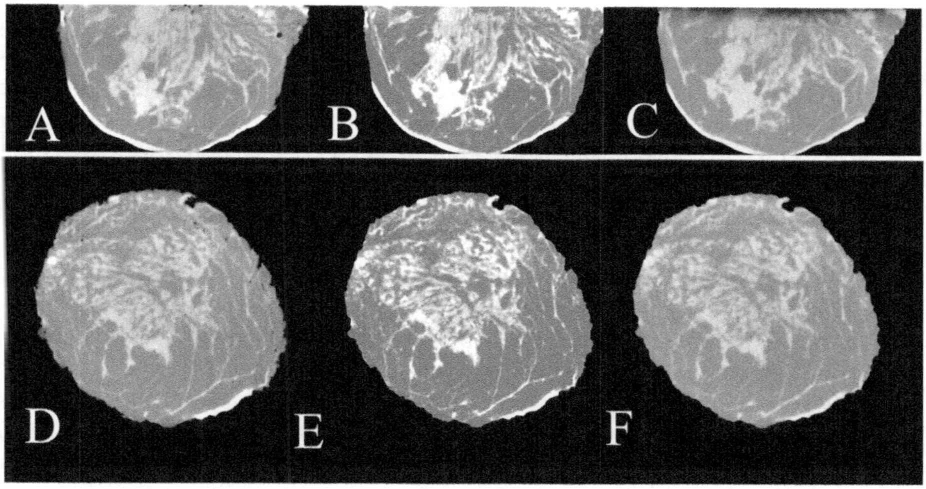

Fig. 1. Specimen 47 A (sagittal) & D (coronal) slices from a mastectomy specimen reconstruction (basis for digital phantom). B & E corresponding slices from resultant digital phantom (multi-value model). C & F slices from a computer simulation made using this digital phantom (i.e., generating simulated projections followed by FBP reconstruction). The simulation technique used was: 40kVp, 4mGy mean glandular dose for 300 simulated projections with simulated detector characteristics similar to prototype detector. Note the similarity between simulated reconstructions (C&F) and the original specimen reconstruction (A&D). A post reconstruction ADF was applied to the simulation. [Note: display window not uniform].

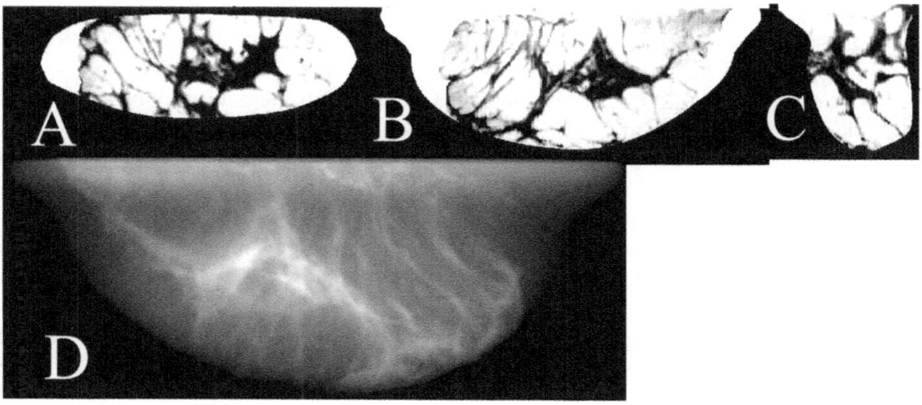

Fig. 2. Specimen 52 A,B,C are coronal, axial and sagittal slices respectively of a compressed phantom. The phantom here is displayed as a 'negative' (adipose tissue values light and glandular values dark). D is a simulated mammogram using this compressed phantom. The simulation technique used here was Mo/Mo 30kVp spectra at 1.5mGy mean glandular dose with simulated detector characteristics of 100 micron^2 pixel and CsI thickness of 100 microns.

4 Discussion

There is similarity between this approach to generate digital breast object models based on images of mastectomy specimens and the previously mentioned approach by another research team that uses clinical breast CT images to generate models [19]. There are advantages and disadvantages of using mastectomy specimens to develop breast object models versus using actual clinical breast CT images. Using mastectomy specimens might have the benefit of improved spatial resolution, lower noise and improved image quality because the x-ray technique is not constrained by patient safety dose limitations. Of course using clinical breast CT images to generate breast object models has the advantage that the breast is intact; therefore the chest wall region should have better anatomical fidelity.

We have created an ensemble of digital breast object models that can be useful for evaluating and optimizing tomographic breast imaging systems. We believe that both phantom types, binary and multi-value, give good results. This is confirmed in that objective analysis shows similar results in comparison of simulations based on either phantom type with specimen. However, the multi-value phantom appears to generate more realistic simulated projections with better similarity to the original specimen based on an observer study. We believe this observer evaluation is the reason for favoring the multi-value as an observer model in future studies.

References

1. Screening for Breast Cancer: U.S. Preventative Services Task Force Recommendation Statement. Ann. Intern. Med. 151(10), 716–726 (2009)
2. Nelson, H.D., et al.: Screening for Breast Cancer: An Update for the U.S. Preventative Services Task Force. Ann. Intern. Med. 151(10), 727–737 (2009)
3. Boone, J., et al.: Dedicated Breast CT: Radiation Dose and Image Quality Evaluation. Radiology 221(3), 657–677 (2001)
4. Chen, B., Ning, R.: Cone-beam volume CT breast imaging: Feasibility study. Med. Phys. 29(5), 755–770 (2002)
5. Chen, L., et al.: Cone-beam CT breast imaging with a flat-panel detector: a simulation study. In: Proceedings of SPIE Medical Imaging 2005: Physics of Medical Imaging, vol. 5745 (2005)
6. McKinley, R., et al.: Investigation of cone-beam acquisitions implemented using a novel dedicated mammotomography system with unique arbitrary orbit capability. In: Proceedings of SPIE, Medical Imaging 2005, p. 9 (2005)
7. Tornai, M., et al.: Design and Development of a Fully-3D Dedicated X-ray Computed Mammotomography System. In: Proceedings of SPIE, Medical Imaging 2005, vol. 9 (2005)
8. Yang, W.T., et al.: Dedicated cone-beam breast CT: Feasibility study with surgical mastectomy specimens. Am. J. Roentgenol. 189(6), 1312–1315 (2007)
9. Barrett, H.H., Myers, K.J.: Foundations of Image Science. John Wiley, Chichester (2003)
10. Hsu, J., et al.: Preclincal ROC Studies of Digital Stereomammography. IEEE Trans. Med. Imag. 14(2), 318–327 (1995)
11. Taylor, P., Owens, R., Ingram, D.: Simulated mammography using synthetic 3D breasts. In: 4th Int. Workshop on Digital Mammography (IWDM). Kluwer, Nijmegen (1998)

12. Bakic, P.R., et al.: Mammogram synthesis using 3D simulation. I. Breast tissue model and image acquisition simulation. Med. Phys. 29(9), 2131–2139 (2002)
13. Bakic, P.R., et al.: Mammogram synthesis using a 3D simulation. II. Evaluation of synthetic mammogram texture. Med. Phys. 29(9), 2140–2151 (2002)
14. Bakic, P.R., et al.: Mammogram synthesis using a three-dimensional simulation. III. Modeling and evaluation of the breast ductal network. Med. Phys. 30(7), 1915–1926 (2003)
15. Bliznakova, K., et al.: A three-dimensional breast software phantom for mammography simulation. Phys. Med. Biol. 48, 3699–3719 (2003)
16. O'Connor, J.M., et al.: Using mastectomy specimens to develop breast models for breast tomosynthesis and CT breast imaging. In: Proceedings of SPIE, Medical Imaging 2008: Physics of Medical Imaging, pp. 691315 (2008)
17. Li, C.M., Segars, W.P.,, L.J.Y.: Three-dimensional computer generated breast phantom based on empirical data. In: Medical Imaging 2008. SPIE, San Diego (2008)
18. Li, C.M., et al.: Computerized 3D breast phantom with enhanced high-resolution detail. In: Medical Imaging 2009. SPIE, Orlando (2009)
19. Li, C.M., et al.: Methodology for generating a 3D computerized breast phantom from empirical data. Med. Phys. 36(7), 3122–3131 (2009)
20. O'Connor, J.M., et al.: Comparison of Two Methods to Develop Breast Models for Simulation of Breast Tomosynthesis and CT. In: International Workshop on Digital Mammography (IWDM). Springer, Tucson (2008)
21. IRB Docket #H-12422, I.R.B.I. University of Massachusetts, Editor (2006)
22. O'Connor, J.M., et al.: Characterization of a prototype table-top x-ray CT breast imaging system. In: Proceedings of SPIE, Medical Imaging 2007: Physics of Medical Imaging. SPIE, San Diego (2007)
23. Vedula, A.A., Glick, S.J., Gong, X.: Computer simulation of CT mammography using a flat-panel imager. In: Proceedings of SPIE Physics of Medical Imaging 2003, Physics of Medical Imaging, vol. 5030, pp. 349–360 (2003)
24. Feldkamp, L.A., Davis, L.C., Kress, J.W.: Practical cone-beam algorithm. J. Opt. Soc. Am. 1(6), 612–619 (1984)
25. Altunbas, M.C., et al.: A post-reconstruction method to correct cupping artifacts in cone beam breast computed tomography. Med. Phys. (2007)
26. Perona, P., Malik, J.: Scale-Space and Edge Detection Using Anisotropic Diffusion. IEEE Transactions on Pattern Analysis and Machine Intelligence 12(7), 11 (1990)
27. Insight Segmentation and Registration Toolkit (ITK), `http://www.itk.org/`
28. Metheany, K.G., et al.: Characterizing anatomical variability in breast CT images. Med. Phys. 35(10), 4685–4694 (2008)
29. Burgess, A.E., Jacobson, F.L., Judy, P.F.: Human observer detection experiments with mammograms and power-law noise. Med. Phys. 28(4), 419–437 (2001)

Analysis of Geometric Accuracy in Digital Breast Tomosynthesis Reconstruction

Predrag R. Bakic[1], Peter Ringer[2], Johnny Kuo[2], Susan Ng[2], and Andrew D.A. Maidment[1]

[1] University of Pennsylvania, Dept. of Radiology, Philadelphia, PA 19104
[2] Real-Time Tomography, LLC, Villanova, PA19085
{Predrag.Bakic,Andrew.Maidment}@uphs.upenn.edu,
{Peter.Ringer,Johnny.Kuo,Susan.Ng}@realtimetomography.com

Abstract. The geometric accuracy of a digital breast tomosynthesis (DBT) reconstruction algorithm was assessed using an anthropomorphic software breast phantom with simulated fiducial markers. The locations of the fiducial markers were measured from supersampled images reconstructed to sub-pixel precision. The measured locations were compared with the known ground truth positions of the simulated markers. The fiducial markers simulate small, attenuating objects within the software phantom. Using reconstructed images with resolution of 0.115 mm, and a total of twelve fiducial markers at three different depths, we determined an average difference of 0.105 mm (st. dev. 0.086 mm) between the estimated and true marker locations.

Keywords: Anthropomorphic breast phantom, tomosynthesis, tomographic reconstruction algorithms, geometric accuracy of tomographic reconstruction.

1 Introduction

DBT is an investigational imaging modality that offers improved visualization of breast tissue by eliminating tissue overlap. Early clinical studies have shown that DBT has increased sensitivity and specificity compared to digital mammography (DM), the current gold standard in breast cancer screening. With a radiation dose comparable to DM, DBT has the potential to replace DM as the standard breast cancer screening modality.

Methods for analyzing DBT systems using physical test objects have been described in the literature; these methods use physical test objects, such as a slanted wire [1], or custom-made calibration phantoms [2]. Currently, however, no generally accepted standards exist for testing DBT acquisition and reconstruction. The three-dimensional nature of DBT limits the use of existing tools for mammography (*e.g.*, mammographic accreditation phantom [3]). Similarly, the limited angular range of DBT acquisition prevents the use of most clinical CT tools. In this paper, we describe preliminary results of estimating the geometric accuracy of a DBT reconstruction algorithm using an anthropomorphic software breast phantom with fiducial markers simulated at known positions.

J. Martí et al. (Eds.): IWDM 2010, LNCS 6136, pp. 62–69, 2010.

2 Materials and Methods

We used an anthropomorphic breast phantom described previously [4,5]. The phantom simulates both a realistic breast outline typical of the geometric extent of the breast, and internal tissue structures (glandular tissue, adipose compartments and Cooper's ligaments). The design of the phantom is flexible, covering anatomical variations in breast composition and size. The phantom deformation during clinical breast compression is simulated using a finite element tissue model.

In this study, the breast phantom was simulated with an isotropic resolution of 200 microns. The phantom volume was 450 ml with a compressed thickness of 5 cm and an overall glandularity of 40%. The fiducial markers simulated small, attenuating objects with a size of one voxel (200 microns) and a linear x-ray attenuation 30 times larger than the attenuation of fibroglandular breast tissue. Four fiducial markers were simulated at three different distances from the breast support (6.4 mm, 25.6 mm and 44.8 mm). At each depth, the four markers were positioned in a 20 mm square. Fig. 1 illustrates simulated fiducial markers at different distances from the breast support.

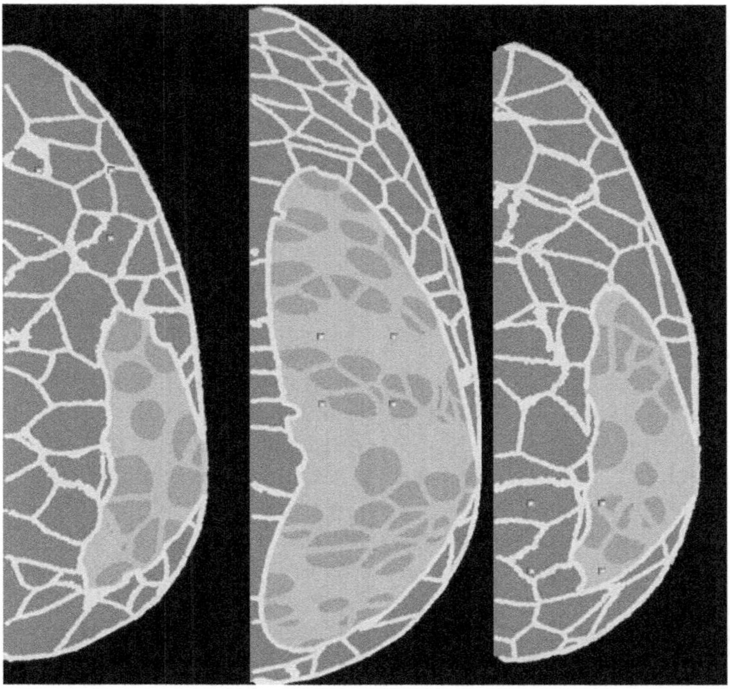

Fig. 1. Cross-sections of the software breast phantom at 6.4 mm (left), 25.6 mm (center), and 44.8 mm (right) from the breast support, containing fiducial markers (in red/white). The markers are shown as magnified 20-fold relative to their simulated size.

Tomosynthesis image acquisition was simulated assuming a monoenergetic x-ray beam without scatter and an ideal x-ray detector. Ray tracing was used to calculate the x-ray attenuation through the phantom. The resolution of the projection images was 100 microns, consistent with the GE Senographe DS tomosynthesis prototype (GE Healthcare, Chalfont St. Giles, UK). The simulated acquisition geometry included 15 projection acquired over a 40 degree arc corresponding to the acquisition protocol of the GE prototype. A commercial back-projection filtered DBT reconstruction algorithm developed by Real Time Tomography, LLC (Villanova, PA) was used. Fig. 2 shows details of a simulated DBT phantom projection (acquired perpendicular to the detector plane) and the reconstructed image plane corresponding to a depth 25.6 mm measured from the breast support.

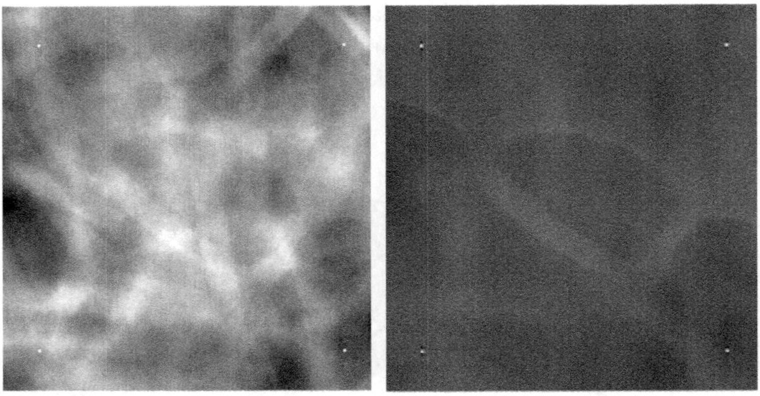

Fig. 2. A detail of a simulated DBT projection image (left) and the reconstructed image plane corresponding to a depth 25.6 mm from the breast support (right) showing four fiducial markers

The reconstruction algorithm used allows us to specify the geometry of the reconstruction plane with arbitrary precision. To measure geometric accuracy, we reconstructed a series of images with sub-pixel shifts within the plane of reconstruction. These images were combined to form a supersampled image. In the current work, 10-fold supersampling was performed in the y-axis (scanning direction) only. In future work, supersampling in the x- and y-axes will be performed. Fig. 3 shows a surface plot of a supersampled image of a fiducial marker.

For our analysis, images of the software phantom with the fiducial markers were reconstructed at a spatial resolution of 0.115 mm, selected to allow display of a 230.4 mm × 192.0 mm reconstructed field-of-view on a 2048×1536 monitor. Image planes were reconstructed every 0.2 mm. For each of the 12 fiducial markers, we identified the distance z_C from the breast support to the reconstructed image plane in which the spatial extent of the objects was smallest. In this plane, we calculated the centroid, i.e., the center of mass, (x_C, y_C) of the marker to determine its location more precisely. The centroid was calculated in a thresholded image of the supersampled objects. The threshold was selected as half of the maximum reconstructed voxel intensity in the region of interest of the fiducial marker.

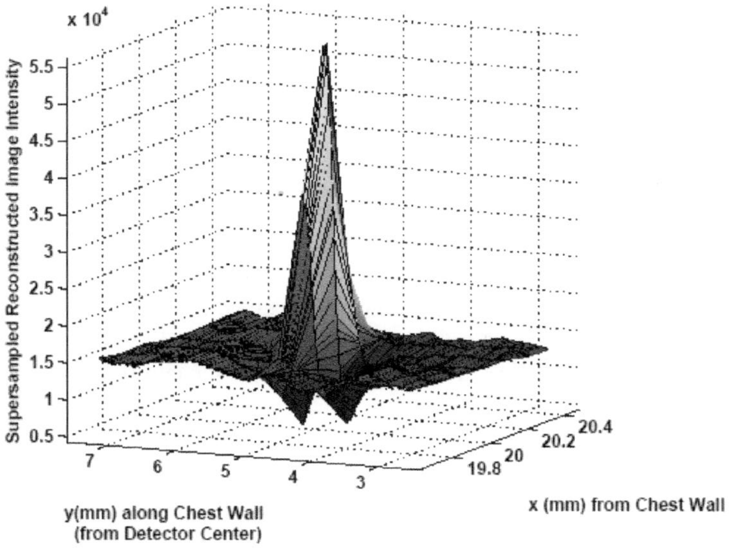

Fig. 3. A surface plot of a supersampled image of a fiducial marker used for estimating the marker location in the reconstructed space with a subpixel precision

The coordinates (x_C, y_C, z_C) were considered as the position of the marker in the three-dimensional reconstructed image space. This position was compared with the known ground truth marker location in the phantom (x_T, y_T, z_T). We defined the error in the position of the fiducial markers E_P as the Euclidean distance between the measured and the true marker position:

$$E_P = \sqrt{(x_C - x_T)^2 + (y_C - y_T)^2 + (z_C - z_T)^2} \; . \tag{1}$$

In addition, we also calculated the error in the distance between a pair of reconstructed fiducial markers, defined as $E_D = d_R - d_T$, where

$$d_R = \sqrt{(x_{C1} - x_{C2})^2 + (y_{C1} - y_{C2})^2 + (z_{C1} - z_{C2})^2} \tag{2}$$

and

$$d_T = \sqrt{(x_{T1} - x_{T2})^2 + (y_{T1} - y_{T2})^2 + (z_{T1} - z_{T2})^2} \; . \tag{3}$$

and indices *C1* and *C2* denote the measured and *T1* and *T2* the true locations for a pair of markers.

Using the supersampled images, we also calculated the relative error in estimating marker size from images reconstructed at a given distance from the plane in which the object is in focus. The estimated marker size was calculated as the full width at half maximum (FWHM) of the reconstructed marker profile. The relative error in marker size was defined as $100\%\times[F(z) - F_{min}]/F_{min}$, where $F(z)$ and F_{min} represent the FWHM estimated at the reconstructed depth z, and the minimum FWHM estimated at the plane of focus, respectively.

3 Results

Fig. 4 shows the error in the measured marker positions E_P as a function of the reconstructed plane depth (*i.e.*, the distance from the breast support). Also shown are the errors calculated separately for each of the three marker coordinates: perpendicular to the chest wall (x_C-x_T), parallel to the chest wall (y_C-y_T), and perpendicular to the detector (z_C-z_T). The overall error E_P averaged over all 12 fiducial markers was 0.105±0.086 mm (average ± st.dev). For the markers located 6.4 mm from the breast support the error E_P was 0.126±0.088 mm; for the markers at 25.6 mm, E_P was 0.073±0.085 mm, and for the markers at 44.8 mm, E_P was 0.116±0.101 mm.

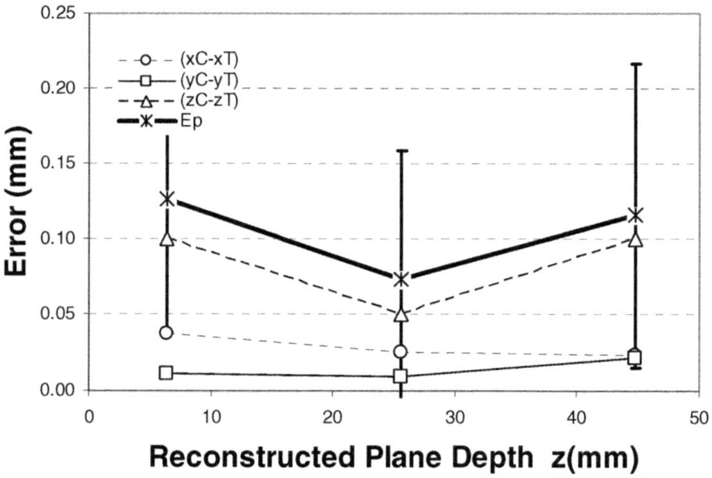

Fig. 4. Average error in the measured 3D position of reconstructed fiducial markers (E_P) as a function of the reconstructed plane depth. (Error bars represent one standard deviation.) Shown also are the errors calculated separately for each marker coordinate.

Fig. 5 shows the error in the reconstructed distance between pairs of fiducial markers E_D. The simulated distance between the markers in the phantom was 20 mm. The distance error is calculated separately for distances measured along the x- and y-axes. The total error E_D averaged over 12 pairs of markers along the two axes was 0.017±0.020 mm. For the markers at a 6.4 mm from the breast support the error E_D was 0.013±0.009 mm; at 25.6 mm, E_D was 0.009±0.007 mm, and at 44.8 mm, E_D was 0.029±0.021 mm.

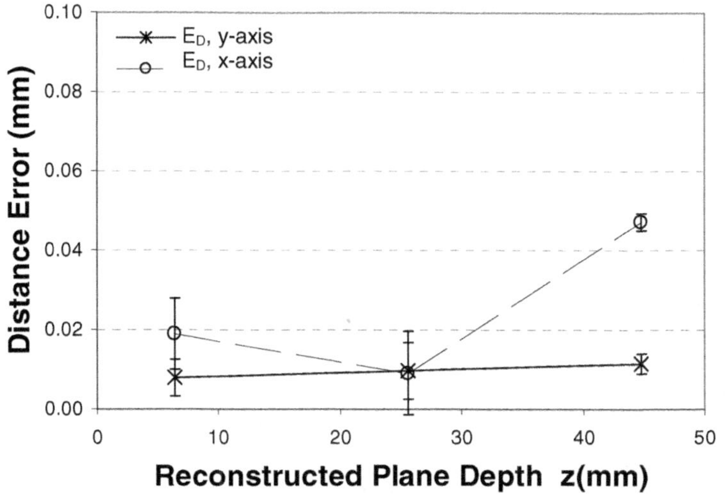

Fig. 5. Error in the reconstructed distance between pairs of fiducial markers (E_D) plotted as a function of the reconstructed plane depth. E_D values were averaged separately over distance measured along the y-axis parallel with tube motion (*) and the x-axis perpendicular to tube motion (○). Error bars represent one standard deviation. Note that supersampling was performed along the y-axis only.

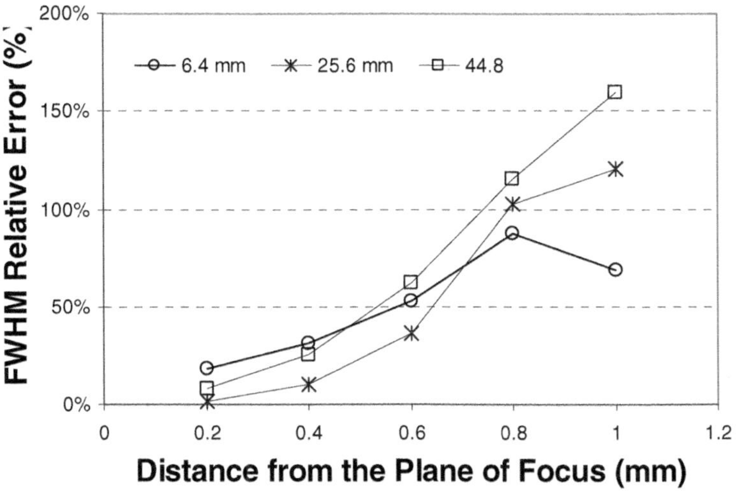

Fig. 6. Relative error in the estimation of the object size at a given distance from the plane of focus, averaged over the fiducial markers positioned at different reconstructed plane depths

Fig. 6 shows the relative error in estimating marker size from images reconstructed at a given distance from the plane of focus. Plotted are the relative errors averaged over the four markers positioned at the same depth. Averaged over all 12 markers, the relative error in estimating marker size at a distance of 0.6 mm away from the plane of focus is equal to 50%; while at a distance of 1.0 mm, the relative error is 116%. The error increases with distance from the detector.

4 Discussion and Conclusions

We have developed a method to assess geometric accuracy with sub-pixel precision. In the example presented, we observed very little dependence of the geometric accuracy on the reconstructed plane; the errors in the measured marker positions E_P were comparable at different distances from the detector. As shown in Fig. 4, the E_P values are on the order of the size of 1 pixel. This is as expected for a well-designed reconstruction system. Note that the error in the z-axis is worst; this arises from the limited angular range of the DBT acquisition geometry.

The observed error in measuring the distance between pairs of fiducial markers E_D, shown in Fig. 5, was a fraction of the pixel size. The E_D values were smaller for distances measured along the y-axis than the x-axis because we performed supersampling along the y-axis only. We plan to extend the presented approach to the analysis of images supersampled in both in-plane directions. The error in distance measurement does not show a strong dependence on the reconstruction plane.

These results have importance for developers of DBT-guided biopsy systems. Based on these data, biopsy systems should be able to localize objects to within one voxel in 3-dimensions, thus providing targeting accuracy comparable to stereotactic biopsy systems.

The presented approach also allowed us to estimate the unsharpness of each object when reconstructed at a given depth from the plane of focus. As shown in Fig. 6, the FWHM error at a distance of 1.0 mm is about twice as large (116% vs. 50%) as at a distance of 0.6 mm. This result is of practical importance as it elucidates limitations of DBT reconstruction in visualizing objects located at a position between two planes of reconstruction. As the analyzed fiducial objects were of size 1-voxel in the phantom, the reconstructed supersampled images of these objects could also be used to calculate spatial dependence of the point spread function.

We are currently performing verification of the presented assessment methodology by analyzing images reconstructed at different magnification. Fig. 7 shows a fiducial marker supersampled 10-times in the y-axis (scan direction), and the same marker from an image reconstructed at 10-times magnification. The images are substantially equivalent; however, accurate side-by-side comparison will require supersampling in both the x- and y-axes, and application of reconstruction filters that are matched in the two magnifications.

In order to validate the effects of simulated image acquisition of the observer accuracy, we also plan to extend the presented approach to include the analysis of physical phantom images. The combined analysis of simulated and physical phantom images offers more flexibility in assessing the impact of the acquisition geometry and system components (e.g., focal spot blurring, scatter, step-and-shoot vs. continuous tube motion, etc.).

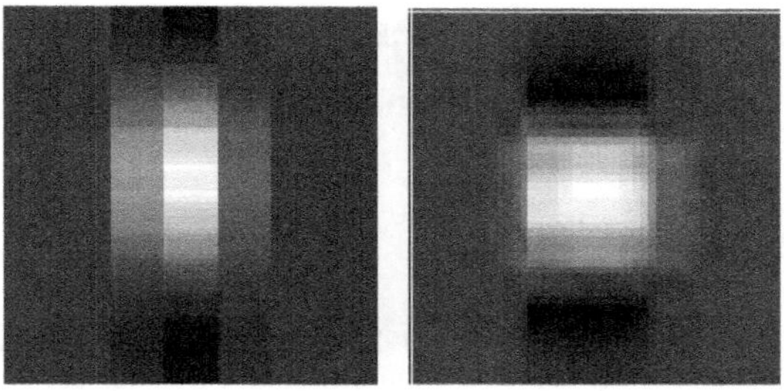

Fig. 7. The same fiducial marker shown in (a) an image supersampled in the (vertical) direction along the chest wall and (b) a 10 times magnified image

In summary, while physical phantoms can exactly measure individual imaging systems, software phantoms offer an advantage of repeated and potentially automated analysis of a large number of acquisition and reconstruction parameter combinations. Moreover, software breast phantoms include simulated anatomical noise. This is particularly relevant in DBT systems, where the reconstruction algorithm must successfully blur out-of-plane objects while preserving in-plane features.

Acknowledgments. This work in progress has been supported by a subcontract from grant R44EB07140 (Real Time Tomography, LLC.) from the National Institutes of Health. P. Ringer, J. Kuo, and S. Ng are employees, and A. Maidment is chair of the Scientific Advisory Board of Real Time Tomography, LLC. The contents of this paper are solely the responsibility of the authors and do not necessarily represent the official views of the funding agencies.

References

1 Hu, Y.-H., Zhao, B., Zhao., W.: Image artifacts in digital breast tomosynthesis: Investigation of the effects of system geometry and reconstruction parameters using a linear system approach. Medica Physics 35, 5242–5252 (2008)
2. Wang, X., Mainprize, J.G., Kempstone, M.P., Mawdsley, G.E., Yaffe, M.J.: Digital Breast Tomosynthesis Geometry Calibration. In: SPIE Medical Imaging 2007: Physics in Medical Imaging (2007)
3. CIRS Tissue Simulation and Phantom Technology: Mammographic Accreditation Phantom, `http://www.cirsinc.com/015_mammo.html` (accessed on January 17, 2009)
4. Bakic, P.R., Albert, M., Brzakovic, D., Maidment, A.D.A.: Mammogram synthesis using a 3D simulation. I. Breast tissue model and image acquisition simulation. Medical Physics 29, 2131–2139 (2002)
5. Zhang, C., Bakic, P.R., Maidment, A.D.A.: Development of an Anthropomorphic Breast Software Phantom Based on Region Growing Algorithm. In: Visualization, Image-guided procedures, and Modeling, Proc. SPIE, vol. 6918 (2008)

Results of Comprehensive Dose Survey for Mammography

Paola Baldelli[1], James B. McCullagh[2], Niall Phelan[1], and Fidelma Flanagan[1]

[1] Breastcheck, The National Breast Screening Programme, 36 Eccles St., Dublin, Ireland
[2] Breastcheck, The National Breast Screening Programme, Infirmary Road, Cork, Ireland
{James.Mccullagh,Paola.Baldelli,Niall.Phelan,
Fidelma.Flanagan}@breastcheck.ie

Abstract. Results of a dose survey carried out at BreastCheck, the Irish Breast Screening Program are presented. The survey includes 12,529 images acquired with GE Essential, Hologic Selenia and Sectra MDM digital mammography systems.

The average MGD per examination was 2.74±0.04mGy, the average compressed breast thickness was 61.4±0.53mm and the average compression force was 10.9±0.7daN. Results for the three different systems are reported and the distribution of the most utilised target/filter combinations are also included.

The results from this survey of a completely digital screening programme reveal a complete reversal of trend in target/filter selection with Mo/Mo usage at 0.3% and Rh/Rh selection at 36.6%. Indeed, the predominant use of novel target/filter combinations is evident from this survey.

A DRL value of 1.75mGy has been established by calculating the 95th percentile of the average MGD for compressed breast thicknesses ranging from 55mm–65mm.

Keywords: Digital Mammography, Breast Screening, Mean Glandular Dose, Dose Reference Level.

1 Introduction

In Ireland, breast cancer is the second most common form of cancer. According to figures from the National Cancer Registry of Ireland [1], breast cancer accounts for 28% of all cancers in women in Ireland, with an average of 1,726 new diagnoses each year.

Mammography is a well-established screening tool and screening has been shown to reduce breast cancer mortality due to earlier detection. As for all practices involving exposure to diagnostic x-rays, mammography should be subject to clinical audit including patient dose measurement at regular intervals.

In order to comply with the European council directive 97/43/Euratom and consequent Irish legislation [S.I.478(2002)], we have measured the Mean Glandular Dose (MGD) to the breast for a representative series of breast examinations and for each mammographic system in service at the national screening program.

Much information can be drawn from a dose audit of this nature. Results were used to define the Dose Reference Level (DRL) and to determine a number of other key

J. Martí et al. (Eds.): IWDM 2010, LNCS 6136, pp. 70–77, 2010.

parameters such as the overall MGD used within the breast screening program, the MGD as a function of the breast thickness and as a function of the breast view as well as the distribution of the breast thickness.

2 Method

Approximately one-hundred patient examinations were acquired for each of the 28 digital systems in operation at the Irish breast screening program. Each examination contains 4 images: a mediolateral oblique (OB) view and a craniocaudal (CC) view for each breast.

The digital systems included in this survey were: 11 *GE Essential,* 10 *Hologic Selenia,* and 7 *Sectra MDM L30*. The three types of systems differ with regard to the target/filter combinations used as well as detector technology and operation of the AEC system.

The GE Essential has a dual track anode composed of Molybdenum (Mo) and Rhodium (Rh) targets in combination with Mo or Rh filters. The GE Essential has an indirect conversion detector which consists of a flat panel amorphous silicon detector with a Cesium Iodide (CsI) needle structure scintillator The system uses a feature called automatic optimisation of parameters (AOP) to optimise image quality under the constraints of breast MGD, exposure time and other technical limitations. It uses a pre-exposure to determine the radiological thickness at the most glandular part of the breast and from this, along with the mechanical compressed breast thickness, the breast composition is estimated. Based on the radiological thickness, composition and system configuration, the final exposure parameters (target, filter, kV and detector dose) are chosen from a pre-computed AOP table. Finally, the mAs value necessary to reach the required detector dose is computed and used to control the exposure [2]. There are three fully automated exposure modes available on the GE Essential; Contrast, Standard and Dose AOP modes.

The Hologic Selenia system is equipped with a Tungsten (W) anode and a choice of Rhodium (Rh) or Silver (Ag) filtration. The detector used in the Selenia is a flat-panel amorphous-Selenium detector which performs a direct conversion of x-rays to a digital signal. It is comprised of a 200μm layer of amorphous selenium bonded to a thin film transistor (TFT) array. The Selenia has four modes of automatic exposure control: Auto-Filter, Auto-kV, Auto-Time and thickness equivalent control (TEC). All the systems included in this survey are operated in the Auto-Filter mode. In this mode the compressed breast thickness is used to determine the kV and filter selection and the required exposure is determined from a pre-exposure of between 2mAs and 10mAs depending on the breast thickness (Hologic, Personal Communication). There are four selectable operating dose levels on the Selenia and all units in the Breast-Check program are operated at the Standard Screening Dose setting.

The Sectra MDM L30 is a multiple slit scanning photon counting system with a W/Al anode-filter combination. It has a pixel pitch of 50μm and field size of 24cm x 26cm. The MDM L30 uses the compressed breast thickness to set the kV and the initial scan speed to select a specific exposure. It then utilises a form of exposure control called *Smart*AEC™ which continuously adjusts the exposure and optimizes the dose level to the local density of the breast throughout the scan. The AEC attempts to maintain, as far as possible, a constant contrast-to-noise ratio (CNR) [3]. There are three dose operating levels on the MDM: C120, C100 and C70 and all units at the national breast screening program are set to the C120 level.

All x-ray units in the BreastCheck screening programme are operated using automated selection of exposure factors. The control mode used depends on the system used and all system types are operated in the same mode. Thus the GE Essential systems are set to STD mode, the Hologic Selenia systems are set to Auto-Filter mode and the Sectra systems operate in its unique automatic exposure control mode.

All of the systems were subject to routine medical physics QA during the period of the survey in accordance with the recommendations of the European guidelines [4] and were found to conform to the specified image quality standards.

Selected examinations for each system were transferred from the PACS archive, the parameters were extracted from the DICOM headers using a locally developed MATLAB routine (Mathworks, Cambridge, UK) and transferred to a Microsoft® Office Excel spreadsheet for further analysis. The parameters extracted from the header for this survey included the patient identification, age, compressed breast thickness, compression force, view position, laterality, exposure kVp, exposure mAs, target material, filter material, organ dose and pre-exposure factors where appropriate.

Dose calculation was carried out according to the method of Dance et al [5] using software developed by KC Young for the National Health Service Breast Screening Program (NHSBSP) [6] with some modification made to take into account new *c*-, *g*- and *s*- factors for HVL values associated with the spectral data for new mammography systems. Further modifications were made to the dose calculations to account for the larger HVL found in the new mammography systems. The s-factor values used in dose calculations with tungsten/aluminum target/filter combinations were obtained from a table of s-factors provided by Dance et al [5].

To determine the dose reference level, the MGD of OB mammograms for breasts with a compressed thickness of 60±5mm was selected for each mammography system. The choice of this thickness was based on the measurement of average compressed breast thickness in this survey. The 95th percentile of the distribution of the MGD was used as DRL.

The dose data from distinct units were then compared to the DRL. The MGD of a particular unit is considered significantly greater than DRL if the MGD, increased by twice the SEM, is higher than the DRL.

3 Results

Data collected from the 28 mammographic systems include 3,016 patients, which result in a total of 12,529 images. The age of the majority (99.7%) of the women was within the standard screening age range with an average age of 57.4 years.

The overall average MGD per examination was 2.74±0.04mGy, the average thickness was 61.4±0.53mm and the average compression force was 10.9±0.7daN. These data are presented in the first column of Table 1 together with average dose, compressed breast thickness and compression force for the OB and CC view.

Table 1. Average MGD, thickness and compression force for the 2 views : CC and OB

		Overall Results	GE	Hologic	Sectra
Exam	Average MGD for exam (mGy)	2.74 ± 0.04	3.03 ± 0.05	2.91 ± 0.06	1.98 ± 0.04
	Average thickness (mm)	61.4 ± 0.53	59.2 ± 0.53	62.5 ± 0.58	64.7 ± 0.64
	Average compression force (daN)	10.9 ± 0.7	10.7 ± 0.1	11.0 ± 0.14	10.9 ± 0.3
CC view	Average MGD for image (mGy)	1.27± 0.01	1.40 ± 0.01	1.36 ± 0.02	0.96 ± 0.01
	Average thickness (mm)	60.0± 0.3	57.1 ± 0.48	61.6 ± 0.53	64.5 ± 0.6
	Average compression force (daN)	10.0± 0.6	10.0 ± 0.09	10.0 ± 0.11	10.2 ± 0.16
OB view	Average MGD for image (mGy)	1.35± 0.01	1.52 ± 0.02	1.44 ± 0.02	0.93 ± 0.01
	Average thickness (mm)	62.5± 0.4	61.3 ± 0.59	63.4 ± 0.62	65 ± 0.7
	Average compression force (daN)	11.9± 0.9	11.8 ± 0.11	11.9 ± 0.15	12.3 ± 0.2

Because of the variation of MGD between systems, we have also reported the results for each of the individual system types included in this survey.

The GE systems, despite having the lowest compressed breast thickness, exhibit an average MGD per exam of 3.03mGy. The Hologic systems had a MGD of 2.91mGy and an average compressed breast thickness of 62.5mm. The Sectra systems had an average MGD of 1.98mGy and an average compressed breast thickness of 64.7mm. The distribution of the MGD for the overall survey and the three types of system are shown in Figure 1.

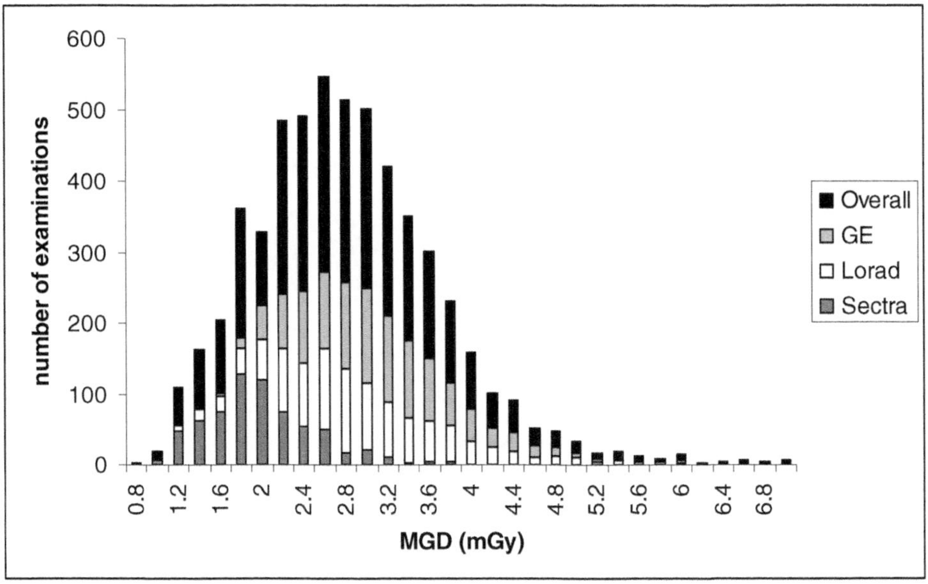

Fig. 1. Distribution of the calculated MGD for examination

Table 2 shows the distribution of target/filter combinations selected in the survey. Rh/Rh (36.6%) is the most commonly used while the Mo/Mo target/filter combination was selected in only 0.3% of cases.

The variation of dose with compressed breast thickness for the overall survey and for the different mammography systems is shown in Figure 2 for the OB view (CC view distribution presents the same trend but is not reported here). Compressed breast thicknesses <40mm and >85mm were neglected as they did not contain a statistically significant number of images.

Finally, Figure 3 shows the MGD calculated for breast thickness between 55mm and 65mm for the OB view images for the 28 systems. The 95th percentile of the average MGD was calculated and reported on the graph to indicate the DRL. This value of 1.75mGy was set as DRL.

Table 2. Proportion of exposures using the different target/filter combinations

Target/filter combination	Proportion of exposures
Mo/Mo	0.3%
Mo/Rh	4.3%
Rh/Rh	36.6%
W/Rh	23.5%
W/Ag	11.2%
W/Al	24.1%

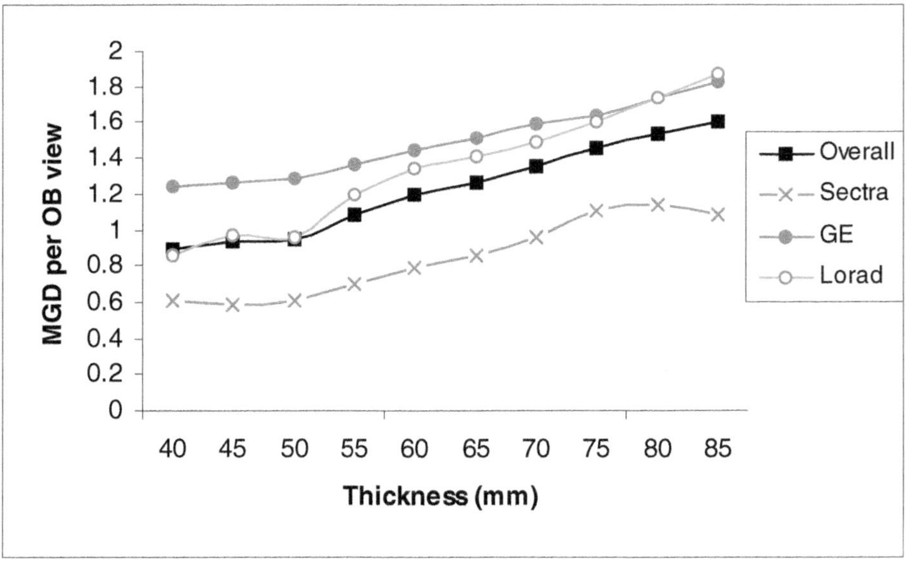

Fig. 2. Average MGD per image main Oblique (OB) as a function of compressed breast thickness

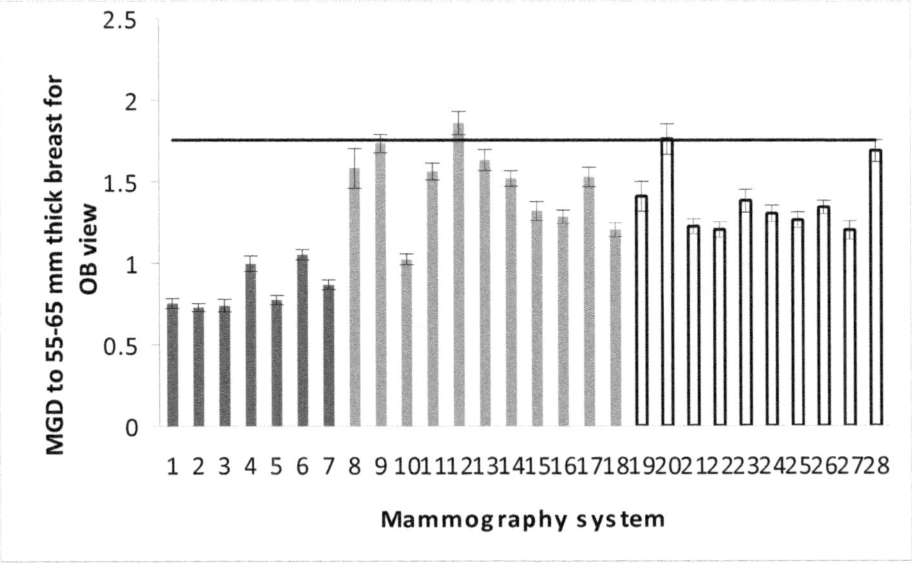

Fig. 3. Distribution of MGD values for OB view images for 55-65 mm thick breasts. Systems 1-7 are Sectra, 8-18 are GE and 19-28 are Hologic. The error bars represent the Standard Error of the Mean (SEM).

4 Discussion

The average MGD per CC view was 6% lower than that of the corresponding OB view. This difference can be accounted for by the equivalent difference in the compressed breast thickness (60mm for the CC view and 62.5mm for the OB view) and the inclusion of denser parts such as the pectoral muscle in the OB view, which can cause an increase in the exposure. The presence of the pectoral muscle in the OB view is also responsible of the higher compression force (11.9daN for the OB view and 10daN for the CC view).

The results from this survey reveal a complete reversal of trend in target/filter selection with Mo/Mo usage at 0.3% and Rh/Rh at 36.6%. Indeed the predominance of new target/filter combinations is evident from this survey (W/Al at 24.1% and W/Rh at 23.5%).

In particular, for the Hologic systems, the W/Rh combination was selected for breast thicknesses less than 70mm and W/Ag was selected for greater breast thicknesses. The selected voltage increased with thickness and ranged between 26kV and 32kV for the W/Rh combination and between 27kV and 36kV for the W/Ag combination.

The majority of exposures on the Sectra units were performed at 35kV (76.7%). For smaller (<45mm) breast thicknesses, 32kV (7.8%) was selected and for the larger (>70mm) breast thicknesses, 38kV (15.3%) was selected.

For the GE systems, most exposures occurred at the single target/filter combination and voltage selection of Rh/Rh 29 kV (71.7%). These factors were selected for breast thicknesses ranging between 30mm and 84mm. The remaining exposures were one of three predominant exposure combinations: 27kV Mo/Rh (8.5%), 30kV Rh/Rh (7%) or 31kV (9.7%).

Regarding the relationship between MGD and breast thickness, as would be expected, the average MGD increases with thickness for all systems included in the survey. The Sectra systems exhibit a significantly lower dose compared with the other two models. Of the other two systems, the Selenia appears to provide a dose advantage, certainly for smaller breasts but less so for larger breasts. The rate of increase in dose appears to be similar for the GE Essential and the Sectra units but the dose level exhibits a plateau for breast thickness >60mm on the Sectra system.

Our results can be compared with results of similar surveys carried on in different countries. In 2001, the results of an Australian survey including 6,000 women was published [7]. The average MGD per view was 2.6mGy and the compressed breast thickness was 5.1cm. In 2003 Jamal et al [8] published results of a small study involving 300 women where the MGD and the compressed thickness were 1.54mGy and 37mm respectively for the CC view and 1.82mGy and 44mm for the OB view. The paper by Young et al [9] reviews a large representative sample of dose measurements collected in the UK screening programme. The MGD was 2.23mGy for the OB view and 1.96mGy for the CC view and the average compressed breast thicknesses were 54.3mm and 51.5mm for the two views respectively. Smans et al [10] in 2006 presented results of a Belgian audit involving 27 centers where the average MGD was 1.76mGy.

Compared with results of previous surveys, the MGD for mammography in the BresatCheck screening programme is low. This is because all imaging equipment in the screening program is digital and 25% of the mammography systems used are low dose systems.

We also compare the DRL value established in this work with those published elsewhere. In particular, the DRL value is significantly lower than the comparative UK value of 3.5mGy [9] and Belgian value of 2.46mGy [10].

Three units were found to have a measured dose slightly higher than the DRL; 2 GE and 1 Hologic system. These systems are currently under investigation to establish the cause of these higher dose measurements.

Acknowledgement. This study was supported by the Eccles Brest Cancer Research Fund.

References

[1] Report of the National Cancer Registry Ireland, Women and cancer in Ireland 1994-2001 (2005)

[2] Shramchenko, N., Blin, P., Mathey, C., Klausz, R.: Optimised Exposure Control in Digital Mammography. In: Yaffe, M.J., Flynn, M.J. (eds.) Proceedings of SPIE, vol. 5368 (2004)

[3] Aslund, M., Cederstrom, B., Lundqvist, M., Danielsson, M.: AEC for scanning digital mammography based on variation of scan velocity. Med. Phys. 32, 3367–3374 (2005)

[4] Perry, N., Broeders, M., de Wolf, C., Tornberg, S., Holland, R., von Karsa, L.: 2006 European Guidelines for quality assurance in breast cancer screening and diagnosis. In: 4th edn. European Commission, Luxemburg (2006), ISBN 92-79-01258-4

[5] Dance, D.R., Young, K.C., van Engen, R.E.: Further factors for the estimation of mean glandular dose using the United Kingdom, European and IAEA breast dosimetry protocols. Phys. Med. Biol. 54, 4361–4372 (2009)

[6] NHSBSP Report 04/05: Breast Dose Surveys in the NSBSP: Software and Instruction Manual. Version 2.0 (2004)

[7] Kruger, R.L., Schueler, B.A.: Survey of doses in screening mammography. Australas. Phys. Eng. Sci. Med. 19, 207–217 (1996)

[8] Jamal, N., Ng, K.H., McLean, D.: A study of mean glandular dose during diagnostic mammography in Malaysia and some of the factors affecting it. Br. J. Radiol. 76(904), 238–245 (2003)

[9] Young, K.C., Burch, A., Oduko, J.M.: Radiation doses received in the UK breast screening programme in 2001 and 2002. Br. J. Radiol. 78, 207–218 (2005)

[10] Smans, K., Bosmans, H., Xiao, M., Carton, A.K., Marchal, G.: Towards a proposition of a diagnostic (dose) reference level for mammographic acquisitions in breast screening measurements in Belgium, Rad. Prot. Dosimetry 117(1–3), 321–326 (2005)

Validation of a Simulated Dose Reduction Methodology Using Digital Mammography CDMAM Images and Mastectomy Images

Mary Yip[1], Federica Zanca[2], Alistair Mackenzie[3], Adam Workman[4], Kenneth C. Young[3], David R. Dance[3], Hilde Bosmans[2], Emma Lewis[1], and Kevin Wells[1]

[1] Centre for Vision, Speech and Signal Processing, Faculty of Engineering and Physical Sciences, University of Surrey, Guildford, GU2 7XH, UK
m.yip@surrey.ac.uk

[2] Department of Radiology, University Hospitals Gasthuisberg, Herestraat 49, B-3000 Leuven, Belgium

[3] National Coordinating Centre for the Physics of Mammography, Royal Surrey County Hospital, Guildford, GU2 7XX, UK

[4] Northern Ireland Regional Medical Physics Agency, Forster Green Hospital, Belfast, BT8 4HD, UK

Abstract. The purpose of this study is to evaluate the effect of simulated dose reduction using CDMAM and mastectomy images acquired on two digital mammography systems. High dose images have been artificially degraded to reduced dose levels by systematically adding filtered noise. Automated scoring has been carried out on the degraded CDMAM images and on experimental CDMAM images, taken at the same corresponding reduced doses. Contrast-detail curves were derived for both, at all doses, and compared. Relative difference in the contrast-detail curves was approximately 5% overall for all four doses.

For the mastectomy images noise power spectra were obtained and the ratio of experimental to synthetic low dose NPS profiles averaged for all doses at 1.04. The largest differences in the NPS profiles were found at the high spatial frequencies, corresponding with the differences in the small discs in the contrast-detail curves.

Keywords: Digital mammography, simulation, CDMAM phantom, validation, mastectomy, dose reduction.

1 Introduction

There is a growing interest in image simulation in digital radiography, and in mammography in particular[1,2,3,4,5,6]. Image simulation provides a means to optimise digital mammography systems for improved breast cancer detection. Image simulation studies could make clinical trials more targeted as simulation studies can be used to study some effects in advance.

Recently, there has been focus on the effect of reducing dose levels in digital mammography images upon detection of mammographic lesions[1,2]. Results

J. Martí et al. (Eds.): IWDM 2010, LNCS 6136, pp. 78–85, 2010.

indicate that dose could be reduced with little detriment to the detection of masses. However, the task of microcalcification detection appeared to have significantly greater dependence on the increased relative noise in the images. This is because breast structure noise dominates when trying to detect relatively large objects such as masses but detector noise is important for the detection of small details such as microcalcifications. In contrast-detail tasks such as scoring of images of the CDMAM phantom, there is a direct link between increasing dose and the detection of small details[7]. This study aimed to validate a method to simulate reduced radiation dose in digital mammograms. The same methodology is applied to mastectomy images for further analysis with use of spectral analysis[8,9].

2 Methodology

2.1 Materials

Sixteen images of the CDMAM (version 3.4) were acquired at each of five doses using the Hologic Selenia system with the CDMAM phantom placed between two slabs of PMMA 2cm thick. The images acquired at the highest dose (31kVp Mo/Rh, 5.91mGy mean glandular dose) were used as the reference images for the subsequent synthesised low dose images (0.34, 0.80, 1.48, 2.95mGy MGD). Flat field images for 5cm of PMMA were acquired at the same beam quality, exposure settings and setup as the CDMAM images. Detector response and NPS measurements were carried out on these flat field images and used for the dose reduction methodology.

Four sets of mastectomy images were acquired on the Siemens Novation system. Each mastectomy sample was placed on the breast support and compressed. The mastectomy sample was exposed with the tube voltage held constant whilst the tube current time product was varied. Care was taken not to move the mastectomy sample after each exposure. The highest dose mastectomy image was used as the reference image for synthesizing subsequent lowered dose images. Table 1 shows the doses used for each mastectomy set as well as their compressed thicknesses. Flat field images using three and five centimetres of PMMA was used in the same setup as the mastectomy samples for detector response and NPS measurements to be used with the dose reduction methodology. Note that all images used in this study are raw and unprocessed.

2.2 Dose Reduction

Båth et al's methodology[10], which had been previously applied to chest X-ray images, has now been implemented for the first time to degrade the following experimentally acquired mammography images: CDMAM and mastectomy images. Furthermore, the NPS was modelled to three noise sources: electronic, quantum and structural. Firstly, the linearised pixel intensity values of the original image, $I_o(x, y)$, were scaled by a ratio of the dose to be simulated, D_{sim}, to the original

Table 1. Beam quality (W = Tungsten, Rh = Rhodium, Mo = Molybdenum), Tube current time product and mean glandular dose settings for each mastectomy sample set. Doses in brackets denote the highest dose image from which the other low dose images were derived.

Set	Beam Quality	Tube Current time product (mAs)	Mean Glandular Dose (mGy)	Compressed Breast Thickness (mm)
1	29 W/Rh	28, 57, 110 (220)	0.3, 0.7, 1.4 (2.7)	50
2	29 W/Rh	10, 20, 50, (68)	0.1, 0.2, 0.4 (0.8)	50
3	25 Mo/Rh	10 (61)	0.2 (1.3)	30
4	25 Mo/Rh	20, 34, 45 (199)	0.4, 0.7, 1.0 (4.3)	30

dose of the image, D_o, to give the original scaled image, $I_{o,sc}$. NPS was measured for each dose level and fitted to each point in the NPS array against detector air kerma, K:

$$NPS_{tot}(u,v) = NPS_e(u,v) + NPS_q(u,v).K + NPS_s(u,v).K^2 \tag{1}$$

where NPS_{tot}, NPS_e, NPS_q and NPS_s are the total, electronic, quantum and structure noise sources, respectively.

Structural noise scales with the dose correction, therefore only electronic and quantum noise images were created. Noise images were created to account for the difference in noise level and frequency content between the $I_{o,sc}$ and that expected for a lower dose image, D_{sim}. This was undertaken through the use of a frequency-based noise transfer function, NTF, equal to the square root of the difference of the NPS between the two images, I_{sim} and $I_{o,sc}$, for each noise source and applied to Gaussian noise images.

$$I_N(x,y) = FT^{-1}(NTF(u,v).(FT(I_G(x,y)))) \tag{2}$$

where

$$NTF(u,v) = \sqrt{\frac{NPS_{sim}(u,v) - NPS_{o,sc}(u,v)}{NPS_{sim}(0,0) - NPS_{o,sc}(0,0)}} \tag{3}$$

For the quantum noise, the difference in NPS_q for the I_o, sc and I_{sim} was applied to equation 2. Both noise sources were scaled on a pixel-by-pixel basis as shown:

$$I_{sim}(x,y) = I_{o,sc}(x,y) + \left[k_e.I_{N,e}(x,y).\left(1 - (D_{sim}/D_{orig})^2\right)\right] + [k_Q.(I_{o,sc}(x,y))^n.I_{N,Q}(x,y)] \tag{4}$$

where k_e, k_Q and n are coefficients to be fitted. n was found to be 0.5 ± 0.03 for both systems and settings implemented in this study.

This method was then applied to derive synthetically dose-reduced CDMAM and mastectomy images. For degrading CDMAM images the highest dose (5.91mGy MGD) CDMAM image was used as the reference image from which low dose images (0.34, 0.80, 1.48, 2.95mGy MGD) were subsequently obtained. All experimentally acquired CDMAM images were scored with an automated scoring tool (CDCOM[11]) and contrast-detail curves derived for each dose as outlined in [12]. The same procedure was repeated for the synthesized dose reduced CDMAM images and compared with the experimentally acquired counterparts.

The dose reduced mastectomy images were compared, quantitatively and visually, with their experimentally acquired counterparts. In addition, noise power spectra[8,9] were measured in the same region of interest for each dose for the simulated and experimental mastectomy images and the profiles compared. To ensure the same region was analysed in each image and that results were not affected by any slight movement of the mastectomy sample, the mastectomy images were checked for alignment. In addition, NPS analysis was undertaken within the central portion of the mastectomy samples to ensure uniform thickness was maintained in the region of interest.

3 Results

Figure 1 show the normalised NPS profiles of high dose flat-field images and the dose reduced flat-field images for the Hologic and Siemens Novation systems. For the Hologic system, there was a relative difference between the experimental and synthetic low dose shown of approximately 4.7% averaged across the radial averaged NNPS. For the Siemens Novation system, there was a relative difference between the experimental and sythetic low dose shown of 2.7%.

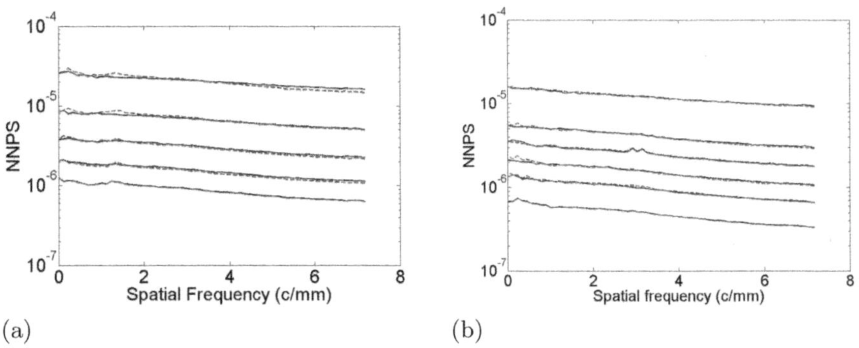

Fig. 1. (a)Normalised NPS profiles for Hologic system. NNPS curves are given for (from top down) 0.34, 0.80, 1.48, 2.95 and 5.91mGy MGD. (b) NNPS profiles for the Siemens Novation. NNPS curves are given for (from top down) 0.2, 0.4, 0.7, 1.0, 1.3 and 4.3mGy MGD. Experimental data is shown in solid blue whilst the synthetic data is shown in the dashed red curve.

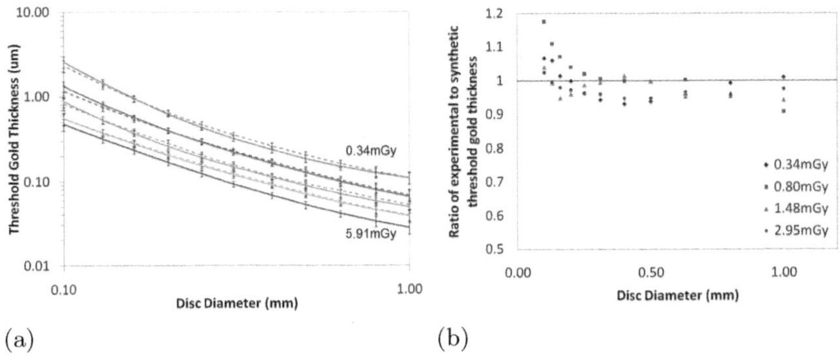

Fig. 2. (a)Contrast detail curves obtained from experimentally acquired CDMAM images at different doses (solid curves) and the simulated reduced dose CDMAM images (dashed curves). Four doses (0.34, 0.80, 1.48, 2.95mGy MGD, from top down) were simulated from the highest dose of 5.91mGy. Error bars denote two standard error means.

3.1 CDMAM Images

Figure 2 shows the contrast detail curves obtained for the experimental and dose-reduced CDMAM images. The dashed curves denote the fitted predicted human threshold contrasts[12] for the synthetic data. The noise added synthetically have led to the contrast details curves to maintain the gradient of the original high dose result, whilst shifting up with increasing relative noise in the image as expected. All the synthetic data curves are similar to that of the experimentally acquired and error bars indicate they are within range of each other. A ratio of the experimental to synthetic results were found to be, across all discs, on average 0.97 for all doses as shown in Figure 2b. This shows that noise may have been overestimated in the synthetic images compared to the experimental CDMAM images.

3.2 Mastectomy Images

Figure 3 shows a region of one set of mastectomy images at 4.3 and 1.0mGy MGD for the experimental (Figure 3b) and synthetic (Figure 3c images, respectively. The appearance of the lesion (a mass with microcalcifications) in the region shown degrades dramatically as the dose was reduced down to one quarter of the high dose image. A profile of the experimental and synthetic images is also shown in Figure 3d, a difference of 1% was found between the profiles. Figure 4 shows the radially averaged noise power spectra for four mastectomy images sets, the settings of which are shown in Table 1. As the dose increased, the system noise increased as expected. NPS of the mastectomy images lead to a relative difference of 7% on average across all doses. The difference increased as the difference between the original dose and the dose to be simulated became greater. The largest difference between experimental and synthetic NPS was found at the higher spatial frequencies.

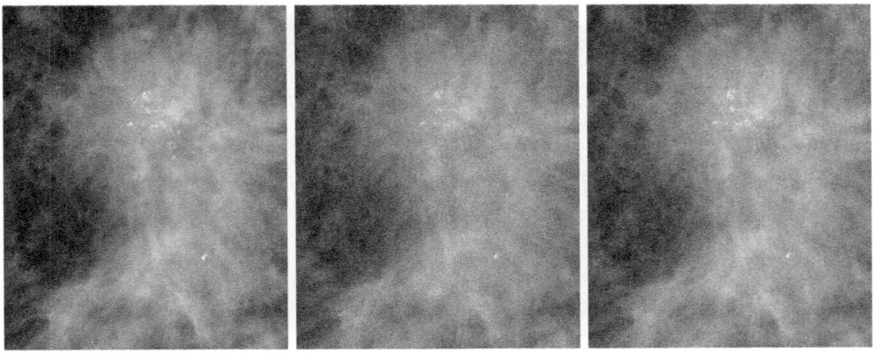

(a) (b) (c)

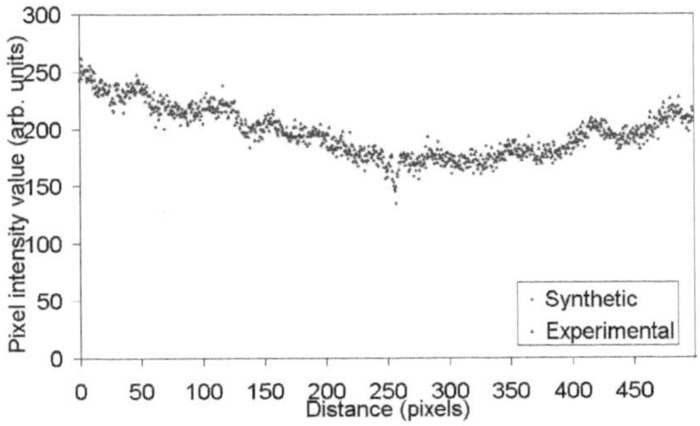

(d)

Fig. 3. (a) and (b) show the same cropped region of a mass with microcalcifications in a mastectomy sample acquired at 4.3 and 1.0mGy (MGD), respectively. (c) shows the mastectomy sample synthetically dose reduced from 4.3 to 1.0mGy. (d) shows the profiles taken across (b) and (c) to compare the experimentally acquired low dose image with the simulated low dose image.

4 Discussion

The contrast detail curves of the simulated dose reduced CDMAM images match well to that of the experimentally acquired images. The detection of smaller discs were inferior for the CDMAM images simulated at a very low dose compared with the CDMAM images experimentally acquired at the low dose. This suggests there may have been some overestimation of the noise to be added as the difference between the original and simulated doses increased.

For the mastectomy work, the simulation has visibly degraded the appearance of the lesions in the mastectomy images, similar to that acquired at lower doses. The noise power spectra measurement of the mastectomy samples at different doses have shown how the system noise increases whilst the shape of the underlying anatomical noise power remained. Larger discrepancies were found in the

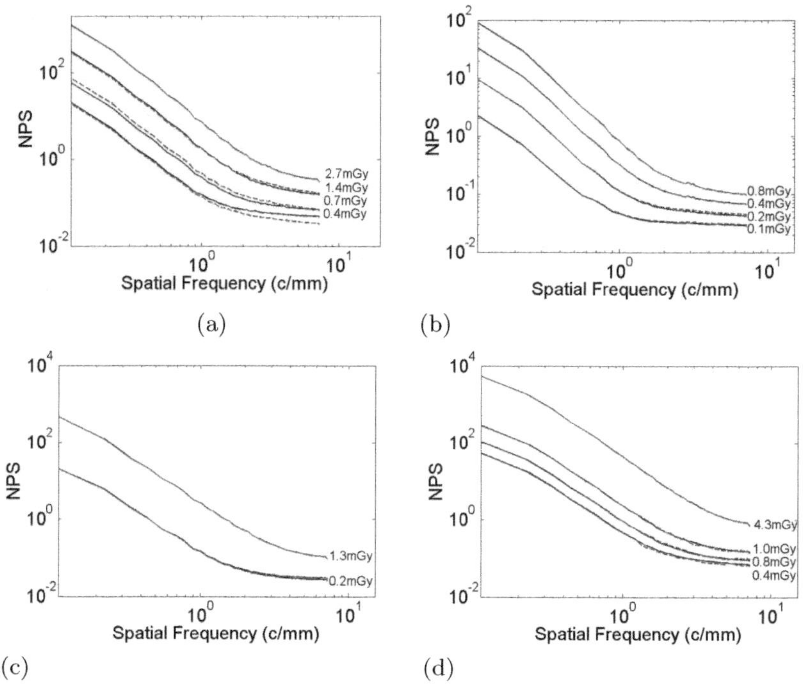

(a) (b) (c) (d)

Fig. 4. (a – d) Radially averaged noise power spectra of a region in sets 1 – 4, respectively, of mastectomy images acquired at different doses and their synthetic counterparts. The dark curve represents the highest dose image from which the other low dose images were degraded from. Red dashed curves illustrates the NPS of synthetically degraded images whilst the blue solid curves show the NPS of their experimentally acquired counterparts.

higher spatial frequencies which corresponds to the poorer detection rates for the small discs in.

Despite reducing the dose to one quarter of the highest dose, the microcalcifications were still visible. However, the very fine calcifications became harder to discern as the relative noise increased in the low dose images. The work presented here shows raw images, thus affects from image processing packages have not been taken into account. Comparison of unprocessed and processed experimentally acquired mastectomy images showed a great difference in the image quality of microcalcifications for the high dose and low dose images. Further application of an image processing package on the synthetic images for further observer studies may be carried out in the future.

In conclusion, this work has validated a method to simulate reduced dose in mammography images, developing on Båth et al's work to incorporate other noise components. Such a method could be applied to study the effect of dose reduction on mass and/or microcalcification detection in clinical mammograms where optimal X-ray exposure factors may be deduced. This work could be developed further to assess the effects of applying alternative beam qualities or even acquiring images of the same mastectomy sample with different detectors.

Acknowledgements. This work is part of the OPTIMAM project and is supported by the CR-UK & EPSRC Cancer Imaging Programme in Surrey, in association with the MRC and Department of Health (England).

References

1. Samei, E., Saunders, R.S., Baker, J.A., Delong, D.M.: Digital Mammography: Effects of Reduced Radiation Dose on Diagnostic Performance. Radiology 243(2), 396–404 (2007)
2. Ruschin, M., Timberg, P., Båth, M., Hemdal, B., Svahn, T., Saunders, R.S., Samei, E., Andersson, I., Mattson, S., Chakraborty, D.P., Tingberg, A.: Dose dependence of mass and microcalcification detection in digital mammography: Free response human observer studies. Med. Phys. 34(2), 400–407 (2007)
3. Saunders, R.S., Samei, E.: A method for modifying the image quality parameters of digital radiographic images. Med. Phys. 30(11), 3006–3017 (2003)
4. Workman, A.: Simulation of Digital Mammography. In: Flynn, M.J. (ed.) Medical Imaging 2005: Physics of Medical Imaging. SPIE, vol. 5745, pp. 933–942 (2005)
5. Zanca, F., Chakraborty, D.P., Van Ongeval, C., Jacobs, J., Claus, F., Marchal, G., Bosmans, H.: An improved method for simulating microcalcifications in digital mammograms. Med. Phys. 35(9), 4012–4018 (2008)
6. Grosjean, B., Muller, S.: Impact of Textured Background on Scoring of Simulated CDMAM Phantom. In: Astley, S.M., Brady, M., Rose, C., Zwiggelaar, R. (eds.) IWDM 2006. LNCS, vol. 4046, pp. 460–467. Springer, Heidelberg (2006)
7. Young, K.C., Oduko, J., Woolley, L.: Technical Evaluation of the Hologic Selenia Full Field Digital Mammography System. Technical Report from National Coordinating Centre for the Physics of Mammography (2007)
8. Bochud, F.O., Valley, J.-F., Verdun, F.R.: Estimation of the noisy component of anatomical backgrounds. Med. Phys. 26(7), 1365–1370 (1999)
9. Heine, J.J.: On the statistical nature of mammograms. Med. Phys. 26(11), 2254–2265 (1999)
10. Båth, M., Håkansson, M., Tingberg, A., Månsson, L.G.: Method of Simulating Dose Reduction for Digital Radiographic Systems. Radiation Protection Dosimetry 114(1-3), 253–259 (2005)
11. Visser, M., Karssemeijer, N.: Manual CDCOM version 1.5: software for automated readout of CDMAM 3.4 images, `http://www.euref.org`
12. Young, K.C., Cook, J.J.H., Oduko, J.M.: Automated and human determination of threshold contrast for digital mammography systems. In: Astley, S.M., Brady, M., Rose, C., Zwiggelaar, R. (eds.) IWDM 2006. LNCS, vol. 4046, pp. 266–272. Springer, Heidelberg (2006)

Measured Dose versus Organ Dose Performance in Digital Mammography Systems

James B. McCullagh[1], Paola Baldelli[2], and Niall Phelan[2]

[1] BreastCheck, National Cancer Screening Service, Infirmary Road, Cork, Ireland

[2] BreastCheck, National Cancer Screening Service, 36 Eccles St., Dublin 7, Ireland

{James.Mccullagh,Paola.Baldelli,Niall.Phelan}@nbsp.ie

Abstract. The results of a dose survey of digital x-ray imaging equipment in a breast screening programme have been used to compare the measured dose values to the dose values displayed by the mammography systems. A total of 12,104 images were used in the survey, acquired on three types of digital mammography system: GE Essential, Hologic Selenia and Sectra MDM L30. The results reveal that estimates of the MGD were largely similar between measured and displayed dose for all models in the survey. It was found that large discrepancies could occur on individual systems which may be attributable to various factors including inaccurate QA measurement data and poorly calibrated mammography systems.

Keywords: Digital Mammography, Radiation Dose, Mean Glandular Dose, Quality Assurance.

1 Introduction

Estimation of the breast Mean Glandular Dose (MGD) is a standard feature on modern full field digital mammography (FFDM) systems [1]. In general the MGD associated with an exposure is displayed on the console and stored in the Organ Dose field of the image DICOM header. An indication of the breast dose is valuable because it provides information about the radiographic technique, the variation in dose with breast size and composition encountered in a given population of screened women and the performance of the imaging system. The breast MGD cannot be measured directly but can be calculated from the breast entrance surface air kerma (ESAK), the compressed breast thickness and x-ray spectral information using appropriate conversion factors [2-4]. Other commonly used conversion factors in modern FFDM systems include those published by Wu et al [5, 6].

Compliance with European and National patient protection legislation requires a regular dose survey of x-ray equipment in the screening programme. The approach currently adopted by our group is automatic extraction of the exposure data from the DICOM image headers and calculation of the breast dose using software which is based on conversion factors compiled by Dance et al [2-4]. The organ dose value stored in the DICOM header presents an alternative approach for the extraction and

J. Martí et al. (Eds.): IWDM 2010, LNCS 6136, pp. 86–91, 2010.

compilation of the dose survey data. There are errors common to both approaches arising from errors in breast composition (~10%), breast thickness measurement (~5%), dosimeter calibration (~3%), HVL measurement (~2%) and dosimeter positioning (~ 1%) which can result in an overall approximate error of 12% in a patient dose calculation [7]. Both approaches also have their own sources of error. For example, one digital mammography system does not include the pre-exposure contribution with the main exposure factors which can be a source of discrepancy. Equally, the organ dose accuracy of a particular system is dependent on accurate field calibration by the manufacturer/service vendor.

There are three types of digital mammography system in use in the Irish National Breast Screening Programme (BreastCheck): General Electric (GE) Essential, Hologic Selenia and Sectra MicroDose L30. The aim of this study was to compare the MGD results obtained by measurement with the organ dose values for each type of system using data collected for a BreastCheck dose survey. This involved comparison of the average MGD per view and the MGD performance over a range of compressed breast thicknesses. Satisfactory comparison would provide assurance of the accuracy of organ dose estimation by digital mammography systems and raise the possibility of its use in future dose surveys.

2 Method

The survey included 28 x-ray systems composed of 11 GE Essential units, 10 Hologic units and 7 Sectra MDM L30 units. The images of at least 100 examinations from each digital mammography system acquired over a three month period between July and September 2009 were used. Each examination contains 4 images: a mediolateral oblique (OB) view and a craniocaudal (CC) view for each breast. Examinations for each system were transferred from the PACS archive, the parameters were extracted from the DICOM headers using a locally developed MATLAB routine (Mathworks, Cambridge, UK) and transferred to a Microsoft® Office Excel spreadsheet for further analysis. The parameters extracted from the header for this survey included the patient identification, age, compressed breast thickness, compression force, view position, laterality, exposure kVp, exposure mAs, target material, filter material, organ dose and pre-exposure factors where appropriate.

The Mean Glandular Dose *(MGD)* for each acquired image was calculated according to Dance et al [2-4] using the formula:

$$MGD = Kgcs\,. \tag{1}$$

where K is the Entrance Surface Air Kerma (*ESAK*) at the upper surface of the breast and g, c and s depend on both x-ray beam and breast characteristics. In particular, the factor s corrects for any difference due to the use of an x-ray spectrum different from Mo/Mo. K was calculated from the tube output a fixed distance above the detector, the tube current exposure time (mAs), the source-to-detector distance and the thickness of the compressed breast. Values for K and the Half Value Layer (HVL) for each unit in the survey were obtained from medical physics quality assurance test results contemporaneous with the survey period. Once the acquisition parameters were extracted from the DICOM header file, dose calculation was carried out using software developed by KC Young for the National Health Service Breast Screening Program (NHSBSP) [8] with some modification made to take into account new *c*-, *g*- and *s*- factors for HVL values associated with the spectral data for new mammography systems. Further modifications were made to the dose calculations to account for the larger HVL found in the new mammography systems. The s-factor values used in dose calculations with tungsten/aluminum target/filter combinations were obtained from a table of s-factors provided by Dance et al [4].

All x-ray units in BreastCheck are operated in fully automatic mode. In the systems which use a pre-exposure, the pre-exposure dose contribution is included in the organ dose estimate. For the Hologic Selenia, the exposure factors stored in the DICOM header include the pre-exposure contribution. For the GE Essential, the pre-exposure factors are not included with the main exposure factors but are stored in a separate field in the DICOM header. The accuracy of GE Essential dose calculations were improved by extraction of the pre-exposure factors, calculation of the pre-exposure dose and adding this to exposure dose for each image acquired.

The accuracy and reproducibility of the breast compressed thickness, tube voltage, x-ray tube outputs and half-value layer measurements for clinical kVp settings and target/filter combinations were periodically checked as part of a routine QA protocol according to European recommendations [9].

3 Results

A total of 2910 examinations were included in the survey. Including extra images, this amounted to 6047 CC and 6057 OB views. The overall average breast thickness was 60.5mm and 63mm, and the median breast thickness was 61mm and 63mm for the CC and OB views respectively. Table 1 shows the average MGD dose obtained by measurement, the average MGD obtained from the organ dose values and the percentage difference, according to model of imaging system. Table 1 demonstrates good correspondence for the GE Essential and Hologic Selenia systems but a larger difference for the Sectra L30 systems. Closer inspection of the results for individual Sectra systems revealed that two systems had differences less than 10%, two systems had differences less than 15% and the remainder had differences of approximately 45%, 37% and 27%.

In Figure 1, the average measured MGD per view was plotted as a function of compressed breast thickness for each type of digital mammography system in the survey. All three models demonstrated increasing mean glandular dose with increasing breast thickness. However, for the Sectra L30 system, the MGD appears to plateau for breast thicknesses greater than 60mm.

The average MGD per view obtained by measurement and from the organ dose field was plotted as a function of compressed breast thickness and the average difference was calculated for each unit in the survey. The system in best agreement (Figure 2) had an average difference of 3.3% and the system in least agreement (Figure 3) had an average difference of 33%. In summary, it was observed that over the entire range of compressed breast thickness, 75% of the units in the survey exhibited an average difference between measured and organ dose of less than 15%.

Table 1. Measured and Organ MGD according to Model (Errors represent 95% confidence limits)

Model	View	Measured Average MGD per Exposure (mGy)	Organ Dose Average MGD per Exposure (mGy)	Percentage Difference
MDM L30	CC	0.96 ± 0.01	0.72 ± 0.01	24.6%
	OB	0.93 ± 0.01	0.71 ± 0.01	24.0%
Selenia	CC	1.36 ± 0.02	1.24 ± 0.02	8.6%
	OB	1.44 ± 0.02	1.31 ± 0.02	8.8%
Essential	CC	1.40 ± 0.01	1.34 ± 0.01	4.2%
	OB	1.52 ± 0.02	1.44 ± 0.02	5.1%

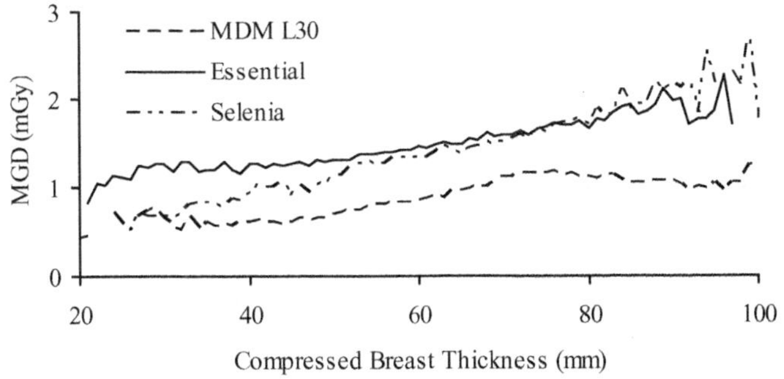

Fig. 1. Average MGD per view as a function of compressed breast thickness

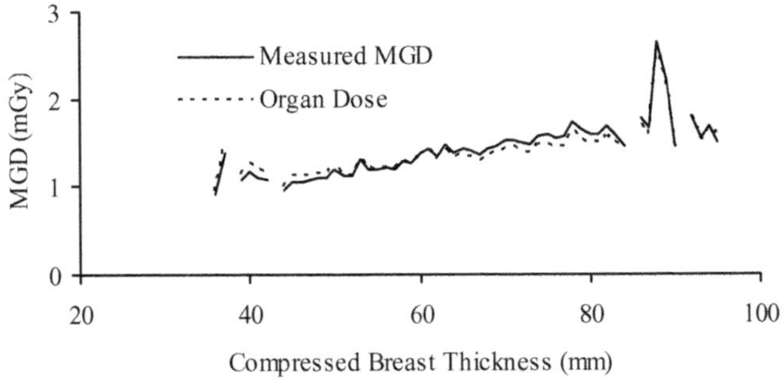

Fig. 2. Comparison of the average MGD values for the unit with the smallest average difference between the measured and organ dose estimates

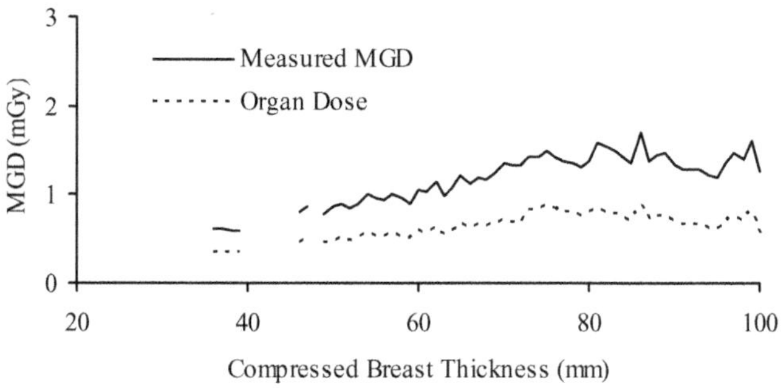

Fig. 3. Comparison of the average MGD values for the unit with the largest average difference between the measured and organ dose estimates

4 Discussion

As shown in Table 1, the GE Essential and Hologic Selenia systems provided similar estimates of measured and organ dose for the average MGD to the breast whereas the Sectra MDM L30 exhibited the greatest difference. This result was contrary to expectation in terms of the conversion factors used by each system. The Sectra systems use the Dance [2-4] conversion factors which we used in our calculation, whereas, the GE and Hologic systems use the conversion factors published by Wu & Barnes [5, 6]. However closer inspection of the results of individual Sectra systems revealed better correspondence between measured and organ dose for four of the Sectra systems. This suggests that the large differences observed for the remaining systems may have been particular to these systems. Possible causes of these large differences include poor system calibration or inaccurate QA measurement data.

As expected, the average MGD increased with compressed breast thickness for all systems included in the survey. Figure 1 clearly demonstrates the lower breast dose of the MDM L30 versus the other two models across the range of breast thicknesses in the survey. It also indicates that the Selenia achieves a lower breast dose for smaller breasts in comparison to the Essential. The rate of increase in dose with breast thickness appears to be similar for all three models until approx 60mm. For greater breast thickness, the breast dose associated with the GE Essential and the Hologic Selenia models continues to increase but appears to plateau for the Sectra systems.

The graphs of measured and organ dose as a function of compressed breast thickness provide a good visual indication of the comparative performance of each system type. As illustrated by Figures 2 and 3, the dose trends were very similar even for the individual units with large differences between measured and organ dose. The graphs also demonstrate that the organ dose underestimated the measured dose for all but two units in the survey.

Cole et al performed a comparative study of the organ dose with established methods of breast dose estimation on a number of screen-film systems and on one full field digital mammography system [10]. They found that once system accuracy was established the organ dose could potentially be used for dose reference level (DRL) comparison. The results from this study also support the use of the organ dose for DRL comparison once system accuracy has been established.

References

1. KCARE: Report 05037 Comparative Specifications of Full Field Digital Mammography Systems. KCARE, Kings College Hospital (2005)
2. Dance, D.R.: Monte Carlo Calculation of Conversion Factors for Estimation of Mean Glandular Breast Dose. Phys. Med. Biol. 35, 1211–1219 (1990)
3. Dance, D.R., Skinner, C.L., Young, K.C., Beckett, J.R., Kotre, C.J.: Additional Factors for the Estimation of Mean Glandular Breast Dose Using the UK Mammography Dosimetry Protocol. Phys. Med. Biol. 45, 3225–3240 (2000)
4. Dance, D.R., Young, K.C., van Engen, R.E.: Further Factors for the Estimation of Mean Glandular Dose using the United Kingdom, European and IAEA Breast Dosimetry Protocols. Phys. Med. Biol. 54, 4361–4372 (2009)
5. Wu, X., Barnes, G.T., Tucker, D.M.: Spectral Dependence of Glandular Tissue Dose in Screen-Film Mammography. Radiology 179, 143–148 (1991)
6. Wu, X., Gingold, E.L., Barnes, G.T., Tucker, D.M.: Normalized Average Glandular Dose in Molybdenum Target-Rhodium Filter and Rhodium Target-Rhodium Filter Mammography. Radiology 193, 83–89 (1994)
7. Burch, A.: Measurement of Mean Glandular Dose. In: Mammography Physics Training, IPEM, Aston University, Birmingham (2009)
8. NHSBSP Report 04/05: Breast Dose Surveys in the NSBSP: Software and Instruction Manual. Version 2.0 (2004)
9. Van Engen, R., van Woudengerg, S., Bosmans, H., Young, K., Thijssen, M.: European Protocol for the Quality Control of the Physical and Technical Aspects of Mammography Screening. In: Perry, N., Broeders, M., de Wolf, C., Törnberg, S., Holland, R., von Karsa, L. (eds.) European guidelines for quality assurance in breast cancer screening and diagnosis, 4th edn., pp. 61–165. European Commission, Luxemburg (2006)
10. Cole, H., Lawinski, C.P., Goodman, D.A.: A Comparative Study of Indicated Breast Dose in Conventional and Digital Mammography Systems. UKRC (2003)

Noise Reduction in Dual-Energy Contrast Enhanced Digital Breast Tomosynthesis Using Regularization

Răzvan Iordache, Maélène Lohezic, Giovanni Palma, Sylvie Puong, and Serge Muller

GE Healthcare, 283 rue de la minière, 78533 Buc, France
{Razvan.Iordache,Maelene.Lohezic,Giovanni.Palma,Sylvie.Puong, Serge.Muller}@ge.com

Abstract. Dual-Energy Contrast Enhanced Digital breast tomosynthesis is an emerging technique for breast cancer detection, which combines the strengths of functional and morphological 3D imaging. The projection images acquired with two energy spectra at several angulations are combined to obtain "iodine" projections. These are then reconstructed to provide 3D iodine images. The combination process significantly increases the noise in the images, which is further amplified by the 3D reconstruction. This paper proposes a regularized reconstruction method based on the simultaneous algebraic reconstruction technique to be used for the reconstruction of the iodine volume. The regularization represents a constraint for the reconstructed volume, which causes the reduction of the noise and preserves the structures of interest. Preliminary results on clinical data demonstrate a significant increase of the visibility of iodine-enhanced regions without affecting their sharpness and morphology.

Keywords: digital breast tomosynthesis, dual energy, iodine imaging, functional imaging.

1 Introduction

During cancer development, angiogenesis occurs in the vicinity of tumors to support their growth [1]. These newly formed blood vessels are characterized by an increased permeability and cause the blood to pool around the tumors.

Contrast-enhanced digital breast tomosynthesis (CE-DBT) takes advantage of this physiological phenomenon to provide 3D functional images of the breast tumor vasculature using an iodinated vascular contrast agent [2, 3, 4]. In dual-energy (DE) CE-DBT, post-contrast projections are acquired at energy levels below and above the K-edge of iodine. Projection images containing only the iodine-uptake are obtained by combining the low-energy and the high-energy projections acquired at the same angle. An iodine volume image is then reconstructed from the iodine projections.

The suppression of anatomical noise in the iodine projections comes at the expense of an increased stochastic noise due to the DE recombination process. The 3D reconstruction further amplifies the noise. The iodine-uptake in tumors is expected to be lower than 4 mg/cm^3, corresponding to a modest increased x-ray attenuation and, consequently, to a low signal difference between the iodine-enhanced regions and the

J. Martí et al. (Eds.): IWDM 2010, LNCS 6136, pp. 92–99, 2010.

background. Therefore the iodine-enhanced regions will exhibit low signal-difference-to-noise (SDNR) levels in the iodine volume.

To address noise reduction, total variation-based noise filtering has been successfully applied on the iodine projections prior to reconstruction, without degrading clinical information [3].

Simultaneous Algebraic Reconstruction Technique (SART) is currently used in both DBT [5] and DE CE-DBT [4] applications. Appropriate a-priori information on the local signal behavior could be integrated into a regularized SART reconstruction to reduce the noise while preserving the morphology and the sharpness of the iodine-enhanced regions.

In this paper, we propose to constrain the noise propagation directly during the reconstruction process by using a regularized SART algorithm.

2 Materials and Methods

2.1 Regularized SART

Consider that the volume to be reconstructed is represented as vector of J elements, $V = [v_j]$, where v_j is the linear attenuation coefficient for the jth voxel. There are I measurements available, represented as a vector $P = [p_i]$. These measurements are logarithmic transforms of the detector pixel values obtained for the different projection views. The geometry of the acquisition system is modeled by the I x J projection matrix $R = [r_{ij}]$, where r_{ij} represents the contribution of the voxel j to the measurement (ray) i. Thus, the tomosynthesis acquisition can be modeled as $P = RV$.

The reconstruction problem consists in inverting the projection matrix R, which is underdetermined and high-dimensioned. Algebraic techniques are iterative methods to solve this problem. An estimate of the solution is projected using the projection matrix and compared to the measurements at each iteration. The resulting error is back-projected to update the estimate.

The SART is an algebraic reconstruction technique where the update is performed after all rays in one projection view have been processed [6]. The estimation of the volume at step $(q + 1)$ is done using the projection $n^{(q)}$:

$$v_j^{(q+1)} = v_j^{(q)} + \lambda^{(q)} \frac{1}{\sum_{i \in \text{projection } n^{(q)}} r_{ij}} \sum_{i \in \text{projection } n^{(q)}} r_{ij} \left(\frac{p_i - \sum_{j=1}^{J} r_{ij} v_j^{(q)}}{\sum_{j=1}^{J} r_{ij} v_j^{(q)}} \right). \tag{1}$$

An iteration is completed when all measurements are used exactly once. Therefore the volume is updated N times during one iteration, where N is the number of projection views. The relaxation factor $\lambda^{(q)}$ is between 0 and 2.

It was proven in [7] that the SART is a particular case of a general iterative scheme for image reconstruction based on Landweber's method, and that it converges to the solution of the following minimum weighted least square estimation problem:

$$\arg\min_v \|p - Rv\|_W^2 = \arg\min_v \{(p - Av)^t W(p - Av)\}, \tag{2}$$

where $W = \mathrm{diag}\{1/\sum_{i=1}^{I} r_i\}$.

The regularized version of equation (2) is:

$$\arg\min_v \{\|p - Rv\|_W^2 + \beta U(v)\}, \tag{3}$$

where β is a positive weighting factor and the regularization term, $U(v) = \sum_{(j,k)\in C} b_{jk}\rho(v_j - v_k)$, is the negative log probability of a prior distribution modeled by a generalized Gaussian Markov random field [8]. C is a set of pairs of neighboring voxels, b_{jk} are directional weighting coefficients, and ρ is a potential function chosen so that it penalizes noise in the reconstructed volume while preserving edges.

Using the preconditioned gradient method employed in [7], it can be shown that the following iterative scheme, called herein regularized SART, converges to the solution of (3):

$$v_j^{(q+1)} = v_j^{(q)} + \lambda^{(q)} \frac{1}{\sum_{i\in \mathrm{proj}\ n^{(q)}} r_{ij}} \left(\sum_{i\in \mathrm{proj}\ n^{(q)}} r_{ij} \left(\frac{p_i - \sum_{j=1}^{J} r_{ij} v_j^{(q)}}{\sum_{j=1}^{J} r_{ij} v_j^{(q)}} \right) - \beta \sum_{\{j,k\}\in C} b_{jk}\rho'\left(v_j^{(q)} - v_k^{(q)}\right) \right) \tag{4}$$

The potential function used in this work was introduced in [9] and has a parameter d that determines the level of smoothing applied:

$$\rho(t) = \delta^2 \left(\frac{|t|}{\delta} - \ln(1 + \frac{|t|}{\delta}) \right), \delta > 0. \tag{5}$$

Voxel value differences below d, which are considered to correspond to noise, are smoothed, whereas differences above d, which correspond to edges, are preserved.

2.2 Evaluation of the Regularized Reconstruction

The evaluation of R-SART was performed both using phantom images and on a clinical case selected from a previously performed pilot trial on DE CE-DBT. Images were acquired with a Senographe DS-based prototype DBT system (GE Healthcare, Chalfont St Giles, UK) modified to allow interleaved LE and HE image acquisitions in a single x-ray tube sweep. Each LE and HE image sequence consisted of 15 projection images acquired over a 40 degree arc. LE images were acquired using acquisition parameters typically employed to retrieve anatomy images, HE images were acquired at 49 kVp using a Mo or Rh target and a Cu filter. The LE and HE projection images were decomposed into iodine equivalent projection images using a previously developed algorithm

[10]. The decomposed projection images were then reconstructed to obtain DE CE-DBT (iodine) volumes using SART and R-SART. Three iterations were used for both SART and R-SART. "Iodine" volumes were reconstructed in planes (slices) parallel to the detector using 1 mm increments with an in-plane pixel size of 0.1 mm^2.

A digital subtraction angiography (DSA) phantom (Nuclear Associates - DSA phantom, linearity insert), composed of a 2.4 cm block of PMMA with embedded iodine inserts at different surfacic concentrations was imaged. Sections of 50% breast-equivalent material (BR12 plates manufactured by CIRS) were added to reach a thickness of 5 cm.

For patient imaging Visipaque-320 (GE Healthcare, Chalfont St Giles, UK) was power-injected at a dosage of 1ml/kg bodyweight prior to image acquisition. Then, the breast was positioned for MLO view image acquisition using normal compression.

The reconstructed volumes were visually inspected to assess an overall impression of vascular enhancement (clinical case only), lesion/iodine inserts morphology, and image noise. The cross-sectional images were reviewed in stack mode on a SenoAdvantage review workstation (GE Healthcare, Chalfont St Giles, UK).

A slice of interest was selected in each "iodine" volume for further analysis. The slice contains in-focus iodine inserts for the phantom dataset, and an in-focus iodine enhanced lesions for the clinical case. The following criteria were used to compare the performance of the two reconstruction algorithms:

(1) signal-difference-to-noise ratio per pixel between an iodine insert/iodine enhanced lesion and non-enhanced background, as a measure for the detectability of iodine in the DE CE-DBT volumes. *SDNR* per pixel was computed as:

$$SDNR = \frac{SI_{iodine} - SI_{background}}{\sigma_{background}}, \tag{6}$$

where SI_{iodine} and $SI_{background}$ are the per-pixel average signal intensities in the iodine insert/iodine enhanced lesion and non-enhanced background and $\sigma_{background}$ is the standard deviation in the background;
(2) in-plane noise power spectrum (NPS) estimated in the central uniform region of the DSA phantom, to evaluate the spectral properties of the noise. Its computation was performed by averaging Fourier power spectra of overlapping, detrending corrected, 256×256 pixel ROIs;
(3) line profiles through the center of iodine inserts/enhanced lesions, as a measure of sharpness of the margins of inserts/lesions;
(4) visual inspection of the pixel-to-pixel difference between the reconstructions of the same region of interest with the two algorithms to evaluate the information loss caused by the regularization.

3 Results

Figure 1 illustrates a region of the slice of interest through the DSA phantom, containing the 0.5 mg/cm^2 insert in focus. The use of the regularization leads to the increase of the *SDNR* per pixel of the insert from 0.52 (SART reconstruction) to 1.24 (R-SART reconstruction). SDNR was computed using (6), where the average signal intensity in the iodine was estimated in ROI 1 and the average signal intensity and the noise in the background were estimated in ROI 2.

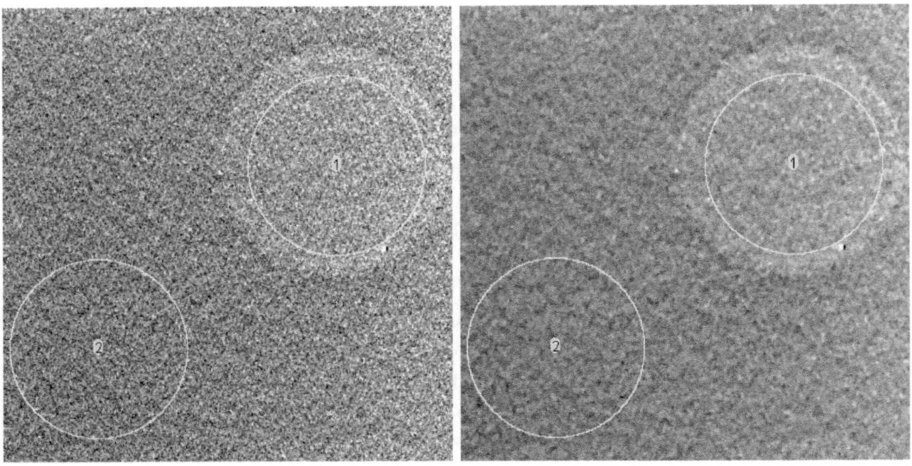

Fig. 1. Detail of the slice of interest through the DSA phantom, containing the 0.5mg/cm^2 iodine insert, reconstructed using SART (left) and R-SART (right). The insert *SDNR* per pixel is 0.85 for SART and 2.02 R-SART.

To facilitate quantitative comparison of the in-plane NPS of the reconstruction schemes, the measured NPS in the x-direction (parallel to the sweep plane) for $f_y = 0$ lp/mm and in the y-direction (perpendicular to the sweep plane) for $f_x = 0$ lp/mm are shown in Figure 2. Due to the high noise on the central axes, seven lines on each side of $f_y = 0$ lp/mm and, respectively, $f_x = 0$ lp/mm were averaged to compute the 1D NPS. The in-plane NPS is anisotropic, which is consistent with previously reported results [11]. The regularization has a low pass filtering effect on the noise that explains the coarser noise "grain" in the R-SART reconstruction (Figure 1 right) than in the SART reconstruction (Figure 1 left).

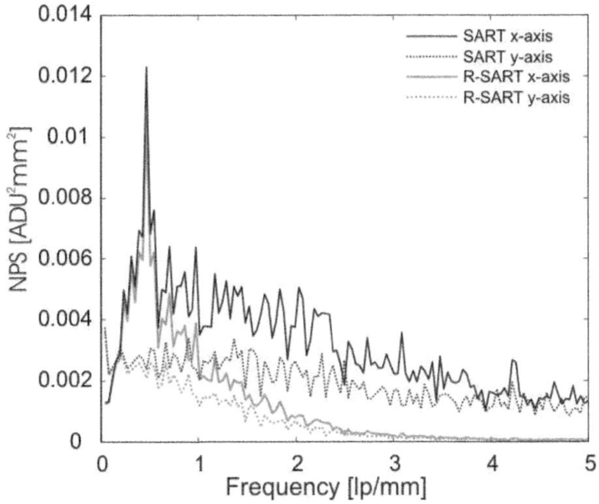

Fig. 2. In-plane NPS in the x-direction and the y-direction computed in the central uniform area on the slice of interest through the DSA phantom

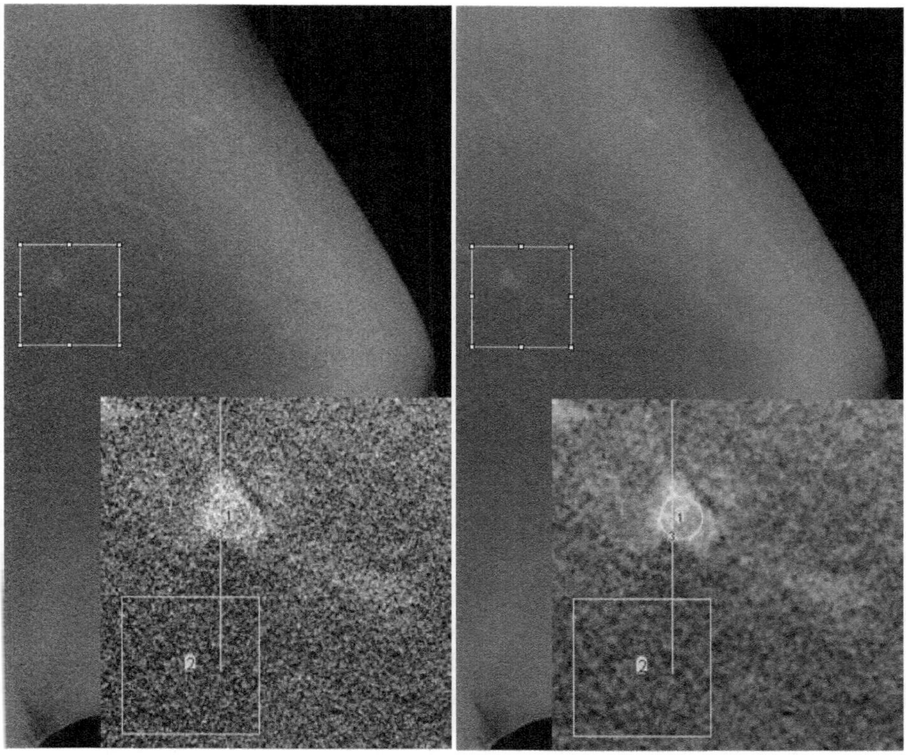

Fig. 3. Slice of interest of a DE CE-DBT clinical case containing an iodine-enhanced lesion (IDC), reconstructed using SART (left) and R-SART (right). The lesion SDNR per pixel is 2.18 for SART and 5.23 for R-SART. Images courtesy of Dr Dromain, Institut Gustave Roussy - Villejuif, France.

Figure 3 illustrates a DE CE-DBT slice of a patient reconstructed with SART (left) and with R-SART (right). This patient presented a 13×6 mm invasive ductal carcinoma (IDC) that exhibits iodine uptake, depicted in the zoomed area. Based upon visual inspection, both reconstruction techniques demonstrate consistent lesion morphology and border sharpness. The enhancement of the index lesion was also qualitatively concordant. The noise in the images reconstructed with R-SART appears to be lower. Reduction of noise in the images results in a superior visualization of the enhancing tumor and blood vessels. The *SDNR* of the mass in the images reconstructed with R-SART (*SDNR* = 5.23) is twice as high than in the image reconstructed with SART (*SDNR* = 2.18). The *SDNR* was computed using (6), where the average signal intensity in the iodine-enhanced region was estimated in the ROI labeled 1 in Figure 3 and the average signal intensity and the noise in the background were estimated in the ROI labeled 2 in Figure 3.

The difference between the zoomed areas in Figure 3 is displayed in Figure 4 (left). Neither morphological structures relevant for iodine uptake nor residual texture are visible in this image.

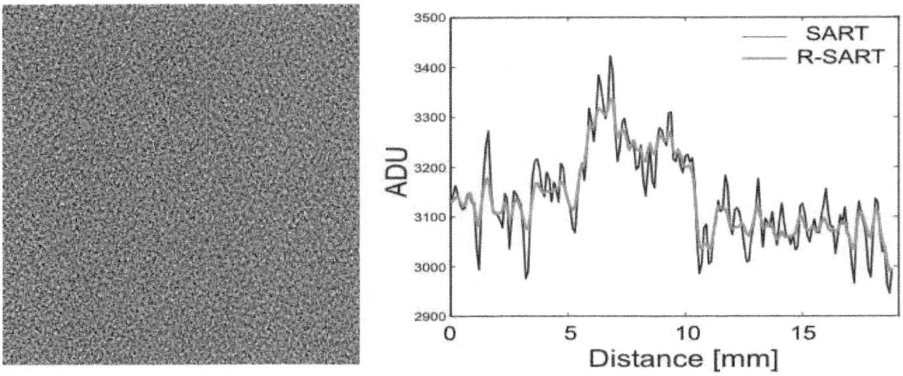

Fig. 4 Difference between the zoomed areas in Figure 3 (left) and vertical profiles along the line labeled 3 in Figure 3 (right)

Vertical line profiles through the iodine-enhanced region are shown in Figure 4 (right). The profiles correspond to the lines labeled 3 in Figure 3. The SART profile looks like a noisy version of the R-SART one. R-SART does not alter the sharpness of the contrast-uptake lesion.

4 Discussion

We showed that the SART algorithm could be regularized to constrain the increase of the noise of the reconstructed "iodine" volume in DE CE DBT. The new algorithm, R-SART, converges to a solution that fits the input data (projections) and has well behaving noise properties.

On a DSA phantom study, it was demonstrated that the *SDNR* of iodine inserts is increased when the regularization is used. This is because the noise is reduced. From a spectral perspective, the reduction of the in-plane NPS is quasi-uniform over the frequency range, and the power spectrum shape is not distorted.

The analysis of an iodine-enhanced region in a clinical DE CE-DBT case demonstrated that the R-SART reconstruction conducted to a better visibility on the lesion (the *SDNR* is more than two times higher) than SART, without modifying its morphology and its sharpness.

The results are consistent between the phantom and clinical acquisitions, in terms of impact of the regularization on the noise, preservation of the morphological features of the iodine insert/iodine enhanced lesion, improvement of the *SDNR*.

The preliminary results presented in this paper illustrate the potential of the regularized SART to improve the iodine-uptake conspicuity and to facilitate the application of automatic algorithms to detect and characterize the contrast-uptake in DE CE-DBT. Of course, the impact on the clinical assessment needs to be evaluated.

Acknowledgments. The authors would like to thank Ann-Katherine Carton for fruitful discussions and knowledgeable advice.

References

1. Frouge, C., Guinebretière, J., Paola, R.D., Contesso, G., Blery, M.: Correlation between contrast enhancement and tumor angiogenesis. Investigative Radiol. 29, 1043–1049 (1994)
2. Lewin, J., Isaacs, P., Vance, V., Larke, F.: Dual-energy contrast enhanced digital subtraction mammography: feasibility. Radiol. 229, 261–268 (2003)
3. Carton, A.-K., Gavenonis, S.C., Currivan, J.A., Conant, E.F., Schnall, M.D., Maidement, A.: Dual-energy contrast-enhanced digital breast tomosynthesis – a feasibility study. Br. J. Radiol., 2009 0: 80279516 (2009)
4. Puong, S., Patoureaux, F., Iordache, R., Bouchevreau, X., Muller, S.: Dual-energy contrast enhanced digital breast tomosynthesis: concept, method, and evaluation on phantoms. In: Proc. Med. Imaging 2007: Physics of Med. Imaging. SPIE, vol. 6510 (2007)
5. Gennaro, G., Toledano, A., di Maggio, C., Baldan, E., Bezzon, E., La Grassa, M., Pescarini, L., Polico, I., Proietti, A., Toffoli, A., Muzzio, P.C.: Digital breast tomosynthesis versus digital mammography: a clinical performance study. European Radiol. (2009) (Published Online)
6. Andersen, A.H.: Algebraic reconstruction in CT from limited views. IEEE Tr. on Med. Imaging 8, 50–55 (1989)
7. Jiang, M., Wang, G.: Convergence studies on iterative algorithms for image reconstruction. IEEE Tr. on Med. Imaging 22, 569–579 (2003)
8. Thibault, J.-B., Sauer, K.D., Bouman, C.A., Hsieh, J.: A three-dimensional statistical approach to improved image quality for multislice helical CT. Med. Phys. 34(11), 4526–4544 (2007)
9. Lange, K.: Convergence of EM image reconstruction algorithm with Gibbs smoothing. IEEE Tr. on Med. Imaging 9, 439–446 (1990)
10. Puong, S., Bouchevreau, X., Patoureaux, F., Iordache, R., Muller, S.: Dual-energy contrast enhanced digital mammography using a new approach for breast tissue canceling. In: Flynn, M., Hsieh, J. (eds.) Proc. Medical Image 2007: Physics of Med. Imaging. SPIE, vol. 6510, p. 65102H (2007)
11. Zhao, B., Zhou, J., Hu, Y.-H., Mertelmeier, T., Ludwig, J., Zhao, W.: Experimental validation of a three-dimensional linear system model for breast tomosynthesis. Med. Phys. 36(1), 240–251 (2009)

Phantom Dose Levels Found at Annual Physics Surveys in a National Mammography Screening Program: Comparison of Doses from Analog and Digital Equipment and from Digital Equipment at Different Points in Time

Kristin Pedersen[1], Ingrid D. Landmark[1], Kirsti Bredholt[1], Ingrid H.R. Hauge[1,2], and Hilde M. Olerud[1,3]

[1] Norwegian Radiation Protection Authority, P.O. Box 55, NO-1332 Østerås, Norway
[2] Oslo University College, Faculty of Health Sciences, P.O. Box 4 St. Olavs plass, NO-0130 Oslo, Norway
[3] University of Oslo, Faculty of Mathematics and Natural Sciences, Department of Physics, P.O. Box 1048 Blindern, NO-0316 Oslo, Norway
kristin.pedersen@nrpa.no

Abstract. In a national breast screening program using analog mammography systems, the dose to a 45 mm phantom was routinely calculated at annual physics surveys. This test procedure was retained through the gradual transition to digital mammography. Here the phantom doses recorded for 50 analog mammography systems, and 24 (2007) and 35 (2008/2009) digital systems are compared. The phantom doses for the digital systems were found to be statistically significantly lower than the doses for the analog systems. The digital doses were essentially unchanged from 2007 to 2008/2009.

Keywords: quality control, phantom doses.

1 Introduction

When the Norwegian Breast Cancer Screening Program (NBCSP) was initiated in 1995, only screen-film based mammography systems were available. A protocol mandating annual physics surveys was developed. This protocol required both regular patient dose surveys and dose monitoring through calculation of the mean glandular dose (MGD) to a specified phantom with thickness 45 mm. For the phantom MGD, required and desirable maximum dose levels were given (2.0 mGy and 1.5 mGy respectively).

Currently, the majority of mammography systems used in the screening program are digital. In several ways, these systems are fundamentally different from their analog counterparts. New test protocols have therefore been developed. Our most recent protocol describes assessment of system dose in accordance with the European Guidelines [1], i.e., the MGD obtained with clinical system settings is evaluated for PMMA phantoms with thicknesses in the range 20 to 70 mm. However, when the first digital

J. Martí et al. (Eds.): IWDM 2010, LNCS 6136, pp. 100–105, 2010.

systems were brought into use, protocols for routine physics testing were not firmly established. Therefore, in order to facilitate comparison between "historic" and current dose levels in a program with both analog and digital systems, and to allow assessment of the dose levels with digital systems in terms of the previously established dose limits, the dose evaluation for the 45 mm phantom was retained for digital systems.

Here we present mean glandular doses calculated for the 45 mm phantom at the last annual physics survey conducted for 50 analog mammography units (screening and assessment), and the MGDs calculated for the same phantom for all digital units surveyed in 2007 and from October 2008 to November 2009 (2008/2009 in the following). This allows both a comparison between "analog" and "digital" dose levels, and an assessment of the development of the dose level with digital systems over time.

2 Materials and Methods

A test phantom of the material BR12 (CIRS, Norfolk, VA, USA) with total thickness 45 mm and surface area 100 mm x 125 mm was placed on the breast support table, centered laterally and with one edge coinciding with the edge of the table. Compression was applied and an exposure made with system settings corresponding to settings that would have been used for a breast with the same thickness. The target and filter material, tube voltage (kV) and tube current-time product (mAs) were recorded. The tube output for the target and filter material and kV employed was measured with an ion chamber assembly (Radcal Corporation, Monrovia, CA, USA). The same instruments were also used in the dose measurements conducted for calculation of half value layer (HVL). The MGD was calculated according to Dance et al. [2] as

$$MGD = Kgcs, \tag{1}$$

using conversion factors published by Dance et al. [2,3].

MGD was calculated for a total of 50 analog mammography units, from the manufacturers General Electric (GE Medical Systems, Buc, France), Instrumentarium (now GE) and Siemens (Siemens, Erlangen, Germany). Data from the last physics survey conducted for each particular unit after the introduction of the first digital system in the screening program in 2000 was used. Due to a very gradual transition to digital mammography, the data collection period extends from 2000 to 2009.

The doses for digital systems in 2007 were calculated for a total of 24 units, in 2008/2009 the number of units was 35. The digital system manufacturers represented were General Electric, Hologic (Hologic, Inc., Danbury, CT, USA), Sectra (Sectra Imtec AB, Linköping, Sweden) and Siemens. No CR systems were included. An overview of the x-ray set models, number of units, available target-filter combinations, and target-filter combination and kVp used for the phantom exposures, is shown in Table 1 for the analog systems, and in Table 2 and Table 3 for the digital systems.

Table 1. Analog x-ray set models included in the study. Mo=molybdenum, Rh= rhodium. "Target-filter used" and "kVp used" refer to the phantom exposures.

Manufacturer	Model	# units	Target- filter available	Target-filter used	kVp used
GE	Senographe 600T	2	MoMo	MoMo	28 29
GE	Senographe 800T	2	MoMo MoRh	MoMo	29
GE	Senographe DMR	5	MoMo MoRh RhRh	MoMo	28 29
				MoRh	26
Instrumentarium	Alpha	1	MoMo MoRh	MoMo	29
Instrumentarium	Diamond	9	MoMo MoRh	MoMo	28 29 30
Siemens	Mammomat 300	6	MoMo	MoMo	29 30
Siemens	Mammomat 3000/ 3000 Nova	25	MoMo MoRh WRh	MoMo	27 29 30

Table 2. Digital x-ray set models surveyed in 2007 included in the study. Mo=molybdenum, Rh= rhodium, W=tungsten, Al=aluminium. "Target-filter used" and "kVp used" refer to the phantom exposures.

Manufacturer	Model	# units	Target- filter available	Target-filter used	kVp used
GE	Senographe 2000D	2	MoMo MoRh RhRh	MoRh	27 28
GE	Senographe DS	10	MoMo MoRh RhRh	RhRh	29
GE	Senographe Essential	4	MoMo MoRh RhRh	RhRh	29
Hologic	Lorad Selenia	2	MoMo MoRh	MoMo	28
Sectra	MicroDose D40	1	WAl	WAl	32
Siemens	Mammomat Novation	5	MoMo MoRh WRh	WRh	26 27

Table 3. Digital x-ray set models surveyed in 2008/2009 included in the study. Mo=molybdenum, Rh= rhodium, W=tungsten, Ag=silver, Al=aluminium. "Target-filter used" and "kVp used" refer to the phantom exposures.

Manufacturer	Model	# units	Target- filter available	Target-filter used	kVp used
GE	Senographe 2000D	2	MoMo MoRh RhRh	MoRh	27 28
GE	Senographe DS	7	MoMo MoRh RhRh	RhRh	29
GE	Senographe Essential	12	MoMo MoRh RhRh	RhRh	29
Hologic	Lorad Selenia	1	WRh WAg	WRh	28
Sectra	MicroDose D40	3	WAl	WAl	26 29 32
Siemens	Mammomat Novation	4	MoMo MoRh WRh	WRh	27 28
Siemens	Mammomat Inspiration	6	MoMo MoRh WRh	WRh	28

3 Results

All phantom MGD values were found to be below the 1.5 mGy limit in the quality assurance protocol. The mean phantom MGD for the analog systems was found to be 1.11 mGy. The MGD values for the digital systems were found to be lower, at 0.98 mGy (a reduction of 11.7 %) and 0.96 mGy (a reduction of 13.5 %) respectively for the 2007 and 2008/2009 digital data sets. The difference between the analog and digital dose levels was found to be statistically significant with a two-sided t-test and a significance level of 0.05. There was no statistically significant difference between the doses in the two digital data sets. The ratio between the highest and lowest MGD were calculated for the three data subsets and found to be lowest for the doses with analog systems, intermediate in the 2007 digital group and highest in the 2008/2009 digital group. A summary of the key dose findings is shown in Table 4. In Table 5, mean MGD values is shown for the digital x-ray set models represented in both digital data sets with three or more unist, GE Senographe DS and Essential, and Siemens Mammomat Novation. For these system models the mean MGD in practice remained unchanged from 2007 to 2008/2009.

Table 4. Various features of the phantom doses found in the three data sets

	# units	MGD mean (standard deviation) [mGy]	MGD median [mGy]	MGD range [mGy]	MGD_{max}/MGD_{min}
Analog	50	1.11 (0.11)	1.13	0.84-1.30	1.55
Digital 2007	24	0.98 (0.18)	1.03	0.61-1.43	2.34
Digital 2008/09	35	0.96 (0.21)	1.01	0.36-1.24	3.44

Table 5. Mean MGD calculated for the digital x-ray set models represented in both digital data sets with three or more units

Manufacturer	Model	# units 2007	MGD mean (standard deviation) [mGy]	# units 2008/09	MGD mean (standard deviation) [mGy]
GE	Senographe DS	10	1.01 (0.05)	7	0.98 (0.05)
GE	Senographe Essential	4	1.10 (0.04)	12	1.07 (0.04)
Siemens	Mammomat Novation	5	0.81 (0.07)	4	0.86 (0.07)

4 Discussion

Image quality and dose are closely connected. Whether the recorded phantom doses were appropriate in terms of giving the necessary image quality is not the subject of this paper. In our experience, the digital systems were used in accordance with the manufacturers' recommendations. The use of the analog systems might to a larger extent have been subject to local custom, for instance in terms of preferred film optical density level.

Data for analog systems was selected from the last physics survey available for each unit rather than from one particular year. The transition to digital systems has been very gradual and we wanted to include analog data from as close as possible to the actual shift of technology at each site. For analog systems, the dose level determines the film optical density. Different film and screen types and generations might require different doses. All units included in our study used film from Kodak (Eastman Kodak Company, Rochester, NY, USA). Only two film types were represented: Kodak Min-R 2000 and Min-R EV. Min-R 2000 was used exclusively in the period 2000 to 2005. While Min-R EV dominated from 2005, Min-R 2000 was still used by some sites in 2009. The conditions in terms of film used were therefore comparable throughout the data collection period.

The doses calculated for the digital systems were found to be statistically siginficantly lower than the doses calculated for the analog systems. It can be seen in Table 1 that, with one exception, the target-filter combination used for the analog systems was molybdenum-molybdenum (MoMo). Table 2 and Table 3 show that with the digital systems, a shift to harder x-ray beams had taken place. Similar findings have recently been reported in a study of patient doses by Hendrick et al. [4]. Harder x-ray beams carry the potential to provide a reduction in dose. However, since digital detectors can provide diagnostically usable images over a much larger dose range than film, including at quite high doses, the shift to harder x-rays does not, in itself, guarantee a reduction in dose. That a dose reduction was actually observed indicates that the dose operating levels chosen by the system manufacturers have been such that at least some of the dose reduction potential of the harder x-ray beams has been realized.

From the first to the second digital data set there seems to be a non-significant trend towards lower doses. This could be partly attributable to the introduction of a new system model from Sectra (representing 8.6 % of the units in the 2008/2009 data set) which in our material was found to operate at approximately 35 % lower dose level than the previous model. In addition, the Hologic Selenia unit represented in both data sets had its original molybdenum-target x-ray tube replaced with a tungsten-target tube, contributing to an approximate halving of the phantom dose for this unit. One can conclude from this that the choice of x-ray set model is among the factors that might strongly influence the "digital" dose level. A larger spread in dose level among the digital systems than the analog is clearly evident in the ratios MGD_{max}/MGD_{min} shown in Table 4. The largest ratio was found for the most recent digital data set. For the digital system models that were represented in both digital data sets with three or more units, the mean MGD essentially remained unchanged from 2007 to 2008/2009 in spite of a change in the number of units for each model.

It was stated earlier that the phantom exposures were made with system settings corresponding to settings that would have been used for a breast with the same thickness. Over time the automatic exposure control systems in digital mammography have grown continuously more sophisticated and fine-tuned to the task to which they are actually designed: ensure optimal exposure of a human breast. While the degree of sophistication might vary from one model to another, this development has still made the choice of how to actually run the machine for phantom exposures considerably more challenging for digital compared to analog systems.

References

1. van Engen, R., Young, K., Bosmans, H.: European protocol for the quality control of the physical and technical aspects of mammography screening: Digital mammography. In: Perry, N., Broeders, M., de Wolf, C., Törnberg, S., Holland, R., von Karsa, L. (eds.) European guidelines for quality assurance in breast cancer screening and diagnosis, 4th edn., Luxembourg: Office for Official Publications of the European Communities, pp. 105–150 (2006)
2. Dance, D.R., Skinner, C.L., Young, K.C., Beckett, J.R., Kotre, C.J.: Additional factors for the estimation of mean glandular breast dose using the UK mammography dosimetry protocol. Phys. Med. Biol. 45, 3225–3240 (2000)
3. Dance, D.R., Young, K.C., van Engen, R.: Further factors for the estimation of mean glandular dose using the Unite Kingdom, European and IAEA breast dosimetry protocols. Phys. Med. Biol. 54, 4361–4372 (2009)
4. Hendrick, R.E., Pisano, E.D., Averbuk, h.A., Moran, C., Berns, E.A., Yaffe, M.J., Herman, B., Acharyya, S., Gatsonis, C.: Comparison of acquisition parameters and breast dose in digital mammography and screen-film mammography in the American College of Radiology Imaging Network digital mammographic imaging screening trial. Am. J. Roentgenol. 194(2), 362–369 (2010)

Task-Based Evaluation of Image Quality of Filtered Back Projection for Breast Tomosynthesis

Xinying Wang[1], James G. Mainprize[1], Gang Wu[1,2], and Martin J.Yaffe[1,2]

[1] Imaging Research, Sunnybrook Health Sciences Centre, 2075 Bayview Ave., Toronto, Ontario, Canada, M4N 3M5
[2] Dept. of Medical Biophysics, University of Toronto, Sunnybrook Health Sciences Centre, 2075 Bayview Ave., Toronto, Ontario, Canada, M4N 3M5
{cindy.wang,james.mainprize,gang.wu,martin.yaffe}@sunnybrook.ca

Abstract. A task-based method for evaluating tomosynthesis reconstruction image quality is developed. An ideal observer model using a signal known exactly/background known exactly (SKE/BKE) is demonstrated for a breast tomosynthesis system. Image quality is evaluated by calculating the detectability (d′) of a spherical lesion. The detectability is calculated in the central slice of the reconstruction. The effect of various filter choices is demonstrated on lesion detectability at various dose levels. This model is extended to backgrounds which simulate anatomic variability in a "background known statistically" model.

Keywords: tomosynthesis, ideal observer, image quality, noise power spectrum, Fourier analysis, filtered back projection.

1 Introduction

Filtered back projection (FBP) is a standard algorithm for reconstructing images in cone-beam computed tomography (CBCT), where a full dataset is available (>180°, and 200+ ~ 500+ projections). Breast tomosynthesis (DBT) is a form of cone-beam CT with limited-angle acquisition and a limited number of views. The goal of DBT is generally to produce slice images with a high in-plane resolution (≈1 mm) but only a modest resolution in the slice thickness (≈1 mm). Reconstruction for tomosynthesis can be performed using FBP, but special filter modifications are required to compensate for the incomplete data and undersampling in the slice thickness direction.[1] These are angle-dependent filters used to compensate for aliasing and to reduce the appearance of high spatial frequency noise. Additional filter modifications may be used to restore low-spatial frequency content lost by a standard ramp reconstruction filter.[2]

Image quality analysis for breast tomosynthesis can be evaluated by determining the noise equivalent quanta (NEQ) following the approach by Siewerdsen *et al.* used for CBCT.[3] Furthermore, the NEQ can be extended to evaluate task-based performance of an imaging system. Here, we will develop a 2D task-based measure for a simple lesion detection task in the central slice. We examine the effect of slice-thickness filters on detectability of lesions of various sizes in both homogeneous and heterogeneous backgrounds.

J. Martí et al. (Eds.): IWDM 2010, LNCS 6136, pp. 106–113, 2010.

2 Method

The task evaluated here is a detection task of a Gaussian lesion embedded in a cubic block of tissue to simulate mass detection in breast tomosynthesis. A computer simulation of projection imaging and a filter back-projection reconstruction algorithm were implemented.

2.1 X-ray Model

The tomosynthesis geometry shown in Fig.1 is modeled. The source to detector distance, *SDD* is 660 mm, the source to pivot distance *SPD* is 620 mm and the detector size is 2048 x 512 with detector element pitch of 100 μm. Image acquisitions consists of 15 projection images acquired with a gantry arc from -20° to 20° and a stationary detector. Projections were simulated by simple ray-based projections through a voxel model with attenuation coefficients equal to 50%/50% fibroglandular/adipose tissue by composition.

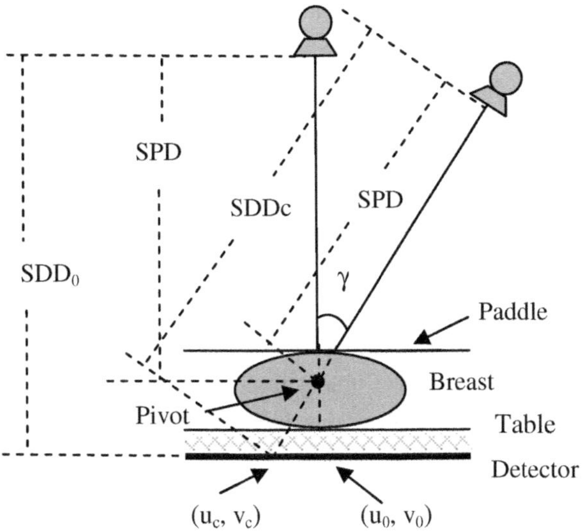

Fig. 1. Partial isocentric breast tomosynthesis geometry used in the simulation

Noise-free projection images are generated assuming 20 keV monoenergetic x-rays. Poisson noise realizations of the projection images are generated at four air kerma levels: 0.17, 0.29, 1.14, 2.86 mGy. An ideal detector with 100% quantum efficiency, zero electronic noise and transparent paddles and tables were assumed. Table height above the detector was assumed to be zero.

2.2 Filtered Back Projection for Breast Tomosynthesis

Following the approach of Mertelmeier, the filtered projection image is calculated as the following in the Fourier domain

$$g^*(f_x, f_y) = g(f_x, f_y)H_{\mathrm{R}}(f_x)H_{\mathrm{ST}}(f_z)H_{\mathrm{SP}}(f_x) \tag{1}$$

where H_{ST} is the slice thickness apodization filter to reduce aliasing and artifacts in the z-direction (perpendicular to the detector), H_{SP} is a Hanning spectral filter, and H_{R} is the reconstruction or ramp filter.H_{ST} is given by:

$$H_{ST}(f_z) = \frac{1}{2}(1 + \cos(\frac{\pi f_z}{\beta f_{z-ny}})) \tag{2}$$

where $0 < \beta \leq 1$ is a filter parameter that can be adjusted to increase the amount of filtering in the slice direction, and $f_{z\text{-ny}}$ is the Nyquist frequency in the z direction .

Back-projection followed with bilinear interpolation on a reconstructed volume size of 120 × 30 × 60 mm^3 and a voxel size of 100 μm × 100 μm × 1 mm.

2.3 Task and NEQ

The detectability, d', of an ideal observer for a 2D image with a signal known exactly/background known exactly (SKE/BKE)[1][1][2] test is related to the NEQ as follows[3],

$$\begin{aligned} d'^2 &= \iint NEQ(f_x, f_y)\left|\overline{\Delta S}(f_x, f_y)\right|^2 df_x df_y \\ &= \iint \frac{K^2 MTF^2(f_x, f_y)\left|\overline{\Delta S}(f_x, f_y)\right|^2}{NPS(f_x, f_y)} df_x df_y, \end{aligned} \tag{3}$$

where K is the gain factor of the system, MTF is the 2D modulation transfer function,, NPS is the 2D noise power spectrum, and $\overline{\Delta S}$ is the expectation of the difference between signal present and signal absent images. For simplicity, we will use the 2D NEQ to evaluate the detectability for the central slice through a lesion in a DBT reconstructed image.

The numerator simplifies to the Fourier transform of the difference between the signal present and absent output image, ΔI, yielding

$$d'^2 = \iint \frac{\left|\Delta I(f_x, f_y)\right|^2}{NPS(f_x, f_y)} df_x f_y. \tag{4}$$

To eliminate any low spatial frequency bias, the NPS was calculated using a difference of images approach.[7] The NPS was evaluated for the ensemble of subregions of size 64×64 voxels in each image.

2.4 SKE/BKE Simulation

For this model, the signal-absent volume consists of a uniform rectangular block of breast tissue equivalent material with 50% fibroglandular tissue and 50% adipose tissue for a breast thickness of 60 mm. For the signal present case, a spherical lesion is inserted at the centre of the volume. The diameter of the lesion was varied from 2 to 8 mm. The mean difference between signal present { $\tilde{V}_S$ } and signal absent { $\tilde{V}_B$ } reconstructed volumes was calculated,

$$\Delta V = \left\{ \frac{1}{n} \sum_{j=1}^{n} \tilde{V}_S^{(j)} - \tilde{V}_B^{(j)} \right\}. \tag{5}$$

and $\Delta I(f_x, f_y)$ was calculated as the 2D Fourier transform of the central-slice image of ΔV.

2.5 Heterogeneous Background

In clinical situations, the background is not homogeneous, nor is it "known exactly." Instead, the background is heterogeneous with tissue structures that tend to have common texture measures. We extend our modeling to include heterogeneous backgrounds in a "background known statistically" (BKS) simulation. Here, 3D "random cloud" backgrounds are generated by filtering uniform random deviates by an inverse power-law to simulate mammographic backgrounds.[8] of the form

$$H(f) = \frac{1}{\left(1 + \frac{f}{f_0}\right)^{b/2}}. \tag{6}$$

where $f_0 = 0.01$ mm^{-1} and $b = 2.5$ for our simulations. The resulting "cloud" volume was converted to attenuation coefficients normalized to range between 100% fat and 100% fibroglandular tissue.

The simulation geometry is a slightly different from the homogeneous case. The volume size was 32x32x40 mm^3, with isotropic voxels 100×100×100 µm^3 to maintain a consistent spatial frequency content in all three directions in the simulated backgrounds. A single lesion size was considered (4 mm diameter) for the signal- present volumes. Twenty realizations of signal-present, $V_S^{(j)}$ volumes were created. Projection images were simulated as above, with quantum Poisson noise added to create, { $I_S^{(j)}$ } for the jth realization. Twenty (n=20) different realizations of random backgrounds with no lesion present, $V_B^{(j)}$, are generated and the same level of quantum Poisson noise is added on the projected images { $I_B^{(j)}$ }. Reconstruction follows to yield the estimated reconstruction volumes $\tilde{V}_S^{(j)}$ and $\tilde{V}_B^{(j)}$.

The signal is estimated from average of the difference between a signal-present and a signal-absent reconstructed volume,

$$\Delta V = \frac{1}{n} \sum_{j=1}^{n} (\tilde{V}_S^{(j)} - \tilde{V}_B^{(j)}) \tag{7}$$

A reliable NPS that captures the quantum noise and the "anatomic noise" cannot be obtained from a difference of images. Instead, the background was corrected for global reconstruction artifacts by calculating an average reconstructed signal-absent volume and calculating the NPS on each realization of the corrected volume

$$\tilde{V}_{Bcorr}^{(i)} = \tilde{V}_{B}^{(i)} - \frac{1}{n}\sum_{j=1}^{n} V_{B}^{(j)} \tag{8}$$

3 Results

Examples of the simulated projection images are shown in Fig. 2 for four different lesion sizes. The smallest lesion is almost invisible at the given noise level. Fig. 3 shows examples of the central slice through reconstruction of the corresponding lesions at the same air kerma (0.019 mGy per projection, or 0.59 mGy per exam).

Fig. 4 shows the effect of varying the slice thickness filter factor, β on the detectability SNR (d') for a range of air kerma and lesion diameters. As expected, the detectability improves with more severe filtering (β) but the improvement appears to be marginal, especially at low doses.

Because only a 2D detectability was evaluated in the central slice, the impact of the slice filter in the third dimension was examined. In Fig. 5, the orthogonal slice in the x-z plane through the lesion is shown. Qualitatively, the slice filter has a strong impact on reducing the appearance of artifacts (note the reduced streaks in the upper half of the slice for $\beta = 0.3$).

Fig. 6 shows examples of the random backgrounds created in simulations for noise-free and noisy projections and central slices. Fig. 7 shows the calculated detectability for the SKE/BKS trials. Again the slice thickness filter parameter has a marginal effect on improving detectability. Interestingly, increasing the air kerma beyond 1.14 mGy did not improve the detectability as was seen in the homogenous cases (Fig. 4). This suggests that in the presence of heterogeneous backgrounds, the detectability is no longer quantum limited for an air kerma greater than 1.14 mGy.

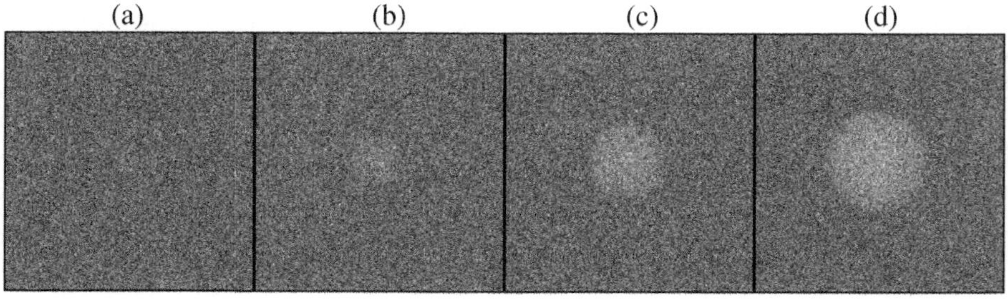

Fig. 2. Different lesion sizes projection images at gantry angle 0°, 20 keV, for a single projection kerma of 0.019 mGy (per projection) for lesion diameters of a) 2 mm, b) 4 mm, c) 6 mm, and d) 8 mm. Contrast settings are identical for each image.

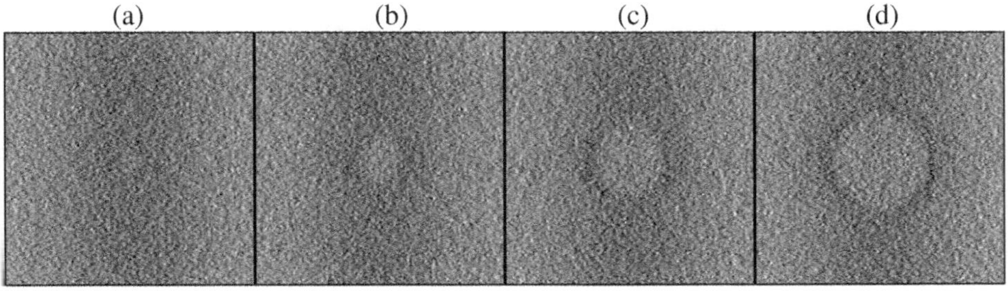

Fig. 3. Central slice through the lesions for reconstruction by filtered back projection using slice thickness filter factor β=0.3 , and total exam air kerma is 0.29 mGy for different lesion size a) 2 mm, b) 4 mm, c) 6 mm, and d) 8 mm. Contrast settings are identical for each image.

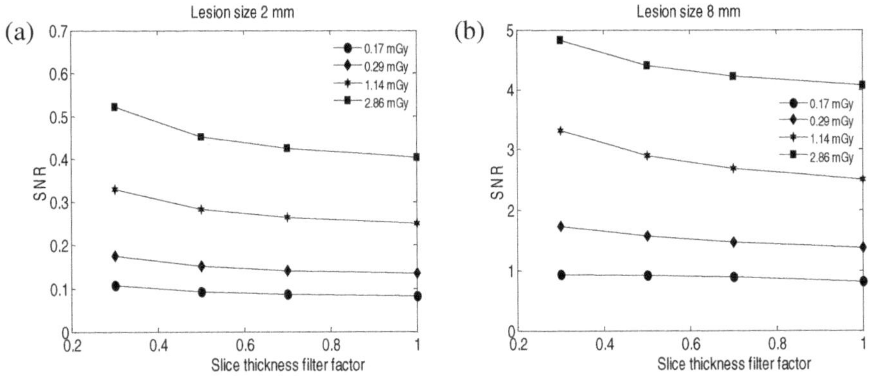

Fig. 4. 3D volume reconstructions from projections with different lesion sizes. Filtered back projection with different slice thickness filter factors was used. SNR was calculated for the corresponding volume slices for lesion diameters of (a) 2 mm and (b) 8 mm.

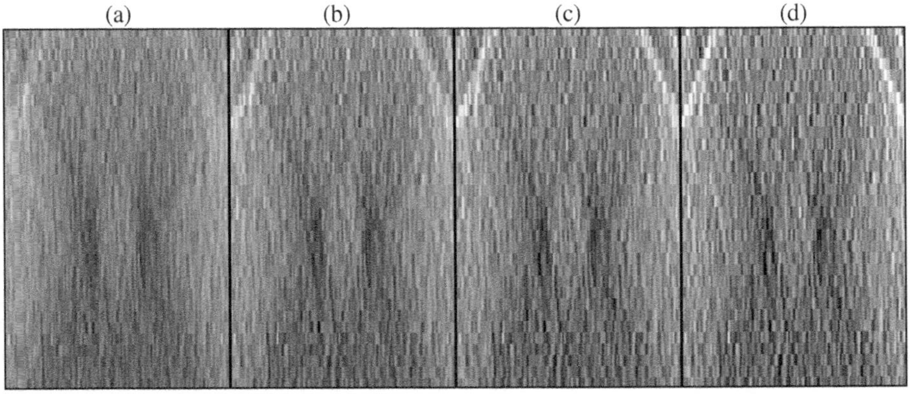

Fig. 5. For the 4 mm lesion, the reconstructed slice in the x-z plane is shown for the 2.86mGy. exam for different slice thickness filters with (a) β=0.3 (b) β=0.5 ,(c) β=0.7 (d) β=1.0. Contrast settings are identical for each image.

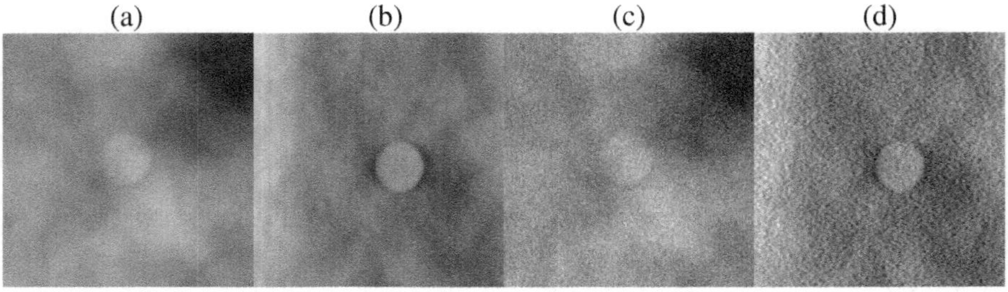

Fig. 6. Examples of lesions in heterogeneous backgrounds generated from power-low filtered random structures. A noise-free projection image is shown in (a) and the corresponding central slice of the reconstruction is shown in (b). Noisy versions of the (c) projection and (d) reconstruction (d) are shown for an incident air kerma of 0.29 mGy for the total exam.

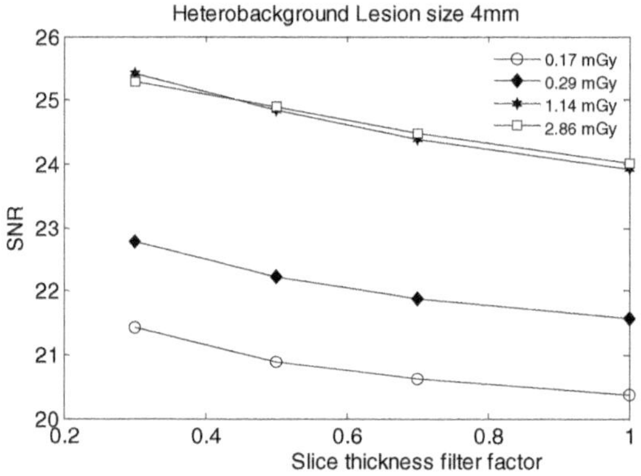

Fig. 7. Lesion size is 4 mm. Heterogeneous backgrounds SNR calculation.

4 Discussion

Note that the detectability SNR is much higher than was calculated for the homogeneous case. Part of this is expected, as the phantom thickness was reduced (40 mm instead of 60 mm) which should lead to a 2× improvement in SNR. In addition, there may be discrepancies in the two different NPS calculation techniques. Future work will include validation of the two approaches to ensure consistent results.

Further work is necessary to extend this to a 3D NEQ model for task-based evaluation. However, the coarseness of the reconstruction (1 mm slices) in the 3rd dimension makes it very difficult to obtain a reliable 3D NPS, in part, because the number of samples is small in the 3rd dimension as well as the fact that assumption of a stationary system is extremely difficult to justify over the entire reconstruction volume.

Note that the effect of slice thickness filtering is believed to be more dramatic for the detectability of microcalcifications but was beyond the scope of this work.

5 Conclusion

An NEQ-based approach calculating the detectability for a SKE/BKE and SKE/BKS task for tomosynthesis is demonstrated for a filtered back-projection technique. The effect of different slice thickness filter factors (β) is evaluated to determine the impact on lesion detectability. For homogeneous or heterogeneous phantoms with a spherical lesion, the results show that strong filtering (small β) results in the best SNR but the improvement is small. However this may have an adverse effect on lesion localization in the z-direction.

References

1. Mertelmeier, T., Orman, J., Haerer, W., Dudam, M.K.: Optimizing filtered backprojection reconstruction for a breast tomosynthesis prototype device. In: Proc. SPIE, vol. 6142, pp. 1–12 (2006)
2. Jun Zhou, J., Zhao, B., Zhao, W.: A computer simulation platform for the optimization of a breast tomoynthesis system. Med. Phys. 34(3), 1098–1109 (2007)
3. Siewerdsen, J.H., Jaffray, D.A.: Three-dimensional NEQ transfer characteristics of volume CT using direct and indirect-detection flat-panel imagers. In: Proc. SPIE, vol. 5030, pp. 92–102 (2003)
4 Emil, Y., Sidky, E.Y., LaRoque, S.J., Pan, X.: Accurate computation of the Hotelling Observer for the evaluation of image reconstruction algorithms in helical, cone-beam CT. In: IEEE Nucl. Sci. Symp., vol. 15, pp. 3233–3236 (2007)
5. Abbey, C.K., Boone, J.M.: An ideal observer for a model of x-ray imaging in breast parenchyma tissue. In: Krupinski, E.A. (ed.) IWDM 2008. LNCS, vol. 5116, pp. 393–400. Springer, Heidelberg (2008)
6. Gallas, B.D., Barrett, H.H.: Validating the use of channels to estimate the ideal linear observer. Opt. Soc. Am. 20(9), 1725–1738 (2003)
7. Aufrichig, R., Su, Y., Cheng, Y., Granfors, P.R.: Measurement of the noise power spectrum in digital x-ray detectors. In: Proc. SPIE, vol. 4320, pp. 362–372 (2001)
8. Burgess, A., Jacobson, F., Judy, P.: Human observer detection experiments with mammograms and power-law noise. Med. Phys. 28(4), 419–437 (2001)

A Comparative Study of the Inter-reader Variability of Breast Percent Density Estimation in Digital Mammography: Potential Effect of Reader's Training and Clinical Experience

Emily F. Conant[1], Diane Li[1], Sara Gavenonis[1], Predrag R. Bakic[2], Ann-Katherine Carton[2], Cuiping Zhang[2], Andrew D.A. Maidment[2], and Despina Kontos[1]

[1] Hospital of the University of Pennsylvania, Breast Imaging Division, 1 Silverstein Building HUP, 3400 Spruce St., Philadelphia PA 19104-4206
[2] University of Pennsylvania, Department of Radiology, Physics Section, 1 Silverstein Building HUP, 3400 Spruce St., Philadelphia PA 19104-4206
{Emily.Conant,Diane.Li,Sara.Gavenonis,Predrag.Bakic, Ann-Katherine.Carton,Cuiping.Zhang, Andrew.Maidment,Despina.Kontos}@uphs.upenn.edu

Abstract. The variability of breast percent density (PD%) estimation from digital mammography (DM) images was evaluated using measurements from readers with different training and clinical experience. Post-processed DM images (*PremiumView*™, GE Healthcare) from 40 women were analyzed. Breast PD% estimation was performed using the *Cumulus* software (*Ver. 4.0, Univ. Toronto*). Two groups of readers were considered, one with clinical (*i.e.,* radiologists) and one with non-clinical training (*i.e.,* physicists). Consistency of PD% was analyzed using the Pearson correlation coefficient (r) and ANOVA. Inter-reader agreement was higher among clinical (r=0.91, p<0.001), than non-clinical readers (r=0.83, p<0.001). Intra-reader consistency after repeated reads was on average equally high for both groups (r=0.91, p<0.001). Our results suggest that the reader's experience and training has an effect on the obtained PD% measures. The higher correlation among the clinically trained readers could be attributed to their extensive exposure to post-processed DM images and their knowledge of breast anatomy.

Keywords: Digital mammography, breast percent density, breast cancer risk estimation.

1 Introduction

It is well known that the sensitivity of mammography in detecting breast cancer decreases as parenchymal density increases. There is also growing evidence that suggests that breast density is an independent risk factor for breast cancer [1]. Breast imaging radiologists incorporate a density estimate in all clinical mammographic reports using the 4-tiered Breast Imaging Reporting and Data System (BI-RADS) system and they are very familiar with the wide variety of breast parenchymal patterns that are

J. Martí et al. (Eds.): IWDM 2010, LNCS 6136, pp. 114–120, 2010.

present in clinical mammographic images. Unfortunately, because the assignment of a BI-RADS density category is subjective, there is a high degree of inter-reader variability [2, 3]. Currently, the most widely used methods to quantify breast density rely on measures derived from mammograms using a semi-automated image thresholding method [4]. Most of the studies published to date have been performed using digitized screen-film mammograms [5]. Digital mammography (DM) is increasingly replacing screen-film mammography in breast cancer screening. DM imaging systems typically produced two-types of images; the "FOR PROCESSING" images, which are proportional to the x-ray attenuation (*i.e.,* raw data), and the "FOR PRESENTATION" images, which are post-processed according to vendor-specific image processing algorithms prior to presentation to the radiologist for diagnostic interpretation. Currently, the strategy adopted by most clinical breast imaging divisions is to archive only the post-processed (*i.e.*, "FOR PRESENTATION") images due to storage and cost constraints. Therefore, studies are needed to determine the optimal approaches for utilizing the digital data for breast density estimation. Investigating the potential use of post-processed images in breast density estimation could result in a more widely-adopted translation of density-based breast cancer risk assessment in clinical practice. Currently, not many studies exist that have investigated this potential [6].

We performed a study to compare the inter-reader variability of area-based breast percent density (PD%) estimation in post-processed DM images performed by groups of readers with different training and clinical experience. Post-processed DM images were used in our study because they are the DM images that clinicians are most familiar with and are the images that best display breast anatomy. In addition, the post-processed images are often more widely available and accessible in clinical practices. The main goals of our study were to investigate *i*) if post-processed DM images can provide viable means for PD% estimation and *ii*) if the obtained measures are affected by the type of training and the clinical experience of the readers.

2 Methods

The MLO DM image from the contralateral (*i.e.,* unaffected) breast of 40 women with recently detected abnormalities and/or previously diagnosed breast cancer, recruited as part of a separate multi-modality imaging clinical trial that has been completed in our department[1], were retrospectively collected and analyzed under HIPAA and IRB approval. All women were study volunteers who had signed informed consent. DM imaging was performed with a GE Healthcare DS full-field DM system (GE Healthcare, Chalfont St. Giles, UK) [7-9]. Images were acquired with 0.1 mm/pixel resolution at 12 bit gray-level. Image post-processing was performed using *PremiumView*™ (GE Healthcare), an embedded adaptive histogram equalization method [8]. Area-based breast percent density (PD%) was estimated using the semi-automated image thresholding technique of *Cumulus* (*Ver. 4.0, Univ. Toronto*) [4]. Using *Cumulus* the image background and the pectoral muscle region are excluded from the breast density calculations and user-defined gray-level intensity thresholds are applied to outline the fibroglandular tissue regions within the breast. PD% is then computed as the percent of the breast occupied by fibroglandular tissue.

[1] Evaluation of Multimodality Breast Imaging, NIH P01 CA85484, PI: M.D. Schnall.

Two groups of readers were considered, one with clinical experience (*i.e.*, breast imaging radiologists) and one with non-clinical training (*i.e.*, medical physicists). Each group included three readers at different levels of experience (*i.e.*, years of training). The clinical readers had varying degrees of clinical experience of breast anatomy and digital mammographic images (*i.e.*, 3 months, 3 years and 20 years of experience). The non-clinical readers had knowledge of breast anatomy and varying degrees of training in medical physics (*i.e.*, 2, 12 and 14 years of experience). Both groups received the training outlined in the manual which accompanies the *Cumulus* software. Each group performed two rounds of readings on the same dataset.

To evaluate the consistency of the obtained PD% measures among readers in the same group (*i.e.*, inter-reader agreement) and between the repeated reads of each reader (*i.e.*, intra-reader agreement), pair-wise Pearson correlation coefficients were computed and analysis of variance (ANOVA) was performed. To compare the PD% measures between the groups of readers the Student's paired t-test was applied on the mean of the PD% estimates of the corresponding groups.

3 Results

Inter-reader correlation was high for both groups (Table 1), with the clinically trained group having a higher average inter-reader Pearson correlation (r=0.91, p<0.001) than the non-clinically trained group (r=0.82, p<0.001). Linear regression analysis also demonstrated lower variability and a stronger association between the PD% estimates obtained by the clinically trained readers than between the non-clinically trained readers (Fig. 1). Overall, the mean of the PD% estimates for the non-clinically trained group was statistically significantly higher (p<0.001) than that of the clinically trained group ($mean_{Read1}$=27%, $mean_{Read2}$=24%), both for the first ($mean_{Read1}$=30%) and the second ($mean_{Read2}$=35%) reads. ANOVA showed that the obtained breast PD% measures were more consistent among the clinically trained readers than among the non-clinically trained readers (Fig. 2). The means of the PD% estimates were not statistically significantly different among the clinically trained readers (p_{ANOVA1}=0.39, p_{ANOVA2}=0.05), but statistically significantly different among the non-clinically trained readers (p_{ANOVA1}<0.001, p_{ANOVA2}<0.001). Intra-reader agreement after repeated reads was on average equally high (r=0.90, p<0.001) for both groups (Table 2).

Table 1. Pair-wise Pearson correlations (r) for inter-reader variability in breast PD% estimates between readers in each group. Readers are ordered in increasing order of experience.

	Pearson correlation coefficients (r) for PD% inter-reader variability							
	Clinically Trained Readers				Non-clinically Trained Readers			
	Read 1		Read 2		Read 1		Read 2	
Reader	2	3	2	3	2	3	2	3
1	0.95	0.90	0.83	0.93	0.83	0.85	0.79	0.84
2		0.91		0.92		0.77		0.86

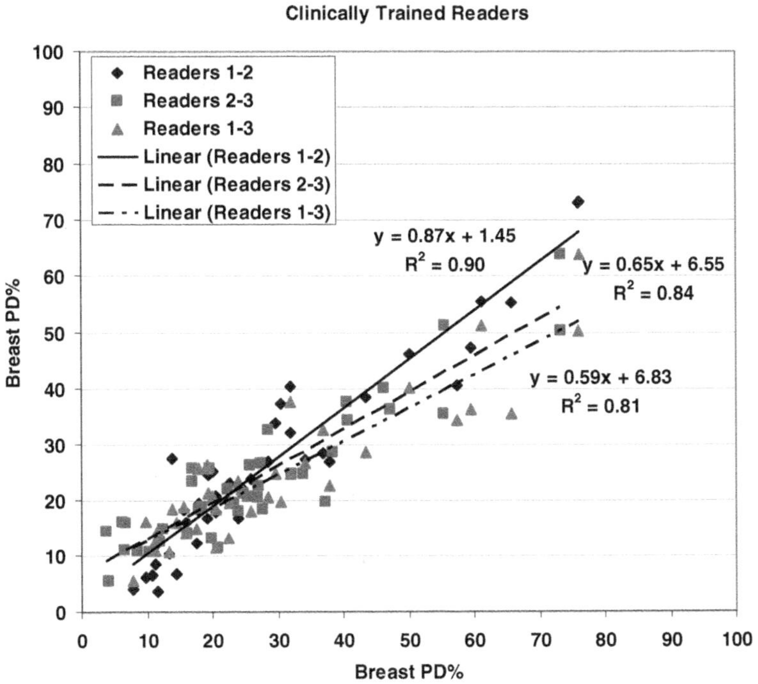

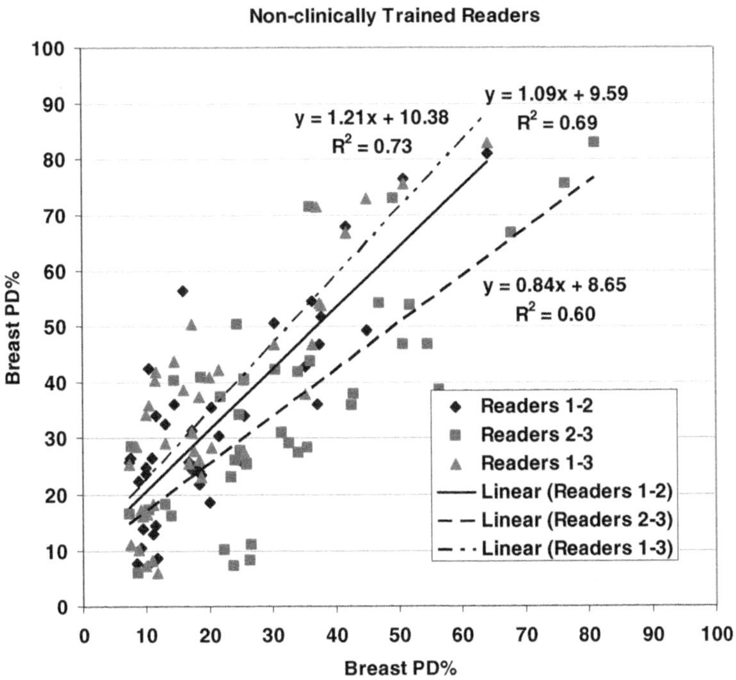

Fig. 1. Linear regression plots for pair-wise comparisons in the breast PD% estimates between the readers in each group (Read 1) with adjusted-R^2 estimates and linear regression equations. Readers in each group are ordered in increasing order of experience.

Table 2. Pearson correlations (r) for intra-reader variability in breast PD% estimates between the two reads of the readers in each group. Readers are ordered in increasing order of experience.

	Pearson Correlations (r) for PD% intra-reader variability					
	Clinically Trained Readers			Non-clinically Trained Readers		
Reader	1	2	3	1	2	3
	0.98	0.79	0.95	0.91	0.91	0.89

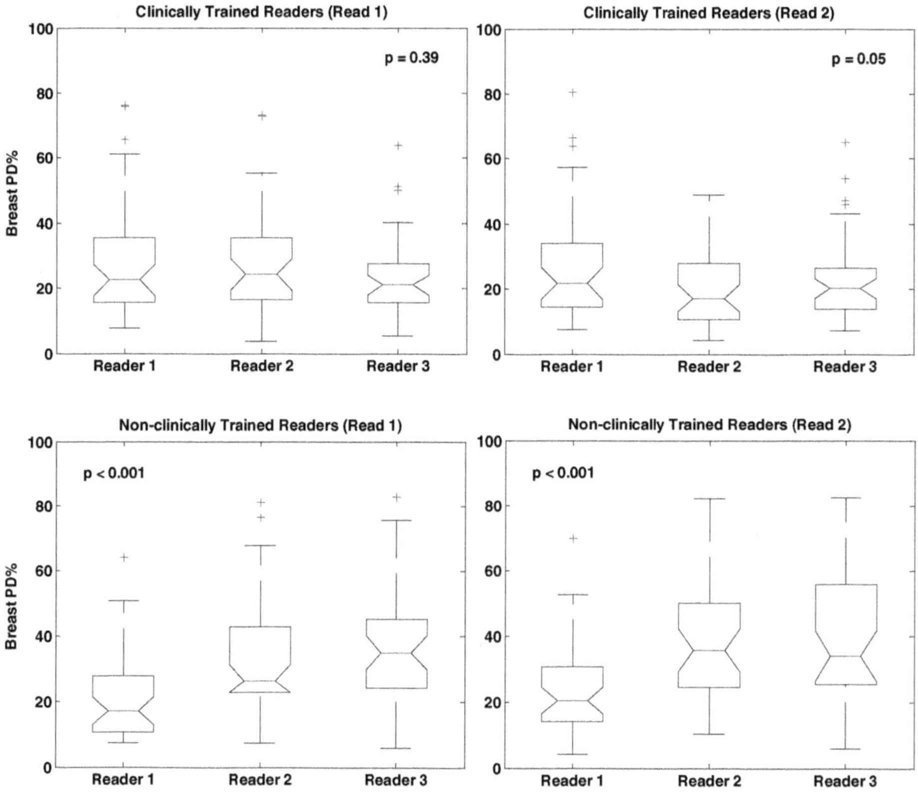

Fig. 2. Box-plots and ANOVA p-values for the breast PD% estimates of the readers in the clinically trained group (up) and the non-clinically trained group (down) after the first and the second breast PD% reads. Readers are ordered in increasing order of experience.

4 Discussion

The observed overall high inter- and intra- reader correlations suggest that area-based breast PD% estimates obtained from post-processed DM images could be a viable means for obtaining breast density measures in DM. However, the level of experience and training of the reader may have an effect on the obtained measures. Our study suggests that clinically trained readers, such as breast imaging radiologists, with minimal training in the use of the semi automated image-thresholding software (*i.e., Cumulus* Ver. 4.0, Univ. Toronto), have an overall higher inter-reader agreement in performing breast PD% estimation on post-processed DM images. The higher correlation among the clinically trained readers may be attributed to their greater clinical experience with the post-processed DM images and their clinical knowledge of breast anatomy.

The statistically significant difference in the means of the breast PD% measures between the two groups suggests that differences in breast imaging training could potentially impact a patient's risk assessment outcome. We attribute the observed differences in breast PD% estimates between the clinically and the non-clinically trained readers in potentially inherent differences in their corresponding visual perception of the dense breast tissue region in the DM image. Figure 3 illustrates an example where the difference in the thresholded dense tissue region resulted in significantly different breast PD% estimates between the two readers.

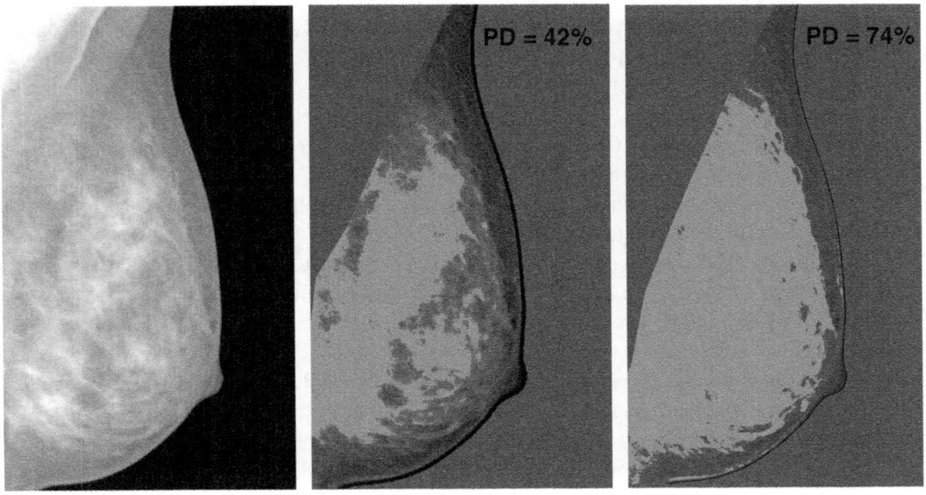

Original DM Clinically-Trained Reader Non-Clinically Trained Reader

Fig. 3. An example of an MLO DM image and the corresponding *Cumulus* dense tissue thresholding for a clinically trained and a non-clinically trained reader with PD% estimates.

Consistent and reproducible measures of breast PD% will become increasingly necessary as breast density measures become more frequently incorportated in breast cancer risk assessment algorithms. In addition, consistent and reproducible breast density measures will continue to be important in understanding the potential sensitivity of mammographic screening for individual women as we move forward towards adopting personalized screening algorithms for breast cancer detection. Fully-automated methods for estimating breast PD% hold the promise to alleviate the subjectivity introduced by individual readers and result in more accurate quantitative measures [10-12]. Further work is underway to compare the breast PD% estimates obtained from the post-processed DM images with the corresponding measures estimated from the raw (*i.e.*, unprocessed) digital mammograms.

Acknowledgments. This work was partially funded by the Susan G. Komen for the Cure Foundation (Postdoctoral Fellowship KG080756), the DOD (Concept Award BC086591), and the National Institutes of Health/National Cancer Institute (Program Project Grant P01-CA85484, PI M.D. Schnall). We would like to thank Dr. Martin Yaffe for providing the *Cumulus* software and Dr. Johnny Kuo for developing and maintaining the RSNA Medical Imaging Resource Center (MIRC) image archive. This manuscript reflects the views of the authors and may not reflect the opinions or views of the NIH.

References

1. Boyd, N.F., Guo, H., Martin, L.J., Sun, L., Stone, J., Fishell, E., Jong, R.A., Hislop, G., Chiarelli, A., Minkin, S., Yaffe, M.J.: Mammographic density and the risk and detection of breast cancer. New England Journal of Medicine 356, 227–236 (2007)
2. Nicholson, B.T., LoRusso, A.P., Smolkin, M., Bovbjerg, V.E., Petroni, G.R., Harvey, J.A.: Accuracy of assigned BI-RADS breast density category definitions. Academic Radiology 13, 1143–1149 (2006)
3. Ooms, E.A., Zonderland, H.M., Eijkemans, M.J., Kriege, M., Mahdavian Delavary, B., Burger, C.W., Ansink, A.C.: Mammography: interobserver variability in breast density assessment. Breast 16, 568–576 (2007)
4. Yaffe, M.J.: Mammographic density. Measurement of mammographic density. Breast Cancer Research 10, 209 (2008)
5. McCormack, V.A., dos Santos Silva, I.: Breast density and parenchymal patterns as markers of breast cancer risk: a meta-analysis. Cancer Epidemiology, Biomarkers & Prevention 15, 1159–1169 (2006)
6. Bakic, P.R., Carton, A.-K., Kontos, D., Zhang, C., Troxel, A.B., Maidment, A.D.A.: Breast percent density estimation from mammograms and central tomosynthesis projections. Radiology 252, 40–49 (2009)
7. Kontos, D., Ikejimba, L., Bakic, P.R., Troxel, A.B., Conant, E.F., Maidment, A.D.A.: Digital breast tomosynthesis parenchymal texture analysis for breast cancer risk estimation: Results from a screening trial. Presented at 95th Scientific Assembly and Annual Meeting of the Radiological Society of North America, Chicago, IL (2009)
8. Kontos, D., Bakic, P.R., Carton, A.K., Troxel, A.B., Conant, E.F., Maidment, A.D.A.: Parenchymal texture analysis in digital breast tomosynthesis for breast cancer risk estimation: A preliminary study. Academic Radiology 16, 283–298 (2009)
9. Bakic, P.R., Carton, A.K., Kontos, D., Zhang, C., Troxel, A.B., Maidment, A.D.: Breast percent density: estimation on digital mammograms and central tomosynthesis projections. Radiology 252, 40–49 (2009)
10. Jeffreys, M., Warren, R., Highnam, R., Smith, G.D.: Initial experiences of using an automated volumetric measure of breast density: the standard mammogram form. Br. J. Radiol. 79, 378–382 (2006)
11. Heine, J.J., Carston, M.J., Scott, C.G., Brandt, K.R., Wu, F.F., Pankratz, V.S., Sellers, T.A., Vachon, C.M.: An automated approach for estimation of breast density. Cancer Epidemiol Biomarkers Prev. 17, 3090–3097 (2008)
12. van Engeland, S., Snoeren, P.R., Huisman, H., Boetes, C., Karssemeijer, N.: Volumetric breast density estimation from full-field digital mammograms. IEEE Trans. Med. Imaging 25, 273–282 (2006)

Concepts for Efficient and Reliable Multi-modal Breast Image Reading

Horst K. Hahn[1], Markus T. Harz[1], Heike Seyffarth[2], Fabian Zöhrer[1], Tobias Böhler[1], Konstantinos Filippatos[2], Lei Wang[1], André Homeyer[1], Felix Ritter[1], Hendrik Laue[1], Matthias Günther[1], Thorsten Twellmann[2], László K. Tabár[3], Ulrich Bick[4], and Kathy J. Schilling[5]

[1] Fraunhofer MEVIS, Institute for Medical Image Computing, Universitaetsallee 29, 28359 Bremen, Germany
horst.hahn@mevis.fraunhofer.de
[2] MeVis Medical Solutions AG, Bremen, Germany
[3] Department of Mammography, Falun Central Hospital, Sweden
[4] Department of Radiology, Charité, Berlin, Germany
[5] Center for Breast Care, Boca Raton Community Hospital, Boca Raton/FL, USA

Abstract. We describe a group of concepts that facilitate reading of multi-modality breast imaging data in a single workplace and discuss their use and limitations. Our concepts comprise intelligent preprocessing, spatial referencing and dedicated workflow tools and aim at homogenizing and simplifying the multi-modality workplace, at improving the standardization across modalities and vendors, at supporting cross-modality information linkage, and at reducing required user interaction and waiting times, all at a high level of flexibility for the user to access the available imaging information at any time required. As a result, many situations where information from multiple modalities and time points must be assessed, both qualitatively and quantitatively, are expected to be handled more efficiently and reliably.

Keywords: multi-modality, image display, human-computer interaction, data fusion, image processing, image registration, computer-aided radiology.

1 Introduction

As new and improved imaging modalities continue to add options for diagnostic and clinical discoveries, the efficient and reliable access and fusion of the available information becomes a key factor for optimizing the reading workflow, for maximizing the overall diagnostic outcome, and for evaluating new versus existing technology. In breast imaging, multi-modal reading is in general required as soon as any single modality is insufficient for a conclusive decision regarding the treatment of women. Specifically, women with mammographic findings detected in screening are recalled for further imaging. Any finding from other modalities in the ipsilateral and contralateral breast then needs to be correlated to the respective mammographic location, often necessitating error-prone manual comparison of modalities. Currently, a number of specialized computer workstations from different vendors are needed to

J. Martí et al. (Eds.): IWDM 2010, LNCS 6136, pp. 121–128, 2010.

take full advantage of combinations of established and new imaging modalities, such as digital mammography (MG), breast MRI, positron emission mammography (PEM), handheld ultrasound (US), automated 3d breast ultrasound (ABUS), or tomosynthesis (DBT). The combined use of several workstations results in inadequate workflows with different user interfaces, lack of synchronization, suboptimal data transfer, multiple keyboards and monitors, etc. Our goal is to describe and discuss a group of novel software based concepts to support multi-modal breast image reading.

2 State of the Art

Concepts for integrated or fused viewing and methods for joint analysis are sparse. The same holds true for methods that make direct use of available data beyond diagnostic imaging, be it histology and pathology data or patient information such as hormone receptor status, family history, etc. This is despite the known correlations of patient risks and outcomes, and despite the amount of work spent on registration approaches for multi-modal combinations of imaging modalities.

In literature, the term *multi-modality* most often refers to the complementary analysis and diagnostic characterization of images of different modalities (MG, MRI, US, PET, etc.), comparing their respective additional cancer yield etc. [1] In general, such comparisons are conducted utilizing software provided by the vendor of the respective modalities. Those publications emphasize clinical aspects, rather than workflow and computer support. Examples are numerous, recent work encompassing the multicenter ACRIN 6666 trial to assess the benefit of ultrasound in addition to screening MG, the controversy regarding the utility of contrast-enhanced MRI, alone or in combinations with other modalities and in women with increased risk of developing cancer [2, 3], and also work on the combination of per-modality CAD systems on MG and US [4].

The necessity of fusion imaging and fused viewing is widely acknowledged [5] but still community efforts to provide integrated viewing of different modalities are few. First multi-modal workplaces do exist from different vendors, but without dedicated cross-modality workflow or reading support or joint analysis of the data. Also, systems exist that are not specifically targeted at breast imaging [6], the free *Fusion-Viewer* is a research tool targeting the evaluation of the underlying MRI-PET registration scheme [7], and most major PACS systems permit multi-modal viewing. Though with the advent of digital image data, the workflow of radiology technologists was examined and changed [8], research on the diagnostic workflow of radiologists using multi-modal, multi-examination breast images is lacking. Hence, most software tools support a limited set of simple diagnostic or procedural aspects, and some are integrated into the hospital PACS.

Also, intra- and inter-modality coregistration has been addressed, providing support for specific clinical scenarios, e.g. easier look-up of prior findings in current mammograms, reduction of motion artifacts in dynamic contrast-enhanced MRI, and more [9, 10]. Many of the current work are founded on methods established with no particular focus to breast imaging, e. g. various standard registration methods [11, 12]. More recently, breast imaging related issues were addressed such as 2d-3d spatial correlation [13], correlation between MG and MRI [14], and correspondence analysis regarding the deformations induced by different breast imaging modalities [15]. Only few of these techniques are in clinical use, mainly because spatial registration algorithms are often computationally slow and also sensitive to parameterization and therefore in a routine workflow only accessible to trained operators who can compensate for lack of accuracy.

Reliable automated advanced preprocessing needs assured quality standards, regarding image quality [16] and acquisition artifacts for all modalities. Quality checks are today only performed after image acquisition, with the patient no longer available for a repeat scan. The only *online* validation of image quality is performed visually by a trained operator, not necessarily guaranteeing quality criteria crucial for computerized processing steps. Automated processing pipelines, such as proposed by Zijdenbos et al. [17], rely on a level of image quality that cannot always be granted.

Regarding digital pathology, open technical issues have been described in detail [18]. Research has provided tools for interactive and automated segmentation of cell nuclei and other structures from digital pathology slides, and also three-dimensional reconstructions of thin-slice sections are being explored [19, 20]. Web-based multiscale access to the enormous amounts of data has also been suggested. However, the issue of storage, exchange, and collaboration on these images in a clinical setting combined with a radiology workplace is still unsolved.

3 Concepts

The integration of different imaging modalities into a single, multi-modality workstation poses significant challenges to user interface design, information standardization, cross-modality information linkage, and workflow adaptation. Consequently, we propose four novel building blocks for an efficient multi-modal workplace as summarized by Figure 1: (i) The *Diagnostic Workflow Gallery* providing comprehensive access to all available information while providing an intuitive means to define and adapt complex multi-modal reading workflows, (ii) An *Intelligent Preprocessing Engine* supporting standardization and speedup of multi-modal reading, (iii) *Advanced Spatial Referencing*, both for efficient prior-current and cross-modality comparisons, and (iv) *Cross-Disciplinary Embedding* comprising real-time access to full-resolution histological data and integration with the required tools for interventional procedures from biopsies to open surgery.

Diagnostic Workflow Gallery. To support efficient reading, the current diagnostic task is analyzed based on the available data combined with user preferences and related information is prefetched and processed (cf. preprocessing engine). Resulting image series are sorted with respect to image modality and acquisition time and represented by miniature images, so-called *thumbnails*, as entry points for the *Diagnostic Workflow Gallery*. The physician may now simply click on one of the thumbnails to activate related diagnostic tasks. From this point, a number of additional workflow options based on pre-analyzed images are added to the gallery, and options that no longer apply are removed. Using this concept, the main functionality of the multi-modal workstation intuitively reveals itself to the physician without requiring multiple menu-based mouse interactions. For each workflow step, the user interface is streamlined to the specific diagnostic task while further advanced functionality is offered and displayed through dynamically generated thumbnails within the gallery.

Subsequently, for example, if the current situation allows reporting of a lesion seen in MRI, the *Diagnostic Workflow Gallery* offers tools for segmentation based kinetic and morphologic lesion characterization, for altering the segmentation, for correlating the lesion location at other time points and in other modalities, and for retrieving similar findings from a case database. Note that, in case of lesion characterization, the gallery shows a preview of the individual lesion segmentation, not only a fixed button or symbol. Also, the *Diagnostic Workflow Gallery* does not force user interaction. Instead, the reader may interact with the main screen and upon parameter or view alterations, the gallery previews update their appearance.

From the history of steps performed, workflow templates can be generated or adapted to the physician's individual requirements. Likewise, all image series are accessible through drag-and-drop for intuitively defining a user's preferred mono- or multi-modal hangings that are also saved as a part of the workflow templates. Through an *adaptive fingerprint* algorithm, hangings are filled with data of new patients in a best-match fashion that robustly tolerates even considerable changes in acquisition protocols [6]. Once findings are defined on any of the images, these are persistently managed within the gallery together with associated bookmarks. Bookmarks permit the immediate retrieval of the full workflow state including all viewers at any time, and a parameter set history to enable full interactivity from the state in which the workflow was left.

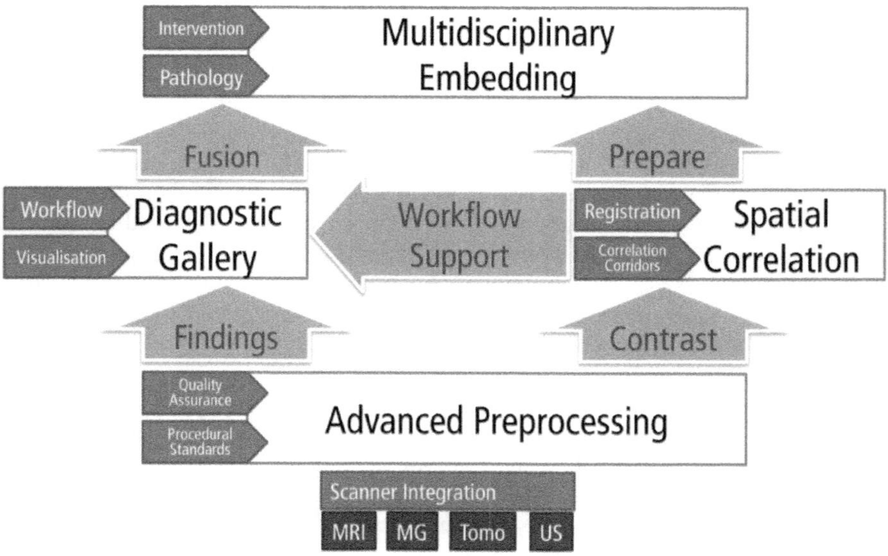

Fig. 1. Conceptual collaboration graph: *Advanced preprocessing* provides the basis of subsequent visualization for the *diagnostic gallery* and to calculate *spatial correlations*. Symbolic visualizations and thumbnails enable intuitive workflow operation, while temporal and cross-modality correlation and registration enable easy navigation and is the foundation of multi-modal analysis tools. Both contribute to efficient diagnosis and serve the *multidisciplinary embedding* of radiology, pathology, and surgery.

Intelligent Preprocessing Engine. In order to reduce interaction times but also to increase the robustness of information extraction, we propose an intelligent preprocessing engine that (i) is modular and extensible and that (ii) can be automated based on the available image data and on the specific requirements of departments and individual clinicians. The basic concept for automation is similar to the *adaptive fingerprint* algorithm, but requires a stricter rule set because preprocessing must be able to run fully automatically and without supervision. After proper configuration, unread patient series are retrieved from PACS and required image analysis methods are applied automatically including lesion segmentation and characterization.

As an important prerequisite for reliable application of post-processing and CAD algorithms, we propose to employ automated content-based quality assurance of 3d and dynamic images prior to processing and visualization. Specifically, dedicated software modules automatically detect imaging artifacts and ensure the defined image acquisition standards with regard to the intended post-processing, CAD tool, or visualization. In particular, artifacts such as improper timing of contrast-agent bolus, patient motion, or adjustment problems may occur during MRI acquisition and can be detected automatically. To some extent, artifacts can be compensated, for example by modeling the specific contrast agent relaxivity or acquisition sequence in order to normalize image intensities.

Further methods within the preprocessing engine are deformable motion correction methods for dynamic imaging and for ABUS to minimize post-processing artifacts. Other tools provide automatic cross-vendor homogenization for cases where only *for presentation* images have been stored and enhancement of mammograms [21], or offer plug-in interfaces to dedicated CAD tools. In order to maximize display speed even on slow network connections, intelligent preprocessing is complemented by progressive on-demand retrieval of large datasets.

Advanced Spatial Referencing. While motion correction in breast MRI is part of most current workstations, an increasing number and diversity of acquired images per patient asks for additional spatial coregistration methods. We focus on two different problems here, namely precise spatial correlation of prior and current imaging series and fast cross-modality spatial alignment for synchronization of imaging findings.

The joint analysis of prior and current images provides additional information on temporal lesion development. However, numerous factors in the context of imaging prohibit a direct comparison of image intensity values, including changes in scanner hardware and software, different acquisition parameters, pathological and physiological breast tissue changes, or geometrically different breasts after intervention. We propose to use deformable image registration methods and an automated slice synchronization tool to allow the direct comparison of breast MRI images over time that performs reasonably well even for extreme deformations [10, 22].

To enable fast cross-modality alignment and CAD, we propose a set of automatic segmentation and localization methods for anatomical structures such as nipple, skin boundary, fatty tissue, fibroglandular tissue, pectoral muscle, chest wall, and possibly vasculature besides lesions. Based on these structures we propose to employ non-linear coordinate system transformations between the respective modalities in both directions including physical deformation modeling. Typically, a trained clinician visually correlates findings between modalities based on spatial sense and experience.

Our approach is intended to perform a similar job even in difficult cases, such that for each position in one modality, an approximate 2d or 3d search region can instantly be displayed in the other modality. If automatic extraction fails for a single anatomical structure, we propose to use an efficient interactive tool as a fallback solution.

Cross-Disciplinary Embedding. Despite the progress of radiology, pathology remains the diagnostic gold standard for breast abnormalities. The result from pathology needs to be correlated to other available information. Such cross-disciplinary correlation provides the basis to clarify the diagnosis in ambiguous cases. Also, the biopsy position within the breast tissue needs to be assured and documented [23]. In MRI, those lesions with diffuse and low contrast uptake, e.g. DCIS, may hardly be visible under MRI-based biopsy guidance. As a solution, we propose to perform a transformation of the target area from a high-resolution pre-interventional MRI to the intra-interventional situation (cf. spatial referencing, Fig. 2).

Any surgery or biopsy specimen is submitted to the pathology department with the ultimate goal of correlation to the imaging findings. Still, only selected, if any, snapshots of histological images are exported to PACS and directly accessible to radiologists.

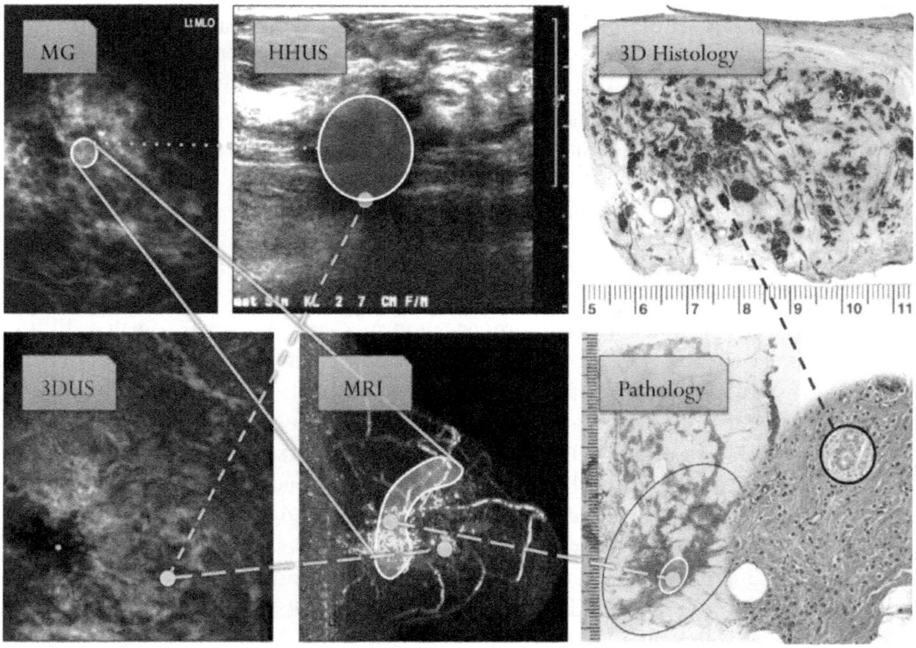

Fig. 2. Advanced spatial referencing: Computations can provide estimations of corresponding locations or search regions, wherever features suffice for the task. From mammography (MG), referencing to MRI can be deducted from combinations of CC and MLO views. Volume ultrasound (3DUS) can be mapped to MRI in a local loose coupling fashion. Solid connections indicate computer-provided spatial referencing; dashed connections indicate interactively established correlations.

Moreover, unlike digital radiology, digital pathology is still in its infancy, with most pathologists relying on conventional microscopes. However, due to the growing availability of automated microscopes that enable batch digitization of histological slides, pathology becomes increasingly computerized. A lack of infrastructural integration combined with insufficient tools for digital pathology is the most likely cause for continuing skepticism towards digital pathology. Additionally, digital whole-slide images cannot easily be integrated into medical workstations. Typical file sizes for such slides are at least one order of magnitude larger than typical radiological images, thus current PACS systems may not be capable of storing these files.

As a possible solution, we propose specialized multi-scale viewers with efficient on-demand data streaming connected to a histological image server, resulting in a seamless integration of full high-resolution histology data for cross-disciplinary evaluation of ambiguous cases. Furthermore, in order to support histological-radiological communication, we propose illustration tools for pathological annotations. We argue that a comprehensive breast imaging workstation should enable cross-disciplinary planning of interventions from biopsy over surgery to radiation therapy, and allow for full cross-disciplinary correlations of radiological and histological findings.

4 Discussion and Perspectives

An integrated software solution for multi-modal reading has the potential to associate corresponding locations by providing visual guidance, e.g. search corridors or co-registration based exact lesion matches. Computer assistance relating a finding seen on MG views to an approximate region of likelihood in breast MRI or ABUS volumes, or vice versa, requires further fundamental research. Also, new image analysis and CAD algorithms, which show potential for advanced diagnostic support, provide a challenge to user interface design due to the increasing complexity of resulting software systems. The proposed extensible *Diagnostic Workflow Gallery* aims at a solution here. Importantly, an integrated workplace will provide a homogenized look-and-feel for basic and advanced display options of all modalities, in contrast to the combination of several proprietary single-modality workplaces. Intuitive customization of cross-modality 2d, 3d, and dynamic views without requiring technical knowledge is facilitated through a hanging editor with *adaptive fingerprinting* of image series. Post-processing and CAD tools often sensitively depend on the compliance with standards for acquisition protocols and image quality. Practically, however, strict standardization across sites, vendors, and clinicians has severe limitations and is currently out of reach. Instead, we propose sophisticated algorithms to estimate levels of artifacts as a basis for automated image quality assurance. Ideally, this would take place during image acquisition to allow reacquisition in case of low-quality data.

Current limitations relate to a lack of standardization in image acquisition and storage, coarseness of cross-modality spatial correlation, availability of digital high-resolution histology data, and availability and specificity of dedicated multi-modal CAD algorithms. Even though digital pathology is still young, an upcoming extension of the DICOM standard will permit the combined management of radiological and pathological images in a single PACS system. Ultimately, making direct use of multi-modal and cross-disciplinary information will facilitate more sensitive detection and more specific characterization of lesions seen in more than one modality.

References

1. Lopez, J.K., Bassett, L.W.: Invasive lobular carcinoma of the breast: spectrum of mammographic, US, and MR imaging findings. Radiographics 29(1), 165–176 (2009)
2. Kuhl, C., Weigel, S., Schrading, S., et al.: Prospective Multicenter Cohort Study to Refine Management Recommendations for Women at Elevated Familial Risk of Breast Cancer: The EVA Trial. J. Clin. Oncology 28(9), 1450–1457 (2010)
3. Sardanelli, F., Podo, F., D'Agnolo, G., et al.: Multicenter comparative multimodality surveillance of women at genetic-familial high risk for breast cancer (HIBCRIT study): Interim results. Radiology 242(3), 698–715 (2007)
4. Horsch, K., Giger, M.L., Vyborny, C.J., et al.: Classification of Breast Lesions with Multimodality Computer-aided Diagnosis: Observer Study on an Independent Clinical Data Set. Radiology 240(2), 357–368 (2006)
5. Zaidi, H., Montandon, M.L.L., Alavi, A.: The clinical role of fusion imaging using PET, CT, and MR imaging. Magn. Reson. Imaging Clin. N. Am. 18(1), 133–149 (2010)
6. Hahn, H.K., Valer, V., Schlemmer, H.P., et al.: Efficient Whole Body MRI and Multimodal Softcopy Reading by Persistent Workflow Modeling. Proc. RSNA: LL-I 3100 (2006)
7. Baum, K.G., Helguera, M., Krol, A.: Fusion viewer: A new tool for fusion and visualization of multimodal medical data sets. J. Digit Imaging 21(Suppl. 1), S59–S68 (2008)
8. Reiner, B., Siegel, E., Carrino, J.A.: Workflow Optimization: Current Trends and Future Directions. J. Dig. Imaging 15(3), 141–152 (2002)
9. Guo, Y., Sivaramakrishna, R., Lu, C.C.C., et al.: Breast image registration techniques: a survey. Med. Biol. Eng. Comput. 44(1-2), 15–26 (2006)
10. Li, X., Dawant, B.M., Welch, E.B., et al.: A nonrigid registration algorithm for longitudinal breast MR images and the analysis of breast tumor response. MRI. 27(9), 1258–1270 (2009)
11. Collignon, A., Maes, F., Delaere, D., et al.: Automated multi-modality image registration based on information theory. In: Proc. IPMI, pp. 263–274 (1995)
12. Davis, M.H., Khotanzad, A., Flemig, D.P., Harms, S.E.: A physics-based coordinate transformation for 3-D image matching. IEEE TMI 16(3), 317–328 (1997)
13. Marti, R., Zwiggelaar, R., Rubin, C., Denton, E.: 2D-3D correspondence in mammography. Cybernetics and Systems 35, 85–105 (2004)
14. Ruiter, N.: Registration of X-ray mammograms and MR-volumes of the female breast based on simulated mammographic deformation. PhD thesis, Univ. Mannheim (2003)
15. Azar, F.S., Lee, K., Khamene, A., et al.: Standardized platform for coregistration of non-concurrent diffuse optical and magnetic resonance breast images obtained in different geometries. J. Biomed. Opt. 12, 051902 (2007)
16. Eskicioglu, A.M., Fisher, P.S.: Image quality measures and their performance. IEEE Trans Communications 43(12), 2959–2965 (1995)
17. Zijdenbos, A.P., Forghani, R., Evans, A.C.: Automatic Pipeline Analysis of 3-D MRI Data for Clinical Trials: Application to Multiple Sclerosis. IEEE TMI 21(10), 1280–1291 (2002)
18. Madabhushi, A.: Digital Pathology Image Analysis: Opportunities and Challenges. Imaging Med. 1(1), 7–10 (2009)
19. Petushi, S., Garcia, F.U., Haber, M.M., et al.: Large-scale computations on histology images reveal grade-differentiating parameters for breast cancer. BMC Med. Imaging 6, 14 (2006)
20. Petushi, S., Zhang, J., Zhu, W., et al.: A visual analytics system for breast tumor evaluation. Anal. Quant. Cytol. Histol 30(5), 279–290 (2008)
21. Zöhrer, F., Harz, M.T., Bödicker, A., et al.: Interactive Multi-Scale Contrast Enhancement of Previously Processed Digital Mammograms. In: Martí, J., et al. (eds.) Proc. IWDM. LNCS, vol. 6136. Springer, Heidelberg (2010)
22. Böhler, T., Schilling, K., Bick, U., Hahn, H.K.: Deformable Image Registration of Follow-up Breast Magnetic-resonance Images. In: Proc. WBIR (in print 2010)
23. Irwin, M.R., Downey, D.B., Gardi, L., Fenster, A.: Registered 3-D ultrasound and digital stereotactic mammography for breast biopsy guidance. IEEE TMI 27(3), 391–401 (2008)

Modeling the Impact of CAD on the Outcomes of the UK Breast Screening Programme

Paul Taylor

Centre for Health Informatics and Multiprofessional Education, University College London, Archway Campus, Highgate Hill, London, N19 5LW, United Kingdom
p.taylor@chime.ucl.ac.uk

Abstract. *Objective:* To assess the impact on breast cancer mortality of improving the sensitivity of breast screening programmes. *Methods:* A Markov Model was populated with data describing the UK screening programme and the incidence and mortality of breast cancer. The model was used to study the impact of CAD on cancer detection rates and mortality for a cohort followed from age 45 to age 90. *Results:* Running the model with values of sensitivity from 75 to 95% shows the proportion of cancers detected at screening increasing as screening improves, and deaths from breast cancer falling. The drop in breast cancer deaths is however modest: increasing sensitivity from 75% to 85% reduces the number of breast cancer deaths from 28 to 27 per thousand. *Conclusions:* The likely improvements in the sensitivity of screening with CAD do not have a marked effect on breast cancer mortality.

1 Introduction

Computer aided detection (CAD) systems analyse digital mammograms and alert the user to regions of the image more likely to contain an abnormality. One question that is often asked of CAD is whether it allows a single reader to achieve the sensitivity normally attained by double reading mammograms. Attempts to compare CAD with double reading have largely failed to achieve either methodological rigour or statistical power with the notable exception of the CADET II trial which suggested that single reading with CAD provides comparable sensitivity to double reading.[1,2]

To assess the importance of sensitivity in screening we need to understand the relationship between sensitivity and outcomes. Clearly, better screening should have a greater impact on mortality, but it is naïve to assume that each extra cancer detected at screening is a life saved. Mathematical simulations have been carried out to assess the financial value of digital mammography and of CAD.[3,4] This paper presents a mathematical modeling exercise which takes values from official UK statistics and the academic literature and simulates the impact of changing the sensitivity of screening on various outcomes.

2 Methods

Cancer in a screened population is modelled as a Markov process. The states and the allowed transitions are shown in Figure 1. Each transition is associated with a probability.

J. Martí et al. (Eds.): IWDM 2010, LNCS 6136, pp. 129–136, 2010.

Examples of age-related transition probabilities for women aged 45, 60 and 75 are shown in Table 1. Using this model, a cohort of women is followed from age 45 to 89. Each year women are moved from state to state according to the transition probabilities. The transition probabilities can be adjusted to model different scenarios.

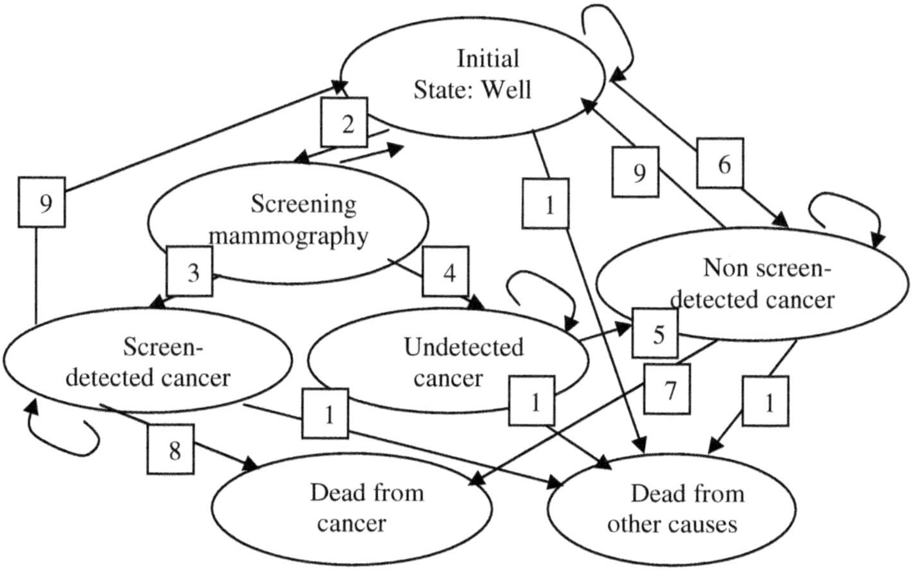

Fig. 1. Markov model of breast cancer screening. The ovals represent states through which women pass. Arrows show transitions between states.

Between the ages of 50 and 64, a proportion of the cohort attends for screening every three years. Two categories of detected cancer are considered: detected at screening and detected outside screening. No distinction is made between cancers detected in the intervals between screening visits and those detected in women who have not attended for screening. The progression of disease is not modelled explicitly, but detected cancers are assigned a prognostic category, according to a profile of prognostic categories for the method of detection. The difference in the profiles of screen-detected and non screen-detected) is the basis, in the model, for the benefit due to screening: women with cancers detected at screening are more likely to survive because these cancers are more likely to be assigned to a category with a better prognosis. All cancer deaths are assumed to occur within 10 years of diagnosis, after which patients are returned to the initial state.

Authoritative estimates of transitions labelled 1,2,3 in Figure 1 (death from 'other causes', screening uptake, proportion of women with cancers detected at screening, were obtained from the UK mortality statistics and the NHS screening programme.[5,6] The approach to modeling the other transitions is outlined here.

Table 1. Age-related transition probabilities for women aged 45, 60 and 75

Age	Other cause mortality rate, per 100,000	Screen detected cancer rate, per 100,000 of population	Non screen detected cancer rate, per 100,000 of population	Mean sojourn time in years
45	124	6	164	3
60	503	207	96	4
75	2507	65	248	6

Transitions 4,5,6: If the model is to be used to assess improvements to the screening programme, it must include a pool of detectable but currently undetected cancers. The model therefore includes an 'undetected cancer' state to which women are transferred with a probability reflecting the sensitivity of the screening programme. These cancers are undetected until the 'sojourn' time elapses, after which they present clinically and the woman transfers to the 'non screen-detected cancer' state. Estimates of sensitivity vary, however the accuracy of the baseline estimate of sensitivity used is not crucial for the model. Consider Figure 1. If an estimate is chosen for *4* and used to calculate a figure for *5*, then so long as the figure for *6* is set so that the total for *5* and *6* is accurate, the accuracy of the estimate at *4* is of no consequence. We followed Elmore et al. and assume 25% of cancers are missed at screening so set the probability of transition *4* to be 1/3 the value of transition *3*.[7]

Transition *5* is the rate at which missed cancers present clinically. Women stay in the missed cancer state – unless they die of 'other causes' – until the sojourn time has elapsed. An estimate of mean sojourn time (MST) at each age is used to generate a Poisson distribution from which the number of missed cancers presenting each year is determined. Undetected cancers are attenuated with the force of the other cause mortality rate (transition *1*) each year that they stay in this state. In order to determine age-related estimates of MST, PubMed was searched and five papers containing seven sets of estimates for the age-related MST obtained. [8-12] Plotting age against MST and fitting a simple exponential curve made it possible to generate a figure for MST at each age. These ranged from 3 yrs at age 50 to 6yrs at age 75. To test the sensitivity of the model to this estimate, the performance of the model was assessed as the fitted curve was varied through plus and minus 40%.

The total of transitions *5* and *6* was calculated by subtracting the figure for screen detected cancer from the total number of cancer registrations.[13] The figure for transition *6* is then determined as the balance required given the calculated value of transition *5*.

Transitions 7 & 8. Profiles for screen-detected and non screen-detected cancers are used to assign each cancer a 'prognostic category'. Each category is associated with a separate survival curve. The modeled impact of screening on mortality will depend very closely on these profiles and associated survival curves. If the profile of cancers is similar for screen detected and non screen detected cancers, screening can have

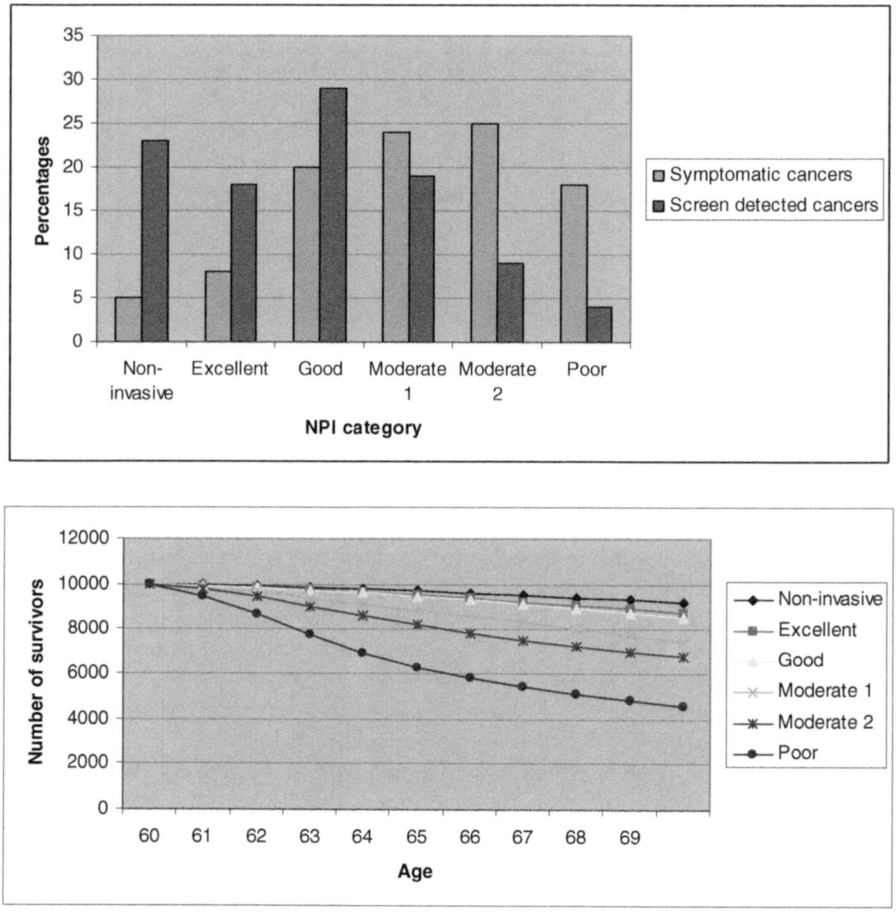

Fig. 2. The prognostic profiles for screen and non-screen detected cancers. The bar chart shows the percentages of non screen-detected and screen detected cancers in the different prognostic categories of the NPI. Cancers in the simulation are assigned to a prognostic category according these frequencies, depending on the method of detection. The graph shows the survival curves, for an imaginary cohort of 10,000 women aged 60 in 2006 in each these categories. The curves are based on data for cancer deaths collated by Nottingham City Hospital combined with 'Other Cause Mortality' rates, for ages 60 to 69, calculated from UK mortality statistics.

little benefit. If the survival curves for the different prognostic categories are not widely separated, screening can have little benefit. Two approaches were compared: one based on the staging of cancers and one on categories of the Nottingham Prognostic Index (NPI), shown in Figure 2. In order to compare the performance of the model with different estimates of these values, two 'figures of merit' were used.

First, the model was used to generate an estimate of the overall impact of the screening programme on the number of breast cancer deaths. This was achieved by running a simulation with sensitivity of 0% (transition *4*), simulating the complete absence of screening. Comparing this to a simulation with the baseline sensitivity of

75% allows an estimate of the overall impact of screening on mortality. Second, age-specific rates of breast cancer mortality were compared to those in the UK mortality statistics and the mean squared error calculated. Note that the two datasets should be similar but there is no a priori reason why they should be identical, the model simulates a cohort aged 45 in 2006 and followed for 45 years, UK mortality statistics are calculated for the women aged between 45 and 90 in 2006.

Two implementations of the model, one using the profiles and survival curves published for categories of the NPI, the other based on stages, were compared on these two 'figures of merit'. The NPI model gives an estimate for the impact of NHSBSP of a 16% reduction in breast cancer deaths. The stage model gives an estimate of 9%. Both sets of data seem consistently to underestimate the number of deaths, compared to the 2006 data, and to markedly underestimate deaths in older women. On balance the NPI data seemed more robust, with a more plausible estimate of the impact of screening and was used in the remaining simulations. The NPI data were however calibrated with the figures for breast cancer deaths in 2006, in order to allow for a possible interaction between age and mortality, resulting in a mean squared error of 1.1% in the central region. The NPI profiles were determined from data published by Wishart et al. and survival curves based on data published by Blamey et al. [14,15] Since the NPI profile is for invasive disease only, estimates of the proportion of detected cancers that are non-invasive in screening and outside screening were obtained from the NHSBSP and the literature.[6,16]

Transition 9. An assumption that survival curves plateau at 10 years after diagnosis is enforced, and surviving patients return to the well state.

The model was used to simulate changes in the cancer detection rate that might follow from changes in the screening protocol. The model was run varying the sensitivity from 75% to 95%. Each simulation was run 1,000,000 times.

3 Results

Running the model from age 45 to 89 with sensitivity set at the baseline value of 75%, predicts the detection at screening of 44 cancers per thousand of population and the detection outside of screening of 82 cancers per thousand of population. Improving screening to a sensitivity of 85% should increase the number of cancers detected at screening to 50 per thousand and decrease the number detected outside screening to 76 per thousand. The number of breast cancer deaths moves only slightly, from 28 to 27 per thousand. Table 2 shows the impact of changing the sensitivity of screening from 75% to 95%. There is a drop in breast cancer deaths as screening improves over this range, but almost no impact on overall survival is observed.

4 Discussion

The original aim of this work was to assess the possible impact of CAD on the outcomes of the screening programme. In order to determine how CAD might alter the sensitivity of screening, a pair of linked systematic reviews was carried out, looking first at comparisons of single reading and CAD and then at comparisons of single and double reading. This work has been described in detail elsewhere.[17] Ten studies were found comparing single reading to single reading with CAD, and seventeen comparing single reading with double reading. Double reading with arbitration (films where the two initial readers disagree are reviewed by a third) shows a significant increase in detection rate (odds ratio: 1.08, 95% CI:1.02-1.15) and a significant decrease in recall rate (odds ratio:0.94, 95% CI:0.92-0.96). CAD studies do not show a significant increase in cancer detection rate (odds ratio: 1.04, 95% CI: 0.95-1.13) and – although there is considerable heterogeneity between studies – show an increased recall rate (odds ratio: 1.10, 95% CI: 1.09-1.12). These differences in cancer detection rate are relatively small (expressed as a risk difference the best estimate of the effect due to CAD is 0.16 extra cancers per 1000 women screened, for double reading 0.44 extra cancers per 1000 women screened). Rather than attempt to compare the model run on such similar estimates, it seemed more illuminating to use the model to simulate larger changes in sensitivity. The conclusion is that the kinds of improvements in sensitivity that we can anticipate are unlikely to have a significant impact on the main outcomes of interest to the screening programme.

Perhaps the best evidence about the impact of CAD is from the CADET II trial. CADET II was designed as an equivalence trial and powered to detect a 10% difference between the intervention and control conditions. No difference was detected so the two are considered equivalent. The systematic review of trials suggests that neither CAD nor double reading are likely to improve screening sensitivity by 10%. The modeling reported here suggests that improvements of less than 10% are not going to have an impact on the chief outcome of interest to screening programmes.

Table 2. Outcomes per 100,000 of population for different levels of sensitivity of screening

Sensitivity of screening	Screen detected cancers	Non screen detected cancers	Breast cancer deaths	Total deaths
95	5534	7143	2647	71317
92.5	5389	7260	2682	71212
90	5255	7374	2696	71291
87.5	5096	7521	2690	71260
85	4953	7642	2728	71333
82.5	4816	7785	2748	71388
80	4694	7879	2751	71308
77.5	4505	8063	2798	71338
75	4370	8246	2816	71353

One conclusion to draw from this is that proposed enhancements of screening should target the specificity of screening rather than the sensitivity, where the scope for real improvement is perhaps limited. By itself, this would argue for the retention of double reading and not single reading with CAD: both the systematic review and the CADET II trial showed a significant increase in recall rate with CAD. CAD can be justified on cost grounds if the saving in radiologist time of not doing double reading is less than the combined cost of CAD and of dealing with the additional recalls generated by CAD. The data suggest that screening is beneficial, but since screen detected cancers do not directly translate into lives saved, improvements in screening have only a small impact on mortality.

Acknowledgments. The study was partly completed as a component of an evaluation of CAD supported by the UK NHSBSP. The help and assistance of Dr Given-Wilson and Dr Potts is gratefully acknowledged.

References

1. Bennett, R.L., Blanks, R.G., Moss, S.M.: Does the accuracy of single reading with CAD compare with that of double reading? Clin. Radiol. 61(12), 1023–1028 (2006)
2. Gilbert, F.J., Astley, S.M., Gillan, M.G., et al.: Single reading with computer-aided detection for screening mammography. N. Engl. J. Med. 359(16), 1675–1684
3. Lindfors, K.K., McGahan, M.C., Rosenquist, C.J., Hurlock, G.S.: Computer-aided detection of breast cancer: a cost-effectiveness study. Radiology 239, 710–717 (2006)
4. Tosteson, A.N., Stout, N.K., Fryback, D.G., et al.: Cost-effectiveness of digital mammography breast cancer screening. Ann. Intern. Med. 148(1), 1–10 (2008)
5. National Statistics Mortality statistics: deaths registered in 2006 Office of National Statistics (2008)
6. NHS Breast Screening Programme Breast Screening Programme, England: 2005-06 The Information Centre (2007)
7. Elmore, J.G., Armstrong, K., Lehman, C.D., Fletcher, S.W.: Screening for breast cancer. JAMA 293(10), 1245–1256 (2005)
8. Weedon-Fekjaer, H., Lindqvist, B.H., Vatten, L.J., Aalen, O.O., Tretli, S.: Estimating mean sojourn time and screening sensitivity using questionnaire data on time since previous screening. J. Med. Screen. 15(2), 83–90 (2008)
9. Mandelblatt, J., Schechter, C., Lawerence, W., Yi, B., Cullen, J.: The SPECTRUM Population Model of the Impact of Screening and Treatment on US Breast Cancer Trends from 1975 to 2000. Journal of the National Cancer Institute Monographs 36, 47–55 (2006)
10. Olsen, A.H., Agbaje, O.F., Myles, J.P., Lynge, E., Duffy, S.W.: Overdiagnosis, sojourn time, and sensitivity in the Copenhagen mammography screening program. Breast J. 12(4), 338–342 (2006)
11. Tabar, L., Vitak, B., Chen, H.H., et al.: The Swedish Two-County Trial twenty years later. Updated mortality results and new insights from long-term follow-up. Radiol. Clin. North Am. 38, 625–651 (2000)
12. Paci, E., Duffy, S.: Modelling the analysis of breast cancer screening programmes: sensitivity, lead time and predictive value in the Florence District Programme (1975-1986). Int. J. Epidemiol. 20(4), 852–858 (1991)
13. National Statistics Cancer statistics registrations: Registrations of cancer diagnosed in 2006, England. Office of National Statistics (2008)
14. Wishart, G.C., Greenberg, D.C., Britton, P.D., et al.: Screen-detected vs symptomatic breast cancer: is improved survival due to stage migration alone? British Journal of Cancer 98, 1741–1744 (2008)

15. Blamey, R.W., Ellis, I.O., Pinder, S.E., et al.: Survival of invasive breast cancer according to the Nottingham Prognostic Index in cases diagnosed in 1990-1999. Eur. J. Cancer 43(10), 1548–1555 (2007)
16. Evans, A.J., Pinder, S.E., Ellis, I.O., Wilson, A.R.: Screen detected ductal carcinoma in situ (DCIS): overdiagnosis or an obligate precursor of invasive disease? J. Med. Screen 8(3), 149–151 (2001)
17. Taylor, P., Potts, H.W.: Computer aids and human second reading as interventions in screening mammography: two systematic reviews to compare effects on cancer detection and recall rate. Eur. J. Cancer 44(6), 798–807 (2008)

Introducing a Novel Image Quality Measure for Digital Phase-Contrast-Image Evaluation

Satoru Matsuo[1], Hiroshi Fujita[2], Junji Morishita[3], and Tetsuro Katafuchi[4]

[1] Department of Radiology, Shiga University of Medical Science
Otsu 520-2192, Japan
matsuo@belle.shiga-med.ac.jp
[2] Department of Intelligent Image Information, Graduate School of Medicine, Gifu University
Gifu 501-1194, Japan
fujita@fjt.info.gifu-u.ac.jp
[3] Department of Health Sciences, Faculty of Medicine Science, Kyushu University
jm@shs.kyusyu-u.ac.jp
[4] Department of Radiological Technology, Gifu University of Medical Science
Seki 501-3892, Japan
katafuchi@u-gifu-ms.ac.jp

Abstract. Recently, detective quantum efficiency (DQE) arising from the concept of signal-to-noise ratio (SNR) has been used for assessing digital x-ray imaging systems. Using a phase-shift of x-rays that occurs when passing through an object, digital phase contrast imaging (herein referred to as "phase imaging"), which involves magnification, can produce images different from those of standard contact imaging (herein referred to as "regular imaging"). For this reason, assessment of the image quality based on DQE which does not include the object information may not be appropriate to compare image quality between the phase images and the regular images. As an alternative method, we proposed a new image quality assessment method based on radial direction distribution function (RDDF) and signal intensity distribution function (SIDF) in two-dimensional power spectra of images that contain information of an object. To evaluate the usefulness of our method based on RDDF and SIDF, we assessed images of different contrast, noise characteristic and sharpness using simple phantoms. Our results showed that the accurate evaluation of these factors was successfully performed. Comparing the image quality of projected plant seeds by phase imaging and regular imaging, we found the phase imaging method provided higher image quality in terms of edge sharpness than that of the regular imaging.

Keywords: phase contrast imaging, radial direction distribution function, signal intensity distribution function, edge enhancement, two-dimensional power spectra.

1 Introduction

For evaluation of digital x-ray imaging systems, detective quantum efficiency (DQE) is sometimes applied based on the concept of signal-to-noise ratio (SNR).[1, 2] This is because DQE has been considered useful for comprehensive evaluation of an x-ray

J. Martí et al. (Eds.): IWDM 2010, LNCS 6136, pp. 137–144, 2010.

detection system, as data calculations of DQE are done using the gradient obtained with input-output conversion characteristics of the system (characteristic curve), modulation transfer function (MTF) for resolution properties, and the Wiener spectrum or noise power spectrum for noise properties [3]

For image evaluation of digital phase contrast imaging developed for mammography examination, the authors have been conducting comparative studies between conventional digital x-ray imaging (herein referred to as "regular imaging") and new phase contrast imaging (herein referred to as "phase imaging") using simple phantoms [4-9]

The feature of the phase imaging is to provide edge-enhancement of an imaged object, utilizing refracted x-rays which occur when x-rays pass through the object. Therefore, phase imaging should be assessed based on the distribution of x-ray intensities after passing the object, including the refracted x-rays. Namely, to assess differences between the phase images and regular images, conventional methods are not appropriate, such as MTF that has been used to evaluate feature of an x-ray detector without including an imaged object, Wiener spectrum, DQE, or noise equivalent quanta (NEQ) that has been used for image quality evaluation. In this study, we proposed a new image evaluation method including an imaged object, utilizing radial direction distribution function (RDDF), which was obtained based on two-dimensional (2D) power spectrum (herein referred to as "power spectrum") [10-14].

2 Materials and Methods

The system used in our experiment (Mermaid, Konica Minolta, Tokyo, Japan) consists of mammography x-ray equipment (MGU-100B, Toshiba, Tokyo, Japan) capable of performing regular imaging and phase imaging and a data reader (REGIUS MODEL190, Konica Minolta, Tokyo, Japan) which includes a photostimulable phosphor plate of Computed Radiography (PM-6M, Konica Minolta, Tokyo, Japan) for the x-ray detector. A molybdenum x-ray tube with a 0.1 mm / 0.3 mm focal spot size was used, in which selection of the size was automatically done (a 0.3 mm size was used for regular imaging and a 0.1 mm size was used for phase imaging). The distance between the focal spot of the x-ray tube and the target object was 65 cm. In regular imaging, the detector was positioned right behind the object; whereas, in phase imaging, the detector was positioned 49 cm apart from the object. Therefore, in phase imaging, the imaged object was magnified by 1.75-times.

The parameter of the system had a sampling pitch of 0.04375 mm, matrix size of 4360 × 5736, and density resolution of 12 bits. Therefore, the effective sampling pitch at 1.75-time magnification imaging was 0.025 mm.

Data analysis was conducted after transferring the raw digital image data of the imaged phantom to a personal computer and post-processing the data. Power spectrum was obtained by 2D Fourier transformation of images.

$p(r)$ and $p(\theta)$ are the sum total of power spectrum in spatial frequency regions; a round-shape and fan-shape, as shown in Fig. 1 (a) and (b), respectively. The $p(r)$ is defined as RDDF that shows the degree of texture roughness. The $p(\theta)$ can be defined as ADDF, which shows the direction of texture.

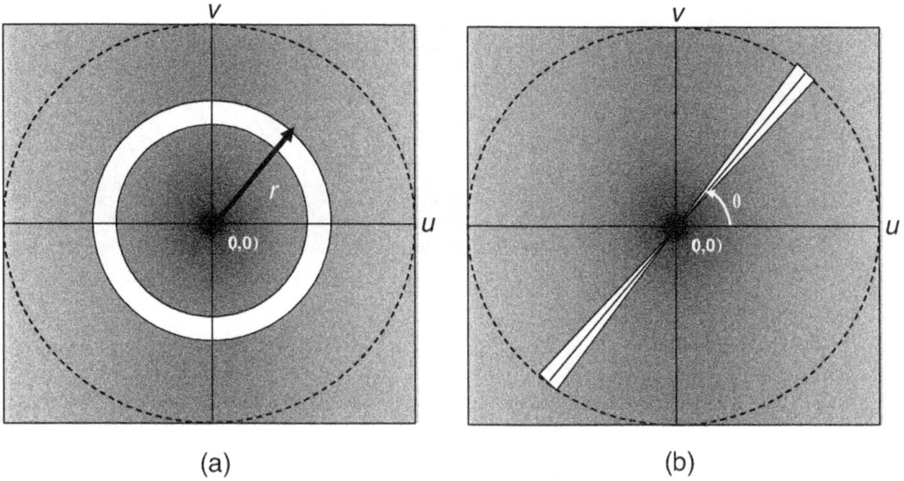

Fig. 1. The concept of (a) radial direction distribution function (RDDF) of the power spectrum, $p\ (r)$ and (b) angle direction distribution function (ADDF) of the power spectrum, $p\ (\theta)$

Next, as shown in the following Eq. (1), by using a logarithm of ratio between the $p_{s+n}\ (r)$ of "signal + noise" image and $p_n\ (r)$ of "noise" image, subtraction of $p_n\ (r)$ from $p_{s+n}\ (r)$ is done, and thus RDDF of only "signals" is obtained. In the present study, we defined this as signal intensity distribution function (SIDF), that is $P_s\ (r)$.

$$P_s(r) = 10\, log_{10} \frac{p_{s+n}(r)}{p_n(r)} \quad [\text{dB}] \tag{1}$$

In this study, as our purpose was to evaluate sharpness of x-ray images, image quality difference between phase images and regular images were evaluated using RDDF $p\ (r)$ (for spatial frequency component evaluation) and SIDF $P_s\ (r)$ obtained from $p\ (r)$.

To evaluate usefulness of our image evaluation method applicable when including the subject information by RDDF and SIDF, we used two types of simple phantoms which provide different levels of contrast, sharpness and noise, independently.

Secondly, image quality of phase images and regular images was compared using plant seeds (including grains of rice embedded in glue) having major axes of 4 -5 mm.

3 Results and Discussion

RDDF and SIDF obtained from spherical phantom images with different contrast are shown in Fig. 2 (a) and (b), respectively. Although the difference of image contrasts cannot alter the shapes of the curves of RDDF or SIDF, it can be shown as different levels of RDDF or SIDF in all frequencies. These facts helped to accurately analyze the difference of contrasts.

RDDF and SIDF obtained from spiked-ball phantom images with different sharpness are shown in Fig. 3 (a) and (b). Difference of image sharpness was clearly shown in the different RDDF or SIDF curve shapes and their levels in the frequency band areas between 0.2 and 2.5 cycles/mm, although those were not different in the low frequency

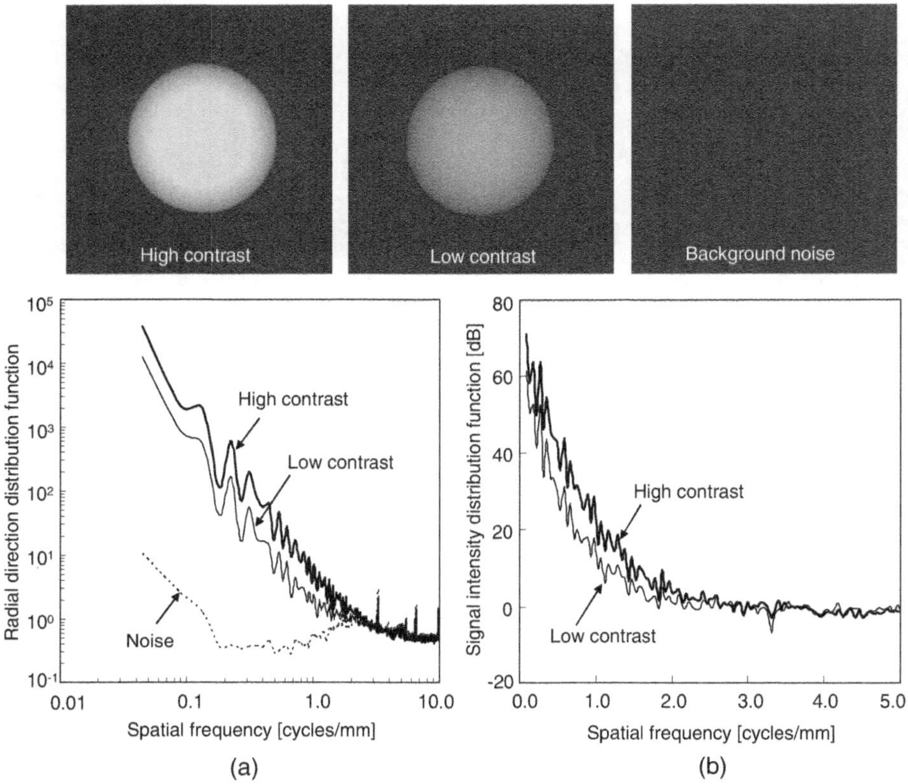

Fig. 2. Radial direction distribution functions (RDDFs) (a) and signal intensity distribution functions (SIDFs) (b) for high and low contrast phantom images as well as for the noise image

band areas. When the sharpness is different, a spatial frequency band area recognizable as signals is altered. These results provide accurate analysis of the sharpness difference.

RDDF obtained from different noise images (Fig. 4 (a), (b)) are shown in Fig. 4 (c). Using the same detector, the difference of noise characteristic of these images occurs only in the low frequency region, as these reflect difference of quantum noise only. The results clearly reflect the noise difference.

RDDF and SIDF of the phase image and regular image are shown in Figs. 5 (a) and (b), respectively. Frequency band of signals is considered different between regular images and phase images, as signal component of phase images contains signals up to 6.0 cycles/mm, though that of regular images becomes 0 dB at 3.0 cycles/mm. These results concur with the difference in sharpness observed with image evaluation of the simple phantom (as shown in Fig. 3), and that phase images provide better sharpness than regular images. Based on the images of plant seeds with RDDF or SIDF, we consider the image quality improvement by phase imaging was resulted from improvement of image sharpness, and the causes must be the scaling effect derived from magnification imaging and edge enhancement effect due to refracted x-rays.

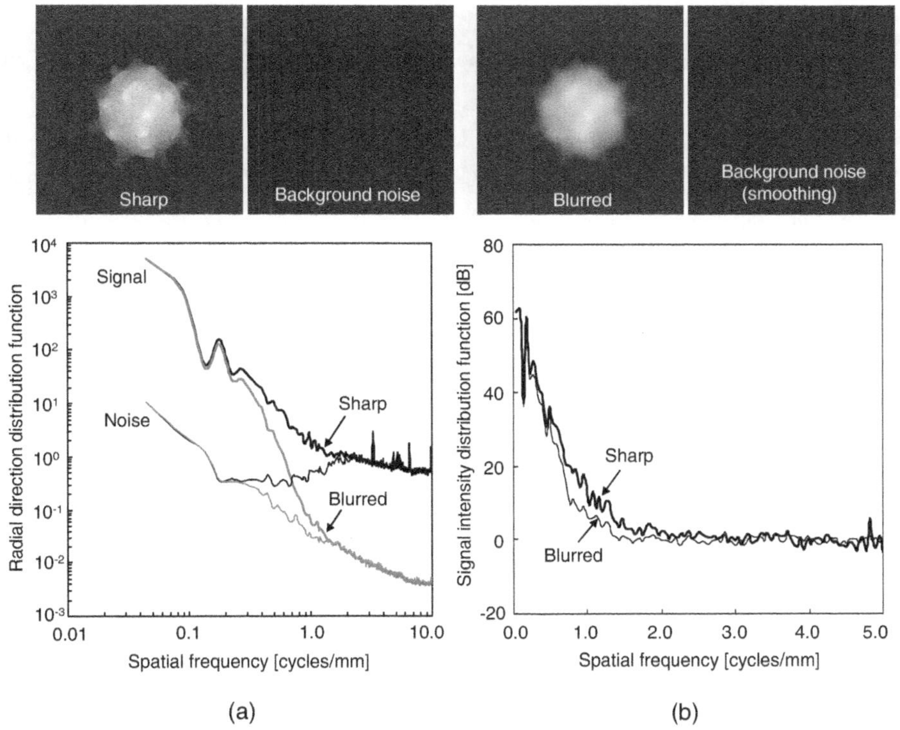

Fig. 3. Radial direction distribution functions (RDDFs) of the spiked spherical phantom images of different sharpness and with the different level of noise (a). Signal intensity distribution functions (SIDFs) of the spiked spherical phantom images having different sharpnesses (b).

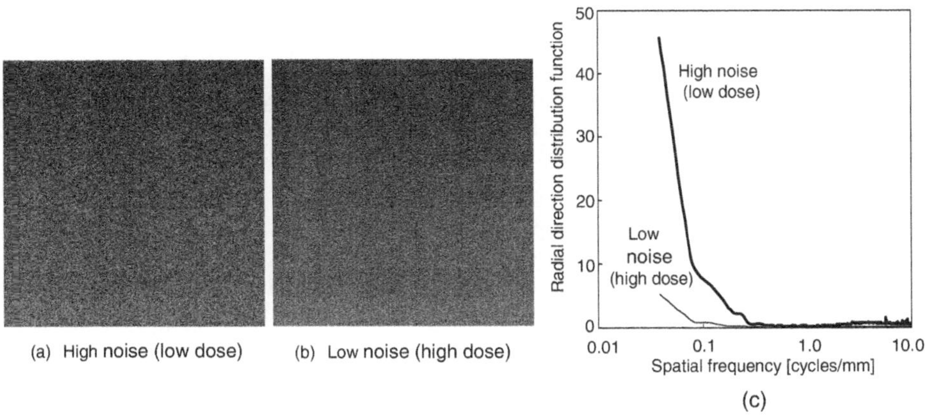

Fig. 4. The noise images by the difference in a dose. (a) high noise image (dose : 5mAs) and (b) low noise image (dose : 16mAs). (c) Radial direction distribution functions for high and low noise images.

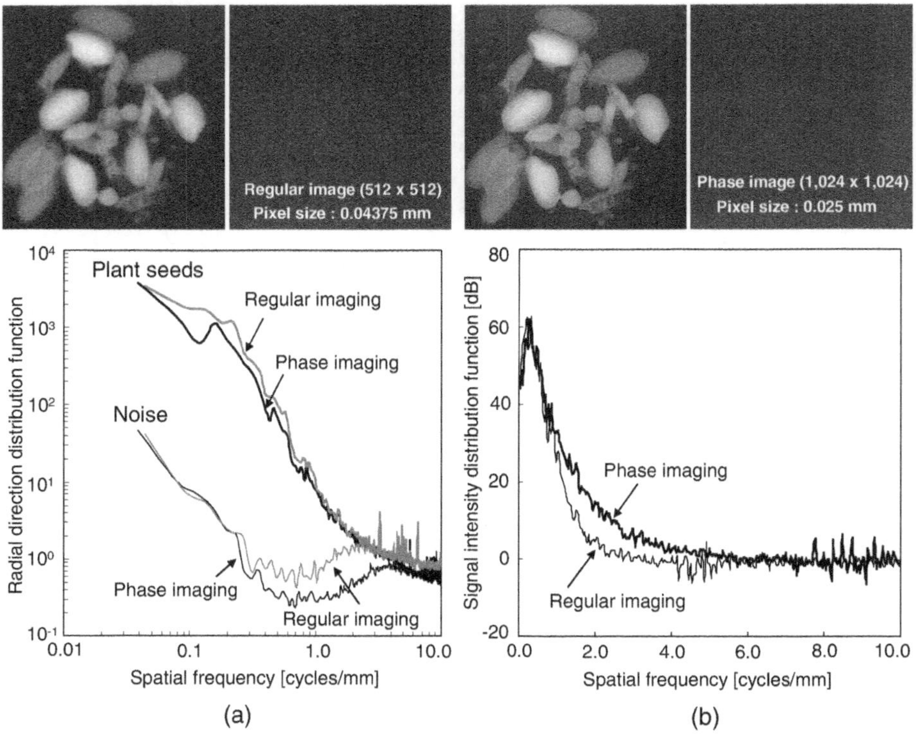

Fig. 5. Radial direction distribution functions (RDDFs) of plant seeds and noise images for the phase imaging and regular imaging (a). Signal intensity distribution functions (SIDFs) of plant seeds images for the phase imaging and regular imaging (b).

4 Conclusion

In this study, using phantoms, we investigated whether image quality evaluation was possible between different images, using RDDF and SIDF which were obtained from power spectrum. The results of our experiment proved that differences of image contrast and sharpness could be correctly evaluated. This method is effective for phase imaging, where evaluation of sharpness is impossible without images containing information of the scanned object.

Acknowledgments

The authors appreciate fruitful discussion with Dr. Chika Honda and Mr. Hiromu Ohara of Konica Minolta Medical and Graphic, Inc. We also thank Prof. Kiyoshi Murata of Shiga University of Medical Science and Mr. Masatake Imai of Shiga University of Medical Science Hospital for their permission to publish this work.

References

1. Dainty, J.C., Shaw, R.: Image science. Academic Press, London (1974)
2. Shaw, R.: Evaluating the efficiency of imaging processes. Rep. Prog. Phys. 41, 1103–1155 (1978)
3. Van Metter, R.L., Beutel, J., Kundel, H.L.: Handbook of medical imaging. Physics and psychophysics, vol. 1. SPIE Press, Washington (2000)
4. Tohyama, K., Katafuchi, T., Matsuo, S., Morishita, J., Yamada, K.: Application of phase contrast imaging to mammography. Jpn. J. Radiol. Technol. 61, 245–252 (2005) (in Japanese)
5. Tohyama, K., Katafuchi, T., Matsuo, S.: Clinical implications of phase-contrast imaging in mammography. Jpn. J. Medical Imaging and Information Sciences 23, 79–84 (2006) (in Japanese)
6. Matauo, S., Katafuchi, T., Tohyama., K., Yamada, K., Fujita, H.: Evaluation of edge effect due to phase contrast imaging for mammography. Med. Phys. 32, 2690–2697 (2005)
7. Honda, C.: Fundamental technical concept of digital phase contrast mammography. Jpn. J. Medical Imaging and Information Sciences 21, 230–238 (2004) (in Japanese)
8. Ohara, H., Honda, C., Ishisaka, A.: Digital phase imaging using an X-ray tube with a small focal spot. Jpn. J. Photographic Science and Technology 65, 501–505 (2002) (in Japanese)
9. Matsuo, S., Fujita, H., Morishita, J., Katafuchi, T., Honda, C., Sugiyama, J.: Evaluation of a phase contrast imaging with digital mammography. In: Krupinski, E.A. (ed.) IWDM 2008. LNCS, vol. 5116, pp. 130–136. Springer, Heidelberg (2008)
10. Onishi, H., Takahashi, M., Matsuo, S., Ushio, N., Noma, K., Masuda, K.: E valuation of a SPECT image using a textual analysis. Jpn. J. Radiol. Technol. 51, 710–716 (1995)
11. Matsuo, S., Komizu, M., Kida, T., Noma, K., Hashimoto, K., Onishi, H., Masuda, K.: Analysis of frequency components of X-ray images. Jpn. J. Radiol. Technol. 53, 1655–1672 (1997) (in Japanese)
12. William, K.P.: Digital Image Processing, pp. 473–477. John Wiley & Sons, Canada (1978)
13. Rafael, C.G., Paul, W.: Digital Image Processing, pp. 418–423. Addison-Wesley Publishing Company Inc., Canada (1987)
14. Rangaraj, M.R.: Biomedical Image Analysis, pp. 612–621. CRC Press, Florida (2000)

Appendix : Clinical Mammograms

The clinical image of the breast are shown in Fig. 6 [9]. The improved sharpness in the phase contrast image is obvious from clinical images due to the edge enhancement effect than that in regular image.

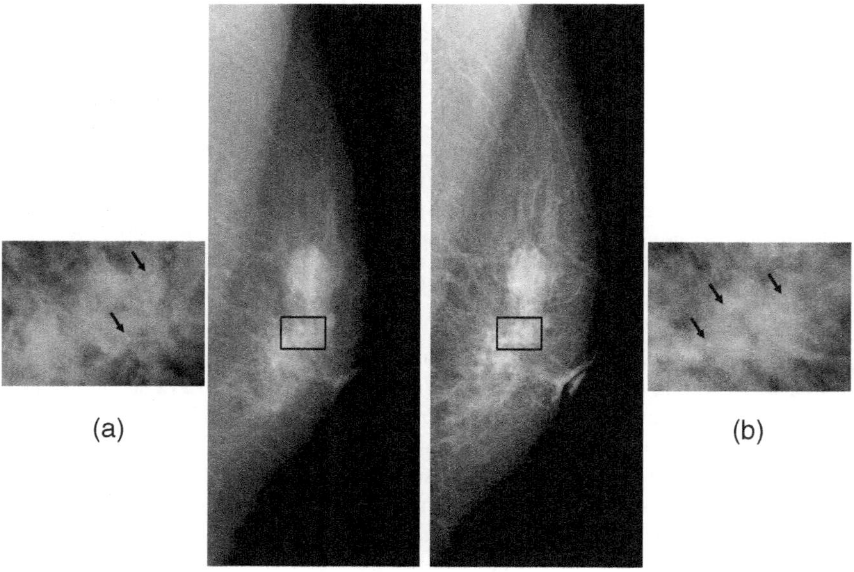

Fig. 6. Clinical image comparison. (a) Regular image (b) Phase image

Modelling Structural Deformations in Mammographic Tissue Using the Dual-Tree Complex Wavelet

Michael Berks[1], Chris Taylor[1], Rumana Rahim[2], Caroline Boggis[3], and Susan Astley[1]

[1] Imaging Science and Biomedical Engineeering, University of Manchester, Oxford Road, Manchester, M13 9PT, UK
[2] Department of Clinical Radiology, University of Manchester, Manchester, UK
[3] Nightingale Breast Centre and Genesis Prevention Centre, University Hospital of South Manchester, Southmoor Road, Manchester M23 9LT, UK
michael.berks@manchester.ac.uk

Abstract. The appearance of breast tissue in mammograms is altered by the presence of a malignant mass. Existing synthesis methods have not addressed this structural deformation. We aim to use a set of mass background images that display altered breast tissue to simulate such deformations in regions of digital mammograms previously showing no signs of disease. Regions are decomposed using the dual-tree complex wavelet transform (DT-CWT) to obtain a richer representation of local structure than provided by image grey-levels alone. Synthesis is achieved by modifying the high-frequency DT-CWT coefficients of normal regions to match those in mass backgrounds. Three methods for completing this task are described. The results, advantages and current limitations of the methods are discussed.

Keywords: Mammography, malignant mass, tissue deformation, lesion synthesis, texture models, dual-tree complex wavelet transform.

1 Background

Several authors have described the benefits of synthesising signs of breast disease in digital mammograms [1-4]. When synthesising masses, a common theme is to generate an image representing the increased attenuation through the dense mass tissue and superimpose this on a real mammogram previously displaying no signs of disease. The appearance of a stellate lesion may be achieved by adding spicules to the mass [1, 3, 4]. However, such approaches treat the mass (and spicules) independently from the tissue to which they are added and fail to account for deformation in existing breast tissue caused by the malignancy. In this paper we present methods for synthesising mammographic background changes associated with the presence of a malignant mass.

The principal purpose of this work is to create regions that can be combined with existing methods for synthesising the central density of masses, thus achieving a more complete simulation of breast disease. We report results in terms of how realistic synthesised regions appeared to expert mammogram readers.

J. Martí et al. (Eds.): IWDM 2010, LNCS 6136, pp. 145–152, 2010.

2 Data and Methods

2.1 Mammogram Data

Two sets of mammographic image data were used in this work: *normal regions* and *mass backgrounds*. Normal regions were extracted from mammograms displaying no signs of disease. Each mass background was obtained by extracting a mammographic region about a biopsy-proven malignant mass and then subtracting the central density from this region using a process described in our earlier work [5]. As a result, a mass background displays both spicules (if present) and the structural deformation of breast tissue in the region. Our aim is to simulate the appearance of a mass background in a normal region.

2.2 Modelling and Synthesising Mammographic Texture Using the Dual-Tree Complex Wavelet Transform

Because mammographic appearance is often rich in linear structures we apply the dual-tree complex wavelet (DT-CWT) [6] to our image data. The DT-CWT decomposes images into 6 oriented subbands localised in scale. The magnitudes of DT-CWT coefficients have the important property of shift invariance and the phase of coefficients can be used to infer information on local structural shape. Thus the DT-CWT provides a richer representation of our data than using the grey-levels alone, whilst the construction of the transform minimises redundancy in the coefficients without introducing inconsistencies associated with shifts in the decomposition frame.

In addition, the decomposition is invertible, allowing the image to be reconstructed from the set of coefficients. We make use of this fact and apply the following three stage process to synthesise mass background appearance in a normal region:

1) Compute the DT-CWT decomposition of a normal region
2) Modify coefficients in the decomposition to match the properties of DT-CWT coefficients in real mass backgrounds
3) Invert the modified DT-CWT to reconstruct a region in which mass background appearance has been synthesised

We now describe three methods of completing the second stage of the process.

2.3 Directly Transferring DT-CWT Coefficients

As an initial proof-of-concept, we transfer DT-CWT coefficients from the high frequency subbands of real mass regions into the decomposition of a normal region. This replicates the appearance of spicules and deformed tissue in the modified region, whilst maintaining the underlying global appearance of the normal region.

However, if in addition to abnormal structures the mass background has a textured appearance with an underlying dominant orientation, the structures that give rise to the orientation will be encoded in the transferred DT-CWT coefficients. As a result the appearance of a dominant orientation will be transferred to the target normal region. If this orientation differs from the existing dominant orientation of texture in the normal region, there will be a mismatch at the borders of the transferred region causing unrealistic appearance. For example, Fig. 1 (a) depicts a normal region containing structures with a primarily vertical orientation, whilst in Fig 1 (b) we show a mass background with approximately horizontal dominant linear structures. When DT-CWT coefficients are transferred from the mass background into the normal region, the mismatch in orientation causes a notably unrealistic texture in the modified region displayed Fig 1 (c).

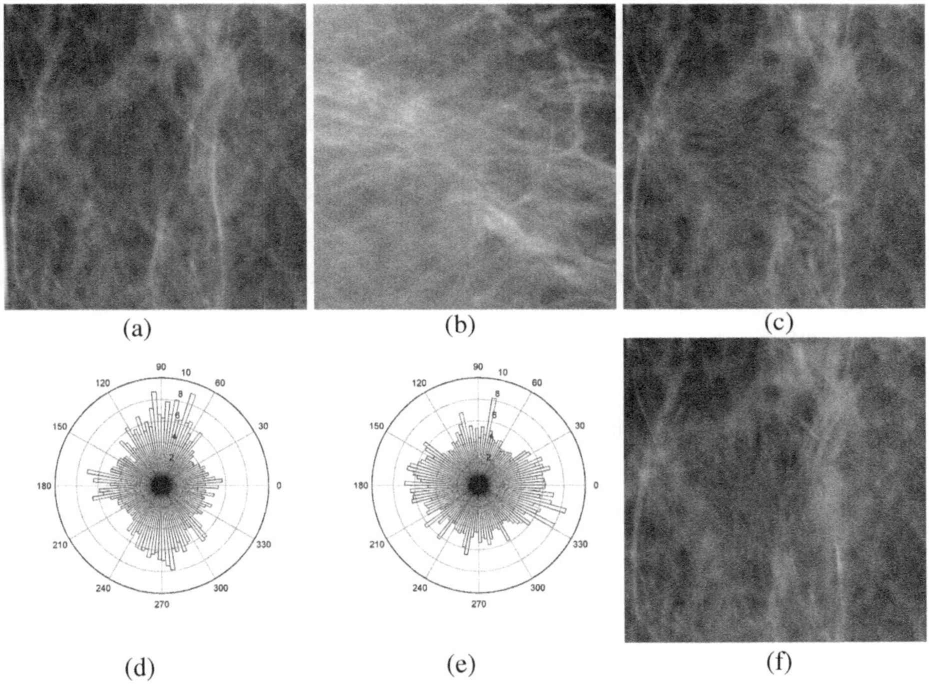

Fig. 1. Aligning orientation in the transfer region *a)* normal region; *b)* mass background; *c)* modified region with unaligned orientations; *d)* orientation histogram for normal region; *e)* orientation histogram for mass background; *f)* modified region following orientation alignment

To prevent this mismatch of orientations, we rotate the mass background prior to computing its DT-CWT so that its dominant orientation matches that of the target region. To find the optimal angle of rotation, we compute orientation histograms for each region (Fig 1 (d) and (e)) using a function of the DT-CWT coefficients [7], and select the angle that would achieve greatest correlation between the two histograms. For these two regions, an angle of 102° was calculated as the best angle by which to rotate the mass background. The result of applying this alignment is depicted in Fig 1 (f). Note how the modified region now appears much more realistic than the case prior to the rotation.

Whilst aligning structure orientation helps to remove some texture artefacts from the modified regions, there will still be some mismatches for particular pairs of normal regions and mass backgrounds (for example, when transferring DT-CWT coefficients from a mass background rich in parenchymal structure to a mainly fatty normal region).

Moreover, because we use a real mass background as a template to modify each normal region, we are limited to sampling from the finite set of mass backgrounds in our training data. To overcome these limitations, in the following two sections we explore methods for sampling new DT-CWT coefficients in normal regions with respect to probabilistic texture models learnt from our training data.

2.4 Sampling DT-CWT Coefficients from Local Texture Models

Rather than using DT-CWT coefficients directly sampled from the decompositions of real mass backgrounds, we can attempt to model the spatial distribution of coefficients in the real data. Due to the heterogeneity of mammographic texture, constructing a single model to describe the variation in coefficients across a whole region is not possible. Instead we model local patches of coefficients.

For a given location (x,y) in the L-th level of a DT-CWT, we define a local patch $\mathbf{x}$ such that it includes coefficients in a 5 by 5 neighbourhood centred on (x,y) from each of the 6 oriented subbands. We also include the coefficient sampled at (x,y) from each of the oriented subbands in the next coarsest level. Each coefficient has a complex magnitude and phase, thus $\mathbf{x}$ is a 312-element feature vector representing the local structure at (x,y).

We extract all such feature vectors from the set of real mass backgrounds to populate a feature space for each DT-CWT level and describe the distribution of vectors within each feature space by fitting a Gaussian Mixture Model (GMM) using the k-means algorithm [8].

As a result of the model fitting we can describe the probability of any feature vector $\mathbf{x}$ as

$$P(\mathbf{x}) = \sum_{i}^{k} P(\mathbf{x} \mid i) P(i)$$

where i indexes the i-th Gaussian component and

$$P(\mathbf{x} \mid i) \sim N\left(\boldsymbol{\mu}_i, \boldsymbol{\Sigma}_i\right).$$

The mean $\boldsymbol{\mu}_i$ and covariance $\boldsymbol{\Sigma}_i$ of each Gaussian component are computed as the sample mean and covariance of each cluster returned from the k-means algorithm, whilst the prior probabilities for each component $P(\mathrm{i})$ are computed as the proportion of the total points assigned to each cluster.

To select k, models were built for each level using increasing values of k (starting at k = 10 and increasing in increments of 10) until at least one cluster was dropped during the algorithm. Because the k-means algorithm is susceptible to getting stuck in local minima, having selected a value of k, we repeated the modelling 100 times at each level, choosing a different randomly selected initialisation of clusters for each repeat. Within each level, the model that returned the most compact clusters was selected. Applying this process to the 5 finest DT-CWT level produces 5 GMMs that can be used to sample new coefficients in a normal region.

The sampling procedure follows a method we have described in earlier work [5]. In summary, in the DT-CWT of a normal region, we discard all existing coefficients from the area in which we wish to synthesise mass background appearance. We then fill in the missing area, working from the outside inwards, at each step sampling new coefficients from the GMMs conditioned on the neighbouring coefficients that have already been filled. The DT-CWT levels are filled from coarse to fine, so that information on long range structural interactions are passed upwards through the decomposition.

2.5 Adding Structure to the Texture Models

Early experiments using the method described in section 2.4 indicated that although general mammographic texture was being generated, it did not include linear structures associated with breast tissue or spicules. To overcome this, we can force the models to generate DT-CWT coefficients associated with structure at specific locations, by applying a line detection algorithm to obtain maps of structure for each real mass background in our training data. Local feature vectors are then constructed as previously; however the set of feature vectors corresponding to structure is modelled separately from feature vectors not corresponding to structure. Thus for each level we have a structure GMM and a non-structure GMM.

At synthesis time, we require a structure map to determine which model should be used to sample coefficients at each location in the normal region we are modifying. Ideally, this map itself would be generated by sampling from a probabilistic model that encapsulates the spatial distribution of linear structure in real mass backgrounds. However as yet, we do not have a method for constructing such a model. Instead, for each normal region we used the structure map from a randomly selected mass background to act as a template. In this way, the method can be seen as a combination of the direct transfer method described in section 2.3 with the probabilistic sampling method described in section 2.4.

3 Results

The three methods described above have been used to synthesise mass background appearance in 89 regions randomly selected from normal mammograms. Fig. 2 depicts synthesis results for three normal regions. On the top row of Fig. 2 the regions are shown in their original form. The second, third and fourth rows show the result of modifying each region using the methods described in section 2.3, 2.4 and 2.5 respectively.

Evaluating the results is difficult because by construction each synthetic mass background will only resemble a region of a real mammogram when a synthetic central density has been added to it (thus reversing the separation process used to generate mass backgrounds, as outlined in section 2.1).

30 mass backgrounds synthesised using the direct transfer of DT-CWT coefficients have been combined with our method for synthesising the central density of malignant mammographic masses ([9]) and quantitatively evaluated by expert mammogram readers. In this study, ten expert readers were shown randomised set of 30 real and 30 synthetic masses, and asked to rate each mass on a scale varying from definitely real to definitely synthetic. ROC curves were fitted to the reader ratings, and a mean area-under-curve of $A_z = 0.70 \pm 0.09$ was computed. This suggests experts could identify synthetic masses at a rate significantly better than chance. Moreover, from feedback provided by the readers during the study, it would appear that in the

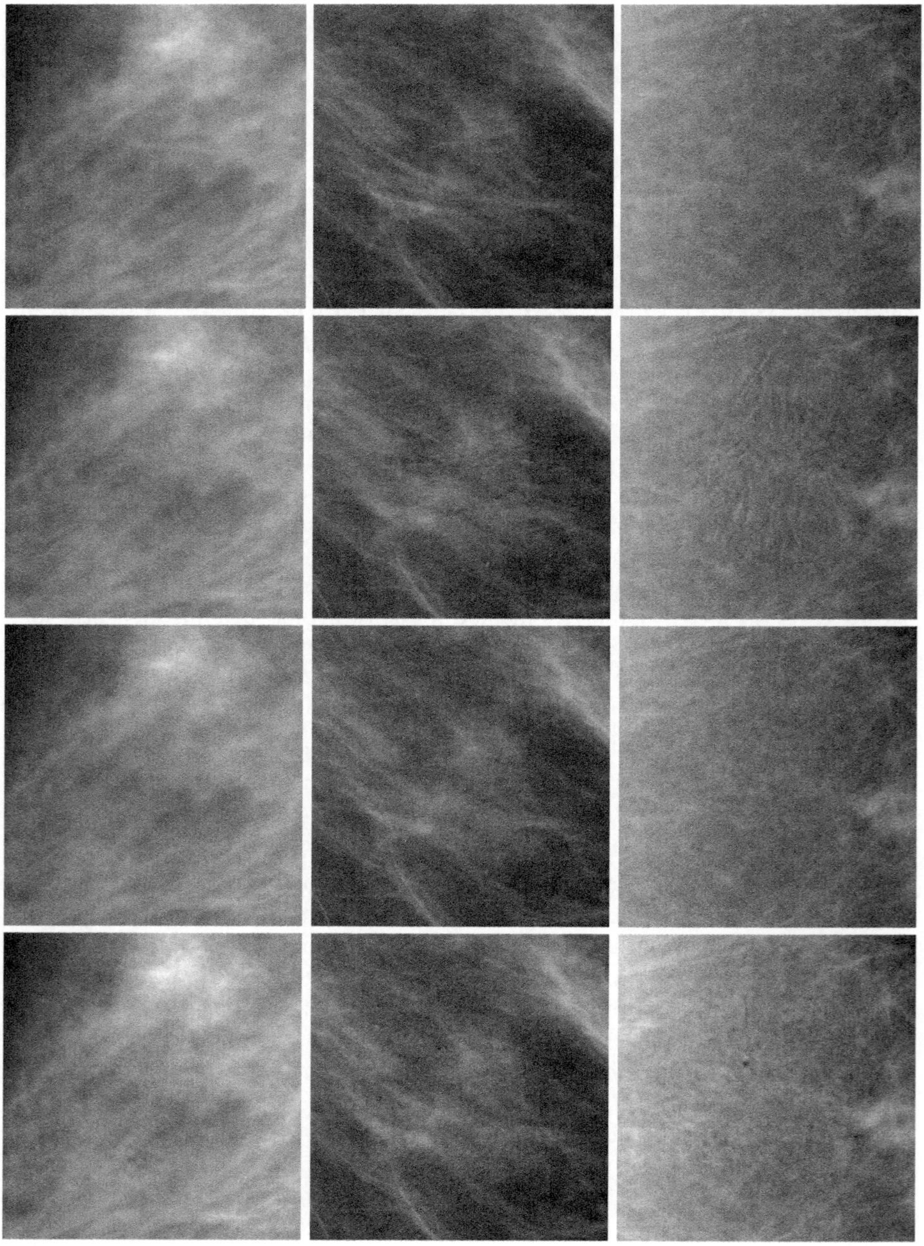

Fig. 2. *Top row:* three normal mammogram regions; Three methods of synthesising mass background appearance: directly transferring DT-CWT coefficients from real data (*2nd row*); probabilistically sampling coefficients from local texture models (*3rd row*); probabilistically sampling with respect to a map of structure (*bottom row*)

majority of correctly identified synthetic masses, unrealistic texture artefacts in the synthesised mass backgrounds were the cause.

This provides further motivation to develop the methods in which new DT-CWT coefficients are sampled from probabilistic models. However as yet, we have not performed a quantitative evaluation in which synthetic mass backgrounds generated using either of the methods described in sections 2.4 and 2.5 are combined with synthetic mass densities. Instead, a qualitative assessment of the methods is given in the next section.

4 Discussion

The image results obtained by directly transferring DT-CWT coefficients (section 2.3) show that mass background appearance can be synthesised within the framework we have specified in section 2.2. However the direct transfer method will only work for certain mass background/normal region pairs. Where the existing mammographic texture in normal regions is not a good match to the texture in the randomly selected mass background, texture artefacts may generated in the modified region that ultimately allow such regions to be identified by experts as synthetic.

In contrast, the two probabilistic methods for sampling new DT-CWT coefficients (sections 2.2 and 2.3) have the potential to generate mass background appearance in any normal region.

These two methods have yet to be assessed by experts. However, a simple visual assessment of the images produced show that the structures present in a typical mass background (such as spicules) are not being generated by the methods. We belive that a significant cause of this failure is that the representation we have used to encode local structure produces a feature space for which we cannot adequately learn a probability distribution. This is evidenced by the fact such structures were still not created despite the explicit structure maps used in the method described in section 2.5. We are currently working on compacting this representation (e.g. by using rotation invariant features) to obtain a feature space that will be easier to model.

5 Conclusion

The results of this work should be considered in the context of previous attempts to synthesise mammographic appearance. Modelling and synthesising distortions in normal breast tissue are extremely challenging tasks that to the best of our knowledge have not been addressed by other authors in previously published literature. Whilst we have not solved the problem, we believe we have at least set it in framework that will enable a successful solution. Obtaining such a solution is crucial to any method attempting a realistic simulation of malignant mammographic masses. We also propose that our models for synthesising tissue deformations associated with a visible mass could be adapted to synthesise, and ultimately detect, more general patterns of architectural distortion in mammograms. This proposal is the subject of ongoing research.

References

1. Caulkin, S.: Generating Synthetic Abnormalities in Digital Mammograms Using Statistical Models. Imaging Science and Biological Engineering, PhD. University of Manchester, UK, Manchester, 268 (2001)
2. Lado, M.J., Tahoces, P.G., Souto, M., et al.: Real and simulated clustered microcalcifications in digital mammograms. ROC study of observer performance. Med. Phys. 24, 1385–1394 (1997)
3. Ruschin, M., Tingberg, A., Bath, M., et al.: Using simple mathematical functions to simulate pathological structures–input for digital mammography clinical trial. Radiat. Prot. Dosimetry 114, 424–431 (2005)
4. Saunders, R., Samei, E., Baker, J., et al.: Simulation of mammographic lesions. Academic radiology 13, 860–870 (2006)
5. (Reference not shown to maintain anonymity) (2008)
6. Kingsbury, N.: Complex wavelets for shift invariant analysis and filtering of signals. Applied and computational harmonic analysis 10, 234 (2001)
7. Anderson, R., Kingsbury, N., Fauqueur, J.: Determining Multiscale Image Feature Angles from Complex Wavelet Phases. In: Kamel, M.S., Campilho, A.C. (eds.) ICIAR 2005. LNCS, vol. 3656, p. 490. Springer, Heidelberg (2005)
8. MacQueen, J.B.: Some Methods for Classification and Analysis of Multivariate Observations. In: 5th Berkeley Symposium on Mathematical Studies and Probability (1967)
9. Berks, M., Taylor, C.J., Rahim, R., et al.: Synthesising Malignant Breast Masses in Normal Mammograms. In: 10th International Workshop on Digital Mammography. Springer, Girona (2010)

Classification of Linear Structures in Mammograms Using Random Forests

Zezhi Chen, Michael Berks, Susan Astley, and Chris Taylor

Imaging Science and Biomedical Engineeering, School of Cancer and Enabling Sciences, University of Manchester, Oxford Road, Manchester, M13 9PT, UK
zezhi.chen@manchester.ac.uk

Abstract. Classification of linear structures, such as blood vessels, milk ducts, spiculations and fibrous tissue can be used to aid the automated detection and diagnosis of mammographic abnormalities. We use a combination of dual-tree complex wavelet coefficients and random forest classification to detect and classify different types of linear structure. Encouraging results are presented for synthetic linear structures added to real mammographic backgrounds, and spicules in real mammograms. For spicule/non-spicule classification in real mammograms we report an area $A_z = 0.764$ under the receiver operating characteristic.

Keywords: mammography, linear structures, classification, random forests, dual-tree complex wavelet.

1 Introduction

It has been reported recently that that current CAD systems do not detect architectural distortion (AD) with adequate sensitivity or specificity [1]. Previous attempts at detecting patterns of distorted breast tissue – including both patterns of spicules associated with malignant masses and more general cases of AD in which no focal mass is visible – have used a two stage approach involving i) detecting linear structures, ii) analysing the orientation patterns of these structures to determine if AD is present [2-4].

We hypothesise that the sensitivity of AD detection algorithms could be improved if different types of linear structure could be labelled automatically, and used selectively in the second stage of the analysis outlined above. Although there is an extensive literature on detecting linear structures in digital mammograms [2-7], less attention has been paid to classifying different structure types [8].

We present a novel method for classifying linear structures, based on the use of a complex wavelet transform to provide a rich representation in which local shape can be inferred from the phase relationships between coefficients (cf [5-7]). We use random forest classification [9] applied to this representation to detect linear structures and classify their types. Results are given for synthetic linear structures added to real mammographic backgrounds, and for spicule detection and classification in real mammograms.

J. Martí et al. (Eds.): IWDM 2010, LNCS 6136, pp. 153–160, 2010.

2 Data and Methods

2.1 Mammogram Data

We used a sequential set of 84 abnormal mammograms with biopsy-proven malignancy, drawn from a screening population (Nightingale Breast Centre, South Manchester University Hospitals Trust, UK), and a set of 89 normal mammograms of the contralateral breasts of the same individuals (where disease was radiologically confirmed to be confined to one breast). All mammograms were digitised to a resolution of 80μm, using a Vidar CADPRO scanner. A 4×4 cm patch was extracted around each abnormality, and a similar patch was sampled randomly from each of the normal mammograms. For each abnormal patch an expert radiologist annotated some (though not necessarily all) of the spicules associated with the abnormality, using in-house software, resulting in a total of 555 spicule annotations.

2.2 Synthetic Data

We generated synthetic images by adding linear structures to 128×128 pixel normal mammogram patches, pre-processed to remove naturally occurring linear structure. 6130 normal mammogram patches were sampled randomly from 185 normal screening mammograms, including the 89 described above. For the experiments in Section 3 we added linear structures with Gaussian or rectangular cross-sections. For the experiments in Section 4.1 we added linear structures with elliptical cross-sections, simulating the x-ray projection of uniformly dense cylindrical structures.

2.3 Representing Local Structure Using the DT-CWT

Wavelet transforms have been used extensively in image processing and analysis to provide a rich description of local structure. The dual-tree complex wavelet transform (DT-CWT) has particular advantages because it provides a directionally selective representation with shift-invariant coefficient magnitudes and local phase information [10]. The DT-CWT combines the outputs of two discrete transforms using real wavelets, differing in phase by 90°, to form the real and imaginary parts of complex coefficients. For 2-D images, the DT-CWT produces 6 directional sub-bands, oriented at ±15°, ±45°, ±75°, at each of a series of scales separated by factors of 2.

In the experiments described in Sections 3 and 4, we used the complex coefficients (in phase/magnitude form) from the 6 oriented sub-bands in each of the s finest decomposition scales from a $w \times w$ neighbourhood centred on each pixel. This produced feature vectors with of length $12sw^2$. In some experiments we formed more compact vectors of length $2sw^2$ by including only the coefficients at each location and scale for the sub-band that gave the largest response.

2.4 Classifying Structure Using Random Forests

Given a set of training data consisting of N samples each of which is a D-dimensional feature vector labelled as belonging to one of C classes, a random forest comprises a set of tree predictors constructed from the training data [9]. Each tree in the forest is built from a bootstrap sample of the training data (that is, a set of N samples chosen randomly, with replacement, from the original data). The trees are built using a standard classification and regression tree (CART) algorithm; however, rather than assessing all D dimensions for the optimal split at each tree node, only a random subset of $d < D$ dimensions are considered. The trees are built to full size (i.e. until a leaf is reached containing samples from only one class) and are not pruned.

During classification, an unseen feature vector is classified independently by each tree in the forest; each tree casts a unit class vote, and the most popular class can be assigned to the input vector. Alternatively, the proportion of votes assigned to each class can be used to provide a probabilistic labelling of the input vector.

Random forests are particularly suited to learning non-linear relationships in high-dimensional multi-class training data, and have been shown to perform as well as classifiers such as Adaboost or support vector machines, whilst being computationally more efficient [9]. For all the experiments described below we followed published guidelines [9], constructing forests containing 200 trees and setting $d = \sqrt{D}$.

3 Experimental Results for Synthetic Data

We conducted initial experiments to test our approach, using synthetic images containing linear structures with Gaussian or rectangular cross-sections superimposed on real mammographic backgrounds, as described in Section 2.2. For each image, the bar type and orientation were selected randomly, whilst the contrast and width were randomly sampled from ranges typical of linear structures in real mammograms: widths [4, 32] pixels (0.3 – 2.5mm), peak contrast [8, 16] grey-levels (relative to images scaled 0 – 255). Fig. 1 shows two synthetic images, and the largest complex coefficient (over the 6 orientation sub-bands) for three different levels in the transform.

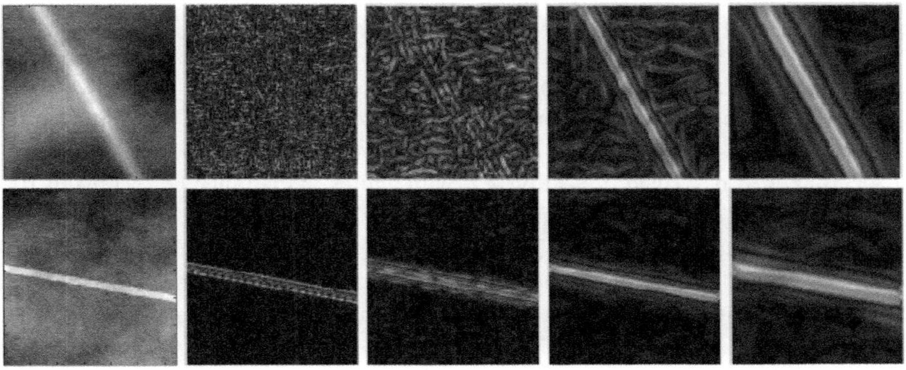

Fig. 1. Synthetic images and their DT-CWT coefficients. Top: Gaussian bar, width (SD) 4.33 pixels, contrast 10.12 grey-levels. Bottom: rectangular bar, width 8 pixels, contrast 9.05 grey-levels. Columns L to R show original and maximum response over orientation sub-bands for the 2nd, 3rd and 4th levels of the DT-CWT, using intensity (magnitude), hue (phase) coding.

We generated training sets containing 10, 20, 40, 80 and 160 images and a test set containing 100 images. The pixels in each image were labelled as belonging to either background or rectangular/Gaussian bar, giving a three-class classification problem. We extracted 432-dimensional feature vectors, using all 6 orientation sub-bands, a neighbourhood size of $w = 3$, and $s = 4$ scales. We sampled 40,000 vectors randomly from each of the training sets, and constructed a random forest classifier as described in Section 2.4. The forest was used to classify all the pixels in the test images and classification error (misclassified pixels / total pixels) was calculated. The results are summarised in Table 1. Classification accuracy improves as the number of training images increases, reaching 97.7% correct classification for the largest training set tested. Given these promising results, we moved on to real data.

Table 1. Random forest classification error rates for 3-class labelling of sythetic images

Number of training images	10	20	40	80	160
Classification error	0.0522 ± 0.0584	0.0354 ± 0.0429	0.0365 ± 0.0425	0.0237 ± 0.0348	0.0231 ± 0.0322

4 Experimental Results for Real Data

To apply our approach to real mammographic data we proceeded in three stages i) detecting the linear structures in a set of normal and abnormal training images, ii) building a spicule/non-spicule classifier using the expert annotations of the training images, iii) using the classifier to label pixels in unseen test images, using expert annotations to evaluate classification accuracy. Because we were working with a limited dataset, we used a cross-validation approach to evaluation.

4.1 Detecting Linear Structures in Mammograms

For line detection in real mammograms, we trained a random forest classifier on synthetic images designed to contain similar structures to those found in mammograms. Linear structures with elliptical cross-sections were added to normal mammogram backgrounds as described in Section 2.2, with widths in the range [2, 32] pixels (0.15 – 2.5 mm), and contrasts in the range [4/256, 16/256] grey-levels. We experimented with neighbourhood sizes $w = 1$, 3 and 5, scales $s = 4$ and 6, and number of sub-bands all or maximum response only, building in each case a random forest classifier using 200,000 training vectors sampled with equal probability from the line and background classes. We also varied the lower and upper bounds $[l, u]$ of the range of widths used during training.

We found that it was possible to achieve close to 100% classification accuracy on unseen synthetic data with virtually all parameter combinations, though all-sub-band classifiers generally outperformed maximum-sub-band classifiers. Example linear structure probability images obtained by applying classifiers built using different training regimes to a real mammogram patch are shown in Fig 2. All the classifiers produce plausible results, but because we do not have ground truth data for the real linear structures it is difficult to draw firm conclusions. Inspecting the results for a large number of images, we made the following observations regarding different training regimes:

- minimum line width $l = 2$ gave more noisy results than $l = 4$;
- maximum line width $u = 32$ gave less sensitivity to subtle lines than $u = 16$;
- neighbourhood size $w = 3$ or 5 gave better signal-to-noise than $w = 1$, probably because information on phase derivatives is captured;
- scales $s = 6$ gave better discrimination between lines and edges than $s = 4$;
- using all orientation sub-bands gave better results near line crossings than the maximum sub-band approach.

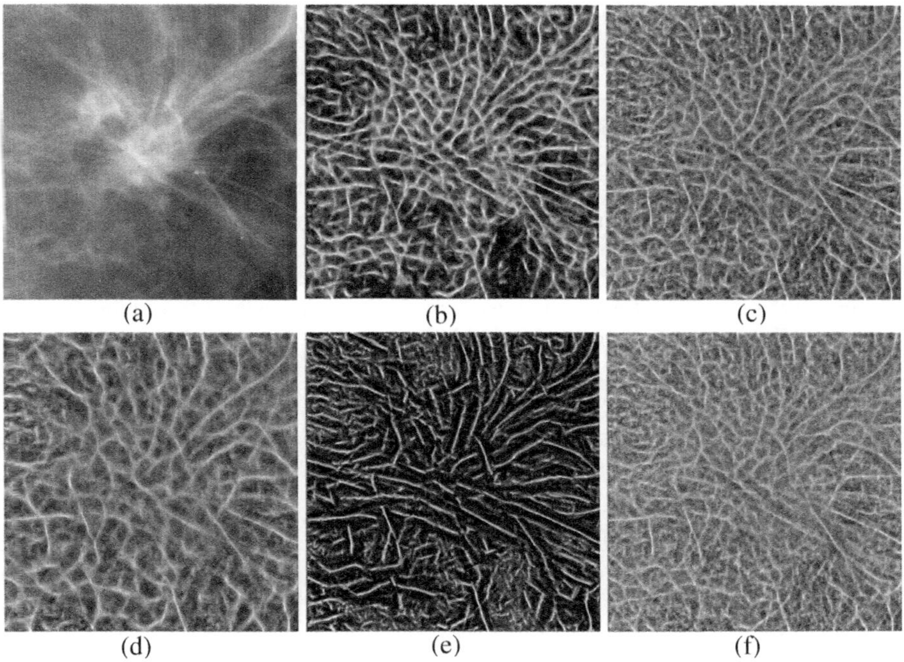

Fig. 2. (a) Original mass region; (b)-(f) Line probability maps using for varying parameter sets: (b) Bar widths = [4, 16], $w = 3$, $s = 6$, all sub-bands; (c) Bar widths = [2, 16], $w = 3$, $s = 6$, all sub-bands; (d) Bar widths = [2, 32], $w = 3$, $s = 6$, all sub-bands; (e) Bar widths = [2, 16], $w = 3$, $s = 5$, maximum sub-band response; (f) Bar widths = [2, 16], $w = 1$, $s = 6$, all sub-bands

Based on these observations, we selected for subsequent experiments a classifier trained using synthetic bars of width [4, 16] pixels, with 648-dimensional feature vectors constructed using neighbourhood size $w = 3$, scales $s = 6$, and all oriented sub-bands (see Fig 2 (b)). This classifier was used to construct linear structure probability images for the 84 abnormal and 89 normal regions.

4.2 Classifying Spicules in Mammograms

We used the linear structure probability images described above to train a random forest classifier to distinguish between spicules and other linear structures in real mammograms, using DT-CWT features.

The expert spicule annotations for the abnormal images were used as a basis for selecting spicule pixels, though they were not sufficiently accurate to be used directly. To refine the annotations, we initialised a snake [11] using each original annotation, and iterated it to convergence, using evidence from the linear structure probability image. The 555 refined spicule annotations identified a set of 36,514 spicule pixels.

We also randomly sampled an equal number of pixel locations from the 89 normal patches such that the distribution of linear structure probabilities in the normal samples matched the distribution of those in the spicule sample.

Random forest classifiers were trained using DT-CWT feature vectors constructed using varying neighbourhood size *w*, scales *s*, and number of sub-bands with the spicule/non-spicule labels, and evaluated using a 10-fold cross-validation design. The set of normal and abnormal regions were divided into 10 groups so that the total number of normal and spicule samples in each group were as close as possible to a 10th of the total. The samples in each group were then classified using a random forest trained on the samples from the remaining 9 groups. The classification results from each group were pooled to generate an unbiased class probability for each sampled pixel. These probabilities were used to compute an ROC curve for each training regime, and the area under the curve (A_z) was computed and used as a measure of classification performance. The results are tabulated in Table 2.

Table 2. Spicule classification results for varying compositions of feature vectors

Composition of feature vectors			Size of feature vectors (*D*)	ROC A_z
Neighbourhood size (*w*)	N$^{o.}$ of decomposition scales (*s*)	N$^{o.}$ of subbands		
3X3	4	All	432	0.693
3X3	5	All	540	0.699
3X3	6	All	648	0.755
3X3	6	Maximum	108	0.752
1X1	6	All	72	**0.764**
5X5	6	All	1800	0.754

From the results we can see the advantage of including more decomposition scales in the feature vectors and using the responses in all oriented subbands as opposed to using only the maximum response. However, somewhat surprisingly, the classification results appear to be slightly better using 1x1 neighbourhoods rather than 3x3 neighbourhoods, in contrast to the trend observed when performing line detection (see Section 4.1).

We also applied the 10-fold cross-validation approach and the best classifier design (neighbourhood size $w = 1$, scales $s = 6$, all sub-bands) to generate unbiased spicule probability images for all 89 normal and 84 abnormal regions. Typical results are shown in Figure 3, where the increased spicule probability in spiculated areas of the abnormal region – relative to both the normal region and the non-spiculated areas of the abnormal region – is clearly visible.

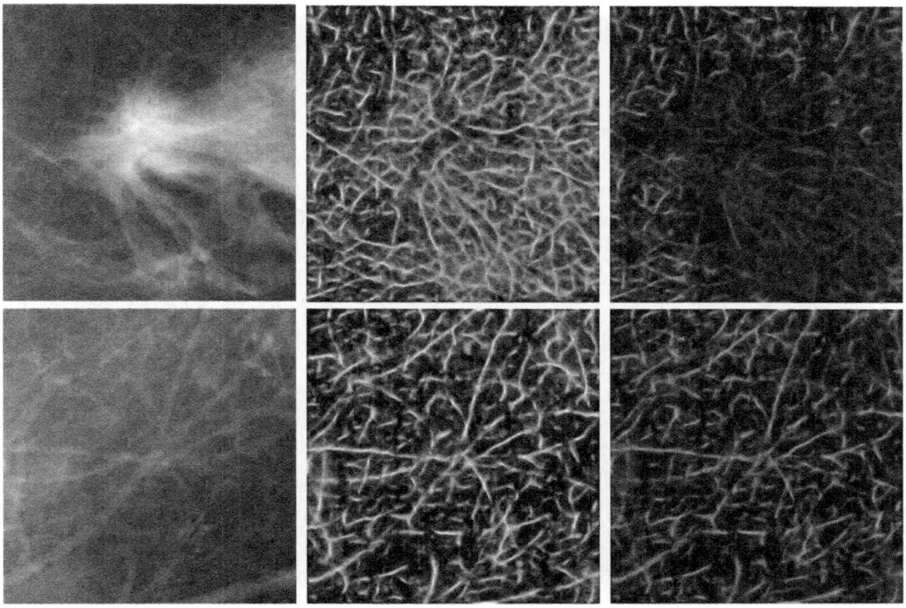

Fig. 3. Left column: a mass region and normal region; centre column: line probability maps of each region; right column: spicule probability depicted as hue from cyan (normal) to pink (spicule), modulated by line strength

5 Conclusion

In this paper we have presented a new method for classifying local structure in mammograms. We have applied the method to detect and differentiate between two types of synthetic linear structures added to real mammographic backgrounds. The accuracy of the classification highlighted the promise of the approach.

For real data, we first trained a classifier on synthetic images to perform line detection. We then used the results of this detection scheme, together with radiologist annotations to perform spicule/non-spicule classification. An ROC A_z of 0.764 suggests that a meaningful differentiation can be made between the two classes. Whilst such a classification may not be strong enough to detect abnormal malignant patterns on its own, the spicule probabilities may allow us to assign a weighting to each pixel when it is included in other measures (for example probability maps) designed to detect such patterns, thus improving the specificity of these measures. This will be the subject of further work.

References

1. Prajna, S., et al.: Detection of Architectural Distortion in Mammograms Acquired Prior to the Detection of Breast Cancer using Texture and Fractal Analysis. In: Giger, M.L., Karssemeijer, N. (eds.) proc. SPIE Medical Imaging, vol. 6915 (2008)

2. Parr, T., et al.: Statistical Modelling of Lines and Structures in Mammograms. In: Duncan, J.S., Gindi, G. (eds.) IPMI 1997. LNCS, vol. 1230, pp. 405–410. Springer, Heidelberg (1997)
3. Karssemeijer, N., te Brake, G.M.: Detection of Stellate Distortions in Mammograms. IEEE Transactions on Medical Imaging 15(5), 611–619 (1996)
4. Bornefalk, H.: Use of Phase and Certainty Information in Automatic Detection of Stellate Patterns in Mammograms. In: Proc. SPIE Medical Imaging, vol. 5370, pp. 97–107 (2004)
5. Wai, L.C., Mellor, M., Brady, M.: A Multi-resolution CLS Detection Algorithm for Mammographic Image Analysis. In: Barillot, C., Haynor, D.R., Hellier, P. (eds.) MICCAI 2004. LNCS, vol. 3217, pp. 865–872. Springer, Heidelberg (2004)
6. Schenk, V., Brady, M.: Finding CLS Using Multiresolution Orientated Local Energy Feature Detection. In: Heitgen, H. (ed.) Proc. 6th International Workshop on Digital Mammography, pp. 64–68 (2002)
7. McLoughlin, et al.: Connective Tissue Representation for Detection of Microcalcifications in Digital Mammograms. In: Sonka, M., Fitzpatrick, J.M. (eds.) Proc. SPIE Medical Imaging, vol. 4684, pp. 1246–1256 (2002)
8. Zwiggelaar, R., Astley, S.M., Boggis, C.R., et al.: Linear Structures in Mammographic Images: Detection and Classification. IEEE Transactions on Medcal Imaging 23(9), 1077–1086 (2004)
9. Breiman, L.: Random Forests. Machine Learning 45(1), 5–32 (2001)
10. Selesnick, I., Baraniuk, R.G., Kingsbury, N.G.: The Dual-Tree Complex Wavelet Transform. IEEE Signal Processing Magazine 22(6), 123–151 (2005)
11. Kass, M., Witkin, A., Terzopoulos, D.: Snakes: Active Contour Models. International Journal of Computer Vision 1(4), 321–331 (1988)

Should CAD Be Used as a Second Reader? – Exploring Two Alternative Reading Modes for CAD in Screening Mammography

Richard Lederman[1], Isaac Leichter[2,3], Eli Ratner[2], Menachem Abramov[2], Alexandra Manevich[2], and Jonathan Stoeckel[2]

[1] Department of Radiology, Hadassah University Hospital, Jerusalem, Israel
[2] Siemens Computer Aided Diagnosis, Jerusalem, Israel
[3] Department of Medical Engineering, Jerusalem College of Technology, Jerusalem, Israel

Abstract. 16963 FFDM cases (280 cancers), were culled retrospectively and run with a CAD algorithm. Instead of using CAD as a "second reader", the study investigates the feasibility of using CAD for prescreening, allowing cases with no CAD prompts to bypass review, thereby decreasing the workload. The study also investigates the outcome of presorting all cases with matching CAD marks of the same type in both views, to enrich the case mix, thereby enhancing the reader's willingness to accept true CAD prompts. The sensitivity of the CAD algorithm was 83.4% and the mean false mark rate generated by CAD per case was 1.15. It was found that prescreening decreases the workload by about 42%, but is not feasible since 6.4% of the cancers would be missed. Using presorting, 73.2 % of the cancers and only 14.2% of the normals would be prioritized for interpretation, enriching the case mix by 5 times.

Keywords: FFDM, Screening mammography, Computer Assisted Detection (CAD), CAD sensitivity, False mark rate, Second reader, Pre-screening.

1 Introduction

True CAD prompts identify cancers, but some CAD marks are false, pointing to structures not related to malignancy. Since in a screening environment almost all cases are normal, almost all the CAD marks presented to the radiologist are false prompts, even though the false mark rate generated by the CAD algorithm is quite low. Thus, when CAD is used in a screening environment as a "second reader", an additional workload is imposed on the radiologist by the necessity to evaluate each of the CAD prompts. On the other hand, if CAD could be used for prescreening, the need to interpret all mammograms with no CAD prompts would be eliminated. Developing prescreening is worthwhile, since it offers a great savings in man- hours, and this could be translated into much wider access to mammography, without adding to the cost of screening programs. However, since CAD algorithms have not yet reached 100% sensitivity, when cases are read in the prescreening mode, malignant cases with no CAD marks will also not be reviewed by the radiologist, thus reducing the cancer detection rate.

Another area of concern when CAD is used as a "second reader" in a screening environment, is the radiologists' reluctance to accept CAD prompts, assuming that almost

J. Martí et al. (Eds.): IWDM 2010, LNCS 6136, pp. 161–167, 2010.

all prompts are false marks [1], [2] since most of the cases are, indeed, normal. However, if the reader would know that the case mix has been enriched with malignant findings rather than a routine screening case mix, the reader's approach would be modified [3]. A novel approach, investigated in the present study, to creating an enriched case mix may be obtained by presorting the cases with matching CAD marks of the same type (i.e. a mass or cluster) in both views. This method could be effective since cancers are usually visible in both views, while false marks tend to occur randomly and would usually not have corresponding marks in both views. Thus, when the cases are read following presorting, true CAD prompts in those cases with matching marks of the same type, are more likely to be accepted by the radiologist.

This study investigates the feasibility of using CAD for prescreening in order to decrease the workload in screening mammography, by allowing all cases with no CAD prompts to bypass review by a radiologist. Furthermore, the study also investigates the outcome of automatically presorting all cases with matching CAD marks, in order to determine by how much such an approach would enrich the case mix.

2 Material and Methods

2.1 CAD Methodology

All the cases were run with a prototype CAD algorithm (Siemens), which is trained to detect suspicious masses and clustered micro-calcifications. In the first step of the detection process, the algorithm identifies all possible candidates for suspicious findings. For each candidate, the algorithm calculates a score based on its characteristics, reflecting the likelihood of malignancy. Only candidates with a score above a certain threshold are displayed on the mammogram. The number of displayed CAD marks depends on this pre-selected operating point of the CAD algorithm. Modifying the operating point affects both the detection sensitivity and false mark rate. High detection sensitivity is associated with a larger number of false marks. The operating point should be set to optimize the balance between sensitivity and false mark rate.

2.2 Case Analysis - Outcome of Prescreening

16963 FFDM cases (280 cancers, 16683 normals), were culled retrospectively in a consecutive manner, from 6 facilities. The normal cases were either negative cases or cases with findings deemed benign by the radiologist, at screening. No follow-up was available to assure that these cases indeed represented "true-normal" cases. For the normal mammograms, the mean rate of CAD false marks per case was calculated. Then, the percentage of normal cases for which the CAD algorithm generated no prompts, was evaluated to assess what percentage of normal cases would bypass interpretation in the prescreening reading mode.

In order to analyze the sensitivity of the CAD algorithm, for the malignant cases the prompts generated by CAD were compared with the mammographic findings that were forwarded for biopsy. For this purpose, a non-blinded radiologist marked on the digital image the location of the biopsied finding (ground truth). Then, the percentage of malignant cases with no CAD prompts was calculated to determine the rate of missed cancers which would result from using CAD for prescreening rather than as a "second reader".

2.3 Case Analysis - Outcome of Presorting

In addition, the percentage of normal cases with matching CAD marks and of malignant cases with matching CAD marks were calculated. Furthermore, the subgroup of malignant cases in which the cancer was detected by CAD was analyzed to determine the percentage of cases with matching CAD marks. First, the cases with matching marks that are indeed corresponding marks correctly identifying the cancer in both views were analyzed. Then, the malignant cases in which the cancers were correctly detected only in one view with a false matching mark of the same type in the other view were analyzed. In the remaining cases, the matching CAD prompts are both false marks, while the true mark does not have a matching mark. These data were used to determine the extent to which the presorting of cases with matching CAD marks will enrich the case mix.

3 Results

3.1 CAD Analysis Using the Prescreening Reading Mode

Figure 1 shows that for almost 43% of the normal cases no false marks were generated. The percentage of cases with false marks generated by the CAD algorithm decreased, as the number of false marks per case increased. The average false mark rate per case was 1.15.

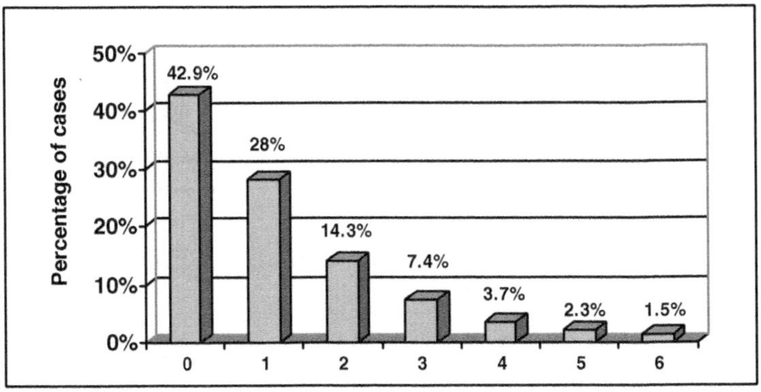

Fig. 1. Distribution of false CAD marks in normal cases

The CAD algorithm detected the cancer in at least one view, in 236 of the malignant cases, yielding a sensitivity of 83.4%. The algorithm did not detect the cancer in 44 cases, but as shown in figure 2, in 26 of these cases false marks were generated by the CAD algorithm. On the other hand, in 18 malignant cases (42% of the cancers missed by CAD) the cancer was missed with no CAD prompts whatsoever. Only these 18 malignant cases would bypass review by the radiologist using the prescreening reading mode.

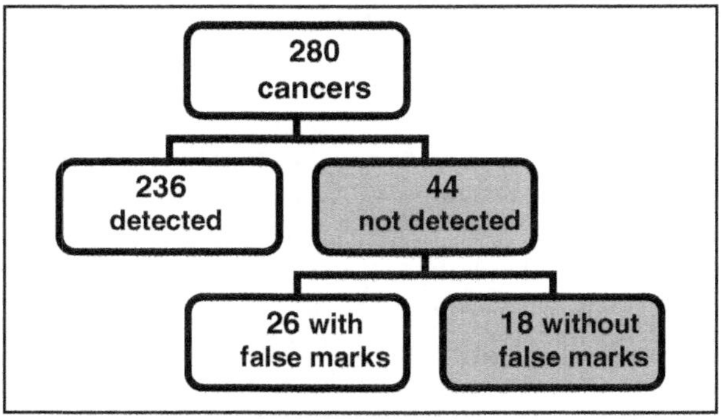

Fig. 2. A detailed scheme of the detection results

As shown in figure 3, the percentage of normal cases with no CAD prompts was 42.9%, while the percentage of malignant cases with no CAD prompts was only 6.4%.

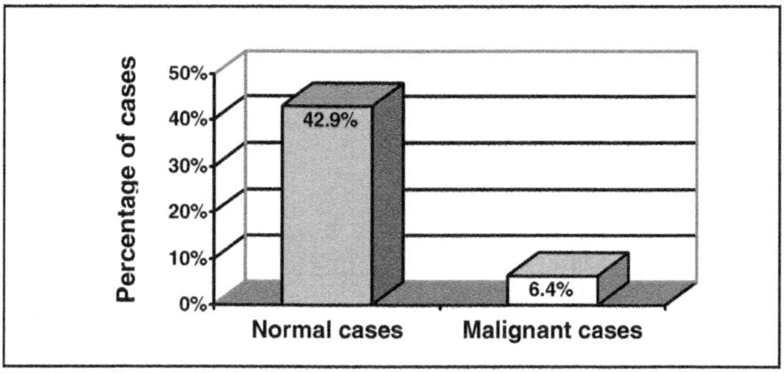

Fig. 3. Percentage of cases without CAD marks by pathology results

Overall, in 7167 of the cases (42.3%) no prompts were generated by the CAD algorithm. Hence, if CAD had been used for prescreening, i.e. all the cases without CAD prompts would have been bypassed reviewed by the radiologist, the workload would be reduced by 42.3%. Thus, with prescreening 42.9% of the normal cases will bypass review by the radiologist, but 57.1% of the cases will still have to be interpreted. The effect of prescreening on the malignant cases would be to present the 236 cases with the true CAD marks and the 26 missed cases with false CAD marks to the radiologist for interpretation. But, unfortunately, 6.4% of the malignant cases (18 missed cases without false marks) would bypass review by a radiologist, reducing the cancer detection rate.

3.2 CAD Analysis Using the Presorting Reading Mode

Figure 4 shows the percentage of normal cases and of malignant cases with matching CAD marks.

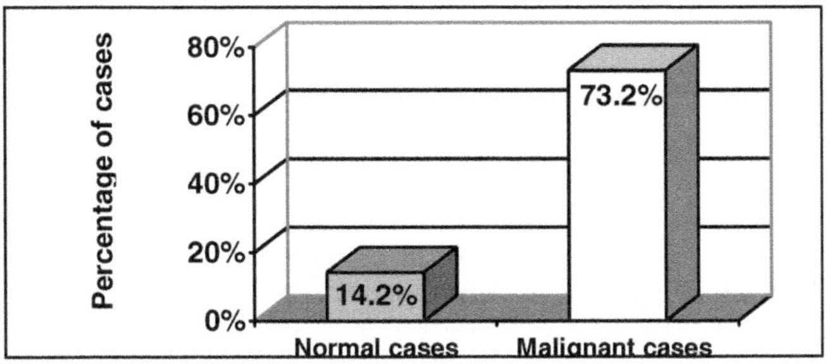

Fig. 4. The percentage of cases with matching CAD marks for normal cases and for malignant cases

This figure shows that of the normal cases, only 14.2 % (2369 cases) had, coincidentally, matching CAD marks of the same type in both views. On the other hand, 73.2% (205) of malignant cases had matching CAD marks of the same type in both views. However, this subgroup included some cases with a true CAD mark in one view, while the matching CAD mark of the same type on the other view was a false mark. As shown in figure 5, further analysis showed that 193 of the 236 cases in which the cancer was detected by CAD (81.8%) had matching CAD marks in both views. In 181 cases (76.7%) of the 236 cases with cancers detected by CAD, corresponding prompts were, indeed, marked by CAD, identifying the cancer correctly in both views.

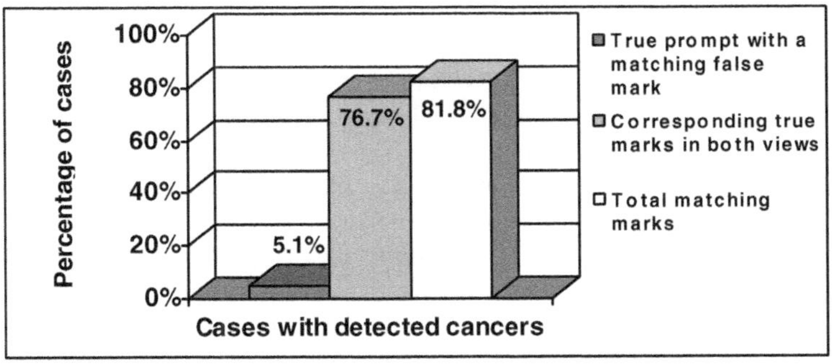

Fig. 5. The percentage of cases with matching CAD marks for malignant cases, in which cancers were detected by CAD

Thus, if presorting would be selected as the reading mode, the radiologist would prioritize the interpretation of all cases with matching marks, i.e. 2369 of the 16683 normal cases, and 205 of the 280 malignant cases. Thus, the case mix in the prioritized subgroup would be enriched 5 times, from 1.6% (280 of 16963) without presorting to 8% (205 of 2574) with presorting.

4 Discussion

Since its inception, CAD in screening mammography has been envisioned as a "second reader", to be used following conventional interpretation, as a means of avoiding oversights. The necessity to evaluate each CAD prompt imposes an additional workload on the radiologist and increases the reading time, since in a screening environment almost all cases are normal. Although the false mark rate of the CAD algorithm is very low, in a screening environment the vast majority of CAD marks presented to the radiologist are false prompts. Therefore, perhaps another approach to the use of the information generated by CAD would be more beneficial. In this study the use of CAD for prescreening was explored. In this scenario, all cases with no CAD marks could bypass review by a radiologist, thus offering a great savings in man-hours, which could be translated into a cost savings for screening programs.

Even though this study shows that prescreening by CAD would decrease the workload by about 42%, the results of this study indicate that it is not feasible to exclude the reading of cases with no CAD prompts, since 6.4% of the cancers would be missed. However, the operating point of the CAD algorithm analyzed in the study was optimized for the use of CAD as a "second reader". Hence, additional operating points, resulting in a different balance between sensitivity and false mark rate, should be investigated to allow prescreening. Prescreening would require an operating point with higher detection sensitivity, in order to minimize the missed cancer rate, but this operating point will result in a concomitant higher false mark rate and, therefore, in a lower decrease in workload.

The use of CAD for presorting in order to enrich the case mix, was also explored in this study. This method of prioritizing the cases with matching marks, could be effective since cancers are usually visible in both views while false marks tend to occur randomly and would usually not have corresponding marks in both views. Based on the results of this study, presorting by matching CAD marks in both views will raise the level of concern of the radiologist regarding 73.2% of the cancers but will prioritize only about 14.2% of the normal cases. Thus, when the cases with matching CAD marks are automatically presorted by the algorithm, the radiologist reads a case mix enriched by 5 times, which may enhance the willingness to accept true CAD prompts in a screening environment. The remaining cases without matching CAD marks would have a much lower incidence of cancer, and furthermore, would include more subtle cancers. A more expert reader could be assigned to interpret these cases, which would be more challenging, with a very low incidence of cancers including many subtle findings.

These results indicate that another approach to the use of CAD could be used to better advantage. Instead of using CAD as a "second reader", CAD could be run in the background just in order to presort cases, and case interpretation will be performed without viewing the CAD prompts, but with a raised level of concern for the prioritized cases, knowing that the case mix is enriched. Likewise, CAD could also be used for prescreening without presenting the prompts to the radiologist, but for this purpose the algorithm must be improved to reduce the missed cancer rate. In both scenarios the CAD prompts could be presented to the reader on demand. Alternatively, following prescreening, the remaining cases with CAD marks could be displayed with CAD prompts, but using the operating point appropriate for the use as a "second reader", rather than the operating point used for prescreening.

The success of CAD as a "second reader" ultimately depends upon the acceptance of true CAD prompts by the radiologist for overlooked cancers. While for medico-legal reasons, most radiologists fear missing a cancer in mammography, most readers find it difficult to identify subtle cancers which do not demonstrate characteristic features of malignancy. These are the very cancers that are likely to be overlooked, and that are most likely to be rejected when presented as CAD prompts. Indeed, in a recent study [4], it was found that performance for radiologists without fellowship training in mammography, improved most during their first 3 years of clinical practice, due to a decrease in false-positive reading but with no associated increase in sensitivity. In a related study [5], it was further stated that fellowship-trained radiologists had significantly higher recall and false-positive rates, but this finding was associated with improved sensitivity. It is possible that fellowship-trained radiologists would be more amenable to the use of CAD as a "second reader". Alternatively, when CAD is used in a prescreening or presorting mode, the cases could be assigned to fellowship trained or non-fellowship trained radiologists for interpretation based on the CAD results, even without ever showing the CAD marks to the readers.

References

1. Paquerault, S., Samuelson, F.W., Petrick, N., Myers, K.J., Smith, R.C.: Investigation of Reading Mode and Relative Sensitivity as Factors that Influence Reader Performance When Using Computer-Aided Detection Software. Acad. Radiol. 16, 1095–1107 (2009)
2. Georgian-Smith, D., Moore, R.H., Halpern, E., Yeh, E.D., Rafferty, E.A., D'Alessandro, H.A., Staffa, M., Hall, D.A., McCarthy, K.A., Kopans, D.B.: Blinded Comparison of Computer-Aided Detection with Human Second Reading in Screening Mammography. Am. J. Roentgenol. 189, 1135–1141 (2007)
3. Warren Burhenne, L.J., Wood, S.A., D'orsi, C.J., Feig, S.A., Kopans, D.B., O' Shaughnessy, K.F., Sickles, E. A., Tabar, L., Wybrony, C.J., Castellino, R.A.: Potential Contribution of Computer-Aided Detection to the Sensitivity of Screening Mammography. Radiology 215, 554–562 (2000)
4. Miglioretti, D.L., Gard, C.C., Carney, P.A., Onega, T.L., Buist, D.S., Sickles, E.A., Kerlikowske, K., Rosenberg, R.D., Yankaskas, B.C., Geller, B.M., Elmore, J.G.: When Radiologists Perform Best: the Learning Curve in Screening Mammogram Interpretation. Radiology 253, 632–640 (2009)
5. Elmore, J.G., Jackson, S.L., Abraham, L., Miglioretti, D.L., Carney, P.A., Geller, B.M., Yankaskas, B.C., Kerlikowske, K., Onega, T., Rosenberg, R.D., Sickles, E.A., Buist, D.S.: Variability in Interpretive Performance at Screening Mammography and Radiologists' Characteristics Associated with Accuracy. Radiology 253, 641–651 (2009)

Automated Assessment of Area of Dense Tissue in the Breast: A Comparison with Human Estimation

Camilla Jeffries-Chung[1], Jennifer Diffey[2], Michael Berks[2], Joanna Morrison[1], Rosanne Verow[1], Julie Morris[3], Mary Wilson[4], Caroline Boggis[4], Nicky Barr[4], Anthony Maxwell[5], Alan Hufton[2], and Susan Astley[2,*]

[1] University of Manchester Medical School, Stopford Building, Oxford Road, Manchester M13 9PT, England
[2] Cancer and Enabling Science, University of Manchester, Stopford Building, Oxford Road, Manchester M13 9PT, England
sue.astley@manchester.ac.uk
[3] Medical Statistics, University Hospital of South Manchester, Wythenshawe Hospital, Southmoor Road, Manchester M23 9LT, England
[4] Nightingale Breast Centre and Genesis Prevention Centre, University Hospital of South Manchester, Southmoor Road, Manchester M23 9LT, England
[5] Royal Bolton Hospital, Bolton, BL4 0JR

Abstract. Both interactive thresholding tools and human visual assessment have been related to the risk of developing breast cancer. In this paper we explore the relationship between human assessment of area of dense tissue and the actual thickness of tissue in the breast by using a volumetric density technique to compute areas of dense tissue, varying the threshold below which areas of low density are discounted and observing the correlation with visual assessment of density at different thresholds. Based on analysis of thresholds used in the automated method, radiologists' definition of a dense pixel is one in which the percentage of glandular tissue is between 10% and 20% of the total thickness of the compressed breast at that point.

Keywords: risk assessment, human perception, computer analysis, breast density.

1 Introduction

The quantity and pattern of dense glandular tissue in the breast has for many years been linked with the risk of developing cancer. Wolfe developed a set of classes into which mammograms could be assigned depending on the appearance of the breast tissue, but despite demonstrating a clear relationship with risk, assignment was subjective [1]. Boyd introduced a more quantitative measure which was more readily understandable, using the percentage of dense tissue as an indication of risk [2]. Subsequent attempts at automation have either used texture measures to try to replicate Wolfe's classes [3], or focused on the quantification of breast tissue to improve the objectivity of Boyd's method [4,5]. A widely used method is Cumulus, an interactive thresholding programme for use with digitised images which requires an operator to

* Corresponding author.

J. Martí et al. (Eds.): IWDM 2010, LNCS 6136, pp. 168–174, 2010.

define thresholds to separate the breast from the background, and dense from fatty tissue [6]. This still has a considerable element of subjectivity and raises the question as to what radiologists consider to be dense tissue when assessing mammograms. In order to gain an insight into this, we have taken a calibration-based volumetric technique which provides the thickness of gland and fat at each pixel in the mammogram and used it to compute the area of gland in the mammogram [7]. By applying a threshold on the thickness of gland that is used to compute the area it is possible to simulate the way in which radiologists decide where to place their internal threshold when assessing density.

2 Materials and Methods

A set of 672 screening mammograms was randomly selected from mammograms obtained during routine film-based mammographic screening of women over the age of 50 in Bolton and Bury. All participants had consented to breast density measurement. Breast density was assessed by two consultant radiologists who marked the proportion of dense tissue on a 10cm Visual Analogue Scale. Both readers are highly experienced in mammographic interpretation, although radiologist 2 had greater experience than radiologist 1 in using Visual Analogue Scales for density estimation.

A subset of 168 mammograms was digitized, and breast density was measured using a calibration-based semi-automated technique [7]. This involves calibrating an aluminium step-wedge by imaging it alongside known thicknesses of tissue equivalent material. The calibrated step-wedge is then imaged next to the breast in each mammogram. Radio-opaque markers are placed on the compression plate. The magnification of these markers in the mammogram enables accurate measurement of compressed breast thickness and tilt of the compression plate. The grey-level of

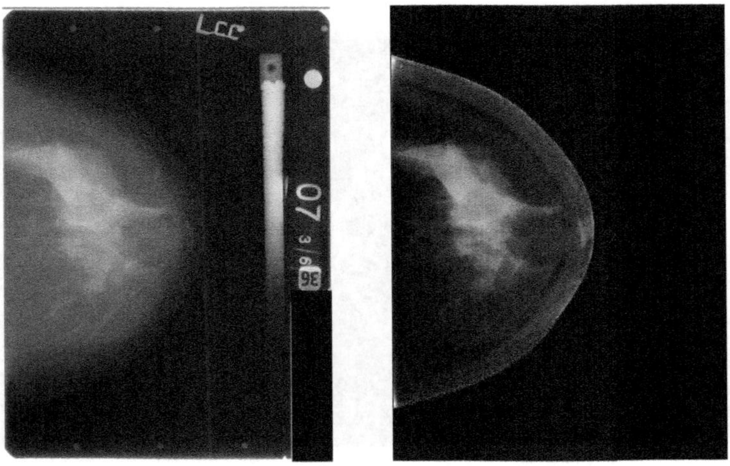

Fig. 1. Left: An original mammogram showing the calibration object and the images of the circular markers on the compression plate used to compute compressed breast thickness. Right: An unthresholded glandular tissue map of the mammogram.

every pixel in the mammogram is then matched to the equivalent grey-level value in the image of the step-wedge. The corresponding thickness of the step-wedge at this point, combined with the breast thickness measurement, allows composition to be uniquely determined at each pixel.

The automated method thus provides an estimate of the thickness of glandular and fatty tissue at every pixel in the image, from which the volumes of fatty and glandular tissue in the breast can be computed. However, for this study we used the technique to produce maps of fatty and glandular tissue in the breast from which we could compute the area of dense tissue in the breast as a percentage to mirror the visual assessment process. Figure 1 shows an example of a glandular tissue map alongside the original mammogram from which it was derived. The calibration object is seen as a vertical step-wedge at the right hand side of the mammogram, and the circular compression plate markers are visible along the top and bottom edges of the mammogram.

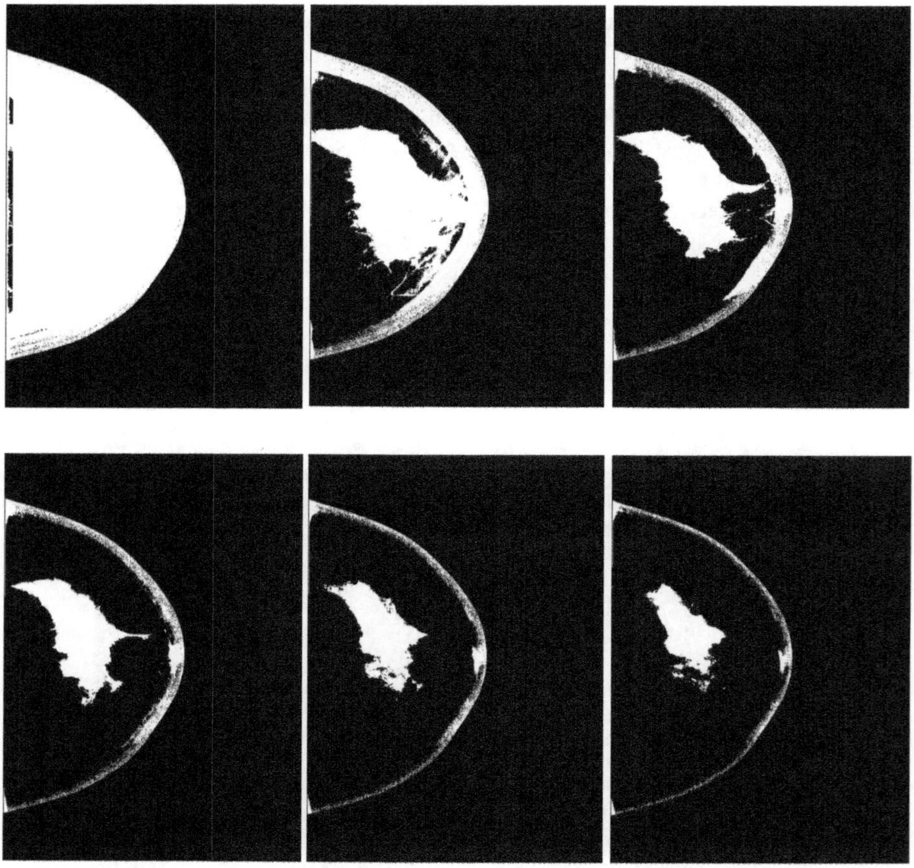

Fig. 2. Thresholded glandular tissue maps for the image shown in figure 1. Dense glandular regions are shown as white. The top row shows the glandular tissue map thresholded at 0%, 5% and 10%, whilst the bottom row shows thresholds of 15%, 20% and 25%.

We produced six such maps for each digitized mammogram, using a threshold on the percentage thickness of glandular tissue at 0, 5, 10, 15, 20 and 25% so for example at the 20% threshold a pixel which had 18% glandular tissue would be considered as being fatty background for the purposes of computing the % dense tissue in the breast by area, whilst at the 15% threshold, the same pixel would have been classed as dense tissue. At the 0% threshold, any pixel which contained even the smallest amount of gland was classed as dense. Figure 2 shows examples of the thresholded density images for the mammogram in Figure 1. At a threshold of 0%, virtually the whole breast appears to be dense. As the threshold increases, the area of dense tissue decreases.

The percentage area of dense tissue was computed for each threshold from the number of pixels above the threshold and the number of pixels within the breast border. These areas were compared with the visual assessment of percentage area of dense tissue.

3 Results

There was a significant difference in density estimation by the two radiologists using the visual analogue scale, with radiologist 1 (R1) giving lower readings for all mammographic projections with an average mean of differences of 15.9% ($p<0.05$). Figure 3 shows a scatter plot of the two radiologists' assessments. They were more likely to agree when estimating densities of very dense or fatty breasts, with the least agreement in fatty-glandular breasts, which fall in the middle of the density range.

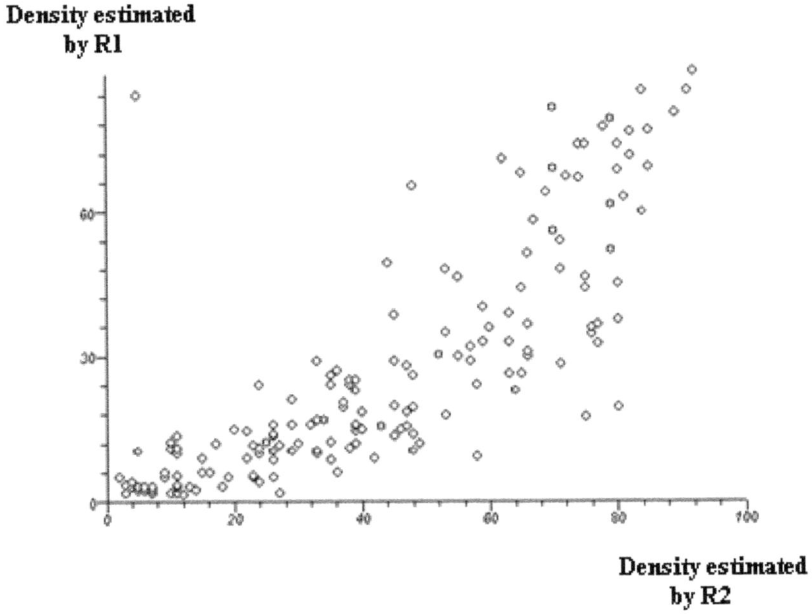

Fig. 3. Densities estimated by two consultant radiologists, R1 and R2

When compared with the automated measure of area using thresholds of 0% and 5%, both radiologists produced values that were significantly lower than the dense area output by the automated method. At thresholds of 10% and 15%, the agreement improved, however, as the threshold increased further, the radiologists' readings were found to be generally higher than those generated by the software. This reached statistical significance at a threshold of 25%. The radiologists' readings of low and high density breasts were found to be closer to the software values than their assessments of breasts of intermediate density. This can be seen in Figure 4, which shows an agreement plot [8] of one radiologist's assessments with the automated measurement at the 10% threshold. A similar pattern was observed for both observers at all thresholds.

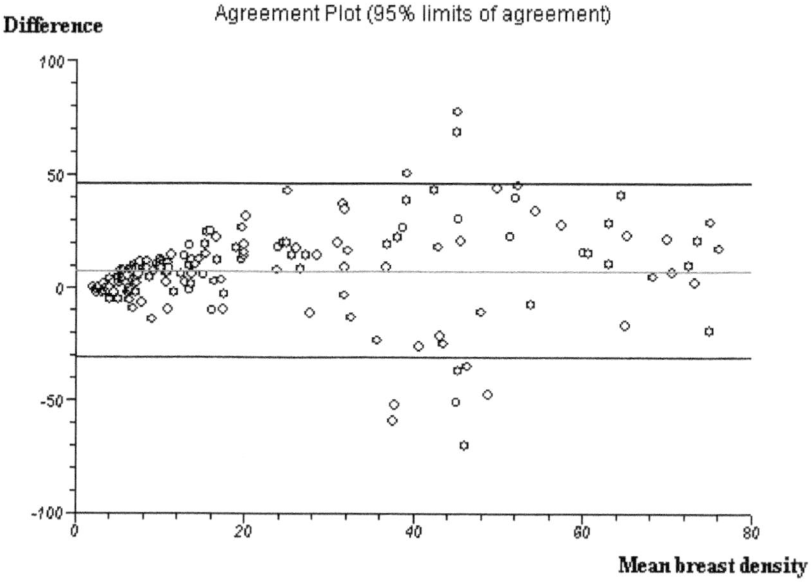

Fig. 4. Agreement plot showing the difference between R1's assessment of percentage dense area and the automated measure plotted against mean density for the RMLO view using a threshold of 10%

4 Discussion

Based on analysis of thresholds used with a volumetric method for computing gland and fat in the breast, the radiologists' definition of a dense pixel is one in which the percentage of glandular tissue is between 10 and 20% of the total thickness of the compressed breast at that point.

This could indicate that when radiologists make a judgement about what part of the breast constitutes 'dense tissue' they ignore pixels in which there is actually a substantial glandular component. However, when the dense tissue maps at different thresholds are examined (Figure 2), we can see that there is actually a significant edge effect near the breast border, particularly noticeable at the lower thresholds (5-10%).

At the border, the estimate of breast thickness is less reliable: in the part of the breast where the compression plate is in contact with the skin we have an accurate measure of thickness of tissue from the markers on the compression plate, but where the breast loses contact with the compression plate we must rely on a model to estimate thickness. The model used in the quantitative method used in this study was that the breast adopts a semicircular profile at the boundary. Clearly, as the estimated breast thickness tails off towards the edge of the breast, the impact of any inaccuracy will have a large effect on the thickness maps thresholded according to percentage of density.

One approach to overcoming this would be to exclude the edge of the breast when computing glandular thickness. We are currently exploring methods by which this could be achieved without impinging on genuine glandular tissue; it would not be appropriate to exclude a uniform band around the edge of the breast, as the glandular tissue in the breast is not uniformly distant from the breast border, being closer at the nipple. Furthermore, the example shown here is a cranio-caudal view in which the area of loss of contact with the compression plate is relatively uniformly distributed around the breast border. In medio-lateral oblique views, this is no longer the case because of the pectoral muscle. Automated methods [9,10] could be used to identify the nipple, and thus facilitate definition of a tapered region to be excluded. The component of non-fatty tissue due to the skin should also be taken into account. A pragmatic approach to correcting images for edge effects would be to use a low threshold on the glandular map as a basis for excluding regions from further analysis.

Another approach would be to produce maps based on actual thickness of gland rather than percentage thickness, with thresholds on thickness of gland in mm. This is more intuitive when compression paddle tilt is taken into account, for example, since in this case a constant thickness of gland would correspond to different percentages across the same breast. However, it is less clearly related to radiologists' perception of density.

It is difficult to establish a gold standard against which methods of density measurement can be assessed. Visual assessment varies from observer to observer, and according to the method used to record it [11]. In this study we used Visual Analogue Scales, as our readers are familiar with this approach and it has been related to risk of developing cancer in our screening centre [12]. Although semi-automated methods have been described as gold-standard approaches, they still rely on subjective judgement [13]. It is possible to measure density using other imaging modalities, but there is no guarantee that the volume of the breast imaged and assessed is the same across modalities. Current breast phantoms allow calibration to a limited extent, but do not provide a useful way of assessing whether methods are dealing with the breast edge in an appropriate fashion.

Area-based methods of visual assessment that express density as a percentage of the whole breast are the standard method of identifying women at risk on the basis of increased breast density. Such methods may not always provide an accurate assessment as they are unduly affected by weight change [14]. For example, when women gain weight, their breasts rapidly gain volume, due to an increase in fat. There is no corresponding rapid gain in dense glandular tissue, so weight gain would be associated with a decrease in the percentage of dense glandular tissue, and hence a decrease in risk, whereas actually post-menopausal weight gain is associated with an increased risk of cancer.

In addition, visual assessment of density requires the radiologist to make a decision about what level of brightness actually corresponds to dense tissue. This is problematic when women are imaged with different parameters as may be the case across populations, or for a given woman, over time. It may also be the case that the perceived brightness depends on specific overlaps of tissue in the breast volume. which could arise or disappear depending on how the breast is positioned for imaging. We believe that this supports the use of quantitative, volumetric methods in which such subjectivity is removed, although it is important to understand how visual assessment relates to volumetric density, given the significant body of knowledge relating visual assessment to risk.

Acknowledgments. The authors are grateful for the support of the John Lewis Partnership and the Genesis breast cancer prevention charity for funding this research.

References

1. Wolfe, J.N.: Risk for breast cancer development determined by mammographic parenchymal pattern. Cancer 37(5), 2486–2492 (1976)
2. Boyd, N.F., Lockwood, G.A., Byng, J.W., Tritchler, D.L., Yaffe, M.J.: Mammographic densities and breast cancer risk. Cancer Epidemiol.Biomarkers Prev. 7(12), 1133–1144 (1998)
3. Li, H., Giger, M.L., Olopade, O.I., Margolis, A., Lan, L., Chinander, M.: Computerized texture analysis of mammographic parenchymal patterns of digitized mammograms. Acad. Radiol. 12(7), 863–873 (2005)
4. van Engeland, S., Snoeren, P.R., Huisman, H., Boetes, C., Karssemeijer, N.: Volumetric breast density estimation from full-field digital mammograms. IEEE Trans. Med. Imaging 25(3), 273–282 (2006)
5. Highnam, R., Pan, X., Warren, R., Jeffreys, M., Davey Smith, G., Brady, M.: Breast composition using retrospective standard mammogram form. Physics in Medicine and Biology 51, 2695–2713 (2006)
6. Boyd, N.F., Byng, J.W., Jong, R.A., Fishell, E.K., Little, L.E., Miller, A.B., et al.: Quantitative classification of mammographic densities and breast cancer risk: results from the Canadian National Breast Screening Study. J.Natl.Cancer Inst. 87(9), 670–675 (1995)
7. Diffey, J.L., Hufton, A.P., Astley, S.M.: A New Stepwedge for the Volumetric Measurement of Breast Density. In: Astley, S.M., Brady, M., Rose, C., Zwiggelaar, R. (eds.) IWDM 2006. LNCS, vol. 4046, pp. 1–9. Springer, Heidelberg (2006)
8. Bland, J.M., Altman, D.: Statistical methods for assessing agreement between two methods of clinical measurement. Lancet 1(8476), 307–310 (1986)
9. Kinoshita, S.K., Azevedo-Marques, P.M., Pereira Jr., R.R., Rodrigues, J.A.H.,, Rangayyan, R.M.: Radon-Domain Detection of the Nipple and the Pectoral Muscle in Mammograms. Journal of Digital Imaging 21(1), 37–49 (2008)
10. Petroudi, S., Brady, J.M.: Automatic Nipple Detection on Mammograms. In: Ellis, R.E., Peters, T.M. (eds.) MICCAI 2003. LNCS, vol. 2879, pp. 971–972. Springer, Heidelberg (2003)
11. Sukha, A., Berks, M., Morris, J., Boggis, C., Wilson, M., Barr, N., Astley, S.: Visual Assessment of Density in Digital Mammograms. In: Marti, J., et al. (eds.) Digital Mammography (2010)
12. Duffy, S., et al.: Visually assessed breast density, breast cancer risk and the importance of the craniocaudal view. Br. Can. Research. 10(4) (2008)
13. Boyd, N.F., Byng, J.W., Jong, R.A., Fishell, E.K., Little, L.E., Miller, A.B., et al.: Quantitative classification of mammographic densities and breast cancer risk: results from the Canadian National Breast Screening Study. J. Natl. Cancer Inst. 87(9), 670–675 (1995)
14. Patel, H.G., Astley, S.M., Hufton, A.P., Harvie, M., Hagan, K., Marchant, T.E., Hillier, V., Howell, A., Warren, R., Boggis, C.R.M.: Automated breast tissue measurement of women at increased risk of breast cancer. In: Astley, S.M., Brady, M., Rose, C., Zwiggelaar, R. (eds.) IWDM 2006. LNCS, vol. 4046, pp. 131–136. Springer, Heidelberg (2006)

Local Greylevel Appearance Histogram Based Texture Segmentation

Reyer Zwiggelaar

Department of Computer Science
Aberystwyth University
Aberystwyth SY23 3DB, UK
rrz@aber.ac.uk

Abstract. We have developed a segmentation approach, which is based on modelling local texture information and incorporates both greylevel and spatial aspects. Variation in local greylevel configuration/appearance is represented in histogram format for which the distribution varies with texture appearance. Segmentation results based on full mammographic images are presented. In addition, the potential use of the segmentation results for mammographic risk assessment and abnormality detection is discussed.

1 Introduction

Over the past years (texture) segmentation approaches have been developed, which have been linked to the detection of abnormalities [1,2] and the automatic estimation of mammographic risk [3,4,5,6,7,8,9]. With respect to the detection of abnormalities, such segmentation approaches have been linked to asymmetry estimation in the comparison of left and right breast images. In addition, local texture aspects can be used to detect subtle abnormalities, such as spiculated lesions [10]. Automatic estimation of mammographic risk approaches are directly linked to the work of Wolfe [11], Byng [12] and Tabar [13], who have used mammographic image information and correlated the appearance of mammographic tissue with mammographic risk assessment.

We present a texture based segmentation approach, which is based on histograms of local greylevel appearance. Each variation in local greylevel configuration/appearance is represented by an unique number. The distribution of local greylevel appearances forms the basis for texture models which are used for segmentation purposes. Various approaches to the modelling and the use of local greylevel appearance histograms are presented and discussed.

The layout of the paper is as follows. In Sec. 2 the local greylevel appearance (LGA) histogram based approach is presented, which covers both the modelling of the data and the use of such models in automatic segmentation of mammographic images. In addition, potential application areas, such as the detection of abnormalities and automated mammographic risk assessment, are discussed. The paper concludes with a discussion and conclusions. As might be expected with a texture based segmentation approach, there are similarities with existing approaches which are covered in the discussion section.

J. Martí et al. (Eds.): IWDM 2010, LNCS 6136, pp. 175–182, 2010.

2 Local Greylevel Appearance (LGA) Histograms

The aspects discussed in the section cover how models can be generated and how these models can be used in segmentation (and classification) applications. Parameters within the models are introduced.

2.1 Building Models

The building of models based on LGA histograms incorporates the following steps (the parameter values that will be used in Sec. 3 will be indicated):

1. Set the size of the local window, which indicates the number of dimensions to represent the local greylevel appearance (a 3×3 local window is used in Sec. 3).
2. Set the greylevel resolution, which determines the number of greylevel bins to be used in the subsequent processing steps. This can be used to reduce the greylevel range (8 bins were used in Sec. 3).
3. Create storage allocation for a number of elements equal to the number of pixel samples from the image(s).
4. Extract at each pixel sample in the image(s) a local window and at the same time reduce the greylevel resolution with the values set under 2).
5. At each pixel transform the local window greylevel information into a unique number by using

$$\sum_{i,j} \#bins^{counter(i,j)} I(i,j), \tag{1}$$

   where $counter(i, j)$ starts at a value of zero and increments at each location in the local window, and $I(i, j)$ is the (reduced resolution) greylevel value at position (i, j) in the local window. Each number is unique in that it represents a specific greylevel configuration/appearance within the local window.
6. Store the unique numbers obtained under 5) as indicated under 3).
7. Sort all the stored values using an appropriate technique.
8. Transform the stored values into a LGA histogram that only contains the combination of unique numbers and their occurrence. At this stage it is possible to remove "noise" with low occurrence from the LGA histograms. The LGA histograms can be normalised with a L_1 metric.

This completes the formation of a LGA histogram model. It is possible to generate a single LGA histogram model from a single image, or a number of LGA histograms models representing various image regions (textures), or a single LGA histogram model from a series of images (all representing similar texture appearances). L_1 normalisation tends to appear after all relevant histograms have been combined.

2.2 Using Models

Using the models is similar in approach as building the models, but only based on the information from a local region and not a full image. This involves the following steps (again the parameter values that will be used in Sec. 3 will be indicated):

1. Set the size of the local region (24 × 24 regions were used in Sec. 3), which will contain a number of local windows (the window is expected to be smaller that the region).
2. For each pixel in the image to be segmented use a local region and extract all the local windows within the local region and use these to form a local LGA histogram of unique numbers and their occurrences. The local LGA histogram are L_1 normalised.
3. Determine the distance between the local LGA histogram and the LGA histogram model(s) using an appropriate distance metric. (e.g. Euclidean, transportation or hybrid transportation [4]). This can use some clever coding to avoid going through all entries in the LGA histogram model and/or local LGA histogram.

This process is repeated for all pixels in the image and the results form either a "probability" or a classification image.

3 Results

The initial results presented in this section are based on the Mammographic Images Analysis Society (MIAS) database [14]. Each mammogram in the MIAS database was labeled according to the four Birads density classes.

3.1 One Model per Birads Density Class

The first results are based on one LGA histogram model for each of the four Birads density classes. These models were obtained by combining the individual LGA histograms from the relevant MIAS images. To achieve this, a non-normalised LGA histogram was generated for each mammogram in the database. Subsequently, histograms belonging to the same Birads density class were combined and the resulting LGA histograms were L_1 normalised.

The four *Birads LGA histogram* models have been used to classify pixels within full mammograms as belonging to one of the four classes. Typical segmentation results, covering the full Birads density range, can be found in Fig. 1. These examples indicate that the various appearances associated with the four Birads classes are represented in the segmented images and result in realistic segmentation results. Birads I associated regions are represented as major areas in the left two mammograms (Birads I and II), but can also be found as smaller regions in the Birads III and IV cases. Similar aspects can be concluded for the regions associated with the Birads IV class, which is represented mainly in the right two images (Birads III and IV). This almost binary segmentation needs further investigation, although it should be noted that there are significant segmented regions associated with Birads classes II and III.

3.2 Four Models per Birads Density Class

Secondly, we have also generated results based on four LGA histogram models for each of the four Birads density classes. These models were obtained by obtaining four cluster centres from the individual LGA histograms from the relevant MIAS images (k-means

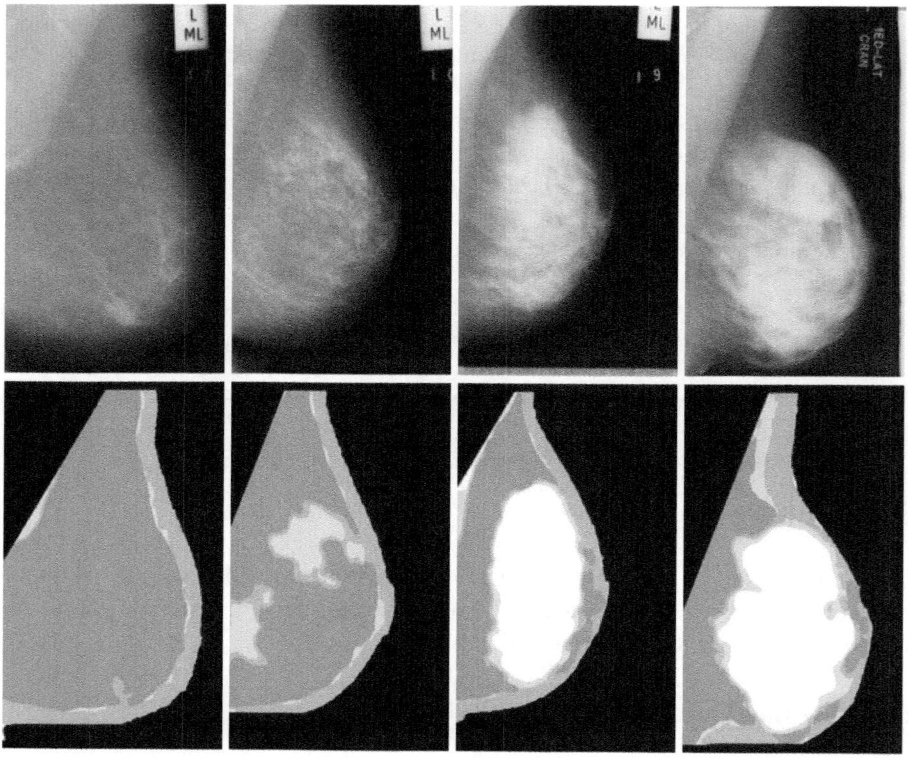

Fig. 1. Example mammograms (top row) and associated segmentation results (bottom row) based on one LGA histogram model per Birads density class. From left to right the mammograms range from Birads I to Birads IV.

was used as a clustering technique). To achieve this, a non-normalised LGA histogram was generated for each mammogram in the database. Subsequently, histograms belonging to the same Birads density class were used clustering and the resulting four LGA histograms were L_1 normalised.

The sixteen *Birads LGA histogram* models have been used to classify pixels within full mammograms as belonging to one of the four classes, and within each class there are four sub-classes. Typical segmentation results, covering the full Birads density range, can be found in Fig. 2. These results indicate overall similar results when compared with Fig. 1 and in general the four regions are sub-divided into the four sub-classes for each Birads density class. However, there is some mixing between classes, but overall the results appear realistic. One aspect that should be noticed is that breast boundary region tends to be associated with the high risk Birads class and this should be further investigated.

3.3 Potential Application Areas

Two potential application areas are discussed where the described segmentation results could be used. This covers automatic mammographic risk assessment and the detection of potential abnormal regions in mammographic images.

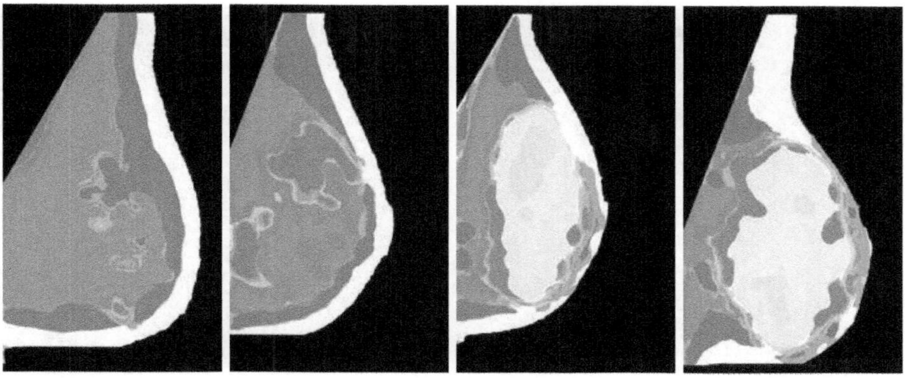

Fig. 2. Segmentation results based on fours LGA histogram model per Birads density class. From left to right the mammograms range from Birads I to Birads IV (see Fig. 1 for the original mammograms).

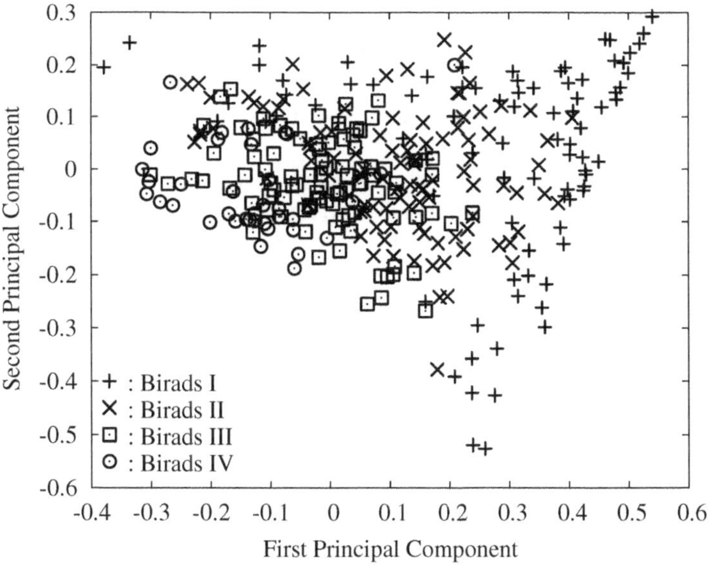

Fig. 3. First two principal components of the 4D feature space. The different markers represent the ground truth for the four Birads density classes represented by sixteen LGA histogram models.

To indicate the potential of the developed segmentation approach for mammographic risk estimation we have preformed a simple experiment. Each mammogram is represented by the relative number of pixels within each of the four Birads classes represented by sixteen LGA histogram models (i.e. the relative areas of the sixteen classes in the segmented images), which results in a sixteen dimensional feature space. The distribution of a set of 322 cases from the MIAS database can be found in Fig. 3, which displays the first two principal components. It should be clear that low/high risk shows

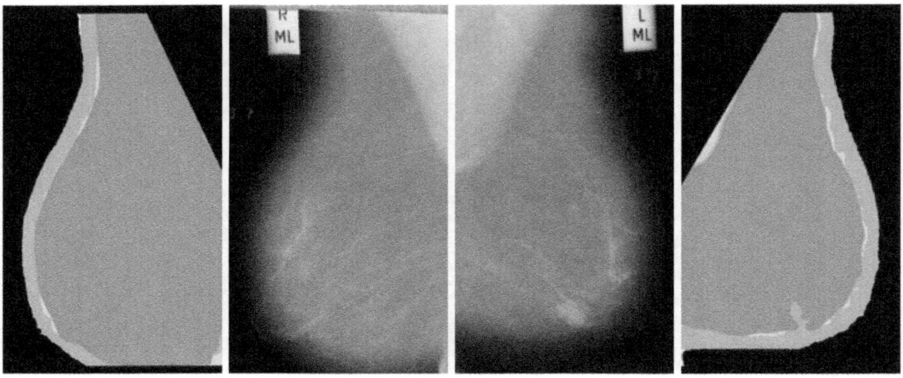

Fig. 4. Example mammograms (middle) and associated segmentation results (outside) based on one LGA histogram model per Birads density class

good discrimination, but at the same time it seems that a more detailed classification might be problematic. The correct classification was 64% (using a simple k-nearest-neighbour classifier) with respect to four Birads density classes, while low/high risk estimation was correct for 98% of the cases. The former is not at the same level as some of the published results (see references in Sec. 1), but could be used as the basis to do subsequent processing.

For the detection of potential abnormalities in mammographic images, a direct comparison between left and right segmentation results could indicate areas of dissimilarity. A typical example of this can be found in Fig. 4, which shows the results of segmentation of two mammograms where the abnormality is clearly indicated as a difference between the two results. A full evaluation is outside the scope for this paper, but the potential of this should be clear from the shown example.

4 Discussion

Clearly, one of the parameters within the developed approach is the local window size. In general, it would be reasonable to expect that a larger window size would contain additional texture information as long as it does not exceed the structural size of the texture under analysis. However, it should also be clear that an increase in the window size is directly linked to an increase in the feature space dimensionality. Without an increase in the number of samples this would mean a sparser populated feature space, which tends to result in a degradation of segmentation/classification results. With only a limited number of samples this might mean that a small local window size provides the most robust results, which is in line with the work of Varma who used a 3×3 window to generate a feature space [15].

The second parameter, which determines the greylevel resolution, can be seen to have at least two effects: a) reducing the greylevel resolution initially results in the removal of noise and subsequently in the loss of information, and b) reducing the greylevel resolution is directly related to the feature space dimensionality and has such has similar effects as the local window size (see above).

4.1 Related Approaches

Up to a certain point, the underlying information within the local window is the same as that being used in the texton based approaches [5]. However, within the texton based approaches the texture models are formed by cluster centre related histograms, while here we model the full feature space.

There are also similarities with local binary pattern (LBP) based texture analysis [16,17]. However, the main differences is that with the developed there is no reduction to binary patterns, but the full greylevel range can be used. This means that the resulting histograms might contain more information, but at the same time would be sparser populated. Closely related to the original LBP work is that of Zhang et al. [18] who use local histograms of LBP results (and call these spatial histograms).

The term spatial histogram is also used by Birchfield et al. [19], but in their case this is used for triple-histograms that contain the number of pixels at a specific greylevel, but also the mean and covariance of the position of all the pixels at that same greylevel.

5 Conclusions

We have investigated a novel texture segmentation methodology based on a concept of LGA histograms. Initial results show realistic segmentation of full mammographic images and the potential of the developed approach for mammographic risk estimation and the detection of potential mammographic abnormalities has been indicated.

References

1. Astley, S.M.: Computer-based detection and prompting of mammographic abnormalities. British Journal of Radiology 77, S194–S200 (2004)
2. Miller, P.I., Astley, S.M.: Detection of breast asymmetries using anatomical features. International Journal of Pattern Recognition and Artificial Intelligence 7(6), 1461–1476 (1993)
3. Miller, P.I., Astley, S.M.: Classification of breast tissue by texture analysis. Image and Vision Computing 10, 277–282 (1993)
4. Zwiggelaar, R., Denton, E.R.E.: Texture based segmentation. In: Astley, S.M., Brady, M., Rose, C., Zwiggelaar, R. (eds.) IWDM 2006. LNCS, vol. 4046, pp. 433–440. Springer, Heidelberg (2006)
5. Petroudi, S., Brady, M.: Breast density segmentation using texture. In: 8th International Workshop on Digital Mammography, pp. 609–615 (2006)
6. Oliver, A., Freixenet, J., Martí, R., Pont, J., Pérez, E., Denton, E.R.E., Zwiggelaar, R.: A novel breast tissue density classification framework. IEEE Transactions on Information Technology in BioMedicine 12, 55–65 (2008)
7. He, W., Muhimmah, I., Denton, E.R.E., Zwiggelaar, R.: Mammographic segmentation based on texture modelling of tabar mammographic building blocks. In: Krupinski, E.A. (ed.) IWDM 2008. LNCS, vol. 5116, pp. 17–24. Springer, Heidelberg (2008)
8. Yaffe, M.J.: Mammographic density - measurement of mammographic density. Breast Cancer Research 10 (2008)
9. Subashini, T.S., Ramalingam, V., Palanivel, S.: Automated assessment of breast tissue density in digital mammograms. Computer Vision and Image Understanding 114, 33–43 (2010)

10. Zwiggelaar, R., Parr, T.C., Schumm, J.E., Hutt, I.W., Astley, S.M., Taylor, C.J., Boggis, C.R.M.: Model-based detection of spiculated lesions in mammograms. Medical Image Analysis 3(1), 39–62 (1999)
11. Wolfe, J.N.: Risk for breast cancer development determined by mammographic parenchymal pattern. Cancer 37(5), 2486–2492 (1976)
12. Byng, J.W., Yaffe, M.J., Lockwood, G.A., Little, L.E., Tritchler, D.L., Boyd, N.F.: Automated analysis of mammographic densities and breast carcinoma risk. Cancer 80(1), 66–74 (1997)
13. Gram, I.T., Funkhouser, E., Tabar, L.: The tabar classification of mammographic parenchymal patterns. European Journal of Radiology 24(2), 131–136 (1997)
14. Suckling, J., Parker, J., Dance, D., Astley, S., Hutt, I., Boggis, C., Ricketts, I., Stamatakis, E., Cerneaz, N., Kok, S., Taylor, P., Betal, D., Savage, J.: The mammographic images analysis society digital mammogram database. In: Gale, D., Astley, Cairns (eds.) Digital Mammography, pp. 375–378. Elsevier, Amsterdam (1994)
15. Varma, M., Zisserman, A.: A statistical approach to material classification using image patch exemplars. IEEE Transactions on Pattern Analysis and Machine Intelligence 31, 2032–2047 (2009)
16. Pietikainen, M.: Image analysis with local binary patterns. In: Kalviainen, H., Parkkinen, J., Kaarna, A. (eds.) SCIA 2005. LNCS, vol. 3540, pp. 115–118. Springer, Heidelberg (2005)
17. Ojala, T., Pietikainen, M., Maenpa, T.: Multiresolution gray-scale and rotation invariant texture classification with local binary patterns. IEEE Transactions on Pattern Analysis and Machine Intelligence 24, 971–987 (2002)
18. Zhang, H., Gao, W., Chen, X., Zhao, D.: Object detection using spatial histogram features. Image and Vision Computing 24, 327–341 (2006)
19. Birchfield, S.T., Rangarajan, S.: Spatiograms versus histograms for region-based tracking. In: Proceedings of the IEEE Conference on Computer Vision and Pattern Recognition (CVPR), pp. 1158–1163 (2005)

Quantification of Vascular Calcifications on Digitized Mammograms

Enrique Nava[1], Irene Barba[1], Francisco Sendra[1], and Celestino Gómez–Rebollo[2]

[1] Laboratory of Digital Radiology, University of Málaga, Campus de Teatinos s/n, Málaga, Spain
en@uma.es
[2] Hospital Universitario Virgen de la Victoria, Málaga, Spain

Abstract. In this paper, it's presented an automatic quantification computer algorithm of breast arterial calcifications in digital mammograms, using image processing techniques, to be used as a quantitative cardiovascular risk or diabetes indicator for women. The proposed image processing algorithm is composed of two main steps: a detection phase, using a combination of the line operator method with edge detection, and a segmentation phase which use a thresholding technique to obtain seeds for a region growing algorithm. Obtained results shows a good agreement with ground truth images of calcified vessels which have been hand–segmented by an experienced radiologist with a high degree of detail and quality .

1 Background

Breast arterial calcifications (BAC) is a consequence of calcium deposition along the inner breast vessel lumen and are a common finding in screening mammograms, with a significant prevalence (7-25%) that increases with age [1]. The relationship between BAC and age is agreed by all researchers in our knowledge. Usually, those findings are often unreported because they are not related with breast cancer.

Some studies have reported the association of BAC seen in mammography with diabetes [2,5,6,7], hypertension [1], coronary artery disease [3,4,5,6,8,9], retinopathy [10], and osteoporosis [11] and they have found a statistical significant association, so screening mammography can be used to identify women with higher risk of these diseases, or as another useful disease risk indicator. Other studies suggest that presence of BAC are not a useful indicator of diabetes [12] or cardiovascular risk [13,14,15]. The relationship between BAC and pregnancy and lactation has been also reported by Maas et al [13], hypothesizing that hormonal changes during pregnancy are important in the deposition of calcium in breast arteries.

In most of the papers, the indicator is the presence or absence of arterial calcification, and, in some cases, the location and severity scored by radiologists are also included. In a recent study, the number of BAC and the longest continuous segment is used as a quantitative measure of severity of arterial calcifications.

J. Martí et al. (Eds.): IWDM 2010, LNCS 6136, pp. 183–190, 2010.

The use of digital image processing techniques can be useful to obtain better quantitative measures of calcium deposition, but to our knowledge, only one study has been reported [16]. Better quantitative measures of calcium deposition can be very useful to improve the discussion about the relationship between BAC and those diseases. Additionally, digital image processing techniques can be used to analyze digital mammograms without any time loss for radiologists, because they can be performed automatically by computers.

The aim of this paper is to present a novel automatic computer algorithm which allows the quantification of calcium deposition on vessels and can be used to estimate the quantity and severity of breast arterial calcifications. This algorithm performs a detection and segmentation of BAC on the mammogram so it can be obtained a collection of BAC indicators, as the number of pixels with high probability of calcification, the number and length of calcified vessels, etc. Obtained results of segmented images are compared with annotated mammograms.

2 Method

We have used three sets of digital mammograms to evaluate the detection and quantification of breast vascular calcifications

1. Standard database: 20 pairs (both projections) of digitized mammographic images with BAC has been selected from University of South Florida Digital Database for Screening Mammography (DDSM)[17]. This database has been selected because it's publicly available and it can be used as a reference to compare different algorithms from different researchers. These images are normalized using the calibration curve of the digitized scanner to reduce distortion, following a detection and removal of artifact objects in the background of the film (e.g. labels).
2. From this set, a small (five) group of mammograms which exhibits BAC has been hand–segmented by an expert radiologist with a very high degree of accuracy to be used as ground truth images. This small set is used as an aid in the development of the quantification algorithm.
3. Full digital database: 20 pairs (both projections) of full–field digital mammograms (FFDM) with BAC has been selected from Laboratory of Digital Radiology at University of Malaga database. This database is chosen as a target because the increasing use of digital mammography systems will obsolete film mammography in the near future.

Before the detection of vascular calcification, the breast contour is detected and the algorithm only needs to perform computations on pixels located inside the breast region.

2.1 Detection of Calcified Vessels

To extract vascular calcifications, we used a similar approach to Zwiggelaar line operator method [18,19]. For each pixel inside the breast, a line-strength S is computed as

$$\mathbf{S} = \alpha(\mathbf{L} - \mathbf{N}) + (1 - \alpha)(\mathbf{I} - \mathbf{G}) \tag{1}$$

where **L** is the average grey level of the pixels that lay on a line passing through the target pixel in the orientation that produces the maximum value, **N** is the average grey level of the pixels inside a square neighborhood with the same orientation, **I** is the original pixel value, **G** is the grey level pixel value obtained with gaussian filtering with a σ value which matches the size of the square neighborhood used to compute **N** and $\alpha \in [0, 1]$ is a parameter. The line-strength **S** is a weighted average of the line operator method, which is able to obtain higher values for linear structures, and a classical edge detection algorithm, which obtains higher values at vessel edges. The combined detection of line structures and edges is used to improve detection of curve-shape calcified vessels.

2.2 Segmentation of Calcified Vessels

In the previously mentioned papers [18,19], the goal is to obtain a skeletonized description of the linear structures. For our purposes, the goal is the detection of pixels with higher probability of calcification, so thinner or connected structures are not required. Therefore, segmentation is performed with a two step algorithm.

1. A simple thresholding technique is used on the line-strength **S** image to obtain an image with the pixels of the vessels with higher probability of calcification.
2. A region-growing algorithm, using the result of the thresholded image as seeds, to include pixels with high intensity gray level pixel value compared with the neighborhood, connected with seed pixels but with not enough line-strength pixel value to be detected in the first step of the algorithm. Usually, added pixels are located inside the calcified vessels.

3 Results

Figure 1 shows a typical result obtained with the proposed algorithm before the region growing phase. In this case, the threshold value has been selected to obtain a very low of false positive BAC, but it can be shown that most of the BAC are detected. The main difference is that BAC in the ground truth image have bigger area that in the detected ones, because the line strength measure exhibits greater values in the boundaries of individual BAC. By the addition of the second term of the equation 1 to the line operator, we obtain a way to balance this behavior using the α parameter. If α has high values (close to 1) linear structure detection increases, it's obtained high values of **S** at pixels located inside vessels, but if α value is low (close to 0) then calcified contours detection is favored, with higher values of **S** located at the vessel boundary pixels.

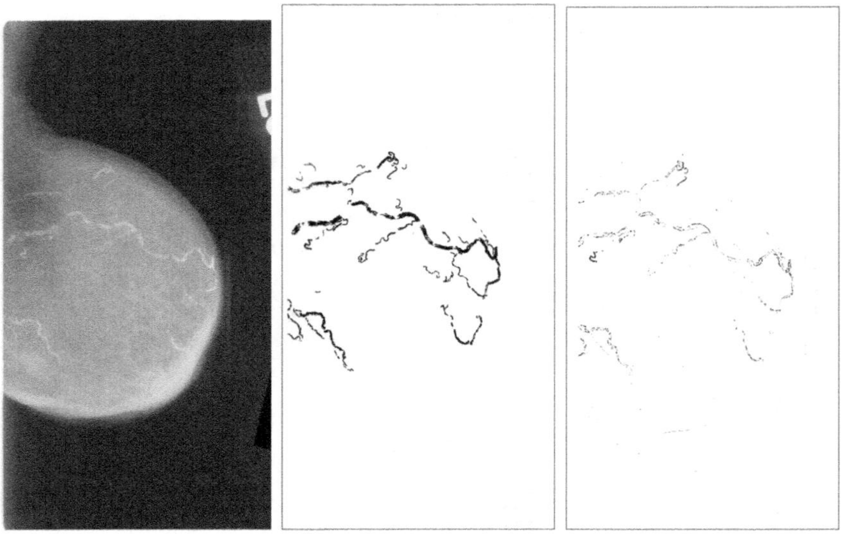

Fig. 1. Segmented BAC with the proposed algorithm. a) Original image (Case 4101, left mammogram, lateral–oblique projection, from USF-DDSM database). b) Ground truth image, segmented by an expert radiographer. c) Obtained result ($\alpha = 0.1$).

Boundary pixels have a higher probability of calcification because the projection of the deposition of calcium in the intima of the vessels produces images with linear or curved parallel lines similar to a railroad, so in our approach we used small α values ($0.05 - 0.20$) to produce this pattern in detection. On the contrary, if we are interested in the deposition of calcium in the media, a higher value of α must be selected.

Figure 2 shows the **S** line strength pixel values in magnified portions of a mammograms which contains arterial calcifications using a value $\alpha = 0.1$ before the thresholding technique. In this case, it can be shown that the operator gives an strong response on the boundaries of the calcified vessels.

In Figure 3, it is shown magnified portions of a BAC where the proposed algorithm illustrates a typical behavior with high positive and negative predictive values (PPV 0.98 and NPV 0.96, respectively in the whole database) and high specificity (0.999) when it's considered a false positive (FP) a pixel that has been detected with the algorithm but the pixel is not included in the ground truth annotated image and it's considered a FN (false negative) when the pixel is plotted in the annotated image but it's not detected by the algorithm. If we consider BAC instead individual pixels, also sensibility is very high (close to 1).

Even if a optimized segmentation algorithm can show a better adjustment with the ground truth images, authors believe that obtained results are better to be used as a quantitative measure of calcium deposition in vessels if they

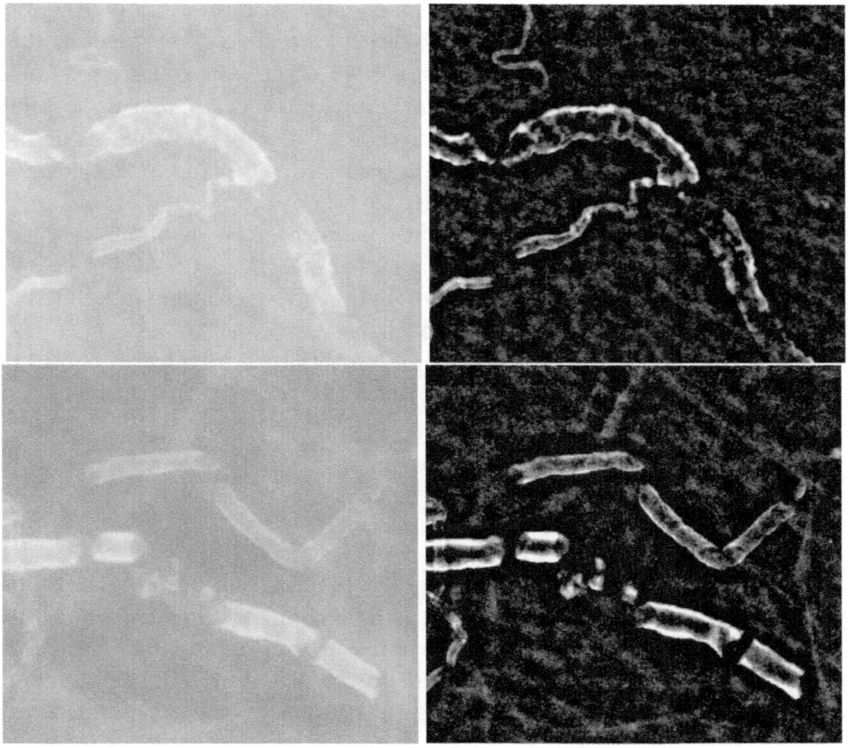

Fig. 2. Typical examples (case 4131 and 3481) of line strength operator ($\alpha = 0.1$). a) Original image. b) Line strength result.

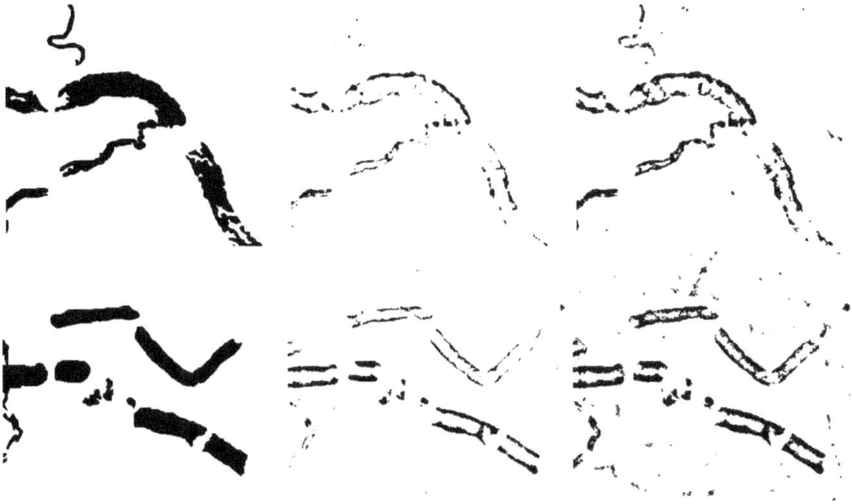

Fig. 3. Typical examples (case 4131 and 3481) of segmentation results for a magnified BAC. a) Ground truth image, segmented by an expert radiographer. b) Obtained result after thresholding. c) Final result after region growing.

don't include the interior part of the calcified vessels, in an attempt to quantify only the calcium deposited in the intima, with a possible exception with highly calcified cases, which exhibits higher grey level pixel values. This argument is also supported by Penugonda et al [15], arguing that a lack of a clear association between BAC seen in mammography and coronary disease may reflect differences in the mechanism of calcium deposition in breast arteries (uniformly located in the media) and coronary arteries (localized to the intima).

The results obtained with the computer algorithm are compared with the small set of hand–segmented BAC images to determine the accuracy of the proposed method. At this point, receiver–operator parameters and ROC curve graphics can be plotted. Finally, using the breast contour and the obtained results, a quantitative measure of calcium deposition in vessels can be obtained as the relation of pixels who exhibits BAC to the total number of pixels inside the breast in the mammogram.

4 Discussion

A automatic computer quantification method to assess cardiovascular risk using screening mammograms has been developed. This method has been tested with digitized images from the USF–DDSM database and compared with ground truth images, segmented with very high degree of detail by an experimented radiologist, and tested with full field digital mammographies. Obtained results shows that BAC are detected and segmented with enough degree of precision to obtain a quantitative measure of calcium deposition.

Cardiovascular diseases as heart stroke or coronary artery disease are one of the leading causes of mortality for women above the age of 45, the same population who is the target of mammographic screening programmes. Screening mammography is widely used and it's considered an excellent and standard method to detect early breast cancer in women with an age range which matches higher coronary disease and diabetes risk. Early diagnosis and therapeutic intervention can significantly reduce the mortality from cardiovascular episodes in women, so if a relationship between with BAC shown in screening mammography can be established, it will very useful.

A hot controversial can be found in the scientific literature about the usefulness of BAC to predict cardiovascular risk. Most of the papers rely their conclusions on the use of very basic indicators of BAC, mostly presence/absence of BAC. The use and research on new quantitative parameters of BAC, as the proposed measure of calcium deposition, can be very useful to elucidate this question.

An automatic detection of BAC and the quantification of calcium deposition can be performed on the computer easily with FFDM images without an increase of the radiologist reading time and can offer an indicator of cardiovascular risk and other diseases risk or can be valuable to improve the discussion about the usefulness of BAC using quantitative improved indicators. Computer aided diagnosis is now widely accepted and used and the proposed quantification measure can be easily implemented and included in the software.

A prospective study is underway at our laboratory to assess the utility of the proposed quantification indicator on screening mammography in clinical practice.

References

1. Çetin, M., Çetin, R., TamerDale, N.: Prevalence of breast arterial calcification in hypertensive patients. Clinical Radiology 59, 92–95 (2004)
2. Baum, J.K., Comstock, C.H., Joseph, L.: Intramammary arterial calcifications associated with diabetes. Radiology 136, 61–62 (1980)
3. Moshyedi, A.C., Puthawala, A.H., Kurland, R.J., O'Leary, D.H.: Breast arterial calcifications: association with coronary artery disease. Radiology 194, 181–183 (1995)
4. Kemmeren, J.M., Beijerinck, D., van Noord, P.A.H., Banga, J.-D., Deurenberg, J.J.M., Pameijer, F.A., van der Graaf, Y.: Breast arterial calcifications: association with diabetes mellitus and cardiovascular mortality. Radiology 201, 75–78 (1996)
5. Adkins, J.R., Gamblin, T.C., Christie, D.B., Collings, C., Dalton, M.L., Dale, P.S.: Mammography as screening for coronary artery disease. Am. Surg. 73(7), 717–721 (2007)
6. Dale, P.S., Richards, M., Mackie, G.C.: Vascular calcifications on screening mammography identify women with increased risk of coronary artery disease and diabetes. Am. J. Surg. 196(4), 537–540 (2008)
7. Dale, P.S., Mascarenhas, C.R., Richards, M., Mackie, G.C.: Mammography as a screening tool for diabetes. J. Surgical. Research. 159, 528–531 (2008)
8. Iribarren, C., Go, A.S., Tolstykh, I., Sidney, S., Johnston, S.C., Spring, D.B.: Breast vascular calcification and risk of coronary heart disease, stroke, and heart failure. J. Women Health 13, 381–389 (2004)
9. Topal, U., Kaderli, A., Topal, N.B., Ozdemir, B., Yesilburşa, D., Cordan, J., Ediz, B., Aydinlar, A.: Relationship between the arterial calcification detected in mammography and coronary artery disease. Eur. J. Radiology 63, 391–395 (2007)
10. Calisir, C., Yavas, U.S., Erol, N.: The Relationship between the Breast Arterial Calcification Detected by Mammography and the Hypertensive Retinopathy in Hypertensive Women. Korean J. Radiology 9, 320–324 (2008)
11. Reddy, J., Bilezikian, J.P., J, Smith, S.J., Mosca, L.: Reduced Bone Mineral Density Is Associated with Breast Arterial Calcification. J. Clin. Endocrinol. Metab. 93, 208–211 (2008)
12. Sickles, E.A., Galvin, H.B.: Breast arterial calcification in association with diabetes mellitus: Too weak a correlation to have clinical utility. Radiology 155, 577–579 (1985)
13. Maas, A.H.E.M., van der Schouw, Y.T., Beijerinck, D., Deurenberg, J.J.M., Mali, W.P.T.M., van der Graaf, Y.: Arterial calcifications seen on mammograms: Cardiovascular risk factors, pregnancy, and lactation. Radiology 240(1), 33–38 (2006)
14. Zgheib, M.H., Buchbinder, S.S., Rafeh, N.A., Elya, M., Raia, C., Ahern, K., Smith, M.C., Costantino, T., Flory, M.J., Lafferty, J.C., Castellanos, M.R.: Breast arterial calcifications on mammograms do not predict coronary heart disease at coronary angiography. Radiology 254(2), 367–372 (2010)
15. Penugonda, N., Billecke, S.S., Yerkey, M.W., Rebner, M., Marcovitz, P.A.: Usefulness of breast arterial calcium detected on mammography for predicting coronary artery disease or cardiovascular events in women with angina pectoris and/or positive stress test. Am. J. Cardiology 105, 359–361 (2010)

16. Cheng, J.-Z., Cole, E.B., Pisano, E.D., Shen, D.: Detection of arterial calcification in mammograms by random walks. In: Prince, J.L., Pham, D.L., Myers, K.J. (eds.) Information Processing in Medical Imaging. LNCS, vol. 5636, pp. 713–724. Springer, Heidelberg (2009)
17. Heath, M., Bowyer, K., Kopans, D., Moore, R., Kegelmeyer, W.P.: The Digital Database for Screening Mammography. In: Yaffe, M.J. (ed.) Proceedings of the Fifth International Workshop on Digital Mammography, pp. 212–218. Medical Physics Publishing (2001)
18. Hadley, E.M., Denton, E.R.E., Zwiggelaar, R.: Mammographic Risk Assessment Based on Anatomical Linear Structures. In: Astley, S.M., Brady, M., Rose, C., Zwiggelaar, R. (eds.) IWDM 2006. LNCS, vol. 4046, pp. 626–633. Springer, Heidelberg (2006)
19. Zwiggelaar, R., Astley, S.M., Boggis, R.M., Taylor, C.J.: Linear Structures in Mammographic Images: Detection and Classification. IEEE Tr. Med. Imaging 23(9), 1077–1086 (2004)

Comparison of Tilt Correction Methods in Full Field Digital Mammograms

Michiel Kallenberg and Nico Karssemeijer

Radboud University Nijmegen Medical Centre, Department of Radiology,
Geert Grooteplein Zuid 18, 6525 GA Nijmegen, The Netherlands
m.kallenberg@rad.umcn.nl

Abstract. During the acquisition of a mammogram the breast is compressed between the compression paddle and the support table. When compression is applied the upper plate is tilted which results in variation in breast thickness from the chest wall to the breast margin. Variation in breast thickness influences the grey level values of the image and hampers image analysis, such as volumetric breast density estimation. In this paper we present and compare two methods that estimate and correct image tilt. The first method estimates tilt from fatty tissue regions. The second method is based on the entropy of the grey level distribution of the image. 1876 images are obtained from relatively young women with a high breast density on average. The tilt correction methods are evaluated by assessing their accuracies in estimating artificial tilts that are added to the images that are expected to have only a small tilt. On average both methods are able to estimate the artificial tilt, although the accuracy is relatively low. To the best of our knowledge this is the first paper that presents and validate tilt correction methods on individual mammograms. We expect that results will be better in screening populations which forms the majority of cases utilised in image analysis.

Keywords: tilt correction, full field digital mammography, breast density.

1 Background

Mammograms are obtained by compressing the breast between two plates of imaging radiation transparent material, and taking an image of the compressed breast tissue. Due to the forces that are applied during compression, the upper plate, the compression paddle, is subject to deformation. This deformation may lead to variation of the breast thickness up to 2 cm from the chest wall to the breast margin. It is seen in almost all mammography systems. [1,2,3].

Variation in breast thickness affects image analysis by its impact on the pixel values. In addition, the knowledge of breast thickness variations is essential in volumetric breast density estimation. It would therefore be of great value if compression paddle tilt can be estimated to accurately correct mammograms.

At our institute we have collected a database of over 100,000 mammogram images to study the relation between breast density and breast cancer. However,

J. Martí et al. (Eds.): IWDM 2010, LNCS 6136, pp. 191–196, 2010.

the images were recorded with a Hologic Selenia FFDM system that contained a paddle that is designed to bend during compression. In order to accurately estimate the volumetric breast density of the images in our database the effect of the paddle tilt needs to be removed.

In this work we present and compare two methods that estimate and correct the tilt on an image based level.

2 Method

2.1 Estimating Tilt From Fatty Tissue Regions

The first tilt correction method is based on estimating the breast thickness along the x direction of the mammogram by measuring the distribution of pixel values of fatty tissue as a function of x, where x runs from the chest wall to the nipple side.

For pure fatty tissue the following equations hold:

$$I_f = I_0 e^{-\mu_{f,eff} h} \tag{1}$$

with I_0 the X ray exposure of the incident beam, h the breast thickness, $\mu_{f,eff}$ the effective attenuation coefficient for fat tissue, and I_f the exposure at the detector of an X ray beam that is exclusively attenuated by fatty tissue.

In raw FFDM images pixel values are proportional to exposure. If we log-transform the image we obtain

$$\begin{aligned} y_f &= \ln g I_f = \ln g + \ln I_0 - \mu_{f,eff} h \\ \frac{dy_f}{dh} &= -\mu_{f,eff} \end{aligned} \tag{2}$$

with g the gain.

Previous research [1,2] showed that the thickness of the breast can be modelled by a linear function of x

$$\begin{aligned} h &= t * x + b \\ t &= \frac{dh}{dx} \end{aligned} \tag{3}$$

in which t is the tilt. (see fig 1)

Combining eq. 2 and eq. 3 yields

$$t = \frac{dy_f}{dx} * \frac{1}{-\mu_{f,eff}} \tag{4}$$

The tilt angle (a) between the compression paddle and the support table can be computed by

$$a = \arctan(\frac{-dy}{dx}) = \arctan(-t) = \arctan(\frac{dy_f}{dx} * \frac{1}{\mu_{f,eff}}) \tag{5}$$

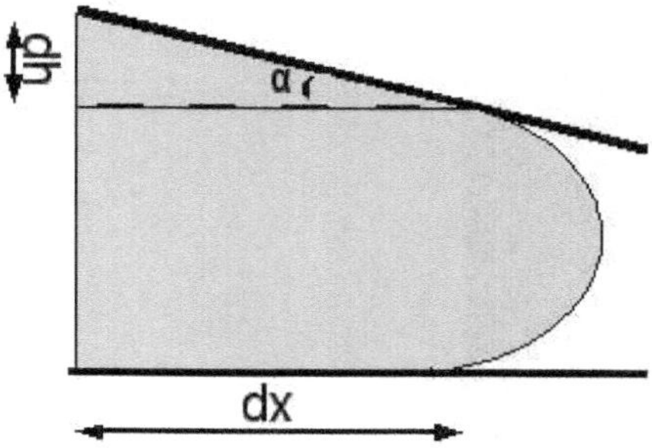

Fig. 1. Schematic representation of the tilt

We try to estimate $\frac{dy_f}{dx}$ by evaluating the histogram of pixel values in fatty tissue regions as a function of x. For each column of 0.2mm width in the image we estimate y_f by computing the 5th percentile of the log transformed pixel values. Next $\frac{dy_f}{dx}$ is obtained by fitting a regression line through (x, y_f), after which the tilt tangle is calculated according to equation 5. (see fig 2)

As in the periphery the breast is not fully compressed, the computation of y_f is restricted to the interior of the breast. In addition, dense columns are excluded as in dense areas a reliable value of a fatty pixel is hard to obtain. The density of a column is assessed by observing the standard deviation of the pixel values in that column. If the standard deviation exceeds a threshold the column is not included in the estimation of the tilt angle. Images that do not have at least 10 fatty columns (i.e. 2mm) are rejected.

2.2 Estimating Tilt by Minimising Entropy

The second tilt correction method is based on the entropy of the grey level distribution of the image. In the histogram of a tilt corrected image the peaks of the fatty and dense tissue are relatively narrow, which is reflected in the entropy computed by

$$H(X) = -\sum_{i=1}^{n} p(x_i) \log_2 p(x_i) \tag{6}$$

with $H(X)$ the entropy of the histogram, $p(x_i)$ the probability that a pixel has a grey level x in bin i, and n the number of bins.

To estimate the tilt of an image, H is computed as a function of (a). For a given value of a H is computed by correcting the image using equations 1-5 and subsequently applying equation 6 as is depicted in figure 3. The tilt that gives the minimum entropy is taken as the estimated tilt of the image.

2.3 Validation

The tilt correction methods are evaluated on images recorded with a GE Senograph 2000D. As this system contains a paddle that is relatively rigid, the images

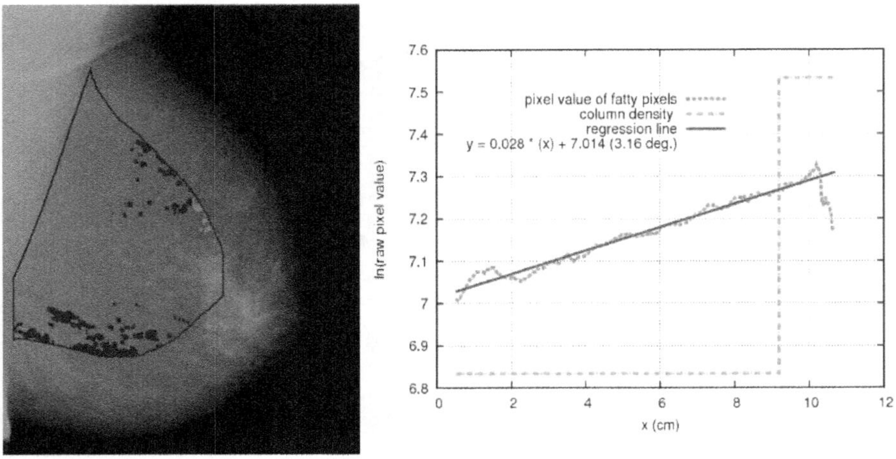

Fig. 2. In the interior of the breast pixel values of fatty pixels are measured. The pixels marked black in the left image are fatty pixels from fatty columns. The pixels marked grey are fatty pixels from dense columns. The right graph shows the fit of the regression line. An artificial tilt of 3.0 deg was added to the image. The tilt that is estimated (3.16 deg.) is close to the artificial tilt of 3.0 deg.

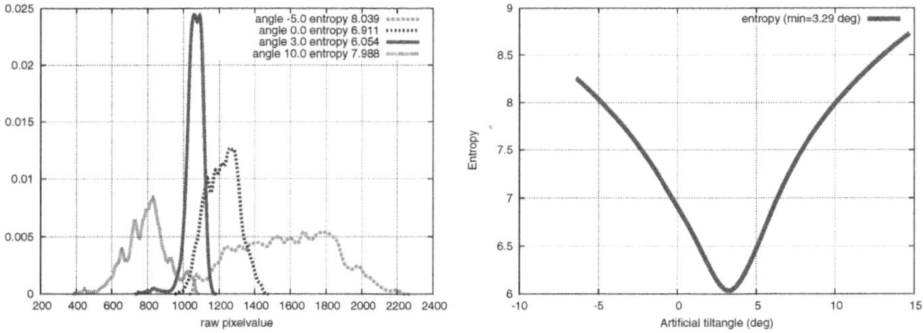

Fig. 3. Grey level distributions and entropy for a series of potential tilts for the mammogram shown in fig. 2. The minimal entropy is found at a tilt of 3.29 deg. which is close to the artificial tilt of 3.0 deg. that was added to the image.

are expected to have only a small tilt with a maximum of 1.0 deg. [1,2]. By adding an artificial tilt to each image according to the tilt model described above we can obtain a series of images with known tilt angles. Consequently we can compare the proposed methods by assessing their accuracies in the estimation of the artificial tilt.

2.4 Database

The database used in this study contained 1876 images of 553 women with a mean age of 51.5 (± 11.1). The BI-RADS density of each image was determined automatically using [4], yielding a mean BI-RADS density of 2.75. The images were acquired on a GE Senograph 2000D using standard clinical settings, including the use of an anti-scatter grid. The tilts that were simulated were homogeneously distributed in the range of 2.8 to 8.5 deg. which is the range that is observed in clinical practice with Hologic.

3 Results

Figure 4 shows the relation between the artificial tilt and the estimated tilt for both tilt correction methods. It can be seen that on average both methods are able to estimate the artificial tilt, although the tilt is slightly underestimated. Accuracy is relatively low as is indicated by the the large inter quartile ranges. Tilt correction method 1 rejected 27,5% of the images because it could not find enough fatty columns in the breast.

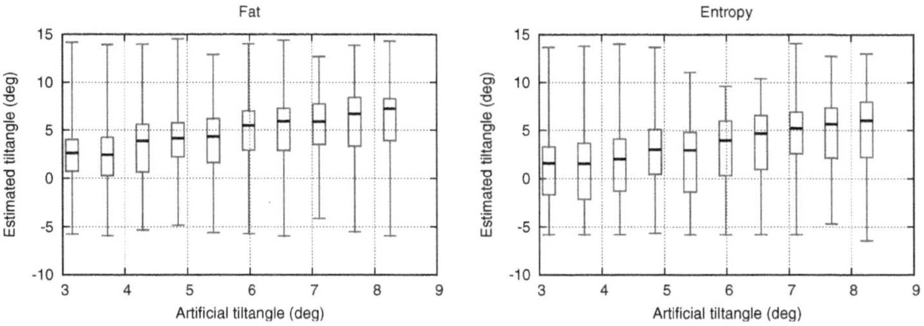

Fig. 4. Box plots for both tilt correction methods, showing the relation between the artificial tilt and the tilt that is estimated. The solid lines represent the median. It can be seen that both methods are able to estimate the tilt to some extent. Accuracy, however, is relatively low.

4 Discussion

To the best of our knowledge this is the first paper that presents and validate tilt correction methods on individual mammograms. Results showed that accuracy of the developed methods is still relatively low. Part of this low accuracy may be attributed to the fact that the GE images do have a small tilt themselves.

The images were obtained from relatively young women with dense breasts, which hampers especially the first tilt correction method. It is expected that results will be better for mammograms from screening, which forms the majority of the cases we want to study.

Acknowledgments. This work was supported by the Dutch Cancer Society, Grant UU 2008-4071.

References

1. Tyson, A.H., Mawdsley, G.E., Yaffe, M.J.: Measurement of compressed breast thickness by optical stereoscopic photogrammetry. Med. Phys. 36(2), 569–576 (2009)
2. Mawdsley, G.E., Tyson, A.H., Peressotti, C.L., Jong, R.A., Yaffe, M.J.: Accurate estimation of compressed breast thickness in mammography. Med. Phys. 36(2), 577–586 (2009)
3. Diffey, J., Hufton, A., Beeston, C., Smith, J., Marchant, T., Astley, S.: Quantifying Breast Thickness for Density Measurement. In: Krupinski, E.A. (ed.) IWDM 2008. LNCS, vol. 5116, pp. 651–658. Springer, Heidelberg (2008)
4. Karssemeijer, N.: Automated classification of parenchymal patterns in mammograms. Phys. Med. Biol. 43(2), 365–378 (1998)

Optimising Beam Quality Selection in Mammographic Acquisition Using the Standard Attenuation Rate

Christopher E. Tromans[1], Jennifer Diffey[2], Mary Cocker[3], and Sir Michael Brady[1]

[1] Wolfson Medical Vision Laboratory, Department of Engineering Science, University of Oxford, Parks Road, Oxford, UK, OX1 3PJ
[2] Imaging Science and Biomedical Engineering, School of Cancer and Imaging Sciences, Stopford Building, University of Manchester, Oxford Road, Manchester, UK, M13 9PT
[3] Radiation Physics and Protection, Oxford Radcliffe Hospitals NHS Trust, Churchill Hospital, Oxford, UK, OX3 7LJ
cet@robots.ox.ac.uk

Abstract. A common metric used to optimise digital mammography image acquisition is contrast-to-noise ratio. Using the standard attenuation rate (SAR), a quantitative normalised representation of breast tissue for image analysis applications, we demonstrate that the image contrast may be completely separated from the acquisition parameters, in particular the beam quality, used for acquisition. Optimising the contrast-to-noise ratio at acquisition is therefore suboptimal, since the contrast may be manipulated by post processing. A tissue equivalent phantom is used to investigate the variation in both signal-to-noise ratio, and image sharpness within the SAR images. The results show that the primary effect of varying the acquisition parameters through the various automated optimisation of parameter modes, and hence the mean glandular dose, is to vary the global contrast of the acquired image, an effect successfully mapped to a common normalised basis using the SAR. The signal-to-noise ratio and image sharpness are second order effects, and are therefore dominated by the global image contrast when image acquisition is optimised using the contrast-to-noise ratio.

Keywords: image acquisition optimisation, quantitative mammography.

1 Introduction

A key benefit of digital mammography is the decoupling of the x-ray image formation and detection process, from that of image display. In practice however, this benefit is seldom fully realised in current clinical workflows. Pisano et al [1] presented a pictorial essay describing the various algorithms employed for the "manipulation of fine differences in image contrast". Such algorithms generally involve generic image enhancement techniques, such as contrast-limited adaptive histogram equalisation and unsharp masking, whose performance leads the paper to conclude "different digital image processing algorithms are likely to be useful for different tasks". In this work we propose a different approach to image processing, namely to use a quantitative, normalised representation of breast tissue, which we call the standard attenuation rate

J. Martí et al. (Eds.): IWDM 2010, LNCS 6136, pp. 197–204, 2010.

(SAR) [2] in which radiodensity is measured at each pixel. In simple terms, SAR provides the sum of the breast composition attenuation values along the path to a given pixel location, including the sums of all forms of adipose tissue, fibroglandular tissues, and calcifications. Its value is smaller if the path contains more adipose tissue, larger for dense tissues, and much larger for calcifications. The SAR thereby removes the dependency upon the acquisition conditions and results in an image that depends solely upon the underlying tissue composition, so, regardless of the beam quality from which the mammogram was acquired, global contrast depends only on the tissues of the breast. One may think of this as being similar to the Hounsfield unit used universally in CT, though specifically optimised for mammography. Computation utilises a model of the physics of image formation to quantify the difference between the primary x-ray photon fluence incident upon the breast and the primary x-ray photon fluence exiting the breast and subsequently recorded by the image detector. The difference between these fluences, which is a result of the radiodensity of the constituent tissues, is expressed per unit distance traversed by the primary beam through the breast relative to a reference material, which, in this work, is a 50/50 mixture of adipose/fibroglandular tissue (though the method can be straightforwardly adapted to any other measure). The physics model includes: photon production within the x-ray tube given the acquisition configuration of the mammography unit; an empirical detector calibration model to calculate the photon fluence exiting the breast; and a scatter model which calculates an estimate of the component of the image signal from scattered photons, enabling the primary component to be estimated.

In this paper, we present an analysis of image quality, more specifically image sharpness, and noise, traded against mean glandular patient dose for a range of image acquisition conditions, using the SAR image.

Much of the work in the literature makes use of the contrast-to-noise ratio (CNR) to quantify image quality. For example, a study by Young et al [3], used a phantom constructed of PMMA (background) and aluminium (foreground), and defined a metric of image quality (which they state "has been adopted in European and UK Guidance on testing AECs") around the CNR, which they applied to the "unprocessed" images acquired on a GE Senographe 2000D. It is worth noting that the pixel intensity in such images is linearly related to photon fluence, and exponentially related to tissue composition (see [2]). The specific measure they adopted is:

$$CNR = \frac{mean(bgd) - mean(Al)}{\sqrt{\frac{[sd(bgd)^2 + sd(Al)^2]}{2}}}$$

Since digital image processing, such as the SAR, can remove the dependency of image contrast on beam quality, and contrast in the SAR may be linearly scaled at display time by the user through the use of window and level controls, or rescaled using a non-linear function should the need arise, the use of a quality metric which includes a contrast measure in the raw acquisition seems sub-optimal, especially since global contrast may be manipulated by algorithmic post processing. Ideally, acquisition is optimised for that which cannot be corrected by signal processing. The noise measure in the denominator is one such property over which we cannot apply post processing correction, and so it should be minimised at image acquisition time, since stochastic noise intrinsically cannot be corrected for algorithmically (since, by definition, no technique exists for identifying the combination of random events which occurred during any given exposure: filters applied to signals containing stochastic noise will have equal effect on both noise and signal). Young et al make the implicit assumption that the MTF of the imaging system is not significantly affected by varying beam quality, an assumption they state may need to be verified for some designs, and one which we investigate in this paper. The MTF describes the frequency response of the imaging system, and whilst it is the low frequency components that describe the underlying foundation of the image, the fine details, such as lesion spicules or microcalcifications, are a result of the presence of high frequencies. It follows that a sharp image containing the fine detail required for clinical diagnosis results from a MTF in which the high frequencies have a high contrast relative to low frequencies: the more extended the frequency response, the finer the detail included and the sharper the image.

2 Materials and Methods

A series of images were acquired of a tissue equivalent phantom, which we designed and constructed. The phantom is 100mm by 100mm, 60mm thick, and consists of a pair of interlocking step wedges (with steps varying in height between 1 and 3mm), one of glandular, the other of adipose equivalent material (supplied by CIRS), with a further two adipose wedges forming the sides (pictured in our companion conference paper [2], together with an experimental study using the phantom to verify the performance of the SAR, specifically its ability to derive an image in which contrast is independent of beam quality). The phantom includes adipose/fibroglandular compositions ranging between 12% and 83%.

Phantom images were acquired using a GE Senographe Essential, and in order that the acquisition parameters adopted were as close as possible to those which would be employed clinically, the built-in automatic exposure control was used. This has three selectable modes of operation for the automated optimisation of parameters (AOP): standard (STD), contrast (CNT) and dose (DOSE). The DOSE mode attempts to limit the glandular dose delivered to the patient, whilst the CNT mode attempts to maximise the global contrast in the acquired image (rather than by software post-processing) at the expense of patient dose. An image was acquired in each mode, and it is upon these three images that the SAR image was computed. The mean glandular dose was calculated for each image using the technique proposed by Dance et al [4-5]. The tube output and half value layer measurements required by the calculation were taken from the routine quality assurance reports undertaken by the NHS radiation protection physicists, and the "breast" composition was taken to be 50/50.

Subtraction images of both the raw "FOR PROCESSING" and the SAR images were computed giving the percentage difference between the DOSE and STD, and CNT and STD images so as to assess the overall effect of the variation in acquisition parameters resulting from the three AOP modes.

The noise in the SAR images was measured by computing the standard deviation in a 42 by 58 square of pixels within each step of the phantom. It should be stressed that pixel intensity in the SAR image is independent of the acquisition beam quality and exposure (within the bounds of modelling and measurement error), and dependant only on the underlying tissue composition, and since the same phantom is used in all images remains constant, and so the standard deviation alone may be used to assess image noise (which is not the case in the raw acquired image).

When developing a metric to assess image sharpness, one should note the subtle distinction between contrast (the magnitude of the difference between two location) and image sharpness (a measure of the distinctness of a features edges). In a noise-free x-ray image of a fine detail arising from a discontinuity in composition, the contrast will depend on the exact location of the two points chosen to measure contrast between, and the resulting variation in contrast between slightly varying locations of the two measurement points, will be governed by the image sharpness. In a technique analogous to anisotropic diffusion, image sharpness is measured by plotting local contrast between two points either side of, and equidistant from, the centre of an image feature, against the distance between the points. The shape of the resulting plot describes image sharpness. Fig. 1 illustrates the method, where the sharpness of the two signals on the left is plotted on the right, and in which each point describes the contrast between the points a distance 'x' either side of the centre of the discontinuity. Measuring image sharpness in the spatial domain in this way allows the MTF of the image system to be indirectly quantified at any image point, in this case for a tissue equivalent phantom and thus the measure of the system response closely corresponds to those found during clinical examination.

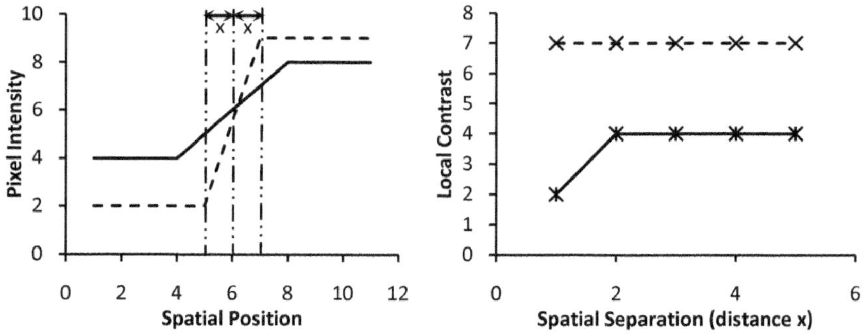

Fig. 1. The computation of the image sharpness metric

3 Results

For each of the three AOP modes, the acquisition parameters selected were: 31kVp Rh-Rh 74mAs, 31kVp Rh-Rh 57.8mAs and 30kVp Rh-Rh 126.8mAs for STD, DOSE and CNT respectively. The resulting mean glandular doses were 1.757mGy, 1.372mGy and 2.656mGy. Subtraction images quantifying the percentage difference between the raw images are shown in fig. 2 and the SAR images in fig. 3. Note the variation in the scales of the "jet" colour map between the various images, depicted by the scale bars immediately to the right of each plot which show the percentage difference corresponding to each colour.

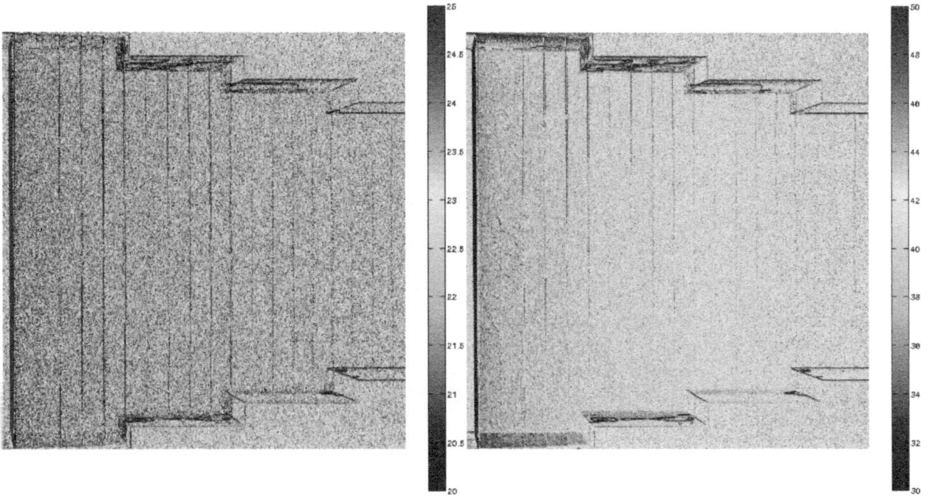

Fig. 2. Subtraction images showing the percentage difference between raw "FOR PROCESSING" images, DOSE-STD (left) and CNT-STD (right)

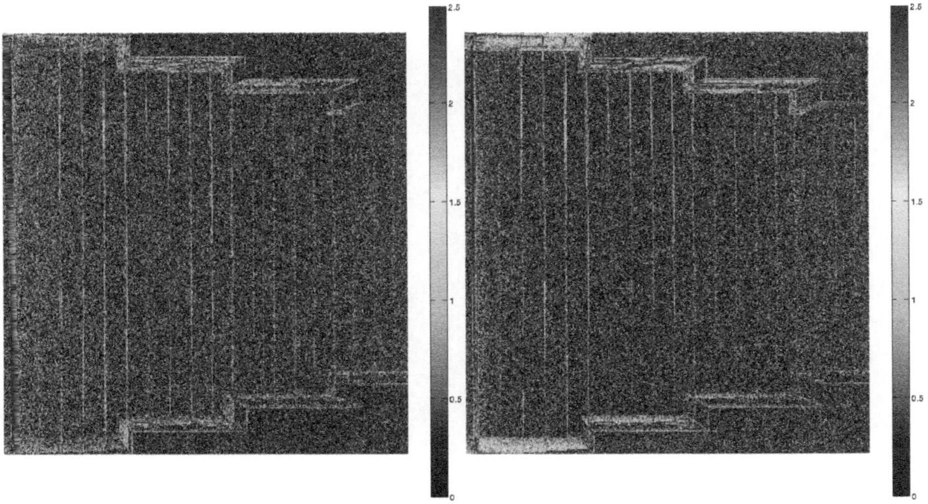

Fig. 3. Subtraction images showing the percentage difference between SAR images, DOSE-STD (left) and CNT-STD (right)

The variation in the noise between the SAR images is shown in fig. 4, and the variation in image sharpness is shown in fig. 5.

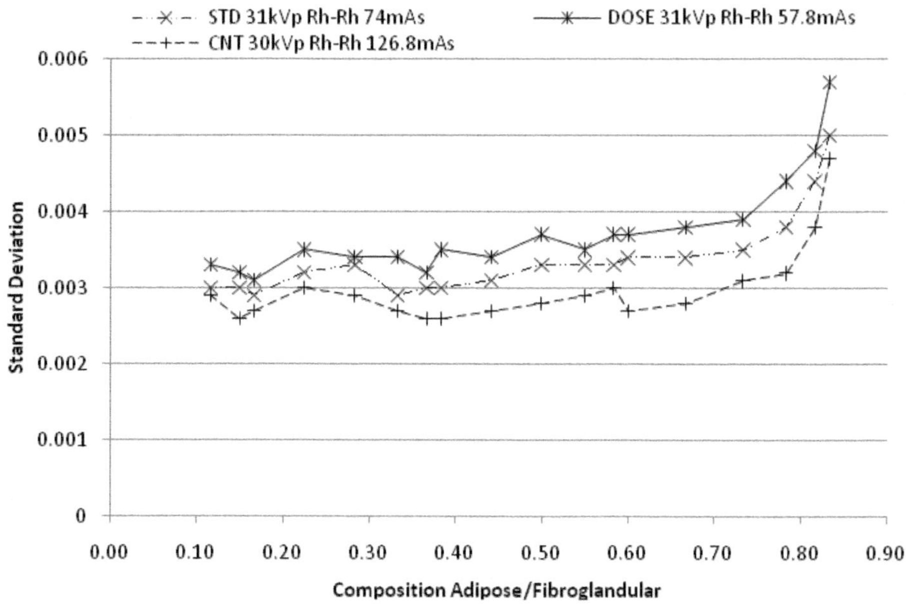

Fig. 4. The variation with composition of the image noise in the SAR images measured by computing the standard deviation

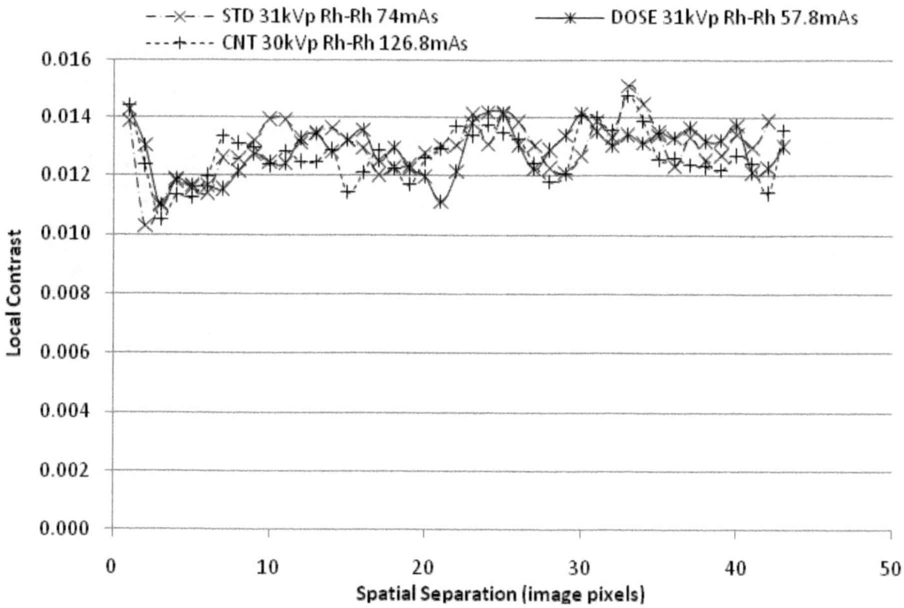

Fig. 5. The local image contrast measure assessing image sharpness computed at across the discontinuity arising over a step at the centre of the step wedge

4 Discussion

The variation in global image contrast arising within the exponential relationship between composition and pixel intensity occurring as a result of the decrease in tube accelerating potential from 31 to 30kVp between the STD and CNT modes is noticeable from the subtraction images on the right of fig. 2 where the percentage difference rises from the high thirties in the highly dense fibroglandular left hand side of the phantom, to the low forties in the mostly adipose material on the right. In the case of the comparison of the differences between DOSE and STD on the left of fig. 2, where the beam quality has remained constant and only the exposure has varied between 57.8 and 74mAs, as one would expect no global contrast variation exists between adipose and fibroglandular, however the dominant effect which may be observed is the increase in quantum noise resulting from the reduction in exposure.

As expected, the high similarity between the normalised SAR images may be seen in fig. 3, where the vast majority of pixels fall into the range of the colour map which only encompasses 0 to 2.5%. The variation in noise may be observed between the various doses, though these generally result in differences confined to below 1%. The largest discrepancies visible occur at the discontinuities between steps, though given the display range finishing at 2.5%, these are only minor.

The noise is quantified using the standard deviation in fig. 4, where the expected relationship due to quantum noise between increased mean glandular dose, and decreased standard deviation/image noise may be observed, though it should be noted these are only small given the range of 0 to 2.5%.

The image sharpness measured via local contrast depicted in fig. 5 shows a very high degree of similarity between the three acquisition conditions images, suggesting that the assumption that the MTF of the imaging system is not significantly affected by varying beam quality made by Young et al [3] is valid.

The British National Health Service remedial dose level for a 60mm thick breast, the thickness of the phantom, is 3.0mGy, and so it may (happily) be concluded that the equipment is performing satisfactorily.

The dose between STD and DOSE mode dropped by 21.9%, whilst the average standard deviation across all the steps of the phantom rose by 10.5% (equivalent to a step in the fibroglandular wedge within the phantom of 0.048mm). Similarly, the dose between CNT and STD mode rose by 33.9%, whilst the average standard deviation dropped 11.68% (equivalent to a step in the fibroglandular wedge within the phantom of 0.064mm). A qualitative visual evaluation between the SAR images revealed very little difference, as is to be expected from the difference image falling almost entirely between 0 and 2.5%. Noise and image sharpness would appear to be second order effects compared to global image contrast, however since this may be algorithmically manipulated, as demonstrated by the SAR computation, the use of 33.9% extra dose to achieve the same primary effect at image acquisition is unnecessary.

Further work is required to relate this approach to the perceptual characteristics of the images, particularly as viewed by highly experienced radiologists. It may, for example, be possible to ascertain a minimum signal-to-noise ratio required for a given specificity and sensitivity of diagnosis. Optimal patient doses and hence acquisition parameters when using the SAR may then be calculated. Work is ongoing to assess diagnostic performance of SAR using both human readers, and for the case of computer aided diagnosis.

5 Conclusion

The primary effect of varying the acquisition parameters through the operating mode of the AOP algorithm, and hence the mean glandular dose delivered to the patient, is a variation in global contrast of the acquired image. The SAR is seen to successfully map between the global contrast variations resulting from the beam quality variations tested to a common normalised basis in which the image is dependant solely on underlying tissue composition. The signal-to-noise ratio and image sharpness are seen to be only second order effects, and are therefore dominated by the global image contrast when image acquisition is optimised using the contrast-to-noise ratio, suggesting this measure is suboptimal for the case of digital mammography where contrast may be manipulated by algorithmic post processing.

Acknowledgments

Thanks to the radiographers at the Breast Imaging Centre in the Churchill Hospital Oxford for their help in facilitating the validation experiments. Christopher Tromans wishes to thank the EPSRC for funding his research fellowship.

References

[1] Pisano, E.D., et al.: Image processing algorithms for digital mammography: a pictorial essay. Radiographics 20, 1479–1491 (2000)
[2] Tromans, C., Brady, M.: The Standard Attenuation Rate for Quantitative Mammography. In: Proceedings of 10th International Workshop on Digital Mammography (IWDM), Girona, Spain (2010)
[3] Young, K.C., et al.: Optimal beam quality selection in digital mammography. Br. J. Radiol. 79, 981–990 (2006)
[4] Dance, D.R., et al.: Additional factors for the estimation of mean glandular breast dose using the UK mammography dosimetry protocol. Physics in Medicine and Biology 45, 3225 (2000)
[5] Dance, D.R., et al.: Further factors for the estimation of mean glandular dose using the United Kingdom, European and IAEA breast dosimetry protocols. Phys. Med. Biol. 54, 4361–4372 (2009)

Investigating the Replacement of the Physical Anti-scatter Grid with Digital Image Processing

Christopher E. Tromans[1], Jennifer Diffey[2], and Sir Michael Brady[1]

[1] Wolfson Medical Vision Laboratory, Department of Engineering Science, University of Oxford, Parks Road, Oxford, UK, OX1 3PJ
[2] Imaging Science and Biomedical Engineering, School of Cancer and Imaging Sciences, Stopford Building, University of Manchester, Oxford Road, Manchester, UK, M13 9PT
cet@robots.ox.ac.uk

Abstract. Scattered photons degrade mammographic image quality, so, almost universally, a physical anti-scatter grid is used to limit their effect. Physical grids are not completely effective in rejecting only scattered photons, so patient dose must be increased in order to maintain low levels of quantum noise. The standard attenuation rate (SAR), a quantitative normalised representation of breast tissue for image analysis applications, incorporates a model of scatter, and a software correction of the image blurring arising from scatter within the image signal. A tissue equivalent phantom is used to investigate the possibility, in terms of both image sharpness and noise, of replacing the physical grid with the software correction in the SAR. Encouraging results are reported, software correction almost matching the performance of the grid, whilst maintaining a superior signal-to-noise ratio.

Keywords: algorithmic scatter correction, physics modelling, acquisition optimisation.

1 Introduction

Scattered photons are those which do not follow a straight path in their traversal through the breast, from focal spot to image detector. As a result of their often complex paths, they convey no useful information, and degrade the mammographic image. Their detrimental effect manifests itself in two forms: the signal-to-noise ratio worsens due to increased quantum noise; and image sharpness is reduced through the introduction of a low frequency blurring. Quantum noise arises due to the counting of a number of random x-ray photon detection events at the detector over a finite time, and may be modelled as the standard deviation of a Poisson distribution, for which the signal-to-noise ratio may be written as:

$$SNR = \frac{Signal}{Noise} = \frac{I_{primary}}{\sqrt{I_{primary} + I_{scatter}}}$$

where the presence of the scattered photon fluence in the denominator, together with its absence in the numerator (since scatter conveys no useful information), illustrates

J. Martí et al. (Eds.): IWDM 2010, LNCS 6136, pp. 205–212, 2010.

the detrimental effect on the signal-to-noise ratio within the image. Quantum noise should be minimised at acquisition time, since stochastic effects intrinsically cannot be corrected by post processing (since, by definition, no technique exists for identifying the combination of random events which occurred during any given exposure: filters applied to signals containing stochastic noise will have an equal effect on both noise and signal). Image blurring may, however, be corrected by digital image processing, for example unsharp masking. Currently, the inclusion of a physical anti-scatter grid between the patient and detector is the technique used almost universally to reduce the image degradation effect of scatter. Anti-scatter grids are not completely effective in rejecting scattered photons, so the image is still degraded, though to a lesser extent than in the absence of a grid. However, anti-scatter grids are also not completely effective in the transmission of primary photons, which necessitates an increase in patient dose in order to maintain a given detector fluence, hence magnitude of quantum noise. Evidently, it would be of interest were one be able to dispense with the anti-scatter grid, not least in the context of digital breast tomosynthesis, which requires a greater number of projection images to be acquired within a strict dose budget. In this paper, we investigate the impact on image quality when acquisition takes place both with and without an anti-scatter grid, and when the acquired image is post processed to calculate the standard attenuation rate (SAR) image [1]. SAR is a quantitative normalised representation of breast tissue for image analysis applications, in which pixel intensity depends only on the underlying tissue radiodensity. SAR is calculated using a model of the physics of image formation to quantify the difference between the primary x-ray photon fluence incident upon the breast and the primary x-ray photon fluence exiting the breast and subsequently recorded by the image detector. In order to calculate the primary fluence it includes a model of scattered photons. Though the technique was originally developed with the aim of quantifying the magnitude of scatter in the image signal to facilitate quantitative mammography, its wider application in the possible replacement of conventional physical grids is presented in what follows.

2 Materials and Methods

Two scattering phenomena occur within the mammographic energy range: coherent scatter, that is the photon changes direction but there is no loss of energy; and incoherent scatter, in which a change in photon direction occurs together with a portion of the photon energy being transferred to an electron. The coherent scattering cross section $d\sigma_T$ arising from a free electron for which photons are scattered in direction ϕ into solid angle $d\Omega$ may be written precisely as:

$$\frac{d\sigma_T}{d\Omega}(\phi) = \frac{r_e^2}{2}\left(1+\cos^2\phi\right)$$

Scaling this free electron case by the atomic/molecular form factor, tabulated in [2], gives the atomic/molecule cross section, σ_{coh}:

$$\sigma_{coh} = \int_{\phi} d\sigma_T(\phi) F_m^2(x) \text{ where } x = \frac{1}{\lambda}\sin\left(\frac{\phi}{2}\right)$$

Similarly, for incoherent scatter, the cross section of the photon scattering arising from an electron (bound electron Compton) using the Klein-Nishina formula is:

$$\frac{d\sigma_{KN}}{d\Omega}(\phi) = \frac{r_e^2}{2}\frac{k(1-\cos\phi)}{[1+k(1-\cos\phi)]^3} \times \left[1+\cos^2\phi + \frac{k^2(1-\cos^2\phi)}{1+k(1-\cos\phi)}\right]$$

Scaling the single electron case by the atomic/molecular scatter function $S_m(x)$, tabulated in [2-3], gives the atomic/molecule cross section, σ_{incoh}:

$$\sigma_{incoh} = \int_{\phi} d\sigma_{KN}(\phi) S_m(x)$$

The remaining photon energy, E', post incoherent scattering, is then given by:

$$E' = \frac{E}{1+\left(\frac{E}{m_0 c^2}\right)\times(1-\cos\phi)}$$

These relations are used to calculate the scatter arising around a primary ray, as depicted in fig. 1, where for each small traversal *dt*, centred around point **P**, along primary ray path **AB**, the scatter fluence adopting path **PC**, and being incident upon the pixel centred at **C**, is calculated for all pixels in the vicinity of **B**. The scatter signal at each image pixel, arising from scattering at **P** and being attenuated by the tissue in path **PC** is given by:

$$I_{scatter}(\varepsilon) = \left(1 - e^{-\rho\left(\frac{\sigma_{coherent,pixel}+\sigma_{incoherent,pixel}}{uA}\right)dt}\right) e^{-\mu|\mathbf{PC}|(\varepsilon)} I_0$$

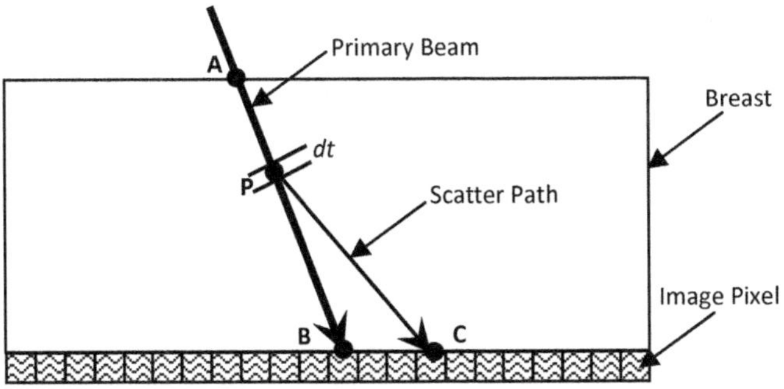

Fig. 1. The calculation of the scatter arising around a primary ray

Fig. 2 shows the result of a typical calculation of the scatter around a primary ray.

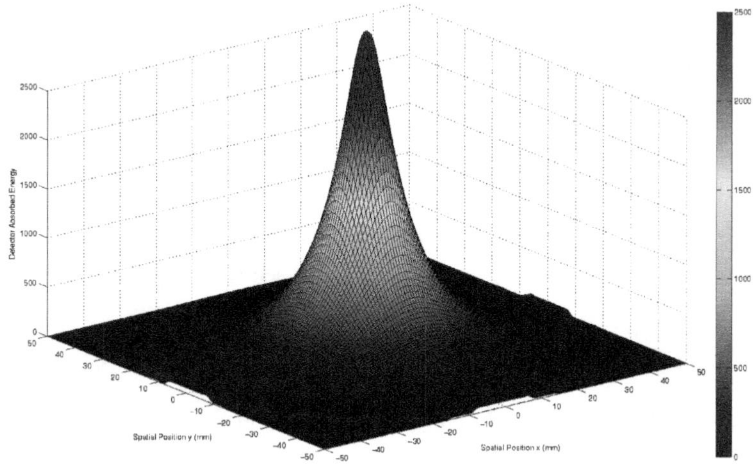

Fig. 2. The calculated scatter point spread function arising around a primary ray traversing 60mm of 50/50 adipose/fibroglandular tissue, in the absence of a grid

The probability of a photon travelling in a given direction successfully traversing the scatter grid is calculated from the grid geometry, and the resulting effect of the grid on the scatter around a primary ray is shown in fig. 3.

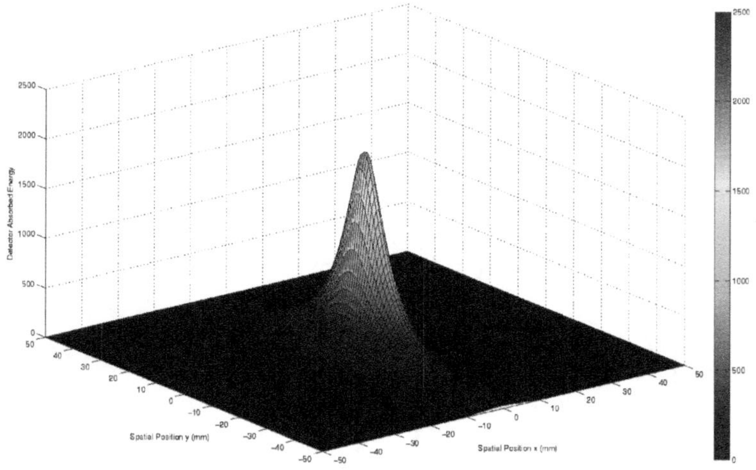

Fig. 3. The calculated scatter point spread function arising around a primary ray traversing 60mm of 50/50 in the presence of a 5:1 31 lp/cm reciprocating focused anti-scatter grid

The scattering arising around a number of "seed" primary rays spaced throughout the image area are calculated (since they are spatially variant), and the intermediaries are found by interpolation. Assuming a 50/50 composition of the entire breast volume, the scatter-to-primary ratio is computed by summing at each pixel the scatter contributions from all primary rays in the vicinity of the pixel. The scatter-to-primary ratio is then used to calculate the scatter component within the acquired image.

The low frequency blurring effect of scattered photons is modelled using a kernel derived from the image acquisition conditions and the underlying scatter relations. Subtraction of the calculated scatter signal from the acquired image yields both an approximation of the magnitude of primary signal, and the removal of the image blurring arising as a result of scatter. This is analogous to unsharp masking. However, in this instance, the subtracted low pass signal is of a magnitude equivalent to that of the scattered photon fluence, and the kernel we use has the shape of the point-spread arising from scatter around a primary beam (for example those shown in fig. 2 or fig. 3), both of which, as discussed earlier, are calculated utilising the fundamental physical relations describing scatter.

Numerous studies have investigated whether or not an anti-scatter grid should be utilised in digital mammography (though none propose a physics based post processing operation as included here). For example, Veldkamp et al [4] use the CDMAM phantom which consists of a matrix of square cells with gold disks of varying sizes, fixed onto a 0.5mm thickness of aluminium, surrounded by PMMA. The choice of phantom is critical since the magnitude of the scattered fluence, and its spatial characteristics (governed by the angular scattering properties) need to be as close as possible to breast tissue, so as to ensure that any optimisation of the imaging system applies to images of the breast, and not just to images of the phantom. At the k-α edge of Molybdenum, which falls at 17.5keV (a peak in the photon energy spectrum due to characteristic emission), the linear attenuation coefficients for photoelectric absorption and scattering (both coherent and incoherent) [5] for adipose tissue are 0.0762 and 0.0596; 0.785 and 0.285 for fibroglandular tissue; and 27.3 and 1.47 for calcium hydroxyapatite (crystalline); whilst those for gold are 2090 and 52.5 (orders of magnitude different from anything found in the breast), and 12.6 and 1.03 for aluminium. To reduce the possibility of optimising to the phantom materials, rather than breast tissue, we have manufactured a phantom from tissue equivalent resins (supplied by CIRS). The phantom comprises of an upper and lower cuboid of adipose material, measuring 100mm by 100mm, with a thickness of 10mm, and a further two cuboids, one of adipose, and one of glandular, measuring 100mm by 50mm, with a 40mm thickness are placed in-between to form a large discontinuity in tissue composition, across which the blurring effect of scatter will be significant. This phantom is an extreme case, such a large discontinuity is highly unlikely to be encounter within a human breast in clinical practise, but the objective is to test the algorithm at its extremities, to allow the quantification of the worst case performance. A second tissue equivalent phantom which far better approximates clinical cases is also investigated. This comprises of interlocking step wedges, and is used for the validation of the SAR computation in our companion paper [1], in which a full description of it design and photographs may be found.

The phantoms were imaged on a Siemens Mammomat Inspiration, both with and without the factory fitted 5:1, 31 lp/cm reciprocating focused anti-scatter grid. The SAR images were computed for each acquired image, both with and without the image blurring due to scatter correction enabled. Since the SAR image is both quantitative and normalised, it provides an ideal basis for comparison between the four images (as discussed in our companion paper [6]).

3 Results

Fig. 4 shows a horizontal profile, spatial averaged vertically over 300 rows, across the SAR images of the phantom, and the variation in sharpness of the discontinuity can be seen to be the underlying phenomena governing the results of the visual assessment.

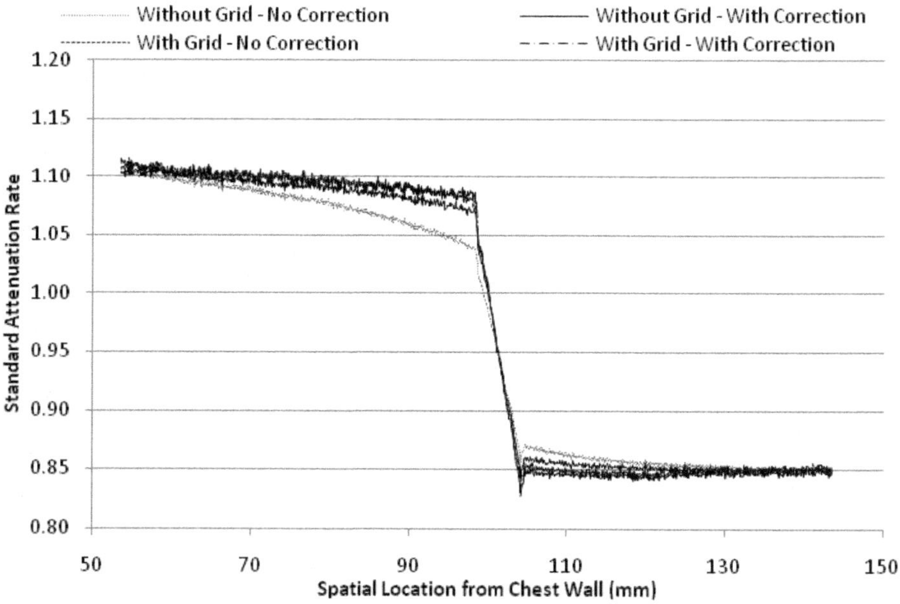

Fig. 4. A horizontal profile, spatial averaged vertical over 300 rows, across the SAR images

The signal-to-noise ratio in each of the SAR images was calculated as the ratio of the median and standard deviation, in the middle of both the high and low density areas of the phantom. Since the SAR is dependant only upon the underlying tissue composition, and the dependency upon the image acquisition parameters is removed (see [1]), the signal-to-noise ratio is the appropriate measure of image quality, since the same phantom is used for the comparisons.

Table 1. The signal-to-noise ratio in the SAR images acquired at 32kVp W-Rh 45mAs

	With Grid		**Without Grid**	
	High Density	**Low Density**	**High Density**	**Low Density**
Blur Corrected	75.2	91.2	97.2	109.9
No Blur Correction	81.1	98.1	120.2	134.8

For the case of the interlocking step wedge phantom, percentage difference images are given in fig. 5.

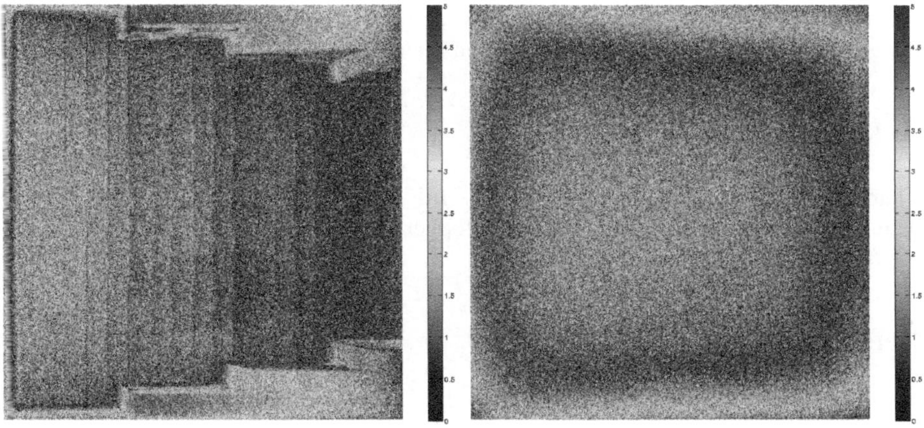

Fig. 5. Percentage difference images between SAR images: acquired with and without grid present, no algorithmic blurring correction (left); and with the grid but no blurring correction, and without the grid but with blurring correction (right)

The scatter model execution time was 81.79 seconds and 75.35 seconds (85 micron image) for the cases with and without the anti-scatter grid respectively.

4 Discussion

The performance of the software blur correction may be seen in fig. 4, where the corrected image sharpness is almost identical to that observed when the physical anti-scatter grid is used. It worthy of note that this test is an extreme case, using a phantom which contains a discontinuity in tissue composition far greater than that ever likely to be found in clinical practice, and thus tests the model to the extremities. Blur correction comes at the cost of a worsening of the signal-to-noise ratio, as may be observed in table 1, since it is not possible to discriminate noise from signal, so the noise is also sharpened.

In the case of the more clinically realistic interlocking step wedge phantom, the results are given as percentage difference images in fig. 5. On the left, the blurring effect of scatter may be observed, since in this image the presence of the grid is compared with its absence, and in the resulting subtraction all the salient features of the phantom can be made out. When the blurring correction is applied to the image acquired with the grid absent, the difference image on the right shows successful correction of the blurring since no features of the phantom can be seen, only the variation in image noise.

5 Conclusion

A technique is proposed for the algorithmic correction of scatter with the aim of replacing the physical grid. Experiments upon phantom images suggest that the blurring effect of scatter is significantly reduced by the technique, even at large discontinuities in tissue composition, to a level very similar to that seen when the grid is included. A superior signal-to-noise ratio is observed when the algorithm is used in the absence of the grid, due to the increased primary photon fluence at the detector.

Acknowledgments

Thanks to the radiographers at Stepping Hill Hospital, Stockport for their help in acquiring the experimental images. Christopher Tromans wishes to thank the EPSRC for funding his research fellowship.

References

[1] Tromans, C., Brady, M.: The Standard Attenuation Rate for Quantitative Mammography. In: Proceedings of 10th International Workshop on Digital Mammography (IWDM), Girona, Spain (2010)

[2] Hubbell, J.H., Veigele, W.M.J., Briggs, E.A., Brown, R.T., Cromer, D.T., Howerton, R.J.: Atomic Form Factors, Incoherent Scattering Functions, and Photon Scattering Cross Sections. Journal of Physical and Chemical Reference Data 4, 471–538 (1975)

[3] Hubbell, J.H., Overbo, I.: Relativistic atomic form factors and photon coherent scattering cross sections. Journal of Physical and Chemical Reference Data 8, 69–106 (1979)

[4] Veldkamp, W.J., et al.: The value of scatter removal by a grid in full field digital mammography. Med. Phys. 30, 1712–1718 (2000)

[5] Berger, M.J., Hubbell, J.H.: XCOM: Photon cross section on a personal computer, US Department of Commerce NBSIR 87-35971987

[6] Tromans, C., et al.: Optimising Beam Quality Selection in Mammographic Acquisition using the Standard Attenuation Rate. In: Proceedings of 10th International Workshop on Digital Mammography (IWDM), Girona, Spain (2010)

Diffusion-Weighted Imaging (DWI) for Breast Cancers; Challenging to Diagnose Ductal Carcinoma in Situ (DCIS) and Invasive Lobular Carcinoma (ILC)

Yoshifumi Kuroki[1], Katsuhiro Nasu[2], Seiko Kuroki[3], Ryuzo Sekiguchi[1], Shigeru Nawano[4], and Mitsuo Satake[5]

[1] Tochigi Cancer Center, Diagnostic imaging, Utsunomiya, Japan
[2] University of Tsukuba, Department of Radiology, Tsukuba, Japan
[3] National Cancer Center, Research Center for Cancer Prevention and Screening, Tokyo, Japan
[4] International University of Health and Welfare, Mita Hospital, Tokyo, Japan
[5] National Cancer Center Hospital East, Department of Radiology, Kashiwa, Japan

Abstract. Diffusion-weighted imaging (DWI) visualizes the thermal motion of protons in tissue, which is less affected by the state of the background mammary gland and therefore has sufficient capability to diagnose DCIS as well as ILC. Of 30 patients with DCIS, sensitivity was 87%, and 4 cases were false negative (FN); of 30 patients with invasive lobular carcinoma ILC, the rate of true positive cases (TP) was 93%, and 2 cases were FN. With an ability to provide steady, high-contrast resolution tissue images, DWI is expected to play a significant role in future breast cancer diagnosis. We should promote the application of DWI in clinical practice, taking advantage of its high-contrast resolution.

Keywords: Diffusion-weighted imaging, Magnetic Resonance Imaging, Breast cancer.

1 Introduction

Diffusion-weighted imaging takes advantage of the slight density discrepancies that can occur even under the same temperature and pressure as surrounding tissue, due to the random thermal movement of molecules. The clinically applied diffusion-weighted imaging (DWI) visualizes the diffusion caused by the random thermal movement of water molecules. In oncology, the diffusion of water molecules is closely related to cellularity, and the diffusion image can reveal the density of the tissue. Therefore, the tissue structure can be inferred from the analysis of DW images. To date, DWI has been developed mainly in neuroradiology, but recent advances in parallel imaging have allowed some clinical studies to report the application of DWI in examinations of the body trunk [1], [2], [3], [4], [5]. With regard to breast tumors, DWI is expected to aid in discriminating between malignancy and benignancy and in evaluating the efficacy of neoadjuvant chemotherapy (NAC) [6].

In this study, we investigated the imaging capacity of non-contrast MRI in combination with DWI and short TI inversion recovery (STIR) for diagnosing DCIS and ILC.

J. Martí et al. (Eds.): IWDM 2010, LNCS 6136, pp. 213–218, 2010.

2 Method and Materials

2.1 Patients

The subjects were 30 cases of DCIS and 30 cases of ILC that had been consecutively diagnosed by post-operative pathology. All patients underwent preoperative MRI with DWI and mammography (MMG), which had been performed within a one-month interval. Cases of a synchronous cancer in an ipsilateral mammary gland and cases with prior breast cancer treatment and/or a history of NAC were excluded.

2.2 Image Protocol of Mammography and MRI

We used a mammography exposure unit, LORAD M-IV (HOLOGIC, USA), and an image reader, FCR PROFECT CS (FUJIFILM Corporation, Japan). The sampling size was 0.05 mm. The recommended characteristic curve (Type T) was applied. Bilateral cranio-caudal view and mediolateral oblique view were obtained. The LCD we used was monochrome with resolution 2048 x 2560. The brightness was 660 cd/m2 and the gray scale resolution was 256 levels.

MR examinations were performed with a 1.5-Tesla MR imager (Gyroscan Intera 1.5-T Master Grade; Philips Medical Systems, the Netherlands). Bilateral breast images were acquired using a SENSE body coil with patients in the prone position. All imaging was performed in the transaxial plane. Subsequent to DWI and STIR, dynamic study was carried out on each image using Gd-DTPA enhancement. Single-shot echo planar imaging was employed for DWI. Employed b factor of DWI was as follows. Single DWI protocol; b factor=1000 sec/mm^2 and Dual DWI protocol; combinations of 750 sec/mm^2 and 1500 sec/mm ^{2}. The details of imaging protocol are shown in Table 1.

Table 1. The details of Image parameters

	FOV (mm)	TR/TE (msec)	Thk/Gap (mm)	Mtrx	NEX	Scan time (sec)
STIR	320	4039/60	5/0	512x269	2	147
Single DWI	320	3100/85	5/0	256x101	1	39
Dual DWI	320	4500/76	5/0	256x102	3	109

2.3 Imaging Analysis

Three radiologists with 15 and 14 years of experience in diagnostic radiology interpreted MR images and MMG by the retrospective consultation. Mammography was interpreted according to category classification of BI-RADS, and surrounding mammary glands of MMG were classified into fatty, scattered fibroglandular tissue; heterogeneously fibroglandular tissue; and mostly fibroglandular tissue and further subdivided into two groups, fatty/scattered fibroglandular tissue and heterogeneous fibroglandular tissue/mostly fibroglandular tissue.

STIR and DWI images were interpreted on the basis of the diagnosis criteria shown in Table 2. The apparent diffusion constant (ADC) was not used for diagnosis, and images were evaluated visually. We classified the signal intensity in DWI and STIR of the area in question as high, iso, or low in comparison with the signal intensity of the corresponding srroundung mammary gland. Areas of high intensity on DWI were located and then investigated on STIR with the help of signal intensity, marginal aspect, and internal structure. Based upon the results, they were classified as malignant, highly suspicious of malignancy and requiring further examination, and benign with low likelihood of malignancy and requiring no further examination. An area of high intensity on DWI and low or iso-intensity on STIR was classified into the malignant group. In the case of high intensity on both DWI and STIR, the area was classified as malignant if there was marginal irregularity and/or internal structure, for example a papillary projection. As an exception, an area of extremely low intensity on STIR was classified as malignant based on morphologic findings, regardless of its signal intensity on DWI.

The true positive rate (TP) and false negative rate (FN) were calculated using the pathology results, analyzing their relationship to MMG categories and surrounding mammary gland types, respectively.

Table 2. Criteria of MR image Interpretation

Intensity of DWI	Intensity of STIR	Irregular Margin	Internal structure	Diagonosis
High	Low – Iso	----------	----------	Malignancy
High	High	Positive	----------	Malignancy
High	High	Negative	Positive	Malignancy
High	High	Negative	Negative	Benign
Low ~ Iso	Low ~ High	----------	----------	Benign
----------	Extremely low	----------	----------	Malignancy

3 Results

Of 30 patients with DCIS, sensitivity was 87%, and 4 cases were FN. The TP rate of patients with fatty/scattered fibroglandular tissue was 90%. For patients with heterogeneous fibroglandular tissue/mostly fibroglandular tissue, the TP rate was 85%. The TP rate of category 1/2 patients was 71%, while the TP rate of category 3/4/5 patients was 91%.

Of 30 patients with ILC, the TP rate was 93%, and 2 cases were FN. The TP rate of patients with fatty/scattered fibroglandular tissue was 100%, while the TP rate of patients with heterogeneous fibroglandular tissue/mostly fibroglandular tissue was 92%. The TP rate of category 1/2 patients was 83%, and the TP rate of category 3/4/5 patients was 96%. Mammography and MR images of a DCIS case were and a ILC case were shown in Fig. 1, Fig. and 2, Fig. 3, and Fig. 4.

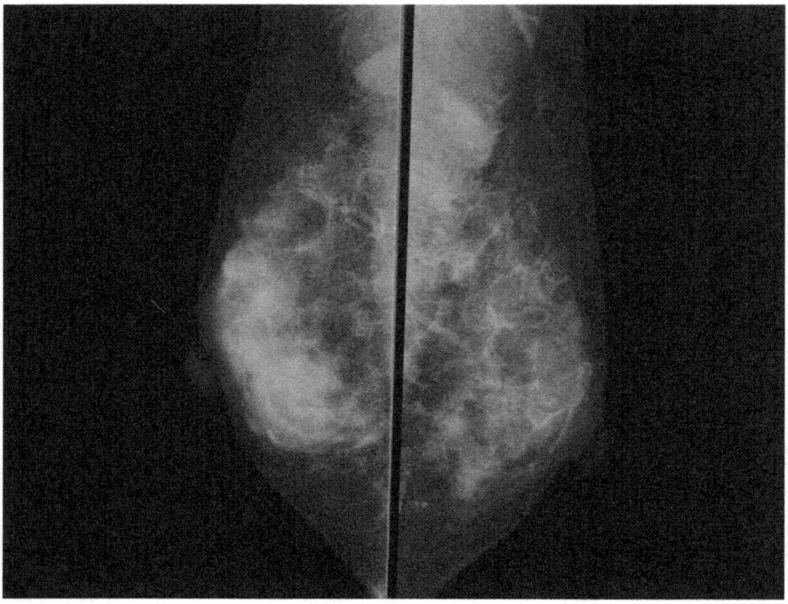

Fig. 1. MLO view of Mammography of a DCIS case

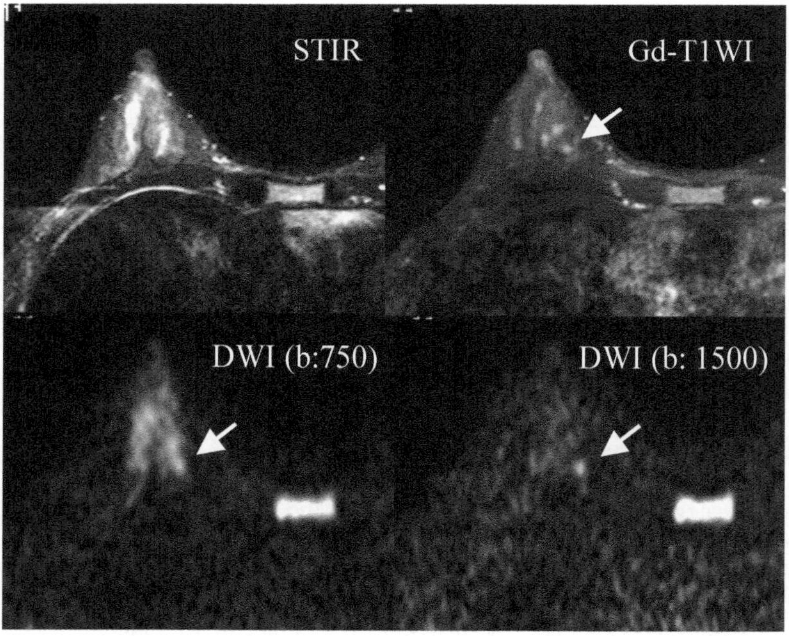

Fig. 2. MR images of a DCIS case

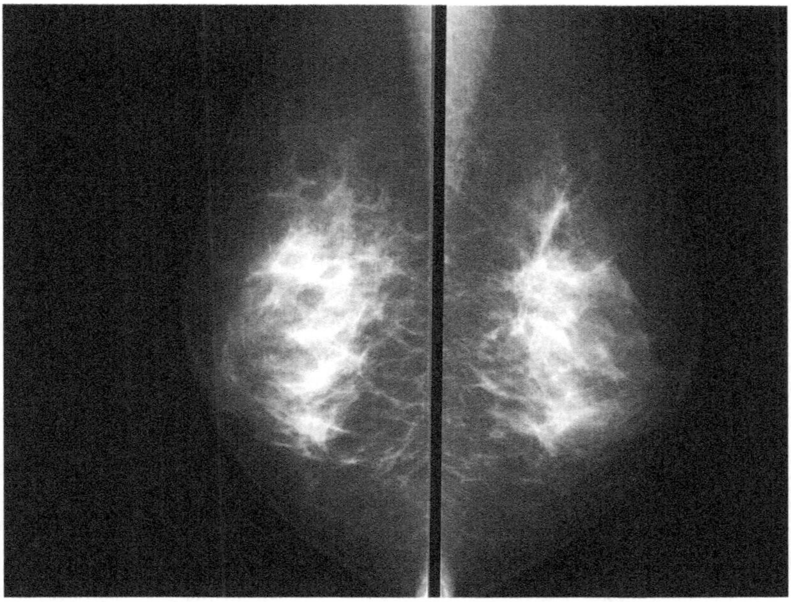

Fig. 3. MLO view of Mammography of a ILC case

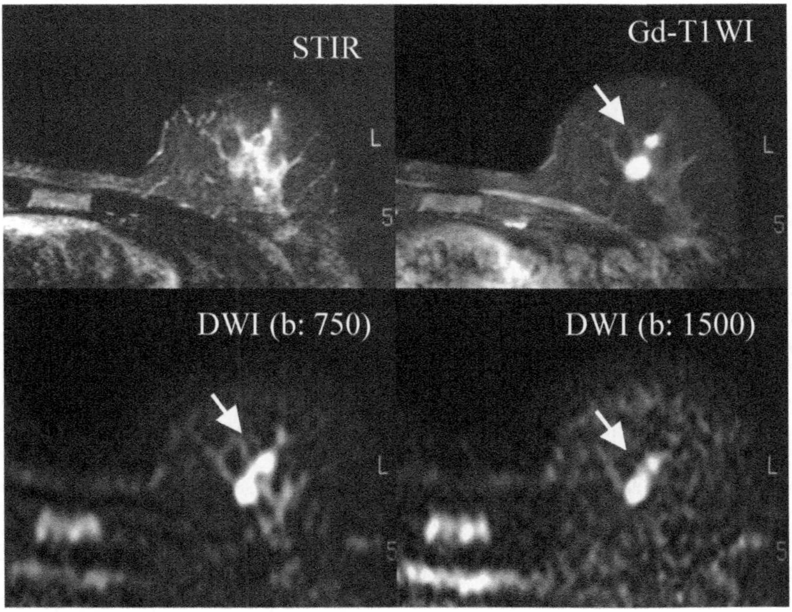

Fig. 4. MR images of a ILC case

4 Discussion

Though Gd-DTPA enhanced dynamic study is required in Breast MRI in some clinical guidelines, we investigated it without contrast medium daringly to confirm potential of the non contrast MRI for breast cancer survey. In this study, the combination of DWI and STIR, using no contrast agent, was shown to have a possibly higher depiction capacity than MMG. The capacity was characterized by the independency of cancer from the surrounding mammary gland that was dense breast type. The reason of detectability was speculated with the effective signal suppression of surrounding mammary gland using DWI. However, some drawbacks were found with DWI: some cases with a high degree of mastopathy demonstrated an insufficient suppression of signal of sourrounding mammary gland, which hindered lesion recognition. Also, fat suppression is indispensable in DWI, so cases with insufficient fat suppression resulted in difficulties in distinguishing between the lesion and the surrounding fat, which lowered the lesion-depicting capacity of DWI. DWI has a high-contrast resolution, but its spatial resolution is low. This latter characteristic together with background glandular suppression means DWI generally cannot provide anatomical information about the structure being imaged, which means supplemental anatomical information must be provided by some other sequence images or modality.

Limitations of this study are as follows. Firstly it is that there is a small number of cases, and secondly that the false positive rate for the benign disease is not considered.

5 Conclusion

DWI has excellent potential to depict covert breast cancer in the dense breast without contrast medium, even if it was hard to detect DCIS and ILC by mammography. DWI may be valuable in close inspection of the patients who cannot be administered the contrast media and survey.

References

1. Padhani, A.R., Liu, G., Mu-Koh, D., et al.: Diffusion-Weighted Magnetic Resonance Imaging as a Cancer Biomarker: Consensus Cancer Biomarker: Consensus and Recommendations. Neoplasia 11, 102–125 (2009)
2. Pruessmann, K.P., Weiger, M., Scheidegger, M.B., et al.: SENSE: sensitivity encoding for fast MRI. Magn. Reson. Med. 42, 952–962 (1999)
3. Kuroki, Y., Nasu, K., Kuroki, S., et al.: Diffusion-weghted imaging of breast cancer with the sensitivity encoding technique: analysis of apparent diffusion coefficient value. Magnetic Resonance in Medical Science 3, 79–85 (2004)
4. Woodhams, R., Matsunaga, K., Iwabuchi, K., et al.: Diffusion-weighted imaging of malignant breast tumor: the usefulness of apparent diffusion coefficient (ADC) value and ADC map for the detection of malignant breast tumors and evaluation of cancer extension. J. Comput. Assist. Tomogr. 29, 644–649 (2005)
5. Kuroki-Suzuki, S., Kuroki, Y., Nasu, K., et al.: Detecting Breast Cancer with Non-contrast MR Imaging: Combining Diffusion-weighted and STIR Imaging. Magn. Reson. Med. Sci. 6(1), 21–27 (2007)
6. Tozaki, M., Sakamoto, M., Oyama, Y., et al.: Monitoring of early response to neoadjuvant chemotherapy in breast cancer with (1)H MR spectroscopy: Comparison to sequential 2-[18F]-fluorodeoxyglucose positron emission tomography. JMRI 28, 420–427 (2008)

CR Mammography: Image Quality Measurement and Model Calculation for Needle vs. Powder Imaging Plate

Dirk A.N. Vandenbroucke and Paul J.R. Leblans

Agfa Healthcare N.V., Medical Imaging, R&D Consumables, Septestraat 27, B.-2640 Mortsel, Belgium
Dirk.Vandenbroucke@agfa.com

Abstract. Computed radiography (CR) is a digital radiography technology in which a storage phosphor plate is used to store a latent X-ray image. The plate is exposed in a light-tight cassette and then read out in a digitizer to create the digital image. Traditionally, CR powder imaging plates (PIP) are used based on $BaFBr_{1-x}I_x:Eu^{2+}$ phosphor. The active layer consists of phosphor micro-crystals in a polymer binder. A needle imaging plate (NIP), created by vapor deposition of needle-shaped phosphor crystals, is expected to lead to better image quality. A first reason is that lateral light spread is less in NIP. Further, the system gain is higher, because more storage centers are created per unit of absorbed X-ray energy, because read-out depth can be higher and because the stimulated light escape efficiency is higher. The more transparent NIP guarantees a more constant image contribution over the thickness of the plate. Finally, the NIP layer is more homogeneous than the PIP layer, which leads to a lower degree of screen-structure noise. Measurements confirm that $CsBr:Eu^{2+}$ NIP's in CR mammography have significantly better image quality (DQE), especially in the high frequency range. A linear-systems approach is used to model signal and noise transfer in a CR system using PIP or NIP. The transfers are described by cascading transfer relationships for each process. The calculated image quality (DQE) is in good agreement with measurement for both the NIP and the PIP systems.

Keywords: digital mammography, computed radiography (CR), detective quantum efficiency, needle imaging plate, image quality measurement, modeling.

1 Introduction

In CR storage phosphors, free electrons and holes released by X-ray quanta are trapped in storage centers, to generate a latent image. In the read-out process in the digitizer, the trapped electrons are stimulated with red light to give rise to a blue luminescence signal, which is transformed in an electronic signal by a light sensor and digitized [1].

Today, storage phosphors of the $BaFBr_{1-x}I_x:Eu^{2+}$ family are used in commercial CR systems. $BaFBr_{1-x}I_x:Eu^{2+}$ is used in the CR Mammography plates of Agfa, Fuji, Carestream and Konica. It is an excellent storage phosphor with a high storage capacity and a relatively high specific X-ray absorption. Since $BaFBr_{1-x}I_x:Eu^{2+}$ does not melt congruently it decomposes upon vaporization, which makes it unfit for NIP manufacturing.

J. Martí et al. (Eds.): IWDM 2010, LNCS 6136, pp. 219–226, 2010.

Agfa discovered an excellent new storage phosphor, $CsBr:Eu^{2+}$, with a chemical composition and density of 4.5 g/cm^3 that lead to a specific X-ray absorption that is similar to that of $BaFBr_{1-x}I_x:Eu^{2+}$ [2]. NIP is expected to lead to better image quality than PIP for several reasons. The needles provide strong forward light scattering, thereby strongly reducing light spread in lateral direction in the phosphor layer. As a consequence, image sharpness is higher at equal thickness. Further, a NIP system's gain is higher, because i) more storage centers are created per unit of absorbed X-ray energy, ii) read-out depth can be higher and iii) the stimulated light escape efficiency is higher. The more transparent needle shaped $CsBr:Eu^{2+}$ layer guarantees a more constant image contribution over the thickness of the layer, which leads to a higher Swank factor. Evaporated CsBr layers are more homogeneous than coated $BaFBr_{1-x}I_x:Eu^{2+}$ layers, ensuring a lower degree of screen-structure noise. Today, $CsBr:Eu^{2+}$ needle image plates are used in commercial CR systems for general radiography (Agfa) and have, more recently, also been introduced in CR Mammography systems (Konica, Agfa). As indicated in [3] the much higher DQE of the NIP CR general radiography system results in a superior performance of the system in observer performance studies [4].

Superior performance of NIP CR systems compared to PIP systems is also expected in mammography. The purpose of the actual study is to confirm the superior image quality by measurement and to illustrate the physical backgrounds based on a simple simulation.

2 Experimental

CR Systems

The reference NIP system is the commercial Agfa DX-M system with Agfa HM5.0 NIP, as presented in [5]. The DX-M is a new generation flying-spot digitizer that supports both standard phosphor plates (PIP) and needle-based detectors (NIP). The HM5.0 NIP is a dedicated high resolution CsBr:Eu2+ Mammo plate.

The Agfa CR MM3.0R plate a rigid BaFBr1-xIx:Eu2+ high resolution PIP was scanned with the same DX-M digitizer, with adjusted scan parameters for optimal image quality.

Measurement Procedures and Exposure Conditions

The technical image quality for both systems was determined in according to the methods of the IEC 62220-1-2 standard [6]. The X-ray spectrum used in the measurements corresponded to the RQA-M2 radiation quality, with an X-ray tube voltage of 28 kVp, an internal Mo filter and an external Al filter of 2.0 mm. The resulting Al half-value layer was ca. 0.61 mm Al. The normalized noise power was determined at a dose of 77 μGy, and the MTF at a dose of 156 μGy. The dose was measured with an ionization chamber dosimeter prior to the exposure. As X-ray source a Siemens Mammomat 1000 X-ray tube with focal spot size of 0.3 mm was used. The tube is mounted on a horizontal bench, allowing easy introduction of extra diaphragms for optimal exposure conditions to guarantee high primary-to-scatter ratio.

Measurement Analysis

DQE Calculation

The 2-dimensional frequency-dependent detective quantum efficiency DQE(u,v) is given by:

$$DQE(u,v). = MTF^2(u,v).W_{in}(u,v)/W_{out}(u,v) \quad (1),$$

where MTF is the pre-sampled modulation transfer function of the imaging device, W_{out} the noise power spectrum of the flat-field image made with the X-ray imaging device and W_{in} the noise power spectrum of the incoming X-radiation at the detector, i.e. phosphor plate surface. The input noise power spectrum is equal to the incoming X-ray quantum fluence and is constant for all spatial frequencies (white noise).

The DQE, calculated as the average of the DQE measured in slow- and fast-scan direction, was used for comparison between the systems and for comparison with model calculations.

3 Model Calculations

Sections 1: X-ray spectrum generation in X-ray tube

The X-ray spectra are calculated using the model of Tucker et al [7]. Using this semi-empirical model, the bremsstrahlung and characteristic X-rays produced at varying depth in the W anode is simulated, resulting in a spectral distribution of X-rays generated in the tube. The attenuation of arbitrary internal filtration on the spectra is calculated using the energy-dependent Lambert-Beer law [8]. Attenuation data used for the materials were obtained from the National Institute of Standards and Technology (NIST) [9].

Section 2: spectral changes by filtration

Section 2 generates the X-ray spectrum after transmission through an external filter, usually present in an X-ray imaging system. The output of section 2 is the X-ray spectrum incident on the phosphor plate.

An X-ray spectrum corresponding to the RQA M2 Al half-value layer of 0.612 mm was generated. To obtain this HVL, the Mo-generator X-ray source high voltage had to be set at 28 kVp for the combination with the internal Mo filter of 0.025 mm and the external 2.0 mm Al filter.

Section 3: Interaction of X-rays with phosphor plate

Section 3 covers the interaction of X-ray quanta with the phosphor layer and the generation of light in the stimulation process. The linear-systems approach [10] is used to describe both signal and noise transfer in the system. The link is made to metrics of image quality and system performance including the modulation transfer function (MTF), noise power spectrum (NPS) and detective quantum efficiency (DQE). The cascaded systems approach represents the imaging system as a series of discrete stages, where each stage represents a process which affects either the mean number of image carriers, a gain stage, or the spatial distribution of the image carriers, a spatial spreading stage. Each of these processes has distinct signal and noise transfer characteristics.

First, the interaction of X-rays with the plate is modeled. The spectrum of the X-rays that deposit energy in the screen is calculated using the energy dependent law of Lambert-Beer and using the mass-energy absorption coefficients of the phosphor material. Since the plate response is not homogeneous, it is split up in 15 virtual layers in the thickness direction and the energy deposited in each layer is calculated. The energy needed to create a storage center in the storage phosphor being known, this allows calculation of the number of storage centers being created in each layer.

Next, stimulation of the phosphor plate by the digitizer in the read-out phase is modeled. The stimulation efficiency is assumed for each virtual layer of the phosphor screen, giving the fraction of storage centers that will give rise to an optical photon. It is assumed that the stimulation efficiency in the top layer is higher than that in the bottom layer. For the intermediate layers, the stimulation efficiency is obtained by linear interpolation. Photons generated in the stimulation process have a certain escape efficiency towards the digitizer optics. Again, each virtual phosphor plate layer is assumed to have a different escape efficiency, the escape efficiency from the top layer being higher than the escape efficiency from the bottom layer. The stimulation and light emission stages give rise to spatial spreading of the signal, caused by lateral light diffusion. Therefore, an MTF is connected to these stages. In our model, the MTF of the stimulation and emission stages is described by a Lorentzian, which is different for each virtual phosphor plate layer:

$$\mathrm{MTF}(u) = [1 + (u/H)^2]^{-1} \tag{2}$$

The Lorentz factor H is normally assumed to be higher for the top than for the bottom layer of the plate. The H values for the intermediate layers are calculated by linear interpolation. The noise transfer of these 3 stages is calculated using the cascaded systems approach. The NPS is calculated for every virtual layer using the input parameters for the individual layers as described. The total NPS is calculated by summing the NPS of the individual layers.

The system MTF at this stage is calculated as the weighted average of the virtual layers' MTF, the weighing factors being the relative contributions of the layer to the signal. The weighing factor, therefore, depends on the X-ray absorption and on the stimulation and escape efficiencies of the virtual layer.

The DQE at this stage is calculated as:

$$\mathrm{DQE}(u) = q_0.G^2.\mathrm{MTF}(u)^2/\mathrm{NPS}(u) \tag{3}$$

where q_0 is the mean number of quanta incident on the screen and G is the gain of the phosphor plate, i.e. the number of photons per incident X-ray quantum.

Table 1 summarizes the parameters used for the simulation of DQE of both systems.

Section 4: Conversion of photons to photoelectrons

The signal conversion stage deals with the transmission of photons to the detector by the digitizer optics, with the passage through the filter for stopping the stimulation light and with the conversion to photoelectrons. The total system gain is calculated as the gain of the previous stage, multiplied by the transmission efficiency of the light guide, the transmission of the filter and the quantum efficiency of the light detector.

The noise transfer is again calculated using the cascaded systems approach.

Table 1. Model parameters for PIP and NIP system DQE calculation

System	PIP	NIP
Phosphor	$BaFBr_{1-x}I_x:Eu^{2+}$	$CsBr:Eu^{2+}$
Coating weight (mg/cm^2)	52	50
Effective thickness (µm)	104	110
Fill factor (%)	75	90
Physical thickness (µm)	138	122
X-ray absorption (%)	85	87
# virtual plate layers	15	15
Storage center generation energy (eV)	100	75
Est. number of F-centers / abs. X-ray	200	265
Stimulation efficiency (X-ray side) (%)	80	75
Stimulation efficiency (bottom) (%)	30	65
Escape efficiency (X-ray side) (%)	70	80
Escape efficiency (bottom) (%)	20	60
MTF Lorentzian (X-ray side)	2.2	1.6
MTF Lorentzian (bottom)	3.5	4.9
Coupling efficiency optics (%)	60	60
Filter transmission (%)	50	45
PMT quantum efficiency (%)	28	25
Pixel aperture (µm)	50	50

The pre-sampled MTF represents the deterministic spreading of the photons by the pixel aperture. It is considered as deterministic since each integrated photon is conceptually redistributed to a single point (the pixel center) at which it will be counted. The pre-sampled MTF is modeled with a sinc function. The total system MTF is calculated as the MTF of the previous stage, multiplied by the sinc function.

Section 5: Addition of screen structure noise
Since it is very difficult to model the screen structure noise, a simple approach is used [11]. Per frequency extra noise is added to the total NPS.

4 Results

4.1 MTF

Figure 1 shows the measured and calculated MTF. The MTF for NIP systems is higher than for PIP systems. The improvement in sharpness is most pronounced at high frequency. The correspondence of simulated MTF and experimental MTF is for both systems very good.

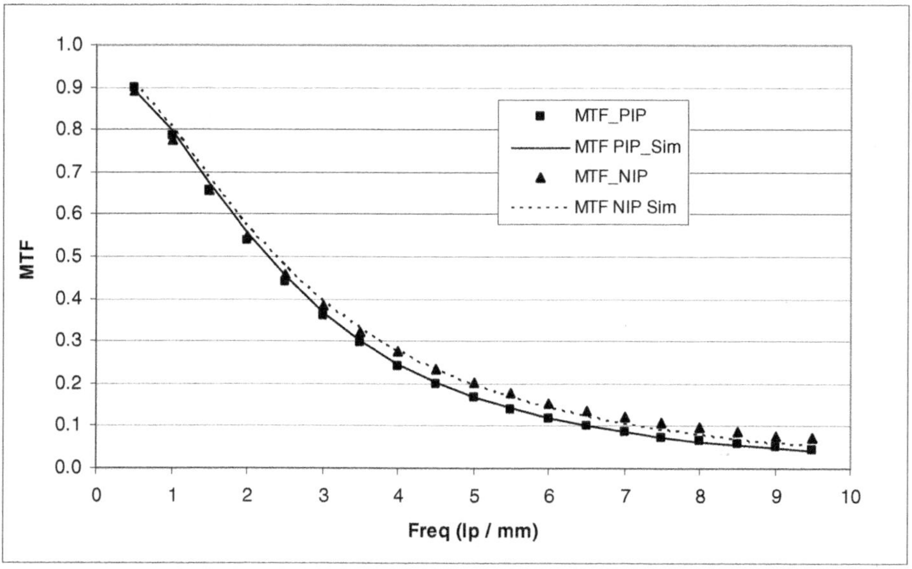

Fig. 1. Measured PIP (squares) and NIP (triangles) CR systems' MTF and fitted values (lines)

4.2 DQE

Figure 2 shows the measured and calculated DQE curves for both systems. In this simulation, the experimental SSN as obtained from the dose dependence of the NPS is used. There is good correspondence between the measured and calculated DQE values for the PIP system. For the NIP system correspondence is perfect when experimental SSN data are used. In trying to optimize the fit for PIP, an optimization of SSN was done. Since only 3 dose levels were tested in dose range is restricted to 38 – 156 µGy, the accuracy of this SSN calculation is not good. Extra dose levels especially at even higher dose levels are needed for a more accurate characterization of the SSN. In fig 2 measured and calculated DQE curves are shown for the PIP systems after optimizing the SSN for optimal fit of calculated to measured data. Compared to the experimental SSN data at low frequencies a 1.5 to ca 2x higher SSN is needed for optimal fit quality. The fact that for NIP no further optimization in SSN is needed to get a perfect match between experimental and simulated DQE is a further illustration of the lower contribution of SSN to the image quality at higher dose levels for NIPs. In the relevant mammographic dose range, screen structure noise is an important contributor to the total noise of the system. The lower SSN for NIP is responsible for a considerable contribution to improvement of the image quality at low frequency.

The experimental DQE is at low frequency at least 30% higher for NIP than for PIP. The relative DQE (NIP / PIP) increases continuously over the entire spatial frequency range, reaching a factor of 2 at 5 lp/mm. As indicated in [5] the higher DQE of the NIP CR system results in a superior performance of the system in observer performance studies, or a potential lower patient dose for the same image quality as compared to the PIP CR systems.

Based on the above estimated number of storage centers generated per absorbed X-ray quantum in the phosphor and the efficiencies to model the various cascades in the imaging process, the number of photon generating photon electrons on the cathode of the PMT that can contribute to the signal is estimated to be 3.5 for PIP and 7.7 for NIP. The high voltage on the PMT is used to amplify the signal to a fixed working point to assure similar grey levels in clinical images. The higher number of photons to generate photo-electron on the PMT cathode for NIP ensures a higher signal to noise level as compared to PIP. Experimentally, at low dose a significantly lower NNPS is measured for NIP as compared to PIP which is more or less constant over the frequency domain (20 – 30%).

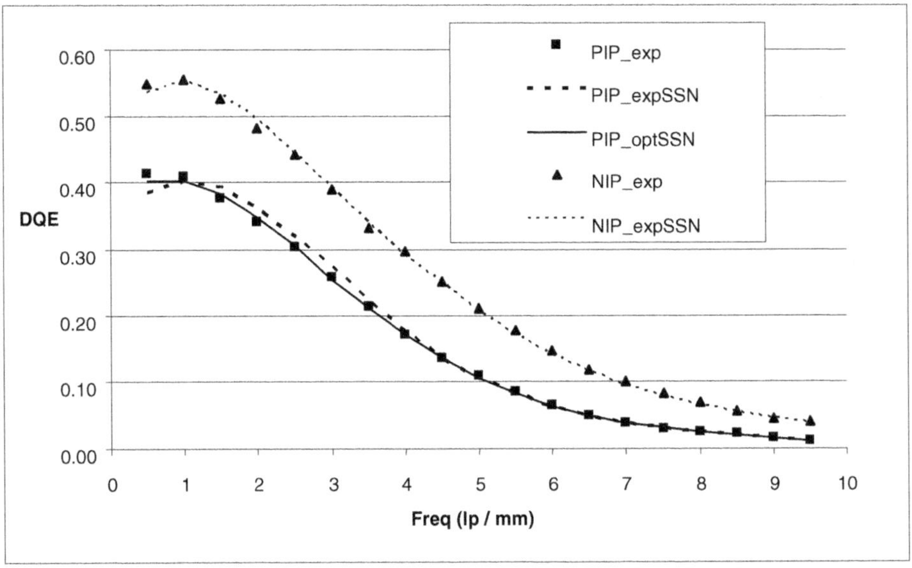

Fig. 2. Measured PIP (squares) and NIP (triangles) CR systems' DQE and calculated values (lines). Model calculations using experimental SSN and model calculations after optimization of SSN for optimal fit to experimental DQE.

References

1. Takabahashi, K., Kohda, K., Shibahara, Y.: J. Electrochem. Soc. Solid State Science and Technology 132, 1492 (1985)
2. Leblans, P.J.R., Struye, L., Willems, P.: Proc. of SPIE, vol. 4320, p. 59 (2001)
3. Schaetzing, R.: RSNA Course in Diagnostic Radiology. Physics 7 (2003)
4. Frankenberger, J., Mair, S., Herrmann, C., Lamotte, J., Fasbender, R.: Proc. of SPIE (2005)
5. Young, K.C., Oduko, J.M., Asad, M.: NHBSP Equipment Report 0905 (November 2009), `http://www.cancerscreening.nhs.uk/breastscreen/publications/nhsbsp-equipment-report-0905.pdf`
6. IEC62220-1-2 ed 1.0 (2007)

7. Tucker, D.M., Barnes, G.T., Chakraborty, D.P.: Med. Phys. 18(2), 211–218 (1991)
8. Beutel, J., Kundel, H.L., Van Metter, R.L.: Handbook of medical imaging, vol. 1, pp. 36–38. SPIE Press (2000), ISBN 0-8194-3621-6
9. `http://physics.nist.gov/PhysRefData/XrayMassCoef/cover.html`
10. Siewerdsen, J.H.: Dissertation submitted in partial fulfullment of the requirements for the degree of Ph D (Physiscs) at the University Of Michigan (1998)
11. Hillen, W., Schiebel, U., Zengel, T.: Med. Phys. 14, 744–751 (1987)

Performance Assessment of Breast Tomosynthesis Systems: Concepts for Two Types of Phantoms

Hilde Bosmans[1], Jurgen Jacobs[1], Eman Shaheen[1], Federica Zanca[1], Lesley Cockmartin[1], Kristina Bliznakova[2], Kim Lemmens[1], Sabrina Vollmar[3], Willi A. Kalender[3], and Nicholas Marshall[1]

[1] Department of Radiology, University Hospitals of Leuven, Herestraat 49, 3000 Leuven, Belgium
[2] University of Patras, Greece
[3] Institute of Medical Physics, University of Erlangen, Germany
hilde.bosmans@uz.kuleuven.ac.be

Abstract. In order to justify the clinical use of a new modality such as digital breast tomosynthesis (DBT), a performance test in terms of a clinical parameter is indicated. For screening applications, lesion detectability is crucial. Whereas a real phantom with typically a homogeneous background that includes a series of details may provide a good indication of such detectability, a more specific approach is needed if the potential advantages of DBT have to be assessed. More specifically, the potential of the reduction of soft tissue superposition has to be covered in the performance test. We propose 2 concepts for a performance evaluation phantom: a hardware phantom with lesion-like inserts and where a background can be simulated on top of the images, and a software phantom with simulated lesions of different size and density (that can originate from theoretical models or from attenuation templates of real lesions). We discuss the need and the use of these phantoms for performance testing.

Keywords: Test objects, breast tomosynthesis, performance, observer studies.

1 Background

Digital breast tomosynthesis (DBT) is a new x-ray breast imaging technique, that can be offered as a separate system or as an option on (2D) mammography systems [1]. In a DBT acquisition, the x-ray tube moves in a small angle around the breast while a series of X-ray images are acquired, either with continuous tube motion or in a step-and-shoot mode. From these projections, images parallel to the detector are reconstructed. Image quality and performance are determined by X-ray tube, detector, beam quality, dose, geometry and motion, image reconstruction and visualisation. There are steps than in 2D full field digital mammography (FFDM).

An absolute performance evaluation of a DBT system should ideally be driven by the clinical task addressed. In this regard, screening needs a high detectability for (subtle) lesions. Justification to implement the new modalities into screening programmes may therefore require evidence that lesion detectability using DBT alone or combined with FFDM, improves compared to FFDM. For purely diagnostic applications, performance

J. Martí et al. (Eds.): IWDM 2010, LNCS 6136, pp. 227–234, 2010.

may relate more to the accurate work-up of a lesion. Today, it is a challenge to understand and assess the performance of a DBT system.

Detectability of lesions in DBT is influenced by three key-issues: (1) the quality of lesion reproduction in a reconstructed plane; (2) the ability of the DBT system to suppress out of plane texture produced by adjacent tissue; (3) the reconstructed image plane thickness presented to the radiologist. A well resolved reconstruction of a plane without any infolding of adjacent tissues may not be the best solution in the clinical practice. For example, there may be a minimum amount of tissue required to help the radiologist recognize patterns like clusters of microcalcifications.

Classical approaches of detectability assessment such as contrast-to-noise ratio (CNR) [2,3], with the contrast created by a homogenous insert in a homogenous background, will not fully assess the possible improvement in detectability produced by a DBT system over FFDM. The fact that the zero spatial frequency component of an object frequency spectrum is largely depressed in the reconstructed planes implies that large area image contrast may be reduced in DBT; the edge enhancement effect often seen in images reconstructed using the filtered back projection method [4] suggests that object edges or perimeter may be determining for detection. There is likely to be a difference between detection of small, irregularly shaped objects with a high contrast (such as clusters of microcalcifications) versus large objects with a more smooth shape. Artefacts may even be of benefit for detection purposes. In practice, benign and speculated masses may end up with different detectability characteristics. It can be anticipated that lesion detectability will be distinctly different, depending on object signal (size) and background structure [2]. Assessment with simple geometrical test objects may not provide a complete answer.

In the present text we discuss concepts for the practical performance testing of DBT systems. Ideally we would need a method (phantom, evaluation strategy, reference values, …) that answers 2 requirements: (1) the results correlate with the clinical performance and (2) the method can be applied on all systems. Several concepts have been simulated with our DBT simulation environment.

2 Material and Methods

In Table 1, we present an overview of parameters with an expected direct impact on the performance of a DBT system and the challenges in assessing them.

We discuss 2 tools to approach a performance test:

1. A 'hardware' phantom. The phantom should include realistic targets (representing microcalcifications and small masses with smooth and irregular borders) with a preference for alternative forced choice (AFC) reader tests (manual reading) or a larger number of objects for computerized reading, and produce a realistic scatter pattern for all competing breast X-ray imaging modalities. It should test at different locations and for different breast thicknesses. Anatomical backgrounds, ranging from fatty to very dense and having different thicknesses, could be foreseen in the phantom base material or added artificially, by multiplying the projection images of the real phantom with 'templates' from a real breast acquisition or from breast models.

Table 1. Overview of system parameters with a possible, direct impact on the performance of a DBT system and the challenges when defining a measurement methodology

Part	Parameter	Challenges
Tube	α, Δα, number of projections	Optimization requires simulation as these parameters cannot be (freely) set
	Geometry	Required measurement accuracy
	Continous vs. step-and-shoot	Include patient motion in the test
Detector	Pixel size (binning), detector material	Define 2D NPS, MTF, lag, homogeneity for low dose oblique projections; Access to raw data of the projection
AEC	Dose (proj, α, thickness, glandularities); Beam quality	Dosimetry for tube in motion; Need new conversion factors; Organize patient dose surveys
Reconstruction & processing	Reconstruction algorithms	Define: 3D NPS, 3D MTF, artefact spread function; Study detectability and characterization of lesions on antropomorphic or anatomical background
	Processing algorithms	
Display	Slabbing, MIP	Evaluation representative for the clinical working condition; evaluation of artefacts
	Hanging protocol' and speed of display	

2. A 'Software' phantom *Method 1:* the use of a specific series of 3D voxel models that represent realistic breasts, both in terms of X-ray attenuation coefficients and visual appearance. Projection images can be calculated from these models and that can then be reconstructed with the DBT reconstruction tools. The model is its own gold standard. Insertion of virtual lesions, either from theoretical models or from templates retrieved from acquisitions of real lesions can improve specificity. *Method 2:* the use of projection data of real patient cases into which templates of attenuation coefficients of real lesions (with given size and position) are simulated. These hybrid images can be reconstructed using routine reconstruction software. Observer studies can be set up.

3 Results

All suggested methods require access to the reconstruction software. In our team, this is performed for the Siemens DBT system using TomoEngine (Siemens, Germany).

1. The 'Hardware' phantom. Fig 1 shows a phantom design for a DBT assessment test and an image acquired from this phantom on a Siemens DBT system. It includes spheres of different glandularities and small particles to represent fibrils and micro-calcifications. The phantom is modular: the objects can be inserted in several configurations. Different types of breast simulating (homogeneous) tissue can be used to overlay the inserts. There are no inhomogeneous layers foreseen. Fig 2 illustrates an approach to create anatomical backgrounds, via the multiplication of the projection images with profiles of real structures. This approach is especially appealing for the study of small details. Larger details may become unrealistic as mass effects cannot be easily simulated. Fig 3 represents a modular phantom concept that can be developed into a performance testing phantom for DBT and breast-CT.

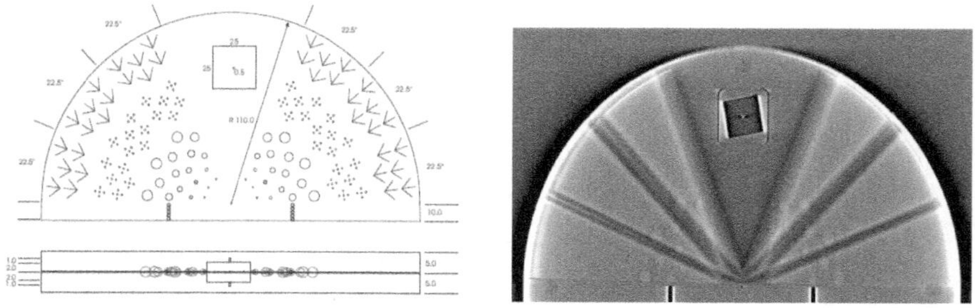

Fig. 1. *Left:* A physical test object for performance testing of DBT with inserts in one plane and triangle segments of glandular tissues in another plane (by A. Walker, Leeds Test Objects, UK). *Right:* DBT acquisition on a Siemens Inspiration DBT system (W/Rh, 26kV, 200mAs).

Fig. 2. *Left:* Reconstructed plane from projections of the phantom (Fig 1) after multiplication with anatomical background. *Middle & Right:* magnified image segments

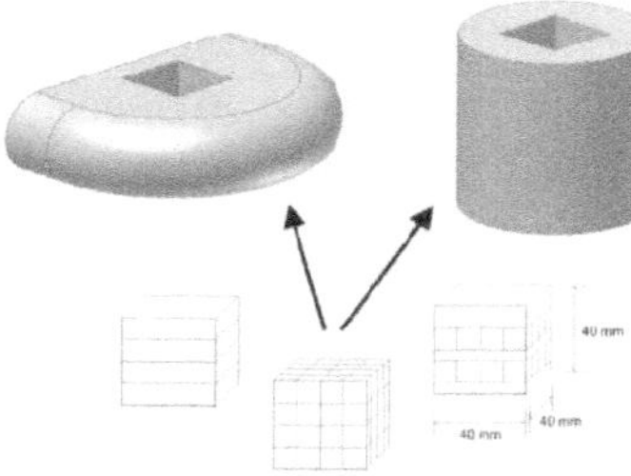

Fig. 3. A modular approach allows the testing of different modalities using the same basic phantoms. Concept developed within the FP7 breastCT project for FFDM, DBT and breastCT.

Breast tissue simulating backgrounds have been used in a commercial phantom (Mammography BR3D phantom), consisting of a set of 6 slabs made of a mixture of breast equivalent material mimicking 100% adipose and glad tissues swirled together in 50/50 ratio by weight. One slab includes a simulation for micro-calcifications, fibrils and masses. AK Carton (Duke conference on tomosynthesis, May 1st, 2009) produced a promising antropomorphic phantom from the Bakic voxel model.

2. The 'software' phantom. *Method 1.* Breast simulating (voxel model) phantoms are available today [5,6]. Fig 4 shows a typical example. Projection data are being calculated from this model using a Monte Carlo detector model [7],[8]. Into such a phantom, lesions with well known contrast, size and 3D position can be simulated [9]. The development of always more sophisticated phantoms is on-going work in several teams. Images thus simulated, with lesions at known locations for the computer but unknown for the human reader, can be input for free response Receiver Operating Characteristic (FROC) studies (human observations).

Fig 5 illustrates the approach (*Method* 2) of the partial simulation, in which lesions are being simulated into the projection images of real patients that are then reconstructed using the clinical tools. Artefacts typically seen in normal DBT images may also be seen and studied in simulated images.

The proposed phantoms are being developed, difficulties are gradually overcome.

4 Discussion and Conclusion

The justification of DBT could be performed via large clinical trials. As with 2D mammography, this may be very expensive and time consuming. Typical occurrence of cancer in screening series is about 6 to 7 cases per 1000 women, and whether a case is indeed normal is only confirmed after the subsequent screening round. The fact that not only global justification is required, but more refined justification that is specific to several types of patients or the use of the technique in comparison to other modalities such as echography and MRI etc, further emphasizes the need to test as many parameters as possible with hardware and/or software phantoms.

With regard to the objective assessment of image quality through the use of calculated parameters, we propose to opt for a DQE that is evaluated from reconstructed data; this would allow a measurement of system efficiency on an absolute scale, even for systems where there is no access to the projection images. However, there are potential problems with this method including the degree to which linear systems analysis can be applied to images with non-linear processing in order to estimate modulation transfer function (MTF), for example. Normalization of the noise power spectrum (NPS) for system gain in the reconstructed image sequence poses a further problem. The effect of anatomical background cannot easily be accounted for.

We have presented two approaches, using hardware and software phantoms, to overcome the limitations of a purely physics metrics. We are convinced that one hardware phantom with a homogenous background will not allow to evaluate the performance of DBT and certainly not the added value of DBT when compared to competing techniques. We hypothesize that the use of software phantoms or the adding of a background structure to phantom images may be a valuable alternative.

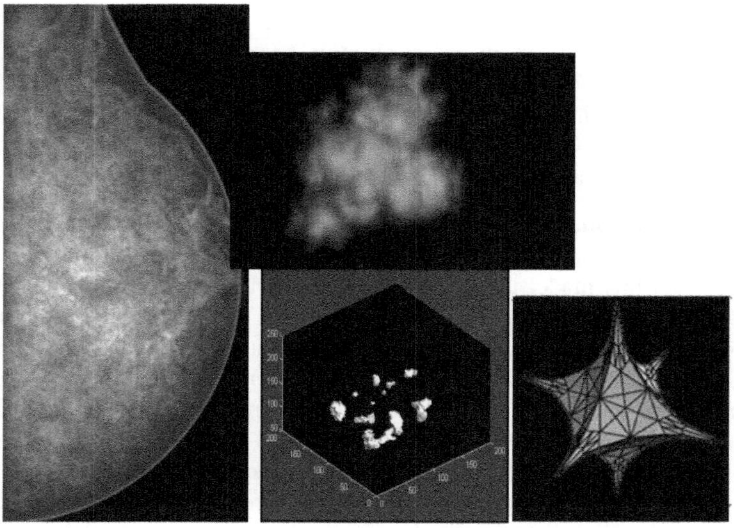

Fig. 4. Software phantom. *Left:* Mammogram calculated from a 3D voxel model of a breast. The model can be foreseen with simulated lesions for performance testing. *Top:* model for a mass based upon a random walk. *Middle:* 3D model of microcalcification cluster, obtained via micro-CT acquisition of a real cluster [9]. *Right:* model for speculated mass.

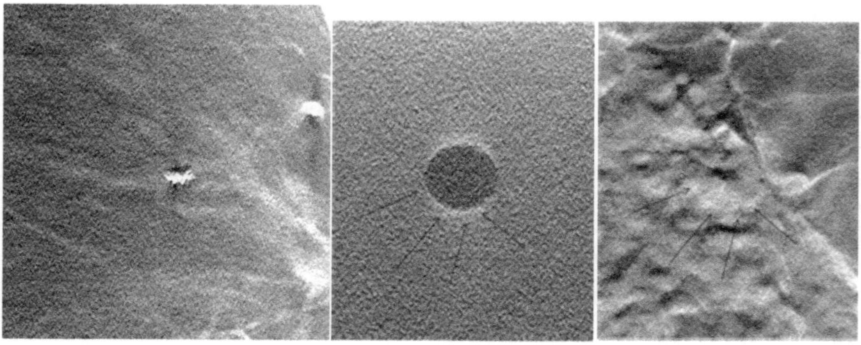

Fig. 5. Simulation of small objects into the projections of real cases or into homogenous PMMA plates. Left: a cluster of microcalcifcations simulated into anatomical backgrounds. Middle: a sphere of 1cm diameter and consisting of glandular tissue has been simulated into PMMA. Right: the same sphere simulated in an anatomical background. Artefacts visible in the PMMA images are faintly visible in the anatomical planes.

CNR evaluations have a crucial role in 2D FFDM, but the predictive power of this metric for system detectability may be weakened for 3D imaging modalities. Appropriate MOs may link better with detectability of objects, although the application of this method is complicated by the presence of structured noise [2]. The 3D detectability could be expressed via the detectability in the 2D tomographic plane with maximal detectability or as a (weighted) sum of detectability in an adjacent series of reconstructed planes. More research, particularly with respect to the validation of MO methods in DBT has high priority.

The software phantoms proposed in this text, can be expanded to breast CT or other X-ray modalities (2D FFDM, phase contrast, dual energy approaches, monoenergetic rays) as long as the detected signal can be simulated. Equally crucial is the access to raw, projection data, the possibility to use the image reconstruction software as a black box and the same with the image processing software.

A good performance test object could test the potential benefits of some particular design features (like geometry, new design of grids, dual energy, ...) or beam quality settings of the systems. Other aspects cannot be easily tested neither with static test objects nor with simulated phantoms: the advantages of step and shoot versus continuous motion, ultra short acquisition times, slot scan devices, etc. For such applications, it may be sufficient to review series of patient cases and calculate rejection rate, the number of obvious motion artefacts, etc... The role of the radiologist and the radiographer should not be underestimated.

The concepts presented here should be further discussed with other groups and compared to other concepts that aim for performance testing.

Acknowledgements

To the EC-EURATOM 7 Framework Program that is part of the project "Dedicated CT of the Female Breast: Feasibility, Optimisation and Comparison to competing imaging modalities", contract no. FP/213153.

References

1. Niklason, L.T., Christian, B.T., Niklason, L.E., Kopans, D.B., Castleberry, D.E., Opsahl-Ong, B.H., Landberg, C.E., Slanetz, P.J., Giardino, A.A., Moore, R., Albagli, D., DeJule, M.C., Fitzgerald, P.F., Fobare, D.F., Giambattista, B.W., Kwasnick, R.F., Liu, J., Lubowski, S.J., Possin, G.E., Richotte, J.F., Wei, C.Y., Wirth, R.F.: Digital Tomosynthesis in Breast Imaging. Radiology 205, 399–406 (1997)
2. Burgess, A.E., Jacobson, F.L., Judy, P.F.: Human observer detection experiments with mammograms and power-law noise. Med. Phys. 28, 419–437 (2001)
3. EC (European Commission) 2006 European guidelines for quality assurance in breast cancer screening and diagnosis, 4th edn. European Commission, Brussels (2006)
4. Mertelmeier, T., Orman, J., Haerer, W., Dudam, M.K.: Optimizing filtered backprojection reconstruction for a breast tomosynthesis prototype device. In: Medical Imaging 2006: Physics of Medical Imaging, Proceedings of SPIE, vol. 6142, pp. 61420F-1–61420F-12 (2006)
5. Bakic, P.R., Albert, M., Brzakovic, D., Maidment, A.D.A.: Mammogram synthesis using a 3D simulation I. Breast tissue model and image acquisition simulation. Med. Phys. 29, 2131–2139 (2002)
6. Bliznakova, K., Bliznakov, Z., Bravou, V., Kolitsi, Z., Pallikarakis, N.: A three-dimensional breast software phantom for mammography simulation. Phys. Med. Biol. 48, 3699–3719 (2003)

7. Smans, K., Carton, A.K., Zoetelief, J., Verbrugge, B., Haek, W., Struelens, L., Vanhavere, F., Bosmans, H.: Med. Phys. Simulation of image detectors in radiology for determination of scatter-to-primary ratios using Monte Carlo radiation transport code MCNP/MCNPX. Accepted for publication in Med. Phys
8. Tward, D.J., Siewerdsen, J.H.: Cascaded systems analysis of the 3D noise transfer characteristics of flat-panel cone-beam CT. Med. Phys. 35, 5510–5529 (2008)
9. Shaheen, E., Van Ongeval, C., Cockmartin, L., Zanca, F., Marshall, N., Jacobs, J., Bosmans, H.: Realistic Simulation of Microcalcifications in Breast Tomosynthesis. In: IWDM (2010)

Realistic Simulation of Microcalcifications in Breast Tomosynthesis

Eman Shaheen, Chantal Van Ongeval, Lesley Cockmartin, Federica Zanca, Nicholas Marshall, Jurgen Jacobs, and Hilde Bosmans

Department of Radiology, University Hospitals of Leuven,
Herestraat 49, 3000 Leuven, Belgium
{eman.shaheen,chantal.vanongeval,lesley.cockmartin,
federica.zanca,nicholas.marshall,jurgen.jacobs,
hilde.bosmans}@uzleuven.be

Abstract. Digital breast tomosynthesis (DBT) provides a possible solution to overcome the problem of overlapping tissue since it provides a 3D volume representation of the imaged object. In order to study the detectability of lesions in DBT, we have developed a simulation tool where objects are simulated into real projection DBT images. The methodology has already been validated for 3D geometrical shapes and has now been extended to irregularly shaped lesions. The work focuses on the simulation of clusters of microcalcifications that are modeled using micro-CT images of biopsy specimens containing such lesions. The compilation of a database of microcalcifications clusters classified following Le Gal nomenclature is ongoing. These extracted model lesions were then simulated into images of biopsy specimens next to the original real cluster in order to confirm the realism of the simulation.

Keywords: breast tomosynthesis, microcalcifications simulation, image quality.

1 Background

Digital tomosynthesis is a form of limited angle tomography that produces section images from a series of projection images acquired as the x-ray tube moves over a prescribed path [1]. The total angular range of the tube movement ranges from 15° to 50° for digital breast tomosynthesis (DBT). The projection images are then reconstructed into planes parallel to the detector. DBT is considered as a potential imaging technique for early detection of breast cancer as it provides volumetric information of the breast that could solve the problem of overlapping tissues hiding pathology of interest [2]. Whether this modality improves the detection of subtle lesions, when compared to 2D mammography, remains to be proven. To study the detectability of lesions, we have developed and validated a software framework that can simulate 3D objects into a DBT image series in a realistic way [3]. The technique of simulation has been applied to both 2D projection mammography [4] and to DBT imaging [5].

In the present work, we report on the simulation of subtle lesions (microcalcifications) into DBT images, including the methodology to produce 3D models of clusters of microcalcifications, the collection of a database of different types of microcalcifications,

J. Martí et al. (Eds.): IWDM 2010, LNCS 6136, pp. 235–242, 2010.

and the procedure to simulate these clusters into DBT projection data. We also report on the steps towards a validation study to verify the realistic appearance of simulated microcalcifications. The goal of the current work is to use the framework described by Shaheen [3] and adapt it to the simulation of clusters of microcalcifications into real patient breast images for later use in observer studies concerning the detectability of such lesions in DBT.

2 Materials and Methods

2.1 Simulation Framework: Theoretical Aspect

A framework was previously designed [3] to simulate small 3D objects into real projection images of breast tomosynthesis systems. Fig. 1 is an overview of the different steps: a 3D object (including its spatial coordinates) is defined and ray traced using the Siddon algorithm [6] to obtain templates representing the total primary X-ray attenuation for the different tomography angles. These ideal templates are then modified to the corresponding detector characteristics by multiplying each template by its corresponding detector modulation transfer function (MTF). Due to the assumption that the simulation object is small, the noise is assumed to be inherited from the background. The scatter fraction is then calculated based on Boone's data [7] and used to estimate the scatter offset and the primary component of the beam. The templates are multiplied with the estimated primary part of the real tomographic input images. These hybrid images are subsequently reconstructed. This same methodology can be used to simulate objects into 2D digital mammography systems by considering the geometry of the system and using correct estimates of scatter fraction for systems with anti-scatter grid. The pre-processing (raw) hybrid image is then processed with the image processing algorithm.

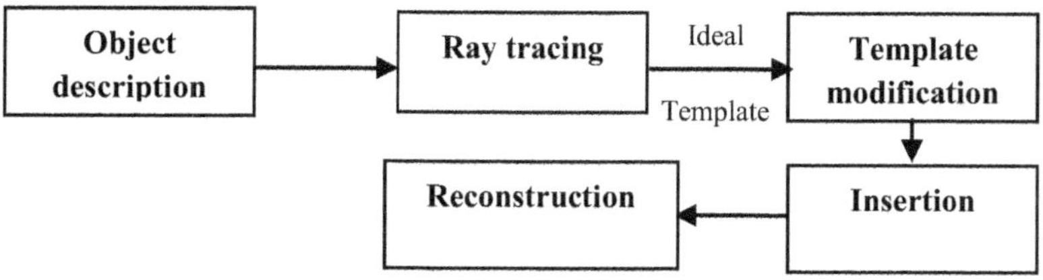

Fig. 1. Simulation process for 3D objects

2.2 3D Model of Microcalcifications

In order to assure a realistic representation of microcalcifications, the simulation starts from biopsy specimens containing clusters of microcalcifications. The specimens are obtained with a guided vacuum large core needle (10 gauge) by means of a digital biopsy table [Mammotest Plus/S (Fisher Imaging, Norderstedt)]. For every patient case, the tissue containing the microcalcifications was preserved in a tube containing Formaldehyde (CH_2O) (for later histological analysis) and was imaged using a cone beam micro-CT scanner (SkyScan 1172) that provides projection and 3D images.

2.2.1 Model Building

To obtain good models of the microcalcifications, the micro-CT exposure settings were adjusted for each sample. Tube voltage was varied between 60 to 80 kVp and the μAs was set to give low noise without saturation. In the simulation framework, the linear attenuation coefficient of the 3D object is assumed to be constant within the object. Therefore, the microcalcification model is obtained by segmenting the micro-CT 3D images to produce a binary 3D model.

A segmentation procedure is applied to every reconstructed image of a biopsy specimen. This included microcalcification edge detection, morphological operations such as dilation and erosion, and the application of a median filter to remove the background noise.

Calcium oxalate (CaC_2O_4) is used as substitute material for microcalcifications since it is considered a close match [8]. The steps used to build the 3D models of clusters of microcalcifications as described in section 2.2 are shown in Fig. 2.

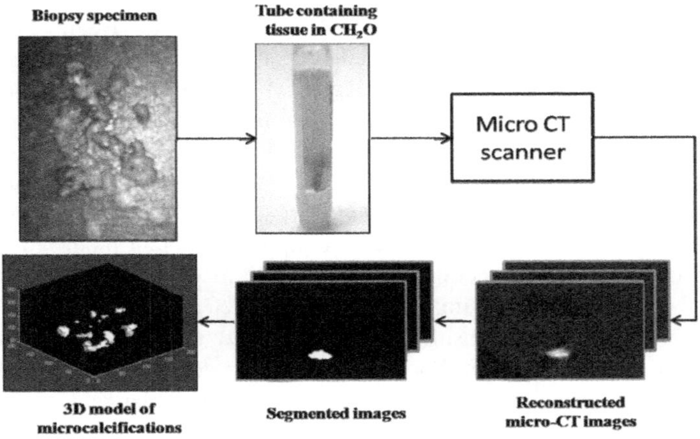

Fig. 2. Steps to create 3D models of clusters of microcalcifications

2.2.2 Database Building

The creation of a database of 3D models of microcalcifications is currently in progress with data acquired on a weekly basis. Since it is important to preserve the shape information of the simulated microcalcifications that is crucial in diagnosis, the database was set up to contain lesions representative for all morphologies found in clinical practice. The morphological characteristics of the microcalcifications were classified using the Le Gal system [9] (widely used in Europe). There are five Le Gal classes (Fig. 3): type 1 calcifications is defined as annular; type 2 is defined as regular punctiform (point-like); type 3 is defined as too fine to characterize the shape (dusty); type 4 is defined as irregularly punctiform; and type 5 is defined as vermicular (worm-like) calcifications. While the goal is to obtain the different Le Gal types of calcifications as described by Zanca [4], this is not a straight forward task and is done by a trained radiologist.

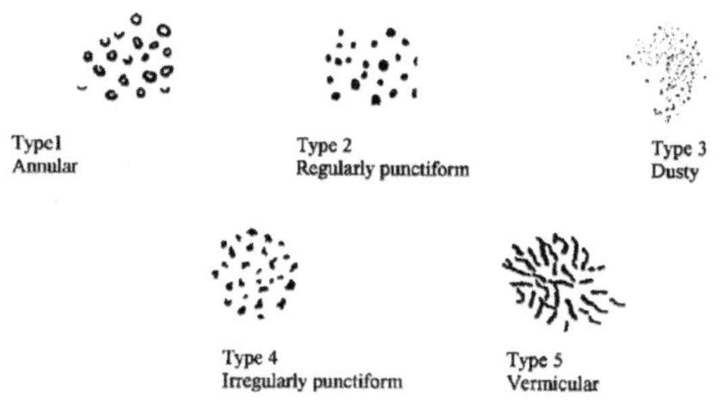

Fig. 3. The classification of microcalcifications by their morphologic characteristics according to Le Gal's classification system.

2.3 Microcalcification Simulation

2.3.1 Simulation Procedure

A DBT system (Siemens Inspiration TOMO, Siemens AG Healthcare, Erlangen, Germany) with an angular range of ±25° that produces 25 projection images is used. The system is equipped with an *a*-Se detector with a pixel size of 85μm. The same system also produces 2D digital mammographic images. In order to simulate microcalcifications into patient images, the 3D models built as described in section 2.2.1 are used as input. The spatial position of the 3D model of a cluster of microcalcifications is defined along with the geometrical information of the DBT system. The simulation framework described in section 2.1 is applied to obtain modified templates to be inserted into real raw projection images prior to reconstruction. In case of tomosynthesis, the raw projection images are reconstructed using the software of the company (TomoEngine, Siemens, Erlangen, Germany) that is based on Filtered Backprojection algorithm (FBP) [10]. For the 2D case, the default image processing algorithm for 2D digital mammography (OpView2, Siemens, Erlangen) is applied to the raw image to obtain a processed image.

2.3.2 Validation Method

The present approach was validated for a selected set of biopsy specimens. These biopsy specimens were put on top of a 4cm PMMA slab and imaged by the DBT system in 2D and in tomosynthesis modes where the dose was set slightly higher for better visualization than for a typical 4.5cm breast in digital mammography (1.3 mGy) and doubled in the tomosynthesis mode (2.6 mGy). In order to study the realism of the simulation of the microcalcifications, the 3D model of the cluster based on the same specimen was simulated into the projection raw biopsy images of the real cluster at a spatial position adjacent to this cluster.

One of the objectives of the presented work is the collection of a database of microcalcifications clusters that are classified according to their morphology. With such a database, it will be possible to study whether DBT preserves the shape information of the simulated calcifications. Our verification of the simulation procedure started with the simulation of the clusters of microcalcifications in the same environment as where they were taken from, in a simple homogeneous background. This would reduce the number of confounding parameters for the analysis, like anatomical background, different breast glandularity, the insertion position into a breast… These will be the topic of future research. Fig. 4 shows an example of a simulated cluster of microcalcifications, using the proposed method, inserted in 2D digital mammography raw image (shown at left the processed version) and in tomosynthesis projections that are reconstructed (shown at right, for the in focus plane), where the region containing both the simulated and real clusters is selected and magnified next to it, to be able to visualize it. The arrow indicates the simulated cluster of Le Gal type 4.

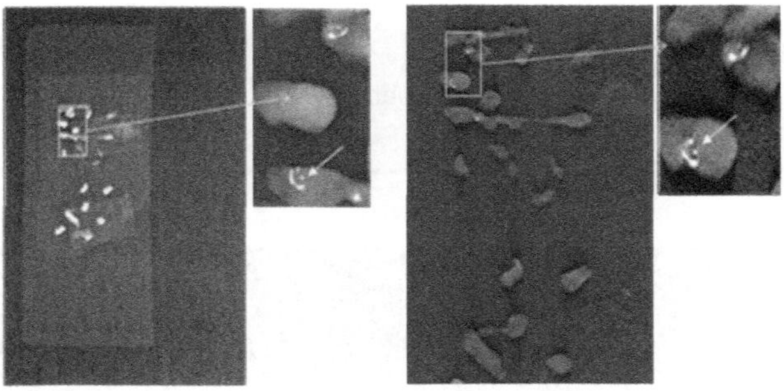

Fig. 4. A simulated cluster of microcalcifications (Le Gal type 4) in 2D processed image (left) and in reconstructed in focus tomographic plane (right). The simulated cluster is inserted adjacent to the real cluster and indicated by the arrow.

3 Results and Discussion

The creation of the database is still in progress. A set of 15 biopsy specimens is collected so far. Every biopsy specimen was scanned by the micro-CT, segmented to form a 3D model, and re-simulated into the biopsy images using the proposed methodology for validation. From the 15 biopsy specimen cases, only 12 could be segmented and modeled. Failure to do so was caused by reconstruction artifacts (beam hardening) of the micro-CT that made the segmentation impossible. An example of these artifacts is shown in Fig. 5.

The simulated clusters in the 2D processed images and in the reconstructed tomosynthesis planes were judged by an expert radiologist who determined whether the simulated and real clusters were distinguishable and classified the clusters to the correct Le Gal type. The classification of the Le Gal types in our database is shown in Table 1. Since some clusters of microcalcifications were not consisting of only one Le Gal type, these clusters were classified to the type dominating the cluster.

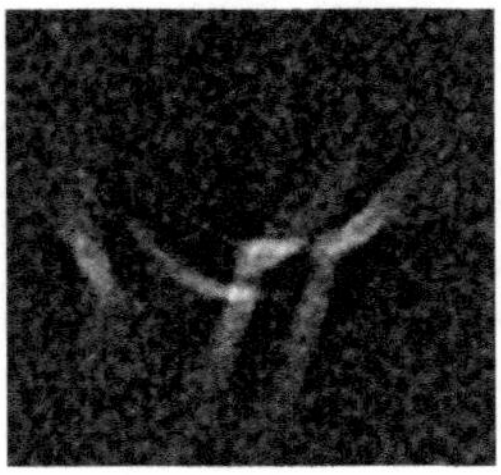

Fig. 5. A reconstructed micro-CT image with artifacts

Table 1. The different Le Gal types currently available in the database

Le Gal type		Number of models
1	(annular)	0
2	(regularly punctiform)	4
3	(dusty)	3
4	(irregularly punctiform)	5
5	(vermicular)	1

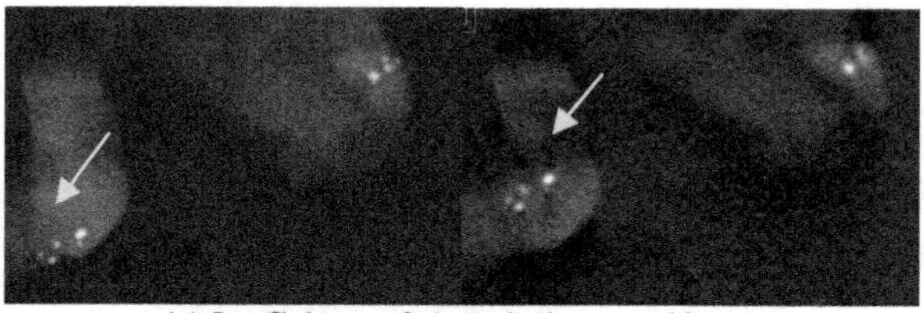

(a) Le Gal type 2 (regularly punctiform)

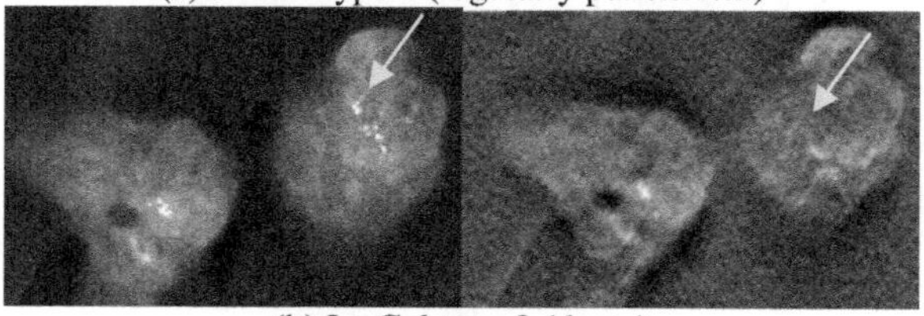

(b) Le Gal type 3 (dusty)

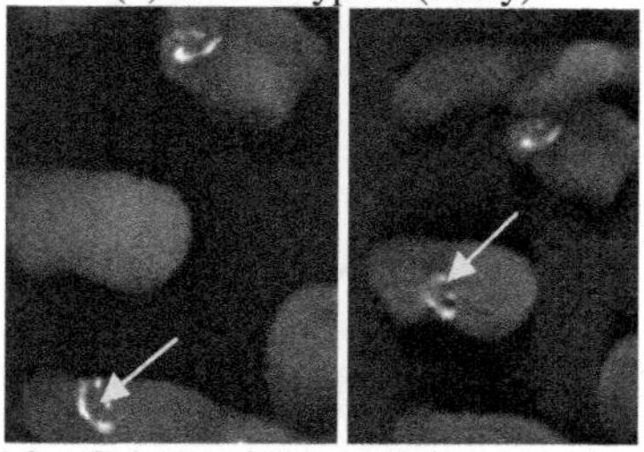

(c) Le Gal type 4 (irregularly punctiform)

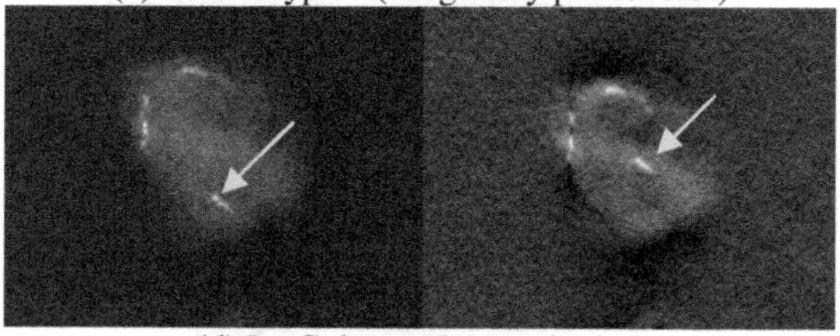

(d) Le Gal type 5 (vermicular)

Fig. 6. Simulated vs Real clusters of microcalcifications of the different Le Gal types. The 2D processed image (left), and the in-focus reconstructed tomographic plane (right). The simulated cluster is indicated by the arrow.

For the distinction between real and simulated clusters of microcalcifications, 10 out of 12 simulations were not distinguished by the expert radiologist, which is considered an encouraging feedback from the radiologist who had the freedom to zoom (in/out), scroll through the different planes, and to change the window level. The 2 cases, which were recognized as simulated, had a slightly different background such that the real cluster was in a glandular tissue and the simulated was inserted into a fatty tissue. Some examples of the simulated and real clusters are shown in Fig. 6. In some cases such as in Fig. 6(b), the visualization and recognition of the Le Gal type of the cluster was better in 2D, this is due to the presence of artifacts around dense microcalcifications.

The visibility of the microcalcifications is preserved throughout the simulation process which is an important aspect in simulating microcalcifications and a crucial requirement for diagnosis.

4 Conclusion

A newly developed methodology is proposed to simulate clusters of microcalcifications into real raw projection DBT images that can be generalized to 2D digital mammography and breast CT. It starts by building 3D models of microcalcifications that are acquired from biopsy specimens and their classification, in terms of morphological characteristics, into the different Le Gal types to form a database. Once an efficient database is established, it can be used to simulate these clusters into patient images using the proposed framework. The future work is to validate this framework with patient data in order to use it in observer studies related to the detectability of such subtle lesions in terms of different aspects. The preliminary results show that it is a promising tool in terms of realism, efficiency and simplicity that can be applied in several applications.

Acknowledgments. To the EC-EURATOM 7 Framework Program that is part of the project "Dedicated CT of the Female Breast: Feasibility, Optimisation and Comparison to competing imaging modalities", contract no. FP/213153.

References

1. Dobbins III, J.T.: Tomosynthesis imaging: At a translational crossroads. Med. Phys. 36(6), 1956–1967 (2009)
2. Wu, T., Stewart, A., Stanton, M., McCauley, T., Phillips, W., Kopans, D.B., Moore, R.H., Eberhard, J.W., Opsahl-Ong, B., Niklason, L., Williams, M.B.: Tomographic mammography using a limited number of lowdose cone-beam projection images. Med. Phys. 30, 365–380 (2003)
3. Shaheen, E., Zanca, F., Sisini, F., Zhang, G., Jacobs, J., Bosmans, H.: Simulation of 3D objects into breast tomosynthesis images. Radiat. Prot. Dosimet. 10 (2010)
4. Zanca, F., Chakraborty, D.P., Van Ongeval, C., Jacobs, J., Claus, F., Marchal, G., Bosmans, H.: An improved method for simulating microcalcifications in digital mammograms. Med. Phys. 35, 4012–4018 (2008)

5. Ruschin, M., et al.: Improved in-plane visibility of tumors using breast tomosynthesis. In: Proc. of SPIE, vol. 6510, pp. 65101J1– 65101J11 (2007)
6. Siddon, R.L.: Fast calculation of the exact radiological path for a three-dimensional CT array. Med. Phys. 12, 252–255 (1985)
7. Boone, J.M., Lindfors, K.K., Cooper III, V.N., Seibert, J.A.: Scatter/primary in mammography: Comprehensive results. Med. Phys. 27, 2408–2416 (2000)
8. Fandos-Morera, A., Prats-Esteve, M., Tura-Soteras, J.M., Traveria-Cros, A.: Breast Tumors: Composition of Microcalcifications. Radiology 169, 325–327 (1988)
9. Le Gal, M., et al.: Diagnostic value of clustered microcalcifications discovered by mammography. Bull. Cancer (Paris) 71, 57–64 (1984)
10. Mertelmeier, T., Orman, J., Haerer, W., Dudam, M.K.: Optimizing filtered backprojection reconstruction for a breast tomosynthesis prototype device. In: Proc. of SPIE, vol. 6142, pp. 61420F1– 61420F12 (2006)

Breast Tomosynthesis Reconstruction Using a Grid of Blobs with Projection Matrices

Gang Wu[1,2], James G. Mainprize[2], and Martin J. Yaffe[1,2]

[1] Department of Medical Biophysics, University of Toronto
Toronto, Ontario, Canada
[2] Imaging Research, Sunnybrook Health Sciences Centre
2075 Bayview Avenue, Toronto, Ontario, Canada
{gang.wu,james.mainprize,martin.yaffe}@sri.utoronto.ca

Abstract. Spherically symmetric basis functions (blobs) are alternatives to the more conventional cubic voxels for image reconstruction in breast tomosynthesis. The volume representation and its projection views (PV) are essential components of iterative algorithms for image reconstruction from data collected from an area detector. This paper addresses the forward projection and backprojection process of three-dimensional (3D) breast reconstruction obtained from cone-beam scans using tomosynthesis imaging equipment. The smoothness of the blob elements allows more realistic modeling of the breast, and the rotational symmetry of the elements leads to more efficient calculation of both directional projection of the represented volume, as required in iterative reconstruction techniques. The combination of blob volume elements and the projection matrix method improves tomosynthesis reconstruction in both accuracy and speed.

Keywords: Breast tomosynthesis, cone-beam projection, image reconstruction, blob, projection matrix.

1 Introduction

Digital breast tomosynthesis (DBT) is an x-ray acquisition and processing technique which is based on a set of projection images acquired over a range of angles. From the reconstruction of the projection images a series of cross-sectional images or slices are obtained. The advantage of DBT over conventional mammography is that the slices do not contain a superposition of the anatomic structures that occur over the thickness of the breast. Resolving the depth in the image to a selected slice eliminates the confounding effect of structures at other depths and can, therefore, enhance the conspicuity of a tumour as well as facilitate spatial localization within the breast. Moreover the patient does not incur additional radiation dose from DBT compared to mammography DBT.

However the reconstruction of a three-dimensional (3D) breast volume is challenging in DBT because the dataset may be sparse and/or noisy: only a limited number of low-dose projections are acquired over an arc. Consequently spatial resolution through the thickness of the breast is typically inferior to the resolution within the plane of the detector. Iterative methods for calculating the reconstruction are preferred

J. Martí et al. (Eds.): IWDM 2010, LNCS 6136, pp. 243–250, 2010.

when projections are sparse, noisy, or when sampling is non-uniform [1]. Iterative reconstruction techniques are based on successive estimation of backprojection and forward projections of the object estimated at intermediate stages of the algorithm [2]. However, they are critical steps in many tomographic image reconstruction techniques. The computed values are based on a model of the data acquisition process which is often simplified or compromised by the need to minimize the reconstruction time. Therefore, much effort has been spent in designing fast projection techniques for reconstruction based on cubic voxels [3]. There remains a need for methods that combine the speed of the fast-projection techniques with more accurate modeling. This is particularly true for DBT, where the data acquisition is inherently incomplete and the breast is under-sampled.

Conventional reconstruction methods generate breast representations that are comprised of volume elements or voxels that are generally cubic or rectangular prisms arranged on a regular Cartesian lattice. These choices, although they simplify the analysis/and are computationally efficient, allow only a crude approximation of the object according to the principles of sampling theory. As an alternative, smooth 3-D, spherically-symmetric volume elements known as 'blobs' are proposed [4]. Blobs, like cubic voxels, are centred on a regular Cartesian grid in 3-D. Unlike the contiguous cubes or prisms, however blobs overlap and have a bell-shaped weighting profile whose value tapers smoothly from unity at the centre of the blob in the radial direction to nil at the surface of the spherical 'blob'.

In this paper we describe the application of smooth blob volume elements to a ray-tracing reconstruction algorithm as well as a projection-matrix, and evaluate the improvements achieved in both computational speed and image fidelity compared to the use of cubic voxels.

2 Method

2.1 Breast Volume Representation by Blobs

Reconstruction in DBT yields a discrete approximation of the (continuous) distribution of breast attenuation on a grid. Image values at intermediate locations (i.e., between the grid points) can be obtained by interpolation. The interpolation process consists of a convolution operation which appropriately weights and shifts copies of a voxel. Therefore, using blobs, in principle a continuous description of the object could (theoretically) be restored by performing these interpolations everywhere in the volume, from the finite set of stored image data. Following the approach of Lewitt [4] and Galigekere, [5] we present a blob-based representation of image reconstruction for DBT.

Let $\boldsymbol{Q} = (x, y, z)^T$ be the point in the object having components x, y and z along the orthogonal X-, Y-and Z-axes. The reconstruction volume $\boldsymbol{V}$ is a Cartesian lattice specified in terms of a reference point $\boldsymbol{Q_0}$ and a transposed array of index vectors $\boldsymbol{J} = (i, j, k)^T$ (all vectors are column vectors). Δx, Δy and Δz describe the grid resolution in the three directions, respectively. In a Cartesian grid $\boldsymbol{G}$, the 3-D coordinates of a grid-point $\boldsymbol{J}$ in $\boldsymbol{V}$ can be expressed as vector:

$$G(J) = G_0 + (i\Delta x, j\Delta y, k\Delta z)^T \quad (1)$$

The image representation denoted by $\bar{f}$ at an arbitrary position Q within the volume is constructed as the superposition of scaled and shifted copies of the local basis function b, as follows:

$$\bar{f}(Q) = \sum_J c(J) b(Q - G(J)) \ . \quad (2)$$

For rotational symmetric basis functions (blobs):

$$b(Q - G(J)) = b(\|Q - G(J)\|) \quad (3)$$

where $\|Q\| = (x^2 + y^2 + z^2)^{1/2}$ denotes the distance of Q from the origin. The general Kaiser-Bessel function is defined in the spatial domain as follows:

$$b^{(m,\alpha)}(r) = \begin{cases} \frac{1}{I_m(\alpha)} (\sqrt{1-(r/a)^2})^m I_m(\alpha\sqrt{1-(r/a)^2}) & \text{for } 0 \le r \le a, \\ 0 & \text{otherwise.} \end{cases} \quad (4)$$

where a is the extent of the blob (set as 2.0), r is the distance from the centre of the blob, and α is the 'taper' parameter (a property of the bell-shaped weighting profile) which should be 10.40 to obtain a high quality low-pass filter performance in the frequency domain of the basis function [2]). The function, I_m. is the modified Bessel function of the first kind with order m (set m = 2 to get a continuous derivative within the blob and at the border [2]).

2.2 Projection Matrix

Regardless of the geometry that is chosen for image acquisition in DBT (e.g., isocentric, partial isocentric or parallel), the image acquisition can be represented (in the homogeneous coordinate system) by a sequence of matrix operations. Following the rigid transformation of gantry and/or detector, the perspective view associated with each angle can be described by a single projection matrix ***P***. The matrix ***P*** can be either estimated by imaging a calibration phantom or constructed from direct specification of the acquisition geometry. We used the following geometry in our simulations: the x-ray source-to-image distance (SID) is 63 cm, and the centre of rotation is 2cm above the detector. The detector remains stationary while the x-ray tube rotates around the pivot, and θ is the rotation angle of the x-ray tube, with zero defined as the perpendicular position of the tube with respect to the detector. The projection matrix can be calculated as:

$$P = \begin{pmatrix} SID & 0 & u_0 \\ 0 & SID & v_0 \\ 0 & 0 & 1 \end{pmatrix} \begin{pmatrix} 1 & 0 & 0 & -SOD\sin\theta \\ 0 & 1 & 0 & 0 \\ 0 & 0 & -1 & SOD\cos\theta \end{pmatrix} \quad (5)$$

where (u_0, v_0) are the coordinates of the perpendicular projection point of the source over the detector plane, and source-to-object distance (SOD) is the source to pivot distance. Deviations from ideal geometry can be incorporated through suitable transformation matrix applied to ***P***. Thus, the projection of an arbitrary grid-point $\boldsymbol{G}(\boldsymbol{J})$ in the volume to be reconstructed ***V*** can be computed by matrix multiplication. Several algorithms have been proposed for voxel-driven projection (VDP), based on the efficiency of matrix multiplication with P [5]. However, as commonly implemented, the quality of ray-driven projection (RDP) is superior to that of the VDP, especially in the case of perspective view or non-isotropic volume elements used in tomosynthesis. Here, we demonstrate the use of matrix ***P*** in ray-driven projection (Fig. 1).

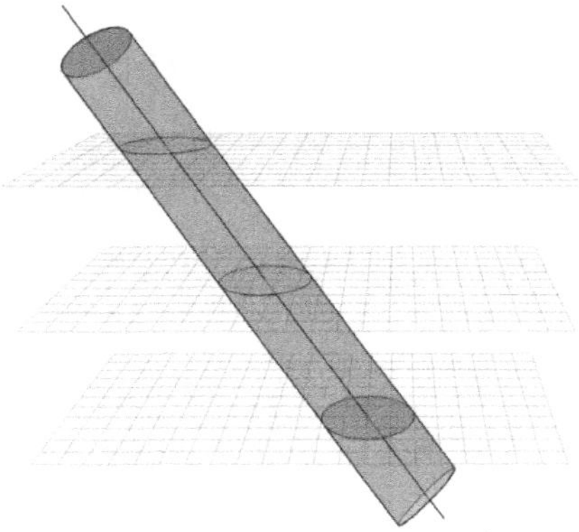

Fig. 1. Ray tracing with computation of distances from the x-ray to centres of blobs encountered in transit determine the active region of neighboring grid points

2.3 Fast Ray Tracing Algorithm

The x-ray passing through source point $\boldsymbol{S} = (S_x, S_y, S_z)^T$ has a unit direction vector $\boldsymbol{U} = (U_x, U_y, U_z)^T$ which is determined by the destination detector element. The equation of the x-ray is specified as: $\boldsymbol{Q}(t) = \boldsymbol{S} + \boldsymbol{U}t$. It is possible to extract source position ***S*** and ray direction ***U*** from ***P*** without matrix decomposition or any explicit knowledge of the detector pixel coordinates [5]. The two most important parts in ray-driven projection (RDP) are finding the interception points of the x-ray with the volume of interest and computing the ray-sum. First, we must choose the most efficient dimension for ray tracing, and this is usually the dimension which has the maximum absolute value in ***U*** components, *i.e.*, $\max(|U_x|, |U_y|, |U_z|)$. For illustration we consider the *Z*- axis; similar equations apply to other directions.

The points intercepted with each (*k*th) grid plane along the chosen direction can be determined by:

$$Q(t_k) = S + Ut_k = \begin{pmatrix} S_x + U_x(G_{0z} - S_z + k\Delta z)/U_z \\ S_y + U_y(G_{0z} - S_z + k\Delta z)/U_z \\ G_{0z} - S_z + k\Delta z \end{pmatrix} \qquad (6)$$

The square distance from any grid point on the kth plane to the x-ray therefore can be expressed as $r^2 = \|(Q(t_k) - G(J)) \times U\|^2$, as shown in Fig. 2, which expands to simpler formula compared to other general ray-tracing algorithms [6]. Let $\boldsymbol{B} = \boldsymbol{Q}(t_k) - \boldsymbol{G}(\boldsymbol{J})$,

$$\boldsymbol{B} = \boldsymbol{Q}(t_k) - \boldsymbol{G}(\boldsymbol{J}) = \begin{pmatrix} S_x - G_{0x} + \frac{(G_{0z} - S_z)U_x}{U_z} + \frac{k\Delta z U_x}{U_z} - i\Delta x \\ S_y - G_{0y} + \frac{(G_{0z} - S_z)U_y}{U_z} + \frac{k\Delta z U_y}{U_z} - j\Delta y \\ 0 \end{pmatrix}. \qquad (7)$$

Because the third component of B in equation 7 equals zero, the time required to compute the square distance is reduced by one-third. Furthermore, other two components of $\boldsymbol{B}$ are comprised of a constant part and an incremental part when moving from one grid position to next along the chosen direction.

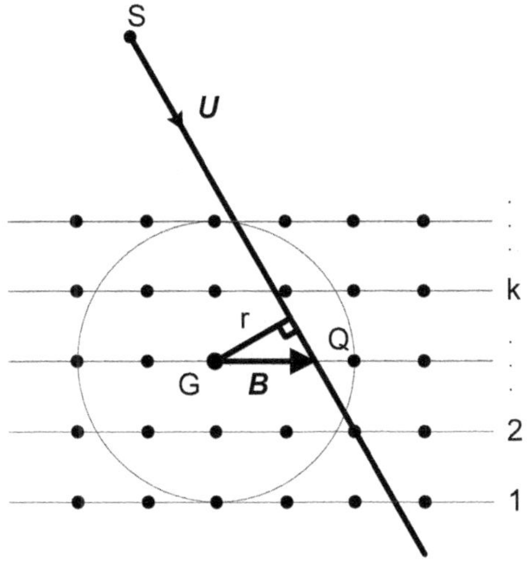

Fig. 2. Distance from an x-ray to the centre of a blob on any grid lattice

The acquisition of the digital breast tomosynthesis was simulated for a partial isocentric geometry. The detector was stationary while the x-ray tube rotated around a pivot point. Eleven (11) projections were taken at different angles from -20° to 20°, with 4° increments, using monoenergetic (20 keV) x-rays. Two mathematical phantoms representing the compressed breast were created: Phantom A and Phantom B. Phantom A has a uniform distribution of 50% fibroglandular and 50% adipose tissue as background. While Phantom A provides a good model of the artifacts inherent in DBT, its uniform background is an oversimplification of the complex detection task in clinical breast screening and diagnosis. Anatomic structures in the normal breast, particularly in the dense breast (which contains a high proportion of fibroglandular tissue), present a 'cluttered' background which may confound image interpretation. Phantom B was designed based on volumetric tissue attenuation data obtained from a clinical breast tomosynthesis examination and, therefore, has a depth-dependent structure. Both mathematical phantoms are rectangular prisms in shape and each contains a simulated small tumor (4 mm × 4 mm × 2 mm) at the centre. The attenuation coefficients of the simulated tumour are equivalent to those measured for infiltrating ductal carcinoma (IDC) [6].

An iterative reconstruction method, simultaneous algebraic reconstruction technique (SART) has been implemented in Matlab and C++ to evaluate the performance of reconstruction algorithm which use blob voxels. The conventional box-voxel reconstructions are also used to compare the resulting image quality.

3 Results

For the uniform phantom (Phantom A), reconstructions based on cubic and blob voxels resulted in almost exactly the same volume, except that the blob-voxel reconstruction tended to show more artifacts near the edge of the phantom if the boundary geometry is not correctly specified.

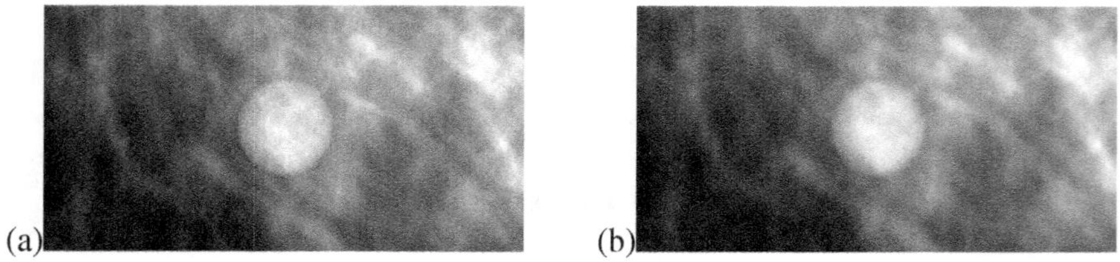

Fig. 3. The central slice of the tumor after two iterations of SART reconstruction for Phantom B, shown in X-Y plane, in comparison of two different voxel types: (a) regular cubic voxels, (b) spherical blob voxels

The central slices of the reconstructed volumes of the phantom B were compared between box and blob RDP methods, in the neighborhood of the simulated tumor as shown in Fig. 3. Fig 3a and 3b are displayed with the same grayscale window and centered at the same level. The blob-voxel reconstruction resulted in a less noisy but more blurry image compared to the cubic voxels. The box-voxel reconstruction is noisier and has more intensity fluctuations in the background tissue and the tumor, which introduce some ring-like artifacts in the tumor area. This is also demonstrated in the profile through the centre of the tumor for Phantom B, as shown in Fig. 4.

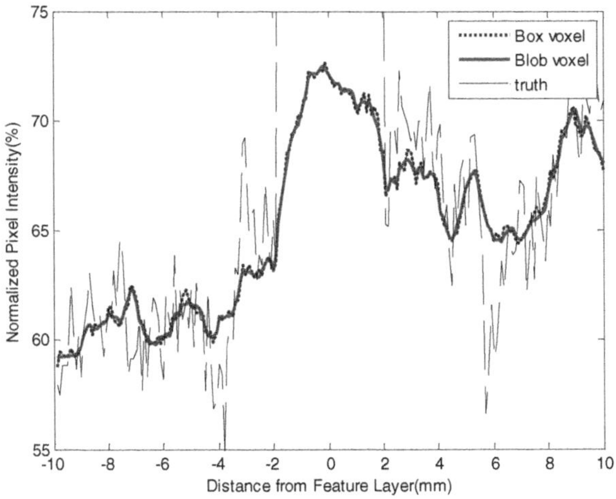

Fig. 4. Profiles through the center of the tumor for Phantom B. Dotted line corresponds to box-voxel and the bold line correspond to blob-voxel. Dash line represents the truth value in the phantom.

With respect to the speed of reconstruction algorithm, we observed computation time using blobs as voxel elements is comparable to that achieved using box voxels, when fast ray-tracing techniques are not applied in both cases. With the projection matrix based ray-tracing, we achieved enhancement in speed for projection simulations by a factor of 6.5, and a factor of 4 for reconstruction.

4 Discussion

The speed of the algorithm is determined by the number of volume elements that are traversed by the ray. When higher accuracy is desired, the size of detector elements must be reduced or the overlapping of the blobs increased and this, increases computation time. As shown by Popescu et al [7], computational speed measured as a function of the number of volume elements traversed per unit time is comparable when ray tracing is performed using blobs or boxes.

Using blobs on a regular grid enables a convenient and flexible model of the forward and backprojection in DBT. The smoother, continuous representation of the sampled signal may yield a more accurate result but this is the subject of future work. Ray-tracing procedures based on the method described here have been implemented and tested, demonstrating flexibility for iterative image reconstruction for DBT. Evaluation of image quality (e.g., edge artifacts and the spread of intensity profiles in a region of interest (e.g., tumour)) will be addressed in further study comparing the performance of blobs and box voxels.

References

1. Wu, T., Moore, R.H., et al.: A comparison of reconstruction algorithms for breast tomosynthesis. Med. Phys. 31(9), 2636–2647 (2004)
2. Mueller, K., Yagel, R., Wheller, J.J.: Fast implementations of algebraic methods for three-dimensional reconstruction from cone-beam data. IEEE Trans Med. Imaging 18(6), 538–548 (1999)
3. Siddon, R.L.: Fast calculation of the exact radiological path for a three-dimensional CT array. Med. Phys. 12(2), 252–255 (1985)
4. Lewitt, R.M.: Alternatives to voxels for image representation in iterative reconstruction algorithms. Phys. Med. Biol. 37(3), 705–716 (1992)
5. Galigekere, R.R., Wiesent, K., Holdsworth, D.W.: Cone-beam reprojection using projection-matrices. IEEE Trans Med. Imaging 22(10), 1202–1214 (2003)
6. Johns, P.C., Yaffe, M.J.: X-ray characterisation of normal and neoplastic breast tissues. Phys. Med. Biol. 32(6), 675–695 (1987)
7. Popescu, L.M., Lewitt, R.M.: Ray tracing through a grid of blobs. In: Nuclear Science Symposium Conference, pp. 3983–3986 (2004)

A Boosting Based Approach for Automatic Micro-calcification Detection

Arnau Oliver[1], Albert Torrent[1], Meritxell Tortajada[1], Xavier Lladó[1], Marta Peracaula[1], Lidia Tortajada[2], Melcior Sentís[2], and Jordi Freixenet[1]

[1] Dept. Computer Architecture and Technology, University of Girona, Girona, Spain
aoliver@atc.udg.edu

[2] Radiology Dept, UDIAT-Centre Diagnòstic, Corporació Parc Taulí, Sabadell, Spain

Abstract. In this paper we present a boosting based approach for automatic detection of micro-calcifications in mammographic images. Our proposal is based on using local features extracted from a bank of filters for obtaining a description of the different micro-calcifications morphology. The approach performs an initial training step in order to automatically learn and select the most salient features, which are subsequently used in a boosting classifier to perform the detection. The validity of our method is demonstrated using 112 mammograms of the well-known digitised MIAS database and 280 mammograms of a full-field digital database. The experimental evaluation is performed in terms of ROC analysis, obtaining $Az = 0.88$ and $Az = 0.90$ respectively, and FROC analysis. The obtained results show the feasibility of our approach for detecting micro-calcifications in both digitised and digital technologies.

1 Introduction

Breast cancer continues to be a significant health problem in the world. It constitutes the most common cancer among women in the European Union [1], and it is estimated that in the United States between one in eight and one in twelve women will develop breast cancer during their lifetime [2,3]. Mammography is the most effective and reliable method for an early detection of breast cancer which is fundamental for improving prognosis [4,5]. Mammographic images are characterised by high spatial resolution allowing the detection of subtle scale signs such as micro-calcifications and masses. In this work, we focused on the detection of micro-calcifications, which are tiny granular deposits of calcium that generally appear in a mammogram as small bright spots within an inhomogeneous background.

The automatic detection of micro-calcifications is a well-known topic in mammography, as can be seen in the different surveys covering this topic [6,7] or the recent works of Chang et al. [8], Nunes et al. [9] and Papadopoulos et al. [10]. The approaches for micro-calcification detection are usually based on two steps. Firstly the detection of suspicious regions is performed, usually tuning the algorithm parameters in order to detect as many suspicious regions as possible (i.e., detecting the largest number of micro-calcifications but also increasing the

J. Martí et al. (Eds.): IWDM 2010, LNCS 6136, pp. 251–258, 2010.

probabilities to detect normal tissue as being a micro-calcification). Secondly, a false positive reduction algorithm is applied in order to find those detected suspicious regions being in fact normal tissue. However, none of these micro-calcification detection approaches have emerged as a standard algorithm.

It is well known that digital mammography allows to improve the detection of micro-calcifications thanks to its superior sensitivity [11]. Unfortunately, this technology is not available in many countries and clinical centres due to its expensive cost. Therefore, reliable automatic approaches able to detect micro-calcifications in film plates are still necessary. In this work we present a boosting based approach for the detection of micro-calcifications in both digital and digitised databases. Our proposal is based on learning the different morphology of the micro-calcifications using local features, which are extracted using a bank of filters. Afterwards, this set of features is used to train a pixel-based boosting classifier which at each round automatically selects the most salient one. Therefore, when a new mammogram is tested only the salient features are computed and used to classify each pixel of the mammogram as being part of a micro-calcification or actually being normal tissue.

The rest of this paper is structured as follows. The following section describes the proposed approach. Section 3 explains the methodology followed to perform the experimental evaluation, which is done using two different databases and ROC and FROC analysis. Finally, the paper ends with the conclusions and further work.

2 Micro-calcification Detection

The presented approach for micro-calcification detection is based on the work of Murphy et al. [12] for object detection using local features and boosting classifier. As is shown in Fig. 1, the proposed approach is divided in three parts. Firstly, we create a visual word dictionary, which is composed by convolving patches containing a micro-calcification with a bank of filters. Afterwards, the training data is found by convolving positive samples (patches containing a micro-calcification) and negative samples (patches of other tissues) with the words of the dictionary defined as the duple patch-filter. Finally, new mammograms are classified pixel-by-pixel by the trained classifier. Hence, the detection problem is translated to a pixel-based classification approach.

In the following subsections we describe in more detail the three parts of our approach.

2.1 Building the Dictionary

The first task of the system consists in building the feature dictionary. This dictionary is similar to an atlas, since it contains samples (patches) of micro-calcifications. However, it also contains the convolution of these patches with a bank of filters, including the delta function (which gives the own patch as a result), 4 Gaussian derivatives, a Laplacian filter, a corner detector, and 2 Sobel

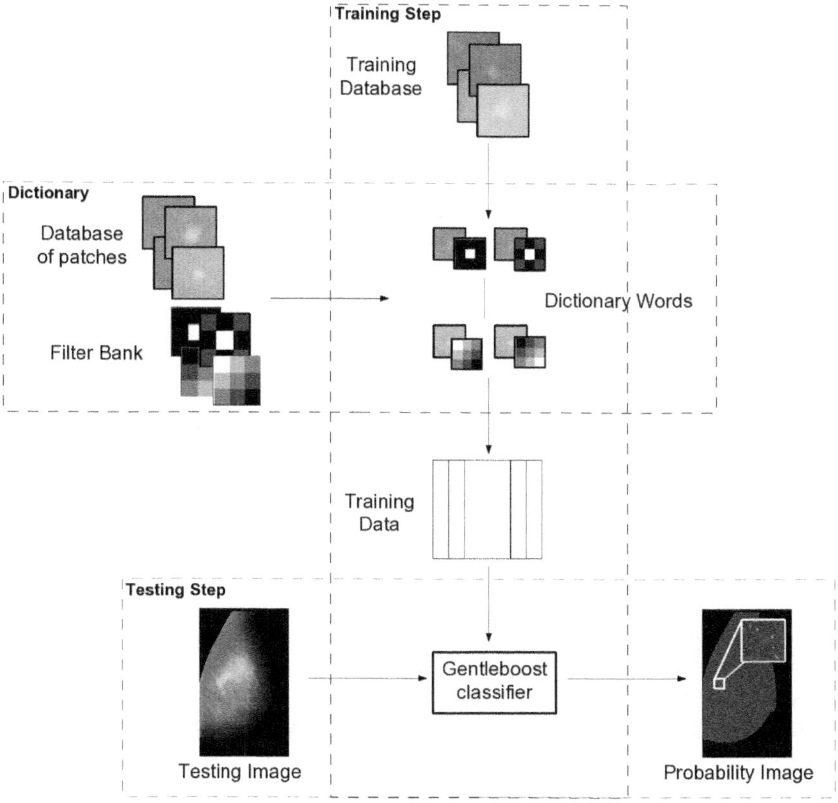

Fig. 1. Schematic representation of our approach. Note that the final result of the algorithm is a probability image where brighter pixels represent more confidence to be a micro-calcification.

filters. Hence, the dictionary contains grey-level and gradient information of the micro-calcifications and their neighbourhood.

Therefore, each dictionary word w_{ij} can be understood as the duple (p_i, f_j), where p_i represents the patch and f_j the filter.

2.2 Training Step

Once the dictionary is built, the words are used to extract the mammographic features that will be used for the micro-calcification detection. Hence, for the training step, we need a different database of patches containing instances of both patches with micro-calcifications and patches from the rest of the mammogram tissues.

The feature extraction of each training image patch consists in two operations. Firstly, it is convolved with all the bank filters, and secondly, the normalised cross-correlation with all the words is computed. Mathematically, both operations can be summarised as:

$$v = (I * f_j) \otimes w_{ij} \quad (1)$$

where I is the training image patch convolved ($*$) with the filter f_j and cross-correlated ($\otimes$) with the word w_{ij} (note that $w_{ij} = p_i * f_j$). The resulting value v represents the similarity of the training patch and the dictionary word. Therefore, for each training image patch, a vector of features v is constructed by cross-correlating all the dictionary words w_{ij} with the convolution of the patch itself with filter f_j. Notice here the necessity of keeping the filter as well as the patch in the dictionary word.

In contrast with the original approach that uses all pixels in the image as the centre of a positive or a negative patch, we manually select these points in each training image. In particular, we select the centre of the micro-calcifications as positive training examples and some random locations of the background containing examples of different tissues as negative training examples. Note that this is necessary in order to reduce the high computational cost due to the large size of the mammograms.

At this point, the positive and negative training examples have been characterised. Therefore, this data can be used to train a classifier. In this work, we have used the Gentleboost algorithm [13].

Boosting algorithms are based on the simple idea that the sum of weak classifiers can produce a strong classifier. In the Gentleboost algorithm, the weak classifiers (h_t) are simple regression stumps with one of the features, so at each round t the feature with less error is selected. The weak classifier used is:

$$h_t(x) = a\delta(x_i > th) + b \quad (2)$$

where th is a threshold that determines if pattern x belongs to the object class, x_i is the i'th dimension of x, and a and b are parameters selected to minimise the error of the classifier (a is the regression slope and b the offset):

$$e = \sum (z(y - (a(x_i > th) + b))^2) \quad (3)$$

At each round the training data weights (z) are updated, increasing in the following round the possibility of classifying correctly the previous incorrectly classified points. In the GentleBoost algorithm the data weights are updated:

$$z_{t+1} = z_t e^{y \cdot h_t(x)} \quad (4)$$

Hence, when testing a new data, the final (strong) classifier is computed using the weak classifier created at each round of the boosting. Therefore, the testing data is classified according to the sign of the sum of weak classifiers:

$$H(x) = \sum h(x) \quad (5)$$

The absolute value of $H(x)$ shows the confidence of the classified data.

2.3 Testing Step

Once the classifier is built, the system is ready for the testing step, where the strong classifier $H(x)$ is applied to new images in order to evaluate the micro-calcifications detection. Note that the classifier is pixel-based, i.e. it is applied one-by-one to all the image pixels. Therefore, the result of our approach after evaluating a mammogram is a probability image, where high values represents more confidence to be a micro-calcification.

Since the classifier is pixel-based, a pre-processing step may be necessary in order to avoid the algorithm detecting micro-calcifications in the background (in digitised images) and in the pectoral muscle. In particular, we used a previous developed algorithm to detect the skin-line border [14] and the approach of Kwok et al. [15] to remove the pectoral muscle.

3 Results

The experimental results were performed using two different subsets of mammograms. The first subset of 112 mammograms was extracted from the MIAS database [16], and contained all the mammograms with micro-calcifications (22 in total) and a set of 90 normal mammograms. In order to train the classifier with positive examples, an expert accurately marked among 5 and 15 micro-calcifications in each mammogram containing micro-calcifications, while the negative examples were randomly obtained from the rest of tissues of normal mammograms. On the other hand, we also used a set of 280 full-field digital mammograms extracted from a non-public database, 90 of them containing micro-calcifications and 190 being normal ones. The mammograms were acquired using a Hologic Selenia mammograph, with resolution 70 micron-pixel, size 4096×3328, and 12-bit depth. The selection of the training points was performed in the same way that using the MIAS subset.

In order to perform the evaluation of our experiments in both databases we used a 10-folder cross-validation methodology. Therefore, we divided both datasets in 10 different groups. One of the groups was used to create the dictionary, eight of them were merged for training the system, while the remaining one was used for testing it. This procedure was repeated until all groups were used for testing. Hence, each mammogram appears in the test set only once.

To perform the quantitative evaluation we used Receiver Operating Characteristic (ROC) and Free-response Receiver Operating Characteristic (FROC) analysis. In ROC analysis, a graphical curve represents the true positive rate (number of detected mammograms with micro-calcifications divided by the total number of mammograms with micro-calcifications) as a function of the false positives rate (number of normal mammograms incorrectly detected as containing micro-calcifications divided by the total number of normal mammograms). Moreover, the percentage value under the curve (A_z) is an indication for the overall performance of the observer, and is typically used to analyse the performance of the algorithms. Note that points in the curve are obtained by thresholding at different levels the result of Eq. 5.

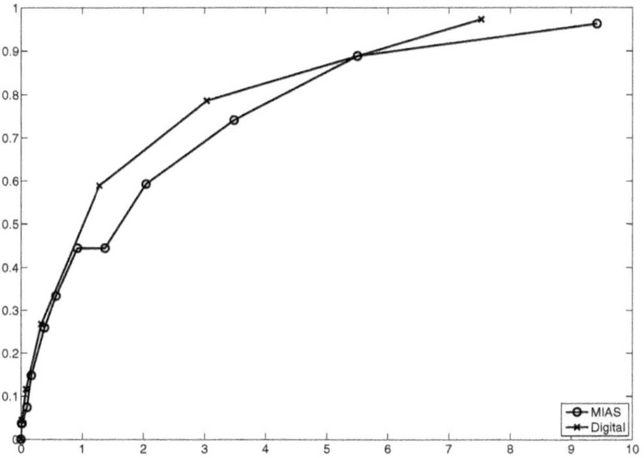

Fig. 2. Obtained FROC when testing the MIAS and the digital database. Note that we obtained better results when testing the latter.

On the other hand, in FROC analysis the Lesion Localisation Fraction (LLF) is obtained as the number of correctly detected lesions relative to the total number of lesions and the Non-Lesion Localisation Fraction as the number of non-correctly detected lesions relative to the total number of images. The FROC curve is the graphical summary of both measures [17]. Note that the definition of a detected region is needed. In this paper we assume that a region is detected if a set of suspicious points are detected inside the region marked by the experts.

Using ROC analysis, we achieved an area under the ROC of $Az = 0.88$ for the MIAS subset and $Az = 0.90$ for the digital database. Note here the benefits of the digital technology, since applying the same method it allows a better detection even when using a bigger number of images for testing.

On the other hand, the obtained FROC curves when testing both databases are shown in Fig. 2. For the MIAS database we obtained 4.3 false positives per image at a sensitivity of 80%, while using the digital database we obtained 3.2 false positives per image at the same sensitivity. Note that these numbers can be decreased using a posterior false positive step. For instance, introducing spatial constraints, and hence assuming that the micro-calcifications appear in the form of clusters. Comparing the performance of the approach using both databases we noticed that again we obtained better results using the digital one. This is the expected behaviour since digital mammography improves the contrast between the different internal structures.

Finally, we compare in Table 1 the results presented in this paper with those obtained by different current state-of-the-art approaches using ROC analysis. Note that each approach used a different subset of images coming also from different databases and hence the comparison is only done in a qualitative way. Note that the obtained results using our approach are of the same order as the

Table 1. Comparison of the obtained experimental results with state-of-the-art algorithms for micro-calcification detection i mammographic images

Authors	Cases	Results (A_z)
Chang et al. (2008) [8]	194	0.90
Nunes et al. (2007) [9]	121	0.93
Papadopoulos et al. (2008) [10]	60	0.92
Our approach – MIAS	**112**	**0.88**
Our approach – Digital	**280**	**0.90**

ones shown in the table. However, we want to stress here that our approach, in contrast with the other ones, does not include a false positive reduction step, and hence they are obtained directly from the detection step. Hence, applying such false positive reduction step would probably increase them.

4 Conclusion

We have presented a new approach for micro-calcification detection based on extracting local features for characterising the morphology of the micro-calcifications. The proposed boosting approach allows the selection of the most salient features at each round, reducing the computational time of the testing step. The performed experiments have shown the validity of our proposal when using either digitised or digital mammograms.

Further work is focused in two directions. Firstly, we would like to integrate a false positive reduction step into the boosting algorithm to improve the results. Secondly, we also want to expand this work for detecting and diagnosing clusters of micro-calcifications.

Acknowledgement

This work was supported in part by Ministerio de Educación y Ciencia of Spain under Grant TIN2007-60553 and by the CIRIT and CUR of DIUiE of Generalitat de Catalunya under Grant 2008SALUT00029.

References

1. Eurostat: Health statistics Atlas on mortality in the European Union. Office for Official Publications of the European Union (2002)
2. Bray, F., McCarron, P., Parkin, D.M.: The changing global patterns of female breast cancer incidence and mortality. Breast Cancer Research 6(6), 229–239 (2004)
3. American Cancer Society: Breast cancer: facts and figures. 2007-08. ACS (2007)
4. Sivaramakrishna, R., Gordon, R.: Detection of breast cancer at a smaller size can reduce the likelihood of metastatic spread: A quantitative analysis. Acad. Radiol. 4(1), 8–12 (1997)

5. Sener, S.F., Winchester, D.J., Winchester, D.P., Barrera, E., Bilimoria, M., Brinkmann, E., Alwawi, E., Rabbitt, S., Schermerhorna, M., Du, H.: Survival rates for breast cancers detected in a community service screening mammogram program. Am. J. Surg. 191(3), 406–409 (2006)
6. Cheng, H.D., Cai, X., Chen, X., Hu, L., Lou, X.: Computer-aided detection and classification of microcalcifications in mammograms: a survey. Pattern Recogn. 36(12), 2967–2991 (2003)
7. Rangayyan, R.M., Ayres, F.J., Desautels, J.E.L.: A review of computer-aided diagnosis of breast cancer: Toward the detection of subtle signs. J. Frankl. Inst. 344(3–4), 312–348 (2007)
8. Chang, T.T., Feng, J., Liu, H.W., Ip, H.H.S.: Clustered microcalcification detection based on a multiple kernel support vector machine with grouped features. In: IAPR Int. Conf. Pattern Recogn., pp. 1–4 (2008)
9. Nunes, F.L.S., Schiabel, H., Goes, C.E.: Contrast enhancement in dense breast images to aid clustered microcalcifications detection. J. Digit. Imaging 1(20), 53–66 (2007)
10. Papadopoulos, A., Fotiadis, D.I., Costaridou, L.: Improvement of microcalcification cluster detection in mammography utilizing image enhancement techniques. Comput. Biol. Med. 10(38), 1045–1055 (2008)
11. Fischer, U., Baum, F., Obenauer, S., Luftner-Nagel, S., von Heyden, D., Vosshenrich, R., Grabbe, E.: Comparative study in patients with microcalcifications: full-field digital mammography vs screen-film mammography. Epidemiol. Rev. 11(12), 2679–2683 (2002)
12. Murphy, K., Torralba, A., Eaton, D., Freeman, W.T.: Object detection and localization using local and global features. In: Sicily W. on Object Recognition (2005)
13. Friedman, J., Hastie, T., Tibshirani, R.: Additive logistic regression: a statistical view of boosting. Ann. Stat. 38(2), 337–374 (2000)
14. Martí, R., Oliver, A., Raba, D., Freixenet, J.: Breast skin-line segmentation using contour growing. In: Martí, J., Benedí, J.M., Mendonça, A.M., Serrat, J. (eds.) IbPRIA 2007. LNCS, vol. 4478, pp. 564–571. Springer, Heidelberg (2007)
15. Kwok, S.M., Chandrasekhar, R., Attikiouzel, Y., Rickard, M.T.: Automatic pectoral muscle segmentation on mediolateral oblique view mammograms. IEEE Trans. Med. Imag. 23(9), 1129–1140 (2004)
16. Suckling, J., Parker, J., Dance, D.R., Astley, S.M., Hutt, I., Boggis, C.R.M., Ricketts, I., Stamatakis, E., Cerneaz, N., Kok, S.L., Taylor, P., Betal, D., Savage, J.: The Mammographic Image Analysis Society digital mammogram database. In: Int. Work. Dig. Mammography, pp. 211–221 (1994)
17. Yoon, H.J., Zheng, B., Sahiner, B., Chakraborty, D.P.: Evaluating computer-aided detection algorithms. Med. Phys. 34(6), 2024–2038 (2007)

Automatic Breast Tissue Classification Based on BIRADS Categories

Gloria Bueno[1], Noelia Vállez[1], Óscar Déniz[1], Pablo Esteve[2], Miguel A. Rienda[2], and Carlos Pastor[2]

[1] Engineering School, Universidad de Castilla-La Mancha, Ciudad Real, Spain
Gloria.Bueno@uclm.es
[2] Hospital General de Ciudad Real, Spain

Abstract. Breast tissue density is an important risk factor in the detection of breast cancer. It is also known that interpretation of mammogram lesions is more difficult in dense tissues. Therefore, getting a preliminary tissue classification may aid in the subsequent process of breast lesion detection and analysis. This article reviews several classification techniques for two datasets, both digitized screen-film (SFM) and full-field digital (FFDM) mammography, classified according to BIRADS categories. It concludes with a hierarchical classification procedure based on k-NN combined with principal component analysis on texture features. The results obtained classifying 1740 mammograms reflect up to 83% of samples correctly classified. The method is being integrated within a CADe system developed by the authors.

1 Introduction

Breast cancer continues to be an important health problem. Early detection is the only way to improve breast cancer prognosis and significantly reduce women mortality. It is by using Computer-Aided Diagnosis systems (CAD) that radiologist can improve their ability to detect and classify breast lesions, [1]. Therefore, the development of reliable CAD systems is an important and challenging task in automated diagnosis. However, automated interpretation of mammogram lesions still remains very difficult. Some of the reasons are the dense tissues. The dense tissues may cause suspicious areas to be almost invisible and may be easily misinterpreted as calcifications or masses [2], [3]. One of the CAD algorithms implemented by the authors which highlights this fact is the adaptive filtering [4]. When processing a mammographic image with dense tissue using this method it is necessary to adjust the input parameters to control the sensitivity of the algorithm in areas of high intensity and thus reduce false positive detections in these areas. Thus, it is possible to design optimal algorithms for a CAD system by previously using breast tissue classification.

Moreover, there is also an interest in investigating the breast tissues since the discovery of the relation between mammographic parenchymal patterns and the

J. Martí et al. (Eds.): IWDM 2010, LNCS 6136, pp. 259–266, 2010.

risk of developing breast cancer in 1976 by Wolfe [5]. A good review of the work on breast tissue classification since then can be found in [6].

This research has been prompted by this need to classify breast tissue and drive the development of CAD algorithms for automatic analysis of breast lesions. This paper is not intended to be applied to the classification of benign or malignant lesion associated with tissue types [7], [8], but as mentioned above, only to drive the CAD algorithms. In our study several classification methods have been compared and a hierarchical classification procedure based on k-nearest neighbors (k-NN) combined with principal component analysis (PCA) on texture features is proposed as the best solution. Experimental results have been given on different mammograms with various densities and abnormalities from both SFM and FFDM.

Section 2 describes the methods and material used for this work. This include the feature extraction procedure applied to the classifiers, the tested classifiers, the data training and testing carried on, as well as the experimental database used. Section 3 describes the results obtained with the proposed method and finally, in Section 5 the main conclusions are drawn.

2 Methods and Materials

There are several classification techniques to classify datasets according mammographic breast density [7]. We use the American College of Radiology BIRADS that has been used in a number of studies and is the most common technique used in the USA [9]. In this classification datasets have been classified according to 4 categories. These are: T.I) fatty, T.II) fatty-glandular or fibroglandular, T.III) heterogeneously dense and T.IV) extremely dense. Figure 1 shows this clasification.

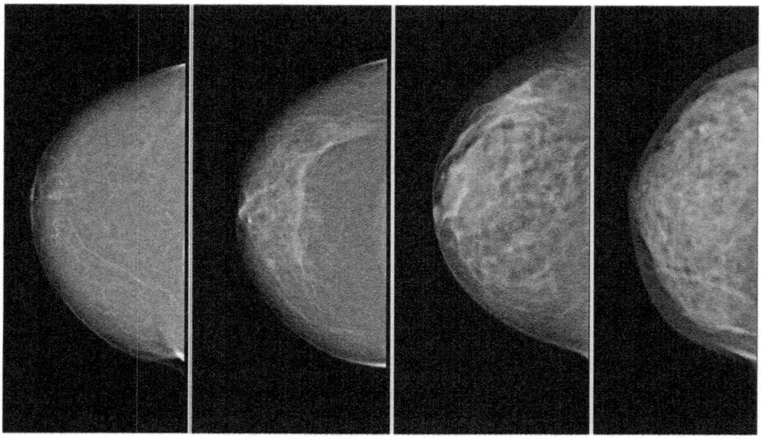

Fig. 1. BIRADS tissue classification. From left to right: T.I) fatty, T.II) fatty-glandular or fibroglandular, T.III) heterogeneously dense and T.IV) extremely dense.

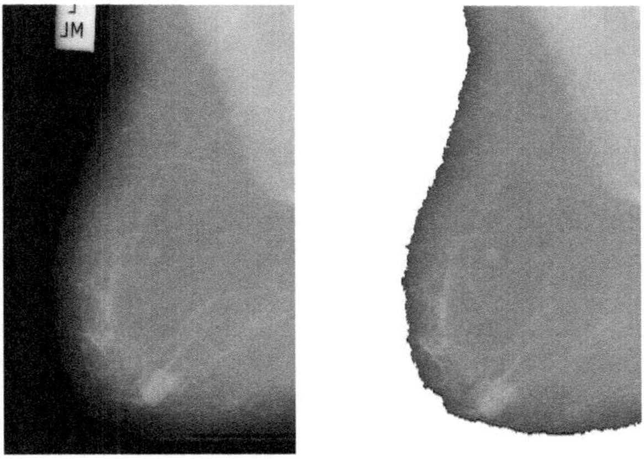

(a) Original image (b) Preprocessed image

Fig. 2. Breast tissue selection

To deal with the breast tissue classification problem several studies have been described in the literature. These studies are based on: a) the use of grey-level histograms and b) texture information extracted from different regions. Our proposal is to apply texture analysis on the whole breast tissue. Thus, all mammograms have been previously preprocessed to identify the breast region and remove the background and possible labels. This is illustrated in Figure 2.

2.1 Feature Extraction

Most studies on texture classification are based on statistical features obtained from the image [10],[11]. Here we analyze 33 features, both 1st and 2nd order texture statistics, obtained from the preprocessed image histogram and the co-occurrence matrix-based features. The latest have been some of the Haralick's coefficients and they have been calculated for a distance parameter equal to 1 at 0, 45, 90 and 135 degrees [12]. These features were:

- (1st order): mean, variance, skewness, kurtosis and entropy.
- (2nd order): energy, variance, contrast, entropy, correlation, homogeneity and the difference statistics: 2nd angular moment, contrast, entropy and mean.

The discrimination power of these features was analyzed with a feature ranking on individual performance for classification method. This evaluation is based on the results from intra-cluster and inter-cluster distances between the four tissue types. These distances measure the variability within and between different classes [13]. This will get the features that maximize these values. The ten most significant features in increasing order of relevance were: 1st order variance, 1st order mean, 1st order entropy, 2nd order and 0 degrees entropy, 1st order and 45 degrees entropy, 1st order and 135 degrees entropy, 1st order and 90 degrees entropy, 1st order asimetry, 1st order kurtosis, 2nd order and 0 degrees correlation.

Moreover, to reduce and select the feature space a principal component analysis (PCA) was applied. This mathematical procedure transforms a number of variables that can be correlated into a smaller set of uncorrelated variables called principal components.

Different tests were performed by varying the number of components from the space reduced by the PCA. This number of components varies between 5 and 30 at intervals of 5. The average errors for all classifiers were measured at each interval. These average errors range between 0.421 and 0.453 with the best performance of 0.12 as shown in Table 1. The minimun error corresponds to the reduction of the space to 10 components by means of the PCA.

In this process, the obtained eigenvectors and eigenvalues were analyzed in order to find the most representative features according to the PCA. The ten most significant features in increasing order of relevance were: 2nd order and 90 degrees variance, 2nd order and 45 degrees entropy, 2nd order and 45 degrees homogeneity, 2nd order and 90 degrees homogeneity, 2nd order and 90 degrees entropy, 2nd order and 45 degrees 2nd angular moment, 2nd order and 45 degrees variance, 2nd order and 135 degrees variance, 2nd order and 0 degrees contrast, 2nd order and 90 degrees correlation.

2.2 Classifiers, Data Training and Testing

Different classification methods from the PRTOOLS Matlab library were tested on the selected features [13]. These methods were: support vector machine (SVM) with polynomial, minkowski distance, exponential, radial_basis and sigmoid kernels, neural networks (feedforward, backpropagation, perceptron and radial basis) (NN), k-NN with k equal to 1, linear bayes normal (LBN), quadratic (QD) and tree-classifier with two layers ({T.I-T.II, T.III-T.IV} and {{T.I, T.II}; {T.III, T.IV}}). The best results obtained for the SVM were with a polynomial kernel and for the NN were with the backpropagation (BPNN), and these are shown here.

To train and test classifiers a combination of hold-out (H-method) and resubstitution (R-method) methods is used [14]. This combination is accomplished through two stages. In the first stage the data is divided into two groups containing the same number of samples. One of these groups is randomly selected to train the classifiers. Once trained the classifiers, the tests were performed on the complete dataset.

The performance of these classifiers are shown and discussed in section 3.

2.3 Experimental Database

Two datasets were considered. One composed of 1418 FFDM provided by the local Hospitals, and the other one composed by 322 SFM obtained from the MIAS public database. The last set was used to compare our results with other authors. Both datasets were labeled according to the BIRADS categories by three expert clinicians from the Hospital General de Ciudad Real. The image sizes are $3328 * 4084$ and $1024 * 1024$ respectively for the FFDM and SFM datasets.

The MIAS database contains images from right and left medial-lateral projections (RMLO, LMLO) of 161 different cases while the FFDM contain in most cases

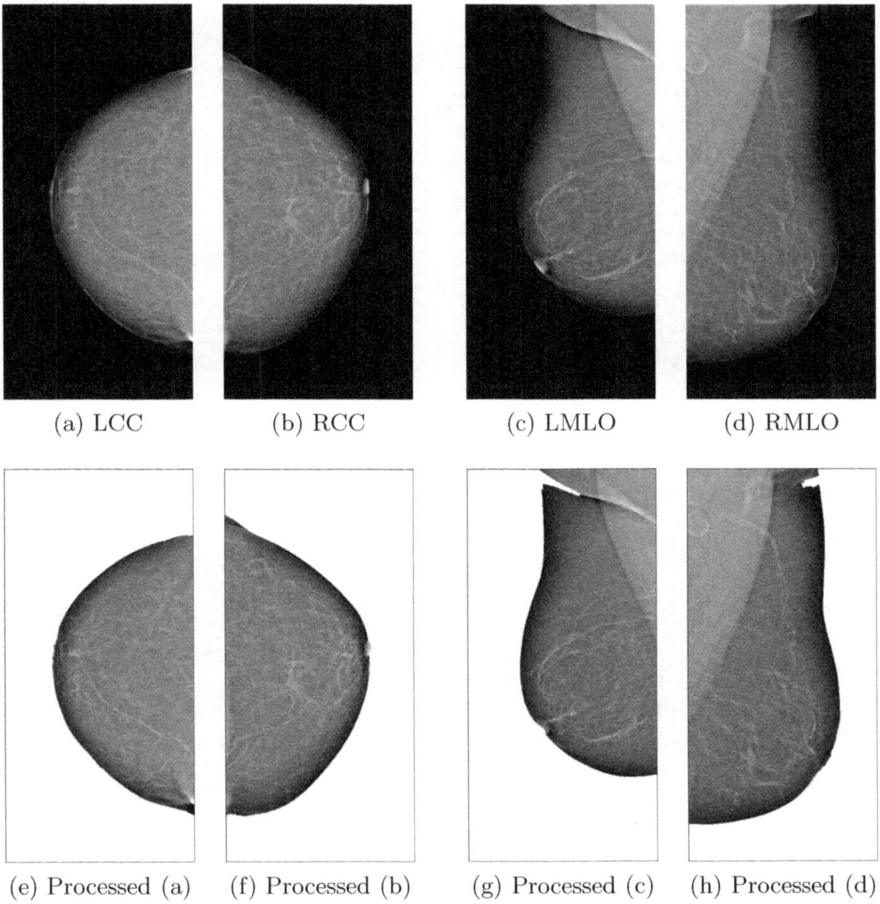

Fig. 3. Breast image projections before and after the preprocessing step

the four projections, right and left medial-lateral as well as right and left cranial-caudal (RCC, LCC). Figure 3 shows the original RMLO, LMLO, RCC and LCC projections and the preprocessed images previous to tissue classification.

3 Results

Table 1 shows the results of the classifiers with and without PCA for the SFM and FFDM datasets. The columns indicate the classifier used while rows indicate the tissue types. The results of Table 1 (c) and (d) are given with 10 features since the PCA obtained the best results with selection of 10 features, as mentioned above.

The best classifiers are shown in bold for those $\geq 80\%$ and light grey for those $\geq 75\%$ and $\leq 79\%$. The best classifier is the SVM for T.I, k-NN for T.II and T.IV and BPNN for T.III. On average and weighted respect to the number of

Table 1. Agreement of Classifiers. % of Mammograms Correctly Classified.

Types	SVM	BPNN	k-NN	LBN	QD
T.I	80%	52%	71%	55%	50%
T.II	69%	49%	71%	55%	45%
T.III	78%	81%	81%	79%	89%
T.IV	56%	23%	58%	42%	31%

(a) FFDM without PCA

Types	SVM	BPNN	k-NN	LBN	QD
T.I	88%	88%	83%	75%	82%
T.II	71%	76%	78%	78%	72%
T.III	42%	82%	77%	77%	67%
T.IV	61%	36%	73%	61%	55%

(b) SFM without PCA

Types	SVM	BPNN	k-NN	LBN	QD
T.I	64%	64%	70%	56%	69%
T.II	70%	38%	72%	49%	55%
T.III	81%	80%	80%	83%	83%
T.IV	60%	41%	58%	40%	55%

(c) FFDM with PCA

Types	SVM	BPNN	k-NN	LBN	QD
T.I	77%	86%	77%	87%	76%
T.II	53%	67%	80%	68%	62%
T.III	61%	62%	80%	71%	49%
T.IV	57%	80%	75%	66%	66%

(d) SFM with PCA

Table 2. Confusion Matrices for the 1-NN Classifier

	Estimated				True
Types	T.I	T.II	T.III	T.IV	Total
T.I	175	37	31	6	249
T.II	28	312	84	9	433
T.III	26	75	484	19	604
T.IV	9	14	32	77	132

(a) FFDM

	Estimated				True
Types	T.I	T.II	T.III	T.IV	Total
T.I	65	12	7	0	84
T.II	5	82	12	3	102
T.III	3	9	74	6	92
T.IV	2	3	6	33	44

(b) SFM

mammograms of each type the k-NN is the best one with similar results with or without PCA (79%). Table 2 shows the confusion matrices for this classifier.

A 2-layer tree classifier was also tested. The results improved upon the previous ones obtaining up to 91% in the 1st layer and 83% in the 2nd layer with k-NN and applying PCA to the MIAS dataset. In the case of the FFDM dataset the results are 83% in the 1st layer and 74% in the 2nd layer. The final results are shown in Table 3.

Examining each tissue: T.I reaches up to 85%, T.II reaches up to 88%, T.III reaches up to 88% and T.IV reaches up to 71%. Therefore the classifiers found the most difficult classification for mammograms of type T.IV.

The difference between the two datasets is due to the use of 4 different projections in the FFDM dataset. While the two CC projections do not contain the pectoral muscle tissue, the MLO projections do contain this muscle. This muscle is usually denser than the rest of the breast tissue and influences the classification because half of the images do not contain it.

In order to improve the results and test the performance and accuracy of the classifiers, we are doing different ongoing tests. These include a cross-validation to train and test the classifiers, different statistical tests for feature selection,

Table 3. 2-Layer k-NN with PCA Tree Classifier for FFDM and SFM

Types	1st Layer	2nd Layer
T.I	81%	68%
T.II		69%
T.III	84%	83%
T.IV		55%

(a) FFDM

Types	1st Layer	2nd Layer
T.I	94%	85%
T.II		88%
T.III	86%	80%
T.IV		71%

(b) SFM

since there is not agreement between the PCA selection and the intra/inter-class feature distances, as above shown, and further selection of the dataset to get more homogeneous set among the tissue types.

4 Discussion and Conclusions

In this work a hierarchical procedure based on k-NN and PCA on texture features has been proposed for breast tissue classification. There are just three works in the literature presenting breast tissue classification according to BIRADS categories on SFM. Their overall correct classification is about 71% [10], 76% [15] without tissue segmentation and 82% [6] with it. Our approach reflect up to 83% of samples correctly classified for the SFM dataset and 74% for the FFDM one.

The method is being integrated within a CADe system developed by the authors and further tests are being carried out to improve the classification results for all tissue types and datasets.

Acknowledgements

The authors acknowledge partial financial support from the Spanish Research Ministry and Junta de Comunidades de Castilla-La Mancha through projects RETIC COMBIOMED and PI-2006/01.1. The authors also thank C. Romero, C. Varela, C. Lozano and M. Arias for providing some of the FFDM dataset.

References

1. Bueno, G.: 10. In: Fuzzy Systems and Deformable Models. Series in Medical Physics and Biomedical Engineering, pp. 305–329. Taylor & Francis Group, London (2008); Book-Title: Intelligent and Adaptive Systems in Medicine
2. Boyd, N., Dite, G., Stone, J., et al.: Realiability of Mammographic Density, a Risk Factor for Breast Cancer. New England Journal of Med. 347(12), 886–894 (2002)
3. Ursin, G., Hovanessian-Larsen, L., Parisky, Y.R., et al.: Greatly increased occurrence of breast cancers in areas of mammographically dense tissue. Breast Cancer Research 7(5), 605–608 (2005)

4. Bueno, G., Ruiz, M., Sánchez, S.: B-spline filtering for automatic detection of calcification lesions in mammograms. In: Proceedings of the Intern. Conference on Information Optics, WIO 2006, pp. 60–70 (2006)
5. Wolfe, J.N.: Risk for breast cancer development determined by mammographic parenchymal pattern. Cancer 37, 2486–2492 (1976)
6. Oliver, A., Freixenet, J., Martí, R., et al.: A novel breast tissue density classification methodology. IEEE Trans. on Inform. Techn. in Biomed. 12, 55–65 (2008)
7. Yafee, M., Boyd, N.: Mammographic breast density and cancer risk: The radiological view. Gynecological Endocrinology 21(suppl. 1), 6–11 (2005)
8. Brem, R., Hoffmeister, J., Rapelyea, J., et al.: Impact of breast density on computer-aided detection for breast cancer. American Journal of Roentgenology 184, 439–444 (2005)
9. Harvey, J.A., Bovbjerg, V.E.: Quantitative Assessment of Mammographic Breast Density: Relationship with Breast Cancer Risk. Radiology 230(1), 29–41 (2004)
10. Bovis, K., Singh, S.: Classification of mammographic breast density using a combined classifier paradigm. In: 4th Intern. Workshop on Digital Mammography, pp. 177–180 (2002)
11. Oliver, A., Lladó, X., Martí, R., Freixenet, J., Zwiggelaar, R.: Classifying mammograms using texture information. In: Proc. Medical Image Understanding and Analysis, July 2007, pp. 223–227 (2007)
12. Haralick, R., Sternberg, S., Zhuang, X.: Image analysis using mathematical morphology. IEEE Transactions on Pattern Analysis and Machine Intelligence 9(4), 532–550 (1987)
13. Heijden, F., Duin, R., Ridder, D., Tax, D.: Classification, parameter estimation and state estimation - an engineering approach using Matlab. John Wiley & Sons, Chichester (2004)
14. Kuncheva, L.I.: Combining Pattern Classifiers. John Wiley & Sons, Inc., Chichester (2004)
15. Petroudi, S., Kadir, T., Brady, M.: Automatic classification of mammographic parenchymal patterns: A statistical approach. In: Proc. IEEE Conf. Eng. Med. Biol. Soc., vol. 1, pp. 798–801 (2003)

Validation of Mammographic Texture Analysis for Assessment of Breast Cancer Risk

Hui Li, Maryellen L. Giger, Olufunmilayo I. Olopade, and Li Lan

Department of Radiology, The University of Chicago, 5841 South Maryland Avenue, Chicago, Illinois 60637
m-giger@uchicago.edu

Abstract. The purpose of this study was to evaluate, using full-field digital mammography (FFDM), our prior computerized texture analysis method that was developed on screen-film mammography method. The evaluation included the analyses on the parenchymal patterns of women with BRCA1 or BRCA2 gene mutations and of women at low risk of developing breast cancer. A total of 180 cases, including 80 women with BRCA1 or BRCA2 gene mutations and 100 low-risk women, were retrospectively collected under an institutional review board approved protocol. Images were obtained with a GE Senographe 2000D FFDM system with 0.1 mm pixel size and 12-bit quantization. Regions-of-interest (ROIs), 256 pixels by 256 pixels in size, were manually selected from the central breast region immediately behind the nipple. The ROIs were used in subsequent computerized feature extraction to assess the mammographic parenchymal patterns in the images. Various mammographic parenchyma features based on local composition, gray-level histogram analysis, spatial relationship among gray-levels, fractal analysis, edge frequency analysis, and Fourier analysis, were automatically extracted from these ROIs. Receiver Operating Characteristic (ROC) analysis was used to assess the performance of the computerized texture features in the task of distinguishing between gene-mutation carriers and low-risk subjects. Computerized texture analysis on digital mammograms demonstrated that gene-mutation carriers and low-risk women have different mammographic parenchymal patterns. In addition, in a round-robin-by-case evaluation with the FFDM dataset with linear discriminant analysis, an AUC value of 0.88 was obtained. Our results indicate the transferability of these radiographic biomarkers for breast cancer risk assessment from SFM to FFDM.

Keywords: Computerized texture analysis, breast cancer risk assessment, mammographic parenchymal patterns, full-field digital mammograms.

1 Introduction

Breast cancer is the most commonly diagnosed cancer among women in the United States, with approximately 192,370 new cases of invasive breast cancer and 62,280 new cases of in situ breast cancer expected to occur among women during 2009. An estimated 40,170 breast cancer deaths are expected in 2009 [1]. Mammographic density and parenchymal patterns have been shown to be associated with the risk of developing breast cancer [2-7]. Computer-extracted mammographic texture features may be useful for identifying women at high risk for breast cancer and for monitoring the treatment of breast cancer patients [8-13].

J. Martí et al. (Eds.): IWDM 2010, LNCS 6136, pp. 267–271, 2010.

We have previously developed computerized texture analysis methods for breast cancer risk assessment using digitized screen-film mammograms (SFM) [8-11]. Our method is schematically shown in Figure 1. Our results showed that women at high risk of developing breast cancer tended to have dense breasts and their mammographic parenchymal patterns were coarse and low in contrast. Thus, the purpose of this study was to validate, using full-field digital mammography (FFDM), our prior screen-film mammography method for the computerized texture analysis. The evaluation included the analyses on the parenchymal patterns of women with BRCA1 or BRCA2 gene mutations and of women at low risk of developing breast cancer.

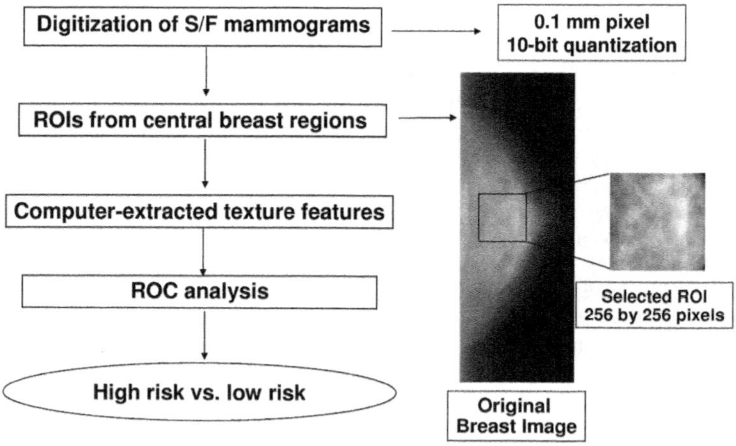

Fig. 1. Computerized texture analysis method for breast cancer risk assessment previously developed for digitized screen-film mammograms

2 Materials and Methods

2.1 Database

In our current study, we extended the evaluation of the performance of our methods to full-field digital mammograms [FFDM] using a database from 180 cases: 80 women with BRCA1 or BRCA2 gene mutations and 100 low-risk women. These full-field digital mammograms were retrospectively collected at the University of Chicago Medical Center under an institutional review board (IRB) approved protocol. All images were obtained with a GE (Waukesha, WI) Senographe 2000D FFDM system. The FFDM images were acquired at 12-bit quantization with a pixel size of 0.1 mm.

The gene mutation carriers were tested at Clinical Laboratory Improvement Amendments (CLIA) approved laboratories under institution review board (IRB) approved protocols. The mammograms of low-risk women were obtained from the screening mammography program. These mammograms were of women who had no family history of breast or ovarian cancer, and no prior history of breast cancer or benign breast disease. These women had a less than 10% lifetime risk of developing breast cancer based on the Gail model.

2.2 Computerized Texture Analysis

Regions-of-interest (ROIs), 256 pixels by 256 pixels in size, were manually selected from the central breast region immediately behind the nipple. Only the left cranial-caudal (CC) view was analyzed. The ROIs were used in subsequent computerized feature extraction to assess the mammographic parenchymal patterns in the images.

A set of computerized texture features was extracted from each ROI to characterize the mammographic parenchymal patterns. These computer-extracted texture features were based on (a) local composition (density related measures), (b) gray-level histogram analysis, (c) spatial relationship among gray-levels, (d) fractal analysis, (e) edge frequency analysis, and (f) Fourier analysis, including RMS variation, first moment of the power spectrum, and power spectral analysis. The detailed descriptions of each individual features can be found elsewhere [8-11].

In order to assess the potential usefulness of these computer-extracted texture features, receiver operating characteristic (ROC) analysis [14-16] was used to determine the performance of each feature in the task of distinguishing mammographic parenchymal patterns from the BRCA1/BRCA2 gene-mutation carriers and those from subjects in the low–risk group. Here, the area under the fitted ROC curve AUC is used as an index to evaluate the inherent discriminant capacity of these texture features in the task of distinguishing between gene-mutation carriers and low-risk subjects.

3 Results

Computerized texture analysis on digital mammograms demonstrated that gene-mutation carriers and low-risk women have different mammographic parenchymal patterns. For gene-mutation carriers, they appear to have dense breast. The skewness,

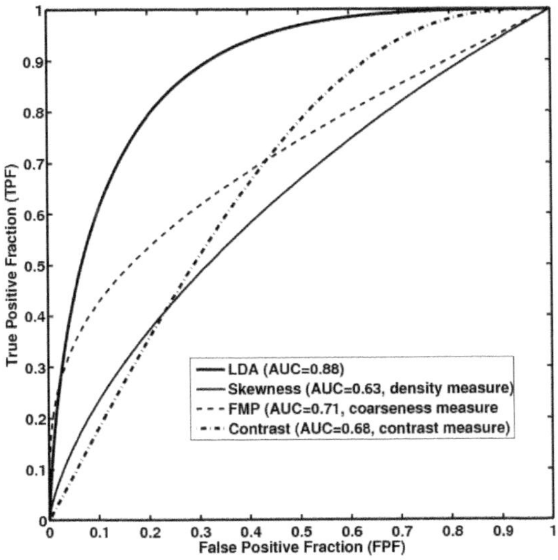

Fig. 2. LDA showed a statistically significant improvement as compared to the individual features in the task of distinguishing between gene-mutation carriers and low-risk women (*P*-value<0.05)

which is related to mammographic density in the breast, yielded an AUC value of 0.63 in distinguishing between gene-mutation carriers and low-risk women from ROC analysis in the entire dataset. Feature values that characterize the texture, i.e., first moment of the power spectrum, yielded AUC values of 0.71 and were found to be similar for high risk subjects imaged with FFDM and, from our earlier studies, with SFM.

In addition, in a round-robin-by-case evaluation with the FFDM dataset with linear discriminant analysis, an AUC value of 0.88 was obtained. (Figure 2) The features merged were those including skewness, first moment and contrast, indicating, as we had found with the prior screen-film mammography dataset, that women at high risk of breast cancer have dense breast and their parenchymal pattern is coarse and of low contrast.

Our results indicate the transferability of these radiographic biomarkers for breast cancer risk assessment from SFM to FFDM. Comparison of the various features between digitized screen-film mammography and FFDM is listed in Table1.

Table 1. Comparison of SFM and FFDM: Performance of texture features in distinguishing between gene-mutation carriers and low-risk groups

Features	FFDM (180 cases) AUC±SE	SFM (172 cases) AUC±SE	p-value
Features related to image local composition (density measure)			
Skewness	0.63 ± 0.04	0.72 ± 0.04	0.1783
Balance	0.62 ± 0.04	0.68 ± 0.04	0.5065
Features related to image coarseness			
Fractal (Box Counting)	0.68 ± 0.04	0.74 ± 0.04	0.1304
MeanGradient	0.67 ± 0.04	0.68 ± 0.05	0.9436
FMP	0.71 ± 0.04	0.75 ± 0.04	0.4191
Feature related to local variation (contrast measure)			
Contrast (Co-occurrence matrices)	0.68 ± 0.04	0.86 ± 0.04	<0.0001

4 Summary

Computerized texture analysis of FFDM provided radiographic descriptors of mammographic parenchymal patterns. Our study on FFDM showed that BRCA1/2 gene-mutation carriers and low-risk women have different mammographic parenchymal patterns. High risk women tend to have dense breast, and their mammographic parenchymal patterns are coarse and low in contrast, which agrees with our previous studies on SFM. The computer-extracted image markers may potentially be used alone or together with clinical measures, as well as biomarkers, for use in identifying women at high risk for breast cancer and for monitoring the treatment of breast cancer patients. Our investigation provided a validation on FFDM of our earlier findings on digitized screen-film mammograms.

Acknowledgements

This research was supported in part by the University of Chicago Breast SPORE P50-CA125183. M. L. Giger is a stockholder in R2 Technology/Hologic and receives royalties from Hologic, GE Medical Systems, MEDIAN Technologies, Riverain Medical, Mitsubishi and Toshiba. It is the University of Chicago Conflict of Interest Policy that investigators disclose publicly actual or potential significant financial interest that would reasonably appear to be directly and significantly affected by the research activities.

References

1. Jemal, A., Siegel, R., Ward, E., Hao, Y., Xu, J., Murray, T., Thun, M.J.: Cancer Statistics, 2009. CA Cancer J. Clin. 59, 225–249 (2009)
2. Wolfe, J.N.: Breast patterns as an index of risk for developing breast cancer. Am. J Roentgenol. 126, 1130–1139 (1976)
3. Boyd, N.F., Martin, L.J., Stone, J., Greenberg, C., Minkin, S., Yaffe, M.J.: Mammographic densities as a marker of human breast cancer risk and their use in chemoprevention. Current Oncology Reports 3, 314–321 (2001)
4. Brisson, J., Diorio, C., Mâsse, B.: "Wolfe's parenchymal pattern and percentage of the breast with mammographic densities: redundant or complementary classifications? Cancer Epidemiol. Biomarkers Prev. 12, 728–732 (2003)
5. Boyd, N.F., Lockwood, G.A., Martin, L.J., Knight, J.A., Byng, J.W., Yaffe, M.J., Tritchler, D.L.: Mammographic densities and breast cancer risk. Breast Disease 10, 113–126 (1998)
6. Boyd, N.F., Martin, L.J., Stone, J., Greenberg, C., Minkin, S., Yaffe, M.J.: Mammographic densities as a marker of human breast cancer risk and their use in chemoprevention. Cancer Prev. 3, 314–321 (2001)
7. Atkinson, C., Warren, R., Bingham, S.A., Day, N.E.: Mammographic patterns as a predictive biomarker of breast cancer risk: effect of tamoxifen. Cancer Epidemiol. Biomarkers Prev. 8, 863–866 (1999)
8. Huo, Z., Giger, M.L., Olopade, O.I., Wolverton, D.E., Weber, B.L., Metz, C.E., Zhong, W., Cummings, S.A.: Computerized analysis of digitized mammograms of BRCA1 and BRCA2 gene mutation carriers. Radiology 225, 519–526 (2002)
9. Li, H., Giger, M.L., Olopade, O.I., Margolis, A., Lan, L., Chinander, M.R.: Computerized texture analysis of mammographic parenchymal patterns of digitized mammograms. Acad. Radiol. 12, 863–873 (2005)
10. Li, H., Giger, M.L., Olopade, O.I., Lan, L.: Fractal analysis of mammographic parenchymal patterns in breast cancer risk assessment. Acad. Radiol. 14, 513–521 (2007)
11. Li, H., Giger, M.L., Olopade, O.I., Chinander, M.R.: Power spectral analysis of mammographic parenchymal patterns for breast cancer risk assessment. J. Digit. Imaging 21, 145–152 (2008)
12. Kontos, D., Bakic, P.R., Carton, A.K., Troxel, A.B., Conant, E.F., Maidment, A.D.: Parenchymal texture analysis in digital breast tomosynthesis for breast cancer risk estimation: a preliminary study. Acad. Radiol. 16, 283–298 (2009)
13. Manduca, A., Carston, M.J., Heine, J.J., Scott, C.G., Pankratz, V.S., Brandt, K.R., Sellers, T.A., Vachon, C.M., Cerhan, J.R.: Texture features from mammographic images and risk of breast cancer. Cancer Epidemiol. Biomarkers Prev. 18, 837–845 (2009)
14. Metz, C.E.: ROC methodology in radiographic imaging. Invest Radiol. 21, 70–733 (1986)
15. Metz, C.E.: Some practical issues of experimental design and data analysis in radiological ROC studies. Invest Radiol. 24, 234–245 (1989)
16. ROC software, http://www-radiology.uchicago.edu/krl/roc_soft6.htm

Towards Learning Spiculation Score of the Masses in Mammography Images

Inna Stainvas*, Jonathan Stoeckel, Eli Ratner, Menachem Abramov, and Richard Lederman

Siemens Computer Aided Diagnosis Ltd., Beck Science Building, Hartom Street 8., Har Hotzvim, PO Box 45202, Jerusalem 91450, Israel
{inna.stainvas,jonathan.stoeckel,eli.ratner, menachem.abramov,richard.lederman}@siemens.com

Abstract. This paper deals with learning spiculation scores of masses in a supervised manner. Three spiculation score prediction models treating the score either as a continuous or ordinary variable are presented. These models were compared on a data-set of 255 masses.

Keywords: spiculation score, variable prediction, automatic feature selection.

1 Introduction

It is well known that lesion spiculation is a very important characteristic that strongly influences the radiologist decision in the malignancy diagnosis of masses in mammography images[1]. There is extensive research devoted to extracting features characterizing the stellate/spiculated structure of the masses (e.g. [2,3,4]).

However, the focus of our paper is learning the spiculation score of lesions and selecting features, which are most required for the spiculation score prediction, in supervised and automatic manner. We use a few novel supervised learning models based on linear regression treating the spiculation score either as a continuous or ordinal variable[1]. Our learning includes automatic feature selection that enables us to validate if heuristically generated spiculation features are indeed selected by the models. The proposed models can be used for learning any other score information. Our intension is to reliably estimate a spiculation score corresponding to the perception of radiologists.

2 Method

We used the following steps to create data for learning the spiculation score. First, in order to create the ground truth, a radiologist is asked to assign the

* Corresponding author.

[1] In statistics, ordinal variables are variables which like categorical variables have a discrete number of values, but are clearly ordered between them.

J. Martí et al. (Eds.): IWDM 2010, LNCS 6136, pp. 272–279, 2010.

spiculation score to the displayed masses; the spiculation score has a number of levels starting from 'non-spiculated' to 'very spiculated' (Section 3). Second, the lesions are described by 231 features among which 20 are specially designed spiculated features. The mass features with the spiculation score attached are considered as training data and are used to learn the system to assign the spiculation score automatically to new unseen findings.

During learning the features are considered as predictor variables, and the score as a response variable. We use two supervised learning approaches to address the problem differing in treating the response score variable either as a continuous or ordinal variable. The first approach treats the score as a continuous variable and includes two models: a Bayesian linear regression model with automatic relevance detection (ARD) ($\mathcal{M}_1$) and nested linear regression ($\mathcal{M}_2$). The first model selects the appropriate features using the ARD approach [5] and the second iteratively adds the best feature one by one using a greedy approach (Chapter 8 of [6]).

The second approach is a novel model introduced by us and is a modified Multiple Instance Learning Relevance Vector model (MIL RVM) that considers the score as an ordinal variable. This model ($\mathcal{M}_3$) is an instantiation of the 'Data Replication Method' framework [7] for the Bayesian MIL RVM binary classifier [8]. This novel classifier automatically selects the relevant features and builds the linear boundaries to assign the spiculation score.

In general all three considered models project data to some 'discriminative' direction in space and thus their number of parameters is of the order of the size of the feature space ($\mathcal{O}(d)$). The models $\mathcal{M}_2$ and $\mathcal{M}_3$ have the same internal feature selection mechanism. The main difference between the two approaches lies in treating the variable either as an ordinal or continuous. The main difficulties of the problem are (i) data sparsity, as the number of lesions is less than the data dimensionality (number of features extracted); (ii) large correlation between the extracted features and (iii) a possible lack of consistency between the scores given by the expert[2].

3 Data Description

Our data is created by using a specially written graphical user interface (GUI) tool which enables medical experts to review the image with the lesion superimposed and assign the spiculation score by clicking the appropriate buttons.

We constrained ourselves to six levels of spiculation from 'non-spiculated' to 'very spiculated' with the values being assigned to $s \in \{0, 0.1, 0.25, 0.5, 0.75, 0.9\}$. (see Fig. 1). Currently, we have a small data base of 255 masses, where each mass is described by the d-dimensional feature vector $\mathbf{x}$ ($d = 231$ for our data-set) and the spiculation score s.

The proportion of the scoring levels is very non-homogeneous[3] and is visually clustered into three groups. Probably, this means that it is more natural and

[2] As well as a subjectivity of the scoring.

[3] The score proportion for our data-set is 4.71%, 9.8%, 7.45%, 13.73%, 27.45%, 36.86% for the score values considered.

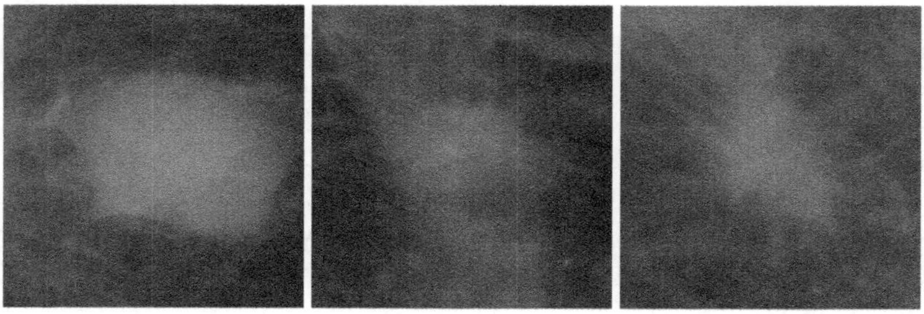

Fig. 1. *S*core of the Masses. The masses with the increasing spiculation score. The spiculation score level is given above each of the image.

easier for radiologists to perceive masses as belonging to three groups 'non-spiculated', 'ambiguous' and 'spiculated', rather than discriminate them in more subtle categories. Therefore, in measuring model performances, we divided the scores into three groups S_1, S_2, S_3 with the score s being $s \leq 0.1,\ 0.1 < s \leq 0.3,\ s > 0.3$, respectively (see also Section 5). This is in agreement with [9], where it was shown that in general the spiculated masses constitute more than 55% of all the masses.

Currently, the responses of a single radiologist are used to learn the spiculation score; however, partial analysis of responses of three radiologists were analyzed. This preliminary analysis show the instability of the spiculation score; we intend to address this issue in the future.

4 Learning Models

This section presents the three learning models to perform the spiculation score prediction.

4.1 Bayesian Linear Regression Model with ARD ($\mathcal{M}_1$)

This section is focused on developing a Bayesian linear regression model for the spiculation score variable y given a training set $\mathcal{D} = \{x_n, y_n\}_{k=1,\ldots,N}$ of N independent samples, where $x_n \in \mathcal{R}^d$ is a d-dimensional predictor variable and y_n is a scalar response variable. The linear regression model assumes linear dependence between the score variable y and the projection of the feature vector onto some unknown direction $\mathbf{w}$ in the space:

$$y = \mathbf{w}^t x + \epsilon, \quad \epsilon \sim \mathcal{N}(0, \sigma^2) \tag{1}$$

where ϵ is a zero-mean Gaussian noise term with variance σ^2.

We use the maximum a-posteriori principle (MAP) to estimate the weight vector $\mathbf{w}$:

$$\mathbf{w}^* = \arg\max_w L(\mathbf{w}) = \arg\max_w [log(p(\mathbf{Y}/\mathbf{w}, \mathbf{X})) + log(p(\mathbf{w}))], \tag{2}$$

where $\mathbf{Y} = \left[y_1\ y_2\ \dots\ y_N\right]^t$ and a design matrix $\mathbf{X} = \left[\mathbf{x_1}\ \mathbf{x_2}\ \dots\ \mathbf{x_n}\right]^t$. The log-likelihood of the data is given by:

$$\log p(\mathbf{Y}/\mathbf{w}, \mathbf{X}, A)) = -N \log \sigma - \frac{N}{2} \log 2\pi - \frac{1}{2\sigma^2} \sum_{i=1}^{N} (y_i - \mathbf{w}^t x_i))^2. \quad (3)$$

The weight prior is given by:

$$p(\mathbf{w}) = (2\pi)^{-d/2} |A|^{1/2} \exp(-\frac{\mathbf{w}^t A \mathbf{w}}{2}), \quad A = diag(\alpha_1, ..., \alpha_d), \quad (4)$$

and states that the weight components are independent and close to zero with a variance parameter $1/\alpha_i$. The parameters α_i are hyper-parameters that are learned from the data to maximize the marginal likelihood $p(\mathcal{D}|A)$. The hyper-parameters provide a mechanism to select effective features [8,5]. As the $\alpha_i \to \infty$ the corresponding weight variance tends to zero (thus concentrating the prior sharply at zero). Hence, regardless of the evidence of the training data, the posterior will be also sharply concentrated on zero, thus corresponding feature can be removed from training.

The MAP estimation for the weight vector and σ^2 lead to a regularized linear regression in the form:

$$\mathbf{w}^* = \frac{1}{\sigma^2} S^{-1} S_{xy}, \quad \sigma^{*2} = \frac{1}{N} \sum_{i=1}^{N} (y_i - \mathbf{w}^t \mathbf{x}_i)^2 \quad (5)$$

$$S = A + \frac{1}{\sigma^2} S_{xx}, \quad S_{xx} = \sum_{i=1}^{N} \mathbf{x}_i \mathbf{x}_i^t, \ S_{xy} = \sum_{i=1}^{N} y_i \mathbf{x}_i. \quad (6)$$

To perform the optimal feature selection, we apply the type-II maximum likelihood [10], which means maximizing the marginal likelihood $p(\mathbf{Y}/\mathbf{X}, A))$ over A.

Though analytical formula for A can be obtained, it is non-linear and requires applying numerical methods; instead we use the same approximation strategy as in [8]. A Hessian of $L(\mathbf{w})$ is easily calculated as $H = -(A + \frac{S_{xx}}{\sigma^2}) = -S$, where S is defined in Eq.6. Using the same evaluation as in Eq.(16-17) of [8], we finally get:

$$-\frac{1}{2} \mathbf{w}_i^{*2} + \frac{1}{2\alpha_i} - \frac{1}{2} H_{ii}^{-1} = 0 \Leftrightarrow \alpha_i = \frac{1}{\mathbf{w}_i^{*2} + H_{ii}^{-1}} \quad (7)$$

The final algorithm iterates recursively between calculating the hyper-parameters α_i by Eq.7 and then the MAP estimation of $\mathbf{w}^*$ is performed according to Eq.5. With the iterations the hyper-parameters some α_i become too large. This means that their corresponding $w_i \to 0$ and the corresponding features are irrelevant and can be eliminated in future iterations.

4.2 Nested Linear Regression ($\mathcal{M}_2$)

We consider the same linear regression model (Eq. 1) and perform a feed-forward model selection based on the χ^2-test on the number of new predictor variables.

A test checks, whether to increase a number of variables by one comparing iteratively on $j = 2, ..., J$ a current null hypothesis:

$$H_0^{j-1} : a_j = 0 \tag{8}$$

against an alternative (that is considered as a current full model):

$$H_0^j : \ arbitrary \ \ a_r, \ r \leq j \tag{9}$$

The statistic of the nested models test is constructed by the method of the "maximum likelihood ratio". The ratio of the maximum likelihood $2 \log \lambda = \frac{p_j}{p_{j-1}}$ is approximately distributed as $\chi^2(r)$ with a degree of freedom $r = 1$ (a number of constraints) [6]; the maximum likelihood p_j is the maximum likelihood of $\mathbf{Y}$ given a truncated set of variables $\mathbf{X}$ with only j first columns.

The $\chi^2(r)$ test with significance level α stops the model growing if

$$\Delta e_{rse} \leq \sigma_j^2 [\chi^2(1)]^{-1}(1 - \alpha),$$

where $[\chi^2(r)]^{-1}$ is an inverse function of the cumulative density function of $\chi^2(r)$ with r degrees of freedom, Δe_{rse} is a difference of residual square errors e_{rse} for the regression variable $y(\cdot)$ with j and $j-1$ variables; $\sigma_j = e_{rse}$ is an estimated variance of an independent Gaussian noise in the full model. We set the significance level to $\alpha = 0.05$ in our experiments.

4.3 MIL RVM with the Ordinal Data ($\mathcal{M}_3$)

This section presents the modified MIL RVM for ordinal data. Below, we briefly present the MIL RVM binary classifier and explains mapping the "data replication (DR) framework" [7] to MIL RVM.

MIL RVM Binary Classifier. In the MIL framework the data is considered to be aggregated into the so called bags $\mathbf{x}_\mu, \mu = 1 \ldots \mathcal{M}$, ($\mathcal{M}$ is the number of bags). All the instances of the bag share the same extra bag-state label being positive or negative. The bag-μ is considered to be negative if all its instances x_μ^s, $s = 1 \ldots S_\mu$ (S_μ is the number of samples in the bag μ) are negative; and positive if at least one its instance is positive. The probability of the μ^{the} bag to be positive ($y_\mu = 1$) is given by:

$$p_\mu = p(y_\mu = 1 | \mathbf{x}_\mu) = 1 - \prod_{s=1}^{S_\mu} (1 - \sigma(w^t x_\mu^s), \quad \sigma(z) = \frac{1}{1 + e^{-z}}$$

The classifier optimization criterion is obtained as the MAP estimator of the weight vector $w \in R^d$ from the observed bag's-data $\mathcal{D}$.

$$w^* = arg \max_w [log(p(\mathcal{D}/w)) + log(p(w))], \tag{10}$$

where the log-likelihood of the bag's data is defined as:

$$\log p(\mathcal{D}|w) = \sum_{\mu=1}^{\mathcal{M}} (y_\mu \log p_\mu + (1 - y_\mu) \log(1 - p_\mu)). \quad (11)$$

The weight prior is the same sparsity prior (Eq. 4) as in the linear regression model with ARD.

Mapping the DR Framework to MIL RVM. The RVM MIL is an ideal binary classifier to be used in the DR framework for ordinal data classification that additionally *allows feature selection* and easily generalized to MIL framework. It is used without any particular changes with the replicated data as described in [7]. The only issues that need attention are:

1. *Adding linear constraints on the biases to be ordered:* Instead we add the extra data to the training data-set in the form proposed in [7]. This addition does not guarantee the bias ordering, but modifies a log-likelihood to try to satisfy it. After learning is completed we check that the biases are correctly ordered.
2. *Replication of the data within the MIL concept:* It is easy to see that during the replication we not only replicate the data, but the bag label as well; all the other considerations remain the same.

Currently, our data is constrained to a case where each instance stands for a unique bag (the lesions from different views were not consolidated to a bag), i.e. we constrain ourselves to using RVM without MIL concept.

5 Results

The spiculation score is a very subjective characteristic that has a large inter and intra expert variability. In general the radiologists are very confident in the spiculation presence or absence, but are very uncertain about its value when spiculation is present. We performed training with the originally assigned score values, however we report the model's performances for the scoring groups S_1, S_2, S_3.

The results are presented as confusion matrices for the train and test sets, which are obtained by splitting all the data-set in half randomly. In addition, we introduce so called migration specificity and sensitivity errors as a percentage of non-spiculated lesions to be classified as spiculated and spiculated as non-spiculated, respectively. This consideration is the simplest variant of penalizing differently the errors of spiculated and non-spiculated masses to migrate to ambiguous and opposite scoring groups; i.e. in such consideration assignment to the ambiguous group is ignored by us. Table 1 presents confusion matrices in percent for all the models and training/testing sets. All three models have a good performance in assigning spiculation masses with a low migration sensitivity errors of 17.65% for the models $\mathcal{M}_2$ and $\mathcal{M}_3$ (Table 1).

Table 1. *Scoring results for three models:* The table presents the confusion matrices in percent. The migration **specificity** errors appear in bold and sensitivity errors are shaded. All three models have a good performance in assigning spiculation masses. The most compact model is $\mathcal{M}_3$. S_1, S_2, S_3 are 'non-spiculated', 'ambiguous' and 'spiculated' score zones, respectively. Their size is indicated as $S_k()$ in brackets.

$\mathcal{M}_1$: Linear Regression with the ARD						
	Training			Testing		
score groups	$S_1(17)$	$S_2(8)$	$S_3(100)$	$S_1(20)$	$S_2(11)$	$S_3(99)$
S_1	29.41	5.88	**64.71**	20.00	10.00	**70.00**
S_2	12.50	25.00	62.50	9.09	9.09	81.82
S_3	3.00	10.00	87.00	15.15	11.11	73.74
selected features 73	summary error % 24.8			summary error % 40		

$\mathcal{M}_2$: Nested Linear Regression						
	Training			Testing		
score groups	$S_1(17)$	$S_2(8)$	$S_3(100)$	$S_1(20)$	$S_2(11)$	$S_3(99)$
S_1	29.41	52.94	**17.65**	30.00	15.00	**55.00**
S_2	0.00	25.00	75.00	9.09	0.00	90.91
S_3	0.00	0.00	100.00	8.08	6.06	85.86
selected features 94	summary error % 14.4			summary error % 30		

$\mathcal{M}_3$: Ordinary MIL RVM						
	Training			Testing		
score groups	$S_1(17)$	$S_2(8)$	$S_3(100)$	$S_1(20)$	$S_2(11)$	$S_3(99)$
S_1	5.88	76.47	**17.65**	5.00	55.00	**40.00**
S_2	0.00	25.00	75.00	0.00	36.36	63.64
S_3	0.00	11.00	89.00	2.02	16.16	81.82
selected features 12	summary error % 26.4			summary error % 33.8		

The most compact model is the ordinal RVM MIL model ($\mathcal{M}_3$) that selects 12 features only, among which only 2 features are heuristically built spiculated features. Models $\mathcal{M}_1$ and $\mathcal{M}_2$ select 73 and 94 features, among which 7 and 12 are the spiculation features, respectively. In selection of spiculated features, all the models agree on a single spiculation feature.

6 Discussion

The spiculation score assignment is a clinically demanding task that influences the final diagnosis given by the radiologists and could be helpful for CAD systems. In this paper, we attempted to solve the score prediction task in the supervised manner by considering the score either as a continuous or ordinary variable and using the simplest models based on the linear regression. We showed on the small data-set of 255 masses that it is possible to get satisfactory results. However, more experiments on a large data set are required.

We also notice that the spiculation score has a large inter and intra expert variability that raises other interesting problems, such as (i) the right score quantization that is perceptually stable, (ii) creating models that are stable to unstable response variables [11]. Another interesting application of the score information is considering the score prediction in a multi-task framework for the mass detection.

Acknowledgments

The authors would like to thank David Leib for his advice and help in the preparation of the paper; and the anonymous reviewers for their helpful comments.

References

1. Tabar, L., Dean, P.B.: Teaching atlas of mammography. Georg Thieme Verlag, Stuttgart (1985)
2. Sampat, M.P., Bovik, A.C., Markey, M.K.: Model-based framework for the detection of spiculated masses on mammography. Medical Physics 35(5), 2110–2123 (2008)
3. Liu, S., Delp, E.J.: Multiresolution detection of stellate lesions in mammograms. In: Proceedings of IEEE International Conference on Image Processing, Santa Barbara, California, vol. 1, pp. 109–112 (1997)
4. Zwiggelaar, R., Astley, S.M., Boggis: Linear structures in mammographic images: detection and classification. IEEE Transactions on Medical Imaging 23(9), 1077–1086 (2004)
5. Tipping, M.E., Smola, A.: Sparse bayesian learning and the relevance vector machine (2001)
6. Silvey, S.D.: Statistical Inference. Chapman and Hall, New York (1975)
7. Cardoso Cardoso, J.S., Pinto da Costa, J.F.: Learning to classify ordinal data: The data replication method. J. Mach. Learn. Res. 8, 1393–1429 (2007)
8. Krishnapuram, B., Stoeckel, J., Raykar, V., Bharat, R., Bamberger, P., Ratner, E., Merlet, N., Stainvas, I., Abramov, M., Manevitch, A.: Multiple instance learning improves cad detection of masses in digital mammography. In: Krupinski, E.A. (ed.) IWDM 2008. LNCS, vol. 5116, pp. 350–357. Springer, Heidelberg (2008)
9. Vyborny, C.J., Doi, T., Stein, A.: Breast cancer: Importance of spiculation in computer aided detection. Radiology (215), 703–707 (2000)
10. MacKay, D.J.C.: A practical Bayesian framework for backprop networks. Neural Computation 4, 448–472 (1992)
11. Raykar, V.C., Yu, S., Zhao, L.H., Jerebko, A., Florin, C., Valadez, G.H., Bogoni, L., Moy, L.: Supervised learning from multiple experts: Whom to trust when everyone lies a bit. In: Bottou, L., Littman, M. (eds.) Proceedings of the 26th International Conference on Machine Learning, Montreal, June 2009, pp. 889–896. Omnipress (2009)

Three Digital Mammography Display Configurations: Observer Performance in a Pilot ROC Study

László K. Tabár[1], Alfredo S. Niño[2], Eric Schreiber[3], Richard Ellis[4], Cynthia Goralnik[5], Steven Scislaw[6], Raja Cheruvu[7], Deepa Masrani[8], Daoxian Zhang[9,*], Steve Muller[10], and Patrick Heffernan[9]

[1] School of Medicine, University of Uppsala, Sweden
[2] Banner Good Samaritan Medical Center, AZ, USA
[3] Cleveland Clinic Foundation, OH, USA
[4] Gundersen Lutheran Medical Center, WI, USA
[5] SimonMed Imaging, AZ, USA
[6] Saint Agnes Medical Center and Physicians Diagnostic Imaging, CA, USA
[7] Williamsville, NY, USA
[8] The University of Texas Medical Branch, TX, USA
[9] Three Palm Software, CA, USA
heidiz@threepalmsoft.com
[10] NDS Surgical Imaging, MA, USA

Abstract. The purpose of this study was to determine if displays that provide more grayscale levels (10bit vs. 8bit) can improve observer performance in breast cancer detection. The study was also designed to determine if 3MP (million-pixel) displays can achieve similar observer performance as compared to 5MP displays. The study was performed using the WorkstationOne mammography software on Dome® E5 and E3 high-resolution displays. Ten radiologists reviewed 33 digital mammography screening studies. On 5MP displays, the average Az value across all observers was 0.7912 using 8bit displays, and 0.8306 using 10bit displays. The difference between 8bit and 10bit displays is statistically significant (F=4.43, p=0.0157). The difference of the average Az value from 8bit 3MP displays is not statistically significant compared to reading with 5MP displays. The implications of the study are that, using appropriate software and hardware, 5MP 10bit displays may improve diagnostic accuracy and 3MP displays may not impact negatively diagnostic accuracy.

Keywords: Digital Mammography, Medical Display, Observer Study.

1 Introduction

Most evidence suggests that the human visual system is able to perceive around 1000 grayscale levels over the luminance range currently used in medical displays [1-4]. However an 8bit display can only provide 256 grayscale levels, whereas most digital mammography acquisition devices (such as FFDM or Mammography CR) produce images at higher bit depths, ranging from 10 to 16 bits. This means that all the acquired

* Corresponding author.

J. Martí et al. (Eds.): IWDM 2010, LNCS 6136, pp. 280–287, 2010.

grayscale levels cannot be displayed at once, or at least not displayed at the contrast sensitivity that human observers can utilize. This could result in a potential loss of information during the diagnostic interpretation, with the implication that the observer may require frequent changes to a localized contrast window and level in order to see all the information, which would impact the observer's efficiency.

This study assessed whether human observer perceptual performance would improve with 10bit 5MP (million-pixel) displays compared to 8bit 5MP displays. This study also assessed whether 3MP displays can achieve similar observer performance to 5MP displays. The reading time was also recorded in order to determine if the three different displays have an impact on the efficiency of the reading workflow.

2 Methods

The study used the WorkstationOne software (Three Palm Software, Los Gatos, CA) on the Dome E5 5MP and E3 3MP displays (NDS Surgical Imaging, San Jose, CA) driven by FX4600 graphic cards (NVIDIA, Santa Clara, CA). Dome E5 and E3 displays are FDA-cleared high-resolution grayscale monitors for displaying medical images (Dome E5 is also FDA-cleared for Mammography use). All displays were calibrated to the DICOM GSDF requirements. WorkstationOne has FDA clearance for use with digital mammography systems to interpret digital mammography images by radiologists. WorkstationOne was designed to maximize the radiologists' efficiency as well as accuracy in reading digital mammograms, specifically; the software is designed to support a mammography specific interpretation workflow. With this feature, the radiologist's reading performance is expected to be similar between 3MP displays and 5MP displays. By providing higher pixel grayscale levels, the radiologists' performance is expected to be improved with 10bit displays.

A set of 33 digital mammography cases were randomly selected from an existing database. Among the 33 cases, 16 cases contain mammography visible cancer lesions confirmed by biopsy reports. Each case consists of the standard four view mammogram images. The pixel depth of the images was 10bit or 12bit, and the pixel size of the images was 50μm, 70μm or 94μm. All patient identifiers were removed from the images.

The study recruited 10 observers, including 9 certified radiologists and 1 fellow. The average digital experience was 16.6 months with a range from 0 months to 4 years. The years reading mammograms were from 8 months to 23 years. All participates were voluntary and were from the teaching course "Multimodality Detection and Diagnosis of Breast Diseases". A written informed consent form was provided and signed by each subject before the study began. Participation in this study posed no risks. There is no foreseen benefit to the individuals in this study, and the subjects were not financially compensated.

The WorkstationOne software was downloaded to 6 computers, each with 2 Dome grayscale displays. Among the 6 computers, 2 were setup with 8bit 3MP displays, 2 with 8bit 5MP displays and 2 with 10bit 5MP displays. The study was conducted in a dark room associated with the Mammography Education course held January 21-24, 2009 at the Phoenix convention center.

Reading of the cases was divided into three sessions. In each session, each observer read all 33 cases using the WorkstationOne software. The observers were blinded to the reading results for all three sessions. For session A, the observer used 2 Dome E3 displays (8bit-3MP session). For session B, the observer used 2 Dome E5 8bit displays (8bit-5MP session). For session C, the observer used 2 Dome E5 10bit displays (10bit-5MP session). The order of these sessions was random for each observer. For each case, the observers marked up any lesion that they found, indicated the type of lesion, and ranked the lesion's suspicion (quasi-continuous) level from 0-100%. 10 cases (5 cancers and 5 normal cases) were provided for training before the study began. The similar cancer and normal case distribution for the 33 cases was known to the observers.

A reading methodology was followed on WorkstationOne, which includes the following viewing workflow steps:

1. Overview viewing to display the standard four views of a mammography case from both current and prior (if exists) cases;
2. Bilateral current viewing with same size fit to the display size;
3. Current and prior (if exists) comparison to enhance the detection of tissue density changes;
4. Systematic comparison of left and right breasts using masking [5] to enhance the detection of structural asymmetries;
5. All-pixel viewing of full resolution images using automatic tracking of the viewing path to ensure that there are no areas in the images that are not viewed;
6. Report of interpretation findings.

The study used a standard methodology for multi-reader multi-case (MRMC) receiver operating characteristic (ROC) observer studies [6-12] with a sequential reading model. The MRMC ROC analysis software (DBM MRMC v2.2) from the Kurt Rossmann Laboratories at the University of Chicago was used to calculate ROC curves and the area under the curve (Az value). The software also provided statistical analysis to compare the Az values between the 8bit 3MP group and the 8bit 5MP group; as well as between the 8bit 5MP group and the 10bit 5MP group. The sensitivity and specificity for each observer were also computed. The t-test was used to compare these three measures pooled over the observers for each pair of reading conditions.

3 Results

The study data pooled from all three display configurations were analyzed using DBM MRMC (v2.2) software (see previous section). The program employed jackknifing and ANOVA (Analysis of Variance) techniques [6-12]. The analysis was reported by treating both observers and cases as random samples, i.e., results apply to the observer and case populations. The null hypothesis of equal "treatments" (display configurations) is tested, and the treatments difference 95% confidence intervals are given.

3.1 ROC Area under the Curve (A_z) Analysis

The data collected from 3 "treatments" (display configurations), 10 readers (observers) and 33 cases (17 normal and 16 abnormal mammogram studies) are loaded into the program DBM MRMC for analysis. The curve fitting methodology was PROPROC [6-12]. A graph of the ROC areas under the curve (Az values) for each observer and for each display configuration is shown in the following figure.

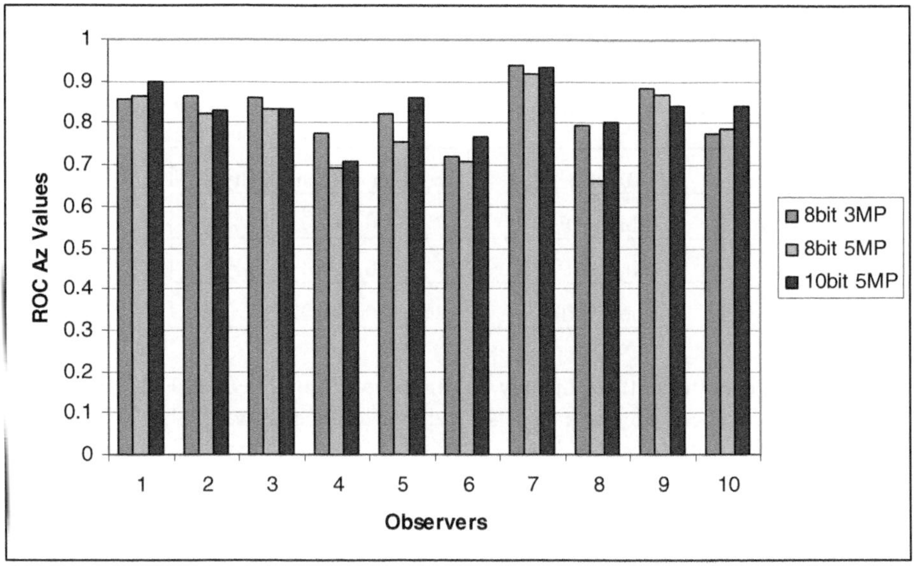

Fig. 1. The Az values for each observer and for each display configuration

The mean values (average across observers) are shown in the following table.

Table 1. The mean values across all observers for each display configuration

Display	Mean Az
8bit 3MP	0.82779296
8bit 5MP	0.79116570
10bit 5MP	0.83063642

The Az values of the three display configurations are not equal, $F(2,18) = 4.43$, $p = 0.0273$. The 95% confidence intervals for the difference between 8bit 3MP and 8bit 5MP is not significant, $p = 0.0235$ (> *0.0167*). The 95% confidence intervals for the difference between 8bit 3MP and 10bit 5MP is not significant, $p = 0.8498$ (> *0.0167*).

The 95% confidence intervals for the difference between 8bit 5MP and 10bit 5MP is significant, $p = 0.0157$ (< 0.0167).

Since the study made comparisons among 3 configurations, the critical p-value (0.05) was adjusted by dividing by 3, so the critical p-value *0.0167* was used.

3.2 Sensitivity Analysis

The sensitivity analysis was also performed at specificity = 0.5 (considering around 50% of normal population of the study data). The mean values (average across observers) are shown in the following table.

Table 2. The mean sensitivities across all observers for each display configuration

Display	Average Sensitivity
8bit 3MP	0.88977372
8bit 5MP	0.83981452
10bit 5MP	0.89634391

The sensitivities of the three display configurations are not equal, $F(2,50) = 4.27$, $p = 0.0194$. The 95% confidence intervals for the difference between 8bit 3MP and 8bit 5MP is not significant ($p = 0.0222$). The 95% confidence intervals for the difference between 8bit 3MP and 10bit 5MP is not significant ($p = 0.7574$). The 95% confidence intervals for the difference between 8bit 5MP and 10bit 5MP is significant ($p = 0.0102$).

3.3 Sequential Reading Analysis

An ROC analysis was also performed for the sessions to determine whether 3 sequential readings in three days would change observer performance. The display configuration was randomly assigned to each session for each observer based on the equipment availability.

Table 3. The mean Az values across all observers for each reading session

Session	Average Az
first	0.83393918
second	0.82404307
third	0.81189700

The Az values of the three sessions are not statistically significantly different, $F(2,62) = 0.74$, $p = 0.4832$. So the session arrangement of 3 sequential readings in three days may not change observer performance.

The reading time was automatically recorded during the study, and the average reading time for the 33 cases is shown in the following table.

Table 4. Reading time for each configuration

Display	Average Time (minutes)
8bit 3MP	34.3
8bit 5MP	31.4
10bit 5MP	35.9

The reading time on 8bit 3MP and 10bit 5MP were longer than 8bit 5MP. But analysis was not performed to determine if the differences between these times are statistically significant.

4 Conclusions

Observer average performance was better on 10bit 5MP displays compared to 8bit 5MP displays, and the difference was statistically significant. Observers also performed at higher detection sensitivity (6% better) on 10bit displays.

Observer performance difference was not statistically significant on 3MP displays compared to 8bit 5MP displays.

The sequential sessions of repeat reading three times of same cases in three days did not change observer performance.

5 Clinical Relevance/Application

The study results suggested that 10bit displays may improve readers' performance for mammography interpretation. If the current recommended 5MP displays are not practical, 3MP displays may provide comparable performance with appropriate software.

6 Discussion

This study results showed that 10bit displays are better than 8bit displays. However another published study (Krupinski chest nodule study) showed no difference [2]. The unique (subtle) image characteristics and features on mammography images and the perceptual reading methodology [5] for mammography may be the factors that enable 10bit displays to improve readers' performance.

3MP displays performed slightly better than 5MP displays, which could be attributed to the perceptual reading methodology [5] provided on WorkstationOne which includes the Tabar masking viewing and all-pixel viewing techniques.

Readers spent slightly more time on 8bit 3MP displays than 8bit 5MP displays (and more time on 10bit 5MP displays than 8bit 3MP displays) which may have contributed to the slightly better performance. More displayed image pixel information (10bit vs. 8bit in depth) might have made the observers involuntarily spend more time to read each case; and more time was required for the observers to scan through all pixels on the lower resolution displays (3MP vs. 5MP), which might explain why observers' performance improved.

7 Limitations

This study used only 33 cases with a variety of image pixel bit depth, a small number which can be a potential bias or imprecision when generalizing the study result. A larger number of cases (>100 cases) with the detailed lesion information (such as lesion type, size and subtlety) and tissue density assessment (especially for normal cases) is planned to be used for a follow-on study.

Validity of whether the reading environments are consistent and whether the time interval between each reading session should be longer or shorter is debatable and needs to be validated.

The study tools (software and hardware) were limited to a small set of manufacturers. Thus it is unknown if the study results apply in general (i.e., to any software and hardware).

References

1. Barten, P.G.J.: Contrast sensitivity of the human eye and its effects on image quality. SPIE Press, Bellingham (1999)
2. Krupinski, E.A., Siddiqui, K., Siegel, E., et al.: Influence of 8-bit vs 11-bit digital displays on observer performance and visual search: a multi-center evaluation. Journal of the Society for Information Display 15, 335–429 (2007)
3. Kimpe, T., Tuytschaever, T.: Increasing the Number of Gray Shades in Medical Display Systems - How Much is Enough? Journal of Digital Imaging 20(4), 422–432 (2007)
4. Samei, E., Ranger, N.T., Delong, D.M.: A Comparative Contrast-Detail Study of Five Medical Displays. Med. Phys. 35(4), 1358–1364 (2008)
5. Tabar, L., et al.: Breast Cancer: The Art and Science of Early Detection with Mammography – Perception, Interpretation, Histopathologic Correlation. Georg Thieme Verlag, New York (2005)
6. Dorfman, D.D., Berbaum, K.S., Metz, C.E.: Receiver operating characteristic rating analysis: Generalization to the population of observers and patients with the jackknife method. Investigative Radiology 27, 723–731 (1992)
7. Dorfman, D.D., Berbaum, K.S., Lenth, R.V., Chen, Y.F., Donaghy, B.A.: Monte Carlo validation of a multiobserver method for receiver operating characteristic discrete rating data: Factorial experimental design. Academic Radiology 5, 591–602 (1998)
8. Hillis, S.L., Berbaum, K.S.: Power estimation for the Dorfman-Berbaum-Metz method. Academic Radiology 11, 1260–1273 (2004)
9. Hillis, S.L., Obuchowski, N.A., Schartz, K.M., Berbaum, K.S.: A comparison of the Dorfman-Berbaum-Metz and Obuchowski-Rockette methods for receiver operating characteristic (ROC) data. Statistics in Medicine 24, 1579–1607 (2005), doi:10.1002/sim.2024

10. Hillis, S.L.: Monte Carlo validation of the Dorfman-Berbaum-Metz method using normalized pseudovalues and less data-based model simplification. Academic Radiology 12, 1534–1541 (2005), doi:10.1016/j.acra.2005.07.012
11. Hillis, S.L.: A comparison of denominator degrees of freedom for multiple observer ROC analysis. Statistics in Medicine 26, 596–619 (2007), doi:10.1002/sim.2532
12. Hillis, S.L., Berbaum, K.S., Metz, C.E.: Recent developments in the Dorfman-Berbaum-Metz procedure for multireader ROC study analysis. Academic Radiology 15, 647–661 (2008)

How Can Image Quality Affect the Detection Performance of Breast CAD (Computer Aided Detection) in FFDM (Full Field Digital Mammography)? – A Comparative Study with Two Different FFDM Systems-

Nachiko Uchiyama[1], Jonathan Stoeckel[2], Kyoichi Otsuka[3], Seiko Kuroki[1], Yukio Muramatsu[1], and Noriyuki Moriyama[1]

[1] Department of Cancer Screening Technology and Development, Research Center for Cancer Screening, National Cancer Center. 5-1-1, Tsukiji, Chuo-ku, Tokyo, Japan, 104-0045
nuchiyam@ncc.go.jp
[2] Siemens Computer Aided Diagnosis Ltd. 8 Hartom St., 91450 Jerusalem, Israel
[3] Siemens-Asahi Medical Technologies Ltd. 3-20-14, Higashi-Gotanda, Shinagawa-ku, Tokyo, Japan

Abstract. This study was conducted to retrospectively evaluate the variation of CAD performance utilizing two different FFDM systems in normal clinical cases.

Keywords: CAD, Digital Mammography, MMG, Digital.

1 Introduction

Computer-aided detection (CAD) has been applied to full field digital mammography (FFDM) for some time. However, unlike the raw imaging data on hard copy, utilizing digitizers for CAD processing in the film-screen system with groups in units of 8-10 bits, the raw imaging data for CAD processing in FFDM are analyzed with groups in units of 12-14 bits, which has a much more dynamic range compared to digitized hard copy data in the film-screen system. According to the background, there are more opportunities to apply a combination of anode/filters such as W/Rh that allows us to decrease the radiation dose while keeping higher image quality compared to the images using Mo/Mo and Mo/Rh in the film-screen system [1]. The raw imaging data for CAD processing in FFDM can be more strongly influenced by the different contrast and image sharpness in clinical images, compared to the CAD dedicated to a film-screen system. From this hypothesis, the detection pattern in CAD could vary even in normal clinical cases.

2 Methods

This study was conducted as part of research that was approved by the IRB at our institute on June 12th in 2007. All patients that were recruited for this study gave informed consent. The clinical cases in this study were selected from screening mammograms

J. Martí et al. (Eds.): IWDM 2010, LNCS 6136, pp. 288–295, 2010.

taken from June 12th in 2007 to December 24th in 2009. Clinical image data were acquired by two different FFDM systems. One was an a-Se FFDM system with a spatial resolution of 70μm (System A) and imaging data were acquired from June 12th in 2007 to November 24th in 2008. Another was an a-Se FFDM system with a spatial resolution of 85μm (System B) and imaging data were acquired from December 7th in 2008 to December 24th in 2009. Mammograms were diagnosed as BI-RADS category 1 or 2 by double-reading and breast ultrasound was performed in each case and diagnosed as a normal or a benign case. The total number of cases was 1140 cases in System A and 1178 cases in System B. The median patient age was 59.8 years old (range 40-75 years old) in System A and 60.0 years old (range 40-88 years old) in System B. To optimize radiation exposure parameters in clinical images, we measured CNR (Contrast to Noise Ratio) in accordance with EUREF (European guidelines for quality assurance in breast cancer screening and diagnosis) guide lines simulating breast thickness, utilizing PMMA phantoms (20-70mm) and radiation exposure parameters, kV (24-34kV) and combinations of anode/filters (Mo/Mo, Mo/ Rh, and W/Rh). In addition, we performed spectral analysis of anode/filters (Mo/Mo, Mo/ Rh, and W/Rh) regarding both FFDM systems. A CAD dedicated to the FFDM systems was applied for purpose of review and was verified, regarding detection areas, with reference to the diagnostic reports of the mammogram and ultrasound. The same CAD algorithm was utilized for the two FFDM systems.

3 Results

We optimized radiation exposure parameters in a clinical setting with reference to the results of the CNR analysis and dosimetry in accordance with EUREF Guidelines [2](Fig.1-2). In System A, under 20mm breast thickness, the combination of 24kV with Mo/Mo was selected; from 21m to 30mm breast thickness, the combination of 26kV with Mo/Mo was selected; from 31mm to 40mm breast thickness, the combination of 28kV with Mo/Mo was selected; from 41to 60mm, the combination of 30kV with W/Rh was selected; from 61mm to 70mm, the combination of 32kV with W/Rh was selected; and above 70mm, the combination of 34kV with W/Rh was selected. CAD detected relatively dense areas as false-positive masses (Fig.3) at a rate of 9.8% (448/4560 images) and fibrous tissue as false-positive microcalcifications (Fig.4) at a rate of 0.7% (34/4560 images). In the cases utilizing 24-28kV Mo/Mo, CAD detected masses as false positives more frequently at a rate of 12.7% (279/2196 images) , compared to the cases utilizing 30-34 kV W/Rh which detected false positives at a rate of 7.1% (169/2364 images). There was a statistically significant difference ($P=0.008< 0.05$) between the two different combinations of anode/filters. CAD detected more false-positive masses in the cases utilizing the combinations with Mo/Mo in comparison with the cases utilizing the combinations with W/Rh. On the other hand, in the cases utilizing 30-34kV W/Rh, CAD detected false-positive microcalcifications more frequently at a rate of 1.1% (26/2364 images), compared to 0.4 % (8/2196 images) detected utilizing 24-28kV Mo/Mo. There was a statistically significant difference ($P=0.022<0.05$) between the two combinations with different anode/filters. CAD detected more false-positive calcifications in the cases utilizing W/Rh in comparison with the cases utilizing Mo/Mo (Table 1a.). In System B,

Table 1. Clinical Radiation Exposure Setting of System A (Table 1a) and System B (Table1b) regarding Frequency of False Positives (FPs) using CAD

Table1a.

	FP Mass		FP Microcalcifications	
24-28kV Mo/Mo	**12.7% (279/2196)**	**P=0.008<0.05**	**0.4% (8/2196)**	**P=0.022<0.05**
30-34kV W/Rh	**7.1% (169/2364)**		**1.1% (26/2364)**	
Total	**9.8% (448/4560)**		**0.7% (34/4560)**	

Table1b.

	FP Mass		FP Microcalcifications	
24kV W/Rh	**1.0% (5/500)**	**P>0.05**	**1.2% (6/500)**	**P>0.05**
26kV W/Rh	**3.3% (32/960)**		**0.7% (7/960)**	
28kV W/Rh	**3.3% (48/1460)**		**1.2% (18/1460)**	
30kV W/Rh	**1.4% (16/1128)**		**0.5% (6/1128)**	
32-34kV W/Rh	**4.2% (28/664)**		**0% (0/664)**	
Total	**2.7% (129/4712)**		**0.8% (37/4712)**	

Table 2. Linear Attenuation Coefficient of Breast Tissue at 20keV [3]

	Linear Attenuation Coefficient(cm^{-1})
Breast Tissue	**0.8**
Fat Tissue	**0.45**
Skin	**0.8**
Mass	**0.85**
Calcification	**12.5**

under 25mm breast thickness, the combination of 24kV with W/Rh was selected; from 26m to 35mm breast thickness, the combination of 26kV with W/Rh was selected; from 36mm to 45mm breast thickness, the combination of 28kV with W/Rh was selected; from 46 to 55mm, the combination of 30kV with W/Rh was selected; and above 56mm, the combination of 32kV-34kV with W/Rh was selected. CAD detected false-positive masses at a rate of 2.7% (129/4712 images) and false-positive microcalcifications at a rate of 0.8% (37/4712 images) in total. With 24kV, CAD detected false-positive masses at a rate of 1.0% (5/500 images) and false-positive microcalcifications at a rate of 1.2% (6/500 images). With 26kV, CAD detected false-positive masses at a rate of 3.3% (32/960 images) and false-positive microcalcifications at a rate of 0.7% (7/960 images). With 28kV, CAD detected false-positive masses at a rate of 3.3% (48/1460 images) and false-positive microcalcifications at a rate of 1.2% (18/1460 images).With 30kV, CAD detected false-positive masses at a rate of 1.4% (16/1128 images) and false-positive microcalcifications at a rate of 0.5% (6/1128 images). With 32-34kV, CAD detected false-positive masses at a rate of 4.2%

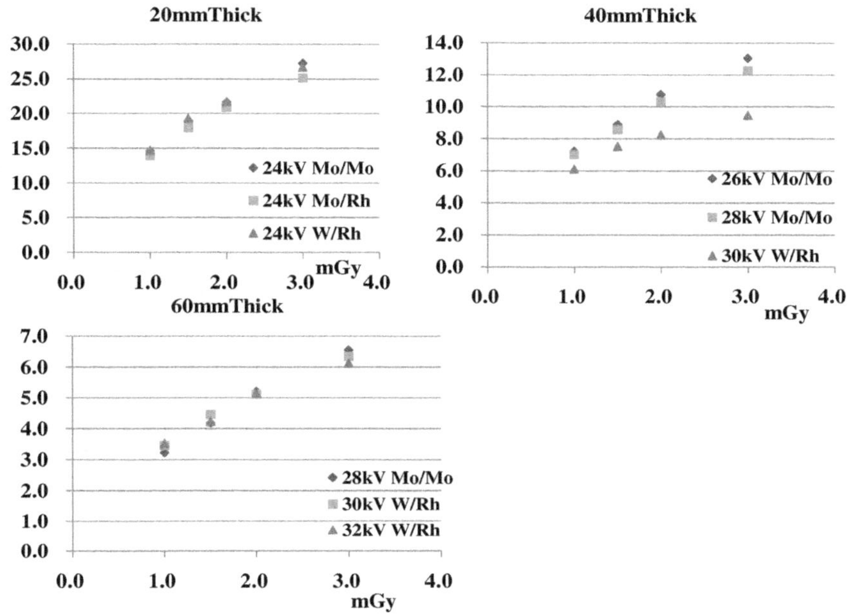

Fig. 1. CNR Analysis: 20mm, 40mm, and 60mmThick PMMA Phantom in System A

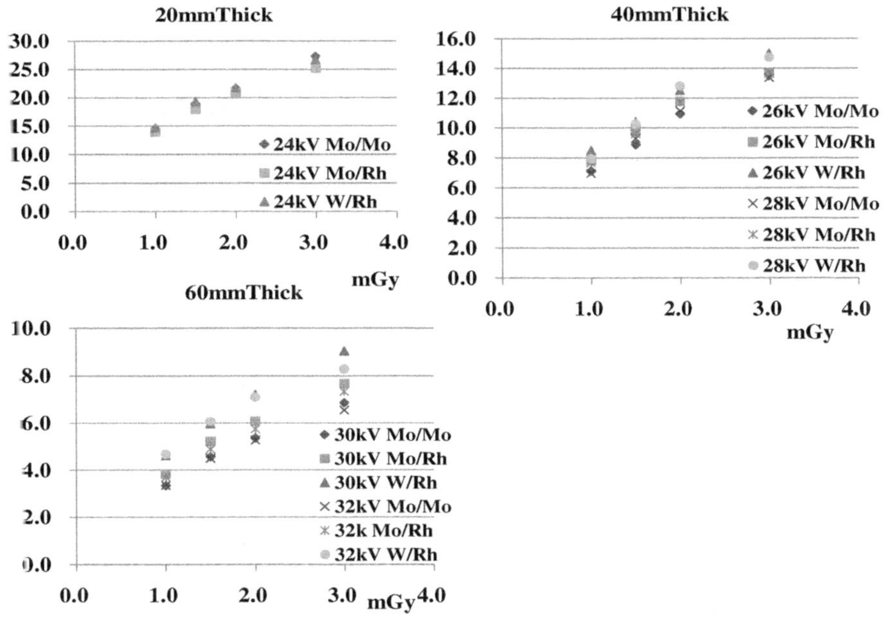

Fig. 2. CNR Analysis: 20mm,40mm,and 60mm Thick PMMA Phantom in System B

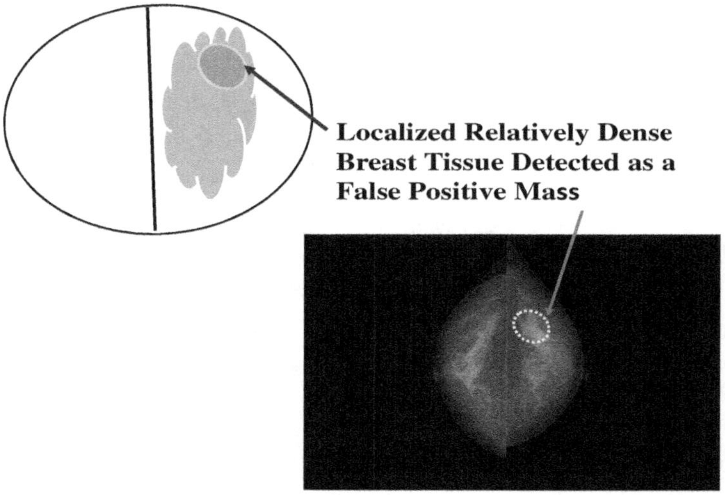

Fig. 3. An Example Case with a False Positive Mass Marked by CAD

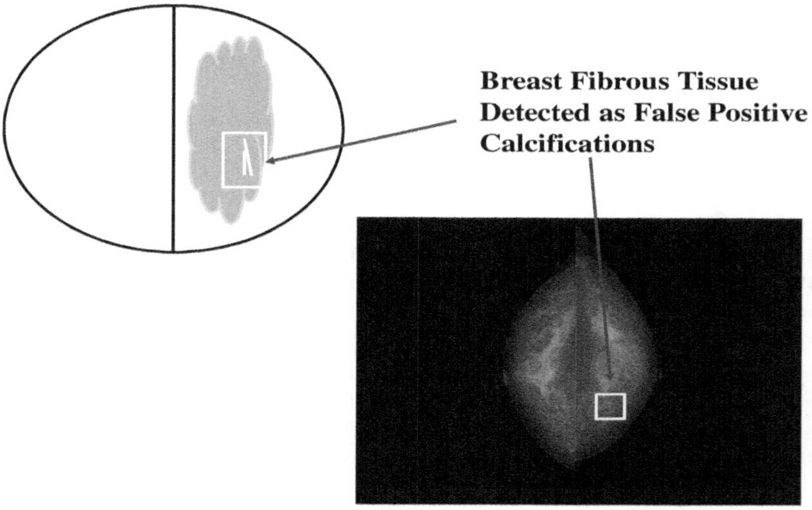

Fig. 4. An Example Case with False Positive Calcifications Marked by CAD

(28/664images) and 0% (0/664 images) false-positive microcalcifications. There was no significant difference among different kV levels with the same combination of anode/filters in System B (P>0.05) (Table 1b.). Regarding spectral analysis of anode/filters, in System A, Mo/Mo and W/Rh demonstrated different spectrum characteristic curves. In addition, the two systems showed different spectrum characteristic curves with W/Rh and the peak value in System B with W/Rh was shown at a higher kV level compared to System A (Fig.5-6).

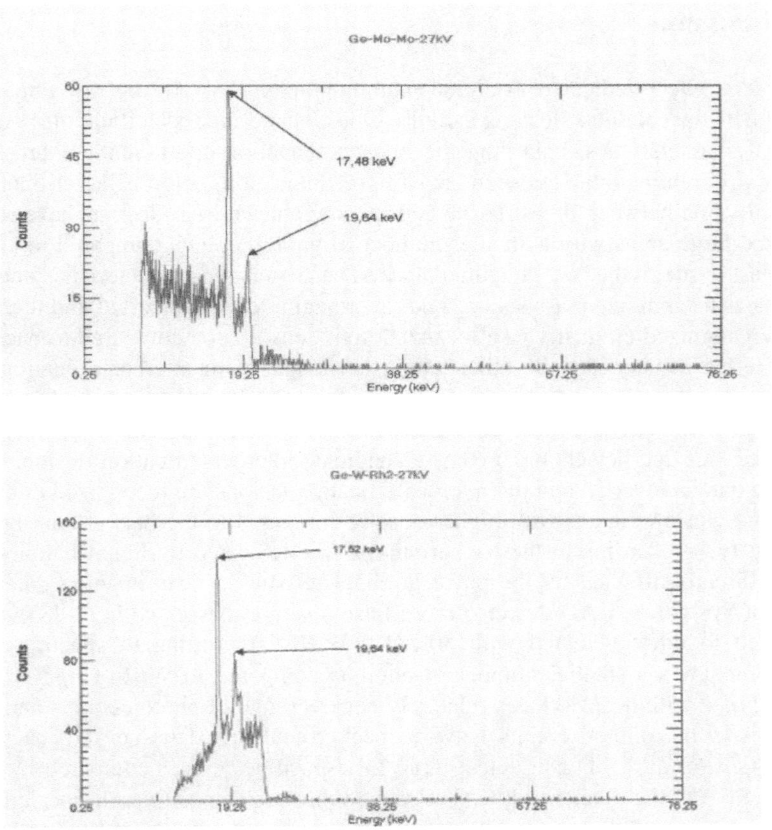

Fig. 5. Spectra of Mo/Mo and W/Rh in System A

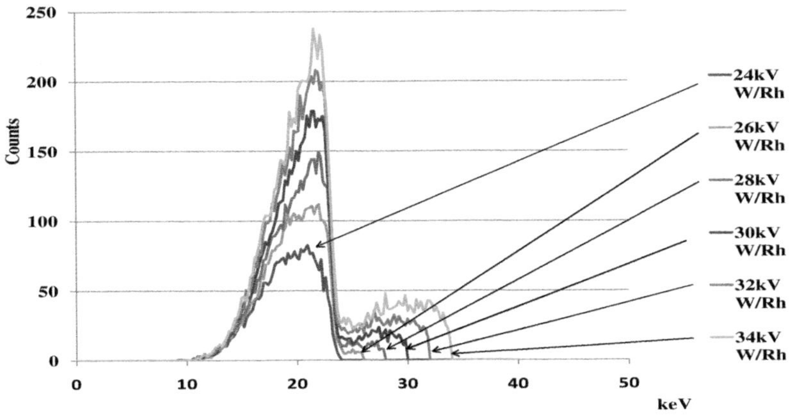

Fig. 6. Spectrum of W/Rh in System B

4 Discussion

At present, CAD dedicated to digital mammography analyzes the raw imaging data and detects the candidate lesions including masses and microcalcifications. As for the physical characteristics, regarding the linear attenuation coefficient for breast tissue [3], the differential value between breast tissue and calcification is larger than the differential value between breast tissue and mass (Table2). Mass lesions have relatively localized large areas with a smaller number of photon counts compared to surrounding breast tissue in the raw imaging data.CAD analyzes the characteristics and detects the area as a candidate mass lesion. The raw imaging data is inverted and the mass lesion is recognized clinically as a localized high density area compared to background breast tissue density. On the other hand, the images with microcalcification lesions have localized small and clustered areas with a smaller number of photon counts compared to the background breast tissue in the raw imaging data. CAD analyzes the characteristics and detects the area as a candidate microcalcification lesion. The raw imaging data is inverted and the microcalcification lesions are recognized clinically as small and clustered areas with higher density compared to the background breast tissue density. According to the background, CAD dedicated to digital mammography can be directly affected by the physical characteristics of raw imaging data. In this study, in System A, CAD detected more false positive masses with 24-28Kv Mo/Mo compared to those detected with 30-34Kv W/Rh. According to spectral analysis, Mo/Mo acquires a smaller number of photons compared to W/Rh (Fig.5). The raw imaging data with Mo/Mo has a relatively narrow range of photon counts and the differentials in the photon counts between background breast tissue and mass can be small. As a result, CAD can detect more false positive masses compared to imaging with W/Rh. On the other hand, CAD detected more false positive microcalcifications with 30-34Kv W/Rh compared to the number detected with 24-28Kv Mo/Mo. This could be a result of the characteristics of W/Rh which can acquire a larger number of photons compared to Mo/Mo. Images with W/Rh have a much wider range of photon counts and the differential value of photon counts between background breast tissue and microcalcifications is large. As a result, imaging data with W/Rh can detect candidate microcalcification lesions with more sensitivity than imaging with Mo/Mo. Even with the same combination of anode/filters, the CAD in System A with 30-34kV W/Rh detected more false positive masses compared to System B with 30-34kV W/Rh.CAD results may differ even when the same system is used, according to which combination of anode/filters is used. On the other hand, CAD results may differ when different systems are used, even though the same combination of anode/filter is used. According to spectral analysis, the spectrum of W/Rh used in System A shows greater similarity to the spectrum of Mo/Mo than the spectrum of W/Rh used in System B (Fig.5-6). As a result, CAD detected more false positives using W/Rh with System A compared to System B.

In conclusion, the CAD performance was affected by the difference in image quality produced by different radiation exposure parameters of the different anode/filters within one system and by differences in the two systems. In FFDM, CAD algorithms should be considered to vary depending on the image acquisition systems.

References

1. Toroi, P., Zanca, F., Young, K.C., et al.: Experimental Investigation on the Choice of the Tungsten/Rhodium Anode/Filter Combination for an Amorphous Selenium-Based Digital Mammography System. Eur. Radiol. 17, 2368–2375 (2007)
2. EUREF; European guidelines for quality assurance in breast cancer screening and diagnosis, 4th edn. (2006)
3. Hammerstein, G.R., Miller, D.W., et al.: Absorbed Radiation Dose in Mammography. Radiol.. 130, 485–491 (1979)

Feature Point Detection for Non-rigid Registration of Digital Breast Tomosynthesis Images

Weiwei Zhang and Sir Michael Brady

Wolfson Medical Vision Lab, Department of Engineering Science
University of Oxford, Parks Road, Oxford, OX1 3PJ, UK
{weiwei,jmb}@robots.ox.ac.uk

Abstract. We present an automatic approach to feature point detection that is well-suited to the non-rigid registration of digital breast tomosynthesis images. The approach combines the scale saliency and the continuous intrinsic dimensionality of image structures in order to detect key feature points along the breast boundary and within the breast. These feature points can be used as the control points for polyaffine transformation regularisation. Experimental results show that non-rigid registration driven by such feature points yields good spatial alignment.

1 Introduction

Digital breast tomosynthesis (DBT) is an emerging imaging modality with the advantage that it delivers 3D information about anatomical structures, potentially improving breast cancer detection and diagnosis at X-ray doses similar to those of conventional mammography. The spatial alignment of DBT images could serve as a computer-aided detection and diagnosis (CAD) tool for the clinician to compare DBT images: two different views of the same breast, left vs right same view, and or to detect significant changes in the"same" view at two different times. For example, bilateral comparison is used primarily to identify breast asymmetry/abnormality, so that an asymmetric area may indicate suspicious regions or an underlying cancer [1]. The image difference analysis used for registered bilateral pairs could be an indicator of a developing mass or future breast disease, such as architectural distortion. As a complementary method to bilateral registration or as an improved means of monitoring pathological changes or growths over time, temporal breast image registration is increasingly recognized as a key component of CAD in mammography community.

DBT image registration poses a difficult and challenging problem using either standard intensity based or feature based methods. Although intensity based registration has the advantage that no prior segmentation is required, it necessitates an expensive computational overhead primarily because DBT images are so large. On the other hand, feature based methods face the fundamental problem of deciding what image features, geometrical or anatomical landmarks, reliably underpin registration. Moreover, the limited-angle acquisition protocol and the

J. Martí et al. (Eds.): IWDM 2010, LNCS 6136, pp. 296–303, 2010.

resulting tilt artifacts, make it difficult to represent and compare DBT images. The potential differences in large volumetric displacements of structures during compression and different imaging parameters or pathological changes over time mean that their appearances are likely to be dissimilar for intra-subject temporal DBT image pairs. Such reasons have led to several registration algorithms relying on the breast boundary, and ignoring internal features that are seriously affected by the dissimilarities of breast image pair. As proposed in Sallam and Bowyer [2], and Karssemeijer and te Brake [3], using the boundary as a basis for registration is based on the observation that the breast tends to be compressed to approximately the same extent in each scan since the breast contents mostly do not vary greatly between scans. However, Marias et al.[4] demonstrate a good registration framework which uses both the breast boundaries and internal landmarks in order to align a pair of temporal mammographic images. They develop a maximum curvature detection algorithm to select boundary landmarks, then a wavelet analysis to separate internal features at different scales and a saliency measure to select certain features as salient internal landmarks.

In this paper, we propose an algorithm that combines scale saliency and continuous intrinsic dimensionality of image structures to automatically select the breast boundary and internal feature points in order to drive the non-rigid registration of DBT images. Although the experiments were done using polyaffine transformation, any warping function, such as those used in a thin-plate spline, b-spline, or more generally, point set based registration regularization can naturally be substituted into our algorithm. In the following sections, the methodology is presented and results of the experiments on DBT image pairs/sequences are shown.

2 Method

Our method consists of five steps:

1. *Weighting the output of the scale saliency algorithm by the continuous intrinsic dimensionality (ci$\mathcal{D}$):* Scale saliency [5] can be considered as a geometrical measure to detect salient regions in an image domain that could be used as matching features. By making use of the scale saliency, we obtain a set of points representing feature regions of different types. However, the feature points obtained in this way correspond to different scales and, potentially, to different image structures. In order to refine the set of feature points to initiate non-rigid registration, we associate the $ci\mathcal{D}$ of image structures [6] with the output of a scale saliency algorithm to assist in determining where the feature points are located. This can be represented as:

$$F(\mathbf{s}, \mathbf{x}) = (\lambda \cdot \mathcal{Y}_D(\mathbf{s}, \mathbf{x}), \beta \cdot ci\mathcal{D}(\mathbf{x})) \quad (1)$$

where $F(\mathbf{s}, \mathbf{x})$ denotes a set of weighted feature points $\mathbf{x}$ at scale range $\mathbf{s}$ by associating the output $\mathcal{Y}_D(\mathbf{s}, \mathbf{x})$ of scale saliency algorithm with the continuously-valued intrinsic dimensionality $ci\mathcal{D}(\mathbf{x})$ (formal definition of $\mathcal{Y}_D$ and $ci\mathcal{D}$ can be found in appendix section). λ and β are weighting parameters for $\mathcal{Y}_D$ and $ci\mathcal{D}$ respectively, and satisfy $\lambda + \beta = 1$. In our case, we use $ci1\mathcal{D}$ and $ci2\mathcal{D}$ to select salient points along the breast boundary or within the breast.

2. *Curve fitting to the breast boundary:* A (spline) parameterised curve $C : t \mapsto \mathbf{x}$, where $t \in \mathbb{R}$ and $\mathbf{x} \in \mathbb{R}^2$, can be fitted to a set of scale salient points identified 'around' the breast boundary in step 1. However, such salient points correspond to a range of scales $\mathbf{s}$ and are located at the centres of different scales. The curve fitted directly to such salient points is usually not a good approximation to the breast boundary. In order to re-direct the fitted curve towards the breast boundary, we re-sample the first fitted curve and for each re-sampled point along the curve we draw a circle taking the mean value of the scales $\mathbf{s}$ as its radius. The final parameterised curve is then re-fitted to the breast boundary by finding the outer envelope of re-sampled and re-scaled circles.

3. *Detection of the maximum curvature in the breast boundary curve and placement of boundary control points:* To register a pair of DBT images, correspondence between the fitted breast boundary curves in step 2 needs to be established. In the model of [2], three points of characteristic geometry in the breast boundary can be used as consistent landmarks. In our DBT case (a similar cranio-caudal view in mammogram), these points could be two invariant points near the chest wall, which are approximated by the beginning and end of the fitted breast boundary curve respectively. The third point (or possibly, the nipple) can be detected by the maximum curvature point (negative curvature by convention) proposed by Marias et al. [4]. Other control points could be placed along the fitted breast boundary curve between the maximum curvature point and one of the chest wall points.

4. *Internal feature point matching:* In this step, we establish internal feature point correspondences. This is done firstly by roughly aligning two DBT images. For example, the floating image is displaced in Cartesian coordinates, such as a small translation in x and y coordinates, in order to get the optimal match against the reference image in terms of intensity difference. Next, the corresponding internal feature points $F_{internal}$ are established by finding the minimum of the Euclidean distance measure between any pair of two internal points in the aligned images.

5. *Using detected feature points to drive polyaffine transformation:* The polyaffine transformation weights the sum of local displacements according to a weight function for each individual image region by solving the following ODE:

$$D(\mathbf{x}) = \dot{x}(t) = \frac{\sum_i w_i(\mathbf{x}) P_i(\mathbf{x}, t)}{\sum_i w_i(\mathbf{x})} \tag{2}$$

where $D(\mathbf{x})$ denotes the globally-weighted transformation for a point $\mathbf{x}$ in the image domain; w_i denotes the weight function and P_i denotes the local displacement for an image region i.

In equation 2, a Gaussian mixture model is used to define the weight function:

$$w_i(\mathbf{x}) = q_i \sum_{i=1} G_{(a_i, \sigma_i)}(\mathbf{x}) \tag{3}$$

where the anchor point a_i describes how each region is geometrically connected. In our case, the anchor points or control points are stored in a location vector using $F_{boundary/internal}$ detected via steps 1 to 4. The standard deviation σ_i specifies the range of each anchor point's effect on the global transformation. σ_i also controls the extent to which each anchor point affects the local deformation around its neighbourhood and hence the extent (or smoothness) of warping that a polyaffine transformation can produce. The relative weight q_i reflects the importance of the individual anchor points. In our case, we assume that each anchor point contributes equally to the global transformation. That is, we set the relative weight to 1.

For a polyaffine transformation, each P_i can be represented as:

$$P_{i(affine)}(\mathbf{x}) = A_i \cdot \mathbf{x} + d_i, \forall \mathbf{x} \in \mathbb{R}^2 \tag{4}$$

where A_i and d_i are respectively the deformation part and translation part. The spatial deformation $D_{spatial}$ is then obtained by associating the $P_{polyaffine}$ with the ODE and integrating equation 2 with respect to time t ($0 \leqslant t \leqslant 1$), such that $D(\mathbf{x}) = \Phi(\mathbf{x}, 1)$, where $\Phi(\mathbf{x}, t)$ isthe solution (i.e. the warp flow) of the ODE.

3 Result

One volunteer DBT dataset with limited-angle projection of $[+12^o, -12^o]$ has been used to date in our experiments.

Feature Point Detection: Figure 1 indicates the result of using our detection algorithm to select breast boundary feature points for a pair of bilateral DBT images. Since the internal structure correspondence of such a pair may not be available, we demonstrate the internal feature point correspondences between the $+12^o$ projection and the 0° projection DBT images from the same side of the breast instead.

Applied to Bilateral DBT Image Registration: Figure 2 is the result of registering a pair of bilateral DBT images. The 0° projection slices are used. Although only nine breast boundary control points are selected by the feature point detection algorithm to drive the polyaffine transformation [7], substantial spatial differences are aligned. The root means square errors (RMSE) of image pair after registration is reduced to 7.48mm (right-to-left registration) and 8.52mm (left-to-right registration), compared with the RMSE of 24.86mm before registration.

Correcting Tilt Artifacts: The maximum $+12^o$ projection and the 0° projection DBT images of the left and right sides of the breast are shown in figure 3 after correcting for tilt artifacts. Both breast boundary and internal feature points are used. The RMSE of a DBT sequence acquisition ($[+12^o, -12^o]$ with small angle increment for each of 13 frames) before registration were $8.61 \pm 2.39mm$ and $6.72 \pm 1.81mm$ for left and right breast respectively; and $1.65 \pm 0.71mm$ and $1.84 \pm 0.89mm$ after registration.

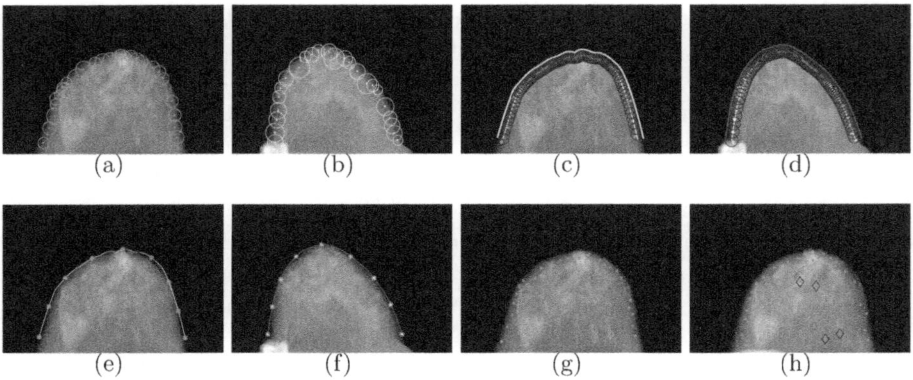

Fig. 1. Result of feature point detection. (a) and (b): associating CID to the output of the scale saliency algorithm to select feature points around the breast boundary for a pair of bilateral DBT images. (c) and (d): parameterised curves fitting to the breast boundary. (e) and (f): using the maximum curvature detection algorithm to place control points. (g) and (h): establishing internal feature point correspondence by the Euclidean distance measure between (g) the 0° projection and (h) the $+12^o$ projection DBT images from the same side of the breast.

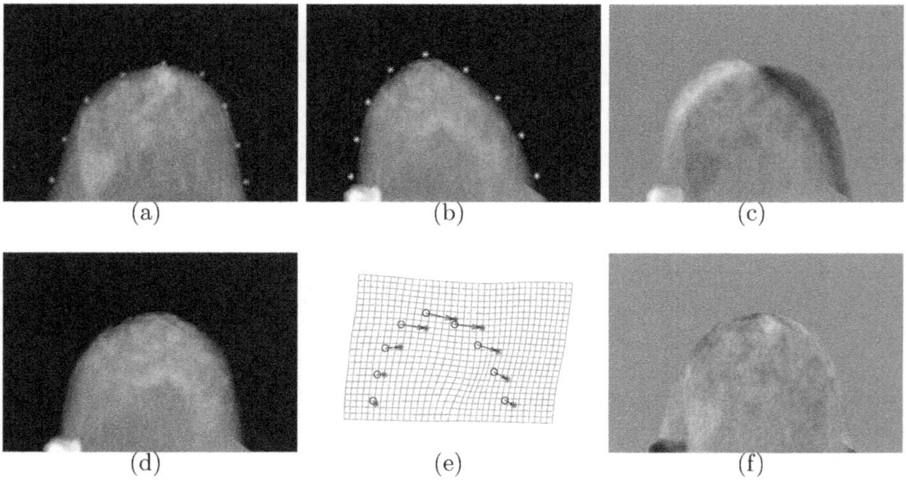

Fig. 2. Result of bilateral DBT image registration. (a) Left and (b) right DBT images with identified breast boundary control points and (c) their difference image before registration. (d): registered right DBT image (right-to-left registration) with (e) its warping grid by polyaffine transformation and (f) difference image between warped right DBT image and left DBT image.

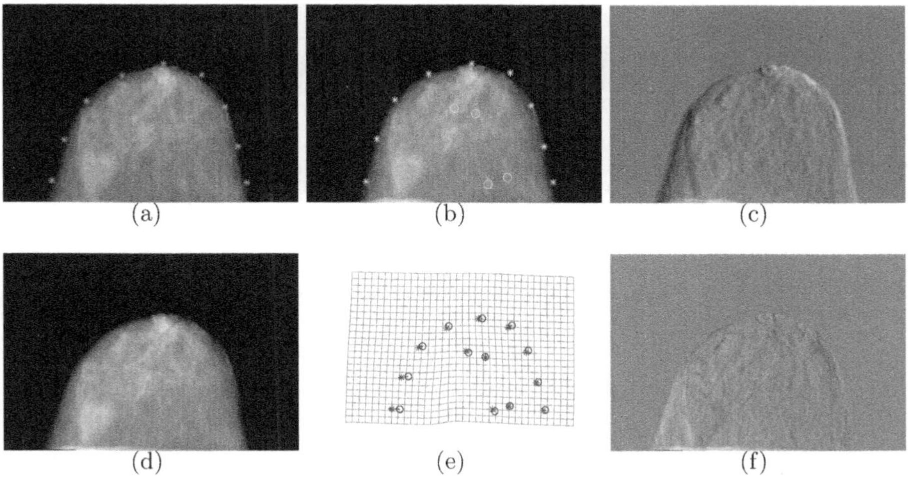

Fig. 3. Result of correcting tilt artifacts. (a) 0° projection and (b) +12° projection left DBT images with (c) identified breast boundary and internal control points and their difference image before registration. (d): registered +12° projection with (e) its warping grid by polyaffine transformation and (f) difference image between warped +12° projection and 0° projection left DBT images.

Table 1. RMSE evaluation on Non-rigid Registration of DBT Images

Application	Feature Point Used	RMSE (before)	RMSE (after)
Bilateral (R to L)	Boundary	$24.86mm^2$	$7.84mm^2$
Bilateral (L to R)	Boundary	$24.86mm^2$	$8.52mm^2$
Artifacts (R seq)	Boundary+Internal	$6.72\pm1.81mm^2$	$1.84\pm0.89mm^2$
Artifacts (L seq)	Boundary+Internal	$8.61\pm2.39mm^2$	$1.65\pm0.71mm^2$

4 Discussion

In this paper, we have presented a feature-based detection algorithm for selecting breast boundary and internal feature points to drive a non-rigid registration algorithm. The key to our algorithm is to associate the CID of image structures to the output of a scale saliency algorithm and separate the breast boundary salient points and internal salient points for further establishment of their individual correspondence. We applied our algorithm to identify the breast boundary points for the polyaffine transformation in order to align a pair of bilateral DBT images and the result shows a good spatial alignment. In the future, we will extend our algorithm to the temporal DBT image registration. Establishing surface correspondence for registering reconstructed DBT volume images is also of our interest.

A Scale Saliency

In equation 1, scale saliency is defined by the weighted entropy across scale and position:

$$\mathcal{Y}_D(\mathbf{s}, \mathbf{x}) \triangleq \mathcal{H}_D(\mathbf{s}, \mathbf{x}) \times \mathcal{W}_D(\mathbf{s}, \mathbf{x}) \tag{5}$$

where $\mathcal{Y}_D(\mathbf{s}, \mathbf{x})$ denotes the out put of a point $\mathbf{x}$ at scale $\mathbf{s}$ from the scale saliency algorithm; $\mathcal{H}_D$ and $\mathcal{W}_D$ denote the entropy and the weighting function respectively, which are defined by:

$$\mathcal{H}_D(\mathbf{s}, \mathbf{x}) \triangleq \int_{i \in D} p_D(s, \mathbf{x}) \log_2 p_D(s, \mathbf{x}) \cdot di \tag{6}$$

$$\mathcal{W}_D(\mathbf{s}, \mathbf{x}) \triangleq s \cdot \int_{i \in D} \left| \frac{\partial}{\partial s} p_D(s, \mathbf{x}) \right| \cdot di \tag{7}$$

and the scale vector is defined by:

$$\mathbf{s} \triangleq \left\{ s : \frac{\partial \mathcal{H}_D(s, \mathbf{x})}{\partial s} = 0, \frac{\partial^2 \mathcal{H}_D(s, \mathbf{x})}{\partial s^2} < 0 \right\} \tag{8}$$

In equation 6 and 7, p_d is the probability density as a function of position x at scale s. By applying the Kadir-Brady scale saliency operator to substantial DBT and mammographic images, it is shown that over 80% of the salient points are located near the breast boundary and up to 20% of the points are located within the breast.

B Continuous Intrinsic Dimensionality

In image processing, the intrinsic dimensionality is formulated as a heuristically discrete distinction between edge-like and corner-like structures. Krüger and Felsberg [8] represent the the intrinsic dimensionality in a continuous and topologically manner as $ci\mathcal{D} = \{ci0\mathcal{D}, ci1\mathcal{D}, ci2\mathcal{D}\}$, for example, in the spectrum of an image patch (i.e., the Fourier transform of a neighbourhood), the $ci0\mathcal{D}$ points, for which their local spectrum is concentrated in the origin, can be used to represent constant image patch. The $ci1\mathcal{D}$ points, for which their local spectrum varies primarily in one direction (i.e., along a line via the origin), represent edge, ridge or boundary-like structures. The $ci2\mathcal{D}$ points where their local spectrum is neither concentrated in the origin nor along a line represent corner or junction-like structures. They further develop a 2D triangular model for $ci\mathcal{D}$. In a polar coordinates $\mathbf{u} \mapsto (q, \theta)$, the radial variance and the angular variance can be defined respectively as:

$$\varepsilon_R^2 = \frac{1}{N'} \int_0^Q q^2 \int_0^{2\pi} |F(q\cos\theta, q\sin\theta)|^2 d\theta dq \tag{9}$$

$$\varepsilon_A^2 = \min_{\theta_0} \frac{1}{N'} \int_{\theta_0}^{\theta_0+\pi} (\theta - \theta_0)^2 \int_Q |F(q\cos(\theta - \theta_0), q\sin(\theta - \theta_0)|^2 dq d\theta \tag{10}$$

where Ω is the region of integration in the Fourier domain, Q denotes radius of Ω, N' is a normalization constant and angle θ_0 represents the local orientation. By multiplying $\tilde{\varepsilon}_A^2$ with $\tilde{\varepsilon}_R^2$, a 2D triangular representation for $ci\mathcal{D}$ is obtained, in which each of the corners of the triangle corresponds to a certain intrinsic dimensionality.

References

1. Miller, P., Astley, S.: Detection of Breast Asymmetry Using Anatomical Features. In: Proceedings of SPIE, vol. 1905, pp. 433–442 (1993)
2. Sallam, M., Bowyer, K.: Registration and difference analysis of corresponding mammogram images. Med. Image Anal. 1(1), 73–91 (1996)
3. Karssemeijer, N., Brake, G.T.: Combining single view features and asymmetry for detection of mass lesions. In: Proc. IWDM, pp. 95–102 (1998)
4. Marias, K., Behrenbruch, C., Parbhoo, S., Seifalian, A., Brady, M.: A Registration Framework for the Comparison of Mammogram Sequences. IEEE. Trans. Medical. Imaging. 24, 782–790 (2005)
5. Kadir, T., Brady, M.: Scale, Saliency and Image Description. International Journal of Computer Vision 45(2), 83–105 (2001)
6. Felsberg, M., Kalkan, S., Krüger, N.: Continuous Dimensionality Characterization of Image Structures. Image and Vision Computing 27, 628–636 (2009)
7. Arsigny, V., Pennec, X., Ayache, N.: Polyrigid and polyaffine trasformations: a novel tool to deal with non-rigid deformation - application to the registration of histological slices. Medial Image Analyis 9, 507–523 (2005)
8. Krüger, N., Felsberg, M.: A continuous formulation of intrinsic dimension. In: British Machine Vision Conference, BMVC (2003)

Contrast Detail Curves on Digital Mammography: Performance Comparison of Raw and Filtered Images

Pedro Collado-Chamorro[1], Camilo Sanz-Freire[1], José Gómez-Amez[2], Alejandro Vázquez-Galiñanes[1], Natalia Lopo-Casqueiro[1], and Maria González-De La Puente[1]

[1] Medical Physics Department, Centro de Investigación Biomédica de La Rioja, Piqueras 98, 26006 Logroño, Spain
[2] Medical Physics Department, Hospital San Pedro, Piqueras 98, 26006 Logroño, Spain

Abstract. Detection of low contrast and very small size objects is of great importance on digital mammography imaging techniques. Hence, when comparing image quality performance for different equipments, it would be desirable to make an objective evaluation primarily based on raw images. In this work we present the results of an image quality performance comparison test for digital mammography systems. Contrast detail curves obtained for raw and filtered images from a CDMAM v3.4 phantom were analyzed. The test resulted in an overall image quality figure index for filtered and raw images of 106.66 and 109.90 respectively for a detection rate threshold of 75%. The largest differences were observed with the small diameter dishes of the phantom.

Keywords: Full field digital mammography, quality assurance, contrast detail curves, CDMAM.

1 Background

Due to the nature of the lesions localized in breast tissue, it is of paramount importance the detection of low contrast and very small size objects. Digital mammography imaging systems should perform well when resolving these objects. Hence, evaluation of the system performance should be objective and independent of the observer.

Image quality may be objectively evaluated by means of the analysis of the Modulation Transfer Function (MTF) of the system as well as contrast detail curves. MTF analysis provides extended information to the spatial resolution test of the system. The contrast detail test adds the information on the contrast performance of the imaging system. This contrast detail test can be seen as a good indicator of the overall quality image performance of the evaluated system.

A major problem when dealing with mammography equipments is the restricted access to raw data. Raw data is normally automatically archived by the image analysis software of the system and most of the times is not accessible without the intervention of the manufacturer. However, to avoid the influence of image processing on the results, the performance tests should also be carried on raw images.

The aim of the present work is to evaluate the influence of image processing on image quality tests results, using raw images as a benchmark. Different filters will be

J. Martí et al. (Eds.): IWDM 2010, LNCS 6136, pp. 304–311, 2010.

applied to these benchmark images. These filters will be firstly characterized by the frequency response obtained from the MTF calculation. The noise behaviour of these filters will be evaluated as well. Results of this pre-analysis will then be confronted to results obtained from the contrast detail analysis as a more general evaluation of image performance.

2 Materials and Methods

The contrast detail tests were conducted on a Seno Essential System flat-panel digital mammography system (GENERAL ELECTRIC). This equipment is based on an amorphous Silicon indirect-detection flat-panel imager: a Cesium Iodine fluorescent layer coupled to an amorphous Silicon photo-detector matrix. The detector features a pixel size of 95 μm and an active detection surface of 24x31 cm^2. In this system a so-called "Premium View" filter is applied by default to all acquired images. However, raw data is stored together with the corresponding processed images and is accessible by the user.

The CDMAM v3.4 phantom (ARTINIS MEDICAL SYSTEM BV) was used in the performance test. The European guidelines for quality assurance in breast cancer screening and diagnosis [1] [2] were followed. The contrast detail phantom was inserted in between of 4 1cm PMMA layers. A total of 6 images of the CD-MAM phantom were taken, using an automatic exposure control in standard operation mode. The automatic selection mode parameters of the exposures were: 29 kV_p, 56 mAs for a combination of anode and filter of Rhodium /Rhodium in all cases.

Images were acquired, stored and then transferred to the analysis software. The CDMAM Analyzer v1.1 image analysis software (ARTINIS MEDICAL SYSTEM BV) was used in the test. For the evaluation, a detection rate threshold of 75% was set as recommended by the Spanish quality assurance protocol in digital mammography [3].

Filter performance was evaluated using the following software applications:

- The application MiQuaela [5] distributed by the Spanish Association of Medical Physics. MiQuaela allows for the calculation of the MTF using an imaged high contrast edge
- The application ImageJ 1.42q [6] distributed by the National Institutes of Health. ImageJ allows for an easy evaluation of image noise

High contrast edge images were acquired using a phantom consistent of 4 PMMA layers 1 cm width each and a 0.5 mm Copper layer interleaved. The Copper layer was inserted slightly rotated in order to obtain an oversampling of the high contrast profile.

These images of the 4 cm PMMA and 0.5 mm Cu phantom were acquire using an automatic exposure control in standard operation mode. The exposure parameters were in this case 29kV_p, 50mAs for a Rhodium /Rhodium anode and filter combination.

Noise images were acquired using a 4 cm width PMMA phantom. Automatic exposure control was employed in this case as well, for the same anode/filter set.

All images were acquired using the bucky compressor assembly, with compression strength of 16 daN.

Vertical MTF values were evaluated from the horizontal edge of the Copper layer and vice versa. For this purpose, 500x250 pixels ROIs were defined centred on these borders (see Figure 1.)

A blank image of the 4 cm PMMA phantom without the Copper layer inserted was also acquired. This is used by MiQuaela to correct the edge images for the actual radiation field profile.

Fig. 1. Defined ROIs for the calculation of horizontal and vertical MTF

3 Results

The contrast detail curve has been first evaluated for the default image filter "Prime View". The obtained values of the contrast detail curve are tabulated below. These are compared with the values for the raw images.

Table 1. Contrast – detail values obtained for the two series of images

Processed series		Raw series	
Diameter (mm)	Thickness (μm)	Diameter (mm)	Thickness (μm)
0.060	2.500	0.060	2.500
0.080	1.917	0.080	1.557
0.100	0.764	0.100	0.769
0.130	0.450	0.130	0.388
0.160	0.401	0.160	0.370
0.200	0.184	0.200	0.185
0.250	0.162	0.250	0.176
0.310	0.110	0.310	0.118
0.400	0.095	0.400	0.095
0.500	0.052	0.500	0.052
0.630	0.048	0.630	0.048
0.800	0.042	0.800	0.042
1.000	0.030	1.000	0.030
1.250	0.030	1.250	0.036
1.600	0.030	1.600	0.030
2.000	0.040	2.000	0.040

The obtained values fulfill both the tolerances and the recommended values as established by the Spanish quality assurance protocol in digital mammography [3]. This is valid for both series of data. The contrast detail curves are depicted in the scatter plot of figure 2.

We obtained the following scores in terms of IQF index value [4]:

- IQF = 106.66 for processed images
- IQF = 109.90 for raw images

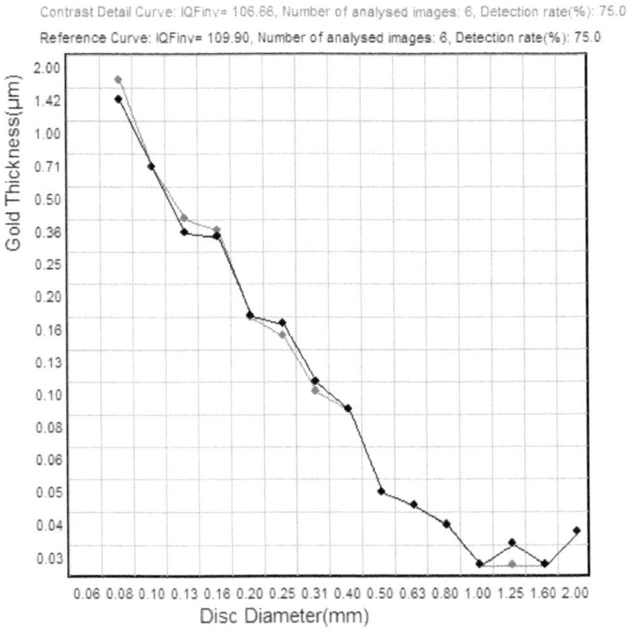

Fig. 2. Contrast detail curves obtained for the two series of images

We considered a number of available filters in order to study the influence on the IQF value. The so-called "Prime View Low", "Prime View Medium" and "Prime View High" filters have been evaluated.

The behavior of these set of filters have been established by the MTF analysis of a high contrast edge processed image. An automatic exposure controlled image of the 4 cm width PMMA phantom was acquire. The automatic exposure X-ray generator parameters were registered. The same parameters (kV_p and mAs) were manually set to acquire the high contrast edge images of the 0.5 mm Copper layer. The mentioned filters were subsequently applied to these images.

The following set of figures was considered from the MTF analysis:

- MTF value at 2cycles/mm.
- MTF value at 4cycles/mm.
- MTF value at the Nyquist frequency.
- The frequency value for the MTF to reach 50% of the its maximum value or f(MTF50).

All MTF curves have been normalized to the 0 cycles/mm MTF value. It is immediate to see that these figures describe edge enhancement filters with different intensities. The corresponding MTF has been represented on figure 3.

The effect of the applied filters on image noise can be evaluated from Figure 4. Here we show a histogram of the values laying in a 328032 pixels homogeneous ROI. The standard deviation for these distributions has a higher value for the filters with higher intensity. Moreover, the minimum values decrease and the maximum values grow, giving a net increase of the range of values. All these data is coherent with an increase of image noise.

Table 2. Evaluation of the MTF for the considered image filters and for raw image

Series	Filters	MTF at 2cycles/mm	MTF at 4cycles/mm	MTF at the Nyquist frequency	f (MTF50) (cycles/mm)
1	Prime View Low	1.28	0.95	0.73	6.02
2	Prime View Medium	1.72	1.11	0.82	6.33
3	Prime View High	2.34	1.53	1.15	7.05
7	Raw	0.77	0.56	0.42	4.39

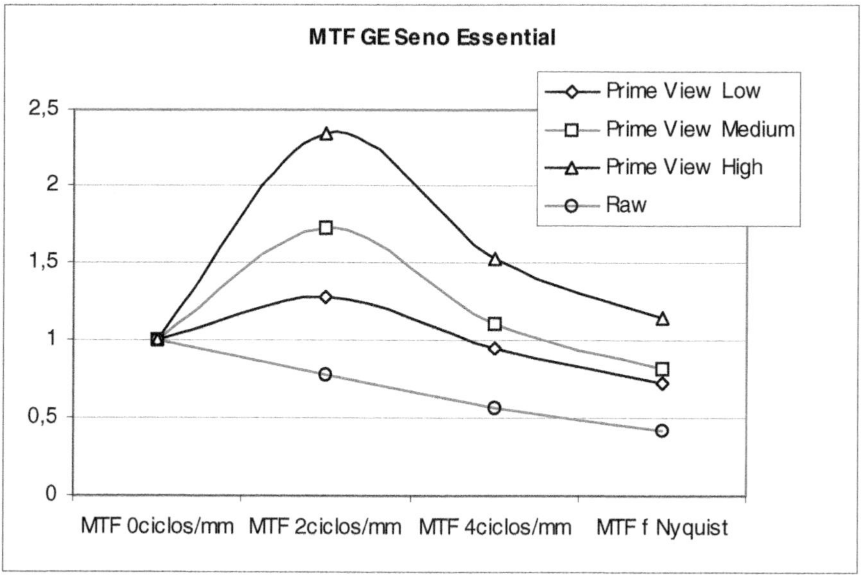

Fig. 3. MTF curves for the different post-processing filters considered and for the non-processed image. The 2 cycles/mm, 4 cycles/mm and Nyquist frequency values have been explicitly plot.

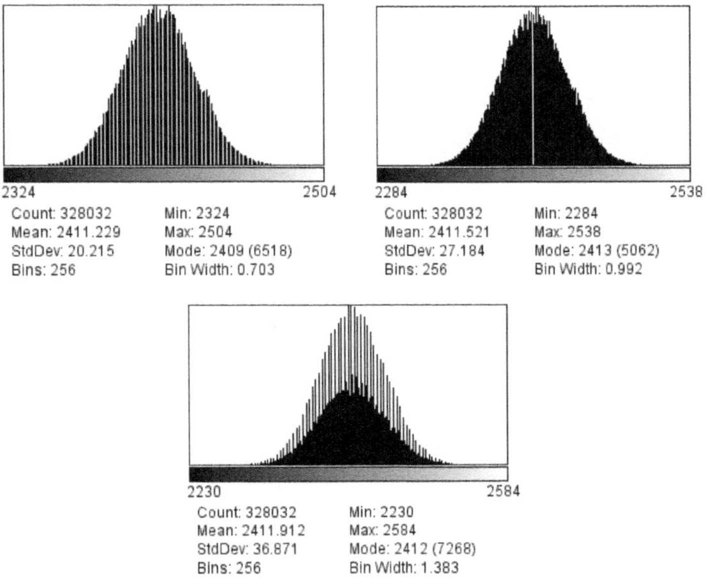

Fig. 4. Homogeneous ROI histograms for the "Prime Viem Low", "Prime View Medium" and "Prime View High" filters

These post-processed images score the following IQF figures:

Table 3. IQF values obtained for postprocesed series of images

Series	Filters	IQF
1	Raw	109,6
2	Prime View Low	107,4
3	Prime View Medium	110,9
4	Prime View High	114,9

Figure 5 shows the contrast detail curve for the "Prime View Filter" and the original raw image. As mentioned, this is the filter that weights the higher frequencies the most. Minimum differences for the object detection rate can be observed, even if this filter has a MTF50 value significantly higher. This can be explained by the fact that the corresponding frequency (7.05 cycles/mm) is above the Nyquist frequency value, thus offering no advantage from the quantitative point of view. The analysis provided by a software application will not show any improvement.

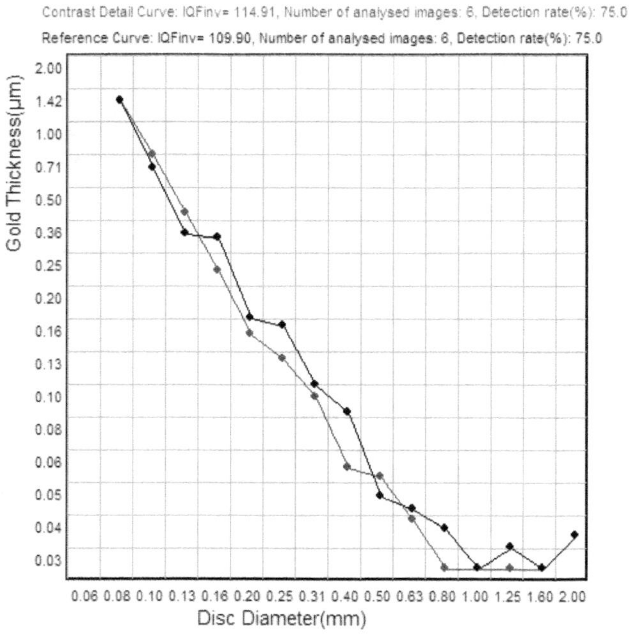

Fig. 5. Contrast detail curves for the raw and post-processed images. Applied filter in this case is "Prime View High".

4 Discussions

The observed differences in terms of IQF value are almost negligible. Nevertheless, the edge filters boost the image noise, especially for the high intensity filters. The detection rate for the small diameter (0.08 mm and 0.1 mm) objects decreases, all this suggesting a net loss of information.

It is important to point out that having access to raw data is always useful. Processed images may improve the ability of the human observer to detect relevant clinical information; however information may be lost during the filtration process. In a general case, this could be reflected on the Image Quality Tests.

References

1. CCE (Comisión de las Comunidades Europeas): European Guidelines for Quality Assurance in Breast Cancer Screening and Diagnosis, 4th edn. European Communities (2006)
2. Mayo, P., Pascual, A., Rodenas, F., Verdú, G., Campayo, J.M., Villaescusa, J.L.: Desarrollo de Procedimientos de Control de Calidad de Equipos de Radiodiagnóstico Digitales. In: XV Congreso Nacional de Física Médica, Pamplona (2005)
3. Chevalier, M., Morán, P., Morant, J.J., Miquélez, S., Torres, R.: Protocolo Español de Control de Calidad en Mamografía Digital. Sociedad Española de Física Médica. Edicomplet, Madrid (2008)

4. Torres, R., Agulla, M., Hernando, I.: Relación Entre Dosis y Calidad de Imagen en Dos Sistemas de Mamografía Digital de Silicio. In: XVII Congreso Nacional de Física Médica, Alicante (2009)
5. Ayala, R., Linares, R., García, R.: MIQuaEla, Servicio de Radiofísica y Radioprotección. Hospital General Universitario Gregorio Marañón (Madrid), http://www.ucm.es/info/fismed/Digital_Imaging/MIQuaEla.rar
6. Rasband, W.: ImageJ, National Institutes of Health (Maryland), http://rsbweb.nih.gov/ij

Reader Fatigue Interpreting Mammograms

Elisabeth A. Krupinski

Dept. of Radiology, University of Arizona
1609 N Warren, Bldg 211, Room 112, Tucson, AZ 85724

Abstract. A number of studies have demonstrated that radiologists can become fatigued after reading cases for a number of hours. Some of these studies have demonstrated that there are associated decreases in observer performance (i.e., reduced accuracy), but mammographic reading has not been a focus of any of these studies. Based on these recent findings regarding decreases in performance as a function of time of day and/or number of hours reading, this retrospective study examined data from a variety of mammography studies in which readers participated in two sessions – once in the morning and once in the afternoon. The ROC Az data from these studies were compared for statistical differences between morning and afternoon reading. Overall there was a small yet significant ($t = 2.365$, $p = 0.0277$) between morning and afternoon diagnostic performance, with performance being degraded in the afternoon. These data suggest that reader fatigue may impact mammography interpretation performance, although more formal studies are required to verify these findings with a prospective study since this retrospective analysis did have limitations.

Keywords: Full field digital mammography, quality assurance, contrast detail curves, CDMAM.

1 Background

Radiologists today read more and more cases with more and more images per case, even in mammography where digital images make it easier to retrieve and view previous exams. This burden is compounded by shortages in radiologists, especially specialty radiologists such as mammographers in rural and medically underserved areas. The result is that radiologists are working longer hours than ever before and some concerns have been raised regarding fatigue and whether it impacts diagnostic accuracy. A more recent problem is the reliance on digital imaging in radiology. Even the best medical-grade displays available have less contrast than traditional radiographic film and they also have reduced spatial resolution. The problem is that it is this information that the visual system uses to regulate image focus, single vision, and direction of gaze. This change to digital displays may have increased strain on radiologists' oculomotor systems, overworking the eyes and resulting in eyestrain (known clinically as asthenopia).2-3 Close work with digital displays may result in oculomotor fatigue, compounding the effects of longer workdays and aging eyes.4 A number of studies have indeed demonstrated that radiologists can become fatigued after reading cases for a number of hours. Some of these studies have demonstrated that there are

J. Martí et al. (Eds.): IWDM 2010, LNCS 6136, pp. 312–318, 2010.

associated decreases in observer performance (i.e., reduced accuracy), but mammographic reading has not been a focus of any of these studies. What has been demonstrated, however, is the fact that FFDM reading still appears to take longer to interpret than screen-film images.5 The potential for increased fatigue due to these longer viewing times would seem to be high.

For example, Gale6 found a significant morning to afternoon drop in sensitivity in detecting pulmonary nodules in chest radiographs. In a more controlled study, Krupinski and Berbaum 7,8 measured visual accommodation before and after test sessions and had subjects complete the Swedish Occupational Fatigue Inventory (SOFI) 9,10 and the oculomotor strain subscale of the Simulator Sickness Questionnaire (SSQ).11-12 Each subject read a series of bone images with fractures before and after a day of clinical reading. Average Receiver Operating Characteristic (ROC) Az (area under the curve) was 0.89 for before work test and 0.85 for the after work test, ($F(1,36) = 4.15$, $p = 0.049 < 0.05$). There was significantly greater error in accommodation after the clinical workday ($F(1,14829) = 7.81$, $p = 0.005 < 0.01$), and after the reading test ($F(1,14829) = 839.33$, $p < 0.0001$). SOFI measures of lack of energy, physical discomfort and sleepiness were higher after a day of clinical reading ($p < 0.05$). The SSQ measure of oculomotor symptoms (i.e., difficulty focusing, blurred vision) was significantly higher after a day of clinical reading ($F(1,75) = 20.38$, $p < 0.0001$).

Based on these findings regarding decreases in performance as a function of time of day and/or number of hours reading, the present retrospective study examined data from a variety of mammography studies in which readers participated in two sessions – once in the morning and once in the afternoon. All of the studies examined another hypothesis than reader fatigue so the present analysis is an ad hoc one that has limitations. The goal however is to provide some preliminary data about reader fatigue in mammography to help determine if a more controlled study is warranted. Therefore, the ROC Az data from these studies were compared for statistical differences between morning and afternoon reading.

2 Method

This retrospective study examined data from four earlier ROC mammography studies [14-17] in which the readers were required to participate in two test sessions. The records were reviewed to identify which of the subjects completed their test sessions once in the morning and once in the afternoon. Those who completed both in the morning or both in the afternoon were not included in the present analysis. Each study originally had 6 radiologists serving as observers.

Study #1 [14] compared screen-film and monitor reading of 20 mammograms (5 mass, 5 microcalcification, 5 mass + microcalcifications, 5 normal) with the lightbox and monitor each at two luminance levels (1100 & 660 ft-L and 140 & 80 ftL respectively). Four of the 6 readers qualified for this analysis (read one condition in the morning and the other in the afternoon) in the screen-film arm and 3 qualified in the monitor arm. Study #2 [15] compared mammograms displays on a monitor calibrated to the DICOM Gray Scale Display Function vs calibrated with the SMPTE standard. Three of the 6 readers from this study qualified. Study #3 [16] compared CRT vs LCD on and off-axis reading of mammograms and 4 of the six qualified for the present analysis. The fourth study [17] compared digital images that were either processed or not processed with a method to correct for the monitor's MTF and 4 of the 6 readers qualified.

The ROC Az data from the qualifying readers were analyzed with a paired t-test to determine if there were statistically significant differences in diagnostic accuracy as a function of whether the cases were read in the morning vs afternoon. In those studies where eye-position was recorded, these data were examined for potential time of day reading effects as well.

3 Results

The ROC Az values from the 4 studies are shown in Table 1. It is important to note that for every experiment about half of the trials for each condition were completed in

Table 1. ROC Az values from the various studies. Am and pm designate the time of day the observer completed the trial. The am trials are in regular font and pm are in bold.

STUDY #1 Film 1100 ftL	***STUDY #1 Film 660 ftL***
.9466 am	**.9398 pm**
.9563 pm	.9517 am
.8850 am	**.8784 pm**
.9772 pm	.9691 am
STUDY #1 Monitor 140 ftL	***STUDY #1 Monitor 80 ftL***
.9655 am	**.9603 pm**
.9843 am	**.9837 pm**
.9663 pm	.9695 am
STUDY #2 DICOM	***STUDY #2 SMPTE***
.9800 pm	.9637 am
.9463 am	**.8704 pm**
.8948 am	**.8560 pm**
STUDY #3 CRT on-axis	***STUDY #3 CRT off-axis***
.9172 am	**.9017 pm**
.8822 am	**.8592 pm**
.9120 pm	.9004 am
.8718 pm	.8783 am
STUDY #3 LCD on-axis	***STUDY #3 off-axis***
.9119 pm	.8632 am
.9079 am	.8544 pm
.9175 am	.9103 pm
.8996 am	.8446 pm
STUDY #4 no correction	***STUDY #4 MTF correction***
.9878 pm	**.9871 am**
.8698 am	**.8301 pm**
.8754 pm	.8927 am
.8967 pm	.9457 am

the morning (am) and about half in the afternoon (pm). Therefore, although the experiments were directed at another hypothesis than fatigue effects, the overall differences between am and pm reading cannot be attributed solely to the experimental condition results.

On average, the ROC Az for morning (am) performance was 0.924 and was 0.910 for afternoon (pm) reading. The difference was statistically significant ($t = 2.365$, $p = 0.0277$). In general, irrespective of the study they were in and the overall results of those studies, reader performance was lower in the afternoon reading sessions than in the morning reading sessions. Interestingly, the study that compared film and digital viewing [14] revealed no significant differences ($F = 1.445$, $p = 0.2876$) in either modality for am (mean Az film = 0.936, mean Az digital = 0.973) vs pm (mean Az fim = 0.938, mean Az digital = 0.970) reading.

Studies 1 and 2 used full images rather than regions of interest so viewing times were compared for these studies. Overall, viewing times were longer (mean am = 39.76 sec, sd = 13.21; mean pm = 46.24 sec, sd = 15.84) during the am trials than the pm trials ($t = 2.637$, $p = 0.0270$). For study 1 am an pm readings were compared for film vs digital viewing and significant differences were found ($F = 43.147$, $p < 0.0001$). The results are shown in Figure 1. For both film and digital images, viewing times were longer in the pm reading sessions than the am reading sessions and the effect was much more pronounced for the digital images overall.

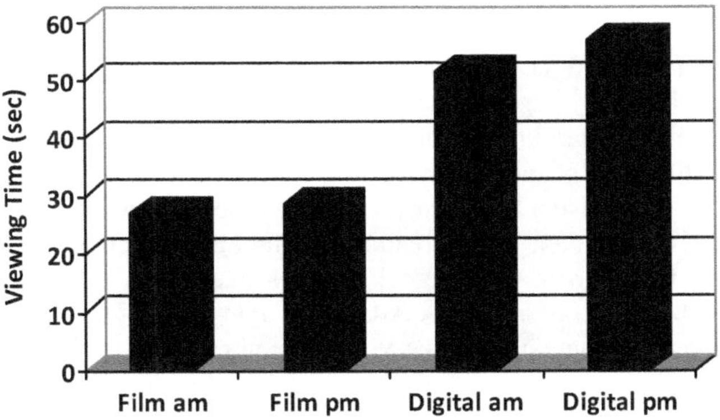

Fig. 1. Mean viewing times for am and pm viewing of film vs digital images in study #1 & study #2

The decision dwell times for study #1 were also analyzed for am and pm readings on film vs digital images. Decision dwell times are derived from the eye-position data recorded as the mammographers search the images for masses and/or microcalcification clusters. Dwell reflects to total amount of time during search that the observer spent on the lesion (true positive if they reported the lesion, false negative if they did not), on a non-lesion area they reported (false positive) or on non-lesion areas they did not report on (true negatives). Overall there were trends to longer dwell times for each decision for digital than film viewing and dwell times tended to be longer in both cases for pm viewing. None of the differences, however, reached statistical significance. The data are shown in Figure 2.

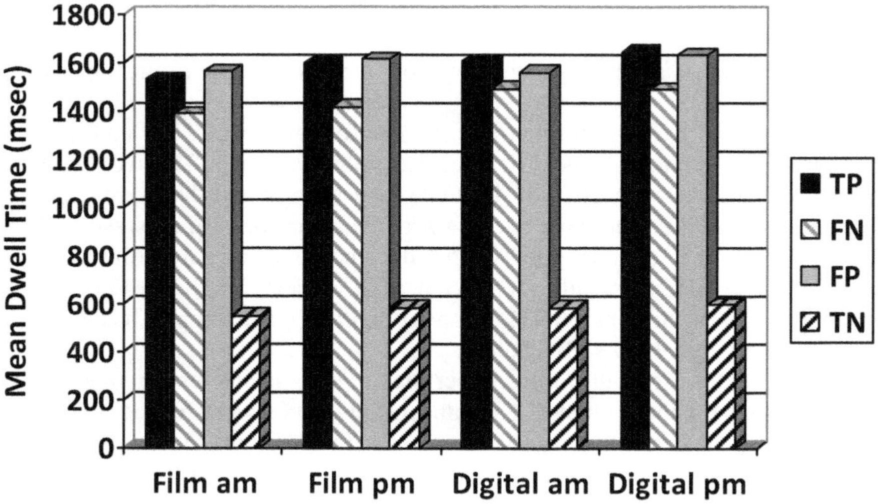

Fig. 2. Mean dwell timers (msec) for true positive (TP), false negative (FN), false positive (FP) and true negative (TN) decisions in study #1 for am and pm reading of film and digital images

4 Discussion

Mammographers are fatigued by their clinical reading workday. Although retrospective in nature, this study suggests that mammographers are less accurate reading images in the afternoon than in the morning. This supports results found by others such as Gale [6] and Krupinski and Berbaum. [7,8] Part of the reason for the degradation in performance is likely visual fatigue resulting from the close nature of mammographic viewing. With both film and digital reading, mammographers tend to work very close to the images. Close work with digital displays may result in oculomotor fatigue, compounding the effects of longer workdays and aging eyes [4].

The viewing time data suggest that viewing times are longer later in the day than earlier and the effect seems to be more pronounced for digital images viewed on a computer monitor. Dwell times for the individual decisions (TP, FN, FP, TN) also show trends towards longer viewing times in the pm vs am for both film and digital reading. Eyestrain has not been very well studied in radiology and never in mammography, but an early self-report study showed that radiologists experience more severe symptoms of eyestrain, blurred vision and difficulty focusing, as they read more imaging studies.[18] Vertinsky and Forster [19] also found that 36% of radiologists reported eyestrain, and the eyestrain could be predicted from the length of work days, the number of breaks, screen flicker, and imaging modality. Goo, *et al.*[20] found that increased ambient light and monitor luminance levels also lead to reports of greater subjective visual fatigue. Eyestrain occurs when oculomotor systems work to maintain accommodation, convergence, and direction of gaze, resulting in physical symptoms such as blurred vision, headaches, and pain in and around the eyes.

The present study does have some limitations. As already noted, the data are retrospective from studies that were not concerned with measuring the impact of fatigue or time of day on performance. However, from Table 1 it is clear that we had a fairly good balance of am vs pm reading across all experimental conditions. Thus the impact of morning vs afternoon reading is likely independent (at least to some degree) of the original experimental conditions. Clearly, however, in order to verify the effects of fatigue on diagnostic accuracy in mammographic reading a prospective study designed to directly address this hypothesis should be conducted.

References

[1] Krupinski, E.A., Berbaum, K.S.: Impact of visual fatigue on observer performance. Proc. SPIE Med. Imag. 7263, 72631O-1 – 9 (2009)

[2] Ebenholtz, S.M.: Oculomotor systems and perception. Cambridge University Press, New York (2001)

[3] MacKenzie, W.: On asthenopia or weak-sightedness. Edinburgh J. Med. & Surg. 60, 73–103 (1843)

[4] Heron, G., Charman, W.N., Gray, L.S.: Accommodation responses and ageing. Investigative Ophthalmology and Visual Science 40, 2872–288 (1999)

[5] Ishiyama, M., Tsunoda-Shimizu, H., Kikuchi, M., Saida, Y., Hiramatsu, S.: Comparison of reading time between screen-film mammography and soft-copied, full-field digital mammography. Breast Cancer 16, 58–61

[6] Gale, A.G., Murray, D., Millar, K., Worthington, B.S.: Circadian variation in radiology. In: Gale, A.G., Johnson, F. (eds.) Theoretical and Applied Aspects of Eye Movement Research, Elsevier Science Publishers, London (1984)

[7] Krupinski, E.A., Berbaum, K.S.: Measurement of visual strain in radiologists. Acad. Radiol. 16, 947–950 (2009)

[8] Krupinski, E.A., Berbaum, K.S.: Does reader visual fatigue impact interpretation accuracy. Proc. SPIE Med. Imag., 7627 (2010)

[9] Ahsberg, E.: Dimensions of fatigue in different workplace populations. Scandinavian J. Psych. 41, 231–241 (2000)

[10] Ahsberg, E., Gamberale, F., Gustafsson, K.: Perceived fatigue after mental work: an experimental evaluation of a fatigue inventory. Ergonomics 43, 252–268 (2000)

[11] Kennedy, R.S., Lane, N.E., Berbaum, K.S., Lilienthal, M.G.: Simulator Sickness Questionnaire: an enhanced method for quantifying simulator sickness. Intl. J. Aviation Psych. 3, 203–220 (1993)

[12] Kennedy, R.S., Lane, N.E., Lilienthal, M.G., Berbaum, K.S., Hettinger, L.J.: Profile analysis of simulator sickness symptoms: application to virtual environment systems. Presence 1, 295–301 (1992)

[13] Kennedy, R.S., Lilienthal, M.G., Berbaum, K.S., Baltzley, D.R., McCauley, M.E.: Simulator sickness in US Navy flight simulators. Aviation, Space, and Environmental Medicine 60, 10–16 (1989)

[14] Krupinski, E.A., Roehrig, H., Furukawa, T.: Influence of film and monitor display luminance on observer performance and visual search. Acad. Radiol. 6, 411–418 (1999)

[15] Krupinski, E.A., Roehrig, H.: The influence of a perceptually linearized display on observer performance and visual search. Acad. Radiol. 7, 8–13 (2000)

[16] Krupinski, E.A., Johnson, J., Roehrig, H., Nafziger, J., Lubin, J.: On-axis and off-axis viewing of images on CRT displays and LCDs: observer performance and vision model predictions. Acad. Radiol. 12, 957–964 (2005)
[17] Krupinski, E.A., Johnson, J., Roehrig, H., Engstrom, M., Fan, J., Nafziger, J., Lubin, J., Dallas, W.J.: Using a human visual system model to optimize soft-copy mammography display: influence of MTF compensation. Acad. Radiol. 10, 1030–1035 (2003)
[18] Krupinski, E.A., Kallergi, M.: Choosing a radiology workstation: technical and clinical considerations. Radiology 242, 671–682 (2007)
[19] Vertinsky, T., Forster, B.: Prevalence of eye strain among radiologists: influence of viewing variables on symptoms. AJR. Am. J. Radiol. 184, 681–686 (2005)
[20] Goo, J.M., Choi, J.Y., Im, J.G., Lee, H.J., et al.: Effect of monitor luminance and ambient light on observer performance in soft-copy reading of digital chest radiographs. Radiology 232, 762–766 (2004)

Adapting Clinical Images to Appear with Different Noise and Sharpness to Model a Different Detector

Alistair Mackenzie[1], Adam Workman[2], David R. Dance[1,3], Mary Yip[4], Kevin Wells[4], and Kenneth C. Young[1,3]

[1] National Coordinating Centre for the Physics of Mammography, Royal Surrey County Hospital, Guildford, GU2 7XX, UK
alistairmackenzie@nhs.net
[2] Northern Ireland Regional Medical Physics Agency, Forster Green Hospital, Belfast, BT8 4HD, UK
[3] Department of physics, University of Surrey, Guildford, GU2 7XH, UK
[4] Centre for Vision, Speech and Signal Processing, Faculty of Engineering and Physical Sciences, University of Surrey, Guildford, GU2 7XH, UK

Abstract. Comparing the clinical performance of digital mammography technologies is challenging. The aim of this work is to develop and test a methodology for adjusting mammographic images taken on a given imaging system to simulate their appearance as if taken on a different system. Such methodology would be very useful for a wide range of system performance and design studies using both phantom and clinical images. The process involves changing the image blurring in accordance with the measured modulation transfer functions and adding noise (electronic, quantum and structure). The method has been tested by adapting flat field images acquired using an amorphous selenium detector and a computed radiography (CR) detector to different dose levels and comparing the resultant simulated NPSs with directly measured NPSs. For the detectors used in this work the NPSs at different dose levels are well predicted. This could be a powerful tool for studies of clinical image quality.

Keywords: mammography, simulation, noise.

1 Introduction

Extensive work has been undertaken on measuring the image quality of digital mammography systems. The physical parameters of images such as noise and sharpness have been measured using quantitative measurements such as modulation transfer function (MTF) and noise power spectra (NPS). Contrast detail measurements are used to ensure that clinical systems meet acceptable standards and have advantages in that they include factors in addition to the detector performance including the visual response of an observer, scatter, scatter rejection and beam quality [1]. Nonetheless, the relationship between contrast detail measurements and measurements of MTF and noise are well understood [2]. However, the relationship between the clinical task of cancer detection and the measured physical characteristics of the detector are less well understood.

J. Martí et al. (Eds.): IWDM 2010, LNCS 6136, pp. 319–326, 2010.

Clinical evaluation of image quality in mammography is expensive and time-consuming. Clinical trials to compare the effectiveness of different systems are rarely conducted as they would require large numbers of patients to achieve both sufficient numbers of detected cancers and statistical significance. In particular it would be desirable to repeat exposures on the same breasts with the same compression to minimise confounding due to differences in breast appearance and compression but this raises ethical issues. Alternative methods involving some degree of image simulation have the potential to enable evaluations at reduced cost and time and without additional radiation exposure. For this purpose it is desirable to be able to acquire images on a given system and to simulate their appearance on a second system, so the performance of the two systems can be compared. This may be possible when the performance of the second system is inferior to that of the second system in terms of unsharpness and or noise, but not conversely. Such a method would enable the background tissue and compression to be the matched in different arms of a study using the insertion of simulated cancers. Work has already been undertaken in correcting for dose and detector used [3-5].

The aim of this work therefore is to set up a framework for adapting images to appear with different imaging characteristics. In particular we are aiming to convert an image acquired using an amorphous selenium detector to appear as a generic CR image. Ultimately, after this has been shown to produce realistic test images, we would apply the methodology to a set of clinical images obtained with DR so that the performance of DR and CR systems for the detection of cancers can be compared at several dose levels.

2 Theory

2.1 Contributions of Electronic, Quantum and Structure Noise to the NPS

There are three different noise sources which contribute to the NPS: electronic, quantum and structure; each has a different relationship with dose and spatial frequency [6, 7]. Equation 1 shows the three components to the total noise power NPS_{tot} and their air kerma dependence. In this expression, NPS_e, NPS_q and NPS_s are the NPS components for the electronic, quantum and structure noise sources respectively at 1 μGy and K is the detector air kerma (DAK) in μGy.

$$NPS_{tot}(u,v) = NPS_e(u,v) + NPS_q(u,v)K + NPS_s(u,v)K^2 \tag{1}$$

2.2 Creation of Noise Images from Measured Noise Coefficients

With the knowledge of the NPS components shown in equation 1, a flat field image can be created. The first stage is to create three noise images for a DAK of 1 μGy based on the NPS components. The separate noise images are then combined (equation 3) to provide a simulation of the linearised flat field image without reference to an initial image.

$$I_M(x, y) = I_e(x, y) + I_q(x, y)\sqrt{K} + I_s(x, y)K + K \qquad (3)$$

where I_M is the modelled image, I_e, I_q and I_s are the flat field images produced from the three noise sources, K is the DAK (μGy) (or the mean pixel value of the modelled image)

2.3 Adapting an Image to Simulate an Image Acquired with a Different Detector and Dose

The methodology is now extended to modify an image acquired with a given image detector with known MTF and noise power coefficients (from equation 1) to simulate images at different dose levels from a different detector.

The image adjustment was undertaken using the following steps:

- Change the original image sharpness to match the model image
- Simulate DAK change (if necessary)
- Adjust the noise components to match the system to be modelled

The original image is blurred by the ratio of the MTFs in frequency space. The three original noise sources are also blurred by the square of the ratio of the two MTFs to give NPS_b to account for the noise in the image being blurred.

Dose change: To simulate dose change in the detector the image is multiplied by a dose factor (R). The difference between the noise expected for the new dose level and the noise in the image which has been modified is shown in equations 5a and 5b. No adjustment is required for the structure noise as this noise source scales correctly with the dose correction.

Detector change: The next stage was to account for differences in noise between the two detectors. The noise differences ($NPS_{\Delta x}$) between the NPS of the three noise sources of the CR and DR systems are calculated (equation 4).

$$NPS_{\Delta x}(u,v) = NPS_{xm}(u,v) - NPS_{bx}(u,v) \qquad (4)$$

$$NPS_{de}(u,v) = NPS_e(u,v)\left(1 - R^2\right) + NPS_{\Delta e}(u,v) \qquad (5a)$$

$$NPS_{dq}(u,v) = NPS_q(u,v)(1 - R) + NPS_{\Delta q}(u,v) \qquad (5b)$$

$$NPS_{ds}(u,v) = NPS_{\Delta s}(u,v) \qquad (5c)$$

where $NPS_{\Delta x}$ is the differences between the detectors' NPS, NPS_{dx} is the total difference in noise between the original image and the image to be simulated.

The NPS_{dx} for the each of the noise source are made into an image using the method described earlier. Each pixel in the blurred image has been linearised to be equivalent to the detector air kerma, therefore an image with the correct magnitude of noise can be created by multiplying the blurred image with the noise on a pixel-by-pixel basis (equation 6). This will ensure that the noise is correct for that dose and will have the correct frequency spectrum and the noise will still be correlated. Finally, the extra noise image (I_N) is added to the blurred original image to obtain the modelled image (equation 7).

$$I_N(x,y)=I_{de}(x,y)+I_{dq}(x,y)\sqrt{I_{blur}(x,y)}+I_{ds}(x,y)I_{blur}(x,y) \tag{6}$$

$$I_M(x,y)=I_{blur}(x,y)+I_N(x,y) \tag{7}$$

I_N is the noise array to be added to the blurred image, I_{blur} is the modelled image with the correct blurring but noise not corrected, I_M is the modelled image, I_R the dose reduced image, I_{blur} the blurred original image, and I_{de} I_{dq} and I_{ds} the images produced from electronic and quantum noise sources and blurred by MTFs.

3 Method

3.1 Measurement of Imaging Characteristics of Digital Mammography Detectors

Two digital mammography imaging systems were used in this study:

DR: Hologic Selenia with an amorphous Selenium detector with pixel pitch of 70 μm.

CR: Carestream CR900 with GE Senographe DMR+ X-ray system with a CR detector with a pixel pitch of 50 μm.

For each system the measurements were made using methods as close as possible to those described by the IEC [8] using 2 mm high purity aluminium at the exit port of the tube, and radiographic factors of 28 kV, and a molybdenum/molybdenum target/filter combination.

The signal transfer function (STP) was measured by acquiring flat field images collimated to about 100 mm × 100 mm field over a range exposure values from a factor of five below to five above the typical dose (147 μGy for DR and 101 μGy for CR). The mean pixel value was measured for each image over 20 mm × 20 mm region of interest, 50 mm from the chest wall side of the image. The mean pixel value was plotted against DAK to give the STP relationship. All images were linearised with the inverse of the STP relationship. The resulting linearised pixel value then represented the DAK which facilitated image comparisons and manipulation. Using the same experimental set up as for the STP a further four images were acquired at normal dose plus quarter, half, double and quadruple typical dose level. The noise power spectra (NPS) were measured at each DAK level of the STP images.

The modulation transfer function (MTF) was measured using a steel edge 0.8 mm thick. The MTF was measured in the two orthogonal directions of the detector using IDL based 'OBJ_IQ_reduced' [9].

Using the NPS and STP images obtained over a range of dose, the separate contributions can be estimated. By fitting equation 1 to the NPS data obtained at the different dose levels, the three noise components are obtained as a function of spatial frequency for both systems. These values are in essence the NPS of the noise source at DAK of 1 μGy.

Using this knowledge it is then possible to predict the difference in the NPS for images taken under different conditions and to adjust the noise in the image on a pixel by pixel basis to model different imaging conditions.

3.2 Simulation of Flat Field Images at Different Dose Level from the Noise Coefficients

To test the image creation process, flat field images were modelled using the fitted noise components. The images produced using these methods were then compared with real images for the same factors. For this purpose, the normalised noise power spectra (NNPS) were calculated for each of the images and compared with the measurement from the original image.

3.3 Adjusting the Imaging Properties of an Image to Appear as if Acquired with a Different Detector

The flat field images used for measurement of NPS and STP were change an acquired image to appear as if acquired using a different detector.

Images acquired on the DR system at high dose were modified to appear as if acquired by the CR system. This was repeated for all five dose levels collected for measurement of the NPS. Images acquired on the DR system were then converted to images of similar dose acquired on the CR system.

A region of interest of 60 x 60 mm with a centre 50 mm from the chest was extracted from each of the DR images. The model images are then created using the method described in the theory section. The NNPS of the model images was measured and compared with those of the real images that were the model.

It was decided not to change the pixel pitch of the image in the conversion and so the modeled CR image had a pixel pitch of 70 μm instead of 50 μm. There is little to be gained in rescaling the image as there is no information in the original image above its Nyquist frequency. In reality the MTF of the CR system above Also the rescaling in itself may create errors [10].

4 Results and Discussion

4.1 Measurement of Signal Transfer Function and Modulation Transfer Function

The STP was measurements showed that the CR system has a logarithmic relationship and the DR system has a straight line relationship with an offset. With this information images from either system were linearised so that the linear pixel value will be equal to the DAK and so the images were comparable.

The MTF of the DR system is greater than the CR at all spatial frequencies. Thus one of the conditions allowing the images to be adapted to simulate another detector has been met for conversion from DR to CR.

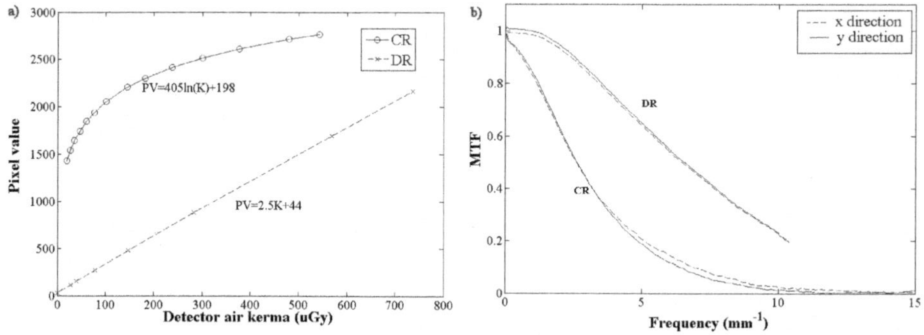

Fig. 1. a) Signal transfer function of DR and CR systems as a function of pixel value against detector air kerma (*K*). b) The modulation transfer function for the CR and DR systems in the x and y axis.

4.2 Simulation of Flat Field Images at Different Dose Levels for a Given Detector

This was undertaken for five dose levels for both CR and DR. Fig. The DR images (fig. 2a) showed the NNPS of the real DR and the modeled image were closely matched over the whole dose range. The NNPS (fig 2b) of CR simulated images and real CR images agreement is slightly poorer than the modeled DR image results, but were still on average within a few percent between the real and modeled image.

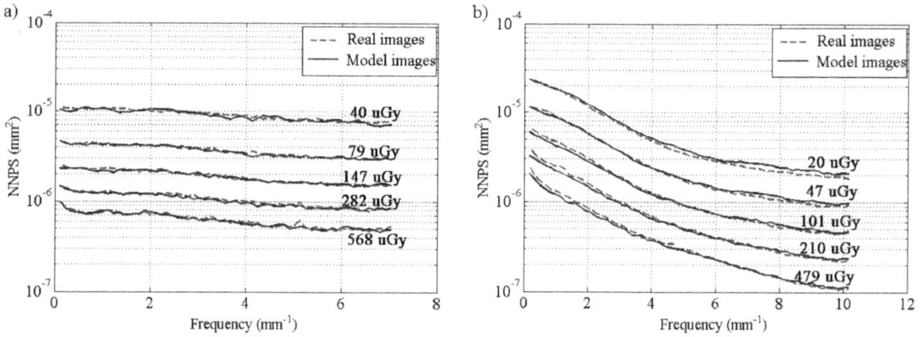

Fig. 2. NNPS of flat field images created from measurements of the noises compared to the equivalent real image for a) DR and c) CR in scan direction

4.3 Adaptation of Image Noise to Simulate Image Obtained with a Different Detector

When converting from a high dose DR images (fig. 3a) to appear as CR images, then the NNPS of the modeled CR images closely matched the real CR image even down to low dose levels. It is easier to convert a large dose change in the image as the original noise in the image is reduced by the square of the dose change. Therefore, the difference between the noise expected and in the dose reduced image is larger and so there is better scope to add the noise into an image with the correct magnitude and shape. However, when the dose change is small as in figure 3b, then there is less scope for accurately inserting noise. The largest difference between the real CR and the simulated CR image was at the conversion from a 28 μGy DR image to 26 μGy CR image. An issue arises at these dose levels as the DR system has more electronic noise than the CR system and at these dose level then electronic noise becomes a significant proportion of the overall noise in the image. The electronic noise cannot be simply removed and its effects are mitigated by a noise correction factor, without this correction the error would be much larger at this low dose level.

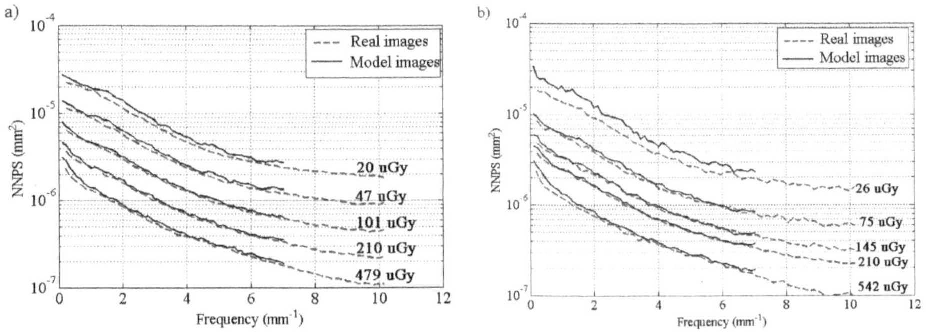

Fig. 3. a) NNPS of simulated CR image produced from DR image at 568 μGy compared to the real CR images; b) NNPS of simulated CR image produced from DR image compared to the real CR images, the dose change was 28 μGy to 26 μGy, 79 μGy to 75 μGy, 147 μGy to 145 μGy, 282 μGy to 210 μGy and 282 μGy to 300 μGy in *x* direction (CR scan direction)

The NNPS of the real CR images extends to higher spatial frequencies than the simulated CR image. The CR system has a Nyquist frequency higher than the simulated images as it was modelled from images obtained from a detector with a larger pixel pitch.

5 Conclusions

This work show shows potential for the conversion of images both in terms of the detector used and the dose acquired. The work requires further improvement and validation. However, if this method is successful it will be a very useful tool for assessing the impact of using detectors with different imaging properties on cancer detection. Clinical trials currently require very large numbers of images to be able to detect differences in acquisition parameters, the reason for the large numbers is the large variability in clinical images due to differences between breasts being imaged. Even if the same breast is imaged using several technologies there will be differences due to the differences in compression/projection. This conversion program can allow the same image to appear as if it was imaged with a different detector and/or different dose. In addition it can remove variability due to differences in scatter rejection systems. This therefore may be a powerful tool for studies of clinical image quality.

Acknowledgments. This work is part of the OPTIMAM project and is supported by CR-UK & EPSRC Cancer Imaging Programme in Surrey, in association with the MRC and Department of Health (England).

We are grateful the help and support of staff from St George's Hospital, London for using their equipment. We acknowledge Hologic and Carestream for their help in access to images from their respective equipments.

References

1. European guidelines for quality assurance in breast cancer screening and diagnosis (2006)
2. Marshall, N.W.: A comparison between objective and subjective image quality measurements for a full field digital mammography system. Phys. Med. Biol. 51, 2441–2463 (2006)
3. Båth, M., Hakansson, M., Tingberg, A., Mansson, L.G.: Method of simulating dose reduction for digital radiographic systems. Radiat. Prot. Dosimetry 114, 253–259 (2005)
4. Svalkvist, A., Båth, M.: Simulation of dose reduction in tomosynthesis. Medical Physics 37, 258 (2010)
5. Workman, A.: Simulation of Digital Mammography. Proc. SPIE Physics of Medical Imaging 5745, 933–942 (2005)
6. Mackenzie, A., Honey, I.D.: Characterization of noise sources for two generations of computed radiography systems using powder and crystalline photostimulable phosphors. Med. Phys. 34, 3345–3357 (2007)
7. Evans, D.S., Workman, A., Payne, M.: A comparison of the imaging properties of CCD-based devices used for small field digital mammography. Phys. Med. Biol. 47, 117–135 (2002)
8. IEC: Medical electrical equipment -Characteristics of digital x-ray imaging devices - Determination of the detective quantum efficiency -Detectors used in mammography. IEC 62220-1-2 , International Electrotechnical Commission, Geneva, Switzerland (2007)
9. NHSBSP: Calculation of Quantitative Image Quality Parameters - Notes Describing the Use of OBJ_IQ_reduced. Equipment report 0902, NHSBSP, Sheffield, UK (2009)
10. Saunders Jr., R.S., Samei, E.: A method for modifying the image quality parameters of digital radiographic images. Med. Phys. 30, 3006–3017 (2003)

Human Observer Performance in a Single Slice or a Volume: Effect of Background Correlation

Ingrid Reiser and Robert M. Nishikawa

Department of Radiology, The University of Chicago, Chicago, IL 60637, USA

Abstract. Human observer performance in a search task was compared for viewing a single 2D image or a 3D image volume in cine mode. The test images consisted of designer nodules added to white noise backgrounds, or filtered noise backgrounds. In white noise backgrounds, performance increased dramatically when the entire volume was provided. In correlated backgrounds, no increase in performance was observed. These results are consistent with our previous findings for signal-known-exactly detection performance in single tomosynthesis slices or the entire reconstructed volume, where the inclusion of adjacent slices in the observer study did not result in a measurable increase in observer performance.

1 Background

With two emerging breast imaging modalities that are generating volume data sets, namely tomosynthesis and breast CT, the goal of this study was to shed light on human readers' performance viewing volume images in slice mode. Tomosynthesis breast images are often reviewed by scrolling through slices, or by viewing the image stack in cine-mode [1].

We have previously presented results from a study that compared tomosynthesis performance when looking at a single reconstructed slice, or a 3D stack of reconstructed slices. The study is summarized briefly: The experiment measured human detection performance for exactly-known signals (SKE) in the tomosynthesis reconstructed volume. Test images consisted of a reconstructed designer nodule [2] with a 8-mm diameter added into reconstructed breast tomosynthesis backgrounds. Two reading conditions were investigated. In the first condition, the 2D slice through the center of the reconstructed signal were shown. In the second condition, all 36 of the 1-mm thick image slices through the volume were shown, and the observers could scroll through slices at their own pace. Reading time was not restricted, and readers rated their confidence of the presence of a signal on a 1-8 scale in each trial. Two readers participated in the study. The results are shown in Fig. 1. No significant difference in area under the ROC curve was found.

This result was not expected and it was not clear whether this finding was caused by errors in the display setup. Conventional wisdom would have suggested that presenting adjacent slices would increasing the information available to the observer and thus improve observer performance. To clarify the cause of these

J. Martí et al. (Eds.): IWDM 2010, LNCS 6136, pp. 327–333, 2010.

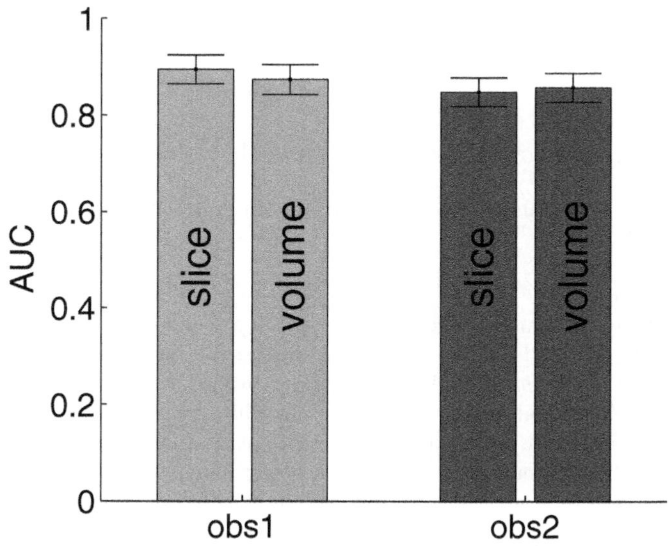

Fig. 1. Observer performance in a SKE detection study in tomosynthesis, for the 2D in-focus slice through the signal, or when scrolling through the 3D volume. We have presented these data at the XIIIth conference of the Medical Imaging Perception Society (MIPS) in Santa Barbara, CA, Oct. 2009.

findings, a comparison of free search in a volume versus free search in a single slice was performed, for signals added to white noise backgrounds, and filtered noise backgrounds.

2 Method

Observer performance was measured by determining the proportion of correct responses in a free-search task. A high-contrast copy of a 2D slice through the center of the signal was shown to the observer, and the observer was asked to indicate the location of the signal in the image. After the observer made his or her decision, the signal location was revealed by a circle cue. A screen shot of the display layout, including the circle cue, is shown in Fig. 2.

First, search experiments were conducted on 2D slices through the volume, located at the center of the signal. Signal amplitude was set so that observers reached a proportion of approximately 0.55 correct answers in the 2D slices. The search experiment was then repeated with the same volume, but this time all volume slices were shown to the observer in an infinite cine-loop. The frame rate was fixed at 20 frames per second.

Search performance was measured in two types of background, uncorrelated and correlated. The uncorrelated backgrounds were uniform with white Gaussian noise. Correlated backgrounds were generated by filtering a random noise volume

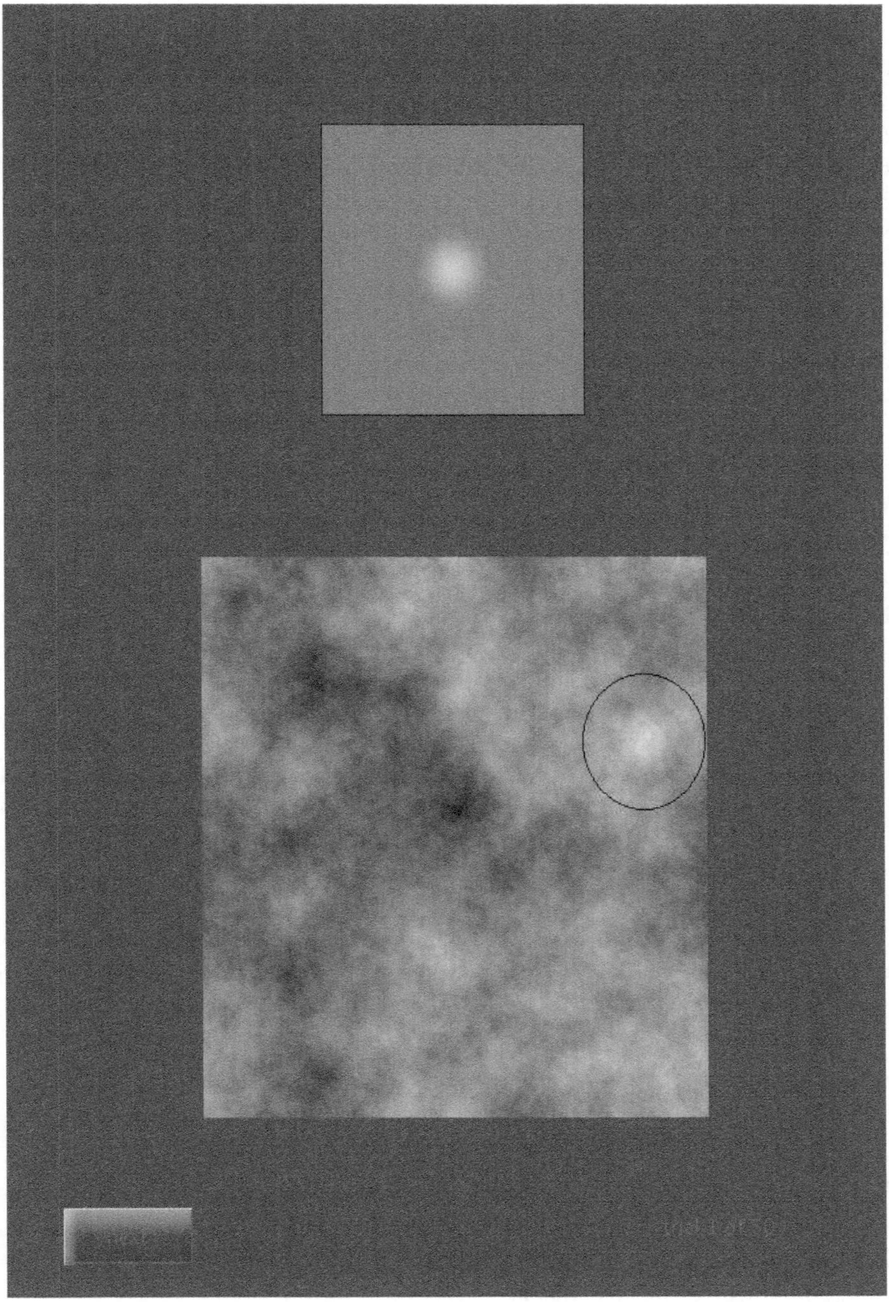

Fig. 2. Screen layout of the observer study. The example shows a filtered noise background. The signal amplitude is higher than was used in the actual experiment for presentation purposes. The circle cue appears after the observer has indicated the signal location.

with a 3D spherically symmetric kernel of the form $k(f) = const/f^2$. This results in volume with a spherically symmetric 3D power spectrum following $P(f) = c/f^4$. In this volume, the correlation is isotropic, i.e., a slice through the volume along the $x - y$ axes looks no different than a slice along the $x - z$ axes. The corresponding power spectrum in a 2D slice is then $P_{2D}(f_2) = c'/f_2^3$, i.e., the texture in the slices is similar to that observed in tomosynthesis [3]. Signals consisted of spherically symmetric designer nodules with a normalized profile of $s(r) = (1-(r/R)^2)^2 \Pi_R$, where Π_R is a rect function. Signal centers were placed randomly within the volume.

Two observers, who were medical physicists (one in training) with experience viewing tomosynthesis images, participated in the reader study. In perception experiments, reader variability tends to be low and typically a small number of observers is sufficient [4,5]. Experiments were conducted with signals of 14, 30 and 60 voxels in diameter. Volume size was 384x384x48 voxels. 64 trials were conducted per experimental condition.

3 Results

Figures 3 and 4 show the proportion of correct responses in white noise backgrounds and filtered noise backgrounds, for signal of 60, 30 and 14 voxels in diameter. Inter-observer variability is within error bars, justifying the use of two observers for this type of experiment. The first column in the figure shows the proportion of correct answers in uncorrelated white-noise backgrounds. For all signal sizes, showing the entire volume increases proportion correct dramatically, from about 0.6 to 0.98. Since the same images were presented to the observers in the 2D and 3D study, a pairwise comparison can be performed for each image, the result of which is shown as "difference per image". As expected, a large increase of proportion correct of 0.4 or larger is found.

The second column shows results obtained in filtered noise backgrounds. In these backgrounds, changes in observer performance are minimal, and not statistically significant in that the change is less than the total length of the error bars.

4 Discussion

For signals added to uniform white-noise backgrounds, this study produced the expected large increase in observer performance. The frame rate of the cine-loop was fast enough to exceed the integration time of the eye, which is about 0.1 sec [6]. This indicates that the display setup was appropriate.

In the second set of experiments, background correlation was introduced at a level similar to that observed in tomosynthesis images. In these backgrounds, differences in human reader performance did not reach statistical significance when the full 3D volume was shown. This could change if the number of trials was increased.

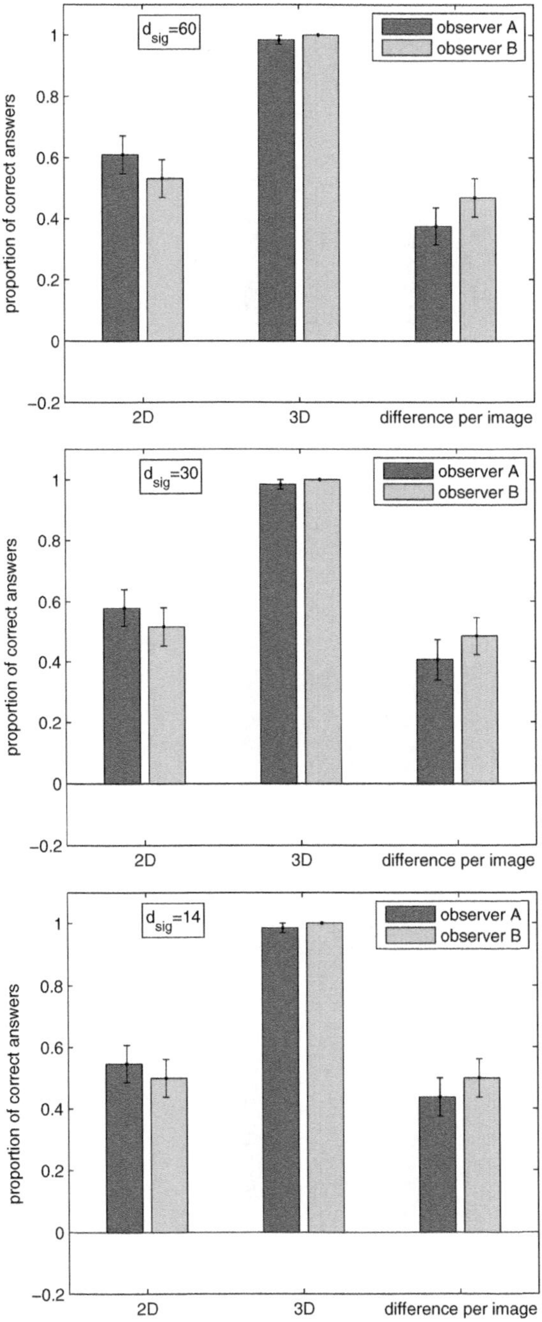

Fig. 3. Proportion of correct answers for two human readers in a search experiment in uncorrelated white noise. First row: 60 voxels signal diameter. Second row: 30 voxels signal diameter. Third row: 14 voxels signal diameter. The total length of the error bars are two standard deviations.

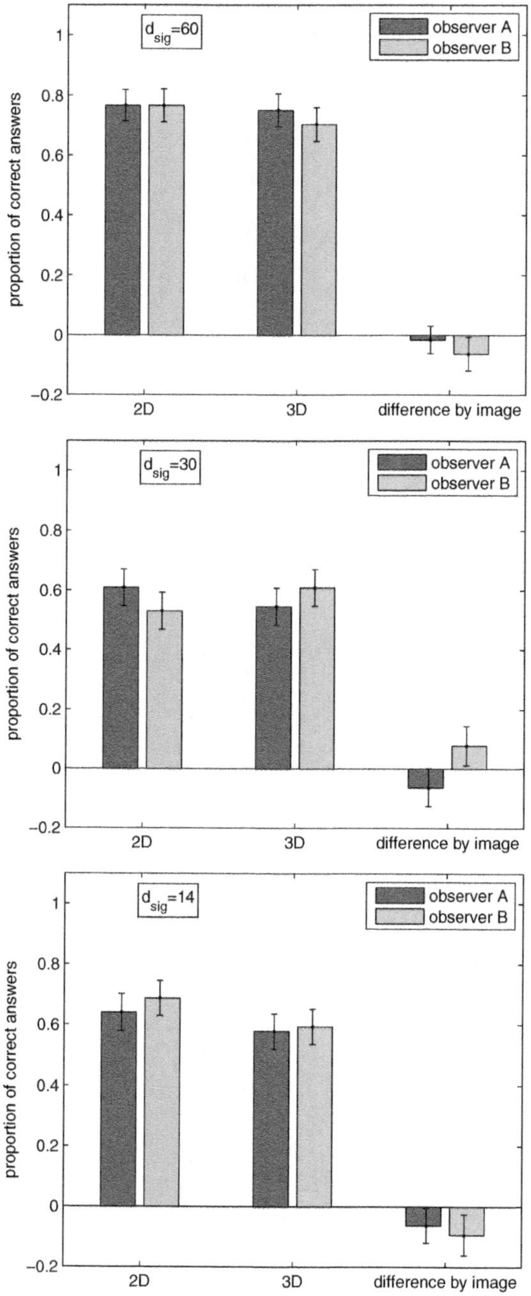

Fig. 4. Proportion of correct answers for two human readers in a search experiment in a correlated background. First row: 60 voxels signal diameter. Second row: 30 voxels signal diameter. Third row: 14 voxels signal diameter. The total length of the error bars are two standard deviations.

A limitation of this study is that there was no participation of a radiologist who routinely reads a 3D imaging modality in slice-viewing mode. Further, the signals as well as the filtered-noise backgrounds did not exhibit any sharp edges, which may not be representative of 3D breast images.

To assess the implication of these results on display modes used in tomosynthesis, we plan to perform 2D and 3D search experiments using clinical tomosynthesis backgrounds and realistic lesions, and we plan to include clinical radiologists in those studies.

5 Conclusion

Our findings indicate that our display setup is capable of measuring differences in performance in slice or volume reading. Our findings also indicate that there may be little benefit of providing adjacent slices when detecting signals in a volume.

Acknowledgments

The authors would like to thank B. Lau for participating in the observer study. RM Nishikawa is a shareholder in and receives royalties from Hologic, Inc. He is a member of the scientific advisory board of Dexela, Ltd. He is a consultant to Carestream Health, Inc. and Siemens Medical Solutions USA, Inc. This project was supported in part by award number R21 EB008801 from the National Institute Of Biomedical Imaging And Bioengineering. The content is solely the responsibility of the authors and does not necessarily represent the official views of the National Institute Of Biomedical Imaging And Bioengineering or the National Institutes of Health.

References

1. Poplack, S.P., Tosteson, T.D., Kogel, C.A., Nagy, H.M.: Digital breast tomosynthesis: initial experience in 98 women with abnormal digital screening mammography. AJR 189, 616–623 (2007)
2. Samei, E., Flynn, M.J., Eyler, W.R.: Detection of subtle lung nodules: Relative influence of quantum and anatomic noise on chest radiographs. Radiology 213, 727–734 (1999)
3. Engstrom, E., Reiser, I., Nishikawa, R.M.: Comparison of power spectra for tomosynthesis projections and reconstructed images. Med. Phys. 36, 1753–1758 (2009)
4. Reiser, I., Nishikawa, R.M.: Identification of simulated microcalcifications in white noise and mammographic backgrounds. Med. Phys. 33, 2905–2911 (2006)
5. Burgess, A.E., Judy, P.F.: Signal detection in power-law noise: Effect of spectrum exponents. J. Opt. Soc. Am. A 24, B52–B60 (2007)
6. Barten, P.G.J.: Contrast Sensitivity of the Human Eye and Its Effects on Image Quality. SPIE press (1999)

Influence of Geometrical Factors on Phase Contrast Fiber Images

Margarita Chevalier[1], Lorena Chanes[2], Eduardo Guibelalde[1], María-Ester Brandan[3], and Tatiana Alieva[2]

[1] Física Médica, Departamento de Radiología, Universidad Complutense de Madrid, 28040 Madrid, Spain
[2] Departamento de Óptica, Universidad Complutense de Madrid, 28040 Madrid, Spain
[3] Instituto de Física, Universidad Nacional Autónoma de México, 04510 Mexico City, Mexico

Abstract. X-ray field phase differences caused by an object can induce edge enhancement in a radiological image. Phase contrast effects have been interpreted using a model based on x-ray Fresnel diffraction. The dependence of edge-enhancement on x-ray energy, object characteristics and magnification has been systematically studied and interpreted by calculations using this model. It was found a good agreement between numerical simulations and experimental results obtained under magnification conditions with a commercial digital mammography unit. The numerical calculations have been extended to the more general situation of a variable total source-to-detector distance. In this paper, we present the results of these calculations as well as the results derived from the analysis of the influence of the fiber radius on the edge enhancement.

Keywords: Phase contrast, Edge enhancement, Digital Mammography, x-ray diffraction.

1 Introduction

In the last years, there has been a growing interest in phase contrast radiography as an alternative method to gain information about low attenuating x-rays objects. Phase contrast is based on the changes experimented by the radiation wave front as a consequence of differences in the real part of the complex index of refraction of different media. For x-rays, the index of refraction can be expressed as $n=1-\delta + i\beta$. In the energy region typical for clinical radiography and for low-atomic-number elements, real coefficients δ are thousands of times larger than imaginary coefficients β which are related with the attenuating properties [1]. Therefore, phase contrast can be of relevance to radiological images of soft tissues, particularly in mammographic images [1,2].

Phase changes cause interference of wave fronts, generating dark and bright fringes around the object's edges -where the refraction index varies abruptly- thus improving the visualization of the borders. To visualize the fine structure of the interference pattern is required to have spatially coherent radiation field at the object plane. In addition, the detector must be placed some distance behind the object to facilitate the Fresnel fringes observation. Different strategies have been designed to accomplish these requirements. The most common one makes use of synchrotron radiation with

J. Martí et al. (Eds.): IWDM 2010, LNCS 6136, pp. 334–341, 2010.

special geometrical conditions to obtain mammographic phase contrast images [3]. In recent years, x-ray tubes with a small focus size (microfocus) have been applied to achieve partially coherent radiation [4,5,6]. Conventional mammography x-ray tubes using the small focus in combination with an object-detector distance greater than the usual (magnification) are commercially available [7,8]. Among the possible modalities to exploit phase observation as image contrast, this work deals with "phase contrast radiography"[1,2], also called "in-line imaging"[9]. Phase contrast images of acrylic fibers with different diameters included into a mammographic phantom were acquired in the magnification mode (i.e. focus size equal to 0.1 mm) of a commercial mammographic unit i.e. total source-to-detector distance equal to 0.66 m. The object-detector distance was varied to obtain images with a series of magnification factors (M), see Fig.2 for definition of M. Enhanced edges can be observed (Fig. 1) in the magnified images of the fibers that were acquired with a detector of 70 μm pixel size. The average of several intensity profiles across the fiber shows the increase of the intensity at the fiber edges.

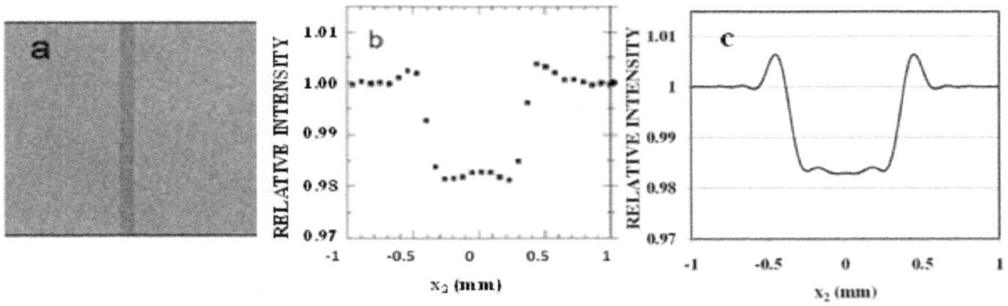

Fig. 1. Image and profiles for a 0.30 mm diameter fiber. a) Image acquired with Mo/Mo at 26 kV, M=2.64 and a 70 μm detector pixel size. b) Average profile for three fibers, including the one shown in the image. c) Simulation of the relative intensity at the detector plane for 15 keV x-rays, M = 2.64 and 70 μm detector pixel size.

Experimental observations were interpreted by a diffraction based simulation. The radiation intensity distribution at the detector plane has been derived from the analytical expression obtained considering the 2-dimensional diffraction of monoenergetic x-rays by a weakly attenuating object. The dependence of the edge-enhancement on energy, object characteristics and magnification has been systematically studied and interpreted by the calculations. It was found a good agreement between experimental results and numerical simulations (Fig. 1)[10]. The numerical calculations have been extended to the more general situation of a variable total source-to-detector distance. In this paper, we present the results of these calculations. In addition, we have analyzed the influence of the fiber radius on the edge enhancement.

2 Theoretical Formalism

The objects considered in this work are long cylinders oriented along the y-axis (see Fig 2). The attenuation and phase shift effects of this object on the radiation field are described by a transmission function $t(x_1)$[6]:

$$t(x_1) = \exp[-\mu(x_1)/2 + i\pi\varphi(x_1)] = A(x_1)\exp[i\pi\varphi(x_1)]\,. \quad (1)$$

where x_1 is the x-coordinate at the object plane that is normal to the beam propagation direction z. $\varphi(x)$ and $\mu(x)$ are the object projected phase shift and linear attenuation coefficients, given in terms of the index of refraction δ and β coefficients, respectively, by

$$\varphi(x_1) = -\frac{2}{\lambda}\int \delta(x_1,z)dz \qquad \mu(x_1) = \frac{4\pi}{\lambda}\int \beta(x_1,z)dz\,. \quad (2)$$

Therefore, the phase contrast image formation can be treated as a two-dimensional (x_1,z) problem. The radiation source focal spot can be considered as a linear, uniform, spatially incoherent monochromatic source of length a and intensity I_0. The object is placed at the distances R_1 and R_2 from the source and detector respectively (Fig. 2).

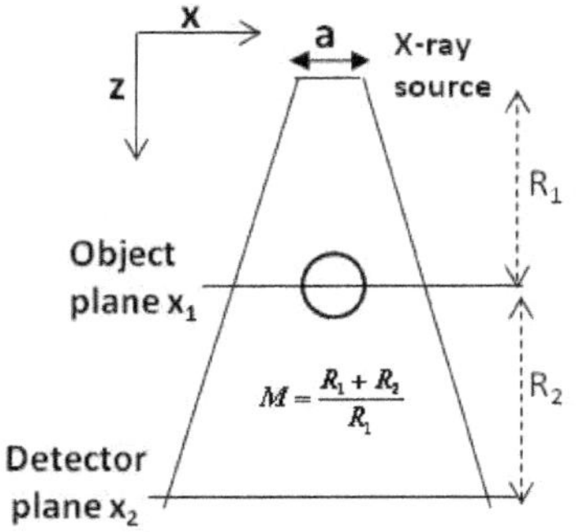

Fig. 2. Geometrical scheme used in the diffraction analysis where R_1 is the source-object distance, R_2 the object-detector distance and M is the magnification factor

Applying the Van Citter-Zernike theorem [11] to relate the mutual intensity $J(x_1,x_1')$ at the object plane x_1 to the intensity of the incoherent source, and after performing some approximations, the intensity distribution at the detector plane x_2 is obtained:

$$I(x_2) = I_0\frac{sR_1}{R_2}\int_{\frac{1}{M}(x_2-\frac{aR_2}{2R_1})}^{\frac{1}{M}(x_2+\frac{aR_2}{2R_1})}|A(X)|^2\,dX - I_0\frac{s\lambda R_1}{2M}\left\{\frac{d\varphi(x)}{dx}|A(X)|^2\Big|_{\frac{1}{M}(x_2+\frac{aR_2}{2R_1})} - \frac{d\varphi(x)}{dx}|A(X)|^2\Big|_{\frac{1}{M}(x_2-\frac{aR_2}{2R_1})}\right\}. \quad (3)$$

where M is the magnification factor. For the cylindrical objects considered in this work, the amplitude and phase of the transmission function obey the following expressions:

$$|A(x)|^2 = \exp\left[-\frac{8\pi r}{\lambda}\beta\sqrt{1-\left(\frac{x-x_C}{r}\right)^2}\right] \qquad \frac{d\varphi(x)}{dx} = \frac{4\delta}{\lambda r}\frac{(x-x_C)}{\sqrt{1-\left(\frac{x-x_C}{r}\right)^2}}\,. \quad (4)$$

where x_C is the displacement of the fiber center from the coordinate origin, which coincides with the center of the radiation source.

3 Results

Eq. (3) was programmed in MATLAB for the specific case of a cylindrical object (Eq. 4). The relative intensity distribution calculated at the image plane shows sharp maxima and minima (Fig. 3a) at the edges of the object that do not appear when only the attenuation effect is considered (Fig. 3b). The internal structure is the combination of attenuation and the minima of the phase oscillation. These effects result in contrast, due to the phase changes, complementary to the contrast produced by absorption alone.

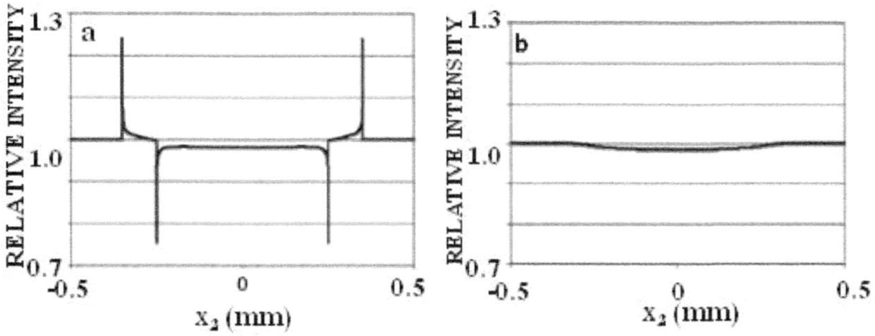

Fig. 3. Relative intensity distribution calculated at the detection plane for the following values of the parameters of interest: effective energy E = 15 keV, source size a = 100 μm, M = 2, fiber diameter d = 0.3 mm in air. (a) Absorption and phase; (b) Pure absorption ($\delta = 0$).

In this work, the detector was characterized by its finite Modulation Transfer Function (MTF) represented in Fig. 4, that was experimentally determined following the methodology described in the IEC standard [12]. Data were processed using the free software MIQuaELa (v.1.0) [13]. The Fourier Transform of the intensity distribution at the image plane was numerically calculated and then multiplied by the detector MTF. The inverse Fourier Transform (FT) of the product of the MTF and the FT of the intensity distribution at the detector plane obtained from Eq. (3) was numerically calculated. The result simulates the intensity distribution measured by the detector, without taking into account the noise.

The study of the dependence of the edge enhancement on fiber size shows that it is slightly more relevant for cylindrical objects with a larger diameter (Fig 5a). The same behavior holds if the detector MTF is taken into account (Fig. 5b). In the latter case, the height of maxima associated with the edge enhancement is significantly reduced and its width is increased. Also it is observed the disappearance of the two sharp minima at the internal part of the fibers as a consequence of both a reduction on its depth and an increasing of its width.

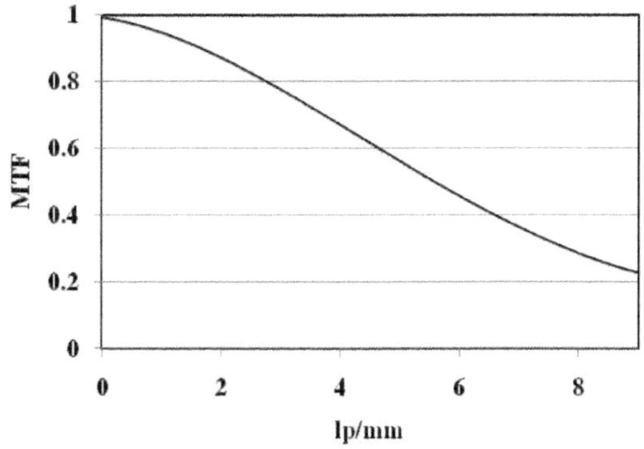

Fig. 4. Presampling MTF of the 70 μm pixel size a-Se detector mounted in the mammography system. The figure shows the average MTF, measured according to IEC.

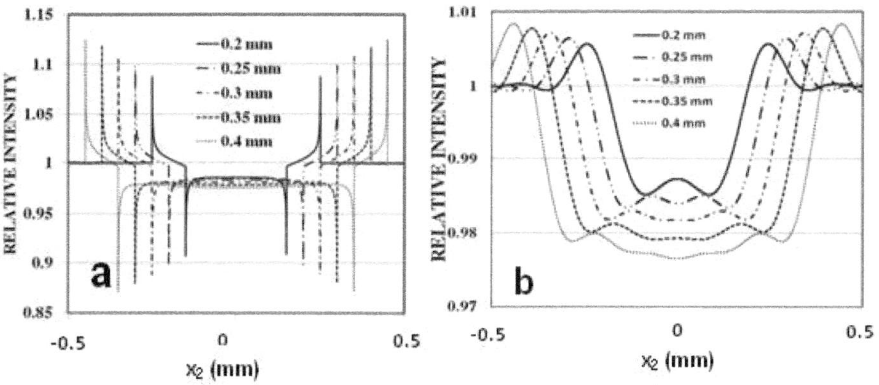

Fig. 5. Relative intensity distribution for several values of the fiber diameter (E = 15 keV, a = 100 μm, M = 2): (a) Simulated profiles at the image plane (b) Simulated profiles at a 70 μm pixel size matrix detector

Figure 6 displays the influence on edge enhancement of both object-detector distance (R_2) and object-source distance (R_1) for the fiber of diameter 0.25 mm. As can be seen in Fig. 6a, edge enhancement does not depend on the increase of R_2 when R_1 is constant. The height of maxima remains the same and their width increases with R_2. The image size also increases due to a major M value. However, a larger edge enhancement is observed for the greater values of R_2 when the detector pixel size is considered (Fig. 6c). This is a consequence of the different contribution to the pixel energy integration due to the different width of the maxima.

As beam spatial coherence increases with distance R_1, greater values of R_1 (lower M values) cause more pronounced edge enhancements i.e. greater values of the height of maxima (Fig. 6b). The detected relative intensity (Fig 6d) exhibits the same behavior.

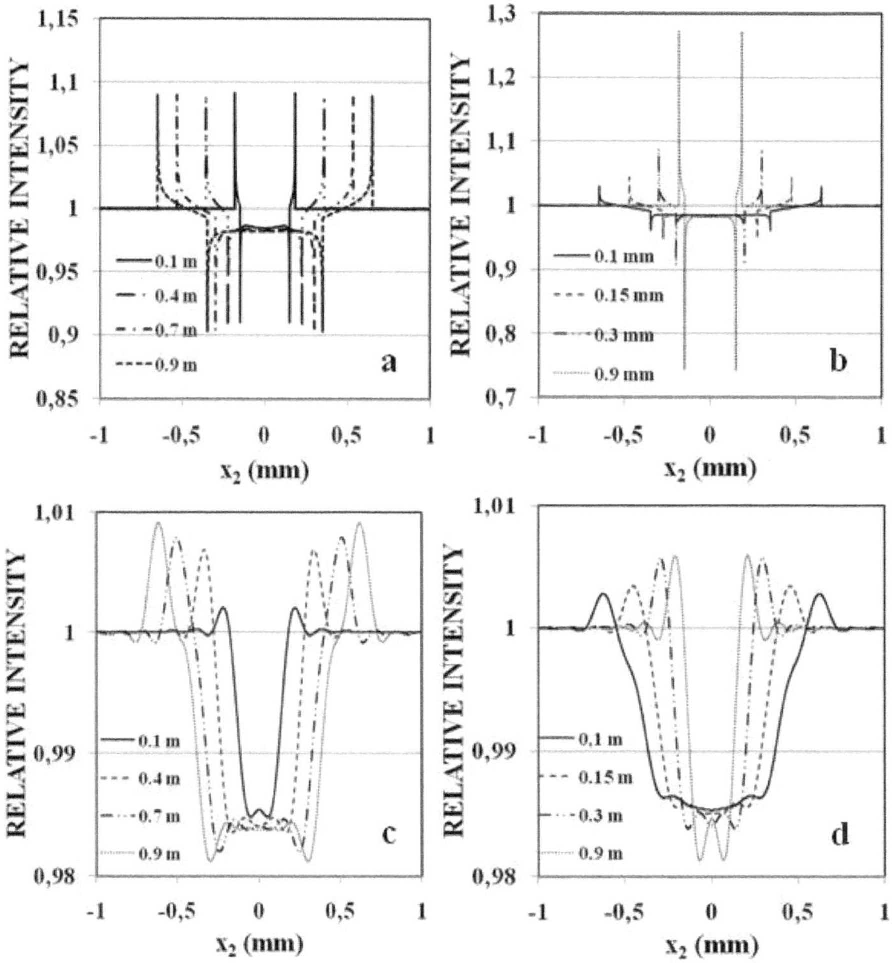

Fig. 6. Relative intensity distribution at the image plane for several values of: (a) object-detector distance R_2 (R_1 fixed at 0.33 m); (b) object- source distance R_1 (R_2 fixed at 0.33 m); (c) & (d) effect of a 70 μm pixel size detector (E = 15 keV, a = 100 μm, fiber diameter = 0.25 mm).

Note that commercial mammographic units have a constant focal spot-detector distance i.e. $R_1 + R_2$ is about 60-66 cm. Under this condition, the effect of phase contrast would be more visible for values of M between 1.6 and 2.8 where the influence of both distances on edge enhancement are positively combined (Fig. 7). The edge enhancement EE was quantified by the simple expression:

$$EE = 100\frac{I_{MAX} - I_0}{I_0} \tag{5}$$

where I_{MAX} is the maximum intensity at the edge and I_0 is the background.

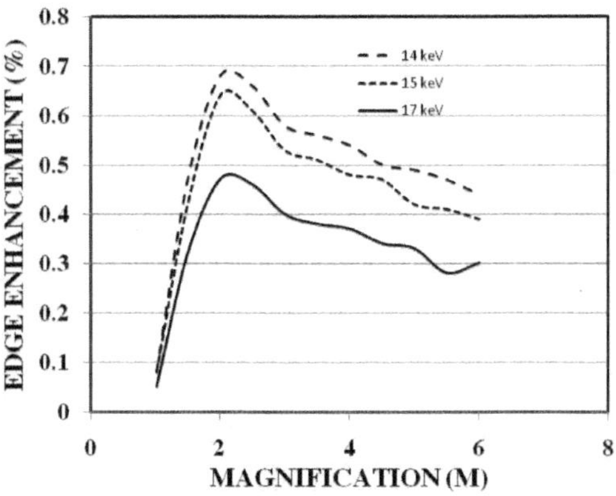

Fig. 7. Edge enhancement as a function of the magnification for three different x-ray effective energies in the range of mammography clinical values. The fiber diameter was 0.25 mm.

The general tendency is for EE to increase rapidly from zero (at contact) up to 0.5-0.7% at about M = 2 and to decrease afterwards. Edge enhancement is larger at the lower energy, about 45% stronger at 14 than at 17 keV.

4 Discussion and Conclusions

A simulation model has been developed to analyze the image formation for in-line phase-contrast imaging. The model has been based on x-ray Fresnel diffraction and the formalism of the mutual coherence function to describe the coherence properties of the incident beam. This model also includes the matrix structure of a detector with a pixel size typical for commercial mammography units. The detector performance has been characterized through its experimentally determined MTF. The sharp structure of maxima and minima that characterizes the theoretical intensity distribution is softened in the detection process and the edge enhancement is reduced by about 10% in the images of the detected fibers. The influence of the geometrical parameters has been systematically analyzed in order to establish the adequate conditions for experimental observation of the edge enhancement associated to the phase contrast effect. Placing the detector away from the object produces an important increasing of the theoretical maxima (about 30%). However, these values are strongly reduced when the detector is considered. For commercial units, intermediate M values (about 2) are demonstrated to be the best conditions to observe these effects. Regarding the geometrical characteristics of the fibers, edge enhancement is more relevant for fibers with greater diameters for relatively small scales (diameter 0.1-0.5mm).

Acknowledgements

Authors acknowledge support from Proyecto Santander-Complutense PR-34/07-15914. Special thanks to Cristina Núñez de Villavicencio and Ángeles Franco (Fundación Jiménez Díaz, Madrid). MEB acknowledges partial support from DGAPA–UNAM and Conacyt-Mexico during sabbatical stay at Universidad Complutense de Madrid.

References

1. Momose, A.: Recent Advances in X-ray Phase Imaging. Jpn. J. Appl. Phys. 44, 6355–6367 (2005)
2. Fitzgerald, R.: Phase-sensitive x-ray imaging. Phys. Today 53(7), 23–26 (2000)
3. Arfelli, F., et al.: Mammography with synchrotron radiation: Phase-detection techniques. Radiology 215, 286–293 (2000)
4. Wilkins, S.W., Gureyev, T.E., Gao, D., Pogany, A., Stevenson, A.W.: Phase-contrast imaging using polychromatic hard X-rays. Nature 384, 335–338 (1996)
5. Kotre, C.J., Birch, I.P.: Phase contrast enhancement of x-ray mammography: a design study. Phys. Med. Biol. 44, 2853–2866 (1999)
6. Pogany, A., Gao, D., Wilkins, S.: Contrast and resolution in imaging with a microfocus x-ray source. Rev. Sci. Instum. 68, 2774–2782 (1997)
7. Honda, C., Ohara, H., Gido, T.: Image Qualities of Phase-Contrast Mammography. In: Astley, S.M., Brady, M., Rose, C., Zwiggelaar, R. (eds.) IWDM 2006. LNCS, vol. 4046, pp. 281–288. Springer, Heidelberg (2006)
8. Yamazaki, A., Ichikawa, K., Kodera, Y.: Investigation of physical image characteristics and phenomenon of edge enhancement by phase contrast using equipment typical for mammography. Med. Phys. 35, 5134–5150 (2008)
9. Wu, X., Liu, H.: Clarification of aspects in in-line phase-sensitive x-ray imaging. Med. Phys. 34, 737–743 (2007)
10. Brandan, M.E., Chevalier, M., Guibelalde, E., Rodrigo, J., Chanes, L., Alieva, T.: Observation of edge-enhancement in digital images obtained with a clinical mammography unit. In: World Congress on Medical Physics and Biomedical Engineering, IFMBE Proceedings, Munich, Germany, September 7 - 12, vol. 25/2, pp. 331–334 (2009)
11. Goodman, J.W.: Introduction to Fourier Optics. Roberts and Company Publishers, Greenwood Village (2005)
12. International Electrotechnical Commission.: Medical electrical equipment -Characteristics of digital X-ray imaging devices – Part 1-2:Determination of the detective quantum efficiency – Detectors used in mammography. IEC Rep. 62220-1-2 Ed. 1.0 (2007)
13. Ayala, R., García, R., Linares. R.: MIQuaELa (2009), `http://www.ucm.es/info/fismed/Digital_Imaging/MIQuaELa.rar`

Robust Breast Composition Measurement - Volpara™

Ralph Highnam[1], Sir Michael Brady[2], Martin J. Yaffe[3], Nico Karssemeijer[4], and Jennifer Harvey[5]

[1] Matakina Technology Limited, 9/2 Hood St., Wellington, 6011, New Zealand
[2] Department of Engineering Science, University of Oxford, UK
[3] Department of Medical Biophysics & Medical Imaging, University of Toronto, Canada
[4] Department of Radiology, University of Nijmegen, Netherlands
[5] Department of Radiology, University of Virginia Health System, USA
ralph.highnam@matakina.com

Abstract. Volumetric breast composition measurements generally require accurate imaging physics data. In this paper we describe a new method (Volpara™) that uses relative (as opposed to absolute) physics modeling together with additional information derived from the image to substantially reduce the dependence on imaging physics data. Results on 2,217 GE digital images, from a diversity of sites, show encouraging agreement with MRI data, as well as robustness to noise and errors in the imaging physics data.

Keywords: Breast density, volumetric, Volpara.

1 Introduction

Area-based breast density measures, both manual and semi-automated, have been shown repeatedly to correlate well with breast cancer risk and with the diagnostic difficulty of a mammogram [1-11]. Such measures are increasingly suggested as a basis for tailoring breast screening for each individual woman. However, such area-based measures are intrinsically subjective, and there is substantial inter- and intra-observer variability [12-14]. Also, such methods require additional decision-time by skilled users. As a result, although the results generated by trained users in research environments are encouraging, the applicability of such methods in the real-world is at best problematic. Unfortunately, to date at least, the automation of area-based density measures, to the point where they can be incorporated into clinical workflow, has proven to be difficult, primarily due to: differences in imaging parameters making the image brighter or darker or affecting the contrast, hence appearing more or less dense; and to textural similarities between very fatty and very dense breasts.

Volumetric breast density measurements, which are, in principle, objective and straightforward to automate, have been proposed, based upon a succession of physics-based models of the x-ray imaging process [15-20]. They have been shown to compare well to visual estimations of breast density. However, to date, the correlation to breast cancer risk has been variable [21]. We contend that this is due primarily to current volumetric models overly relying on imaging physics data that is assumed to be accurate. In this paper, we build upon the work in [16] and describe a "relative" (as opposed

J. Martí et al. (Eds.): IWDM 2010, LNCS 6136, pp. 342–349, 2010.

to absolute) physics model which dramatically reduces the need for accurate imaging physics data by optimizing the information extracted from the image itself.

2 Method

Previously reported volumetric techniques use some form of calibration object or find a position in the breast image that corresponds to a column of tissue that is entirely fat, give or take a thin layer of skin. Of course, the use of calibration objects should, in principle, lead to highly accurate modeling, and should perhaps be the basis for a research-based "gold standard". However, the use of such methods within a clinical environment for routine clinical use poses a number of practical challenges such as the need on numerous occasions to remove the calibration object, for example when imaging large breasts. For this reason, we have focused our efforts on finding an area of the breast that corresponds to entirely fatty tissue, then using that as a reference level (P_{FAT}) to find the thickness of dense tissue (h_d) at each pixel (x,y) as shown by the following equation from [16] where it is simply assumed that the pixel value (P) is linearly related to the energy imparted to the x-ray detector (all direct digital images match this criteria):

$$h_d(x,y) = \frac{\ln\left(P(x,y)/P_{Fat}\right)}{\mu_{\text{fat}} - \mu_{\text{dense}}} \tag{1}$$

The values in the denominator are the effective x-ray linear attenuation coefficients for fat and dense tissue at the particular target, filter, tube voltage and recorded breast thickness combination. This formulation is intrinsically robust to errors in time of exposure, detector gain and other multiplicative variations, since those values appear both in the reference level and in the actual pixel values, so cancel out.

Of course, the difficulty, as pointed out by [17,21], lies in finding an area of the breast which is entirely fat, especially when the breast in question is very dense (which is the category of highest risk and thus of greatest interest). We have overcome this difficulty, while retaining a relative physics model approach by (i) using phase congruency [23], which is invariant to imaging conditions, and (ii) an iterative approach to finding the fatty, uncompressed breast edge as documented in [17] along with realistic, relative, breast edge models [16,17]. With an accurate, breast edge found, we can always find P_{FAT}.

We included a scatter removal process based around the algorithm reported in [16] but adjusted it to again make it work in a relative manner by making simplifications around the variations in scatter-to-primary due to different breast tissue types.

We correct for compression plate slant by assuming a fixed slant [16], an assumption which appears valid for most x-ray systems but not for those systems which purposely have a slanted top compression paddle. For the latter, we are developing a more sophisticated method.

The volume of dense tissue is found by integration of the $h_d(x,y)$ values over the image, while the volume of the breast is determined by multiplying the area of the breast by the recorded breast thickness; the breast density is then the ratio of the two. Evidently, errors in recorded breast thickness remain important; we return to this issue below, where we demonstrate Volpara's robustness.

3 Results

We have collected 2,217 GE digital mammograms from Oslo, Nijmegen and the University of Virginia, and have imaged a range of phantoms in order to validate the software, which we have named Volpara™. The results presented in this paper are for version 1.2.1 of that software.

The performance of our new breast edge detection algorithm, and thus the robustness of finding P_{FAT} on some very dense breasts can be illustrated by visualization of the inner and outer limits of the uncompressed, fatty breast edge within which we search for P_{FAT} – see Figure 1. Critically, note that the inner limit of the determined breast edge does not overlay any dense tissue:

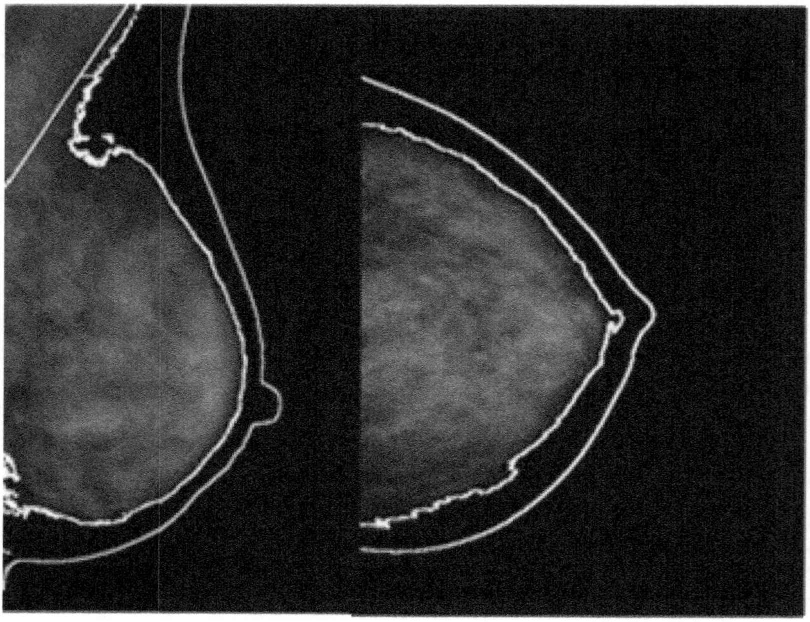

Fig. 1. Dense breasts and the inner edge which Volpara finds

3.1 Accuracy

To demonstrate that Volpara is indeed measuring breast density, we used it to analyze a set of 5 images of a test phantom acquired with different imaging combinations, kindly supplied by the University of Toronto. Each image had 5 "plugs" inserted (labeled A-E below) with different densities and the average error between actual and estimated densities is 1.11%. See Table 1.

Further, we manually measured fibroglandular volume, consequently breast density, from breast MRI volumes for 26 younger women and found a correlation of 0.94 with the Volpara™ fibroglandular volume, and a close relationship between the breast densities as shown in Figure 2, the breast densities will be offset due to the difficulty of measuring total breast volume in the MRI images [16]:

Table 1. Phantom Data

Image	A	B	C	D	E	Imaging Factors
Toronto Plug Phantoms Actual Plug Densities						
#1	0.0	25.0	25.0	12.5	25.0	MoMo26 76mAs
#2	0.0	25.0	25.0	12.5	25.0	MoMo28 51mAs
#3	0.0	20.0	20.0	10.0	20.0	MoMo26 110mAs
#4	0.0	16.7	16.7	8.3	16.7	MoRh26 155mAs
#5	12.5	37.5	37.5	25.0	37.5	MoMo28 55mAs
VolparaV1.2.1 Estimated Plug Densities						
#1	0.43	23.9	24.3	9.9	23.0	
#2	0.16	23.8	24.0	9.7	22.9	
#3	0.41	20.1	20.2	8.6	18.0	
#4	0.64	17.8	18.0	7.8	15.8	
#5	13.6	39.2	39.2	25.2	37.6	

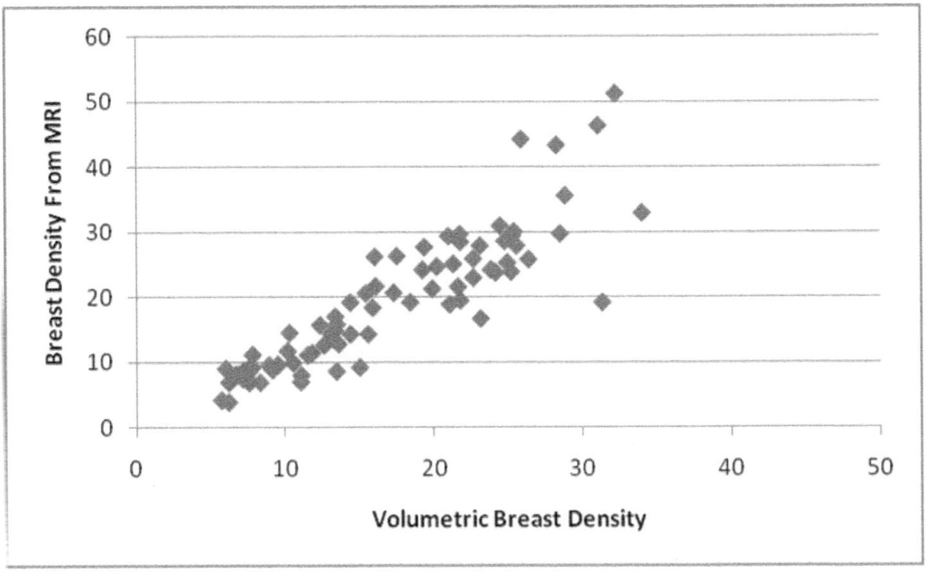

Fig. 2. Volpara breast density versus that derived from breast MRI

3.2 Consistency and Robustness

We analyzed the results over the different detectors in our database; Table 2 shows the median breast density for each detector, along with the Pearson Correlation Coefficient for L/R and CC/MLO along with numbers of images. This demonstrates that we achieve consistent results across detectors. An exception is the PM34_05 detector; but that is the detector on which we have collected the images of young women who have been imaged prior to having breast MRI, thus the high breast density.

Table 2. Comparisons across detectors

	Median	PCC L/R	PCC CC/MLO	#Images
ALL	8.8	0.923	0.915	2217
PM54_01	8.9	0.930	0.908	937
PM460_2	7.8	0.914	0.878	290
PM34_05	16.1	0.915	0.931	104
PM079_04	8.0	0.911	0.916	881

Next, to investigate the robustness of the results to imaging conditions, we edited the mAs in an image (Mo/Mo, 29kVp) by +/- 20% and then again ran Volpara™, as shown in Table 3. As that Table shows, and as expected from Equation (1) ,we found identical results which we also obtained when multiplying the pixel values in the image by various factors to simulate variations in detector gain such as you might expect between detectors, over time, and between manufacturers.

Table 3. Effect of errors in mAs on Volpara's Composition Estimates

	-20% (80mAs)	0% (100mAs)	+20% (120mAs)
Volume of Dense Tissue	114cm^3	114cm^3	114cm^3
Volume of Breast	575cm^3	575cm^3	575cm^3
Breast Density	19.8%	19.8%	19.8%

For a further demonstration, we introduced extra noise into a set of images by randomly adding or subtracting up to 5% and 10% of the pixel value and found that, as expected, noise has limited effect apart from at low breast density. See Table 4.

Table 4. Effect of varying level of random noise on Volpara's Density Estimate

Volpara1.2.1 Breast Density Results Running On:		
Original Image	Original Image ± 5% Random Noise	Original Image ± 10% Random Noise
2.3%	3.4%	5.1%
12.4%	12.5%	13.5%
19.6%	19.7%	17.1%
28.1%	28.2%	28.2%

As noted earlier, breast thickness's major influence is in the breast volume but it is also present to a lesser degree in the dense tissue volume when we work out the effective energy. Fortunately, whereas previous implementations [16] had seen these two factors act in different ways so as to amplify the errors; Volpara™'s implementation has the factors acting in the same way – so, if breast thickness rises, then both breast volume rises and the volume of dense tissue rises so that the overall ratio does not vary widely. Table 5 shows the results from Volpara over four different images when the breast thickness was varied by 20% up and down. Clearly, breast thickness errors will inevitably introduce small errors into the breast density measurement, but thanks to quality requirements in the field, they should rarely exceed 10%, and so, as can be seen, the resulting estimate of breast density remains remarkably accurate. Furthermore, because breast thickness is almost always under-estimated by the x-ray machine, the woman's density assessment will rise, not lower and thus the woman will never be treated in a lower density and thus risk category.

Table 5. Effect of errors in breast thickness on Volpara's Density Estimate

	Variation in Breast Thickness (H)				
H mm	**-20%**	**-10%**	**0%**	**+10%**	**+20%**
86	2.5%	2.4%	**2.3%**	2.2%	2.1%
32	15.1%	13.6%	**12.4%**	11.4%	10.6%
20	23.7%	21.4%	**19.6%**	18.0%	16.8%
40	33.6%	30.5%	**28.1%**	26.2%	24.4%

4 Discussion

The use of relative physics models throughout our new software has produced a robust breast composition measurement tool, Volpara™ and this has uses for a wide range of clinical work, including tailoring screening, but also for temporal comparison of images.

Entirely reasonably and realistically, given the quality regulations that apply to mammography, the algorithm still requires relatively accurate information on kV, target, filter and compressed breast thickness. Inevitably small errors will be present which will have an effect on breast density. Quality control and simple practical measures can be implemented to alert the user as to when their x-ray system or compression paddle needs recalibration.

Acknowledgments

We wish to thank Professor Per Skaane for his valuable insights and provision of some of the data used in this study.

References

1. Boyd, N., Guo, H., Martin, L., Sun, L., Stone, J., Fishell, E., Jong, R., Hislop, G., Chiarelli, A., Minkin, S., Yaffe, M.: Mammographic density and the risk and detection of breast cancer. New England Journal of Medicine 356(3), 227–236 (2007)
2. Boyd, N., Lockwood, G., Byng, J., Tritchler, D., Yaffe, M.: Mammographic densities and breast cancer risk. Cancer Epidemiol. Biomarkers Prev. 7, 1133–1144 (1998)
3. Harvey, J., Bovbjerg, V.: Quantitative assessment of mammographic breast density: relationships with breast cancer risk. Radiology 230, 29–41 (2004)
4. Wolfe, J.: Breast patterns as an index of risk of developing breast cancer. Am. J. Roentgenol. 126, 1130–1139 (1976)
5. Wolfe, J.: Risk for breast cancer development determined by mammographic parenchymal patterns. Cancer 37, 2486–2492 (1976)
6. Saftlas, A., Szklo, M.: Mammographic parenchymal patterns and breast cancer risk. Epidemiol. Rev. 9, 146–174 (1987)
7. Goodwin, P., Boyd, N.: Mammographic parenchymal patterns and breast cancer risk: a critical appraisal of the evidence. Am. J. Epidemiol. 127, 1097–1108 (1988)
8. Carney, P., Miglioretti, D., Yankaskas, B., Kerlikowske, K., Rosenberg, R., Rutter, C., Geller, B., Abraham, L., Taplin, S., Dignan, M., Cutter, G., Ballard-Barbash, R.: Individual and combined effects of age, breast density, and hormonal replacement therapy use on the accuracy of screening mammography. Ann. Intern. Med. 138, 168–175 (2003)
9. Buist, D., Porter, P., Lehman, C., Taplin, S., White, E.: Factors contributing to mammography failure in women aged 40-49 years. J. Nat. Cancer. Inst. 96, 1432–1440 (2004)
10. Saftlas, A., Hoover, R., Brinton, L., Szlo, M., Olson, D., Salane, M., Wolfe, J.: Mammographic densities and risk of breast cancer. Cancer 67, 2833–2838 (1991)
11. Wolfe, J., Saftlas, A., Salane, M.: Mammographic parenchymal patterns and quantitative evaluation of mammographic densities: a case-control study. Am. J. Roentgenol. 148, 1087–1092 (1987)
12. American College of Radiology, Breast imaging reporting and data systems (BIRADS). American College of Radiology, Reston, Va (1993)
13. Martin, et al.: Mammographic density measured with quantitative computer-aided method: comparison with radiologists' estimates and BI-RADS categories. Radiology 240(3), 656 (2006)
14. Nicholson, B., LoRusso, A., Smolkin, M., Bovbjerg, V., Petroni, G., Harvey, J.: Accuracy of assigned BIRADS breast density category definitions. Academic Radiology 13(9), 1143–1149
15. Diffey, J., Hufton, A., Astley, S.: A new step-wedge for volumetric measurement of mammographic density. In: Astley, S.M., Brady, M., Rose, C., Zwiggelaar, R. (eds.) IWDM 2006. LNCS, vol. 4046, pp. 1–9. Springer, Heidelberg (2006)
16. van Engeland, S., Snoeren, P., Huisman, H., Boetes, C., Karssemeijer, N.: Volumetric breast density estimation from full-field digital mammograms. IEEE Medical Imaging 25(3), 273–282 (2006)
17. Highnam, R., Brady, M.: Mammographic Image Analysis. Kluwer Academic, Dordrecht (1999)
18. Shepherd, J.: Automated volumetric density. In: 3rd International Workshop on Breast Density (2008)

19. Tromans, C., Brady, M., Van de Sompel, D., Lorenzon, M., Bazzocchi, M., Zuiani, C.: Progress toward a quantitative scale for describing radiodensity in mammographic images. In: Krupinski, E.A. (ed.) IWDM 2008. LNCS, vol. 5116, pp. 553–560. Springer, Heidelberg (2008)
20. Pawluczyk, O., Yaffe, M., Boyd, N., Jong, R.: Volumetric method for estimation of breast density on digitized screen-film mammograms. Medical Physics 30(3), 352 (2003)
21. McCormack, V., Highnam, R., Perry, N., dos Santos Silva, I.: Comparison of a new and existing method of mammographic density measurement. Cancer, Epidemiol. Biomarkers Prev. 16(16), 1148–1154 (2007)
22. Yaffe, M.: Measurement of mammographic density. Breast Cancer Research 10, 209 (2008)
23. Kovesi, P.: Image Features from Phase Congruency, Videre. MIT Press, Cambridge (1999)

Technical Evaluation of a Digital Breast Tomosynthesis System

Nicholas Marshall, Jurgen Jacobs, Lesley Cockmartin, and Hilde Bosmans

Department of Radiology, University Hospitals Leuven,
Herestraat 49, 3000 Leuven, Belgium
nicholas.marshall@uz.kuleuven.ac.be

Abstract. This paper presents results of a technical evaluation of a digital breast tomosynthesis (DBT) system. Projection images were used to assess x-ray focal spot size and lag as a function of detector air kerma (K) during a tomographic scan. Modulation transfer function (MTF), normalized noise power spectrum (NNPS) and detective quantum efficiency (DQE) were used to obtain a quantitative assessment of detector operation, also from the projection image data. Maximum 1st image lag was 4% at 20 mAs/projection. Both edge and wire methods were used to assess MTF; in the direction of tube travel, MTF was reduced by a factor of 0.33 compared to 2D FFDM (static tube). Detector response results showed that detector gain is increased by a factor of 3.6 for DBT operation compared to 2D FFDM mode. DQE measured at 28 kV W/Rh and 2 mm Al added filtration and a typical detector K of 23.8 μGy per projection was 0.5 at 0.5 mm^{-1}, indicating quantum limited performance of the detector for DBT acquisitions. The availability of projection image data enables the evaluation of important aspects of DBT detector performance in the field.

1 Introduction

Digital breast tomosynthesis (DBT) is a promising technique that hopes to improve the detectability of breast lesions compared to standard projection radiography by removing or at least suppressing the influence of overlying anatomical structure [1]. Overall acceptance of DBT systems into routine clinical service will depend on whether these systems can match or improve on the cancer detection rates of current projection mammography systems for the same patient dose burden. This information is usually provided by clinical trials and is very costly and time consuming to produce. Although far more limited in scope, technical evaluation provides data for comparing different DBT systems and for the eventual generation of Quality Assurance protocols. At the detector acquisition stage, we expect low lag for the detector, only a small effect on the system MTF from focal spot motion and quantum limited operation of the detector, especially at the low exposure per image used in DBT. For the tomographic images, we might expect good localization of contrast for small plane spacing (typically 1mm separation used) good low contrast transfer with regard to the MTF, and quantum limited statistics for the noise in tomographic volume.

J. Martí et al. (Eds.): IWDM 2010, LNCS 6136, pp. 350–356, 2010.

2 Materials and Methods

The system evaluated was a Siemens Mammomat InspirationTomo prototype (Siemens, Erlangen, DE) capable of performing standard projection mammography with/without antiscatter grid and tomosynthesis acquisition without antiscatter grid. This unit is based around an amorphous selenium (a-Se) detector of pixel pitch 0.085 mm, with a 24 cm x 30 cm field of view acquired into image matrix of 2816 x 3584 pixels. In tomographic mode, 25 projection images are acquired in a period of approximately 23s over an angular range of ±25°. Centre of rotation for the tube is 47 mm above the detector; focus detector distance is 65 cm.

2.1 Focal Spot Dimensions and Detector Lag

Dimension of the focal spot was assessed from the projection images using a multiple pinhole test object placed at the tube exit port. This consisted of a matrix of 15 x 0.050 mm diameter holes, each hole separated by 1 cm, drilled into a 0.3 mm thick brass plate. Detector lag was assessed from the projection images using two test objects; a) a 2mm thick lead ball and b) a lead sheet of dimension 24 cm x 30 cm and lead equivalent thickness of 2 mm with a 2 mm square hole cut at the centre. The compression plate tray (upper surface) was used to support each test object at a height of 7 cm above the breast support platform. The hole in the lead sheet generates a high signal in the x-ray detector and the tomographic motion causes the position of this signal to move across the detector during the scan.

2.2 Detector Response, MTF, NNPS and DQE from Projection Images

Radiation output was measured using a calibrated 1 cm^3 air chamber (RTI Ortigo, Sweden) as a function of mAs for a beam quality of 28 kV, W/Rh target filter setting and 2 mm Al added filtration. Pixel value (PV) from a 2 x 2 cm region of interest (ROI) positioned 6cm from the chest wall was plotted against detector air kerma (K); this was done for 2D projection mode and for the 0° projection images from a tomography sequence. In order to examine quantum limited operation of the detector, standard deviation measured for this ROI was plotted as a function of K and a power function fitted ($PV = a.K^b$).

Two methods were used to assess detector MTF, a version of the edge method [2] and the wire method. A 1 mm thick edge of dimension 12 cm x 6 cm was placed at the detector centre on the breast platform, angled at approximately 3° to the pixel matrix and an image was acquired at 28 kV, W/Rh with no additional filtration and grid removed. MTF was calculated from PV data extracted using a 5 cm ROI placed across the edge [3]. For the wire technique, a 20 µm thick tungsten wire was stretched across a wooden frame and placed on the breast platform. With this method applied to the projection image data, the tails of the line spread function (LSF) were extrapolated beyond the 2% point using an exponential curve fit applied to the LSF.

NNPS of the projection images was calculated using a standard method [3] from five flood images taken around the 0° projection (-3.64° to +4.18°), acquired at 28 kV, W/Rh and 2 mm added Al filtration. NNPS was assessed at a typical clinical operating level by using the AEC to set the total tomographic scan mAs. DQE was calculated from MTF, NNPS and number of x-ray photons using a standard formula [4].

2.3 MTF, z-Plane Contrast Sensitivity and NPS for the Reconstructed Tomographic Planes

The in-plane MTF for the tomographic planes was measured with a 20 μm thick tungsten wire, stretched across a wooden frame and placed at various positions above the breast support table. No extrapolation of the LSF tails was performed for these reconstruction images. Rather than measure MTF in the z-direction (vertically through the stack of planes), contrast sensitivity was measured using a 20 μm thick W wire positioned orthogonally to the direction of tube travel and at an angle of approximately 10.1° to the breast support table. Voxel variance for the stack of flood images was measured with a 2cm x 2cm ROI acquired through all the planes at the x-y centre. The in-plane NPS was then calculated from 128 ROIs of dimension 128 x 128 pixels taken from the centre plane from the stack; the NPS was not normalized by the input signal as a logarithmic transform is applied to the projection images before the reconstruction is performed [5]. This transformation yields the linear attenuation coefficients of the imaged objects used in the reconstruction; the NPS of the images after transformation is then inversely proportional to the mean of the projection image (*i.e.* the air kerma or mAs per image).

3 Results and Discussion

Focal spot dimension for the DBT mode measured at a point 6 cm from the chest wall was 0.32 mm in the direction of tube motion and 0.37 mm in the front-back direction. Exposure time per projection for the measurement was 154 ms. Tube travel is approximately 20 mm/s leading to 3.17 mm focus travel in one exposure and hence an effective focus length of 1.01 mm. By varying mAs, it was found that the system changed tube current in order to maintain an exposure time of 155 ms; this will ensure consistent blurring from the focus motion for every DBT scan.

Decay lag and build-up lag were measured as a function of detector exposure, ranging from 56 mAs total (2.2 mAs per projection) to 400 mAs (16 mAs per projection) and corresponding to an approximate range of 30 to 220 μGy per projection at the detector. Figure 1 shows relative signal for the decay lag (2 mm hole cut in lead sheet to generate signal - see Figure 1a(i)). Some slight overshoot is seen; there is little exposure dependency with a measured lag of approximately 3%, in agreement with published data [5]. Note that this method measures the largest exposure range, from direct exposure of the detector to a completely shielded region of the detector. The low lag demonstrated here is important if artifacts are to be avoided in the reconstructed images [6].

Detector response for both 2D FFDM and DBT mode was linear. Both systems have a PV offset of approximately 45; gradient of the detector response for DBT mode was a factor of 3.6 higher, at 12.1 compared to 3.5 for 2D FFDM. Fitting a power function to standard deviation plotted as a function of K for DBT mode gave a *b* coefficient of 0.51, indicating quantum noise limited operation of the detector for the projection images.

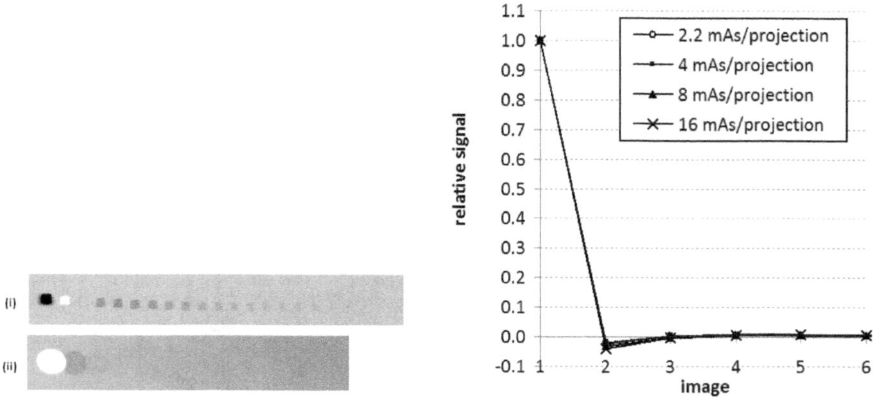

Fig. 1. a) Images used to evaluate lag: i) 2 mm hole in lead sheet and ii) lead ball placed in compression paddle. b) Graph of relative signal plotted as a function of image for the decay lag, with mAs per projection as a parameter.

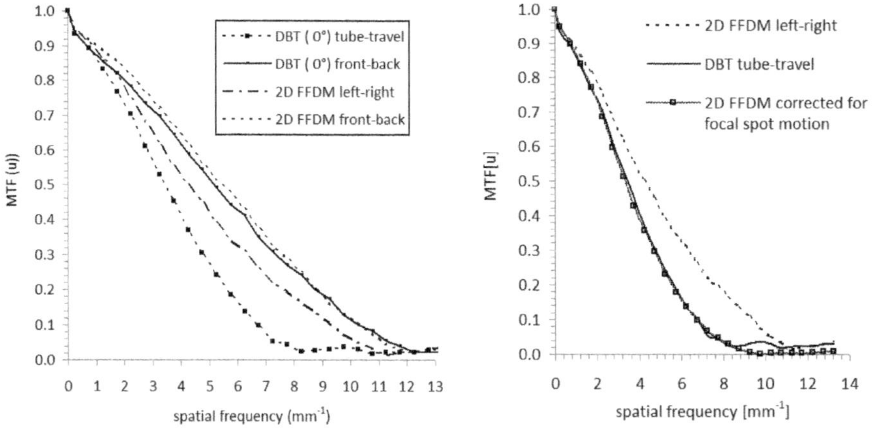

Fig. 2. a) MTF measured for 2D FFDM and DBT mode in the tube-travel and front-back directions across the detector. b) 2D FFDM MTF corrected for tube motion blurring along with the measured DBT MTF in the tube travel direction.

Figure 2a shows that there is some anisotropy of detector MTF calculated from the projection data, even in 2D FFDM mode. MTF is similar in the front back direction for both DBT and 2D FFDM modes, while some reduction is seen in the direction of tube travel for the DBT MTF. Figure 2b shows close agreement between the predicted effect of focal spot motion on the static 2D MTF and the measured result. Figure 3a plots the in-plane tomography MTFs measured with the W wire for the tube-travel and front-back directions, normalized to 1.0 at the respective peak values. These results are consistent with DBT MTF results for an earlier prototype of this system measured with a thin aluminium edge [7]. As expected, the MTFs are significantly different from the projection (detector) MTFs, with a strong reduction in MTF at low spatial frequencies due to the reconstruction algorithm and the limited angular range of the tomographic scan [7]. Contrast sensitivity in the z-direction (normal to the detector) measured with the 20 μm wire for a 1mm reconstructed plane spacing is plotted in figure 3b. Full width half maximum (FWHM) for this system is approximately 3.0 mm. This shows clearly that there is some spread of contrast between planes; small high contrast objects (typically <1mm) such as microcalcifications will appear in more than one plane.

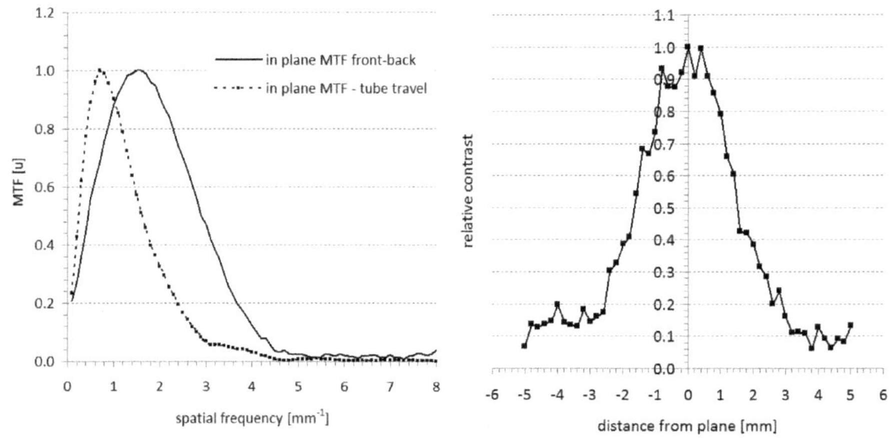

Fig. 3. a) In-plane MTF measured for the tomographic plane in the tube-travel and front-back directions. b) tomographic contrast sensitivity in the z-direction (normal to the detector) measured with a 20 μm W wire.

The AEC selected a total 147 mAs for 2 mm Al beam load, which is equivalent to 5.9 mAs and 24 μGy per projection. As might be expected, this is a significantly lower detector *K* than the range of 50 to 100 μGy typically used in 2D FFDM. At 24 μGy per image (projection), NNPS in 2D FFDM mode is a factor of 1.85 greater than for DBT mode, indicating the presence of electronic noise. This is shown in figure 4a.

Figure 4b plots DQE for 2D FFDM mode and DBT mode, calculated for the projection images. At a detector *K* of approximately 24 μGy, DQE in DBT mode is a factor of 1.8 greater than for 2D FFDM. In fact, DQE in DBT mode at 23.8 μGy is close to the value for 2D FFDM with the detector operated at 98 μGy, a result which confirms quantum limited operation of the DBT system even at low *K*. This is a result of the increased gain for DBT mode reducing the influence of electronic noise at a low detector *K*.

Finally, we can examine the in-plane NPS calculated from the central tomographic plane. Figure 5a plots a greyscale representation of the NPS with the vertical direction (90°) in this figure being the tube travel direction. The greyscale NPS shows the influence of the reconstruction filters on the noise; the noise power is not isotropic and shows clear reduction in noise at low spatial frequency. Figure 5b plots the NPS sectioned at 0° (front back direction) and 90° (tube travel direction).

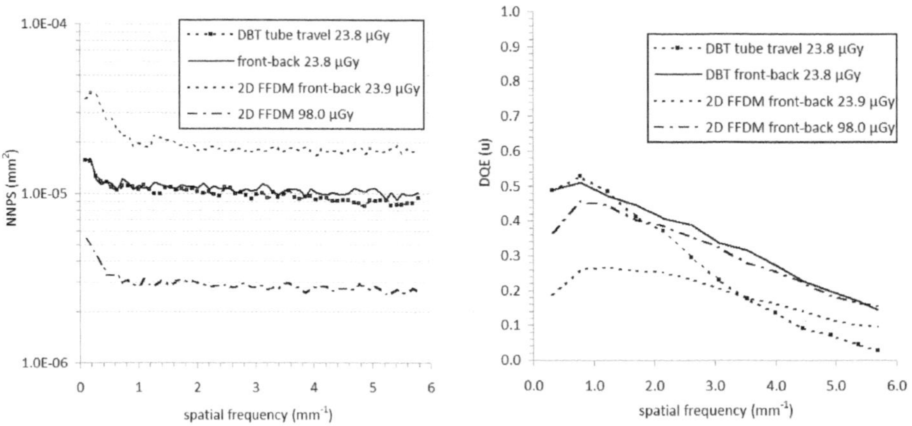

Fig. 4. a) NNPS measured for 2D FFDM mode and for the DBT system projection images acquired at 28 kV W/Rh and 2 mm Al added filtration b) DQE for 2D FFDM mode and for the DBT mode calculated from the projection images; 28 kV W/Rh and 2 mm Al added filtration

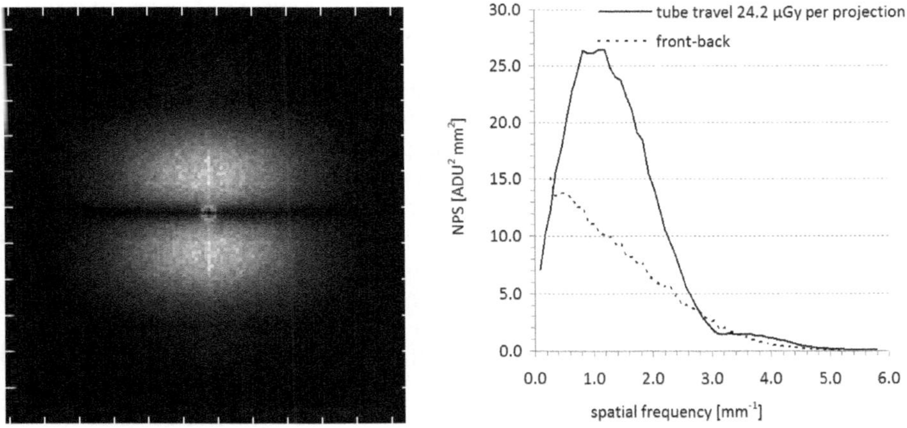

Fig. 5. a) Greyscale figure of the NPS measured for DBT tomographic images acquired at 28 kV W/Rh and 2 mm Al added filtration and 147 mAs for total scan b) Sectioned NPS at 28 kV W/Rh and 2 mm Al added filtration and 147 mAs for total scan

The influence of the tomographic MTF in the tube travel direction is clearly seen in the NPS results. Note that proper normalization of these results will be a further topic of study; this will enable comparison of NPS results acquired on different systems at different exposure levels [8]. The voxel variance offers a means of checking whether the noise in the tomographic image volume is quantum limited. We expect a reciprocal relationship between the noise from logarithmically transformed images and the exposure; alternatively, the exposure multiplied by the noise should be constant for quantum limited operation. The voxel variance for the flood images studied in this work did not follow this relationship – this was not expected as the noise in projection images from which the tomographic planes were reconstructed was shown to be quantum limited. This is a subject of further study.

Access to the projection data enables the measurement of focal spot dimension and focal spot motion blur, and detector lag. These results demonstrate the value of access to the projection image data in a standardized format, a position not currently guaranteed by all manufacturers. While good performance for these parameters is clearly desirable, it is less clear how we should evaluate the results for the reconstruction planes. One means might be via the visibility of different object types. Extension of the analysis presented here to other DBT systems should also help us to see the influence of these parameters for different designs and ultimately implement a practical protocol that usefully characterizes the imaging performance of these systems.

Acknowledgments. This work was partially funded by the BreastCT project.

References

1. Niklason, L.T., Christian, B.T., Niklason, L.E., Kopans, D.B., Castleberry, D.E., Opsahl-Ong, B.H., Landberg, C.E., Slanetz, P.J., Giardino, A.A., Moore, R., Albagli, D., DeJule, M.C., Fitzgerald, P.F., Fobare, D.F., Giambattista, B.W., Kwasnick, R.F., Liu, J., Lubowski, S.J., Possin, G.E., Richotte, J.F., Wei, C.Y., Wirth, R.F.: Digital Tomosynthesis in Breast Imaging. Radiology 205, 399–406 (1997)
2. Samei, E., Flynn, M.J., Riemann, D.A.: A method for measuring the presampled MTF of digital radiographic systems using an edge test device. Med. Phys. 25, 102–113 (1998)
3. NHSBSP (National Health Service Breast Screening Programme) 2009 "Calculation of Quantitative Image Quality Parameters - Notes Describing the Use of OBJ_IQ_reduced" Equipment Report 0902 ((NHSBSP Publications, Sheffield)
4. Dainty, J.C., Shaw, R.: Image Science: Principles, Analysis and Evaluation of Photographic-Type Imaging Processes. Academic Press, London (1974)
5. Zhao, B., Zhao, W.: Imaging performance of an amorphous selenium digital mammography detector in a breast tomosynthesis system. Med. Phys. 35, 1978–1987 (2008)
6. Mainprize, J.G., Wang, X., Yaffe, M.J.: The Effect of Lag on Image Quality for a Digital Breast Tomosynthesis System. In: Proc. of SPIE, vol. 7258, 72580R1–72580R12 (2009)
7. Zhao, B., Zhou, J., Hu, Y.-H., Mertelmeier, T., Ludwig, J., Zhao, W.: Experimental validation of a three-dimensional linear system model for breast tomosynthesis. Med. Phys. 36, 240–251 (2008)
8. Tward, D.J., Siewerdsen, J.H.: Cascaded systems analysis of the 3D noise transfer characteristics of flat-panel cone-beam CT. Med. Phys. 35, 5510–5529 (2008)

Mammographic Interpretation Training: How Could Low Cost Display Devices Help?

Yan Chen and Alastair Gale

Applied Vision Research Centre, Loughborough University, UK
y.chen3@lboro.ac.uk, a.g.gale@lboro.ac.uk

Abstract. This study examined how experienced and less experienced breast screening personnel performed when they examined DICOM screening cases in three conditions: on digital mammography workstations, a LCD monitor and an iPhone. In each condition they either viewed the full images unaided or used post-processing manipulations (HCI). For each case they reported features, rated their confidence on abnormality presence and classified the case. Their visual search behaviour was recorded as well as behavioural data. Additionally, their screening experience was derived from data on a national scheme as well as actual screening information. Both experienced and less experienced screeners performed best on the clinical workstation, however good performance was also demonstrated on the monitor using HCI, with iPhone performance being poor. Overall, results indicate that low cost devices could be used to provide additional tailored training as long as device resolution and HCI aspects are carefully considered.

Keywords: mammographic interpretation training, eye movements, image manipulation, iPhone.

1 Introduction

Breast screening has been undertaken across the UK for over 20 years using mammographic film as the imaging medium [1]. Ongoing developments will see the age range being increased from the current range of 50 – 70 years [2] to encompass women aged 47 - 73 years by 2012 [3]. Both to cope with the additional workload resulting from this increase in screening age range as well as to aid in screening younger women who tend to have denser breasts then Full Field Digital Mammographic (FFDM) imaging is being rolled out nationally. It is planned that in 2010 all UK screening centres will have some digital imaging ability with full digital imaging gradually ensuing as screening using a film medium eventually ceases [3]. Such a drastic change in imaging technique introduces several related changes concerning image acquisition, image storage, and image viewing. Of importance here is the image interpretation as this is a key issue as the soft copy reporting of digital images requires current screening personnel to be trained further in examining these images as their appearances differ to mammographic film. Digitally displayed images have

J. Martí et al. (Eds.): IWDM 2010, LNCS 6136, pp. 357–364, 2010.

limited resolution which is offset by the ability of the health professional to use post processing techniques (such as adjusting window, contrast level and zoom) – these tools are here termed HCI. Furthermore, the expanded screening age range means an increased workload potentially necessitating further new screening staff being trained.

1.1 Training

Current training in image interpretation in breast screening comprises the availability of a number of training courses run by UK breast screening training centres coupled with other approaches such as the PERFORMS scheme [4]. A recent survey of screening centres [5] elucidated that over 80% of radiologists and technologists reported limited time for training with over 20% citing limited access to roller viewers (for mammographic film interpretation training) and almost a similar percentage reporting limited access to a digital workstation (where digital imaging was available). Overwhelmingly, individuals positively reported a need for dedicated training when and where it suited them. Therefore if other image display systems can be used to present mammographic images appropriately for training purposes then it would appear that there is a demand for this from health professionals.

Training for current and future reporting staff should ideally be undertaken on the clinical workstations themselves but this is not always possible as such workstations are in demand for clinical usage. Consequently this study is part of a series of investigations which examine the use of other display devices for training purposes in breast screening. It is accepted that for making screening or diagnostic decisions then mammographic workstations are de rigour.

Specifically, how screeners performed with smaller lower resolution displays was examined as compared to their performance on a clinical workstation. Furthermore their ability in doing this was studied by monitoring their visual search behaviour and interaction behaviour with the display devices. Additionally, task performance was related to participants' data from actual breast screening as well as a national self assessment scheme to examine the role of screening experience.

2 Method

An expert radiologist selected two sets of recent screening cases (20 cases in each set including both MLO and CC views) which demonstrated challenging examples of normal, benign and malignant appearances (features included: masses; calcification, and architectural distortion). Then 14 radiologists and advanced practitioners (technologists who read screening cases) from two UK screening centres undertook three rounds of trials (the whole process spanned over eight months) and examined these DICOM cases on GE digital mammography workstations (dual-monitors), a single standard LCD monitor (using a DICOM viewer, screen size: 21.5", resolution: 1680 x 1050) and an iPhone (using Osirix software, screen size: 3.5", resolution: 480 x 320) respectively (Fig 1).

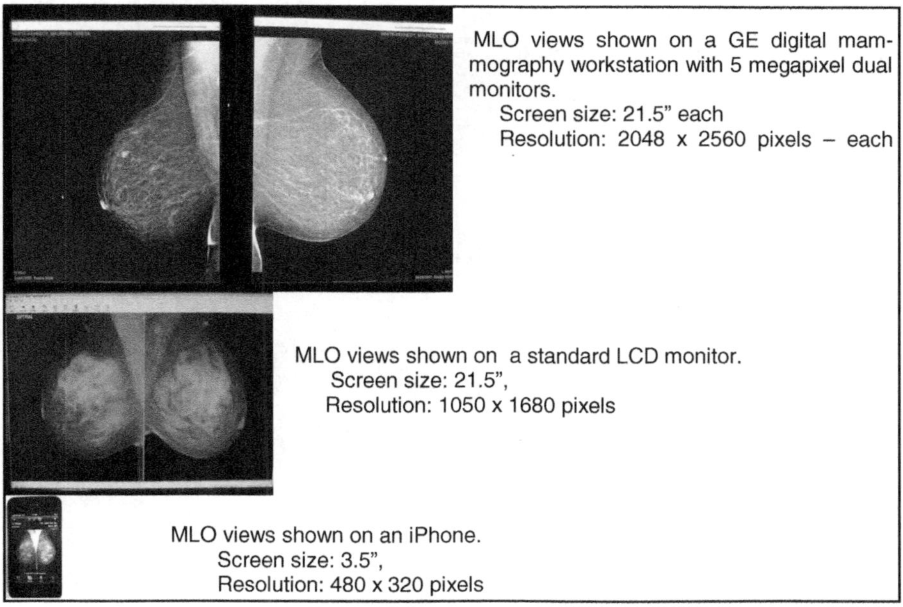

Fig. 1. Illustration of the respective sizes of the three display systems

For half the cases each individual was only allowed to view the full two view cases unaided and for the other half they could use post-processing image manipulations (HCI) – namely zoom, pan and window level/width adjustment. The order of examining images across the centres was count-balanced and there was a gap of at least a month between each trial at each centre. Ambient lighting levels were controlled.

A head mounted eye tracker (ASL 504) was used to record their visual search behaviour throughout (Fig. 2). The viewing distance was 55-65 cm depends on each individual. They were additionally videotaped using a camera to monitor interaction usability. For each case, participants reported if it was normal or abnormal, specified mammographic features, rated their confidence of abnormality presence, classified the case and reported its density.

The performance of each individual was treated anonymously and then related to their known recent performance in the PERFORMS self assessment scheme (where each UK screener reports on a set of difficult exemplar screening images) as well as their known real life performance data from everyday clinical screening to ascertain their screening experience. This was so that the study utilised individuals with various experience levels as one aim was to determine if different display devices were useful for a range of experienced and less experienced screeners [7].

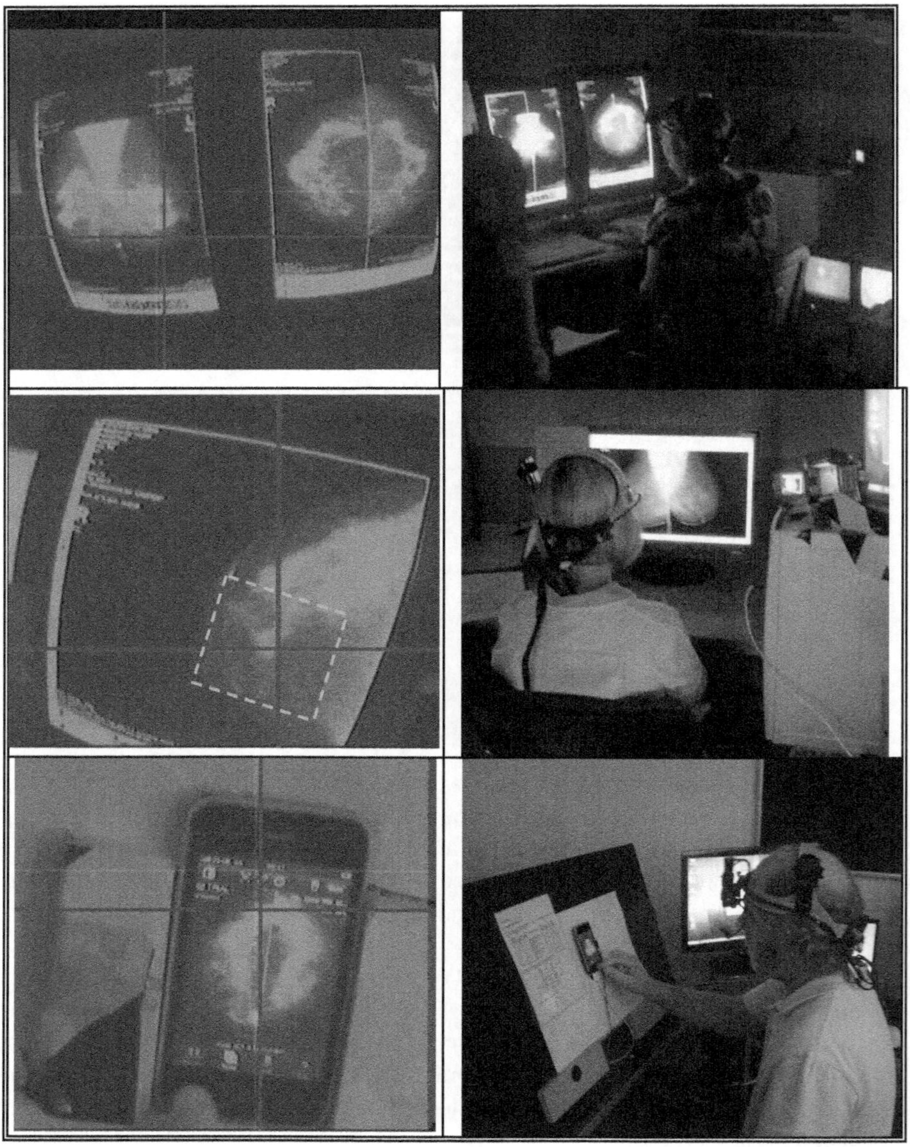

Fig. 2. top row: eye movement record, examples of magnification on the right breast of the MLO views and the experimental setting for the workstation task; Middle row: eye movement record, examples of magnification on the right breast of the CC view and the experimental setting for the LCD monitor task; Bottom row: eye movement record, examples of changing window level/width on the left breast of the CC view and the experimental setting for the iPhone task. (Apparent different ambient lighting conditions are for photographic purposes).

3 Results

In terms of overall performance, clearly the workstation proved to be better, however participants were able to identify abnormalities on the monitor with the support of image manipulation tools just as well as on the workstation. Data were analysed using JAFROC (6) and the Figure-Of-Merit (FOM) values of the workstation with no image manipulation, workstation with image manipulation, and standard monitor with image manipulation were found to be comparable; 0.838, 0.816, and 0.827 respectively. Performance on the iPhone was generally poor, with the FOM of with/without image manipulation being 0.529 and 0.42 respectively (see Fig. 3).

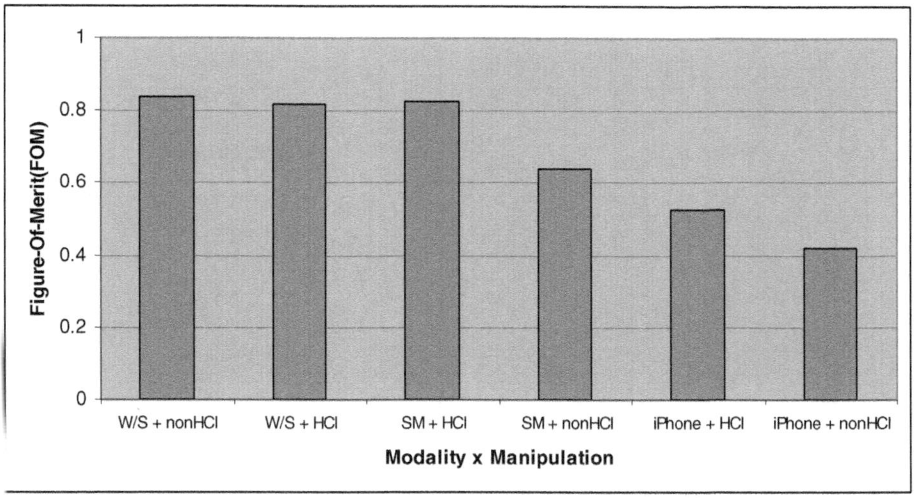

Fig. 3. Figure-Of-Merit (FOM) values on workstation (W/S), standard monitor (SM) and iPhone with (HCI) and without HCI (nonHCI)

On the workstation, performance was not significantly different whether participants manipulated the images or simply viewed them without making any alterations to zoom, window level, etc. Data were evaluated using Receiver Operating Characteristic (ROC) analysis to compare participants' average performance between examining images with, and without, using the image manipulation tools. The area under the ROC curve (A_z) and the trend of the results (Fig. 4) shows that participants performed better when using image manipulation tools (A_z = 0.981) than when examining images without these tools (A_z = 0.951).

The workstation data were also compared between different features types (namely:- Architectural Distortion [AD], Masses, Calcification and Normal cases containing no such features - Fig. 5) in the two conditions. Performance between using or not using the image manipulation tools to identify Masses and AD was about the same. Although there was no significant difference between with/without image manipulation tools to identify micro-calcification, manipulation was found to be slightly better even though it was not significantly so (p>.05, r=.125). Furthermore, using image manipulation led to raising the recall rate, although this was not significant (p>.05, r=.100).

Performance on the standard LCD monitor and iPhone was different to this for certain mammographic features in that whilst masses could be identified sometimes without manipulation, calcifications could generally only be identified when the images were manipulated. Altering the images on the iPhone using the Osirix software was perfectly feasible but due to the large image sizes and the need to move between the four images of each case this proved to be troublesome to some participants and took time.

Additionally, a relationship was found between experience and visual search behaviour with the experienced participants using longer saccadic eye movements and longer fixations in fewer image locations which mirrors the findings in other medical image domains. This also illustrates the potential of utilising visual search data in mammographic interpretation training.

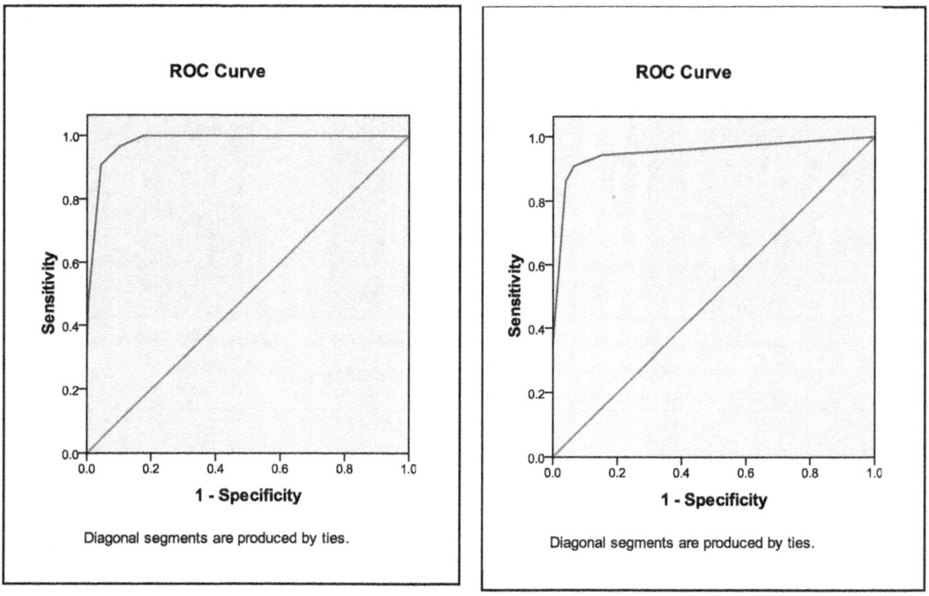

Fig. 4. ROC curves of workstation performance with HCI (left) and without (right)

Recording visual search behaviour on the workstation and monitor was a feasible and useful technique which yielded detailed information on how participants went about assessing the images to find abnormalities. Whilst this was also possible on the iPhone the actual resolution of the ASL system coupled with the small image size meant that the resulting data had to be interpreted with considerable caution.

3.1 Clinical Applications

Overall, the superior workstation performance findings were as expected although it was predicted that using HCI tools would have improved performance. The main finding that using an office monitor with HCI tools could produce similar performance to a

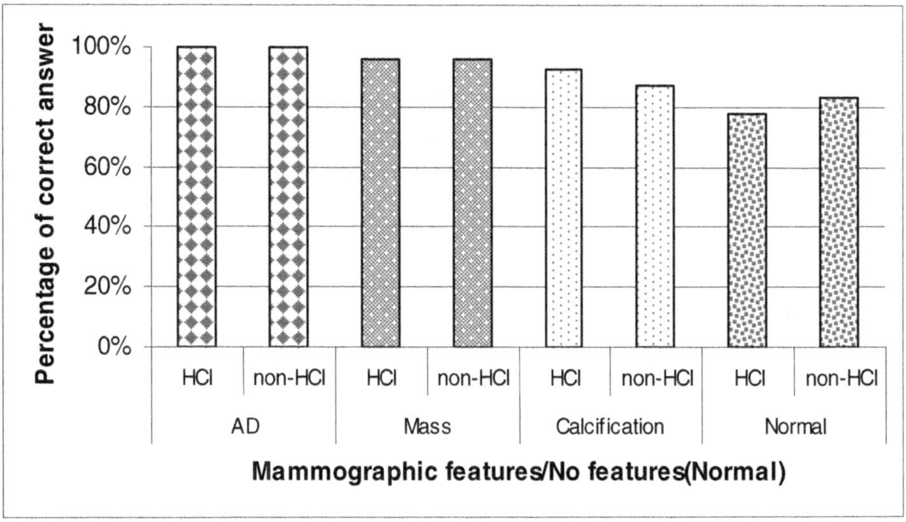

Fig. 5. Workstation performance in correctly identifying key features, with and without HCI usage

clinical workstation was somewhat surprising. These test images were chosen as difficult exemplars demonstrating a range of mammographic features and were thought would be challenging for examination on a monitor. Whilst we would strongly not advocate translating this finding to mean that such monitors could be used for clinical reporting it is evidence to support their use in tele-radiology applications and also as a 'second read' where a clinical workstation is not available. It is also strong support for their use to deliver training in interpreting such digital images. The participants here were all familiar with reporting FFDM images on workstations using DICOM viewers but were unfamiliar with the particular DICOM viewer used here yet they quickly and easily mastered its use.

Small hand held devices such as the iPhone are increasingly being found to be useful in reporting relatively small radiological images (e.g. in CT). Its use here was found to be poor, mainly due to the need to zoom in to the image in order to perceive small mammographic features which then led to participants having difficulty both in encompassing the whole image and remembering where they were within the image.

4 Conclusions

Superior performance was attained using the clinical workstations, however participants were able to identify abnormal features on both the LCD monitor and iPhone, particularly when they were able to manipulate the images. Using the lower resolution monitor with HCI was surprisingly good, equivalent here to the workstations, but the iPhone and its software implementation examined here entailed extensive image manipulation for these large images which inevitably took time and produced poor results.

It is argued that lower resolution displays, such as the LCD display, are therefore potentially very useful for training purposes as long as the user is aware of the limitations of the display and appropriate manipulation software is used. Improved software (tailored specifically for digital mammographic images) for the iPhone would render it much more useful in examining such large images.

Although high-performance display devices are clearly needed diagnostically, it is then possible to use lower quality devices for training purposes. Whilst such usage will not meet every training need they do provide an additional adjunct.

Acknowledgements

This work is partly supported by the UK National Health Service Breast Screening Programme.

References

1. Patnick, J. (ed.): Celebrating 20 years of screening. NHS Cancer Screening Programmes, Sheffield (2008)
2. Patnick, J. (ed.): NHS Breast Screening Programme Annual Review 2009. NHSBSP, Sheffield (2009)
3. Department of Health, Cancer Reform Strategy, Department of Health, London (2007)
4. Chen, Y., Gale, A.G., Scott, H.J.: Mammographic interpretation training in the UK: current difficulties and future outlook. In: Sahiner, B., Manning, D.J. (eds.) SPIE Medical Imaging 2009: Image Perception, Observer Performance, and Technology Assessment, vol. 7263, pp. C1 – C10 (2009)
5. Gale, A.G., Scott, H.: Measuring Radiology Performance in Breast Screening. In: Michell, M. (ed.) Contemporary Issues in Cancer Imaging – Breast Cancer. Cambridge University Press, Cambridge (2010)
6. Chakraborty, D.P.: Jackknife Free-Response Receiver Operating Characteristic Analysis Software, [computer software], Available at: `http://www.devchakraborty.com` [accessed March 10, 2010]
7. Chen, Y., Gale, A.G., et al.: Breast Screening: visual search as an aid for digital mammographic interpretation training. In: Proceeding of SPIE (in press, 2010)

A Survey of Patient Doses from Digital Mammography Systems in the UK in 2007 to 2009

Jennifer M. Oduko[1], Kenneth C. Young[1,2], and Anna Burch[3]

[1] NCCPM, Medical Physics Department, Royal Surrey County Hospital, Egerton Road, Guildford GU2 7XX, UK
[2] Physics Department, University of Surrey, Guildford GU2 7XH, UK
[3] Medical Physics Department, Breast Test Wales, 18 Cathedral Road, Cardiff CF11 9LJ, UK
jenny.oduko@nhs.net, ken.young@nhs.net, anna.burch@velindre-tr.wales.nhs.uk

Abstract. Patient dose data from across the UK were collated, and the information relating to full-field digital mammography systems were analysed, and compared with overall results for film-screen systems. For CR systems, the average mean glandular doses was 2.19 ± 0.07 mGy which was similar to the average for film screen systems 2.15 ± 0.01 mGy. The average patient dose for DR systems was 1.46 ± 0.02 mGy approximately 32% lower than for film screen systems. When different DR systems were compared, the Sectra MDM L30 and the Siemens Novation and Inspiration had the lowest average mean glandular dose, at 0.95 ± 0.02 mGy, 1.16 ± 0.05 mGy and 1.21 ± 0.07 mGy, respectively. It was shown that for DR systems the MGD to the standard breast was broadly correlated with the average MGD for oblique views of 50-60mm thick breasts, while the correlation for CR systems was much lower.

Keywords: patient dose, digital mammography.

1 Introduction

In the UK a nation-wide breast screening programme (NHSBSP) is run by the National Health Service (NHS), and since its inception it has operated a quality system, to ensure optimum performance. As part of the physicists' and radiographers' quality control programmes, patient dose data is collected at least once every three years, for each X-ray set, for a sample of fifty or more women. Although not part of the NHSBSP, many NHS and private hospitals also record the same patient dose data, which increases the pool of information available on doses in both digital and film-screen mammography. The data have been analysed and the results published at intervals [1, 2]. About 20% of the mammography X-ray systems in the NHSBSP are digital at present. The data for digital systems are analysed and presented in this paper, and compared to the average for film-screen systems.

J. Martí et al. (Eds.): IWDM 2010, LNCS 6136, pp. 365–370, 2010.

2 Method

Patient dose data from physics services working with the NHSBSP were gathered, and combined into a single database for analysis. The dose surveys had been carried out in the years 2007, 2008 and 2009. The exposure parameters for individual patients were supplied to the physics services by the radiographers who performed the mammography examinations. The physics services provided technical information from their own regular equipment surveys, (including mean glandular dose (MGD) to the standard breast for each system, and X-ray output and half-value layer at each kV, target, filter combination occurring in the patient dose data). Each physics service used the same programme to record the data and to calculate the MGD for real breasts, according to Dance et al [3].

The data were analysed in detail to provide information about patient doses in direct digital radiography (DR) and computed radiography (CR) systems in use in the UK.

3 Results

3.1 MGD for Oblique Views, for All Breasts, for Different Systems

The numbers of images and digital systems for which data was recorded, and the mean MGD, for oblique (OB) views, and mean compressed breast thickness, are shown in Table 1 for DR systems and Table 2 for CR systems. The data in these tables are averages for the oblique films for all breast thicknesses.

Table 1. MGD and thickness for OB views, for all breasts, for different types of DR system and film-screen for comparison

Manufacturer and model	Number of systems	Number of main images	Mean MGD to breast (mGy) ± 2 SEM	Mean thickness (mm) ± 2 SEM
GE 2000D	2	200	1.30 ± 0.04	57.0 ± 0.8
GE DS	7	1139	1.59 ± 0.03	53.9 ± 0.3
GE Essential	4	805	1.44 ± 0.03	58.1 ± 0.4
Hologic Selenia	3	356	2.00 ± 0.07	53.1 ± 0.5
Hologic Selenia W	3	616	1.44 ± 0.04	52.2 ± 0.8
IMS Giotto	1	118	1.78 ± 0.10	55.5 ± 0.9
Sectra MDM L30	4	316	0.95 ± 0.02	63.5 ± 0.6
Siemens Inspiration	1	128	1.21 ± 0.07	58.8 ± 1.1
Siemens Novation	3	483	1.16 ± 0.05	56.9 ± 0.6
Total/mean (DR)	28	4161	1.46 ± 0.02	55.9 ± 0.5
Average film-screen	454	36814	2.15 ± 0.01	56.7 ± 0.14

Table 2. MGD and thickness for OB views, for all breasts, for different types of CR system and film-screen for comparison

Manufacturer and model	Number of systems	Number of main images	Mean MGD to breast (mGy) ± 2 SEM	Mean thickness (mm) ± 2 SEM
Fuji Profect	5	713	2.20 ± 0.10	54.3 ± 1.1
Agfa CR 35-X	3	208	2.20 ± 0.12	56.6 ± 1.7
Konica CP-1M	1	190	2.08 ± 0.09	55.4 ± 1.8
Kodak EHR-M2	2	68	2.33 ± 0.24	49.3 ± 3.4
Total/mean (CR)	11	1179	2.19 ± 0.07	54.1 ± 0.8
Average film-screen	454	36814	2.15 ± 0.01	56.7 ± 0.14

3.2 MGD for Oblique Views of 50-60mm Thick Breasts for Different Systems

Calculating the average MGD for oblique views, for compressed breast thickness 50-60mm, facilitates comparison between different digital systems. Results are shown (and compared with results for film-screen systems) for DR systems in Table 3 and for CR systems in Table 4. The percentage distribution of mean MGD values (for all breasts) with breast thickness is shown in Figure 1, for both DR and CR systems.

Table 3. MGD and thickness for OB views, for 50-60mm breasts, for different types of DR system and film- screen for comparison

Manufacturer and model	Number of systems	Number of main images	Mean MGD to breast (mGy) ± 2 SEM	Mean thickness (mm) ± 2 SEM
GE 2000D	2	50	1.28 ± 0.06	55.0 ± 0.8
GE DS	7	331	1.62 ± 0.06	55.4 ± 0.3
GE Essential	4	212	1.37 ± 0.04	55.6 ± 0.4
Hologic Selenia	3	103	1.88 ± 0.10	54.5 ± 0.7
Hologic Selenia W	3	215	1.52 ± 0.07	55.0 ± 0.5
IMS Giotto	1	41	1.73 ± 0.08	56.3 ± 0.9
Sectra MDM L30	4	101	0.89 ± 0.03	55.5 ± 0.6
Siemens Inspiration	1	39	1.06 ± 0.07	55.3 ± 1.1
Siemens Novation	3	132	1.16 ± 0.07	55.3 ± 0.5
Total/mean (DR)	28	1224	1.44 ± 0.03	55.3 ± 0.2
Average film-screen	454	11141	1.98 ± 0.01	55.3 ± 0.06

Table 4. MGD and thickness for OB views, for 50-60mm breasts, for different types of CR system and film- screen for comparison

Manufacturer and model	Number of systems	Number of main images	Mean MGD to breast (mGy) ± 2 SEM	Mean thickness (mm) ± 2 SEM
Fuji Profect	5	215	2.10 ± 0.13	55.6 ± 0.5
Agfa CR 35-X	3	70	2.28 ± 0.21	56.1 ± 0.8
Konica CP-1M	1	76	1.99 ± 0.06	55.7 ± 0.7
Kodak EHR-M2	2	24	2.51 ± 0.26	53.5 ± 1.4
Total/mean (CR)	11	385	2.14 ± 0.08	55.4 ± 0.3
Average film-screen	454	11141	1.98 ± 0.01	55.3 ± 0.06

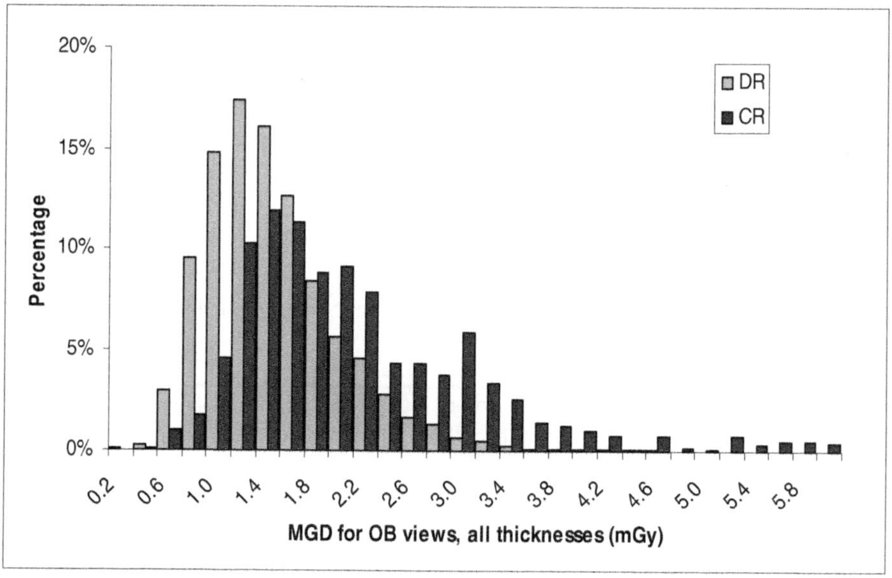

Fig. 1. Distribution of MGDs for oblique views, for all breast thicknesses, for DR and CR systems

3.3 Comparison of Average MGD for 50-60mm Breasts with MGD to the Standard Breast

For each DR and CR system, the average MGD for oblique views of breasts of compressed thickness 50-60mm was plotted against the MGD to the standard breast for that system. The standard breast is represented in measurements by a 45mm thickness of PMMA, equivalent to 53mm thickness of typical breast tissue. The results are shown in Figures 2 and 3 for DR and CR systems respectively. Also shown on Figures 2 and 3 are the national diagnostic reference level (NDRL) of 3.5mGy for the average MGD for 50-60mm breasts, and the remedial value of 2.5mGy for dose to the standard breast.

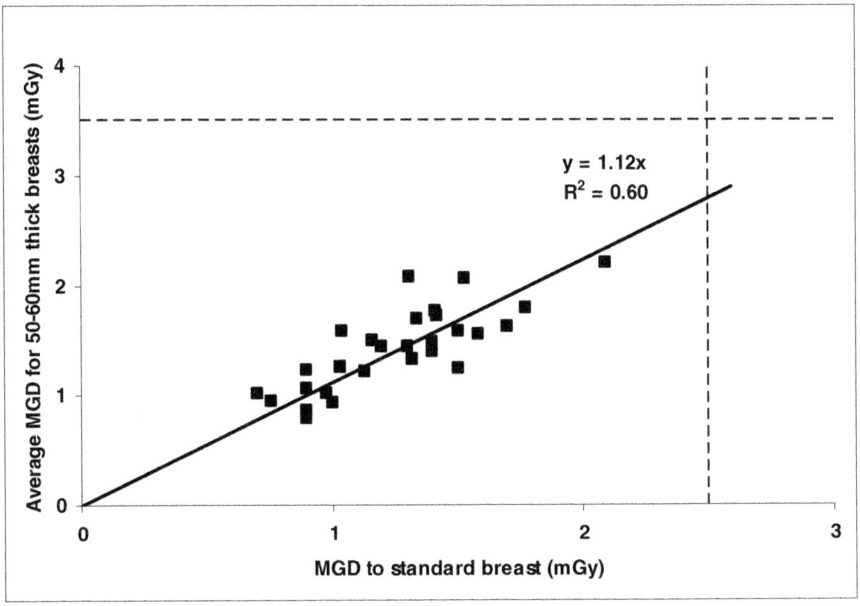

Fig. 2. Average MGD for 50-60mm breasts (oblique view) plotted against MGD to the standard breast for each DR system

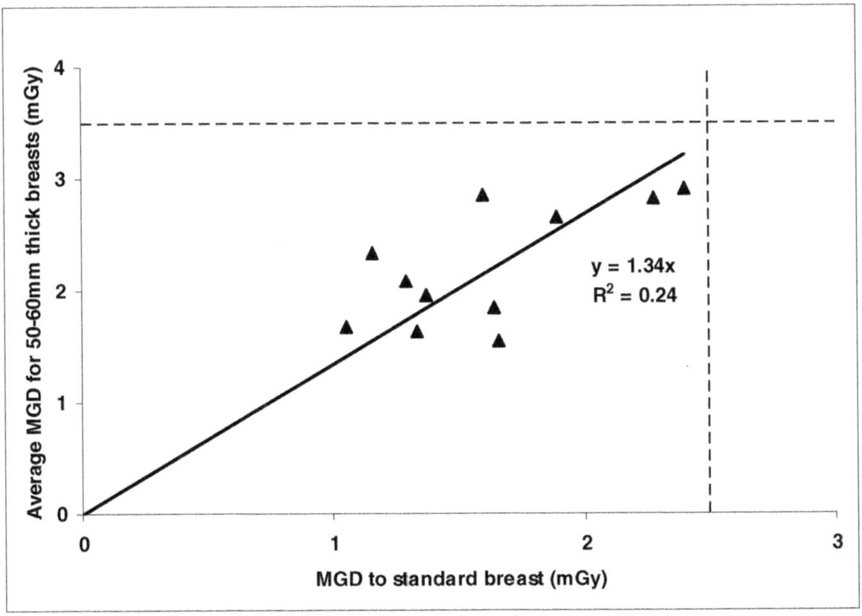

Fig. 3. Average MGD for 50-60mm breasts (oblique view) plotted against MGD to the standard breast for each CR system

4 Discussion

The average MGD for oblique views is about 1.5mGy for breasts imaged with DR systems, and approximately 50% higher at 2.1mGy for breasts imaged with CR systems (Tables 1 to 4). These values are almost the same for the whole population studied and for the selected thickness range of 50-60mm. The average MGD for DR systems is about 30% lower than the average MGD for film-screen mammography, while for CR systems it is 2% and 8% higher, for all breasts and 50-60mm thick breasts respectively. This compares with a 22% average dose saving for digital mammography, reported for the DMIST trial [4]. As expected, the mean compressed breast thickness is close to 55mm for all groups.

Tables 1 and 3 compare the average MGD for a range of DR systems. For 50-60mm thick breasts, the lowest MGDs are 0.89mGy for the Sectra MDM L30, and 1.06mGy for the Siemens Inspiration and Novation (the corresponding figures for all breasts are 0.95mGy for Sectra MDM L30 1.2mGy for the Siemens Inspiration and Novation. The average MGDs for the CR systems presented in Tables 2 and 4 are similar to those found for film screen systems. Systems cannot be ranked on dose alone without considering image quality, so that a comparison based on, for example, dose required to reach achievable image quality [5] is also relevant. Figure 1 shows the distribution of MGDs for DR and CR systems, with the CR results generally at higher values, as expected from the average MGDs noted above.

Figure 2 shows that for DR systems, MGD to the standard breast and average MGD to 50-60mm breasts (oblique views) are broadly correlated, with a line of slope 1.12 fitted to the points. For the CR systems, shown in Figure 3, the correlation is poorer. None of the systems exceed the national DRL or the remedial value shown in these figures.

Acknowledgments. The data collection was conducted under the auspices of the UK National Breast Screening Quality Assurance Coordinating Group for Physics. The authors are grateful to the NHSBSP radiographers who recorded data in the breast screening centres, and the physicists involved in the NHSBSP who provided technical data relating to the systems.

References

1. Young, K.C., Burch, A.: Radiation doses received in the UK Breast Screening Programme in 1997 and 1998. Br. J. Radiol. 73, 278–287 (2000)
2. Young, K.C., Burch, A., Oduko, J.M.: Radiation doses received in the UK Breast Screening Programme in 2001 and 2002. Br. J. Radiol. 78, 207–218 (2005)
3. Dance, D.R., Skinner, C.L., Young, K.C., Beckett, J.R.: Additional factors for the estimation of mean glandular breast dose using the UK mammography dosimetry protocol. Phys. Med. Biol. 45, 3225–3240 (2000)
4. Hendrick, R.E., Pisano, E.D., Averbukh, A., Moran, C.: Comparison of acquisition parameters and breast dose in digital mammography and screen-film mammography in the American College of Radiology Imaging Network digital mammographic imaging screening trial. Am. J. Roentgenol. 194, 362–369 (2010)
5. Young, K.C., Oduko, J.M.: Technical Evaluation of the Hologic Selenia Full Field Digital Mammography System with a Tungsten Tube. NHSBSP, Sheffield (2008)

Analysis of Mammography Quality Control Results: Evidence for a Change in Test Frequency?

Koen Michielsen, Jurgen Jacobs, Nicholas Marshall, and Hilde Bosmans

Department of Radiology, University Hospitals Leuven,
Herestraat 49, 3000 Leuven, Belgium
koen.michielsen@uzleuven.be

Abstract. We investigated the link between the results from daily quality control (QC) data, on the one hand, and from successive half-yearly medical physics QC visits on the other hand. To do so, criteria defining 'unchanged results' were set for both the daily QC data and the half-yearly QC results. Out of the results of 181 QC visits and daily QC at the time of the QC visit, we found that systems that showed little or no deviation in daily constancy performance were likely to show basically unchanged results in the half-yearly medical physics test. The specificity of the daily QC data for a change in results at QC visits was 93% for the dose and SDNR test and 88% for a change in contrast threshold values. Daily QC data could be used to prioritize the work load of the medical physicist.

Keywords: Mammography, Quality control, Daily, Half-yearly.

1 Background

Screening mammography is the subject of substantial quality control procedures. In Europe most are based on the European protocol for the quality control of the physical and technical aspects of mammography screening [1]. These guidelines contain procedures to ensure the quality for both radiological equipment and viewing devices on two levels: (1) at acceptance, the physicist verifies whether a minimum quality level is achieved and this is repeated at half-yearly intervals; (2) constancy tests: regular quality control procedures designed to guarantee that quality is maintained at every moment during the screening actions. The latter tests should be performed by the local personnel and supervision by the physicist is encouraged.

In Flanders, mammography screening is decentralized, allowing each participating center to buy any approved mammography system of their choice, and free to choose which certified quality control reference center to work with. The result of this situation is that we have a large variety of mammography systems in our network, spread over the whole of Flanders.

In our medical physics practice, we opted for a rigorous implementation and centralized supervision of the local QC procedures: every day, radiographers acquire 2 flat-field images using the automatic exposure control (AEC), after which an automatic procedure retrieves and analyses these images and sends the results to our center via e-mail or ftp. The methods and analyses are described in more detail by Jacobs et al [2]. Viewing stations are tested using the variable MoniQA pattern [3].

J. Martí et al. (Eds.): IWDM 2010, LNCS 6136, pp. 371–377, 2010.

This work investigates the link between the results from daily QC data, on the one hand, and from successive half yearly medical physics QC visits on the other hand. More specifically, the hypothesis that systems that show little or no deviation during constancy checks would also show basically unchanged results in the half-yearly quality control was tested. The present study was performed in the frame of optimization of the physics effort to guarantee quality: could the frequency of visits be reduced if the physics team decided to put effort in rigorous daily QC supervision? Can daily QC be used to prioritize the work of the physicist?

2 Method

Results of medical physics QC visits of all digital mammography systems in our network were collected into a database for present analysis. From all available data, the mean glandular dose (MGD) and signal-difference to noise ratio (SDNR) for exposures of 2, 3, 4, 5, 6 and 7 cm of polymethyl-methacrylate (PMMA) with a small aluminum disk of 0.2 mm thickness were selected, along with the contrast-detail results. Results from tests of the x-ray tube, the detector response curve and the short term reproducibility were not included in the further analysis because there is not a single instance of these tests failing in any of our QC visits and the values have never given rise to any discussion or optimization action. Therefore adding them would have no influence on the results of this study.

The first QC acceptance tests were performed in November 2006. At present 63 digital mammography units are included in our network. A total of 50 systems from 6 vendors were included in the study. These are 16 computed radiography (CR) systems, all Profect CS systems by Fuji, and 34 direct radiography (DR) systems. These last were divided between 16 Siemens Mammomat Inspiration systems, 4 Siemens Mammomat Novation, 7 GE Senographe Essential, 2 GE Senographe DS, 6 Hologic Lorad Selenia, 2 Sectra L-30 MDM and 1 IMS Giotto Raffaello system.

The first 3 systems were installed and accepted for screening in 2006, 20 in 2007, 18 in 2008 and 9 in the first half of 2009 (systems installed after this time had the second test too late to be included in present study). One of the systems is located in a mobile screening unit, 13 are located in private radiological practices, 4 in polyclinics, 22 in regional hospitals and 10 in tertiary care centers, as shown in figure 1.

For each physics QC visit, two weeks of daily quality control (DQC) results (one week either side of the QC visit) were summarized into a representative value for further comparison. To do this, the kVp, mAs and SNR in the reference point were averaged. The DQC results were declared as 'unchanged' between these two points if the following criteria were met: no change in anode/filter combination, a maximum change of the average nominal kVp of 1.0, less than 5% change in SNR, less than 10% change in mAs and no change of detector ID.

To compare the results of successive QC visits, the relative change in the parameters from the previous QC visit was calculated. For MGD and SDNR the median value of the changes over the complete thickness range (2 to 7 cm of PMMA) was used. System performance between two half-yearly QC tests was declared 'unchanged' if both SDNR did not decrease by more than 10% and MGD did not increase by more than 10%.

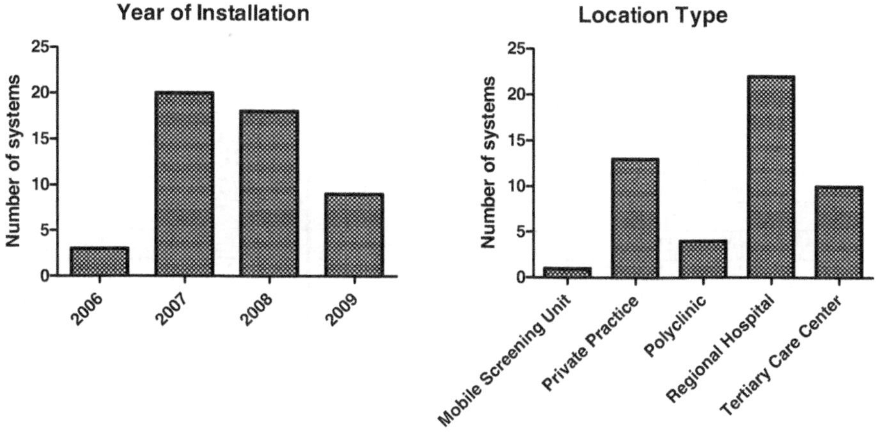

Fig. 1. Overview of installation date and location type

From the contrast-detail analysis, threshold gold thickness for the 0.1 mm detail was retrieved, as this is the crucial factor in determining the limits for the SDNR. These values were obtained with a computerized method applied on sets of at least 8 images [4]. When comparing contrast-detail results, only data acquired with the same phantom in the two consecutive tests were used. A system was declared as unchanged between two half-yearly QC tests if the contrast threshold did not increase by more than 15%.

QC-visit/DQC pairs for which one of the data sets was incomplete were withheld from the analysis. The remaining data pairs were compared with the previous results in order to ascertain whether 'unchanged' data from DQC would give rise to an 'unchanged' result from QC visits based upon SDNR and MGD on the one hand and contrast threshold gold thicknesses on the other hand.

3 Results

There are 129 complete combinations of half-yearly QC visit results and DQC parameters from all mammography systems. There are 74 combinations where the same contrast-detail phantom has been used in both QC visits. Table 1 shows the results for two such combinations, the first a Fuji Profect CS system combined with a Siemens Mammomat 3000 x-ray unit, and the second a Siemens Mammomat Novation system.

For the first system, both the results from the DQC and the half-yearly QC visit stayed the same after 6 months. For the second system, the results from the DQC show that the exposure of the phantom was reduced by 30%, a reduction for which the consequences can be clearly seen in the half-yearly tests: lowered MGD, lowered CNR and an increase in contrast-threshold. The reduction in dose was the consequence of a new calibration for the system after the x-ray tube was replaced.

Out of the initial 129 combinations, 68 (53%) show a change in the DQC parameters, using the above criteria. From this subset of 68 systems, there are 27 systems where the results of the half-yearly visit change more than prescribed in the methods

Table 1. Overview of the complete results for 2 systems

System:	Fuji Profect CS with Siemens Mammomat 3000		Siemens Mammomat Novation	
Date:	Dec/2008	Jun/2009	Nov/2008	Aug/2009
DQC results				
Detector ID:	N/A	N/A	MM12835	MM12835
Anode/Filter:	Mo/Mo	Mo/Mo	W/Rh	W/Rh
kVp:	27	27	27	27
Average mAs ± SD:	78.9 ± 0.9	79.0 ± 0.6	92.0 ± 1.1	64.5 ± 0.7
Av. MPV ± SD:	256.9 ± 4.9	246.1 ± 3.7	351.2 ± 4.4	248.7 ± 3.2
Av. St.Dev. ± SD:	6.55 ± 0.11	6.35 ± 0.12	6.26 ± 0.09	4.91 ± 0.05
Av. SNR ± SD:	39.2 ± 0.7	38.8 ± 0.6	48.2 ± 0.4	40.5 ± 0.5
HY QC results				
MGD 2cm PMMA	0.54	0.53	0.47	0.38
MGD 3cm PMMA	0.91	0.86	0.58	0.46
MGD 4cm PMMA	1.57	1.58	0.81	0.64
MGD 5cm PMMA	2.11	2.05	1.14	0.88
MGD 6cm PMMA	1.88	1.93	1.29	1.00
MGD 7cm PMMA	3.37	3.12	1.77	1.35
CNR 2cm PMMA	9.54	9.31	8.84	7.56
CNR 3cm PMMA	8.63	8.22	7.64	6.64
CNR 4cm PMMA	8.19	8.00	7.16	6.07
CNR 5cm PMMA	6.89	6.74	6.41	5.53
CNR 6cm PMMA	5.73	5.64	4.96	4.35
CNR 7cm PMMA	5.78	5.45	4.50	3.98
Contrast threshold				
A/F & kVp:	Mo/Rh 27	Mo/Rh 27	W/Rh 28	W/Rh 28
mAs:	125	125	125	80
Limit (μm Au) for				
2.00 mm detail:	0.043	0.045	0.047	0.051
1.60 mm detail:	0.046	0.048	0.049	0.054
1.25 mm detail:	0.051	0.053	0.052	0.058
1.00 mm detail:	0.058	0.060	0.056	0.064
0.80 mm detail:	0.069	0.072	0.063	0.073
0.63 mm detail:	0.087	0.090	0.073	0.087
0.50 mm detail:	0.115	0.119	0.089	0.108
0.40 mm detail:	0.156	0.161	0.113	0.137
0.31 mm detail:	0.230	0.237	0.156	0.188
0.25 mm detail:	0.326	0.336	0.211	0.254
0.20 mm detail:	0.475	0.490	0.298	0.354
0.16 mm detail:	0.700	0.721	0.429	0.504
0.13 mm detail:	1.011	1.040	0.610	0.710
0.10 mm detail:	1.618	1.665	0.963	1.108
0.08 mm detail:	2.422	2.490	1.432	1.633
0.06 mm detail:	4.085	4.201	2.403	2.714

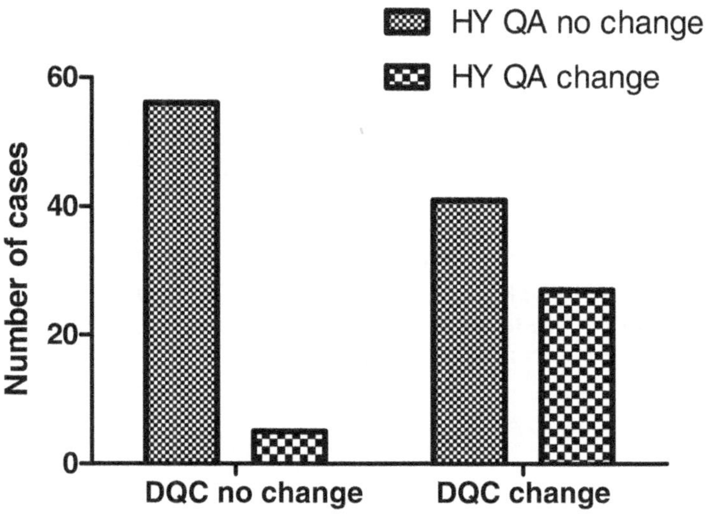

Fig. 2. Results without contrast-threshold data

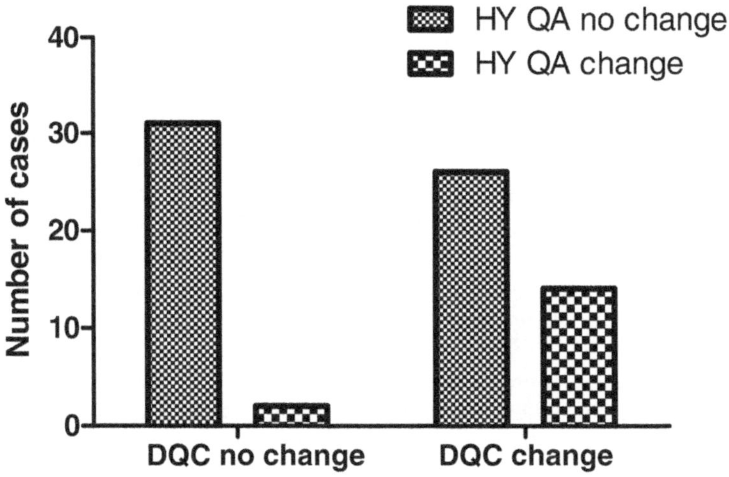

Fig. 3. Results for contrast-threshold data

section (true positives) and 41 where the change is less (false positives). Of the 61 systems where the parameters of the DQC are unchanged following our calculation method, there are 56 for which the results of the half-yearly visit do not change more than prescribed (true negatives), and 5 where the 'system showed a change' following our definition (false negatives). This gives the threshold that we defined for the change in DQC parameters a specificity of 84% for changes in results at the half-yearly QC visits.

If we have a more detailed look at the 5 false negative values, 3 of these cases can be explained by the use of a different AEC exposure mode at the half-yearly test which was not mirrored in the DQC parameters. Both the half-yearly test and the DQC should be performed with the clinical exposure condition. Therefore this difference reflects DQC or half-yearly tests being carried out incorrect, rather than a system failure. If we take this into account, the specificity of our test increases to 93%. These results are shown in figure 2.

A parallel analysis on the contrast threshold results shows that in these 74 cases, we find 14 true positives, 26 false positives, 31 true negatives and 3 false negatives, which gives a specificity of 82%. A closer look at the false negatives shows the same deviating QC visit test due to the other AEC mode. Removing this case increases the specificity to 88%. These results are shown in figure 3.

4 Discussion

The results show that for systems where the change of parameters measured in the DQC remains below our chosen threshold, it is unlikely that results of the half yearly QC visit will show a change. The use of this DQC parameter could be a base to prioritize QC efforts and could have enabled a 40% reduction in the number of half yearly QC visits, leading to a reduction of more than 20% in the total number of QC visits. The principle of a yearly QC visit would be maintained for verification purposes.

It is important to take note of some of the limitations of our proposal. The basic limitation is that this can only be applied if there is a daily routine QC check in place. It should only be applied to systems that have a certain initial quality margin above the acceptable limits. For a system where, for example SDNR is less than 10% above the acceptable limit, the half-yearly test should be performed, even if our method would conclude from the DQC data that the half-yearly test would remain unchanged. A less tangible disadvantage is that for the centers that a physicist would only visit once a year, there is less contact, which may be increasing the barrier for questions and remarks about the QC procedures.

Because we used results from over a period of 3 years, starting from the first introduction of digital mammography in screening in our network, experience has grown substantially, different colleagues performed the QC visits and practical procedures have been gradually introduced. Therefore it is possible that some of the variation in the QC results is caused by a change in procedure and not by system change. Therefore, the positive predictive value of the DQC threshold is likely to be higher in practice than the figure found here. Note that there has been no change in the method used to analyze DQC over this period.

During a normal half-yearly QC visit, the viewing devices are also subject to a test procedure including a visual inspection of the general purpose AAPMtg18-QC pattern [5] and of low contrast resolution patterns, a uniformity measurement and a conformance test to the grayscale standard display function (GSDF) [6]. So in principle, even if we decided the half-yearly test of the mammography system was not needed, we should still go to the site for the monitor QC. Our current experience is that with the exception of acceptance tests, we have never seen anything more serious than small deviations from the GSDF calibration. If there are problems at the acceptance test of the monitors, this usually means that they were installed without applying a calibration. In the daily follow-up we have only seen two real problems (loss in perceived contrast, once due to a loss of calibration tables and once due to incorrect placement of viewing devices) out of the 68 diagnostic workstations included in our network. All other lower scores, obtained by using the MoniQA test pattern [3], can be attributed to environmental conditions like ambient light or the daily tests being performed without due care. This could indicate that a daily evaluation of an extended pattern like MoniQA is redundant and perhaps should be replaced by a simpler and faster test optimized to verify correct viewing conditions.

Finally we would like to point out that instead of checking the DQC whether a half-yearly test is necessary, we could trigger an additional test the moment these results start deviating from the baseline established at the previous QC visit. With this approach, the effective testing frequency would increase for problem sites and decrease for problem-free sites. In addition, other parameters followed in DQC but not used in this study, like detector inhomogeneities or modality software upgrades, can also trigger additional QC visits or actions.

References

1. Perry, N., Broeders, M., de Wolf, C., Törnberg, S., Holland, R., von Karsa, L., Puthaar, E. (eds.): European protocol for the quality control of the physical and technical aspects of mammography screening. Office for Official Publications of the European Communities (2006)
2. Jacobs, J., Lemmens, K., Nens, J., Michielsen, K., Marchal, G., Bosmans, H.: One year of experience with remote quality control of digital mammography systems in the Flemish breast cancer screening program. In: Krupinski, E.A. (ed.) IWDM 2008. LNCS, vol. 5116, pp. 703–710. Springer, Heidelberg (2008)
3. Jacobs, J., Rogge, F., Kotre, J., Marchal, G., Bosmans, H.: Preliminary validation of a new variable pattern for daily quality control of medical image display devices. Med. Phys. 34(7), 2744–2758 (2007)
4. Young, K.C., Cook, J.J.H., Oduko, J.M., Bosmans, H.: Comparison of software and human observers in reading images of the CDMAM test object to assess digital mammography systems. In: Proc. SPIE, vol. 6142, pp. 614206-1–614206-13 (2006)
5. Samei, E., Badano, A., Chakraborty, D., Compton, K., Cornelius, C., Corrigan, K., Flynn, M.J., Hemminger, B., Hangiandreou, N., Johnson, J., Moxley-Stevens, D.M., Pavlicek, W., Roehrig, H., Rutz, L., Shepard, J., Uzenoff, R.A., Wang, J., Willis, C.E.: AAPM TG18. Assessment of display performance for medical imaging systems: executive summary of AAPM TG18 report. Med. Phys. 32(4), 1205–1225 (2005)
6. Digital Imaging and Communications in Medicine (DICOM) part 14: grayscale standard display function, NEMA PS 3.14, Rosslyn, VA: National Electrical Manufacturers Association (2000)

A Comparative Study of Volumetric and Area-Based Breast Density Estimation in Digital Mammography: Results from a Screening Population

Despina Kontos, Predrag R. Bakic, Raymond J. Acciavatti, Emily F. Conant, and Andrew D.A. Maidment

Department of Radiology, University of Pennsylvania, 1 Silverstein Building HUP, 3400 Spruce Street, Philadelphia, PA 19104-4206
{Despina.Kontos,Predrag.Bakic,Emily.Conant, Andrew.Maidment}@uphs.upenn.edu, racci@seas.upenn.edu

Abstract. We compare a volumetric versus an area-based breast density estimation method in digital mammography. Bilateral images from 71 asymptomatic women were analyzed. Volumetric density was measured using *Quantra*^{TM} (*Hologic Inc.*). Area-based density was estimated using *Cumulus* (*Ver. 4.0, Univ. Toronto*). Correlation and regression analysis was performed to determine the association between *i*) density from left versus right breasts and *ii*) volumetric versus the area-based measures. Volumetric breast density measures are strongly correlated but statistically significantly different than the area-based measures (r=0.79, p<0.001). Regression demonstrates a significant non-linear association (R^2=0.70, p<0.001). The density correlation between right and left breasts is also strong for both methods, ($r \geq 0.95$, p<0.001). The strong association with the area-based density measures suggests that volumetric breast density could potentially also aid in breast cancer risk estimation. The observed non-linear association between volumetric and area-based estimates may have implications for risk stratification in clinical practice.

Keywords: Volumetric breast density, digital mammography, breast cancer risk.

1 Introduction

Growing evidence suggests that breast density is an independent risk factor for breast cancer, the strongest known attributable risk factor after age [1, 2]. Currently, the most commonly used methods to quantify breast density rely on semi-automated image thresholding techniques to segment the area of the dense tissue in mammograms [3]. Mammographic breast density is then estimated as the percent of dense tissue area within the entire breast [1]. Although useful for breast cancer risk estimation, these methods are highly subjective and difficult to standardize [1, 3-5], a factor limiting their translation for breast cancer risk assessment in the general population. In addition, they do not provide an estimate of true volumetric breast density but a rather rough area-based estimate measured from the mammographic projection image

J. Martí et al. (Eds.): IWDM 2010, LNCS 6136, pp. 378–385, 2010.

of the breast. Methods are now under development to estimate volumetric breast density from mammograms by incorporating image acquisition physics and breast thickness information [6-10]. Knowing that the risk of breast cancer is mainly associated with the total amount of fibroglandular tissue in the breast (where cancer generally originates), volumetric measures of breast density hold the promise to also provide more accurate measures for breast cancer risk estimation [11].

Studies have demonstrated the reproducibility of different volumetric breast density methods and strong associations with known breast cancer risk factors [7, 12]. However, most studies published to date have been performed using digitized screen-film mammograms and have not demonstrated a clear advantage of the volumetric versus the area-based density measures in breast cancer risk estimation [10, 13]. Methods applied directly to digital mammographic images hold the promise to provide more accurate quantitative measures and fully-explore the potential role of using volumetric breast density assessment in breast cancer risk estimation [14, 15].

We performed a study to evaluate a new volumetric method for breast density estimation in digital mammography (DM) in comparison to the commonly used area-based density estimation method for a screening population of women. Our goals were to evaluate the consistency of the volumetric method, compare it to the commonly used area-based approach, and investigate the nature of the association between the volumetric and the corresponding area-based breast density estimates. The results of this investigation could have significant implications on the implementation of density-based breast cancer risk stratification in clinical practice.

2 Methods

Bilateral DM images from 71 asymptomatic women (age 34-75 yrs, mean 54 yrs) presenting for annual screening mammography were retrospectively collected and analyzed under HIPAA and IRB approval from a separate IRB-approved breast cancer screening clinical trial that has been completed in our department[1] [16]. All women were study volunteers who signed informed consent. Digital mammography imaging was performed with a GE DS FFDM system (GE Healthcare, Chalfont St. Giles, UK) at 0.1 mm/pixel resolution and 12 bit gray-levels. Image post-processing was performed with the GE *PremiumView*™ algorithm [17].

Volumetric breast density (VD%) estimation was performed using *Quantra*™ (*Hologic Inc.*), an FDA approved and commercially available fully-automated software based on an extension of the Highnam & Brady method [6] for digital mammography [14]. Briefly, *Quantra*™ estimates the thickness of the fibroglandular breast tissue above each pixel in the image and aggregates these pixel-wise estimates to compute the total volume of fibroglandular tissue in the breast (Fig. 1). Through a similar process, *Quantra*™ considers the entire imaged breast outline, compensating for those portions of the breast that were not uniformly compressed, to estimate the entire volume of the breast. The estimated fibroglandular tissue volume is then divided by the total breast volume to calculate the volumetric percentage of fibroglandular tissue in the breast (*i.e.*, VD%) [18].

[1] GE Healthcare Protocol Number 804380, Penn PI: E.F. Conant.

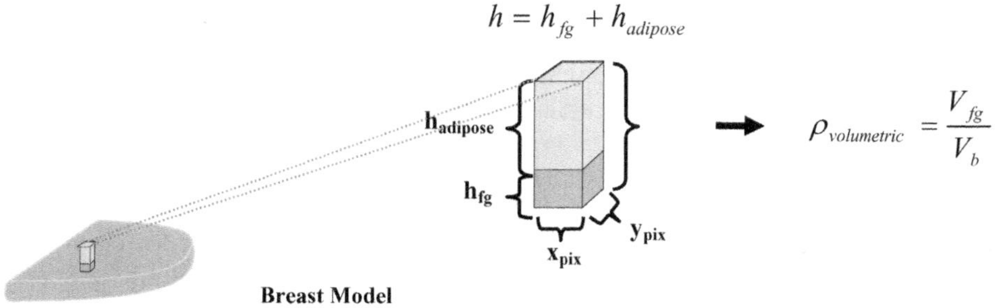

Fig. 1. The main idea of the *Quantra*™ (*Hologic Inc.*) method for estimating pixel-wise breast thickness and deriving a volumetric density measure.

To compare to the commonly used area-based breast density measures, breast percent density (PD%) was estimated in the *PremiumView*™ post-processed DM images by an experienced reader, using the semi-automated image thresholding technique of *Cumulus* (*Ver. 4.0, Univ. Toronto*) [3].

The Student's pair-wise t-test was applied to compare the means of the breast density distributions obtained by the volumetric and the area based methods. The Pearson correlation coefficient (*r*) was computed and linear regression analysis was performed to determine the degree of association between the density estimates from left and right breasts. In addition, both linear and non-linear regression was performed to model the association between the volumetric and the area-based measures.

3 Results

The volumetric breast density (VD%) measures obtained with *Quantra*™ are strongly correlated (r=0.79, p<0.001) but statistically significantly different than the

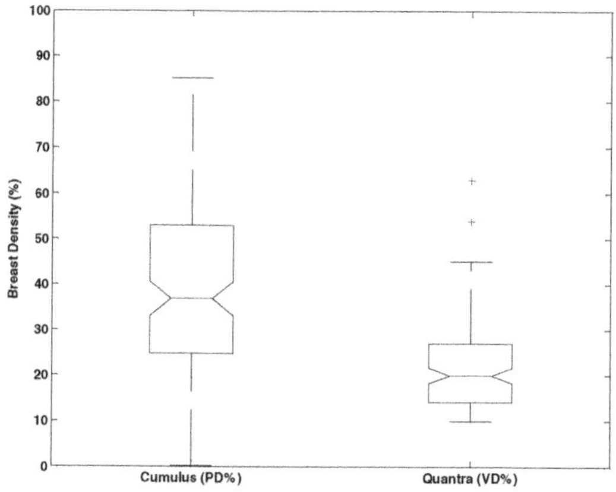

Fig. 2. Box-plots for the distributions of the area-based *Cumulus* percent density (PD%) measures and the *Quantra*™ volumetric density (VD%) measures

corresponding area-based breast percent density (PD%) measures obtained with *Cumulus* ($p<0.001$). As expected, volumetric density estimates (mean=21.94%) are lower than the corresponding area-based estimates (mean=37.97%) (Fig. 2).

When investigating separately for each method the breast density correlations between left and right breasts, both methods are highly consistent, as evidenced by the strong and statistically significant correlations and the linear regression fits (Fig. 3).

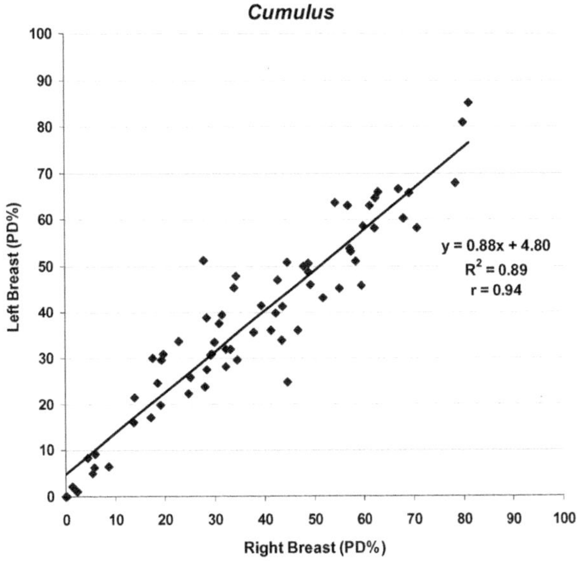

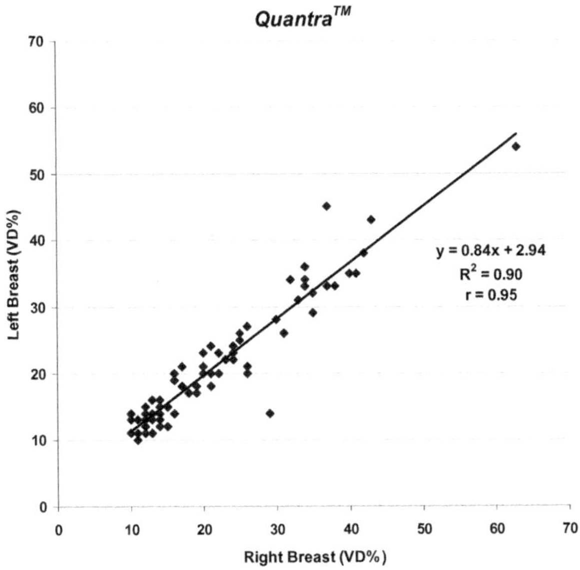

Fig. 3. Linear regression fits between right and left breasts for the *Cumulus* area-based breast percent density (PD%) (up) and the *Quantra*^{TM} volumetric density (VD%) (down)

The *Quantra*™ method had a slightly higher, but significantly different, correlation coefficient than the *Cumulus* method for the volumetric breast density correlation between the right and left breasts ($r = 0.95$, $p<0.001$).

To model the association between the volumetric and the corresponding area-based density measures, both linear and non-linear regression analysis was performed (Fig. 4). Both models show statistically significant associations ($p<0.001$), with the non-linear regression indicating a stronger second-degree polynomial fit (R^2=0.70).

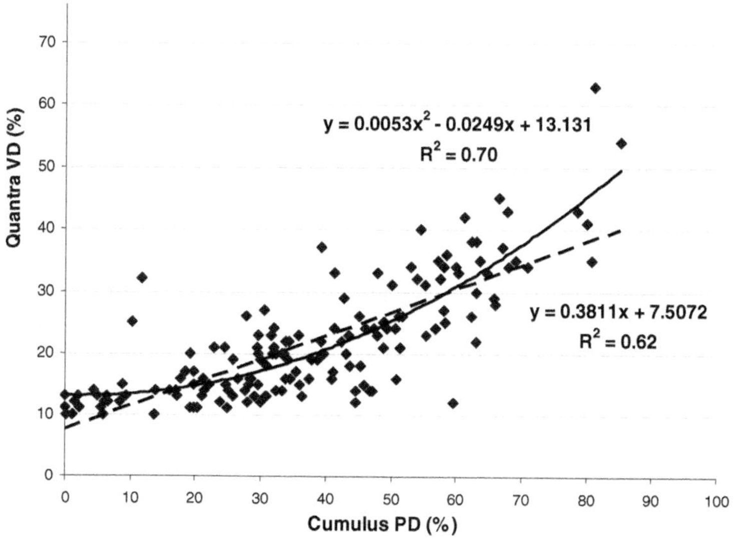

Fig. 4. Linear and non-linear regression fits between the *Quantra™* volumetric (VD%) breast density measures and the *Cumulus* area-based percent density (PD%) estimates.

4 Discussion and Conclusions

Our study performs a comparative evaluation of the first FDA approved fully-automated software for volumetric breast density estimation in digital mammography (*Quantra™, Hologic Inc.*) versus the current gold-standard area-based breast density estimation method (*Cumulus, Ver. 4.0, Univ. Toronto*) for a screening population of women. The strong correlation observed between right and left breasts indicates that volumetric breast density measures computed with *Quantra™* can provide consistent fully-automated measures of breast density for women undergoing mammographic screening. The strong association observed between the volumetric and the corresponding area-based density measures, shown by several studies to correlate with breast cancer risk [1], supports the hypothesis that volumetric breast density measures could also aid in breast cancer risk assessment.

However, the observed non-linear association in our study (Fig. 4) also suggests a potential non-linear relationship between the corresponding volumetric and the area-based risk stratification levels, which could ultimately have implications for risk stratification in clinical practice. This concept is illustrated in Figure 5. Further work is underway to validate these findings with larger clinical studies and to fully investigate the association between volumetric breast density measures and breast cancer risk. Such larger studies will also help to determine more accurately the risk stratification levels using the volumetric versus the area-based density estimates. Our long term hypothesis is that quantitative methods for measuring volumetric breast density can provide more accurate measures of density and ultimately result in more accurate measures to assess breast cancer risk. In addition, fully-automated methods can alleviate the subjectivity of the currently used semi-automated techniques and accelerate the translation of density-based risk stratification in clinical practice.

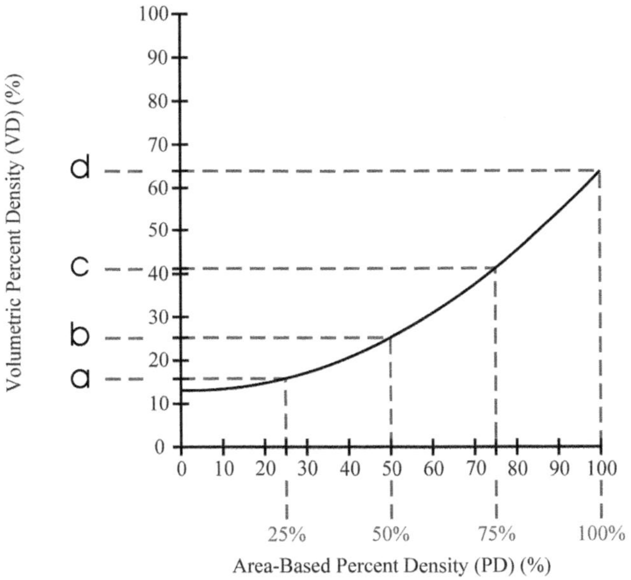

BIRADS
Type 1: Almost entirely fat (<25% glandular).
Type 2: Scattered fibroglandular densities (25-50% glandular).
Type 3: Heterogeneously dense (51-75% glandular).
Type 4: Extremely dense (>75% glandular).

Fig. 5. An illustrative example demonstrating a non-linear association between the established area-based ACR BIRADS density categories for breast cancer risk stratification [19] and the corresponding volumetric density categories based on the specific non-linear trend observed in our study. Larger clinical studies will determine more accurately the corresponding volumetric density risk stratification levels, denoted here for illustration purposes as "a", "b", "c", and "d".

Acknowledgments. This work was partially funded by the Susan G. Komen for the Cure Foundation (Postdoctoral Fellowship KG080756), the DOD (Concept Award BC086591), and GE Healthcare (Clinical trial protocol number 804380). We would like to thank *Hologic Inc.* for providing the *Quantra*^{TM} software and Dr. Martin Yaffe for providing the *Cumulus* software. We would also like to thank Dr. Johnny Kuo for developing and maintaining the RSNA Medical Imaging Resource Center (MIRC) image archive and Julian Marshall for the useful feedback and discussions.

References

1. Boyd, N.F., Guo, H., Martin, L.J., Sun, L., Stone, J., Fishell, E., Jong, R.A., Hislop, G., Chiarelli, A., Minkin, S., Yaffe, M.J.: Mammographic density and the risk and detection of breast cancer. New England Journal of Medicine 356, 227–236 (2007)
2. Tice, J.A., Kerlikowske, K.: Screening and prevention of breast cancer in primary care. Prim. Care 36, 533–558 (2009)
3. Yaffe, M.J.: Mammographic density. Measurement of mammographic density. Breast Cancer Research 10, 209 (2008)
4. Martin, K.E., Helvie, M.A., Zhou, C., Roubidoux, M.A., Bailey, J.E., Paramagul, C., Blane, C.E., Klein, K.A., Sonnad, S.S., Chan, H.P.: Mammographic density measured with quantitative computer-aided method: comparison with radiologists' estimates and BI-RADS categories. Radiology 240, 656–665 (2006)
5. Harvey, J.A., Bovbjerg, V.E.: Quantitative Assessment of Mammographic Breast Density: Relationship with Breast Cancer Risk. Radiology 230, 29–41 (2004)
6. Highnam, R., Jeffreys, M., McCormack, V., Warren, R., Davey Smith, G., Brady, M.: Comparing measurements of breast density. Phys. Med. Biol. 52, 5881–5895 (2007)
7. McCormack, V.A., Highnam, R., Perry, N., dos Santos Silva, I.: Comparison of a new and existing method of mammographic density measurement: intramethod reliability and associations with known risk factors. Cancer Epidemiol Biomarkers Prev. 16, 1148–1154 (2007)
8. Jeffreys, M., Warren, R., Highnam, R., Smith, G.D.: Initial experiences of using an automated volumetric measure of breast density: the standard mammogram form. Br. J. Radiol. 79, 378–382 (2006)
9. Pawluczyk, O., Augustine, B.J., Yaffe, M.J., Rico, D., Yang, J., Mawdsley, G.E., Boyd, N.F.: A volumetric method for estimation of breast density on digitized screen-film mammograms. Medical Physics 30, 352–364 (2003)
10. Boyd, N., Martin, L., Gunasekara, A., Melnichouk, O., Maudsley, G., Peressotti, C., Yaffe, M., Minkin, S.: Mammographic density and breast cancer risk: evaluation of a novel method of measuring breast tissue volumes. Cancer Epidemiol Biomarkers Prev. 18, 1754–1762 (2009)
11. Kopans, D.B.: Basic physics and doubts about relationship between mammographically determined tissue density and breast cancer risk. Radiology 246, 348–353 (2008)
12. Jeffreys, M., Warren, R., Highnam, R., Davey Smith, G.: Breast cancer risk factors and a novel measure of volumetric breast density: cross-sectional study. Br. J. Cancer 98, 210–216 (2008)
13. Ding, J., Warren, R., Warsi, I., Day, N., Thompson, D., Brady, M., Tromans, C., Highnam, R., Easton, D.: Evaluating the effectiveness of using standard mammogram form to predict breast cancer risk: case-control study. Cancer Epidemiol Biomarkers Prev. 17, 1074–1081 (2008)
14. Hartman, K., Highnam, R., Warren, R., Jackson, V.: Volumetric Assessment of Breast Tissue Composition from FFDM Images. In: Krupinski, E.A. (ed.) IWDM 2008. LNCS, vol. 5116, pp. 33–39. Springer, Heidelberg (2008)
15. van Engeland, S., Snoeren, P.R., Huisman, H., Boetes, C., Karssemeijer, N.: Volumetric breast density estimation from full-field digital mammograms. IEEE Trans. Med. Imaging 25, 273–282 (2006)

16. Kontos, D., Ikejimba, L., Bakic, P.R., Troxel, A.B., Conant, E.F., Maidment, A.D.A.: Digital breast tomosynthesis parenchymal texture analysis for breast cancer risk estimation: Results from a screening trial. Presented at 95th Scientific Assembly and Annual Meeting of the Radiological Society of North America, Chicago, IL (2009)
17. Kontos, D., Bakic, P.R., Carton, A.K., Troxel, A.B., Conant, E.F., Maidment, A.D.A.: Parenchymal texture analysis in digital breast tomosynthesis for breast cancer risk estimation: A preliminary study. Academic Radiology 16, 283–298 (2009)
18. Understanding R2 Quantra 1.3, Hologic Inc. (Software Manual) (2009)
19. The ACR Breast Imaging Reporting and Data System (BI-RADS) Atlas. American College of Radiology, Reston (2003)

Volumetric Breast Density and Breast Cancer Risk Factors in a Screening Population

Jennifer Diffey[1], Joanna Morrison[1], Michael Berks[1], Joanne Greene[1], Julie Morris[2], Anthony Maxwell[3], Caroline Boggis[4], Alan Hufton[1], and Susan Astley[1]

[1] Imaging Science and Biomedical Engineering, School of Cancer and Imaging Sciences, Stopford Building, University of Manchester, Oxford Road, Manchester M13 9PT
[2] Department of Medical Statistics, University Hospital of South Manchester NHS Foundation Trust, Southmoor Road, Wythenshawe, Manchester, United Kingdom, M23 9LT
[3] Bolton, Bury and Rochdale Breast Screening Service, Royal Bolton Hospital, Bolton, Lancashire, BL4 OJR
[4] The Nightingale Centre and Genesis Breast Cancer Prevention Centre, University Hospital of South Manchester NHS Foundation Trust, Southmoor Road, Wythenshawe, Manchester, United Kingdom, M23 9LT
jennifer.diffey@postgrad.manchester.ac.uk

Abstract. Breast density is positively linked to the risk of developing breast cancer. Furthermore, the addition of breast density as an input to breast cancer risk prediction models has been shown to improve their predictive power. Such models are used in the management of women at high risk but could potentially be used to determine screening strategy. A stepwedge-based technique has been used to measure volumetric density from the mammograms of 1,289 women in the UK screening programme who additionally completed a questionnaire on risk-related factors. The sample had a mean age of 60.1 (range 48.0 – 78.0), a mean breast thickness of 59mm (range 21 – 102mm) and a mean volumetric breast density of 11% (range 0.5 – 58%). Using Pearson's correlation coefficient, breast density was found to be significantly correlated with weight (r = -0.45), body mass index (r = -0.48), age (r = -0.13) and breast thickness (r =-0.65) at the p = 0.01 level. Absolute glandular volume was also found to be significantly correlated with these parameters although the extent of correlation was weaker.

Keywords: breast density, volumetric technique, risk factors.

1 Introduction

The amount of dense (non-fatty) tissue within the breast is strongly linked to the risk of developing breast cancer [1 – 3]. A number of techniques exist for the measurement of breast density, either by area [3 – 6] or volume [7 – 11]. Although area-based techniques have shown strong correlations between breast density and risk [1 – 3], results are generally based on an interpretation of the mammogram which equates brightness of the image with density; the validity of this assumption has, however, been questioned [12]. Volumetric methods seek to provide a quantitative and more objective approach which takes into account the true three-dimensional nature of the breast and its component tissues.

J. Martí et al. (Eds.): IWDM 2010, LNCS 6136, pp. 386–393, 2010.

2 Materials and Methods

6,000 women attending routine breast screening were invited to participate in a feasibility study which aimed to provide descriptive statistics on breast density distribution in the screening population and to examine the relationship of volumetric breast density to other breast cancer risk factors, collected via questionnaire. The inclusion of breast density in risk prediction models has been shown to offer improved accuracy in the identification of women at high risk of developing breast cancer [13]. Questionnaire data included age at examination, height, weight (and hence body mass index, BMI), date of first pregnancy, ages of menarche and menopause, ethnicity and family history of breast cancer (mother or sister only) including the age at which breast cancer was diagnosed in this relative. Information regarding previous breast disease and use of hormone replacement therapy (HRT) was available in the patients' notes.

Mammograms were taken as usual. A stepwedge calibrated against glandular and adipose tissue equivalent material was placed on the breast support platform and radio-opaque magnification markers were placed on the compression paddle, to facilitate accurate determination of breast thickness across the mammogram [14]. The images were anonymised and digitised and a semi-automated method was used to assess breast density [11]. An operator was prompted to select the radio-opaque markers and define two steps on the stepwedge. An approximate location for the breast edge was determined by i) applying a global threshold based on analysis of the grey-level histogram in each mammogram ii) applying morphological operators to the resulting binary image to isolate the main breast region. The approximate breast edge location was used to initialise an adaptation of active contour algorithm which then computed a more precise (and locally smooth) demarcation of the breast edge. A thickness of glandular tissue was determined at each pixel, allowing breast density to be expressed in terms of the absolute glandular volume and the percentage breast density (defined as glandular volume / total breast volume). An example of the resulting glandular thickness map is shown in Figure 1.

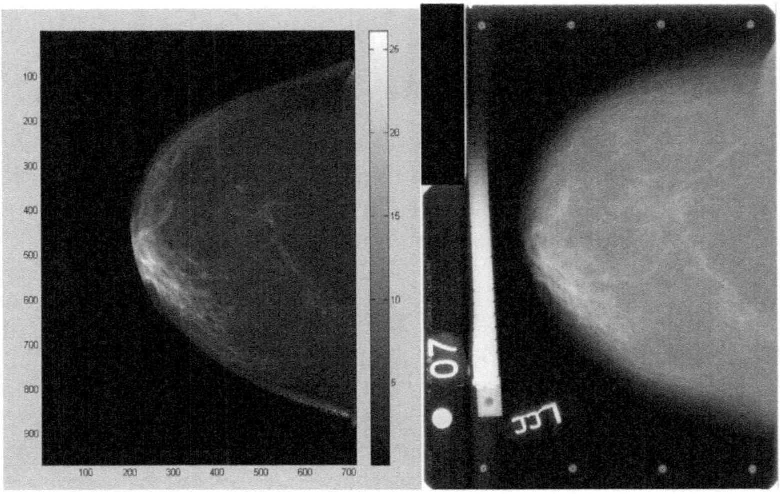

Fig. 1. Glandular thickness map shown alongside an original mammogram

The absolute glandular volume and percentage breast density were calculated for each view. The differences between breast density in the left / right and cranio-caudal (CC) / medio-lateral oblique (MLO) views were examined using paired t-tests. Correlations between breast density and a number of risk-related parameters have been assessed using Pearson's correlation co-efficient.

Results have been compared to those from previous work, which examined the correlation of these parameters with radiologist-assessed percentage breast density for a subset of 294 women from this sample [15].

3 Results

3.1 Population Demographics

1,289 women consented to take part in the study and had their breast density calculated using the method described above. The demographics of the population sample are shown in Table 1. The values quoted for breast thickness, glandular volume and breast density are the average of all four views. The maximum measure of breast thickness is used (i.e. that measured at the chest wall).

Table 1. Demographics of the population sample

Parameter	Mean	Minimum	Maximum	Standard deviation
Age (years)	60.1	48.0	78.0	5.7
Weight (kg)	69.7	32.3	178.1	13.8
Body Mass Index (kgm^{-2})	26.9	13.8	74.1	5.2
Breast thickness (mm)	58.9	20.9	102.3	12.1
Glandular volume (cm^3)	72	8	626	49
Breast density: mean of four views (%)	11.0	0.5	58.0	8.8

3.2 Variation in Breast Density by View

The variation in breast density by view was assessed using paired t-tests. The results are shown in Table 2. Glandular volume and breast density were found to vary with mammographic view. The difference between left and right breasts was not significant but interestingly, the difference between the CC and the MLO view was significant.

Table 2. Variation in volumetric breast density by view

	Mean difference	95% Confidence Interval		Significant difference	p-value
		Lower	Upper		
LMLO – LCC	-1.70	-1.89	-1.50	Yes	<0.01
RMLO – RCC	-1.39	-1.59	-1.20	Yes	<0.01
RCC – LCC	-0.02	-0.21	0.17	No	0.41
RMLO – LMLO	-0.01	-0.19	0.18	No	0.47

3.3 Correlation of Breast Density with Other Risk-Related Factors

The correlations of breast density with age and BMI are shown in Figures 2 and 3 respectively. The correlation of absolute glandular volume and percentage breast density with these and other parameters is compared in Table 3. All correlations were significant at the p=0.01 level.

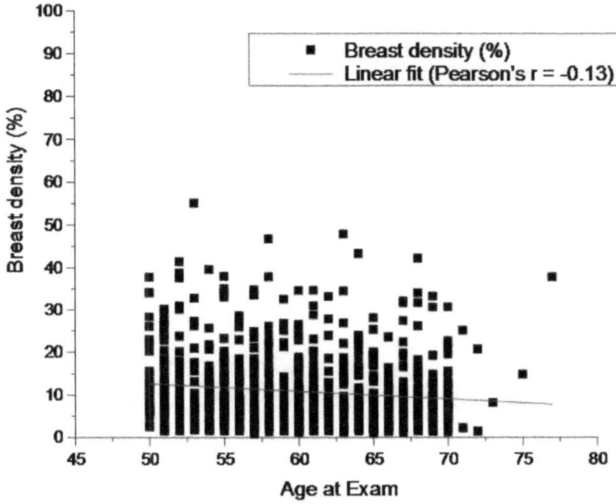

Fig. 2. Breast density versus age at examination

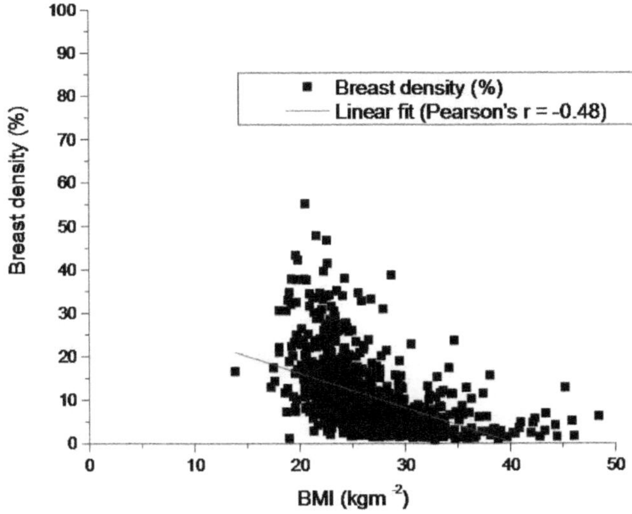

Fig. 3. Breast density versus body mass index (BMI)

As expected, breast density was found to decrease with age, weight, BMI and breast thickness. Glandular volume also decreased with these parameters although the strength of correlation was weaker. The stronger association with percent density can be explained by the evidence that breast volumes increase with weight but there is not a large change in gland volume. The association with BMI for both density variables reflected the association and strength of association with weight.

3.4 Comparison with Radiologist Visual Assessment

294 women from this sample had their breast density measured by a radiologist using a visual analogue scale. The mean age of this sample was 60.2 (range 50 – 72) and the mean weight was 69.3kg (range 41.3 – 139.9kg). Within this sample, the mean breast density was 27% (range 2 – 88%). It is interesting to note that the CC view was denser than the MLO view by an average of 0.5% (left) and 0.8% (right), a trend that agrees with the results in Table 2 for volumetric breast density. However, the right breast density was found to be greater than the left breast density by an average of 1.1%; no significant difference was observed between left and right when using the volumetric method.

The correlations between visually-assessed breast density with risk-related factors are shown in Table 3. Although the trends are similar to those for volumetric breast density, the strength of correlation is generally weaker for visually-assessed density. The relationship with BMI was significant at the p=0.01 level; the relationship with weight was significant at the p=0.05 level.

Table 3. Correlation of glandular volume and breast density with risk-related factors

	Pearson's correlation co-efficient (r)		
	Glandular volume (cm^3)	Volumetric Breast density (%)	Visually-Assessed Breast density (%)
Breast thickness (mm)	-0.11	-0.65	-0.30
Age (years)	-0.18	-0.13	-0.18
Weight (kg)	-0.11	-0.45	-0.29
BMI (kgm^{-2})	-0.16	-0.48	-0.41

4 Conclusion

All women taking part in the study had their breast density measured for each mammographic view using a semi-automated volumetric method. The average breast density of the sample was 11% (range 0.5 – 58%). As expected, breast density was found to decrease as breast thickness increased (r = -0.65). There were also strong negative correlations (p = 0.01) with weight (r = -0.45) and BMI (r = -0.48). Significant correlations were also observed between these parameters and absolute glandular volume although the strength of the correlation was lower. This was expected as although breast volume increases with weight, weight gain is associated with an increase in adipose, rather than glandular tissue.

Our previous work examined the relationship between visually-assessed breast density and risk-related factors. The mean breast density using this method was found to be higher than the software-based measure of volumetric breast density. The visual method was only applied to a subset of the women taking part in the study but the mean age and weight were similar for both samples, as were the ranges for these parameters. Other studies comparing area-based and volumetric techniques have shown that the measure of breast density is higher using area-based techniques [16, 17]. It is interesting to note that the strength of association between visually-assessed density with weight, BMI and breast thickness was lower than that for volumetric density.

Glandular volume and volumetric breast density were found to vary with mammographic view. The difference between left and right breasts was not significant but the difference between the CC and the MLO view was significant. On average, the CC view was found to be denser than the MLO view, a result also supported by radiologist visual-assessment. This suggests that it may not be adequate to measure density for one view only. There is evidence to suggest that the strength of the association between breast density and risk increases when considering both the CC and MLO views compared to the MLO view only [18]. However, this paper was based on a study using radiologist visual assessment of density and other studies using automated techniques have shown that it is possible to use one view only [19].

This study examined the relationship between volumetric breast density and other risk factors for breast cancer. Ideally, we would have liked to examine the relationship with breast cancer risk itself. Independent t-tests were used to compare women who had previously had breast cancer with those who had not. Unfortunately, the analysis into this variable was hampered by the small sample size and the difficulty in accurately matching controls. Meaningful comparison was therefore impossible. Our semi-automated volumetric method is currently being used in a large-scale multidisciplinary study which aims to develop risk-prediction models for the screening population. A number of different methods are being applied and their relationship with risk will be compared. Work is underway to adapt our method to make it suitable for full-field digital mammography. Although it will still be a calibration-based technique, the stepwedge will no longer be required.

Acknowledgements

We are very grateful to John Lewis plc and the breast cancer prevention charity Genesis for funding this study. Many thanks go to Tom Hamnett from the mechanical workshop at The Christie Hospital, Manchester for constructing the stepwedges and markers.

References

1. Wolfe, J.N.: Breast patterns as an index of risk for developing breast cancer. Am. J. Roentgen. 126(6), 1130–1137 (1976)
2. Boyd, N.F., O'Sullivan, B., Fishell, E., Simor, I., Cooke, G.: Mammographic signs as risk factors for breast cancer. Br. J. Cancer. 45(2), 185–193 (1982)
3. Boyd, N.F., Guo, H., Martin, L.J., Sun, L., Stone, J., Fishell, E., Jong, R.A., Hislop, G., Chiarelli, A., Minkin, S., Yaffe, M.J.: Mammographic density and the risk and detection of breast cancer. New Eng. J. Med. 356(3), 227–236 (2007)
4. Byng, J.W., Boyd, N.F., Fishell, E., Jong, R.A., Yaffe, M.J.: The quantitative analysis of mammographic densities. Phys. Med. Biol. 39(10), 1629–1638 (1994)
5. Byng, J.W., Boyd, N.F., Fishell, E., Jong, R.A., Yaffe, M.J.: Automated analysis of mammographic densities. Phys. Med. Biol. 41(5), 909–923 (1996)
6. Zhou, C., Chan, H.P., Petrick, N., Helvie, M.A., Goodsitt, M.M., Sahiner, B., Hadjiiski, L.M.: Computerized image analysis: estimation of breast density on mammograms. Med. Phys. 28(6), 1056–1069 (2001)
7. Pawluczyk, O., Augustine, B.J., Yaffe, M.J., Rico, D., Yang, J., Mawdsley, G.E.: A volumetric method for estimation of breast density on digitized screen-film mammograms. Med. Phys. 30(3), 352–364 (2003)
8. Boyd, N., Martin, L., Gunasekara, A., Melnichouk, O., Maudsley, G., Peressotti, C., Yaffe, M., Minkin, S.: Mammographic density and breast cancer risk: Evaluation of a novel method of measuring breast tissue volumes. Cancer Epidemiol Biomarkers Prev. 18(6) (2009)
9. Highnam, R., Pan, X., Warren, R., Jeffreys, M., Davey-Smith, G., Brady, M.: Breast composition using retrospective standard mammogram form. Phys. Med. Biol. 51, 2695–2713 (2006)
10. Kaufhold, J., Thomas, J.A., Eberhard, J.W., Galbo, C.E., Gonzalez Trotter, D.E.: A calibration approach to glandular tissue composition estimation in digital mammography. Med. Phys. 29(8), 1867–1880 (2002)
11. Hufton, A.P., Astley, S.M., Marchant, T.E., Patel, H.G.: A method for the quantification of dense breast tissue from digitised mammograms. In: Proceedings of the 7th IWDM (2006)
12. Kopans, D.B.: Basic physics and doubts about relationship between mammographically determined tissue density and breast cancer risk. Radiology 246(2) (2008)
13. Barlow, W.E., White, E., Ballard-Barbash, R., Vacek, P.M., Titus-Ernstoff, L., Carney, P.A., Tice, J.A., Buist, D.S.M., Geller, B.M., Rosenberg, R., Yankaskas, B.C., Kerlikowske, K.: Prospective Breast Cancer Risk Prediction Model for Women Undergoing Screening Mammography. J. Nat. Can. Inst. 98, 1204–1212 (2006)
14. Diffey, J., Hufton, A., Beeston, C., Marchant, T., Astley, S.: Quantifying Breast Thickness for Density Measurement. In: Krupinski, E.A. (ed.) IWDM 2008. LNCS, vol. 5116, pp. 651–658. Springer, Heidelberg (2008)
15. Diffey, J., Hufton, A., Astley, S., Mercer, C., Maxwell, A.: Estimating individual cancer risks in the UK national breast screening programme: a feasibility study. In: Krupinski, E.A. (ed.) IWDM 2008. LNCS, vol. 5116, pp. 651–658. Springer, Heidelberg (2008)
16. McCormack, V.A., Highnam, R., Perry, N., dos Santos Silva, I.: Comparison of a New and Existing Method of Mammographic Density Measurement: Intramethod Reliability and Associations with Known Risk Factors. Cancer Epidemiol Biomarkers Prev. 16(6) (2007)

17. Chung, C., Diffey, J., Berks, M., Morrison, J., Verow, R., Morris, J., Wilson, M., Boggis, C., Barr, N., Hufton, A., Astley, S.: Automated assessment of area of dense tissue in the breast: a comparison with visual assessment
18. Duffy, S.W., Nagtegaal, I.D., Astley, S.M., Gillan, M.G.C., McGee, M.A., Boggis, C.R.M., Wilson, M., Beetles, U.M., Griffiths, M.A., Jain, A.K., Johnson, J., Roberts, R., Deans, H., Duncan, K.A., Iyengar, G., Griffiths, P.M., Warwick, J., Cuzick, J., Gilbert, F.J.: Visually assessed breast density, breast cancer risk and the importance of the cranio-caudal view. Br. Can. Research. 10(4) (2008)
19. Byng, J.W., Yaffe, M.J., Lockwood, G.A., Little, L.E., Tritchler, D.L., Boyd, N.F.: Automated analysis of mammographic densities and breast carcinoma risk. Cancer 80(1) (1997)

Determinants and Consequences of Change in Breast Density

Mona Jeffreys[1], Ruth Warren[2], Ralph Highnam[3], and George Davey Smith[4]

[1] Department of Social Medicine, University of Bristol, Canynge Hall, 39 Whatley Rd, Bristol BS8 2PS, UK
[2] Department of Radiology, Addenbrooke's Hospital, Cambridge University Hospitals NHS Foundation Trust, Hills Road, Cambridge, CB2 0QQ, UK
[3] Matakina Technology Limited, Level 9, 50 Manners Street, Wellington, NZ
[4] MRC Centre for Causal Analyses in Translational Epidemiology, Department of Social Medicine, University of Bristol, Oakfield House, Oakfield Grove, Bristol, BS8 2BN , UK
Mona.jeffreys@bristol.ac.uk

Abstract. Single measures of breast density have been consistently related to breast cancer risk, but the role of changes in breast density over the early menopausal period is not clear. We investigated determinants and consequences of change in breast density among 493 women in Scotland. Using simple measures of change, only hormone replacement therapy use and current body mass index were consistently predictive of breast density change. Increases in area-based percent breast density were related to higher breast cancer risk, although the analysis was based on very few women. Modeling of change and rate of change in larger datasets is warranted.

Keywords: Breast density, trajectories, lifecourse epidemiology, breast cancer.

1 Introduction

High breast density is recognised as a strong risk factor for breast density [1]. Understanding determinants of change in breast density not only contributes to our knowledge of the aetiology of breast cancer, but also has public health implications. The identification of women who are at an increased risk of breast cancer, based on their breast density, could provide an opportunity for primary prevention of breast cancer. If breast density is causally related to breast cancer, it is likely that efforts to reduce breast density could result in reductions in breast cancer risk [2].

The role that changes in breast density over the menopause and beyond play in determining future breast cancer risk has received little attention. Key papers in this area are analyses of the Minnesota Breast Cancer Family cohort [3], an East London Screening cohort [2] and the Hawaiian Multi-ethnic cohort [4]. Each of these studies has demonstrated a reduction in breast density with age, but the determinants of this change are not consistent between studies. For example, two studies have found that BMI predicts change in breast density [3,4], whereas this was not found in a third study [2].

We have previously reported on determinants of breast density using area-based and volumetric measures of density [5,6]. The aim of the current study was to identify

J. Martí et al. (Eds.): IWDM 2010, LNCS 6136, pp. 394–401, 2010.

whether risk factors in early or later adulthood affect change in later mammographic breast density.

2 Methods

The women included in the study are part of the Glasgow Alumni Cohort [7]. Students who were registered at the University of Glasgow at some time between 1948 and 1968 were invited to an annual medical examination at the Student Health Service. Approximately 50% of students attended, including 3,584 women. We have previously shown that those who attended were broadly representative of the total student population [7]. Substantial medical data were collected at this examination, including age at menarche (self reported approximately 6 years after the event). The examining physician measured height and weight from which we calculated body mass index (BMI), in kg/m2. Women were traced through the NHS Central Register and those who were successfully traced and still alive (n: 2,169, 61%) were contacted by postal questionnaire in 2001. The response rate was 59% (n: 1,285). Women provided self-reported information on details of all pregnancies, lifetime use of oral contraceptive (OC) and hormone replacement therapy (HRT), physical activity at age 20, 40 and 60 years (categorised as: very, fairly, not very or not at all physically active), body weight at age 40 and 60 years, lifetime smoking habits and current fruit and vegetable intake.

Of the questionnaire respondents who were living in Scotland (n: 935, 73%), the 70% who had ever had at least one screening mammogram were asked for consent to access mammograms taken under the Scottish Breast Screening Programme (1989–2002). Follow-up of the cohort, the questionnaire survey and the mammogram study each received full ethical approval.

Mammograms were retrieved from the relevant screening centre and were digitised on site with a Canon FS 300 digitiser scanner at a resolution of 100 μm with 8 bit precision by a single radiographer. Scanned images were displayed at 300 micron resolution on a flat-panel display system. Mammograms were visually classified using i) Wolfe categories [8] and ii) a six-category classification (SCC) of visually estimated breast density percent [9] by a single experienced radiologist from the digitized images. The SCC categories used were 0%, 1–10%, 11–24%, 25–49%, 50–74% and >75%. We also used an automated estimation of volumetric percent and absolute breast density, the Standard Mammogram Form (SMF), version 2.2, [10]. Cranio-caudal (CC) views were excluded as they resulted in high average density images than medio-lateral oblique (MLO) views, and for most women, they were only available at the first visit. Left and right density measures taken on the same day were averaged; for SMF and SMF%, the median of all values was used; for Wolfe and SCC, categories were given a score of 1-4 and 1-6 respectively, and the mean score for each day was used.

Change on the Wolfe or SCC scale was defined as an increase one or more categories, and was compared to no change or a decrease in density from the first to the last mammogram. Change in absolute and percentage SMF was defined as the absolute difference between the first and last values. Breast cancer was defined as a breast cancer registration on or before 1st June 2008, or breast cancer being recorded on a death certificate on or before 31st August 2009.

Logistic regression (for change in Wolfe and SCC categories) and linear regression (for change in SMF and SMF%) and associated 95% confidence intervals (CI) were used to model determinants of change. Logistic regression was used to estimate odds ratios (OR) and CIs for the association between change in breast density and subsequent breast cancer. All models were adjusted for the age of the woman at the time of the first mammogram, the number of years between the first and last mammogram and for baseline measures of density.

3 Results

A total of 657 women who answered the 2001 questionnaire had had at least one screening mammogram. Of these, 151 women had attended only one screening round, so were excluded. There remained 2,953 mammograms from 506 women. For 19 films (of 15 women) the SMF algorithm produced too poor an estimate of density for analysis. The digitised images of 64 films from 33 women were too pale to assign visual density. One woman who had her first mammogram at age 40 and another who had had a previous single mastectomy were also excluded. A total of 493 women were included in the analyses.

The median age at first mammography was 54.7 years (range 48.6 to 69.9), and the median time between first and last mammogram was 6.2 years (range 0.17 to 12.87). There were reductions in all measures of breast density over this time period (mean change in Wolfe: -0.26 of a category (range -2 to +2), SCC: -0.45 of a category (range -3 to +2.5), SMF: -5.16cm^3 (range -78 to +87), SMF%: -3.2% (range -22.8 to 18.5). The annual rate of change was 0.003 of a category for Wolfe (95%CI -0.24 to 0.24); 0.06 of a category for SCC (95%CI -0.44 to 0.56); -10.5cm^3 for SMF (95%CI -28.3 to 7.3) and -3.5% for SMF% (95%CI -10.4 to 3.4).

Determinants of reduction in breast density were broadly similar for all measures. There was no evidence that age at menarche, ever having been pregnant, fruit intake or exercise or BMI in early and mid-adulthood was related to change in breast density. Those women who had ever taken hormone replacement therapy (HRT) were less likely to have increases in density across all four measures. Higher vegetable intake was strongly protective against increase in Wolfe and SCC pattern, but this was not evident for the SMF measures. Women who had ever smoked were more likely to have increases in Wolfe patterns. Frequent recent exercise patterns were related to lower levels of adverse change in absolute SMF levels. BMI reported at an average age of 60 predicted a lower chance of adverse density change for three density measures (Wolfe, SCC and SMF), whereas higher BMI was related to a higher change in SMF%.

Only six women have developed breast cancer since the date of the final mammogram. Despite this tiny sample, increases in Wolfe (OR 19.5, 95%CI 2.61 to 146) and SCC (OR 11.8, 95%CI 2.4 to 57.4) density were strongly related to breast cancer risk, P=0.003 and P=0.002 respectively), whereas increases in SMF and SMF% appeared not to be; comparing the upper to the lower half of the change distribution: (OR 1.09, 95%CI 0.29 to 4.13 for SMF; 0.88, 95%CI 0.23 to 3.41 for SMF%).

4 Discussion

Using simple measures of change, we have demonstrated that HRT use, current BMI and possibly vegetable intake appear to be predictive of breast density change, whereas more distal measures of lifestyle were not related to change in density.

There are numerous issues which need to be considered when analysing breast density, which have been clearly summarised previously [11]. These include reliability and reproducibility of measurement, the differentiation between age- and menopausal-effects, and time-varying exposures. In the current study, we have addressed the issue of reproducibility of measurement through the use of a single machine being used to digitise all mammograms and an experienced radiologist (RW), performing all of the subjective density analyses. Although in this study we did not incorporate repeat-reading of mammograms to estimate intra-reader reproducibility, she has previously demonstrated high intra-individual reproducibility of density readings on the Wolfe scale, with a correlation coefficient of 0.88 between the two assessments [12], and very good agreement with another radiologist, i.e. high inter- individual reproducibility [13]. A further strength of the current study was the use of a fully automated method to estimate the volumetric breast density.

However, the main limitation of this study is the measurement of the exposure data collection, which occurred at a single point in time. For some women, the exposure data were collected after one or both mammograms were taken. Therefore, it is possible that some of the results that we present could be explained by time-varying exposures. In particular, the HRT effect on breast density may be solely due to the fact, that some women were on HRT at the time of the first mammogram and no longer received HRT at the time of their subsequent mammogram, thus artefactually appearing to show that HRT is related to a reduction in breast density over time. A further limitation of the data is our lack of knowledge of the timing of menopause in relation to the timing of the two mammograms. We were also limited by the relatively small sample size, particularly in relation to subsequent breast cancer risk.

Previous studies have differed in their findings of the associations between recent and past exposures and change in breast density. This is particularly noticeable for BMI, which was found to predict a slower decline in breast density over the menopause in the Minnesota Breast Cancer Family cohort [3] but not in the East London Screening cohort [2]. In our study, BMI measured at age 20, and recalled at age 40 was not predictive of change. However, higher BMI at age 60 (i.e. relatively close to the timing of the mammograms), predicted a lower chance of adverse density change Wolfe and SCC measures, whereas higher BMI was related to a higher adverse change in SMF%. Since the area-based measures of density that we used were not computer-assisted, we could not investigate the effect of BMI on the individual components of area-based breast density (i.e. dense and non-dense area). The reduction in breast density over time is driven by an increase in total breast area and a concomitant decrease in dense area [2,4]. Further investigations of the differential effects of determinants of change on the two components of breast density are therefore warranted.

Our results for reproductive factors, that are known to be related to breast cancer risk, are consistent with those of McCormack and colleagues [2]. This indicates that the mode of action through which these factors affect breast cancer risk appears not to be through their effect on adverse change in breast density over time.

It remains unclear whether the detrimental effect of high breast density on breast cancer risk is due to tracking over time [2], conceptualised as a sensitive period model [14], or due to the cumulative exposure over many years of high density [4]. In the Mayo Clinic mammography screening cohort, no association between change in breast density and subsequent breast cancer risk was seen [15]. Conversely, using pooled data from the Breast Cancer Surveillance Consortium, an increase in breast density among women with originally low density (BI-RADS 1 or 2) was associated with a higher risk of breast cancer, whereas women with originally high breast density retained their higher level of breast cancer risk, irrespective of their subsequent density measure [16]. The numbers of women in the current dataset precluded any precise estimation of the magnitude of the increase in risk associated with a positive change in breast density. Future follow-up of the cohort will allow us to address this.

In summary, we have demonstrated that factors measured in early life, including those that are known to have an effect on a single measure of breast density and breast cancer risk, do not appear to be related to change in breast density. More proximal factors, such as current BMI and HRT use, are modifiable determinants of breast density change.

Table 1. Determinants of change in breast density for four different measures of density

	Change in Wolfe OR (95% CI)	Change in SCC OR (95% CI)	Change in SMF β (95% CI)	Change in SMF% β (95% CI)
Age at menarche				
<=12	1.00	1.00	0.00	0.00
13	1.21	2.18	-0.78	0.85
	(0.41 to 3.56)	(0.84 to 5.65)	(-4.88 to 3.32)	(-0.30 to 2.00)
>=14	0.58	1.78	-3.09	0.52
	(0.14 to 2.43)	(0.62 to 5.08)	(-7.36 to 1.19)	(-0.68 to 1.72)
Ever pregnant				
No	1.00	1.00	0.00	0.00
Yes	0.70	0.82	-3.54	-0.14
	(0.25 to 1.97)	(0.36 to 1.89)	(-7.38 to 0.30)	(-1.22 to 0.93)
HRT use				
Never	1.00	1.00	0.00	0.00
Ever	0.34	0.51	-5.47	-1.03
	(0.12 to 0.99)	(0.23 to 1.12)	(-9.15 to -1.79)	(-2.08 to 0.02)

Table 1. (*Continued*)

Ever smoked				
No	1.00	1.00	0.00	0.00
Yes	3.82	1.98	1.37	0.15
	(1.06 to 13.78)	(0.85 to 4.62)	(-2.15 to 4.89)	(-0.84 to 1.15)
Fruit intake				
Under 2 portions per day	1.00	1.00	0.00	0.00
2 or more portions per day	0.47	1.40	0.51	0.71
	(0.17 to 1.34)	(0.65 to 3.01)	(-2.94 to 3.97)	(-0.26 to 1.68)
Vegetable intake				
Under 2 portions per day	1.00	1.00	0.00	0.00
2 or more portions per day	0.09	0.35	-0.03	-0.28
	(0.01 to 0.71)	(0.13 to 0.96)	(-3.66 to 3.61)	(-1.31 to 0.74)
Exercise @ 20y				
Not /not very active	1.00	1.00	0.00	0.00
Fairly / very active	0.61	0.54	0.97	-0.21
	(0.22 to 1.71)	(0.24 to 1.20)	(-3.14 to 5.07)	(-1.37 to 0.95)
Exercise @ 40y				
Not /not very active	1.00	1.00	0.00	0.00
Fairly / very active	0.88	1.39	-3.61	0.32
	(0.31 to 2.51)	(0.56 to 3.44)	(-7.64 to 0.41)	(-0.83 to 1.46)
Exercise @ 60y				
Not /not very active	1.00	1.00	0.00	0.00
Fairly / very active	0.44	0.77	-3.80	1.08
	(0.16 to 1.21)	(0.35 to 1.69)	(-7.53 to -0.07)	(0.04 to 2.12)
BMI @ 20y				
>=22	1.00	1.00	0.00	0.00
22.1-24.9	0.66	1.01	-1.14	-0.77
	(0.21 to 2.12)	(0.41 to 2.48)	(-5.15 to 2.87)	(-1.89 to 0.35)
>=25	0.82	1.25	0.15	-1.15
	(0.20 to 3.39)	(0.40 to 3.93)	(-5.66 to 5.96)	(-2.77 to 0.48)
Per kg/m^2	0.94	1.10	0.16	-0.13
	(0.77 to 1.15)	(0.95 to 1.27)	(-0.53 to 0.86)	(-0.33 to 0.06)
BMI @ 40y				
>25	1.00	1.00	0.00	0.00
25-29.9	0.23	0.46	-2.30	-1.96
	(0.05 to 1.15)	(0.14 to 1.52)	(-7.14 to 2.52	(-3.35 to -0.58)

Table 1. (*Continued*)

>=30	0.64 (0.07 to 6.25)	0.64 (0.07 to 5.55)	10.27 (-1.34 to 21.9)	-1.27 (-4.61 to 2.07)
Per kg/m^2	0.90 (0.74 to 1.09)	0.91 (0.77 to 1.07)	0.10 (-0.60 to 0.81)	-0.46 (-0.65 to -0.26)
BMI @ 60y				
>25	1.00	1.00	0.00	0.00
25-29.9	0.52 (0.15 to 1.79)	0.56 (0.22 to 1.49)	0.64 (-3.39 to 4.67)	-2.78 (-3.94 to -1.62)
>=30	0.34 (0.06 to 1.86)	0.24 (0.05 to 1.17)	9.44 (3.52 to 15.35)	-3.19 (-4.90 to -1.49)
Per kg/m^2	0.90 (0.77 to 1.04)	0.88 (0.78 to 1.00)	0.80 (0.33 to 1.26)	-0.40 (-0.53 to -0.26)

Note: change in density is coded so that a positive change means an increase in density.

References

1. McCormack, V., dos Santos Silva, I.: Breast density and parenchymal patterns as markers of breast cancer risk: a meta-analysis. Cancer Epidemiol Biomarkers Prev. 15, 1159–1169 (2006)
2. McCormack, V.A., Perry, N.M., Vinnicombe, S.J., dos Santos Silva, I.: Changes and tracking of mammographic density in relation to Pike's model of breast tissue aging: a UK longitudinal study. Int. J. Cancer (2009)
3. Kelemen, L.E., Pankratz, V.S., Sellers, T.A., Brandt, K.R., Wang, A., Janney, C., Fredericksen, Z.S., Cerhan, J.R., Vachon, C.M.: Age-specific trends in mammographic density: the Minnesota Breast Cancer Family Study. Am. J. Epidemiol. 167, 1027–1036 (2008)
4. Maskarinec, G., Pagano, I., Lurie, G., Kolonel, L.N.: A longitudinal investigation of mammographic density: the multiethnic cohort. Cancer Epidemiol Biomarkers Prev. 15, 732–739 (2006)
5. Jeffreys, M., Warren, R., Gunnell, D., McCarron, P., Davey Smith, G.: Life course breast cancer risk factors and adult breast density (United Kingdom). Cancer Causes Control 15, 947–955 (2004)
6. Jeffreys, M., Warren, R., Highnam, R., Davey Smith, G.: Breast cancer risk factors and a novel measure of volumetric breast density: cross-sectional study. Br. J. Cancer 98, 210–216 (2008)
7. McCarron, P., Smith, G.D., Okasha, M., McEwen, J.: Life course exposure and later disease: a follow-up study based on medical examinations carried out in Glasgow University (1948-68). Public Health 113, 265–271 (1999)
8. Wolfe, J.N.: Breast patterns as an index of risk for developing breast cancer. AJR Am. J. Roentgenol. 126, 1130–1137 (1976)
9. Jeffreys, M., Warren, R., Highnam, R., Davey Smith, G.: Initial experiences of using an automated volumetric measure of breast density: the standard mammogram form. Br. J. Radiol. 79, 378–382 (2006)
10. Highnam, R., Pan, X., Warren, R., Jeffreys, M., Davey Smith, G., Brady, M.: Breast composition measurements using retrospective standard mammogram form (SMF). Phys. Med. Biol. 51, 2695–2713 (2006)

11. Byrne, C.: Invited commentary: assessing breast density change–lessons for future studies. Am. J. Epidemiol. 167, 1037–1040 (2008)
12. Atkinson, C., Warren, R., Bingham, S.A., Day, N.E.: Mammographic patterns as a predictive biomarker of breast cancer risk: effect of tamoxifen. Cancer Epidemiol Biomarkers Prev. 8, 863–866 (1999)
13. Sala, E., Warren, R., McCann, J., Duffy, S., Day, N., Luben, R.: Mammographic parenchymal patterns and mode of detection: implications for the breast screening programme. J. Med. Screen. 5, 207–212 (1998)
14. Kuh, D., Ben-Shlomo, Y., Lynch, J., Hallqvist, J., Power, C.: Life course epidemiology. J. Epidemiol Community Health 57, 778–783 (2003)
15. Vachon, C.M., Pankratz, V.S., Scott, C.G., Maloney, S.D., Ghosh, K., Brandt, K.R., Milanese, T., Carston, M.J., Sellers, T.A.: Longitudinal trends in mammographic percent density and breast cancer risk. Cancer Epidemiol Biomarkers Prev. 16, 921–928 (2007)
16. Kerlikowske, K., Ichikawa, L., Miglioretti, D.L., Buist, D.S., Vacek, P.M., Smith-Bindman, R., Yankaskas, B., Carney, P.A., Ballard-Barbash, R.: Longitudinal measurement of clinical mammographic breast density to improve estimation of breast cancer risk. J. Natl. Cancer Inst. 99, 386–395 (2007)

Comparison of Subregional Breast Density with Whole Breast Density

Serghei Malkov[1], Lin Ma[2], Karla Kerlikowske[2], Jeff Wang[1], Steve Cummings[2,3], and John Shepherd[1]

[1] Department of Radiology and Bioimaging, University California at San Francisco, 1635 Divisadero Street, Suite 525, San Francisco, CA 94115 USA
[2] Departments of Medicine and Epidemiology and Biostatistics, University California at San Francisco, San Francisco, CA, USA
[3] San Francisco Coordinating Center, California Pacific Medical Center, San Francisco, CA, USA
John.shepherd@radiology.ucsf.edu

Abstract. The purpose of this paper is to compare subregional breast density and whole breast density and their association with breast cancer risk. The film mammograms of 278 cases and 834 age and ethnicity-matched controls were digitized and analyzed using single-energy x-ray absorptiometry (SXA). The subregion was a 3-cm diameter circle centered in the breast. The whole and subregional densities are found to be highly correlated (r^2=0.7). The 4:1 quartile odds ratio after controlling for other significant risk factors (age, BMI, family history and age at first live birth) was 3.6 (95 CI 2.1-5.4) and the 2.4 (95 CI 1.5-3.7) for the whole and subregional breast density, respectively. Further studies are underway to optimize the placement of the ROI and combined multiple regions.

Keywords: mammographic density, quantitative mammography.

1 Introduction

Breast density, a strong risk factor for breast cancer, is quantified from mammograms as the dense area (or dense volume) normalized to the total breast area (or total breast volume). Little is known regarding the accuracy or predictive power of quantifying breast density in subregions. According to the spatial distribution of dense tissue studies [1] high-density areas are clustered at the central part of the breast. Breast density in subregions is of interest because other technologies, such as optical spectroscopy [2, 3] and quantitative ultrasound [4], may be useful to quantify breast density but not practical for whole breast measures. A centralized subregion has also been used to measure parenchymal patterns as a risk factor for breast cancer. However, little is known regarding the correlation of the density and features to whole breast measures.

Only three papers are known to have addressed any questions regarding local breast density distributions. In a study of the association of quadrant density to the location of DCIS, Ursin et al. used a subjective breast density determination by a radiologist of the fraction of dense tissue area in each breast quadrant [5]. Vachon et al. used areal percent

J. Martí et al. (Eds.): IWDM 2010, LNCS 6136, pp. 402–407, 2010.

mammographic density in breast quadrants to local associations of density to local cancer [6]. Pinto Pereira et al. used areal percent mammographic density subdivided on the mammogram into 48 rectangles to look at the density spatial autocorrelation [1]. All of these investigators were limited in their ability to spatially describe breast density since only regions on the boundary of a dense area would report a region as something different than 0 (total fat) or 1 (total density). However, there are techniques available that measure the volume of dense tissue on a pixel-by-pixel scale [7] that could quantify small subregional densities.

We hypothesize that the volumetric breast density measured in a localized region of interest has good correlation and a similar risk association to that measured from the whole breast. In this study, we compared subregional percent fibroglandular dense volume (sub_%FGV) to the total percent fibroglandular dense volume (%FGV) in an established cohort of women with breast cancer.

2 Methods

The film mammograms of 278 cases and 834 age- and ethnicity-matched controls were digitized and analyzed for both sub_%FGV and %FGV using single-energy x-ray absorptiometry technique (SXA) [7, 8]. The mammograms were acquired prospectively as part of the Breast Cancer Cohort Study, a prospective collection of mammograms, serum, and breast health questions in the San Francisco Bay area. Only the CC-view images were analyzed for this study. The SXA method modeled breasts as a tissue mass of two materials, fat and fibroglandular tissue. SXA uses a calibration phantom in the corner of each mammogram to convert grey-scale values to the associated fat and fibroglandular tissue components. The calibration phantom is an in-image bi-wedge phantom that created a compressible thickness between the compression surfaces to image two reference materials with the same thickness as the breast. The wires imbedded in the phantom allowed for compression thickness and paddle tilt estimates. To determine breast volume, breast thickness was estimated for all pixels in contact with both compression surfaces and those out of contact in the periphery. Knowing the compression thickness under the phantom and tilt angle we estimated breast compression area as a tilted plane. Breast shape in the periphery region was approximated as having a semicircular curvature in cross section. The end results are %FGV and thickness maps for each participant. The total breast %FGV is just the sum of the FGV for each pixel divided by the total breast volume. Subregional %FGV values can be defined as contours drawn on the digitized image. The subregional %FGV studied in this paper (sub_%FGV) was measured in a 3-cm diameter region centered in the breast. A line was drawn from the nipple to the center chest wall that divided the breast into two equal areas. The center of the 3-cm region was placed at the midpoint of this line. This placement was picked for convenience and for easy automation. The size of the region was chosen to insure that the entire region was within the breast tissue in contact with the compression surfaces. Larger regions were found to go into the periphery region of the smaller breast participants. The fibroglandular dense tissue volumes of the whole breast (FGV) and subregion (sub_FGV) were calculated as a product of %FGV and breast volume and as a product of %FGV and subregion breast volume, respectively. Fig. 1 (left) shows the biwedge SXA phantom in the background conforming to the tilted compression angle generated by a quality control phantom in the foreground. Fig. 1 (right) shows the %FGV density map of a LCC-view clinical screen-film mammogram with the subregion circle ROI. The image of the biwedge phantom has been excluded.

The whole breast %FGV and sub_%FGV were compared using linear regression for all participants. The mean values for the cases and controls for each measure were also compared. The significance of the difference was determined using the Pearson's p-value. Lastly, the cases were compared to quartiles of the controls and odds ratios calculated after controlling for BMI, family history, and age at first live birth using logistic regression.

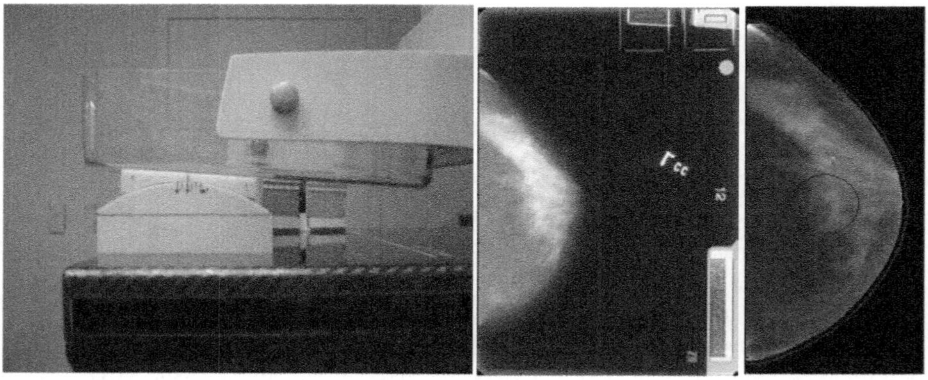

Fig. 1. Compressed wedge phantom (brown) with calibration phantom (left), clinical LCC mammogram showing calibration phantom (center), and SXA density map with the subregional circle ROI (right)

3 Results

A comparison of subregional and whole breast density measures is presented in Fig. 2. The %FGV and sub_%FGV were highly correlated with r^2=0.7.

Using Bland-Altman analysis presented in Fig. 3, we observed positive bias trend in the difference between the sub_%ROI and %FGV when compare to the average values. The sub_%FGV has lower values than the total %FGV for low density breasts and higher values for high density breasts.

The results of case and control breast and population parameters are presented in Table 1. One can see lower mean values for the case and control subregional density distribution in comparison to mean values for the total breast density distribution. In addition, the sub_%FGV case and control distribution is about 6% broader in comparison to the %FGV distribution.

Quartile odds ratios were calculated for both %FGV and sub_%FGV. The 4:1 quartile odds ratio for %FGV was 3.6 (95 CI 2.1-5.4) after controlling for other significant risk factors (age, BMI, family history, and age at first live birth). The odds ratio for the sub_%FGV at similar conditions was 2.4 (95 CI 1.5-3.7).

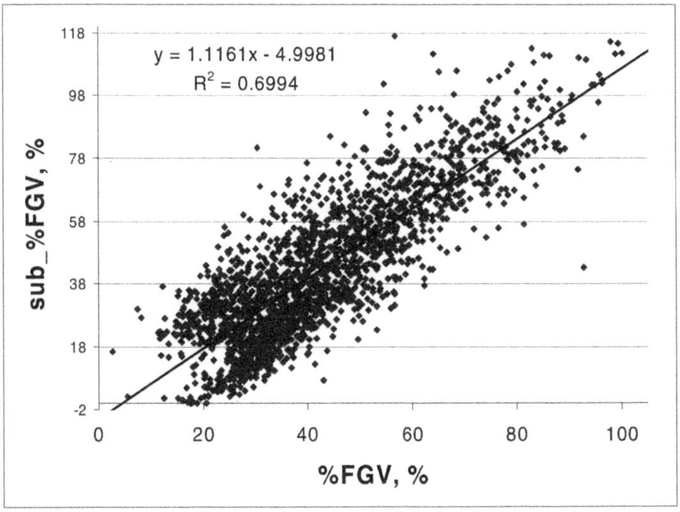

Fig. 2. Sub_%ROI versus %FGV dependence

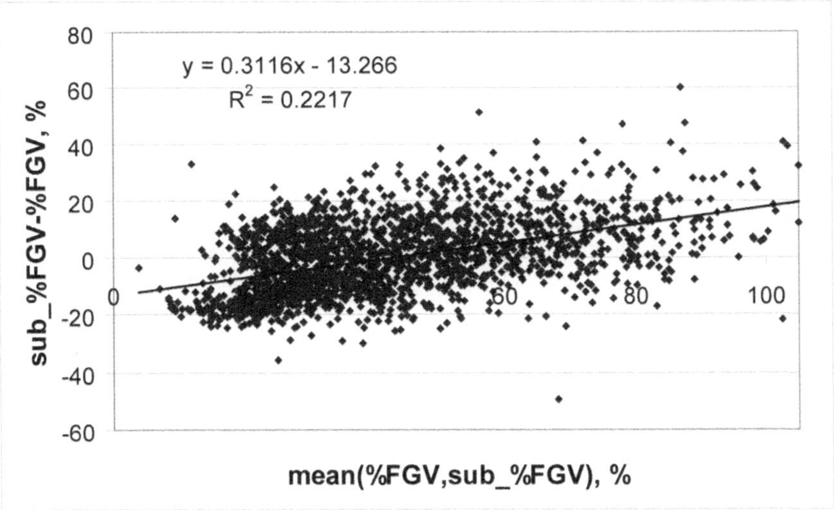

Fig. 3. Bland-Altman plot of difference between the sub_%FGV and %FGV

Table 1. Parameters of case and control distributions

Variable	*P-value (2 sided)*	*Control Mean (Std. Dev.)*	**Case Mean(Std. Dev)**
%FGV, %	0.0007	44.95 (16.95)	48.45 (17.08)
FGV, cm^3	<0.0001	194.46 (99.32)	241.72 (135.28)
Sub_%FGV, %	0.046	42.12 (22.92)	45.31 (23.23)
Sub_FGV, cm^3	0.003	14.95 (7.15)	16.43 (7.18)
BMI, kg/m^2	0.005	24.1 (4.0)	25.0 (5.0)
Age, years	0.95	57.2 (11.3)	57.2 (11.4)

4 Discussion

We found that sub_%FGV was highly correlated to %FGV as well has having a similar relationship to breast cancer risk. Thus, our hypothesis was correct. However, there were systematic differences between the values that ranged from approximately -10% to 20% %FGV over the entire range of values. Individual women had an even higher disagreement with a few differing by as much as 60%. It could be that the subregional density value was more predictive and the preferred density value. Yet, we found that the odds ratio values for sub_%FGV were lower than for whole breast %FGV although both were statistically and clinically significant. Our conclusion for this subregion is that it is a strong predictor of cancer risk but not superior to whole breast density, and that it is unlikely that the women with disparate subregional values are better classified with subregional values than with whole breast values.

The region we picked for this study may not be representative of all subregions. It was picked for convenience. The lower odds ratios were related to the large population standard deviations for sub_%FGV when compared to %FGV. It is known that density is not evenly distributed in breasts. Ursin [5] found that the highest density quadrant was the upper/outer with density values of 56% versus the lowest density quadrant (38%). The %FGV of the projected upper/outer quadrant may be a superior region to sub_%FGV and should also be investigated with volumetric methods.

Our findings to date seem to indicate that methods using subregional values may not be as strong of risk indicators as whole breast %FGV. But there still may be advantages to using technologies that look at subregions. For example, many women choose not to have mammograms for either discomfort or safety reasons. Nonionizing subregional technologies such as optical spectroscopic or ultrasound probes may offer a way to utilize breast density in risk models without the patient having to have a mammogram. In addition, there is controversy surrounding the use of screening mammography for women younger than 50 because of the potential harms of false positives and accumulated radiation dose. The U.S. Preventive Services Task Force (USPSTF) recently recommended against screening mammography for women under 50 years old [9]. Furthermore, the USPSTF stated that starting biennial mammography screening in these women should be up to the individual and take into account their specific benefits and harms. For these women, a nonionizing measurement of breast cancer risk is needed, and this most likely would be a regional measure.

Our study was limited in the following ways. First, we used digitized mammograms versus the current state-of-the-art Full Field Digital Mammography (FFDM). Second, we only used one region when other regions or regions used together may show additional benefit. This study shows that measuring subregional density holds promise as a surrogate measure of whole breast density especially in situations when whole breast density is not available.

References

1. Pinto Pereira, S.M., et al.: The spatial distribution of radiodense breast tissue: a longitudinal study. Breast Cancer Res. 11(3), R33 (2009)
2. Taroni, P., et al.: Seven-wavelength time-resolved optical mammography extending beyond 1000 nm for breast collagen quantification. Opt. Express. 17(18), 15932–15946 (2009)

3. Stromberg, C.N., et al.: Radiographic risk signs for loosening after cemented THA: 61 loose stems and 23 loose sockets compared with 42 controls. Acta Orthop. Scand. 67(1), 43–48 (1996)
4. Glide-Hurst, C.K., Duric, N., Littrup, P.: Volumetric breast density evaluation from ultrasound tomography images. Med. Phys. 35(9), 3988–3997 (2008)
5. Ursin, G., et al.: Greatly increased occurrence of breast cancers in areas of mammographically dense tissue. Breast Cancer Res. 7(5), R605–R608 (2005)
6. Vachon, C.M., et al.: Mammographic breast density as a general marker of breast cancer risk. Cancer Epidemiol Biomarkers Prev. 16(1), 43–49 (2007)
7. Shepherd, J.A., et al.: Novel use of single X-ray absorptiometry for measuring breast density. Technol. Cancer Res. Treat. 4(2), 173–182 (2005)
8. Shepherd, J.A., et al.: Clinical comparison of a novel breast DXA technique to mammographic density. Med. Phys. 33(5), 1490–1498 (2006)
9. Mandelblatt, J.S., et al.: Effects of mammography screening under different screening schedules: model estimates of potential benefits and harms. Ann. Intern. Med. 151(10), 738–747 (2009)

Comparing a New Volumetric Breast Density Method (Volpara™) to Cumulus

Mona Jeffreys[1], Jennifer Harvey[2], and Ralph Highnam[3]

[1] Department of Social Medicine, University of Bristol, Canynge Hall, 39 Whatley Rd, Bristol BS8 2PS, UK
[2] Department of Radiology, University of Virginia Health System, Charlottesville, VA 22908, US
[3] Matakina Technology Limited, Level 9, 50 Manners Street, Wellington, NZ
Mona.jeffreys@bristol.ac.uk

Abstract. Computer-aided thresholding programs, such as Cumulus, are seen as the gold standard for breast density measurement. In this paper we compare a new volumetric breast density software package, Volpara™ to an expert's BI-RADS visual assessment and Cumulus and show that all are closely related, whilst there is a less close relationship between Cumulus percent breast density and absolute volume of dense tissue. These results support the further validation of this new method against breast cancer outcomes.

Keywords: Breast density, volumetric, Volpara.

1 Introduction

Cumulus [1] and similar mammographic density estimator programs are widely considered to be the gold standard for breast density work. They are based on a user-defined thresholding method, and the density calculation is area-based, i.e. calculated from the area of the projected image. These measures have been shown in a meta-analysis to correlate well with breast cancer risk [2]. Nonetheless, the method is subjective and the density measures that results from its use are highly dependent on the observer, with substantial inter- and intra-observer variability, although training does appear to reduce this variability [3,4].

There is increasing interest in the potential for fully automated volumetric measures of breast density. The advantages of these methods include an elimination of user-variability, elimination of the time-consuming density estimation and consideration of the breast as a 3-D organ.

In this paper we compare a new quantitative, user-independent, volumetric breast density method, Volpara™, to visual assessment and to Cumulus. Volpara™ is based around an entirely relative physics model, and is an extension of work described previously [5,6]. The model was designed from the outset with a focus on temporal comparison and for working on images from all digital detectors. A full description can be found in [7]. Key differences with SMF are in the robustness and reliability of the results, especially in dense breasts, and not including skin in the volume of dense tissue.

J. Martí et al. (Eds.): IWDM 2010, LNCS 6136, pp. 408–413, 2010.

2 Methods

As part of the American College of Radiology Imaging Network (ACRIN) Digital Mammography (DMIST) trial [8], the University of Virginia recruited approximately 1,300 women, most of whom had both a film-screen mammogram and a GE digital mammogram on the same day performed by the same technologist.

Cumulus density estimation was performed by one radiologist (JAH) on the left CC image of the digitized film-screen images, and on the "for processing" (i.e. raw) digital image. The BI-RADS breast composition visual assessment (1 to 4) was also performed. From that dataset, we selected the first 105 cases from each of the BI-RADS breast composition categories to investigate the performance of Volpara™, a new volumetric breast density method. Statistical methods used were the correlation coefficient, regression coefficient and Bland-Altman analyses for agreement. 95% confidence intervals (CI) were calculated for each estimation.

The study was approved by the University of Virginia Institutional Review Boards (IRB) and was Health Insurance Portability and Accountability Act (HIPAA) compliant.

3 Results

Out of the 420 cases selected, only 324 cases had readings available for both film screen mammograms and digital mammograms. The BI-RADS density was fatty in 82 patients, scattered in 68 patients, heterogeneous in 102 patients, and extremely dense in 72 patients.

Cumulus breast density percentage (BD%) from the raw digital image and from the films were closely related to each other, Figure 1; the correlation coefficient was 0.95. However, the results from the raw digital image were systematically over-estimated compared to from the digitised film (regression coefficient 0.89, 95% CI 0.86 to 0.92, i.e. a 1% increase in the digital image was associated with a 0.89% increase in the analogue image). Using a Bland-Altman analysis to measure agreement, the mean difference between the two measures was -3.96% (95% CI -4.93 to -2.99).

There was a strong relationship between Volpara™ BD% and BI-RADS categories, with the median Volpara™ percent density rising linearly from 4.0% in the lowest BI-RADS category to 18.9% in the top category (Figure 2).

There was also a strong relationship between Volpara™ BD% and Cumulus BD%, see Figure 3. The correlation coefficient was 0.85. However, as can be seen from a comparison of the scales of the x and y axes in Figure 3, the proportion of the breast that is considered dense is much smaller when measured on a volumetric compared to an areal scale. This explains the rather low regression coefficient; a 1% increase in BD% is associated with an 0.20% increase in Volpara™ BD%, 95% CI 0.18% to 0.21%).

The relationship between the area-based percentage breast density measures and absolute volume of dense tissue was, as expected, weaker than with the volumetric percent density. The correlation coefficient between the absolute volume of dense tissue and the Cumulus BD% was 0.45, see Figure 4. A 1% increase in Cumulus BD% was associated with a 0.56cm^3 increase in absolute volume of dense tissue (95% CI: 0.44 to 0.68).

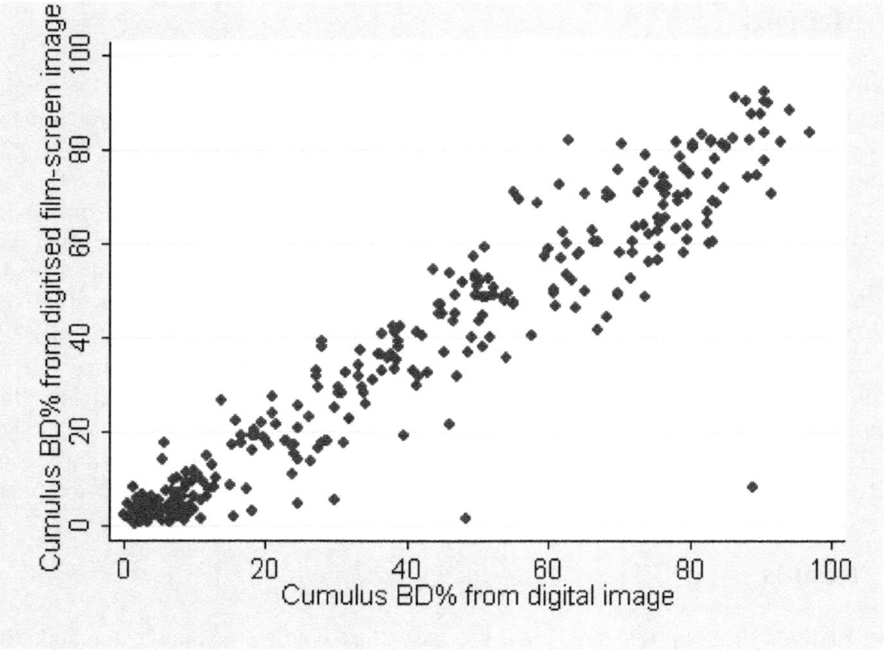

Fig. 1. Relationship between the Cumulus BD% measured on the raw digital (x-axis) and digitized film-screen (y-axis) image

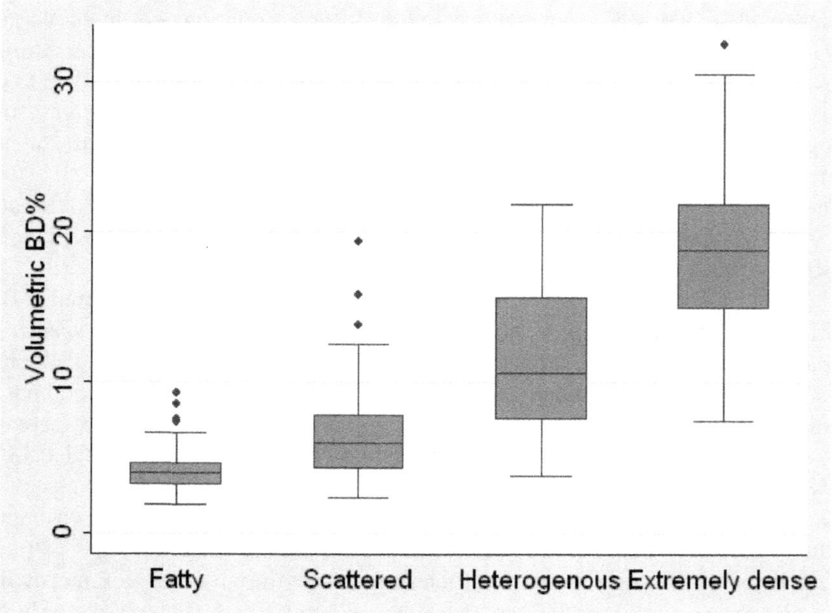

Fig. 2. Association between Volpara™ BD% and BIRADS categories. For each category, the horizontal line shows the median, the box contains the 25th to 75th percentiles of data.

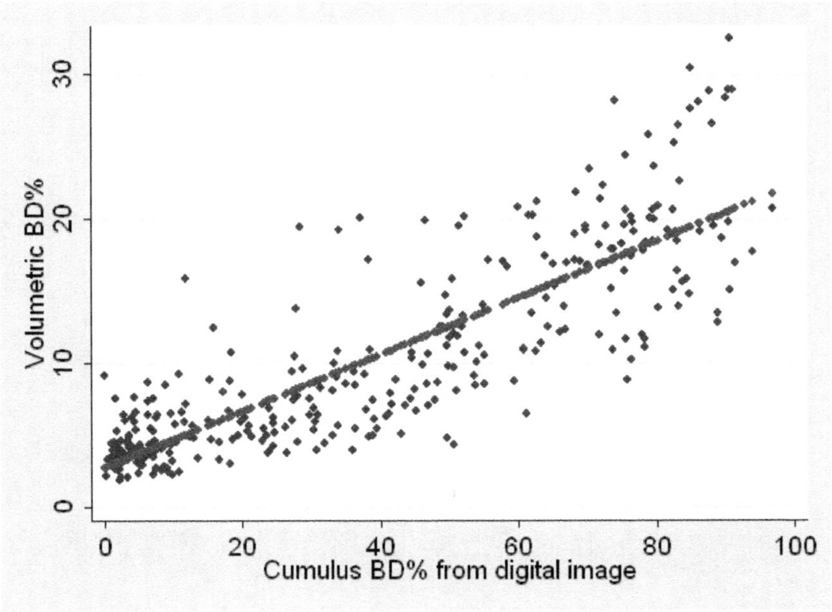

Fig. 3. Association between Cumulus BD% and Volpara™ BD%

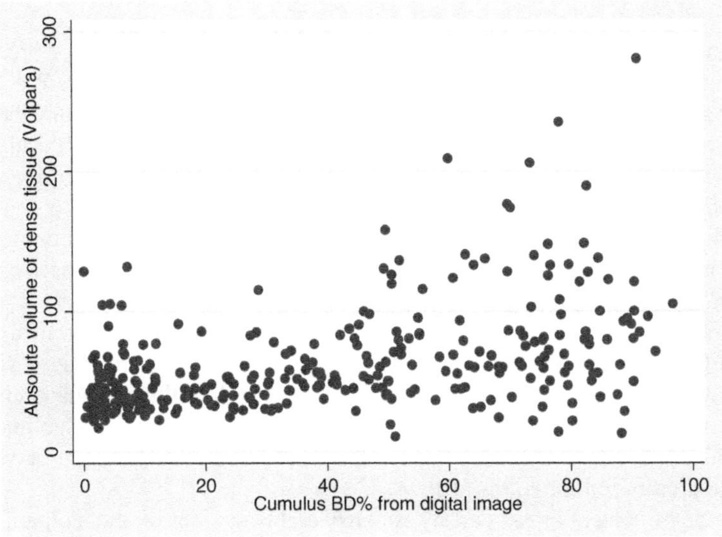

Fig. 4. Association between Cumulus BD% and absolute Volpara™ dense volume

There was a small increase in the absolute dense tissue volume across BI-RADS and Cumulus categories, with the median Volpara™ volume rising from 38cm^3 in the lowest BI-RADS category to 80cm^3 in the highest, and from 41cm^3 in the lowest Cumulus category (<10%) to 80cm^3 in the highest (>75%), see Figure 5.

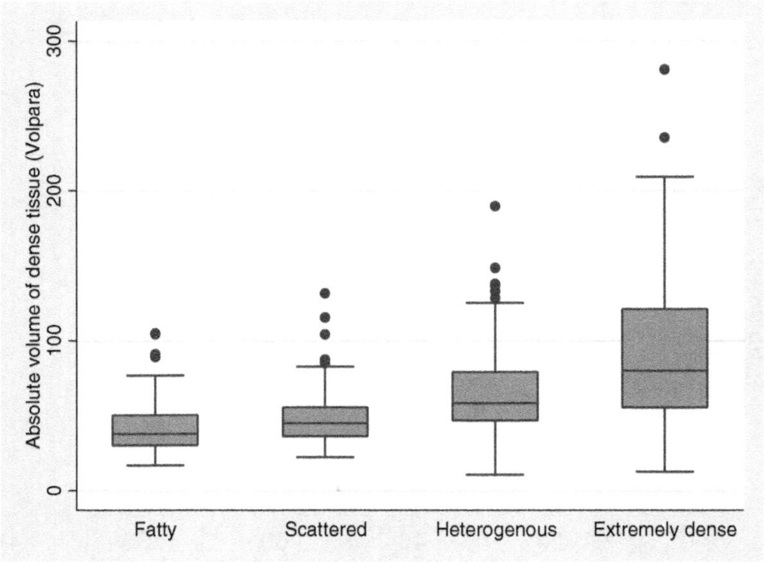

Fig. 5. Association between BI-RADS categories and absolute Volpara™ dense volume

4 Discussion

As far as we are aware, this is the first analysis to compare measures of breast density from digitized film-screen and digital images taken on the same day. We found that BD% from the raw digital image was systematically higher than from the digitised film. A previous analysis suggested the opposite association [9], with a higher mean density when estimated from analogue compared to digital films. However, because the current study was based on films taken on the same day, it is likely that these results are more reliable. As increasingly more studies of breast density are based on FFDM images, these results have important implications for the comparisons of studies of breast density using digital or digitised measures.

Volpara™, the new volumetric method of assessing breast density, designed to be run on FFDM images, is closely related to breast density from area-based visual techniques (BI-RADS) and a semi-automated technique, Cumulus. Unsurprisingly, there is only a weak linear relationship between the volume of dense tissue and the area based density measure, as has been found with previous comparisons between volumetric and area-based measures [10].

The range of percent breast density is considerably smaller on the Volpara compared to the Cumulus scale. The results are not fully consistent with previously reported measures of volumetric breast density. For example, among women without breast cancer, the inter-quartile range of absolute breast volume when measured by SMF v2.2β run on digitised images was 20.8% to 33.5% [11], compared to 4.6% to 15.6% in this study; this is explained by Volpara™ not including skin in its calculation.

In conclusion, since Volpara™ correlates well with the gold standard measure of breast density, we expect that there should also be a strong relationship between Volpara™ and breast cancer risk. The results presented here support further detailed analysis of the potential of this novel measure in relation to breast cancer outcomes. The most useful study design for this will be a case-control analysis.

References

1. Byng, J.W., Boyd, N.F., Fishell, E., Jong, R.A., Yaffe, M.J.: The quantitative analysis of mammographic densities. Phys. Med. Biol. 39, 1629–1638 (1994)
2. McCormack, V., dos Santos Silva, I.: Breast density and parenchymal patterns as markers of breast cancer risk: a meta-analysis. Cancer Epidemiol Biomarkers Prev. 15, 1159–1169 (2006)
3. Prevrhal, S., Shepherd, J., Smith-Bindman, R., Cummings, S., Kerlikowske, K.: Accuracy of Mammographic Breast Density Analysis: Results of Formal Operator Training. Cancer Epidemiol. Biomarkers Prev. 11, 1389–1393 (2002)
4. Martin, K.E., Helvie, M.A., Zhou, C., Roubidoux, M.A., Bailey, J.E., Paramagul, C., Blane, C.E., Klein, K.A., Sonnad, S.S., Chan, H.P.: Mammographic density measured with quantitative computer-aided method: comparison with radiologists' estimates and BI-RADS categories. Radiology 240, 656–665 (2006)
5. Highnam, R., Brady, M.: Mammographic Image Analysis. Kluwer Academic Publishers, Dortrecht (1999)
6. Highnam, R., Pan, X., Warren, R., Jeffreys, M., Davey Smith, G., Brady, M.: Breast composition measurements using retrospective standard mammogram form (SMF). Phys. Med. Biol. 51, 2695–2713 (2006)
7. Highnam, R., Brady, M., Yaffe, M., Karssemeijer, N., Harvey, J.: Robust Breast Composition Measurements - Volpara. In: IWDM Proceedings (in press, 2010)
8. Pisano, E., Gatsonis, C., Yaffe, M., Hendrick, R., Tosteson, A., Fryback, D., Bassett, L., Baum, J., Conant, E., Jong, R., Rebner, M., D'Orsi, C.: American College of Radiology Imaging Network digital mammographic imaging screening trial: objectives and methodology. Radiology 236, 404–412 (2005)
9. Harvey, J.A.: Quantitative assessment of percent breast density: analog versus digital acquisition. Technol. Cancer Res. Treat. 3, 611–616 (2004)
10. Jeffreys, M., Warren, R., Highnam, R., Davey Smith, G.: Initial experiences of using an automated volumetric measure of breast density: the standard mammogram form. Br. J. Radiol. 79, 378–382 (2006)
11. Aitken, Z., McCormack, V.A., Highnam, R.P., Martin, L., Gunasekara, A., Melnichouk, O., Mawdsley, G., Peressotti, C., Yaffe, M., Boyd, N.F., Dos Santos, S.I.: Screen-film mammographic density and breast cancer risk: a comparison of the volumetric standard mammogram form and the interactive threshold measurement methods. Cancer Epidemiol. Biomarkers Prev. 19, 418–428 (2010)

Visual Assessment of Density in Digital Mammograms

Anisha Sukha[1], Michael Berks[2], Julie Morris[3], Caroline Boggis[4], Mary Wilson[4], Nicky Barr[4], and Susan Astley[2,*]

[1] University of Manchester Medical School, Stopford Building, Oxford Road, Manchester M13 9PT, England
[2] Cancer and Enabling Science, University of Manchester, Stopford Building, Oxford Road, Manchester M13 9PT, England
sue.astley@manchester.ac.uk
[3] Medical Statistics, University Hospital of South Manchester, Wythenshawe Hospital, Southmoor Road, Manchester M23 9LT, England
[4] Nightingale Breast Centre and Genesis Prevention Centre, University Hospital of South Manchester, Southmoor Road, Manchester M23 9LT, England

Abstract. This study compares visual assessment of density on full field digital mammograms using visual analogue scales (VAS) and written percentages. Fifty normal digital screening mammograms were selected at random. Nine readers viewed the images on two occasions, firstly indicating density on a VAS and then estimating the percentage of dense tissue in the breast. Although the two methods were correlated, the degree of agreement between the density estimates varied considerably from reader to reader. More experienced readers used a wider range of values, and inter-observer variability for both methods was higher for these readers. The greatest difference between the methods was in mammograms with a mixed fatty-glandular appearance (density between 55% and 75%). Both methods are quick and convenient, although these results demonstrate a need for training to ensure they are used consistently by readers of different degrees of experience.

Keywords: breast screening, risk estimation, breast density, digital mammography.

1 Introduction

Visual assessment of breast density was first related to breast cancer risk more than a quarter of a century ago [1]. Wolfe's classification scheme described both the quantity of dense tissue visible in the mammographic image and the nature of the mammographic pattern, but assignment of images to classes is subjective. Boyd introduced a six class scheme based only on density which was conceptually simpler but still required subjective assessment of density [2]. Despite this, it is possible to obtain consistent estimates of density using both Wolfe's classes and Boyd's, provided that the reader is sufficiently experienced and well-trained [3]. This result has not, however, been replicated using a large number of readers of varying profession and experience.

* Corresponding author.

J. Martí et al. (Eds.): IWDM 2010, LNCS 6136, pp. 414–420, 2010.

More recently, researchers have attempted to automate the process of measuring density, notably using interactive thresholding [4], calibration-based methods [5, 6] and model-based approaches [7,8]. Of these methods, the interactive thresholding method known as Cumulus is the most widely used. This requires a trained observer to select image thresholds that best delineate the breast edge and glandular tissue, and to interactively mark out the edge of the image and pectoral muscle. The vast majority of studies using Cumulus have operated with digitised films. The calibration-based and model-based methods have been facilitated by the availability of digital images, which provide a more straightforward platform on which to work. These techniques can estimate the volumes of fat and gland in the breast, using knowledge of the imaging process to associate pixel values with thicknesses of dense tissue. The relationship of volumetric density to risk has not yet been as thoroughly explored as that of the percentage estimates introduced by Boyd, or the subjective pattern classification used by Wolfe. It is, however, worth noting that volumetric measures enable the separation of the fat and gland components of the breast, and this will overcome one of the major disadvantages of methods based on percentage area. It has been shown that weight loss (or gain) could seriously confound risk studies that use percentage breast density because such measures are dominated by breast fat, which can change rapidly with change in a woman's body weight. As women lose weight, their percentage breast density rises indicating an increase in risk, when lower weight should correspond to lower risk [9]. Volumetric methods will allow the investigation of fat and gland as separate risk factors.

The introduction of full field digital mammography, whilst facilitating the introduction of quantitative volumetric techniques, has proved a complicating factor for breast density measurement, since the majority of work relating density to risk was performed using film-screen mammography and hardcopy or digitised images. With full field digital mammography there is more scope for image processing to optimise image display so that subtle abnormalities can be detected. Automated methods for estimating density usually use the raw (unprocessed) data and thus avoid the influence of image processing algorithms, but many breast imaging centres currently only save the processed images, as these are the images from which a diagnosis or screening outcome has been determined. In longitudinal studies looking at the change in risk over time, women may have had earlier mammograms taken using film-based systems and later ones using full field digital mammography. Here, visual assessment may provide a method of bridging the gap, although a systematic comparison of readers' assessments across the two modalities would be necessary. An accurate measure of change in density could prove beneficial in a number of applications such as identifying women whose density is not reducing as expected with age, or determining whether treatments such as Tamoxifen are effective in reducing density.

There are several different ways in which visual assessment can be achieved including writing down a percentage of dense tissue; classifying according to Wolfe or Boyd's groups; or marking a Visual Analogue Scale (VAS). In our breast centre, many of our readers have extensive experience in using the latter method, and results have shown a clear correlation with risk, especially when density is estimated in both mammographic projections [10]. In this study we compare two methods of visual assessment to determine which might be most appropriate to use either as a baseline against which to compare automated and semi-automated methods, to indicate 'difficulty' of cases when selecting test sets for evaluation of computer-aided detection algorithms, or to provide an indication of risk in large scale longitudinal research. The study uses a wide range of mammogram readers to provide an indication of how practical it would be to use the methods in a typical screening centre.

2 Materials and Methods

Fifty normal full field digital mammograms were selected at random from screening mammograms taken in 2009 as part of the UK National Health Service Breast Screening Programme (NHSBSP) [11]. They were made available to readers on their usual reporting workstation. Nine mammogram readers of varying experience were recruited (table 1). In Manchester, screening mammograms are read independently by pairs of readers drawn from a pool of experienced radiologists, radiographers and breast physicians all trained in mammogram reading and meeting the NHSBSP guidelines for annual reading workloads. The majority of the readers who participated in this study had prior experience in the visual assessment of breast density using Visual Analogue Scales due to participation in a previous research study [12]. In table 1, the readers are ranked in order of experience based on the number of years of film-reading experience multiplied by the approximate number of mammograms read per year, as it proved impractical to obtain a more accurate assessment of lifetime experience.

The set of fifty mammograms, each of which comprised four images (Cranio-Caudal and Medio-Lateral Oblique views of both breasts), was read by each reader on two different days. On the first day the proportion of breast density in each mammographic view was recorded on a 10cm Visual Analogue Scale on a paper pro-forma, and on the second day, the reader simply wrote down the estimated percentage on dense tissue in the breast in each view. The Visual Analogue Scales were processed automatically to extract percentage densities based on the relative positions of readers' marks across each scale line. The results from the four mammographic views were analysed separately; in this paper, results from a single view (the Right Cranio-Caudal, or RCC) are shown, as similar results were found with the other views.

Table 1. Experience of mammogram readers. RG=Radiographer; CR= Consultant Radiologist; BP= Breast Physician.

Reader	*Profession*	*Years of Experience*	*Approximate number of mammograms read per year*
1	RG	1.0	5,000
2	BP	3.0	5,500
3	RG	4.0	5,000
4	CR	3.5	7,000
5	RG	7.0	7,000
6	BP	8.0	7,000
7	CR	15.0	7,000
8	CR	20.0	7,000
9	CR	23.0	7,000

3 Results

The correlation between estimates of density made using the VAS and percentage varied from 0.95 for a relatively inexperienced radiographer reader (Reader 3) down to 0.72 for one of the experienced consultant radiologists (Reader 7). Examples of the results for these two readers are shown in Figures 1 and 2 as Bland-Altman plots [13]. In Figure 1 the difference between the estimates made using the two techniques is plotted against the mean of the two estimates for Reader 3, a relatively inexperienced radiographer with little prior experience of density estimation (95% limits of agreement in this example are -12.71 - 8.51). At the other end of the spectrum, the results for Reader 7, the experienced radiologist, are shown in figure 2. Here there is poor agreement between the two methods of assessing density (95% limits of agreement - 16.90 - 32.46). This reader had a discrepancy in excess of 15% in more than a third of the mammograms viewed, whereas all but one pair of Reader 3's estimates were within 15% of each other. Six of the nine readers had large discrepancies between estimates for the same mammogram, which was in the middle of the density range with an average density assessed as 50%.

Although the two methods were found to be correlated, the degree of agreement between VAS and percentage estimates varied considerably from reader to reader (table 2). More experienced readers used a wider range of values for both methods, and variation in discrepancy was higher for these readers. The greatest difference between the methods was in mammograms with a mixed fatty-glandular appearance (density between 55% and 75%).

There was no evidence from these data that professional group had a significant effect on the discrepancy between measures made by the two techniques. Similarly, reader experience was not systematically related to discrepancy.

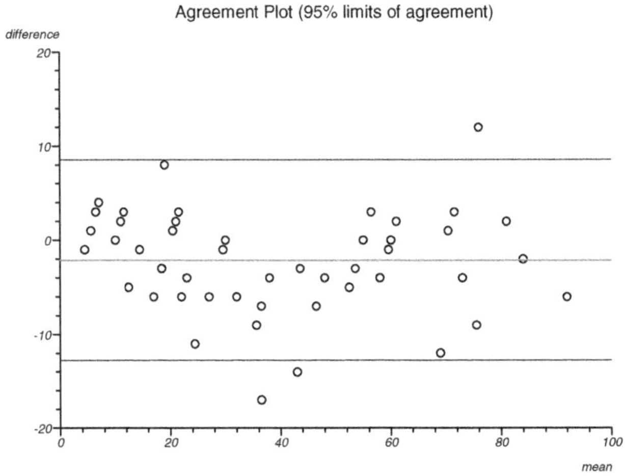

Fig. 1. Bland-Altman plot of the difference between the two estimates of density against the mean. Results are shown for Reader 3 (a relatively inexperienced radiographer).

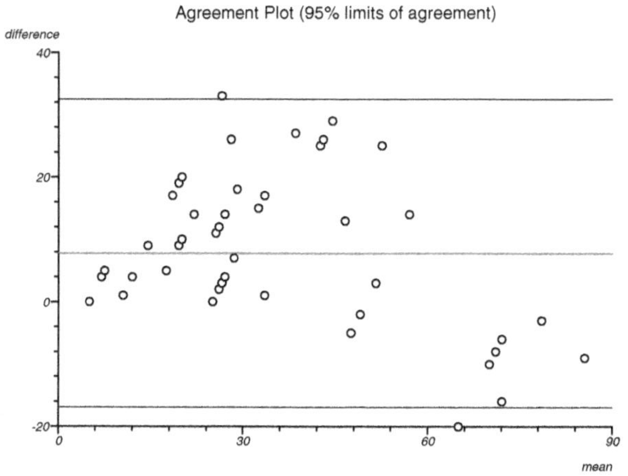

Fig. 2. Bland-Altman plot of the difference between the two estimates of density against the mean. Results are shown for Reader 7 (an experienced consultant radiologist).

Table 2. 95% Limits of agreement ranked by reader experience. Readers 1, 3 and 5 are radiographers, 4, 7, 8 and 9 are consultant radiologists and 2 and 6 are breast physicians.

Reader	*95% limits of agreement (2dp)*	
1	-15.61	15.41
2	-9.28	22.32
3	-12.71	8.51
4	-16.70	5.94
5	-17.67	22.63
6	-22.85	21.69
7	-16.90	32.46
8	-20.14	20.94
9	-16.17	8.09

Figure 3 shows the VAS and percentage estimates assigned by the most experienced reader for all 50 mammograms. Although the images were presented in random order, here we have arranged them in increasing density as assigned by VAS. It is noticeable that this reader has rounded all of the percentage estimates to the nearest 5%. Many of the readers adopted a similar strategy.

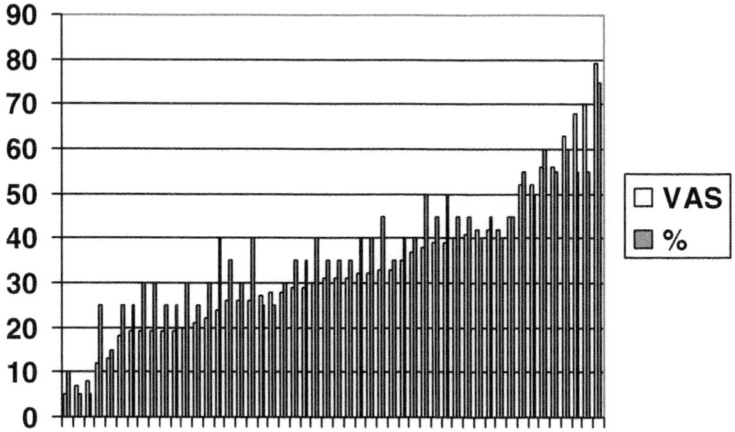

Fig. 3. VAS and % estimates of density for Reader 9 (the most experienced reader, a consultant radiologist). The data is ordered with increasing VAS estimate.

4 Discussion

Both methods of density estimation are quick and convenient, and as expected the estimates of density provided by the two methods show a high degree of correlation. However, for some mammograms there was a large difference between the two estimates and this is a concern. The variation between readers indicates that if the methods are to be used successfully, there is a need for training to ensure that they are used consistently by readers of different degrees of experience. In this study, the readers with least experience in using Visual Analogue Scales were amongst the most consistent.

The discrepancy between estimates made by the two methods was not found to depend systematically on breast density, although the greatest differences between the methods were found in mammograms of mixed fatty-glandular appearance. A larger study with mammograms selected to cover the density range would be necessary to investigate this further; in our data set, which was selected randomly from screening cases, the majority of mammograms were from post-menopausal women, and hence there were only a small number of images showing a high proportion of dense tissue.

Subjective assessment is unlikely to provide sufficient accuracy and reliability to enable subtle changes in density to be monitored, for example to facilitate decisions about the continuation of risk prevention intervention. Whilst previous research has concluded that visual assessment was reproducible, results from only a small number of highly trained readers were investigated [3]. Our results show that different readers behave in different ways when confronted with an image and a scoring system: some use the full range of values available to them, whilst others select particular values that they find conceptually easy (e.g. recording values of 25%, or 50%). The tendency of the majority of readers to round values to the nearest 5 or 10 percent suggests that it may be better to stipulate in advance that this precision is sufficient.

Acknowledgments. The authors would like to acknowledge the contribution of the readers who participated in this study, and of the administrative staff of the Nightingale Breast Centre for their assistance.

References

1. Wolfe, J.N.: Risk for breast cancer development determined by mammographic parenchymal pattern. Cancer 37(5), 2486–2492 (1976)
2. Boyd, N.F., Lockwood, G.A., Byng, J.W., Tritchler, D.L., Yaffe, M.J.: Mammographic densities and breast cancer risk. Cancer Epidemiol. Biomarkers Prev. 7(12), 1133–1144 (1998)
3. Gao, J.N., Warren, R., Warren-Forward, H., et al.: Reproducibility of visual assessment on mammographic density. Breast cancer research and treatment 108(1), 121–127 (2008)
4. Boyd, N.F., Byng, J.W., Jong, R.A., Fishell, E.K., Little, L.E., Miller, A.B., et al.: Quantitative classification of mammographic densities and breast cancer risk: results from the Canadian National Breast Screening Study. J. Natl. Cancer Inst. 87(9), 670–675 (1995)
5. Boyd, N., Martin, L., Gunasekara, A., Melnichouk, O., Maudsley, G., Peressotti, C., Yaffe, M., Minkin, S.: Mammographic density and breast cancer risk: Evaluation of a novel method of measuring breast tissue volumes. Cancer Epidemiol. Biomarkers Prev. 18(6) (2009)
6. Diffey, J., Hufton, A., Beeston, C., Marchant, T., Astley, S.: Quantifying Breast Thickness for Density Measurement. In: Krupinski, E.A. (ed.) IWDM 2008. LNCS, vol. 5116, pp. 651–658. Springer, Heidelberg (2008)
7. Highnam, R., Pan, X., Warren, R., Jeffreys, M., Davey Smith, G., Brady, M.: Breast composition using retrospective standard mammogram form. Physics in Medicine and Biology 51, 2695–2713 (2006)
8. Kaufhold, J., et al.: A calibration approach to glandular tissue composition estimation in digital mammography. Med. Phys. 29(8), 1867–1880 (2002)
9. Patel, H.G., Astley, S.M., Hufton, A.P., Harvie, M., Hagan, K., Marchant, T.E., Hillier, V., Howell, A., Warren, R., Boggis, C.R.M.: Automated breast tissue measurement of women at increased risk of breast cancer. In: Astley, S.M., Brady, M., Rose, C., Zwiggelaar, R. (eds.) IWDM 2006. LNCS, vol. 4046, pp. 131–136. Springer, Heidelberg (2006)
10. Duffy, S.W., et al.: Visually assessed breast density, breast cancer risk and the importance of the craniocaudal view. Br. Can. Research 10(4) (2008)
11. `http://www.cancerscreening.nhs.uk/breastscreen/index.html` (accessed January 2010)
12. Gilbert, F., Astley, S., Gillan, M.G.C., Agbaje, O.F., Wallis, M., James, J., Boggis, C.R.M., Duffy, S.W.: For the CADET II Group. A Comparison of Single Reading with Computer Aided Detection and Double Reading for Screening Mammography. New England Journal of Medicine 359, 1675–1684 (2008)
13. Bland, J.M., Altman, D.G.: Statistical methods for assessing agreement between two methods of clinical measurement. Lancet 1(8476), 307–310 (1986)

Concepts for High-Resolution CT of the Breast

Willi A. Kalender

Institute of Medical Physics (IMP), Henkestr. 91, 91052 Erlangen, Germany
willi.kalender@imp.uni-erlangen.de

Abstract. X-ray computed tomography (CT) has been proposed and evaluated recently as a potential alternative method for breast imaging. Efforts so far showed success with respect to contrast-enhanced dynamic imaging, but suffered from limited spatial resolution. The concept presented here builds upon micro-CT scanning approaches and aims at providing both high spatial resolution at around 100 µm for micro-calcification imaging and advanced dynamic scan capabilities with continuous acquisition and scan times of about 10 seconds for differential diagnosis of lesions. To achieve this, spiral scan modes, slipring technology, high-resolution detectors and high-power micro-focus X-ray tubes are demanded. The concept has been evaluated and confirmed by simulations and basic experiments; first clinical results are expected by the end of 2011.

Keywords: computed tomography, breast, calcifications, image quality, dose.

Full field digital mammography (FFDM) [1–3] represents today's standard and the most widely applied imaging modality for the early detection of breast cancer. However, severe limitations with respect to its sensitivity and specificity are to be acknowledged. In consequence, many alternative approaches are under investigation at present with magnetic resonance imaging (MRI) as one of the promising candidates [4; 5]. For a number of good reasons, X-ray computed tomography (CT) has also been proposed and is presently under investigation as an alternative method.

1 Demands on Breast Imaging

There is a general consensus that 2D projection imaging has limitations; this is especially the case for FFDM when dense breasts are to be examined. The superpositioning of structures can obscure the details of interest to a degree that a diagnostic finding goes undetected. Therefore, full 3D capabilities are an essential demand. Traditionally the detection and diagnosis of microcalcification clusters is one of the most important tasks; high spatial resolution on the order of 100 µm in all three dimensions appears necessary for this. Soft tissue structures shall be discerned just the same, which implies moderate demands regarding low-contrast resolution. In addition, specific information about lesion characteristics is desired; the recording of contrast medium kinetics, that is dynamic imaging capabilities, are of considerable interest as has

J. Martí et al. (Eds.): IWDM 2010, LNCS 6136, pp. 421–427, 2010.

been shown in several studies [6; 7]. Also, any modality under consideration should support intervention and therapy efforts. Patient comfort and safety are a further demand which cannot be neglected. Safety relates both to the radiation dose delivered when X-rays are involved and to the contrast medium dose being applied. Comfort mostly relates to the examination time, patient positioning and to the question if the breast is compressed or not. It is also of importance to image the full breast including all regions close to the chest wall and close to the axilla. These demands are summarized in table 1. It is the purpose of this proposal to show that a dedicated breast CT scanner may fulfill all these demands.

Table 1. General requirements for breast imaging

Full 3D capability
High isotropic spatial resolution of about 100 μm
Dynamic imaging capabilities (sub-minute temporal resolution)
Soft-tissue differentiation
Low dose of X-rays and contrast medium to the patient
Patient comfort, no breast compression

2 Prior Art

Efforts at building a dedicated CT scanner for the breast date back to the 1970s. A so-called CT mammography (CTM) scanner was promoted at the time [8; 9]. The woman was lying prone on the table with one breast at a time hanging through a hole in the table. Spatial resolution was considered inadequate for the mammographic task, and the concept was discontinued.

Renewed efforts started in the early 2000s in the United States [10–13]. Standard components were mostly used for the respective prototype setups. Several limitations resulted from this. E.g., with X-ray tubes and detectors as commonly used in angiography or flat-detector CT with effective pixel sizes of 150 to 400 μm, insufficient spatial resolution was obtained.

However, some of these scanners provide dynamic CT capabilities, i.e. the possibility for differentiating lesions which are enhancing either fast (malignant) or slow (benign). This feature was also shown for standard clinical CT imaging of complete thorax cross-sections with the patient lying prone. Perrone et al. reported [7] remarkably good results for the differential diagnosis of benign and malignant lesions (figure 1). While efforts regarding dynamic CT are being continued, there is a general consensus that dedicated breast CT up to now does not allow sufficient resolution for an adequate diagnosis of microcalcifications to the same degree that mammography does [14].

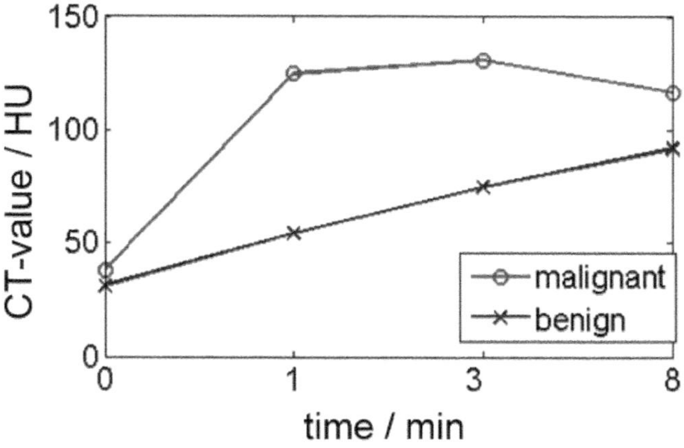

Fig. 1. Dynamic CT allows for differential diagnosis of benign and malignant lesions (Data reproduced from [7]). However, in clinical CT spatial resolution and patient dose values are of the same order as in standard thoracic CT and not acceptable for dedicated breast CT.

3 Proposed Concept

A new effort at breast CT was started by W. Kalender in 2007 [15]. The respective European Union project aimed at investigating the potential of a dedicated breast CT scanner which should be capable of achieving all goals stated in table 1. A sketch of a possible scanner setup proposed at that time is shown in figure 2. It is similar in principle to the setup developed by J. Boone [12] and is used in patient studies [14]. A decisive advantage is expected by providing higher spatial resolution. Special data processing approaches will allow providing different images from a single scan: a high-resolution high-noise image for assessing microcalcification clusters and a low-resolution low-noise scan for viewing soft-tissue lesions [16].

Preliminary proof of evidence for the expected imaging performance of a respective scanner was obtained in first practical test measurements on an experimental micro CT scanner using the parameters as outlined above, in this case a focal spot size of 10 µm and a detector pixel size of 100 µm. Spatial resolution in this case was about 50 µm. Measurement results for a surgical resection specimen are shown in figure 3 to indicate the potential. Microcalcifications in a great range of sizes are displayed with high clarity; to indicate the 3D nature of the findings, three views of the volume are shown side by side from different perspectives.

Essential parts of the necessary basic measurements and simulation studies have been completed; the results confirm that the stated goals can be achieved. However, the concept necessitates the development of new technological solutions for most scanner components. For example, as an X-ray source, a tube with focus size down to 100 µm is required with the focus spot very close to one end of the tube to allow having the focus directly under the table and thereby to include the anatomy up to the chest wall.

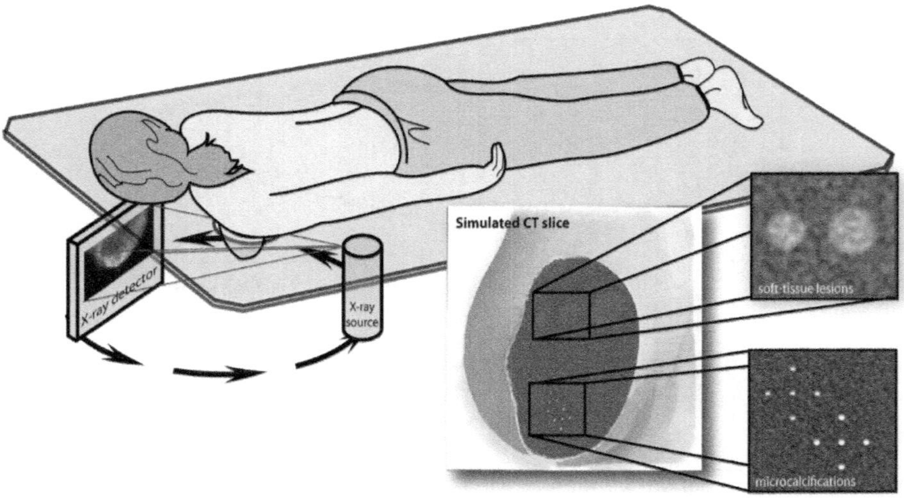

Fig. 2. In dedicated breast CT, the patient will be scanned with low X-ray dose in prone position with only one breast exposed at a time. Calculations and preliminary measurements indicate that we will be able to display microcalcifications as well as or even better than in standard mammography. Soft-tissue lesions which are often obscured in mammography by overlying tissue will be displayed clearly without an increase in dose at decreased spatial resolution. Note: The two separate inserted displays of soft tissue lesions and of microcalcifi-cations are derived from the same, single measurement by different processing.

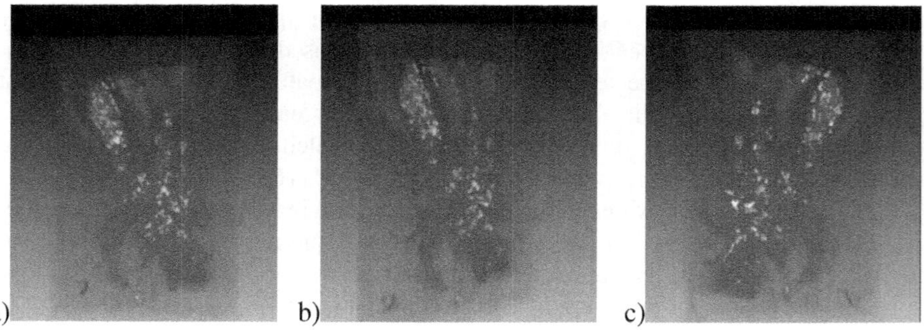

Fig. 3. Experimental micro-CT allows modeling the imaging task. Scans of a breast surgical specimen at about 50 μm isotropic spatial resolution display the 3D microcalcification clusters in a quality not available in clinical imaging so far. Fig. 3a–c display three different perspectives to indicate the 3D nature of the data.

As a detector, a highly efficient design with detector pixel sizes of 100 μm or less is required to achieve the desired high spatial resolution. Also, the number of readings per second has to be increased to allow for a high number of projections being measured during a 360° rotation in very short time. The geometry should be similar to the typical biopsy tables which are in use today, i.e. the scanner should be small and neither demand much space nor intimidate the patient. A proposed setup which shall fulfill these expectations is shown in figure 4. An important step will be the transition from single-circle scans using a flat detector (figure 2) to spiral CT using a dedicated CdTe detector (figure 4a) and a new scanner design allowing for spiral scans (figure 4b).

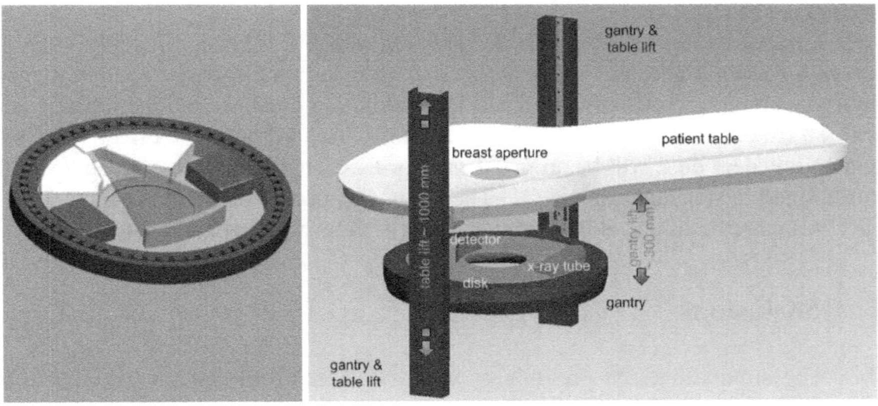

Fig. 4. Proposed CT concept. a) Source-detector design similar to that of modern clinical CT. b) Novel patient-friendly scanner concept which also allows for biopsy taking.

4 Patient Dose Considerations

It is considered essential that patient dose associated with the proposed CT scanner does not exceed the levels necessary and accepted for FFDM. Therefore, the dose situation has been assessed in detail since the very start. In a manner similar to the assessment of image quality, the approach was based both on calculations [17] and on measurements. Respective 3D dose distributions are shown in figure 5. It should be

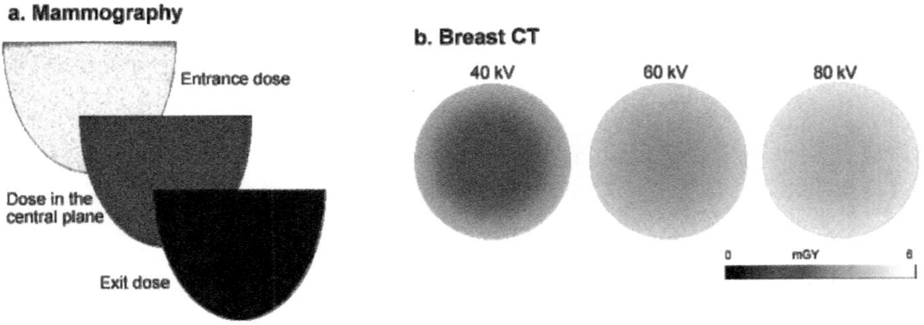

Fig. 5. 3D dose distribution. a) Dose in mammography falls nearly exponentially from entrance to exit plane. Average glandular dose is below 3 mSv. b) Dose in CT is more homogeneous and can be kept to a similar level.

noted that CT delivers a much more homogeneous dose pattern, especially at higher voltage values, and does not reach the high peak values found for FFDM and tomosynthesis on the beam entrance side. Average glandular dose values in the range of 2 to 6 mGy appear feasible for CT [18; 19], which amount to the values typically found in bi-plane FFDM.

It is important to note that dose for dynamic scans will be significantly lower than for high-resolution scans so that the total dose of a complete examination including high-resolution microcalcification and low-resolution dynamic or functional scanning will not exceed the above estimates significantly. Accordingly, the effective or whole body-equivalent dose will be around 1 mSv or less. These values are much lower than those found in standard clinical CT. They may also be compared to the typical natural background radiation level of 3 mSv per year with its range given at 1 to 10 mSv.

5 Conclusions

Based on simulations, on preliminary experiments and on general experience with high-resolution flat-detector micro-CT it appears feasible to build a dedicated CT scanner of the breast fulfilling all the demands given in table 1. In particular, this includes the demand that dose has to be limited to the levels known and accepted for mammography. Necessary developments of respective new technological components, of algorithms and software will take time, but we expect practical solutions for clinical testing to be available by the end of 2011.

Acknowledgement. The author acknowledges support from the European Union Framework Programme 7 [15] for the basic studies on breast CT reported here. The successful cooperation with the project partners from the Universities of Leuven and Rotterdam, respectively, is acknowledged in particular. Special thanks go to Prof. Matthias Beckmann and his team from the Department of Gynecology of the University of Erlangen for very helpful discussions and for providing the specimen shown in figure 3 and to Dipl.-Ing. Felix Althoff for providing the design studies shown in figure 4.

Note: To view the figures in colour, please view the manuscript on http://www.springerlink.com online.

References

[1] Pisano, E.D., Yaffe, M.J.: Digital mammography. Radiology 234, 353–362 (2005)
[2] Fischer, U., Hermann, K., Baum, F.: Digital mammography: current state and future aspects. Eur. Radiol. 16, 38–44 (2006)
[3] Schulz-Wendtland, R., Hermann, K.-P., Wacker, T., Bautz, W.: Aktueller Stand und weitere Perspektiven der digitalen Mammographie. Der. Radiologe 48, 324–334 (2008)
[4] Kuhl, C.K., Schrading, S., Bieling, H.B., et al.: MRI for diagnosis of pure ductal carcinoma in situ: a prospective observational study. The Lancet 370, 485–492 (2007)
[5] Hall, F.M.: The rise and impending decline of screening mammography. Radiology 247(3), 597–601 (2008)

[6] Kuhl, C.K., Schild, H.H.: Dynamic image interpretation of MRI of the breast. J. of Magnetic Res. Imaging 12, 965–974 (2000)
[7] Perrone, A., Lo Mele, L., Sassi, S., et al.: MDCT of the breast. Am. J. Roentgenol. 190(6), 1644–1651 (2008)
[8] Chang, C., Sibala, J., Fritz, S., Gallagher, J., Dwyer, S., Templeton, A.: Computed tomography evaluation of the breast. Am. J. Roentgenol. 131, 459–464 (1978)
[9] Gisvold, J.J., Reese, D.F., Karsell, P.R.: Computed tomographic mammography (CTM). Am. J. Roentgenol. 133(6), 1143–1149 (1979)
[10] Chen, B., Ning, R.: Cone-beam volume CT breast imaging: feasibility study. Med. Phys. 29(5), 755–770 (2002)
[11] Shaw, C.C., Chen, L., Altunbas, M.C., et al.: Cone beam breast CT with a flat panel detector-simulation, implementation and demonstration. In: 27th Annual International Conference 2005. IEEE-EMBS 2005. Engineering in Medicine and Biology Society, pp. 4461–4464 (2005)
[12] Boone, J.M., Lindfors, K.K.: Breast CT: potential for breast cancer screening and diagnosis. Future Oncology 2(3), 351–356 (2006)
[13] Glick, S.J.: Breast CT. Annu. Rev. Biomed. Eng. 9(1), 501–526 (2007)
[14] Lindfors, K.K., Boone, J.M., Nelson, T.R., Yang, K., Kwan, A.L.C., Miller, D.F.: Dedicated breast CT: Initial clinical experience. Radiology 246(3), 725–733 (2008)
[15] Kalender, W.A.: Dedicated CT of the female breast: Feasibility, optimisation and comparison to competing imaging modalities. In: Project in the 7th Framework Research Project of the European Commission: Enhanced safety and efficacy in the medical uses of radiation (2008–2010); FP7-213153
[16] Lapp, R.M., Kyriakou, Y., Kachelrieß, M., Wilharm, S., Kalender, W.A.: Interactively variable isotropic resolution in computed tomography. Phys. Med. Biol. 53, 2693–2713 (2008)
[17] Deak, P., van Straten, M., Shrimpton, P.C., Zankl, M., Kalender, W.A.: Validation of a Monte Carlo tool for patient-specific dose simulations in multi-slice computed tomography. Eur. Radiol. 18, 759–772 (2008)
[18] Vollmar, S., Kalender, W.A.: Dedicated breast CT at the same exposure level as two-view mammography: Influence of detector pixel size and focal spot size on image quality. Eur. Radiol. 18(2)(suppl. 1), 208 (2008)
[19] Vollmar, S., Weigel, M., Kalender, W.A.: Dose distribution and image quality in CT of the breast as a function of projection angle range. In: Radiological Society of North America scientific assembly and annual meeting program, p. 544. Radiological Society of North America, Oak Brook, Ill (2008)

Optimised Breast Tomosynthesis with a Novel CMOS Flat Panel Detector

Steve Naday[1], Edward F. Bullard[1], Spencer Gunn[1], James E. Brodrick[1], Eoin O. O'Tuairisg[1], Alan McArthur[1], Hamdan Amin[1], Mark B. Williams[2], Patricia G. Judy[2], and Anastasios Konstantinidis[3]

[1] Dexela Limited, Wenlock Business Centre, 50/52 Wharf Road, London, N1 7EU, UK
[2] Department of Radiology, University of Virginia, Charlottesville, VA, USA
[3] Department of Medical Physics and Bioengineering, University College London, Malet Place Engineering Building, Gower Street, London, WC1E 6BT, UK

Abstract. Breast tomosynthesis is a promising technology for breast imaging. Although existing tomosynthesis systems using detector technology developed for FFDM and uniform acquisition parameters have demonstrated the potential to improve the effectiveness of breast screening, the full potential of tomosynthesis is yet to be realised. The effectiveness of tomosynthesis depends on multiple factors, including acquisition geometry, number of projections, reconstruction software and X-ray detector performance. In this study, the authors investigated the use of a specially designed 29 cm x 23 cm CMOS flat panel X-ray detector with a novel Active Pixel Sensor with high spatial resolution, high speed read-out, low noise, negligible image lag and a unique ability to reconfigure imaging parameters such as resolution and gain during an acquisition. Advanced tomosynthesis acquisition methods were used with the new detector including non-uniform spacing of projection views. This combination of optimised X-ray detector and optimised acquisition methods provides enhanced imaging performance.

Keywords: Breast imaging, tomosynthesis, CMOS X-ray detector.

1 Introduction

Breast tomosynthesis is a promising technology for breast imaging and especially for breast screening. Studies have shown that tomosynthesis has the potential to significantly enhance specificity in comparison with FFDM [1, 2]. Although tomosynthesis may also bring benefits to diagnostic breast imaging, it is in screening that greatest value can be realised. Breast tomosynthesis promises solutions to many of the problems currently associated with screening mammography, including the reduction of false-positive results caused by the superposition of normal tissues and making lesions that are masked by superimposed breast tissue more conspicuous, especially in the dense breast. In order to be employed as an effective breast cancer screening modality, tomosynthesis must not require long acquisition times, which would increase the risk of patient motion during acquisition.

Tomosynthesis products developed to date have used detector technology originally developed for FFDM (amorphous Silicon readout arrays with and amorphous

J. Martí et al. (Eds.): IWDM 2010, LNCS 6136, pp. 428–435, 2010.

Selenium and CsI converters). Tomosynthesis systems based on such detectors suffer from a number of limitations. First, low frame rate limits the number of images acquired, which can results in image artefacts. Second, read noise, ghosting and lag affect the reconstruction process, reducing the contrast of the reconstructed images. Third, the relatively low frame rate of TFT-based detectors can result in image blurring, both from patient motion and from the motion of the X-ray source tube.

One approach is to use a large number of projection views (PV) over a wide angular range (e.g. 25 PVs over 50 degrees). This provides good image resolution, but results in a slow scan time when performed with a TFT-based detector with a low frame rate. Long acquisition times increase the likelihood of patient motion during acquisition and may increase the level of discomfort felt by patients. A second approach is to reduce the acquisition time by reducing the number of PVs acquired and the angular range (e.g. 11 projections over 15 degrees), but this may have a negative impact on the image quality of the resulting reconstruction (especially in the Z dimension). A third approach is to use multiple scanning linear detectors [3]. This approach avoids the correlation between number of projections and acquisition time found in flat-panel tomosynthesis. However, the scanning linear detector systems developed to date have suffered from slow acquisition times related to the limited output of available mammography X-ray tubes and the low X-ray efficiency of slot scan systems, in which the majority of the X-ray photons output by the X-ray source are absorbed by collimators above the breast. These systems are also complex mechanically and expensive to manufacture, both of which are undesirable in a product intended for use in screening.

In this work, the authors have adopted an alternative approach: using a faster flat panel detector with low read noise and acquiring sufficient projections to endure good image quality. This novel CMOS X-ray detector is inexpensive, robust and offers significant technical advantages for breast tomosynthesis.

The CMOS X-ray detector uses custom CMOS sensors, specifically designed for use in breast tomosynthesis, with an active imaging area of 1536 x 1944 pixels, each 74.8 μm x 74.8 μm, read out through 6 parallel off-chip ADCs. In order to produce a detector suitable for mammography and breast tomosynthesis, the sensors are tiled into a 2 x 2 array bonded to a fibre optic plate (FOP) giving a total active imaging area of 3072 x 3888 pixels. A 150 μm columnar CsI scintillator is coupled to the FOP. The binning modes and associated framing rates are shown in Table 1 below.

Table 1. CMOS detector binning modes and associated framing rates

Binning	1x1	1x2	2x2	1x4	2x4	4x4
Framing rate(frames/s)	26	53	70	72	81	86

The binning mode can be changed instantaneously, allowing multiple detector resolutions to be used in the same acquisition. For example, a central projection with higher dose and 74.8 μm resolution can be incorporated into an acquisition sequence in which the other PVs are acquired at 149.6 μm. The detector also has the capability to change the gain at the pixel level instantaneously, providing both low noise and high saturation modes for each of the binning modes above.

Existing tomosynthesis systems typically use uniform acquisition methodologies, with a uniform distribution of projections and a uniform radiation dose across projections. In this study, non-uniform parameters have been used with the CMOS X-ray detector and compared with uniform parameters [4, 5]. Table 2 shows tomosynthesis parameters used in this study (A) and selected sets of parameters used in three other studies.

The objectives of this study were twofold: (a) To characterise the imaging properties of a high speed, low noise CMOS X-ray detector and (b) to investigate novel variations in tomosynthesis acquisition parameters using a high speed, low noise CMOS X-ray detector and their effect on image quality in breast tomosynthesis.

Table 2. Breast tomosynthesis acquisition parameters

System	Detector pixel size (μm)	Projection Views (PVs)	Angular range (degrees)	Uniform or non-uniform sampling	Detector technology
A	74.8x74.8	13	24	NU	CsI / CMOS
B [9]	139x139	11	15	U	aSe / TFT
C [10]	85x85 or 85x170	25	50	U	aSe / TFT
D [11]	100x100	11-21	30-60	U	CsI / TFT

2 Method

The spatial resolution (MTF), saturation charge and read noise of the CMOS X-ray detector were measured in the laboratory. MTF data were acquired using a spectrum of 25 kVp and a W/Rh Target/Filter combination without the use of additional filtration. The pre-sampling MTF was measured using the edge method [6]. The saturation charge and read noise of the CMOS X-ray detector were measured using standard photon transfer curve (PTC) method described by Janesick [7].

A breast phantom study was performed at dose levels selected to be equivalent to a typical single-view FFDM breast glandular dose of 1.5 mGy. This study was performed on a laboratory system incorporating the Dexela CMOS detector, Sedecal mammography X-ray generator and Varian RAD70 X-ray tube. The X-ray tube was moved using a specially constructed C-arm gantry. The X-ray source was calibrated using a dosimeter.

For the breast phantom study, sets of projections were acquired with different combinations of uniform and non-uniform parameters and different numbers of projections. Both sets had the same aggregate dose. To compare the volumes reconstructed from the different sets of projections these performance measures were used: contrast-to-noise ratio and normalized line profiles of test objects, following the method used by Zhang et al [5]. An iterative method was used to reconstruct the volume in each case, similar to that described in [9].

A clinical breast tomosynthesis case was also acquired using the non-uniform PV angles distribution shown in Table 3. This was acquired using the native 74.8 μm unbinned pixel in all PVs and the detector's high saturation mode.

3 Results

The read noise of the detector was measured as ~148 –e rms in the low noise mode 325 –e rms in the high saturation mode. Modulation Transfer Function (MTF) has been measured as a function of spatial fequency and is shown in Figure 1.

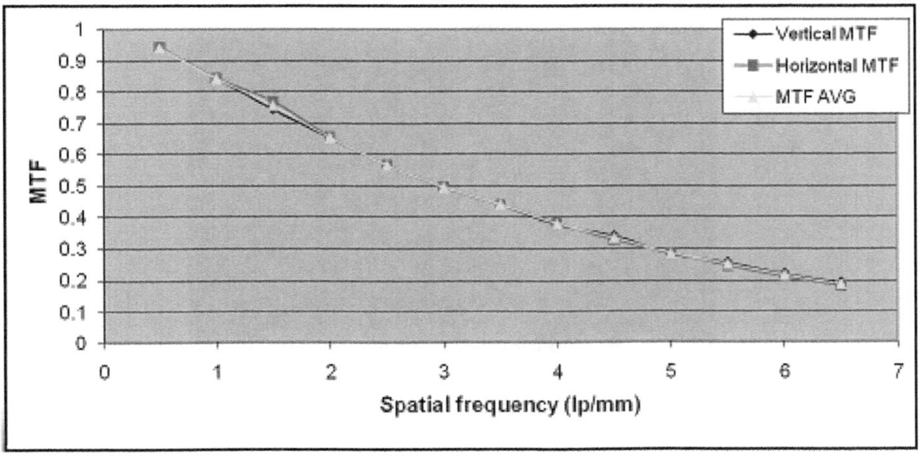

Fig. 1. MTF

Initial results demonstrate fast scan time and excellent image quality can be combined with high in-plane spatial resolution. A clinical trial is planned to determine quantitative measures of clinical performance.

To investigate the effect of the distribution of the PV angles on the reconstructed image, three sets of the projection angle geometries were selected, 1) Uniform angles (U) containing 13 PVs, 2) Non-Uniform angles (NU) containing 13 PVs and 3) Wide Non-Uniform angles which is similar to NU with an extra added angle to contain 15 PVs. The projection angles are specified in Table 3. It should be noted that a limited number of geometries are used here due to the lack of data at the time of submission. More geometries will be evaluated in the near future.

Table 3. Three sets of projection angles in degrees

Uniform Angles (U)		±12	±10	±8	±6	±4	±2	0
Non-Uniform Angles (NU)		±12	±8	±5	±3	±2	±1	0
Wide Non-Uniform Angles (WNU)	±18	±12	±8	±5	±3	±2	±1	0

To compare the volumes reconstructed from the different sets of projections these performance measures were used: contrast-to-noise ratio (CNR) and normalized line profiles (NLP), following the methods used by Zhang et al [5]. An iterative method based on Total Variation regularization was used to reconstruct the volume in each case, similar to that described in [8].

Two test sets were used for this investigation. One was acquired using a phantom containing metal wires to evaluate the blurring in the reconstructed image. The metal wires where placed perpendicular to the path of X-ray source movement and the NLP was obtained. For this test, the PV angles of U and NU were employed. The second test contained the reconstructed images of dead, detail from which is shown in Figure 2. The geometries used for this test were NU and WNU.

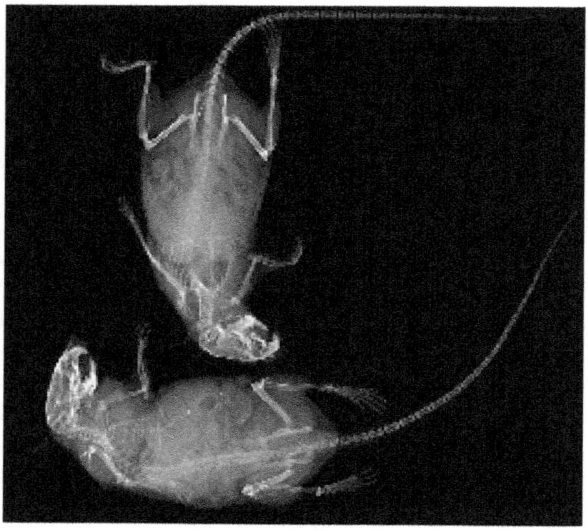

Fig. 2. Detail from a slice taken from the reconstructed image of mice using the WNU distribution

The CNR values of the uniform (U) and non-uniform (NU) projection angles are given in Table 3. Based on the current limited set of geometries used, the results show that the uniform distribution gives better CNR in X-Y plain compared to the NU geometry evaluated.

Table 4. Contrast to noise ratio (CNR) of selected low and high contrast mass applied on phantom

	U	NU
Low contrast mass	5.30	3.65
High contrast mass	15.32	13.84

The results of the CNR of the test mice images are provided in Table 5. The results indicate that the introduction of a pair of extra PV (in WNU geometry) has improved the CNR of both low and high contrast mass region.

Table 5. Contrast to noise ratio (CNR) of selected low and high contrast mass applied on the mice

	NU	WNU
Low contrast mass	2.09	2.19
High contrast mass	8.04	8.90

The in-plane NLP (i.e. the line profile in x-y plane) for the given projection angle distribution did not show significant differences. However, the inter-plane NLP (i.e. the line profile along Z axis) of a high contrast object produced a different outcome, as shown in Figure 3. Here, the uniform distribution produced lower inter-plane blurring than the non-uniform distribution. As indicated in [5], the centrally concentrated PVs (as for the NU geometry) contribute more blurring to the out-of-focus planes.

Figure 4 shows a series of slices from a reconstructed clinical data set acquired at University of Virginia Hospital, Charlottesville. The acquisition used the non-uniform (NU) distribution of thirteen PVs shown in Table 3. The MGD employed was 1.5

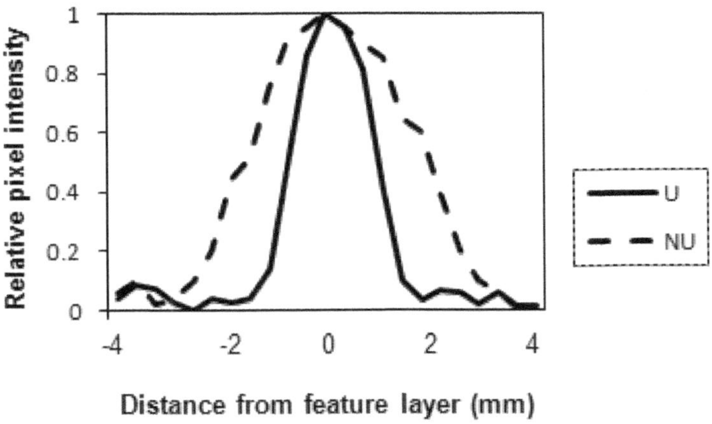

Fig. 3. Normalized inter-plane line profiles along Z axis of a high contrast object

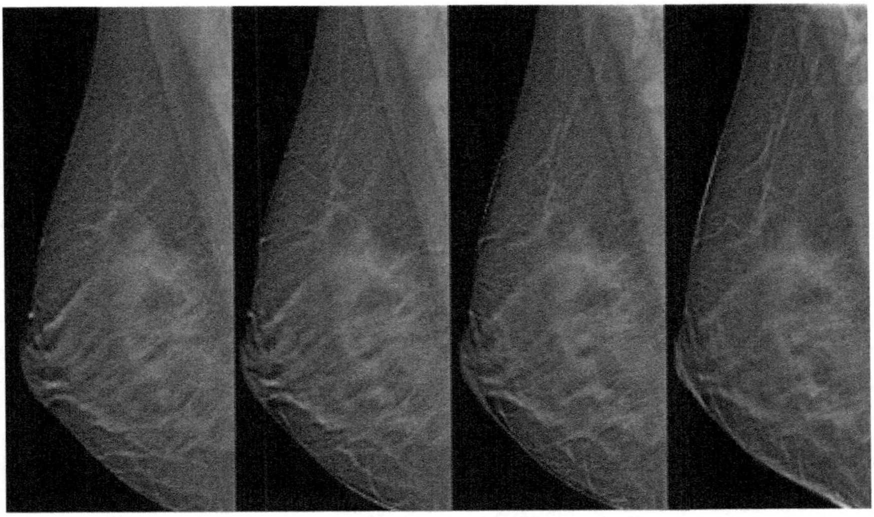

Fig. 4. Details from slices of a reconstruction of a breast acquired with non-uniform PV angles

mGy and the compressed breast thickness was 80 mm. The same breast had previously been imaged on a commercial FFDM system and in that case compressed breast thickness was = 73 mm. The images presented in Figure 4 are a series of four slices separated by 10mm. As can be seen, high-quality images are achieved with the NU distribution, although the results presented in Table 4 favour the uniform distribution method.

The data was acquired using a pixel resolution of 74.8 μm and this spatial resolution was used in the X-Y plane of the reconstruction.

4 Discussion

A novel breast tomosynthesis system has been developed using a fast, low noise CMOS X-ray detector. The CMOS detector is the first to have been developed specifically for breast tomosynthesis. As such, it offers advantages over tomosynthesis systems that use TFT-based detectors designed primarily for FFDM. The flexibility of the CMOS detector allows optimisations of variable binning and variable gain to be incorporated in the acquisition protocol.

The low read-noise and high frame rate of the detector allows breast tomosynthesis to be performed at a higher spatial resolution (74.8 μm in the X-Y plain) than has been achieved with some other breast tomosynthesis systems without lengthening the acquisition time or increasing the mean glandular dose.

Further research is required to determine whether breast tomosynthesis image quality is improved by using more PVs acquired with the CMOS detector's low noise mode and what the optimum tomosynthesis geometry is for this detector.

References

1. Niklason, L., et al.: Digital tomosynthesis in breast imaging. Radiology 205, 399–406 (1997)
2. Dobbins, J.T., Godfrey, D.J.: Digital x-ray tomosynthesis: Current state of the art and clinical potential. Phys. Med. Biol. 48, R65–R106 (2003)
3. Maidment, A.D.A., Ullberg, C., Francke, T., Lindqvist, L., Sokolov, S., Lindman, K., Adelow, L., Sunden, P.: Clinical evaluation of a photon-counting tomosynthesis mammography system. In: Astley, S.M., Brady, M., Rose, C., Zwiggelaar, R. (eds.) IWDM 2006. LNCS, vol. 4046, pp. 144–151. Springer, Heidelberg (2006)
4. Wu, T., et al.: Tomographic mammography using a limited number of low-dose cone-beam projection images. Med. Phys. 30(3), 365–380 (2003)
5. Zhang, Y., et al.: Investigation of Different PV Distributions in Digital Breast Tomosynthesis (DBT) Mammography. In: Krupinski, E.A. (ed.) IWDM 2008. LNCS, vol. 5116, pp. 593–600. Springer, Heidelberg (2008)
6. Marshall, N.W.: A comparison between objective and subjective image quality measurements for a full field digital mammography system. Phys. Med. Biol. 51, 2441–2463 (2006)
7. Janesick, J.: Photon Transfer. SPIE Press Book (2007)

8. Kastanis, I., Arridge, S., Stewart, A., Gunn, S., Ullberg, C., Francke, T.: 3D Digital Breast Tomosynthesis Using Total Variation Regularization. In: Krupinski, E.A. (ed.) IWDM 2008. LNCS, vol. 5116, pp. 621–662. Springer, Heidelberg (2008)
9. Ren, B., Ruth, C., Stein, J., Smith, A., Shaw, I., Jing, Z.: Design and performance of the prototype full field breast tomosynthesis system with selenium-based flat panel detector. In: Proceedings of the SPIE, vol. 5745, pp. 550–561 (2005)
10. Bissonnette, M., Hansroul, M., Masson, E., Savard, S., Cadieux, S., Warmoes, P., Gravel, D., Agopyan, J., Polischuk, B., Haerer, W., Mertelmeier, T., Lo, J.Y., Chen, Y., Dobbins III, J.T., Jesneck, J.L., Singh, S.: Digital breast tomosynthesis using an amorphous selenium flat panel detector. In: Proceedings of the SPIE, vol. 5745, pp. 529–540 (2005)
11. Eberhard, J.W., Albagli, D., Schmitz, A., Claus, B.E.H., Carson, P., Goodsitt, M., Chan, H.-P., Roubidoux, M., Thomas, J.A., Osland, J.: Mammography Tomosynthesis System for High Performance 3D Imaging. In: Astley, S.M., Brady, M., Rose, C., Zwiggelaar, R. (eds.) IWDM 2006. LNCS, vol. 4046, pp. 137–143. Springer, Heidelberg (2006)

Calculation of OTF, NPS, and DQE for Oblique X-Ray Incidence on Turbid Granular Phosphors

Raymond J. Acciavatti and Andrew D.A. Maidment

University of Pennsylvania, Department of Radiology, Physics Section, 1 Silverstein Building, 3400 Spruce St., Philadelphia PA 19104-4206
racci@seas.upenn.edu, Andrew.Maidment@uphs.upenn.edu

Abstract. Digital breast tomosynthesis (DBT) is an imaging modality in which tomographic sections of the breast are generated from a limited range of x-ray tube angles. One drawback of DBT is resolution loss in the oblique projection images. The purpose of this work is to extend Swank's formulation of the transfer functions of turbid granular phosphors to oblique x-ray incidence, using the diffusion approximation to the Boltzmann equation to model the spread of light in the phosphor. As expected, the modulation transfer function (MTF) and noise power spectra (NPS) are found to decrease with projection angle regardless of frequency. By contrast, the dependence of detective quantum efficiency (DQE) on projection angle is frequency dependent. DQE increases with projection angle at low frequencies, and only decreases with projection angle at high frequencies. Importantly, the x-ray quantum detection efficiency (A_Q) and the Swank information factor (A_S) are also found to be angularly dependent.

Keywords: Digital breast tomosynthesis (DBT), oblique x-ray incidence, turbid granular phosphor, optical transfer function (OTF), modulation transfer function (MTF), noise power spectra (NPS), detective quantum efficiency (DQE), Swank information factor (A_S).

1 Introduction

Digital breast tomosynthesis (DBT) is an emerging 3D imaging modality in which x-ray images of the compressed breast are acquired over a limited range of projection angles. Using digital image reconstruction techniques, tomographic sections at all depths of the breast volume can then be generated. Preliminary studies indicate that DBT provides increased sensitivity and specificity for the early detection of breast cancer in women relative to conventional 2D digital mammography [1].

One trade-off of DBT is resolution loss in the projection images as a result of oblique x-ray incidence. Although the degradation in image quality due to oblique x-ray incidence has been studied in columnar cesium iodide phosphors doped with thallium (CsI:Tl) with empirical data [2] and amorphous selenium (*a*-Se) direct converting detectors using Monte Carlo simulations [3], to our knowledge no one has performed a theoretical analysis of the consequences of oblique x-ray incidence. The purpose of this work is to extend Swank's analytical formulation [4] of the transfer functions of x-ray fluorescent screens to oblique x-ray incidence. To this end, we have considered a non-structured turbid granular phosphor such as gadolinium oxysulfide

J. Martí et al. (Eds.): IWDM 2010, LNCS 6136, pp. 436–443, 2010.

doped with terbium (Gd_2O_2S:Tb), which is commonly used in breast imaging and which can reasonably approximate other detector materials.

2 Methods

The optical transfer function (OTF), noise power spectra (NPS), and detective quantum efficiency (DQE) of a turbid granular phosphor irradiated obliquely are now derived from first principles. The spread of visible light in a scintillator can be described by the Boltzmann transport equation. A first-order, steady state solution to the Boltzmann transport equation is a diffusion equation of the form [5]

$$-\nabla^2\phi(\mathbf{r})+\sigma^2\phi(\mathbf{r})=S(\mathbf{r}) \ , \tag{1}$$

where $\phi(\mathbf{r})$ is the product of the density of the secondary carriers (*i.e.*, the optical photons) with the diffusion constant, σ is the reciprocal of the mean diffusion length of the secondary carriers, and $S(\mathbf{r})$ is the source function. The source function $S(\mathbf{r})$ may be modeled as point-like and positioned on ($z_0\tan\theta$, 0, z_0), where z_0 is the depth of the phosphor of total thickness T and where θ is the projection angle relative to normal x-ray incidence. In terms of Dirac delta functions, $S(\mathbf{r})$ can be written as

$$S(\mathbf{r})=\delta(x-z_0\tan\theta)\delta(y)\delta(z-z_0) \ . \tag{2}$$

In the Fourier domain, the source function can be written as the integral

$$S(\mathbf{r})=\delta(z-z_0)\int_{-\infty}^{\infty}\int_{-\infty}^{\infty}e^{2\pi i[(x-z_0\tan\theta)\nu_x+y\nu_y]}d\nu_x d\nu_y \ . \tag{3}$$

Assuming solutions to Eq. (1) of the form

$$\phi(x,y,z)=\int_{-\infty}^{\infty}\int_{-\infty}^{\infty}\psi_{\mathbf{k}}(z)e^{2\pi i(x\nu_x+y\nu_y)}d\nu_x d\nu_y \ , \tag{4}$$

one finds

$$-\frac{d^2\psi_{\mathbf{k}}}{dz^2}+q^2\psi_{\mathbf{k}}=e^{-ik_x z_0\tan\theta}\delta(z-z_0) \ , \qquad q^2=\sigma^2+k_x^2+k_y^2 \ , \tag{5}$$

where $\mathbf{v}$ is the 2D spatial frequency vector and $\mathbf{k}=2\pi\mathbf{v}$. To solve this differential equation, one can apply integral transform techniques. Denoting the Laplace transform of $\psi_{\mathbf{k}}(z)$ as $\Psi_{\mathbf{k}}(p)$, the transform of the differential equation is

$$(-p^2+q^2)\Psi_{\mathbf{k}}(p)+C_1p+C_2=e^{-ik_x z_0\tan\theta}e^{-pz_0} \ , \tag{6}$$

where C_1 and C_2 are the constants of integration. Solving for $\Psi_{\mathbf{k}}(p)$ and taking the inverse transform generates the following piece-wise expression for $\psi_{\mathbf{k}}(z)$.

$$\psi_{\mathbf{k}}(z)=\begin{cases} C_1\cosh(qz)+\dfrac{C_2}{q}\sinh(qz), & 0\le z\le z_0 \\ C_1\cosh(qz)+\dfrac{C_2}{q}\sinh(qz)-\dfrac{e^{-ik_x z_0\tan\theta}}{q}\sinh\big(q(z-z_0)\big), & z_0<z\le T \end{cases} \tag{7}$$

The constants C_1 and C_2 can now be determined from boundary conditions concerning secondary carrier currents directed toward the planes at $z = 0$ and $z = T$. In terms of the inverse relaxation length τ, the secondary carrier currents across any plane of constant z are

$$j_{\text{left}}(z) = \frac{1}{2}\left[\phi\tau + \frac{d\phi}{dz}\right], \quad j_{\text{right}}(z) = \frac{1}{2}\left[\phi\tau - \frac{d\phi}{dz}\right]. \tag{8}$$

In the right-hand side of the two equations, the first term models the effusion current, while the second term comes from Fick's law. The first boundary condition is determined by the reflectivity r_0 of the plane at $z = 0$. Noting that $j_{\text{right}}(0) = r_0\, j_{\text{left}}(0)$, one finds

$$\left.\frac{d\phi}{dz}\right|_{z=0} = \tau\rho_0 \phi\big|_{z=0}, \quad \rho_0 \equiv \frac{1-r_0}{1+r_0}. \tag{9}$$

The second boundary condition is determined from the reflectivity r_1 of the boundary at $z = T$, as stipulated by the expression $j_{\text{left}}(T) = r_1\, j_{\text{right}}(T)$. Defining ρ_1 similar to ρ_0 and noting that the boundary conditions hold for each Fourier component $\psi_{\mathbf{k}}$ of ϕ, it can be shown that

$$C_1 = \left[\frac{(q+\tau\rho_1)e^{q(T-z_0)} + (q-\tau\rho_1)e^{-q(T-z_0)}}{(q+\tau\rho_0)(q+\tau\rho_1)e^{qT} - (q-\tau\rho_0)(q-\tau\rho_1)e^{-qT}}\right] e^{-ik_x z_0 \tan\theta}, \tag{10}$$

$$C_2 = \tau\rho_0 C_1 . \tag{11}$$

Defining $z = T$ as the plane of the photocathode, the OTF of the scattering process, $T_s(\mathbf{v}, z_0)$, can now be determined for a point source from the expression

$$T_s(\mathbf{v}, z_0) = \left[\frac{\rho_1}{1+\rho_1}\right]\left[\psi_{\mathbf{k}}\tau - \frac{d\psi_{\mathbf{k}}}{dz}\right]_{z=T}, \tag{12}$$

giving

$$T_s(\mathbf{v}, z_0) = \tau\rho_1 \left[\frac{(q+\tau\rho_0)e^{(q-ik_x \tan\theta)z_0} + (q-\tau\rho_0)e^{-(q+ik_x \tan\theta)z_0}}{(q+\tau\rho_0)(q+\tau\rho_1)e^{qT} - (q-\tau\rho_0)(q-\tau\rho_1)e^{-qT}}\right]. \tag{13}$$

To calculate the OTF of the entire phosphor, one multiplies Eq. (13) by the relative number of x-ray absorptions as a function of the depth z_0

$$N(z_0) = \frac{\mu e^{-\mu z_0 \sec\theta} \sec\theta}{1 - e^{-\mu T \sec\theta}}, \tag{14}$$

where μ is the x-ray linear attenuation coefficient of the phosphor, and then integrates from $z_0 = 0$ to $z_0 = T$. The OTF is thus

$$T_s(\mathbf{v}) = \frac{\beta\mu \sec\theta}{1 - e^{-\mu T \sec\theta}}\left[\frac{(q+\tau\rho_0)(e^{(\gamma_- - ik_x \tan\theta)T} - 1)}{\gamma_- - ik_x \tan\theta} - \frac{(q-\tau\rho_0)(e^{-(\gamma_+ + ik_x \tan\theta)T} - 1)}{\gamma_+ + ik_x \tan\theta}\right], \tag{15}$$

where

$$\beta \equiv \frac{\tau\rho_1}{(q+\tau\rho_0)(q+\tau\rho_1)e^{qT}-(q-\tau\rho_0)(q-\tau\rho_1)e^{-qT}}\ , \qquad \gamma_\pm \equiv q \pm \mu \sec\theta\ . \tag{16}$$

The normalized modulus of the OTF of Eq. (15) gives the modulation transfer function (MTF). In the absence of other noise sources, the quantum NPS or $W_Q(\mathbf{v})$ is calculated by integrating the product of $N(z_0)$ with $|T_s(\mathbf{v}, z_0)|^2$ from $z_0 = 0$ to $z_0 = T$.

$$W_Q(\mathbf{v}) = \frac{\beta^2\mu\sec\theta}{1-e^{-\mu T\sec\theta}}\left[\frac{(q+\tau\rho_0)^2(e^{(q+\gamma_-)T}-1)}{q+\gamma_-}+\frac{2(q^2-\tau^2\rho_0^2)(1-e^{-\mu T\sec\theta})}{\mu\sec\theta} + \frac{(q-\tau\rho_0)^2(1-e^{-(q+\gamma_+)T})}{q+\gamma_+}\right] \tag{17}$$

With Eqs. (15)-(17), one now has all the tools required for determining the DQE of the phosphor. From the work of Nishikawa, DQE can be formulated as the product of four terms [6]

$$\mathrm{DQE}(\mathbf{v}) = A_Q A_S R_C(\mathbf{v}) R_N(\mathbf{v})\ , \tag{18}$$

where A_Q is the x-ray quantum detection efficiency determined from the Lambert-Beer Law as $1 - e^{-\mu T\sec\theta}$, A_S is the Swank information factor given by $|T_s(\mathbf{0})|^2/W_Q(\mathbf{0})$, $R_C(\mathbf{v})$ is the Lubberts fraction found by normalizing the quotient $|T_s(\mathbf{v})|^2/W_Q(\mathbf{v})$ to unity at $\mathbf{v} = \mathbf{0}$, and $R_N(\mathbf{v})$ is the ratio of the x-ray quantum noise power $W_Q(\mathbf{v})$ to the total noise power $W_T(\mathbf{v})$. Outside of x-ray quantum noise, additional sources of noise which contribute to $W_T(\mathbf{v})$ include optical-detector noise due to silver granules in the phosphor or thermal noise in the photocathode, secondary quantum noise arising from stochastic variation in the number of secondary carriers produced for each incident x-ray, and screen-structure noise [7]. Assuming a quantum-limited imaging system, we treat $R_N(\mathbf{v})$ as unity in this work.

3 Results

The preceding results are now illustrated for a phosphor possessing 90% x-ray quantum detection efficiency at normal incidence, a reflective backing, a non-reflective photocathode, and optical scatter at the diffusion limit ($\tau \to \infty$). Assuming that the frequency vector is oriented along the x direction ($v_y = 0$), Figures 1A-1B show MTF and normalized NPS (NNPS) versus frequency at multiple angles of x-ray incidence for a phosphor with no optical absorption ($\sigma = 0$) and a phosphor with high optical absorption ($\sigma = 2T^{-1}$). Figures 1A-1B demonstrate that increasing the optical absorption increases both MTF and NNPS, which is consistent with Swank's prior work at normal incidence. In addition, Figures 1A-1B indicate that increasing the projection angle decreases both MTF and NNPS, with the relative decrease as a function of projection angle being most predominate at high frequencies. The projection angle dependence of the NNPS is slightly less pronounced than the projection angle dependence of the MTF. For example, comparing 30° incidence to normal incidence at 5

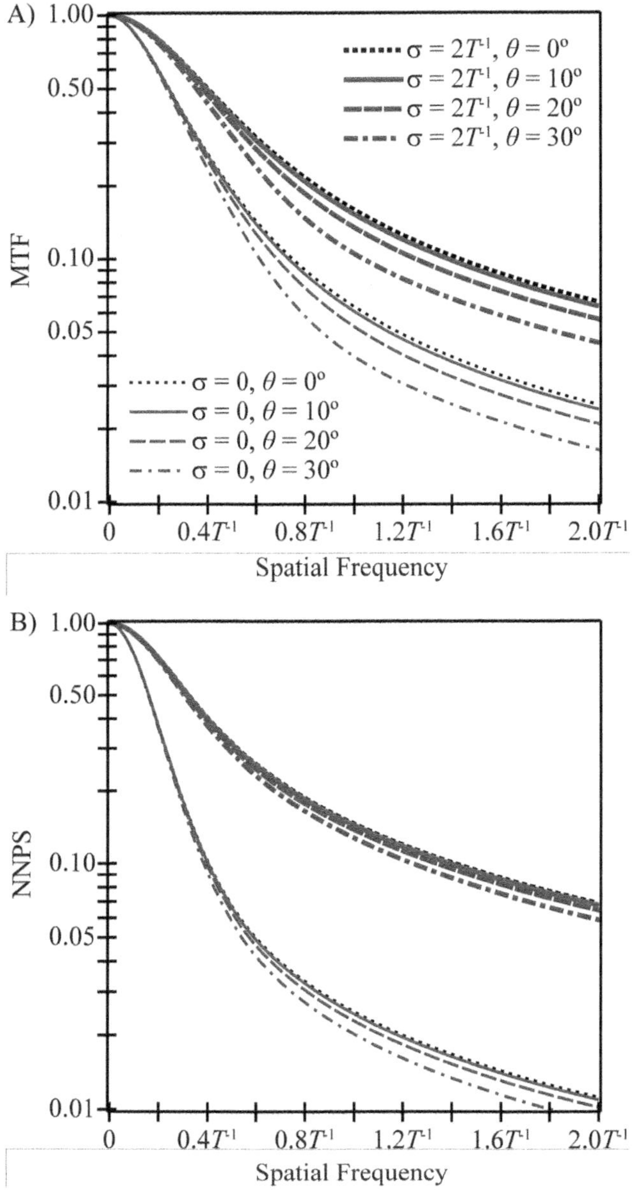

Fig. 1. Assuming that the frequency vector is oriented along the x direction, the modulation transfer function (MTF), normalized noise power spectra (NNPS), and detective quantum efficiency (DQE) are plotted versus frequency in units of inverse phosphor thickness (T^{-1}) in subplots (A)-(C) for multiple projection angles ($\theta = 0°, 10°, 20°, 30°$) and two optical absorption parameters ($\sigma = 0, 2T^{-1}$). The phosphor possesses 90% x-ray quantum detection efficiency at normal incidence, a reflective backing, a non-reflective photocathode, optical scatter at the diffusion limit, and quantum-limited noise. Subplots (A)-(C) implicitly share a common legend. In (D), the explicit dependence of the Swank information factor (A_S) and DQE($\mathbf{0}$) on the angle of x-ray incidence is studied for the two optical absorption parameters.

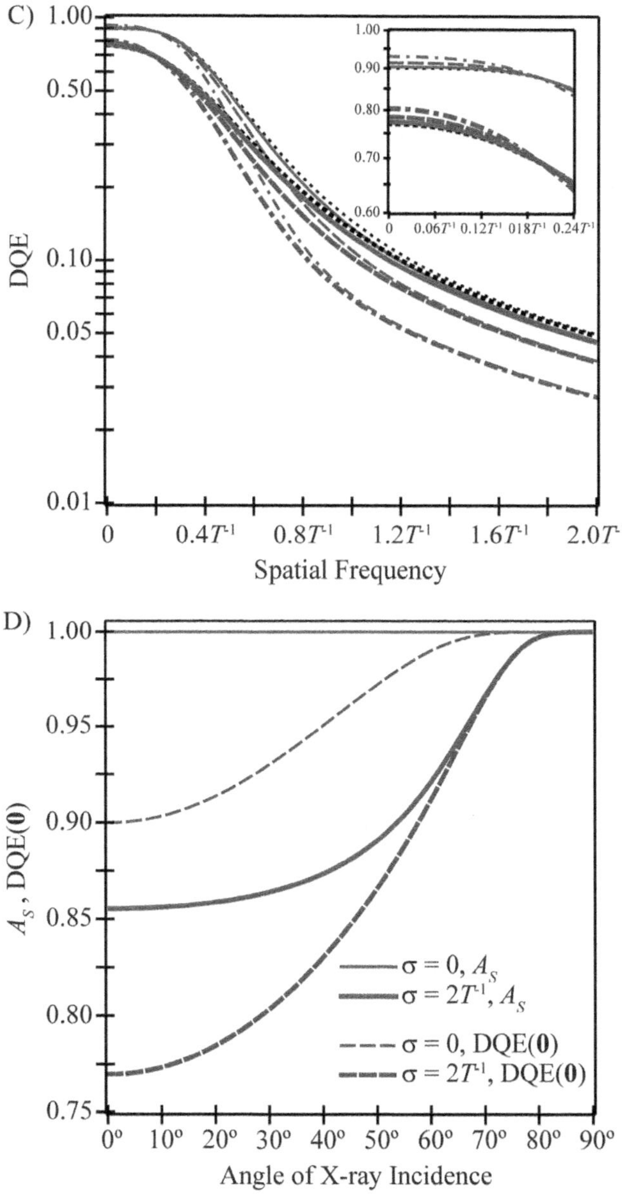

Fig. 1 *(continued)*

line pairs per millimeter (lp/mm) in a scintillator with 100 μm thickness and no optical absorption, the MTF decreases by 20% whereas the NNPS decreases by only 14% (19% and 8.9%, respectively, with high optical absorption).

Figure 1C shows DQE versus frequency for the same phosphor. Consistent with Swank's previous observations, Figure 1C indicates that increasing the optical absorption reduces the DQE at low frequencies but has a smaller effect on the DQE at high frequencies. In addition, Figure 1C demonstrates that unlike MTF and NNPS which decrease with projection angle at all frequencies, DQE actually increases with projection angle at low frequencies and only decreases with projection angle at high frequencies. The DQE degradation with projection angle at high frequencies is slightly more pronounced than either the MTF or NNPS degradation. For example, at 5 lp/mm in a scintillator of 100 μm thickness irradiated at a 30° angle, the DQE decreases by 24% relative to normal incidence for both optical absorption parameters shown in the figure.

In Figure 1D, the Swank information factor (A_S) and DQE($\mathbf{0}$) are plotted versus the angle of x-ray incidence. Swank has shown that A_S provides a measure of the fluctuation in signal generated from each x-ray due to variability in the absorbed energy of each interacting x-ray and in the number of optical photons generated from each interacting x-ray [8]. Figure 1D demonstrates that in a phosphor with no optical absorption, A_S is unity for all projection angles, but in a phosphor with high optical absorption, A_S increases with projection angle from 0.86 at normal incidence to unity at shearing incidence. In the typical range of projection angles used in DBT, the projection angle dependence of A_S is slight. For example, comparing 30° incidence to normal incidence in a phosphor with high optical absorption, A_S increases by merely 1.0%. Unlike A_S, DQE($\mathbf{0}$) is projection angle dependent for all possible optical absorption parameters. The projection angle dependence of DQE($\mathbf{0}$) is more pronounced than the projection angle dependence of A_S. In Figure 1D, DQE($\mathbf{0}$) increases from 0.90 (no optical absorption) and 0.77 (high optical absorption) at normal incidence to unity at shearing incidence. Comparing 30° incidence to normal incidence, the relative increase in DQE($\mathbf{0}$) is 3.3% in a phosphor with no optical absorption and 4.4% in a phosphor with high optical absorption.

4 Discussion

This work develops analytical models of OTF, NPS, and DQE for a turbid granular phosphor irradiated obliquely. In agreement with Mainprize *et al.* who studied CsI:Tl phosphors experimentally [2], we show that at high frequencies, oblique x-ray incidence gives rise to considerable degradation in MTF and hence poorer resolution. We have also observed that NPS is degraded with projection angle for all frequencies, where the NPS degradation is much less pronounced than the MTF degradation. Although Mainprize *et al.* did not study the dependency of NPS on projection angle, our finding of small changes in NPS with increasing projection angle is qualitatively concordant with the prior work of Hajdok and Cunningham in their Monte Carlo simulations of *a*-Se detectors [3]. As a final point, we have demonstrated that DQE increases with projection angle at low frequencies but decreases with projection angle at high frequencies. Consistent with the observations of Hajdok and Cunningham, the DQE degradation with projection angle at high frequencies is more pronounced than the MTF degradation, reflecting the dependency of DQE on the square of MTF.

In this work, it has been observed that the Swank information factor (A_S) is angularly dependent, but its dependency is small over the angles used in DBT. This observation is consistent with Monte Carlo simulations of columnar CsI:Tl phosphors conducted by Badano *et al.*, who show that the variation in A_S over projection angles typical of DBT is minimal [9]. While the relative change in A_S with angle is slim, the relative increase in DQE(**0**) is more substantial, as it includes the influence of increasing x-ray quantum detection efficiency (A_Q) with increasing angle.

Acknowledgments. The project described was supported by Grant T32EB009321 from the National Institute of Biomedical Imaging and Bioengineering (NIBIB). The content is solely the responsibility of the authors and does not necessarily represent the official views of the NIBIB or the National Institutes of Health.

References

1. Rafferty, E.: Tomosynthesis: New Weapon in Breast Cancer Fight. Imaging Economics 17(4) (2004)
2. Mainprize, J.G., Bloomquist, A.K., Kempston, M.P., Yaffe, M.J.: Resolution at oblique incidence angles of a flat panel imager for breast tomosynthesis. Med. Phys. 33(9), 3159–3164 (2006)
3. Hajdok, G., Cunningham, I.A.: Penalty on the detective quantum efficiency from off-axis incident x rays. In: Yaffe, M.J., Flynn, M.J. (eds.) Proc. of SPIE, Medical Imaging 2004: Physics of Medical Imaging, vol. 5368, pp. 109–118. SPIE, Bellingham (2004)
4. Swank, R.K.: Calculation of Modulation Transfer Functions of X-Ray Fluorescent Screens. Appl. Opt. 12(8), 1865–1870 (1973)
5. Marshak, R.E., Brooks, H., Hurwitz Jr., H.: Introduction to the Theory of Diffusion and Slowing Down of Neutrons – I. Nucleonics 4, 10–22 (1949)
6. Nishikawa, R.M., Yaffe, M.J.: Model of the spatial-frequency-dependent detective quantum efficiency of phosphor screens. Med. Phys. 17(5), 894–904 (1990)
7. Nishikawa, R.M., Yaffe, M.J.: Effect of various noise sources on the detective quantum efficiency of phosphor screens. Med. Phys. 17(5), 887–893 (1990)
8. Swank, R.K.: Absorption and noise in x-ray phosphors. J. Appl. Phys. 44(9), 4199–4203 (1973)
9. Badano, A., Kyprianou, I.S., Sempau, J.: Anisotropic imaging performance in indirect x-ray imaging detectors. Med. Phys. 33(8), 2698–2713 (2006)

Scanner for Integrated X-Ray Breast Tomosynthesis and Molecular Breast Imaging Tomosynthesis

Mark B. Williams[1], Patricia G. Judy[1], Zongyi Gong[1], Audrey E. Graham[1], Stan Majewski[2], and Spencer Gunn[3]

[1] University of Virginia, Department of Radiology, Charlottesville, VA, USA
mbwilliams@virginia.edu
[2] West Virginia University, Department of Radiology, Morgantown, WV, USA
[3] Dexela, Ltd., London, UK

Abstract. A dual modality tomosynthesis (DMT) breast scanner has been developed that combines x-ray breast tomosynthesis (XBT) and molecular breast imaging tomosynthesis (MBIT) on a common upright gantry to obtain co-registered structural and functional tomosynthesis images. This paper describes the scanner's design and operation, and summarizes the results of a pilot clinical evaluation using the tracer ^{99m}Tc-sestamibi. The pilot study results suggest that DMT breast scanning is feasible and provides improved specificity and positive predictive value compared to XBT alone. Potential clinical roles for DMT scanning include problem solving for equivocal mammographic/ultrasound studies; as an aid in biopsy target selection following a positive mammogram with multiple suspicious areas; cancer surveillance in patients with a personal history of breast cancer; pre-surgical planning for determination of disease extent; as an alternative for women for whom MRI is impossible; and for monitoring response to neoadjuvant therapy.

Keywords: breast tomosynthesis, molecular breast imaging, limited angle SPECT, multimodality imaging, hybrid scanners.

1 Background

Digital x-ray breast tomosynthesis is under study by a number of groups (1-11). Also, during the past decade, the development of high spatial resolution, compact gamma cameras for single gamma breast scintigraphy (molecular breast imaging (MBI), or breast specific gamma imaging (BSGI)) has permitted visualization of lesions less than 10 mm in size throughout the breast, including regions near the chest wall (12-14). At the same time, hybrid imaging, in which modalities offering complementary types of information are paired, has become increasingly prevalent. Integrated hybrid scanners containing a nuclear medicine modality and an anatomic modality permit accurate co-registration of functional and structural images, and have been shown to reduce ambiguities in tracer uptake distribution and to improve attenuation correction in the nuclear medicine image (15-22). Recently, investigators wishing to overcome the limitations of using whole-body hybrid scanners for breast imaging have begun development of dedicated PET-CT (23) and SPECT-CT breast scanners (24;25).

J. Martí et al. (Eds.): IWDM 2010, LNCS 6136, pp. 444–451, 2010.

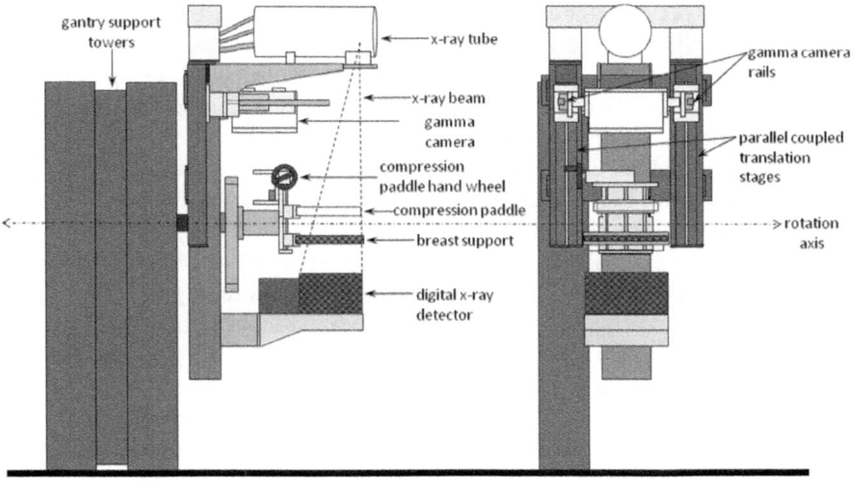

Fig. 1. Scale diagram of the dual modality tomosynthesis scanner at UVa as configured with the CCD x-ray detector

In a collaboration of academic and industrial partners, we have developed a hybrid scanner for dual modality breast tomosynthesis. A schematic of the dual modality tomosynthesis (DMT) scanner is shown in Figure 1. The scanner acquires 3-dimensional dual modality breast images by performing sequential x-ray breast tomosynthesis (XBT) and molecular breast imaging tomosynthesis (MBIT) scans with the breast in a single configuration under mild breast compression. The integration of XBT with MBIT has the potential to improve breast cancer detection, characterization, and management by providing co-registered structural and functional image data. The DMT scanner has recently undergone preliminary clinical evaluation in a pilot study among women scheduled for breast biopsy. Details of that study have been recently reported (26). Here we describe the scanner's operation and summarize the results of the pilot human study.

2 Methods

2.1 Image Acquisition

The x-ray component of the DMT scanner uses a fully isocentric image acquisition geometry in which the tube and detector are rotated about a common axis. The x-ray tube has a tungsten target and uses 50 microns of external rhodium filtration. The source-to-detector distance is 80 cm. Until recently, the x-ray detector was a device developed under NCI funding that incorporates a 2x3 array of CCDs, each coupled to a common gadolinium oxysulfide phosphor via demagnifying fiber optic tapers (27). The detector's field of view (FOV) is approximately 20 cm x 30 cm, and it was typically operated in 2x2 binning mode for an effective detector element size of 90 microns. The CCD detector was used in the pilot clinical study described below. However in light of its relatively long readout time (~2 seconds) and large depth

(16.5 cm) the CCD detector has now been replaced with a CMOS-based detector manufactured by Dexela, Ltd. The model 2923 CMOS detector uses a columnar CsI(Tl) converter optically coupled to a 3072 x 3888 matrix of 75 micron detector elements for a FOV of 23 cm x 29 cm. It can be read out at up to 26 frames per second, and has a depth of only 4.3 cm.

XBT image reconstruction is performed using an algorithm developed by Dexela. The Dexela algorithm is a multi-threaded iterative process that employs anisotropic diffusion-based regularization.

The gamma camera, built at Jefferson Lab (Newport News, VA) includes a tiled 3x4 array of Hamamatsu H8500 position sensitive photomultiplier tubes optically coupled to a pixellated NaI(Tl) crystal array from Saint-Gobain Crystals. Each crystal in the array is 2 mm x 2 mm x 6 mm thick and the crystal pitch is 2.2 mm. The overall gamma camera active area is approximately 15 cm x 20 cm. The camera uses a lead foil parallel hole collimator and has an overall sensitivity of 450 cpm/μCi.

MBIT image reconstruction is performed using an expectation maximization (EM) parallel-beam algorithm developed at UVa. During each iteration a mask based on the prior knowledge of the breast surface location gained from the XBT scan is used as a constraint to compensate for projection data incompleteness by preventing detected gamma events from being attributed to voxels lying outside the breast surface. An inverse cylindrical ray-driven projector-backprojector is used along with the voxel-specific system point spread function (PSF) to avoid overweighting voxels distal to the camera while still permitting resolution recovery. This approach results in reconstructed spatial resolution that is relatively uniform throughout the imaged volume despite the asymmetrical nature of the image acquisition process. Under typical imaging conditions (i.e. 5 – 6 cm compressed thickness, 5 evenly spaced views spanning 40 degrees) the in-plane resolution in the MBIT reconstructions (FWHM of the PSF), averaged at 5 locations within the imaged volume, is 7.6 ± 0.9 mm. The average spatial resolution in the direction of compression is 34.1± 3.1 mm. Phantom studies have shown that compared to 2-D MBI, MBIT provides improved lesion image contrast and signal-to-noise ratio (28).

2.2 Human Study

The pilot study recruited 18 women (a total of 21 biopsied lesions) who were scheduled for breast biopsy at the University of Virginia's Breast Care Center. Eligible participants were identified by monitoring the list of patients scheduled for core or excisional biopsy. Participating volunteers were scanned prior to their biopsy. Approximately 25 mCi (925 MBq) of ^{99m}Tc-sestamibi was injected intravenously. Mild medio-lateral breast compression was used, and subjects were seated throughout the imaging procedure. The breast compression level was determined based on discussion between the subject and the mammography technologist with the goal of achieving the maximum compressive force that could be comfortably tolerated during the ~11 minute total scan time per breast.

Both breasts were scanned. XBT was performed first; 13 unevenly spaced views were obtained distributed over an angular range of ± 12° from the direction

Table 1. kVps used for various breast thicknesses

3 cm	5 cm	7 cm
23 kV	27 kV	29 kV

of breast compression. The tube kVp was chosen based on the results of phantom testing designed to maximize the signal-to-noise ratio relative to the mean glandular dose for this particular target-filter-detector combination for a range of breast compositions and thicknesses (29). Table 1 lists the kVp values used for three different compressed thicknesses. Tube voltages for other thicknesses were determined via interpolation or extrapolation. The total exposure to the breast during the XBT scan was chosen to result in a mean glandular dose that was approximately equal to that in the subject's most recent 2-view mammographic exam.

Immediately following x-ray image acquisition without uncompressing the breast, five 120 second MBIT views were obtained, distributed evenly over an angular range of ± 20° away from the direction of breast compression. For each view the gamma camera collimator was positioned as close as possible to the breast compression paddle. Total MBIT scan time was 10 minutes per breast, chosen to be consistent with current practice in clinical dedicated planar MBI, which is also performed under mild compression. The total acquisition time for both modalities was approximately 11 minutes per breast.

2.3 Reader Study

All images were viewed by a board-certified mammography radiologist (AEG) who was blinded to the biopsy results and to all prior images (mammography, ultrasound, MRI, etc.). Images were viewed on a Dexela dual monitor (5 megapixels per monitor) workstation, which is FDA approved for primary mammographic interpretation. XBT images alone were read first. Each finding identified was scored using a 5-point scale: 1, definitely benign; 2, probably benign; 3, indeterminate; 4, suggestive of malignancy; 5 highly suggestive of malignancy. Next, the MBIT images were viewed alone, but in a sequence that was randomized relative to the sequence in which the XBT images were viewed. The

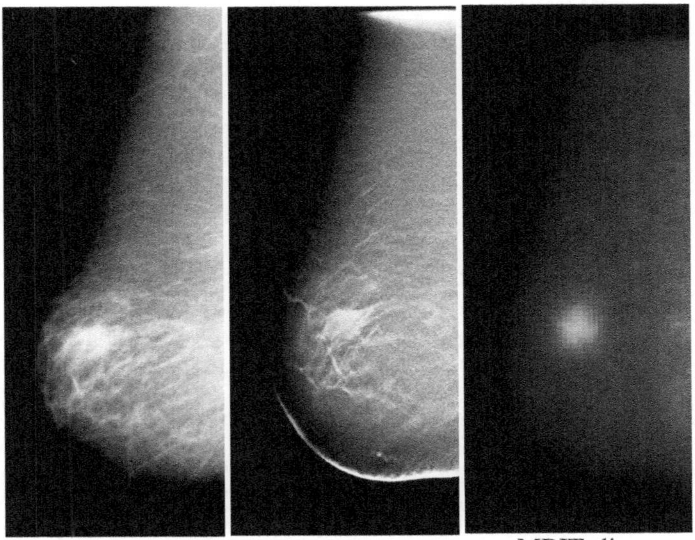

Fig. 2. Planar mammographic image, XBT slice, and corresponding MBIT slice from an example case from the pilot study. In this example the biopsy result indicated poorly differentiated carcinoma.

same 5-point scale was used to score each finding. Finally, the paired XBT and MBIT images were presented together, and each finding was scored.

3 Results

The biopsy results for the 21 lesions showed 7 malignant lesions and 14 benign ones. Figure 2 shows, for one of the malignant cases, the subject's 2-D clinical mammogram, a slice from the 3-D XBT image, and the corresponding slice from the 3-D MBIT image. Based on the results of the reader study, and taking the biopsy results as ground truth, Table 2 shows the sensitivity, specificity, positive predictive value (PPV), negative predictive value (NPV), and accuracy for XBT alone (2nd column), MBIT alone (3rd column), and DMT (4th column). For these calculations, reader scores of 1, 2, or 3 were considered negative while scores of 4 or 5 were considered positive.

Table 2. Results of DMT pilot study, taking reader scores of 1 – 3 as negative and 4 – 5 as positive. PPV=positive predictive value; NPV=negative predictive value.

	XBT Only	MBIT Only	DMT
Sensitivity	86% (6/7)	86% (6/7)	86% (6/7)
Specificity	57% (8/14)	100% (14/14)	100% (14/14)
PPV	50% (6/12)	100% (6/6)	100% (6/6)
NPV	89% (8/9)	93% (14/15)	93% (14/15)
Accuracy	67% (14/21)	95% (20/21)	95% (20/21)

4 Discussion

The results of the pilot study demonstrate that dual modality tomosynthesis is feasible, and suggest that the addition of MBIT to XBT could improve clinical performance, especially in terms of specificity and positive predictive value. In one case DMT scanning detected high-grade ductal carcinoma in situ that was occult on the subject's recent clinical mammographic exam (26). Notably, in this small study MBIT scanning alone performed as well as DMT. More data are needed to determine the added value of DMT relative to MBIT alone.

Retrospective analysis of the data from the pilot study showed that while microcalcifications are well visualized on some XBT scans (based on comparison with the clinical mammograms) they are not clear on others because of subject motion during the ~30 second XBT scan time resulting from the slow CCD detector readout. However with the newly installed CMOS detector the scan time will be ~4 – 8 seconds, which is well within the breath-hold time of most women. The higher likelihood of breathing-induced motion during the longer MBIT scan is considered acceptable given the very different projection data spatial resolutions of XBT (< 80 microns FWHM using the CMOS detector) and MBIT (4 – 11 mm FWHM depending on the viewing angle).

Constant breast configuration during XBT and MBIT scanning is essential for precise co-registration of the two volumetric data sets. The level of breast compression typically used in planar mammography is too high to be comfortably tolerated during the substantially longer MBIT scan. Therefore a lower compressive force is required that satisfies the requirements of both XBT and MBIT. The motivations for breast compression in XBT and MBIT are different. Compression in digital mammography is necessary to reduce scatter, dose, structure superposition, and patient motion while in MBI its principal role is to improve lesion conspicuity by reducing attenuation of gamma rays from the lesion (30-32). Duke investigators have shown that the reduction in structure superposition inherent in XBT permits some increase in compressed thickness (from reduction in compressive force) with negligible impact on conspicuity of microcalcifications or masses under isodose conditions (33). Recent results from that group (2010 SPIE Medical Imaging Conference) have shown that the thickness of a 6 cm breast can be increased by at least 17% without compromising conspicuity, with larger increases possible for thinner breasts. A retrospective analysis of the compressed thicknesses used in the pilot study showed that on average the 'comfortable' compressed thickness was 17% greater during DMT than for the same breast during the preceding clinical mammogram. Thus breast compression consistent with both high quality XBT and comfortable MBIT appears possible.

Like all imaging approaches that use injected tracers, the performance of DMT will depend on the choice of imaging agent. Our DMT studies thus far have used ^{99m}Tc-sestamibi, which is FDA approved for breast cancer imaging and is the most widely used MBI agent when performing 2-D MBI. However, several other agents including bombesin and tetrofosmin have also been successfully used to image breast cancer. These agents are expected to be joined by more specific agents that are now under development; for example those targeting HER2-, ER-, or VEGFR-positive tumors. The sensitivity and specificity of DMT can be expected to vary depending on the particular tracer used.

Potential clinical roles for DMT breast imaging include problem solving for women with equivocal mammographic/ultrasound studies, especially those with radiodense breasts; screening high-risk women, as an aid in biopsy target selection following a positive mammogram, particularly in patients with multiple suspicious areas; workup of palpable masses not demonstrated on mammography or ultrasound; cancer surveillance in patients with a personal history of breast cancer; pre-surgical planning for women with known cancer for determination of disease extent (multi-focal, contralateral breast); as an alternative to breast MRI for women for whom MRI is impossible because of NSF/NFD, obesity, claustrophobia, or who have cardiac pacemakers or other types of implants containing ferromagnetic components; and for monitoring response to neoadjuvant therapy (34;35).

References

1. Badano, A., Kyprianou, I.S., Jennings, R.J., Sempau, J.: Anisotropic imaging performance in breast tomosynthesis. Med. Phys. 34(11), 4076–4091 (2007)
2. Chen, Y., Lo, J.Y., Dobbins III, J.T.: Importance of point-by-point back projection correction for isocentric motion in digital breast tomosynthesis: relevance to morphology of structures such as microcalcifications. Med. Phys. 34(10), 3885–3892 (2007)

3. Ma, A.K., Darambara, D.G., Stewart, A., Gunn, S., Bullard, E., Ma, A.K.W., Darambara, D.G., Stewart, A., Gunn, S., Bullard, E.: Mean glandular dose estimation using MCNPX for a digital breast tomosynthesis system with tungsten/aluminum and tungsten/aluminum+silver x-ray anode-filter combinations. Med. Phys. 35(12), 5278–5289 (2008)
4. Sechopoulos, I., Ghetti, C.: Optimization of the acquisition geometry in digital tomosynthesis of the breast. Med. Phys. 36(4), 1199–1207 (2009)
5. Van de Sompel, D., Brady, M.: A systematic performance analysis of the simultaneous algebraic reconstruction technique (SART) for limited angle tomography. In: Conference Proceedings: Annual International Conference of the IEEE Engineering in Medicine & Biology Society 2008, pp. 2729–2732 (2008)
6. Zeng, K., Yu, H., Zhao, S., Fajardo, L.L., Ruth, C., Jing, Z., Wang, G.: Digital tomosynthesis aided by low-resolution exact computed tomography. Journal of Computer Assisted Tomography 31(6), 976–983 (2007)
7. Zhao, B., Zhao, W.: Three-dimensional linear system analysis for breast tomosynthesis. Med. Phys. 35(12), 5219–5232 (2008)
8. Mainprize, J.G., Bloomquist, A.K., Kempston, M.P., Yaffe, M.J.: Resolution at oblique incidence angles of a flat panel imager for breast tomosynthesis. Med. Phys. 33(9), 3159–3164 (2006)
9. Niklason, L.T., Christian, B.T., Niklason, L.E., Kopans, D.B., Castleberry, D.E., Opsahl-Ong, B.H., Landberg, C.E., Slanetz, P.J., Giardino, A.A., Moore, R., et al.: Digital tomosynthesis in breast imaging. Radiology 205(2), 399–406 (1997)
10. Sechopoulos, I., Suryanarayanan, S., Vedantham, S., D'Orsi, C., Karellas, A.: Computation of the glandular radiation dose in digital tomosynthesis of the breast. Medical Physics 34(1), 221–232 (2007)
11. Wu, T., Stewart, A., Stanton, M., McCauley, T., Phillips, W., Kopans, D.B., Moore, R.H., Eberhard, J.W., Opsahl-Ong, B., Niklason, L., et al.: Tomographic mammography using a limited number of low-dose cone-beam projection images. Medical Physics 30(3), 365–380 (2003)
12. Brem, R.F., Rapelyea, J.A., Zisman, G., Mohtashemi, K., Raub, J., Teal, C.B., Majewski, S., Welch, B.L.: Occult breast cancer: scintimammography with high-resolution breast-specific gamma camera in women at high risk for breast cancer. Radiology 237(1), 274–280 (2005)
13. Brem, R.F., Fishman, M., Rapelyea, J.A., Brem, R.F., Fishman, M., Rapelyea, J.A.: Detection of ductal carcinoma in situ with mammography, breast specific gamma imaging, and magnetic resonance imaging: a comparative study. Academic Radiology 14(8), 945–950 (2007)
14. Hruska, C.B., Phillips, S.W., Whaley, D.H., Rhodes, D.J., O'Connor, M.K.: Molecular breast imaging: use of a dual-head dedicated gamma camera to detect small breast tumors. AJR 2008 American Journal of Roentgenology 191(6), 1805–1815 (2008)
15. Even-Sapir, E., Keidar, Z., Bar-Shalom, R.: Hybrid imaging (SPECT/CT and PET/CT)–improving the diagnostic accuracy of functional/metabolic and anatomic imaging. [Review] [132 refs]. Seminars in Nuclear Medicine 39(4), 264–275 (2009)
16. Magne, N., Chargari, C., Vicenzi, L., Gillion, N., Messai, T., Magne, J., Bonardel, G., Haie-Meder, C.: New trends in the evaluation and treatment of cervix cancer: the role of FDG-PET. [Review] [102 refs]. Cancer Treatment Reviews 34(8), 671–681 (2008)
17. Kuo, P.H., Carlson, K.R., Christensen, I., Girardi, M., Heald, P.W.: FDG-PET/CT for the evaluation of response to therapy of cutaneous T-cell lymphoma to vorinostat (suberoylanilide hydroxamic acid, SAHA) in a phase II trial. Molecular Imaging & Biology 10(6), 306–314 (2008)
18. Delbeke, D., Schoder, H., Martin, W.H., Wahl, R.L.: Hybrid imaging (SPECT/CT and PET/CT): improving therapeutic decisions. [Review] [228 refs]. Seminars in Nuclear Medicine 39(5), 308–340 (2009)

19. Cherry, S.R.: Multimodality imaging: beyond PET/CT and SPECT/CT. [Review] [59 refs]. Seminars in Nuclear Medicine 39(5), 348–353 (2009)
20. Bockisch, A., Freudenberg, L.S., Schmidt, D., Kuwert, T.: Hybrid imaging by SPECT/CT and PET/CT: proven outcomes in cancer imaging. [Review] [120 refs]. Seminars in Nuclear Medicine 39(4), 276–289 (2009)
21. Sung, Y.M., Lee, K.S., Kim, B.T., Choi, J.Y., Chung, M.J., Shim, Y.M., Yi, C.A., Kim, T.S.: (18)F-fdg pet versus (18)F-fdg pet/ct for adrenal gland lesion characterization: A comparison of diagnostic efficacy in lung cancer patients. Korean Journal of Radiology 9(1), 19–28 (2008)
22. Tatsumi, M., Miller, J.H., Wahl, R.L.: 18F-FDG PET/CT in evaluating non-CNS pediatric malignancies.[see comment]. Journal of Nuclear Medicine 48(12), 1923–1931 (2007)
23. Bowen, S.L., Wu, Y., Chaudhari, A.J., Fu, L., Packard, N.J., Burkett, G.W., Yang, K., Lindfors, K.K., Shelton, D.K., Hagge, R., et al.: Initial Characterization of a Dedicated Breast PET/CT Scanner During Human Imaging. J. Nucl. Med. 50(9), 1401–1408 (2009)
24. Crotty, D.J., McKinley, R.L., Tornai, M.P.: Experimental spectral measurements of heavy K-edge filtered beams for x-ray computed mammotomography. Physics in Medicine & Biology 52(3), 603–616 (2007)
25. Brzymialkiewicz, C.N., Tornai, M.P., McKinley, R.L., Cutler, S.J., Bowsher, J.E.: Performance of dedicated emission mammotomography for various breast shapes and sizes. Physics in Medicine & Biology 51(19), 5051–5064 (2006)
26. Williams, M.B., Judy, P.G., Gunn, S., Majewski, S.: Dual modality breast tomosynthesis. Radiology 255(1), 191–198 (2010)
27. Williams, M.B., Simoni, P.U., Smilowitz, L., Stanton, M., Phillips, W.: Analysis of the detective quantum efficiency of a developmental detector for digital mammography. Med. Phys. 26(11), 2273–2285 (1999)
28. More, M.J., Heng, L., Goodale, P.J., Yibin, Z., Majewski, S., Popov, V., Welch, B., Williams, M.B.: Limited Angle Dual Modality Breast Imaging. IEEE Transactions on Nuclear Science 54(3), 504–513 (2007)
29. Williams, M.B., Raghunathan, P., More, M.J., Seibert, J.A., Kwan, A., Lo, J.Y., Samei, E., Ranger, N.T., Fajardo, L.L., McGruder, A., et al.: Optimization of exposure parameters in full field digital mammography. Medical Physics 35(6), 2414–2423 (2008)
30. Pani, R., Scopinaro, F., Pellegrini, R., Soluri, A., Weinberg, I.N., De Vincentis, G.: The role of Compton background and breast compression on cancer detection in scintimammography. Anticancer Research 17, 1645–1650 (1997)
31. Kieper, D., Majewski, S., Goodale, P., Kross, B., Kundu, B., Li, H., More, M.J., Narayanan, D., Popov, V., Smith, M.F., et al.: Improved lesion visibility in a dedicated dual head scintimammography system - phantom results, pp. 1344–1346 (2002)
32. Majewski, S., Kieper, D., Curran, E., Keppel, C., Kross, B., Palumbo, A., Popov, V., Weisenberger, A.G., Welch, R., Wojcik, R., et al.: Optimization of dedicated scintimammography procedure using detector prototypes and compressible phantoms. IEEE Trans. Nucl. Sci. 3, 822–829 (2001)
33. Saunders, R.S.J., Samei, E., Lo, J.Y., Baker, J.A.: Can Compression Be Reduced for Breast Tomosynthesis? Monte Carlo Study on Mass and Microcalcification Conspicuity in Tomosynthesis. Radiology 251(3), 673–682 (2009)
34. Dunnwald, L.K., Gralow, J.R., Ellis, G.K., Livingston, R.B., Linden, H.M., Lawton, T.J., Barlow, W.E., Schubert, E.K., Mankoff, D.A.: Residual tumor uptake of [99mTc]-sestamibi after neoadjuvant chemotherapy for locally advanced breast carcinoma predicts survival. Cancer 103(4), 680–688 (2005)
35. Chuthapisith, S., Eremin, J.M., Eremin, O.: Predicting response to neoadjuvant chemotherapy in breast cancer: molecular imaging, systemic biomarkers and the cancer metabolome (Review). [45 refs]. Oncology Reports 20(4), 699–703 (2008)

An Anthropomorphic Software Breast Phantom for Tomosynthesis Simulation: Power Spectrum Analysis of Phantom Projections

Predrag R. Bakic[1], Beverly Lau[2], Ann-Katherine Carton[1], Ingrid Reiser[2], Andrew D.A. Maidment[1], and Robert M. Nishikawa[2]

[1] University of Pennsylvania, Dept. of Radiology, Philadelphia, PA 19104 ,
[2] University of Chicago, Dept. of Radiology, Chicago, IL 60637
{Predrag.Bakic,Ann-Katherine.Carton,
Andrew.Maidment}@uphs.upenn.edu
{Beverly,IReiser,R-Nishikawa}@uchicago.edu

Abstract. We have performed spectral analysis of simulated tomosynthesis projections generated using an anthropomorphic software breast phantom. Twenty phantoms were generated: ten 450 ml phantoms with 40% glandular fraction and ten 1500 ml phantoms with 20% glandular fraction. Simulated mammographic compression and acquisition was performed using monoenergetic ray-tracing. ROIs were extracted and the modulus-squared 2D FFT was applied to each ROI to obtain periodograms. Radially-averaged periodograms were compared between phantom and clinical images. We observed a good agreement between the spectral power law exponents (β) calculated from phantom projections and clinical images.

Keywords: breast anthropomorphic phantom, digital breast tomosynthesis, power law, power spectrum.

1 Introduction

Clinical validation of novel breast imaging systems is largely unfeasible today as it requires long and expensive clinical trials. On the other hand, physical characteristics of imaging systems such as the MTF, NNPS and NEQ, do not necessarily predict the behavior of the human observer scrutinizing complex mammographic backgrounds and do not take into account clinical processing or display. In an alternative approach, a voxelized anthropomorphic software breast phantom has been developed for use in pre-clinical validation of breast imaging modalities. The phantom realistically simulates the spatial distribution of adipose and fibroglandular tissues with known ground truth in simulated images. Projections of simulated tissue structures generate realistic parenchymal pattern, called anatomical noise. The anatomical noise is known to affect visibility of breast lesions. [1].

A frequently used descriptor of the anatomical noise is the power law exponent (β) of the radially-averaged periodogram. A periodogram is the modulus-squared 2D discrete Fourier transform of a region of interest (ROI). Burgess *et al.* [2] demonstrated

J. Martí et al. (Eds.): IWDM 2010, LNCS 6136, pp. 452–458, 2010.

that periodograms of digitized clinical mammograms follow a power law, $P(f) = B / f^{\beta}$. They performed spectral analysis of 46×46 mm regions in 213 mammograms and calculated the average power law exponent β to be 2.83 with a standard deviation σ=0.35. Engstrom *et al.* [3] analyzed the periodograms from tomosynthesis projection views and reconstructed images. Using 12.8×12.8 mm regions in 55 cases, they found the mean β to be 3.06 (σ=0.21) in the projection view images and 2.87 (σ=0.24) in the reconstructed images. In this paper we report results of estimating β in simulated tomosynthesis images of a software breast phantom.

2 Materials and Methods

We used an anthropomorphic breast phantom described by Bakic *et al.* [4,5], designed based upon the analysis of a large number of clinical breast images and histological slides. The phantom simulates the ellipsoidal shape of the breast outline. The phantom interior includes regions of predominantly adipose and predominantly fibroglandular tissues, and internal tissue structures (adipose compartments and Cooper's ligaments), as illustrated in Fig. 1. The design of the phantom is flexible to cover anatomical variations in breast composition and size.

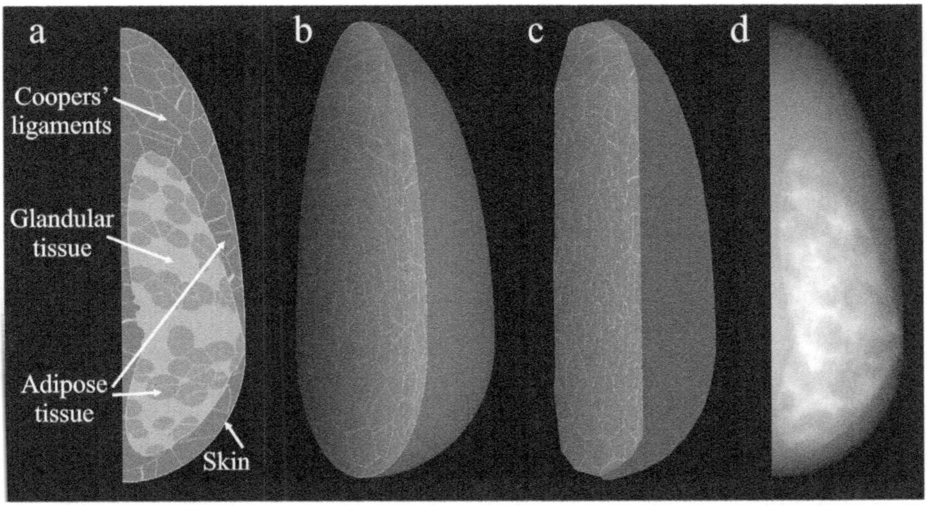

Fig. 1. (a) Orthogonal slice through a software breast phantom. Phantom outlines prior (b) and after (b) simulated mammographic compression. (d) Simulated x-ray projection image of the phantom.

For this study, we used ten 450 ml phantoms with 5 cm compressed thickness and ten 1500 ml, 7.5-cm thick phantoms. The 450 ml phantoms were generated with 40 percent by volume of fibroglandular tissue, while the 1500 ml phantoms were generated with 20 percent by volume of fibroglandular tissue. These glandularities were chosen in light of recent studies that have revealed that 95% of women have breast glandularity below 45% [6].

We simulated mammographic breast compression using a finite element model of tissue deformation. Simulated tomosynthesis projection images of the breast phantom were acquired using a monoenergetic x-ray beam without scatter or quantum noise by a detector which did not introduce any noise or blurring. Ray-tracing was used to calculate the x-ray attenuation through the phantom. The detector pixel resolution was matched with the voxel resolution of the phantom (500 micron). The acquisition geometry included 11 views over a 50-degree arc, with the center of rotation 15 cm above the breast [7].

β values were calculated on the tomosynthesis projection images using the method described by Engstrom *et al.* [3]. For this study, we selected 32x32 mm ROIs that were completely contained within the projected phantom region. ROIs were allowed to overlap by 50%; this is consistent with previous reports in the literature [2]. To reduce spectral leakage, we used a radial Hanning window on the mean-subtracted ROIs. The modulus-squared 2D FFT of each filtered ROI was calculated, and the obtained periodograms were each radially-averaged to reduce noise. β values were calculated as the slope of the linear portion of the log-log plot of the radially-averaged periodogram; these β values were evaluated from 0.15 to 0.7 cycles/mm. β values from all ROIs extracted from one breast and one projection view were averaged, and this average is denoted by β'_{breast}. β_{breast} indicates that the average was taken across one breast and all projection views.

3 Results

Fig. 2 shows examples of ROIs from simulated tomosynthesis phantom projections with corresponding periodograms and 1-D power spectra approximations (radially averaged periodograms). β values for individual ROIs in a single phantom projection image were found to range between 1.5 and 3.5. Fig. 3 shows β'_{breast} values as a function of tomosynthesis projection angle. Tables 1 and 2 show β_{breast} averaged over projection angles and averaged over breast phantoms. The average of β_{breast} over all the phantoms of the same size were 2.65 (σ= 0.318) for 450ml phantoms and 2.62 (σ=0.412) for 1500ml phantoms. Note that σ is the average standard deviation of β values found in a breast.

Table 1. A summary of β_{breast} values and standard deviations for ten 450 ml phantoms and ten 1500 ml analyzed breast phantoms. Mean values and standard deviation were calculated for all ROIs from all projection views for each phantom.

450 ml phantoms			*1500 ml phantoms*		
Phantom ID	β_{breast}	σ_β	***Phantom ID***	β_{breast}	σ_β
B1	2.604	0.305	DD1	2.648	0.400
B2	2.640	0.316	DD2	2.654	0.463
B3	2.621	0.318	DD3	2.618	0.430
B4	2.673	0.255	DD4	2.569	0.380
B5	2.664	0.360	DD5	2.650	0.347
B6	2.665	0.282	DD6	2.544	0.395
B7	2.693	0.355	DD7	2.603	0.369
B8	2.703	0.356	DD8	2.585	0.449
B9	2.651	0.289	DD9	2.582	0.434
B10	2.626	0.348	DD10	2.765	0.451
Average	2.654	0.318	***Average***	2.622	0.412

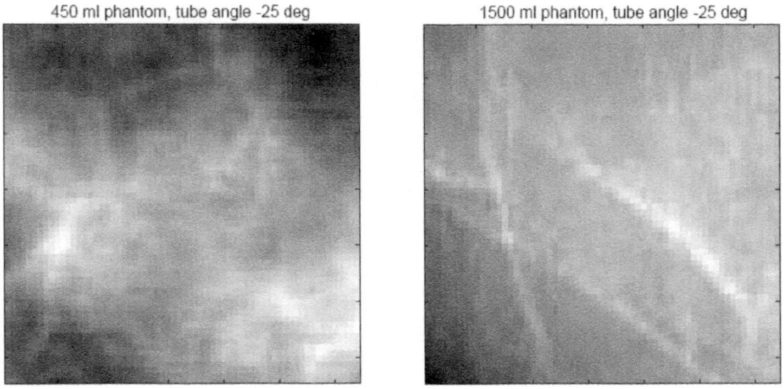

(a)

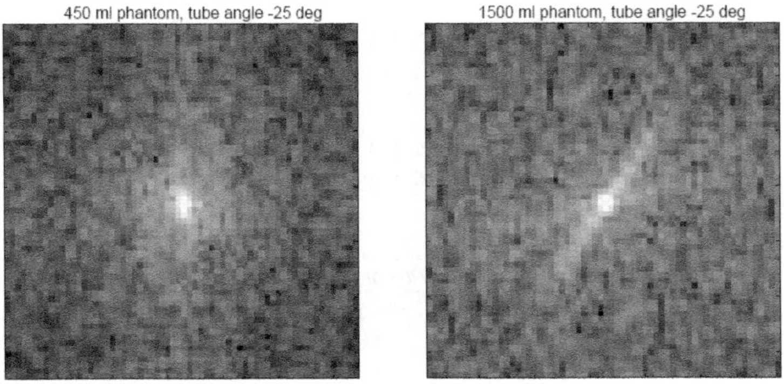

(b)

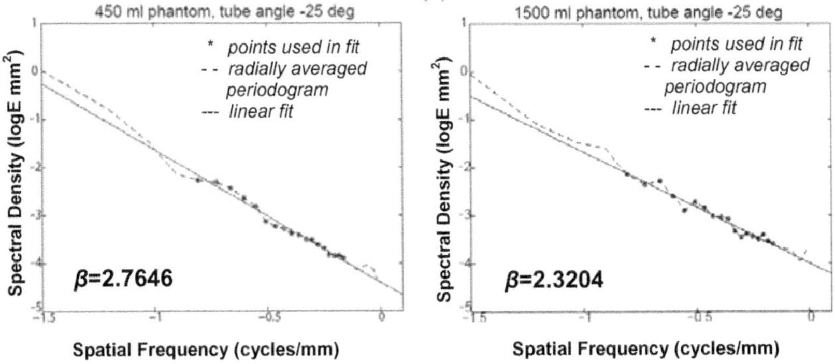

(c)

Fig. 2. Estimation of β values. (a) 32 mm × 32 mm ROIs from phantom projections. (b) Corresponding periodograms. (c) 1-D power spectra for a 450 ml (left column) and a 1500 ml phantom (right column).

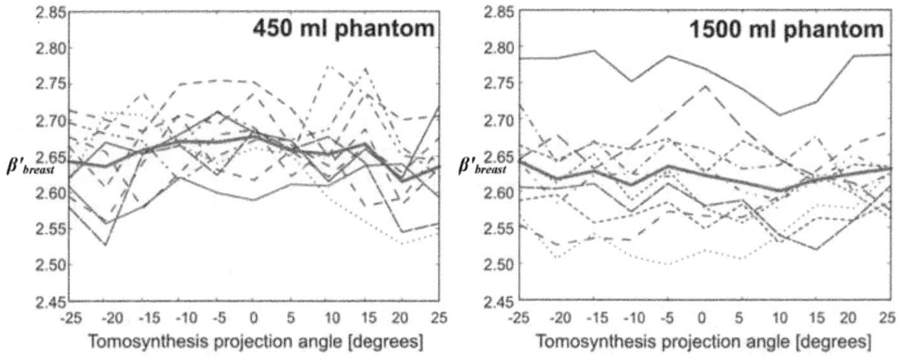

Fig. 3. Dependence of β'_{breast} values on the tomosynthesis projection angle for ten 450 ml software breast phantoms (left) and ten 1500 ml phantoms (right). Blue (dashed or solid) lines are β'_{breast} values for individual phantoms, and red (bold) lines are β'_{breast} values averaged over all phantoms of the same size.

Table 2. Average β'_{breast} values for projection views averaged over ten 450 ml phantoms and ten 1500 ml analyzed phantoms. Standard deviation was calculated across the β'_{breast} values for ten phantoms at each tube angle.

Tube angle	*450 ml phantoms*		*1500 ml phantoms*	
	$<\beta'_{\text{breast}}>$	$\sigma(\beta'_{\text{breast}})$	$<\beta'_{\text{breast}}>$	$\sigma(\beta'_{\text{breast}})$
-25°	2.644	0.043	2.641	0.066
-20°	2.637	0.064	2.617	0.074
-15°	2.659	0.048	2.628	0.073
-10°	2.671	0.041	2.609	0.067
-5°	2.670	0.046	2.635	0.075
0°	2.678	0.046	2.622	0.078
5°	2.658	0.027	2.613	0.066
10°	2.653	0.050	2.601	0.054
15°	2.668	0.064	2.616	0.054
20°	2.616	0.050	2.625	0.064
25°	2.636	0.056	2.632	0.062

4 Discussion

The range of β values estimated from phantom projections (1.5-3.5) is consistent with the range reported in the literature (1.5-4.5) [3]. The average β_{breast} values (2.65 for 450 ml phantoms and 2.62 for 1500 ml phantoms) are comparable with clinical measurements from digitized mammograms (average $\beta_{\text{breast}} = 2.83$) [2] and tomosynthesis projection view images (average $\beta_{\text{breast}} = 3.06$) [3]. The average β_{breast} values from the phantom are slightly lower than those reported in the literature. This might be due to the fact that the β_{breast} values reported in the literature were estimated from contralateral breast images of women with a known breast lesion; parenchymal properties of contralateral breasts are known to be correlated with risk of cancer. The phantom images used in our study did not include any lesions nor did they simulate high-risk parenchymal patterns.

The tomosynthesis projection angle has almost no effect on the measurement of β'_{breast}. In this preliminary study of two groups of phantoms we did not observe a significant effect of the breast volume or glandularity on the estimated β_{breast} values. Further analysis will include more groups of phantoms with different volume, thickness, and glandularity.

We identified limitations of our acquisition simulation approach. We have not used any quantum or detector noise and have not included scatter or detector blur. The inclusion of stochastic noise would affect the power spectrum at high frequencies (above approximately 1.0 cycles/mm [3]). The range of frequencies we have used to estimate β in this study will likely not be affected by the addition of quantum and detector noise. Scatter may increase the entire signal by a uniform amount with potentially a smaller increase at the edges of the image. Since we are using ROIs that do not encompass the entire image, we do not expect a large increase in the low frequency component of the approximated 1D power spectrum.

5 Conclusions

Power-law exponents calculated from simulated tomosynthesis projections through an anthropomorphic software breast phantom are comparable with clinically estimated values.

Acknowledgments. Funding for this research was sponsored in part by DOD BCRP predoctoral fellowship W81XWH-080-1-0353. The project described was supported in part by award R21 EB008801 from the National Institute of Biomedical Imaging and Bioengineering. The content is solely the responsibility of the authors and does not necessarily represent the official views of the National Institute of Biomedical Imaging and Bioengineering or the National Institutes of Health. RM Nishikawa is a shareholder in and receives royalties from Hologic, Inc. He is a member of the scientific advisory board of Dexela, Ltd. He is a consultant to Siemens Medical Solutions USA and Carestream Health, Inc.

References

1. Burgess, A.E., Jacobson, F.L., Judy, P.F.: Human observer detection experiments with mammograms and power-law noise. Medical physics 28, 419–437 (2001)
2. Burgess, A.E.: Mammographic structure: data preparation and spatial statistics analysis. In: Proc. SPIE, vol. 3661, pp. 642–653 (1999)
3. Engstrom, E., Reiser, I., Nishikawa, R.: Comparison of power spectra for tomosynthesis projections and reconstructed images. Medical Physics 36, 1753–1758 (2009)
4. Bakic, P.R., Albert, M., Brzakovic, D., Maidment, A.D.A.: Mammogram synthesis using a 3D simulation. I. Breast tissue model and image acquisition simulation. Medical Physics 29, 2131–2139 (2002)

5. Zhang, C., Bakic, P.R., Maidment, A.D.A.: Development of an Anthropomorphic Breast Software Phantom Based on Region Growing Algorithm. In: Proc. SPIE, Visualization, Image-guided procedures, and Modeling, vol. 6918 (2008)
6. Wu, T., Moore, R., Rafferty, E., Kopans, D.: A comparison of reconstruction algorithms for breast tomosynthesis. Medical Physics 31, 2636–2647 (2004)
7. Yaffe, M.J., Boone, J.M., Packard, N., Alonzo-Prouix, O., Huang, S.Y., Peressotti, C.L., Al-Mayah, A., Brock, K.: The myth of the 50-50 breast. Medical Physics 36, 5437–5443 (2009)

Comparison of Two Novel FFDM Systems with Different a-Se Detector Technology: Physical Characterization and Phantom Contrast Detail Evaluation in Clinical Conditions

Nadia Oberhofer[1,*], Alessandro Fracchetti[1], Elvina Nassivera[2], Aldo Valentini[2], and Ehrenfried Moroder[1]

[1] Azienda Sanitaria Alto Adige, Ospedale Bolzano, Servizio Aziendale di Fisica Sanitaria, Via L, Böhler 5, 39100 Bolzano, Italy
{nadia.oberhofer}@asbz.it

[2] Azienda Provinciale per i Servizi Sanitari di Trento, Ospedale S. Chiara, Servizio di Fisica Sanitaria, L. go Medaglie d'Oro, 38100 Trento, Italy

Abstract. Two new full field digital mammography (FFDM) equipments are compared, Fujifilm Amulet and Siemens Mammomat Inspiration. The systems are identical apart from the detector. Both use a-Se for photon conversion, but differ in detector readout technology: innovative photoconductive switching readout in the first system (pixel pitch 50 μm) and conventional a-Si TFT-matrix readout in the second (pixel pitch 85 μm). Comparison of the physical characteristics in terms of MTF, NNPS and DQE gives similar results for peak-DQE (both > 70% for Mo/Mo anode/target combination at 120 μGy detector entrance $kerma_{air}$), but a slower DQE decline at higher frequencies in the first system. Image quality assessment was based on contrast to noise (CNR) measurement and automated contrast detail (CD) curve analysis (phantom CDMAM ver. 3.4 and CDMAM Analyser ver. 1.1, both Artinis). It revealed a superior performance of the new detector equivalent to an average glandular dose (AGD) reduction comprised between 8% and 28% at equal image quality.

Keywords: digital mammography, direct conversion detector, a-Se detector, photoconductive switching readout.

1 Introduction

In FFDM most commercial systems use a-Se direct conversion Active Matrix Flat Panel Imagers (AMFPI) mainly because of their very high spatial resolution. However, this technology suffers from several drawbacks, first the high noise components if compared to indirect conversion CsI AMFPI, which diminish the detection efficiency. Further, the a-Si TFT matrix readout assembly reduces the fill factor, limiting detector efficiency and pixel size. A pixel pitch not much less than 100 μm is generally considered the limit compromise between image quality, resolution and dose [1].

* Corresponding author.

J. Martí et al. (Eds.): IWDM 2010, LNCS 6136, pp. 459–466, 2010.

Besides, conventional a-Se based detectors are susceptible to ghost artifacts and need to be flushed with an intensive light source before a new exposure to erase residual charge. The time interval between subsequent images is in the order of 30 s for systems on the market. Additionally, a-Se detectors are very temperature sensitive.

Recently a novel a-Se detector has been proposed [2]. The device combines a double layer of a-Se with photoconductive switching readout technology, boasting a 100% fill factor. This allows a smaller pixel pitch of 50 μm, reduced ghost susceptibility and fast readout.

In this work we compare two FFDM equipments with different a-Se direct conversion detectors, but otherwise identical units, only recently brought on the market: Siemens Mammomat Inspiration (System B) and Fujifilm Amulet (System A). The Siemens system uses a detector with the established TFT matrix readout and has previously been shown to exhibit a very good performance with respect to other FFDM equipments [3]. The Fujifilm system employs the innovative photoconductive switching readout detector. Its operation has been described elsewhere [4,5,6]. Comparison is carried out in terms of physical characterization (MTF, NNPS, DQE) and image quality assessment by contrast to noise ratio (CNR) evaluation and contrast detail (CD) analysis.

2 Methods

Table 1 summarizes the main characteristics of the two equipments which are compared in this study. Although the readout data of both detector types follow a linear response function, in System A output data are delivered after a logarithmic transformation to allow the same image processing as in the Fujifilm CR systems. To the purpose of MTF, NNPS and DQE calculation particular attention was posed on the linearization of the image data to minimize uncertainty deriving from truncation errors. All analyzed images were dicom images "for processing". In system A the processing modality FIX-mode was used with sensitivity (S) and latitude (L) set to 121 and 2, respectively. Measurements were taken with calibrated detectors.

Table 1. Systems' characteristics

	System A	System B
Manufacturer	FujiFilm	Siemens
Model	Amulet	Mammomat Inspiration
Target/filter combinations	Mo/30 μm Mo, Mo/25 μm Rh, W/50 μm Rh	
Focus detector distance/grid	650 mm / ratio 5:1, 31 lines/cm	
Detector manufacturer	FujiFilm	Anrad (LMAM)
Detector size	177 mm x 237 mm	240 mm x 300 mm
Pixel number	3540 x 4740	2816 x 3584
Pixel pitch	50 μm	85 μm
Photon conversion layer	a-Se, 187 μm	a-Se, 193 μm
Readout	a-Se, 18 μm, photoconductive switching readout with scanning blue LED array	a-Si, < 5 μm, electronic sequential readout with TFT
Image cycle time	15 s	< 30 s
Detector response function	Linear, 16 bit	Linear
Output data format	Logarithmic, 12 bit	Linear

The physical characterization has been carried out according to IEC 62220-1-2. To permit comparison with previous studies exposures were performed with radiation quality RQA-M2 (Mo target, 30 μm Mo filtration, 2 mm Al additional filtration at tube exit), with neither antiscatter grid nor compression paddle in place. The response curves were determined from a wide range of uniform exposures on the base of the mean pixel value (MPV) of a 1 cm^2 region of interest (ROI,) centered midline 6 cm from the chest wall edge. The same ROIs were considered for noise components analysis using the fit:

$$SD^2 = \alpha x + \beta x^2 + \gamma , \tag{1}$$

where SD is the standard deviation in pixel values within the ROIs, x is the exposure and α, β and γ represent the contributions of the quantum-statistical (Poisson) noise source, structural noise and dose independent electronic noise, respectively.

Evaluation of the physical parameters was performed with the open source Java plug-in "qa-distri/DQE Panel v7" for ImageJ, extended for DQE calculation. MTF was determinated with a tungsten edge test object placed directly on the detector cover with the edge centered midline, 6 cm from the chest wall edge, slightly tilted with respect to the pixel lines or rows. NNPS analysis was always carried out on 3 subsequent flat images and the average value considered for DQE calculation. MTF, NNPS and DQE were determined for two orthogonal directions, perpendicular and parallel to the chest wall edge, which we shall denote briefly as perpendicular and parallel. Correctness of calculation routine has been verified previously by comparison with literature. Dose measurements were taken as $kerma_{air}$ with a regularly calibrated ionisation chamber (Radcal 9010 and 10X6-6M, Radcal Corporation, USA).

Images for MTF determination were acquired with a detector entrance $kerma_{air}$ slightly higher than the values used by each unit in clinical automatic exposure control (AEC) modality for a 4.5 cm polimethylmetacrylate (PMMA) block, which we will denote as E_0. In System B measurements at additional dose values were made. Flat images for NNPS analysis were acquired at $0.5E_0$, E_0 and $2E_0$.

For image quality assessment two methods have been taken into account, CNR measurement and CD evaluation. The former is a straightforward tool, but its numerical value is a relative quantity which is meaningful only if referred to the same unit. On the contrary, the latter represents an objective method suitable for comparison of different equipments, provided the same phantom is used.

CNR has been measured according to the European Guidelines [7]. The contrast object (0.2 mm Al, 1.5 cm x 1.5 cm) was always positioned at a fixed high above a 2 cm homogenous polymethylmetacrylate (PMMA) block covering the whole detector area, with one edge on the midline 6 cm from the chest wall edge. For CD evaluation the same phantom CDMAM ver. 3.4 was used, combined with the commercial software CDMAM Analyser, ver. 1.1 (both Artinis, The Netherlands) for automated image analysis. To simulate clinical exposure conditions, the detail tablet (thickness 0.5 cm, absorption corresponds to 1 cm PMMA) was embedded between PMMA layers, with the smaller details towards the chest wall side. Except for the smallest total phantom thickness of 2 cm, the detail tablet layed on 2 cm of PMMA. For both phantom set-ups additional PMMA layers were put on top of the contrast object/detail tablet, to achieve the desired total phantom thickness.

All reported results represent, in order to reduce uncertainty, the curve fitted value of the analysis (75% detection rate) of 8 raw images for each exposure technique.

Between successive images the detail tablet was slightly moved. Results are expressed in terms of a group contrast detail curve and, additionally, of an overall quality index, the Image Quality Figure Inverse (IQF_{inv}), defined as

$$IQF_{inv} = \sum_{i=1}^{16} \frac{100}{Diameter_i * Depth_i}, \qquad (2)$$

which increases when detail detection rises. This figure has prevously been shown to represent an objective and absolute measure of the image quality [8], suitable for comparison of different equipments.

CNR and IQF_{inv} have been measured for a very wide set of exposure techniques, including very unusual ones, although in both systems the clinical setups used only the target/filter combination W/Rh. For each phantom thickness (from 2 cm to 7 cm) the exposure load was varied at least from a factor 0.5 to 2 relative to the average glandular dose (AGD) value used by the systems in automatic exposure control (AEC) mode. In System A the normal dose Mode L was taken as reference (see Fig. 1, left). The relation CNR $\propto$ IQF_{inv} was verified in both systems. Under the hypothesis that image noise is dominated by quantum noise, image quality improves applying a higher dose according to $CNR^2 \propto$ dose. That means CNR^2/dose is constant for each kV/phantom setting. In analogy we defined the dose independent figure of merit IQF_{inv}^2/AGD to compare systems performances in various conditions.

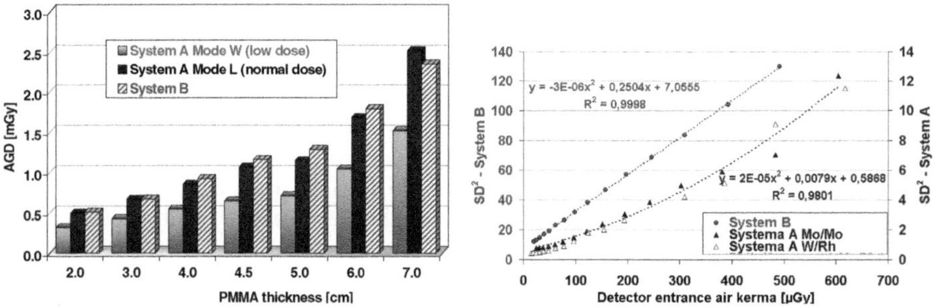

Fig. 1. Comparison of dose levels applied in clinical AEC modalities at various PMMA phantom thicknesses (*left*) and analysis of the noise components (*right*) in both systems

3 Results

Physical Characterization: The results of the noise components analysis are displayed in Fig. 1 (right). The curve related to System B corresponded to an almost perfect straight line, whereas the detector with photoconductive switching readout of System A deviated from the ideal behavior.

Fig. 2 reassumes the results for the presampled MTF up to $f_{Nyquist}$. In both Systems the values showed a marked direction dependance, being higher in the perpendicular direction. In System A the difference increased with increasing exposure, while the MTF value itself reduced (Fig. 2, left). Despite the smaller detector pixel pitch, System A revealed a lower mean MTF value compared to System B (Fig. 2, right).

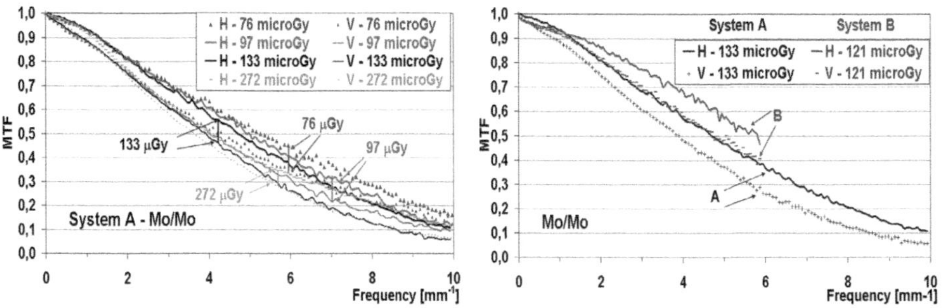

Fig. 2. Presampled MTF. Effect of exposure on calculated values for System A (*left*) and comparison between Systems A and B at similar exposure values (*right*). Both systems show a marked direction dependance. H = perpendicular to chest wall edge, V = parallel. Error: < 2%.

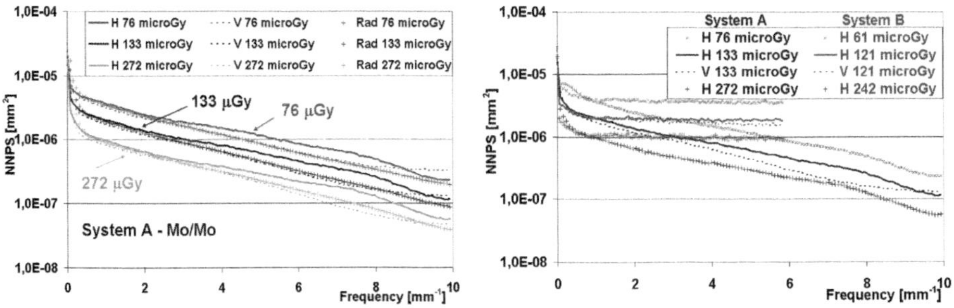

Fig. 3. NNPS. Effect of sampling direction on the calulated values for System A (*left*) and comparison between Systems A and B at similar exposure values (*right*). H = perpendicular to chest wall edge, V = parallel, Rad = radial. Relative error: < 4%.

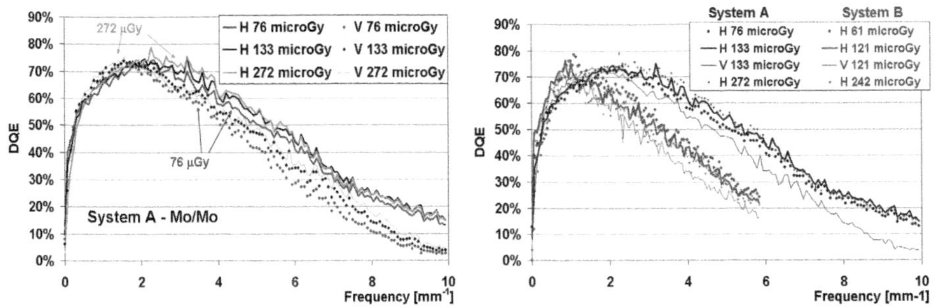

Fig. 4. DQE. Effect of sampling direction on the calulated values for System A (*left*) and comparison between Systems A and B at similar exposure values (*right*). H = perpendicular to chest wall edge, V = parallel. Relative error: 8%.

NNPS results are summarized in Fig 3. In both units the noise diminished with increasing exposure (Fig. 3, right). In System A the NNPS value exhibited a strong direction dependence, with higher noise in the perpendicular direction (Fig. 3, left).

For System B the calculated values evidenced some smaller peaks in the perpendicular direction; beside that the variations linked to the sampling direction were small. With increasing frequency System B exhibited an almost constant NNPS; in System A, with the new detector, the calculated NNPS value decreased, being at comparable detector entrance $kerma_{air}$ on average lower by a factor 1.6 at a frequency $f = 2\ mm^{-1}$, 2.1 at $f = 3\ mm^{-1}$, 2.6 at $f = 4\ mm^{-1}$ and 3.6 at $f = 5\ mm^{-1}$.

Resulting DQE values are shown in Fig. 4. In the investigated dose range both detector types featured a peak DQE value above 70% (Fig. 4, right): in System A it occurred at a frequency of 1 mm^{-1}, whereas in System A the peak was broader and shifted towards 2.5 mm^{-1}. As the frequency increased, in System A the DQE value of the perpendicular direction dropped significantly below the value of the parallel direction. The difference between the DQE values in the two orthogonal directions was in the order of 10% (Fig. 4, left). Nevertheless both values kept high as frequency increased, if compared to System B. For $f > 2.5\ mm^{-1}$ the calculated DQE value in System A exceeded that for System B by an amount between 15% and 25%. To a minor extent also the latter system suffered from direction dipendent behavior, which here affected the peak value, too. The DQE performance was inferior in the perpendicular direction with localized small drops. With respect to dose, in System A the computed DQE values were comparable at normal and high exposure levels and slightly inferior at low dose, but showed constant peak DQE. In system B the DQE turned out to have a slight, opposite dose dependence: DQE decreased with increasing exposure.

Image Quality Assessment: The AEC devices of the two units were calibrated very similarly in the normal dose mode (maximum AGD difference at various phantom thicknesses 10%, Fig. 1, left). In both systems CNR and IQF_{inv} correlated strongly over the wide technique range investigated, as illustrated in Fig.5 . The linear correlation coefficient r resulted 0.9533 and 0.9655 for Systems A and B, respectively.

Fig. 6 reassumes the outcomes of CD analysis. For a given AGD, System A scored a higher IQF_{inv} value for every phantom thickness, as figured in Fig. 6, left. The sign test confirmed the difference being significant ($p < 0.01$). The dose saving permitted by System A with the new detector whilst achieving a specific IQF_{inv} value was compromised between 8% and 28%, depending on the phantom thickness. Performances are compared in Fig. 6 (right) by means of the figure of merit IQF_{inv}^2/AGD. Each bar represents a mean value of 2-4 datapoints corresponding to different AGD values.

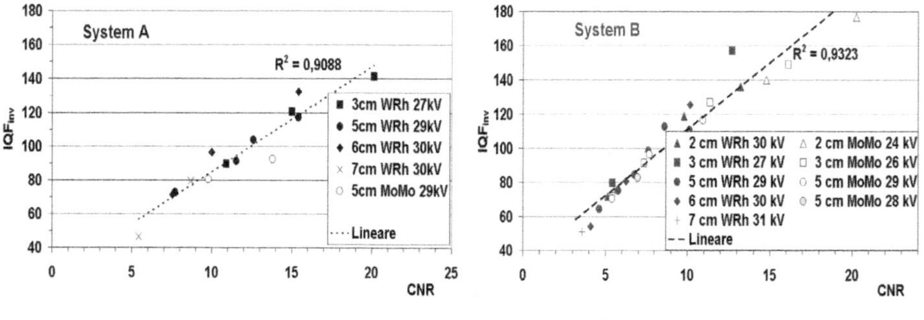

Fig. 5. Study of the correlation between CNR and IQF_{inv}. Relative error: CNR 3%, IQF_{inv} 5%.

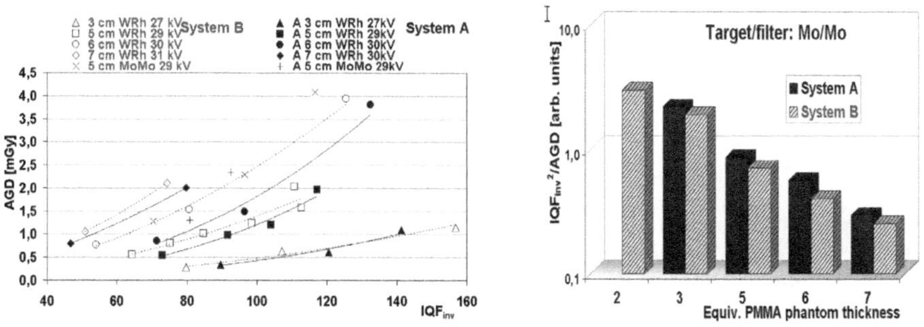

Fig. 6. Comparison of image quality performance of the systems A and B in terms of the overall image quality index IQF_{inv}, AGD (*left*) and the figure of merit IQF_{inv}^2/AGD (*right, System A = full symbols, System B = empty symbols*). Bars represent a mean value. Relative error: IQF_{inv} < 4%, AGD 10% (precision: 3%), mean IQF_{inv}/AGD 5 % - 7%.

4 Discussion

In clinical normal dose mode the AEC devices of the two systems applied similar exposure parameters for every phantom thickness, leading to comparable AGD values.

In system B the noise had substantially poissonian origin, with a small additive (electronic) component and negligible structural noise. In System A the deviation from the ideal linear line confirmed the presence of a structural component, as pointed out by other authors [5,6]. Both detectors exhibited a clearly better MTF-performance in the parallel direction. This may have different causes in the two system, related to the different detector readout technologies.

The most striking difference in the performances of the two detectors regarded the NNPS. In System B the noise contributions were almost constant over the whole spatial frequency range and direction indipendent, as usual for a-Se detectors with TFT-matrix readout. On the contrary System A revealed a decreasing noise with increasing frequency. Although in the parallel direction the performance was markedly poorer, the noise level was generally very low, being on average lower by a factor of almost 2.5 with respect to System B at comparable exposure levels.

Peak DQE values resulted similarly high in both systems (> 70%), but with increasing frequency in System A the DQE value declined much more slowly, owing to the decrease in NNPS. At a frequency of 5 mm^{-1} the mean DQE value of System A was more than twofold the value of System B. In both systems the directional dependence of DQE was governed by the MTF, exhibiting higher values in the parallel direction. In System B the peaks in NNPS caused drops in DQE.

The MTF, NNPS and DQE curves presented in this study for System A demonstrate some variations from literature [5], where data refer mostly to the clinically used target/filter combination W/Rh. Measurements on System A, not discussed in this work, indicate that for the latter tecnique the DQE behaviour is different, with a higher peak value and a quicker decline. The curves seem to be more similar to the reported ones.

In both systems the correlation between CNR and IQF_{inv} was confirmed and could regarded as linear for all practical purposes. Contrast detail evaluation brought out a statistically significant better performance of the new detector at comparable AGD levels, which was summarized by a higher score in the dose independent figure of merit IQF_{inv}^2/AGD. System A allowed on average a dose saving of 20% (min 8%, max 28) at equal image quality.

5 Conclusions

Compared to System B with a conventional a-Se detector, System A exhibited a slightly lower spatial resolution but a definitely better noise behavior. These resulted in a comparable peak DQE value for both detector types at low frequency. With increasing frequency differences between the two systems became more evident. Owing to the decrease in NNPS, in System A the DQE value declined much more slowly. This reflected into a better performance in terms of small detail detection in contrast detail curve analysis. The new detector seamed to allow average dose savings from 8% to 28 at equal image quality.

Acknowledgments

We would like to thank Dr. Mertelmeier from Siemens Women's Health for helpful discussions and Fujifilm Italia for access and assistance at the test facility.

References

1. Neitzel, U.: Status and Prospects of Digital Detector Technology for CR and DR. Rad. Prot. Dosimetry 114, 32–38 (2005)
2. Irisawa, K., Yamane, K., Imai, S., Ogawa, M., Shouji, T., Agano, T., Hosoi, Y., Hayakawa, T.: Direct-Conversion 50 μm pixel-pitch Detector for Digital Mammography using Amorphous Selenium as a Photoconductive Switching Layer for Signal Charge Readout. In: Proc SPIE, vol. 7258, pp. 72581I–1 (2009)
3. Oberhofer, N., Fracchetti, A., Springeth, M., Moroder, E.: Digital Mammography: DQE versus Optimized Image Quality in Clinical Environment – an on Site Study. In: Samei, E., Pelc, N.J. (eds.) Proc. SPIE Medical Imaging 2010, vol. 7622 (in press, 2010), ISBN 9780819480231
4. Kuwabara, T., Iwasaki, N., Sendai, T., Furue, R., Agano, T.: Image Quality Evaluation of Direct-Conversion Digital Mammography System with a New Dual a-Se Layer Detector. In: Proc. SPIE, vol. 7258, pp. 72584P–1 (2009)
5. Rivetti, S., Lanconelli, N., Bertolini, M., Borasi, G., Golinelli, P., Acchiappati, D., Gallo, E.: Physical and psychophysical characterisation of a novel clinical system for digital mammography. Med. Phys. 36(11), 5139–5148 (2009)
6. Young, K.C., Oduko, J.M., Asad, M.: Technical Evaluation of Fuji Amulet Full Field Digtal Mammography System. NHSBSP Equipment Report 0907 (2009), http://www.cancerscreening.nhs.uk
7. Van Engen, R., Young, K., Bosmans, H., Thijssen, M.: European Protocol for the Quality Control of the Physical and Technical Aspects of Mammography Screening. Part B: Digital Mammography. In: European Guidelines for Quality Assurance in Breast Cancer Screening and Diagnosis, 4th edn., European Communities, Luxembourg (2006)
8. Oberhofer, N., Parruccini, N., Moroder, E.: Image Quality Assessment and Equipment Optimisation with Automated Phantom Evaluation in Full Field Digital Mammography (FFDM). In: Krupinski, E.A. (ed.) IWDM 2008. LNCS, vol. 5116, pp. 235–242. Springer, Heidelberg (2008)

Development of a Peripheral Thickness Estimation Method for Volumetric Breast Density Measurements in Mammography Using a 3D Finite Element Breast Model

Olivier Alonzo-Proulx[1], James G. Mainprize[1], Nathan J. Packard[2], John M. Boone[2], Adil Al-Mayah[3], Kristy Brock[3], and Martin J. Yaffe[1]

[1] Sunnybrook Health Sciences Centre, University of Toronto, Toronto, Ontario M4N 3M5, Canada
[2] UC Davis Medical Center, University of California-Davis, Sacramento, California 95817
[3] University Health Network, University of Toronto, Toronto, Ontario M5G 2M9, Canada
oliviera@sri.utronto.ca

Abstract. A method was developed to determine the area in a mammogram where the breast is not in contact with the compression paddle (the periphery), and to predict the breast thickness in that peripheral region. The periphery is determined by evaluating the variation of the signal intensity along radial lines, and the peripheral thickness is modeled assuming the breast has a semi-circular shape. The method was tested on 26 simulated mammograms for which the volumetric information was available. The mammograms were obtained using CT data that were deformed to simulate mammographic compression and then projected using a physical model. The method predicted the thickness in the periphery to within 3.3 mm of the actual value and the volumetric breast density within 4.3 percentage points. The method was also tested on 209 digital mammograms, and on average it was estimated that thickness errors occurred on 9% of the breast image, and the average absolute thickness error on those points was estimated to be approximately 2.0 mm in the periphery and central region of the breast but as large as 10.5 mm in the extreme periphery where the thickness is small.

Keywords: volumetric breast density, digital mammography, periphery detection.

1 Introduction

The estimation of volumetric breast density (VBD) from mammograms is highly sensitive to the breast thickness [1,2]. Several factors can be responsible for uncertainties in thickness. The breast thickness may be reported incorrectly by the mammography unit. In addition, the breast thickness varies due to a slant or curvature in the compression paddle, and at the natural outer bulge, *i.e.* from the point where the breast loses contact with the paddle at the peripheral region of the breast [3,4,5]. Algorithms have been developed to identify the peripheral region on the mammogram and to model the breast shape in the periphery [4,5]. However, the accuracy of such methods cannot be easily estimated without direct measurement of the breast thickness. Furthermore, the effect of these peripheral correction methods on the measurement of the total VBD has not

J. Martí et al. (Eds.): IWDM 2010, LNCS 6136, pp. 467–473, 2010.

been reported. In this work, we tested the validity of a peripheral thickness correction method in mammography using CT data that were deformed to simulate the effects of compression, and thus for which the compressed thickness data were available. The method was also tested on 209 digital mammograms, and an estimation of the thickness error from the algorithm was performed.

2 Methods

2.1 Peripheral Detection Algorithm

The algorithm is similar to ones proposed by others [4,5]. The inner edge of the peripheral region is found using the variation of image intensity along radial lines. The negative logarithm of the pixel value of the digital mammogram, $L(x,y) = -\ln I(x,y)$ is taken. The variation in L represents changes in attenuation, due to changes in tissue thickness and composition. The outer breast edge, **E**, of the breast (see Fig. 1a)) is determined by threshold. For each $\varphi \in [0,\pi]$, the intensity range $\Delta(\varphi) = \max(L)-\min(L)$ is taken from the maximum and minimum intensities in the image. Plots of $L(r)$ are generated, where r is the distance on the image from a point (x,y) to **M** (see figure 1a)), the point midway between the breast edges on the chest wall edge of the image. An example of a function $L(r)$ is shown in Fig. 1b). We can see that for r close to 0, there are relatively small changes in intensity, presumably due to changes in tissue composition rather than from changes in thickness. Further away from **M**, we note a sudden change in intensity, which arises from the reduction in thickness that occurs at the breast periphery. To identify the onset of that sudden change, various linear fits are performed on a progressively larger subset, spanning a distance from **M** of pR_C in $L(r)$ for $p = (0.1, 0.2, \ldots, 1)$ for each φ, and the linear fit with the lowest absolute slope is recorded, yielding the linear function $ar + b$, where a is the slope and b the offset of the fit. The multiple fits are performed to obtain the best estimate of the baseline constant-thickness intensity in the image, by avoiding localized pockets of adipose or fibroglandular tissue. The distance, $R_{C,}$ was chosen so that $R_C = \alpha R_\varphi$, where R_φ is the distance from **M** to the outer breast edge **E** at the angle φ, and α varies from 0.6 to 0.8 for small to large breasts respectively. The maximum distance, r_0, along r in which $(ar + b) - L(r)$ is smaller than the threshold intensity $\beta\Delta(\varphi)$, is determined, where β was determined empirically to be 0.07. That point r_0 denotes the approximate location of the inner edge of the periphery. By repeating the procedure at each angle φ, the approximate inner periphery contour **C'** is found, and is then smoothed using a low-pass filter.

The location of the inner periphery contour **C** is then determined using the thickness profile to be applied in the periphery. The empirical semi-circular thickness profile $T(r)$ is obtained from Rico *et al* [5], and is scaled such that $T(0) = 1$ and $T(1) = 0$, where r is the normalized distance form the inner periphery edge to the breast edge. Assuming that the thickness at **C′** corresponds to $(1 - \beta)$ of the compressed breast thickness T_0, the inner periphery contour **C** is obtained by applying a morphological operation of erosion **C′** by a distance $r' = \delta \times T_0/2$, with δ such that $T(\delta) = 1 - \beta$. For $\beta = 0.07$, $\delta = 0.49$. Finally, the thickness map is applied along radial lines (centred on **M**) in the peripheral region P, such that $T(\mathbf{C}) = T_0$ and $T(\mathbf{E}) = 0$.

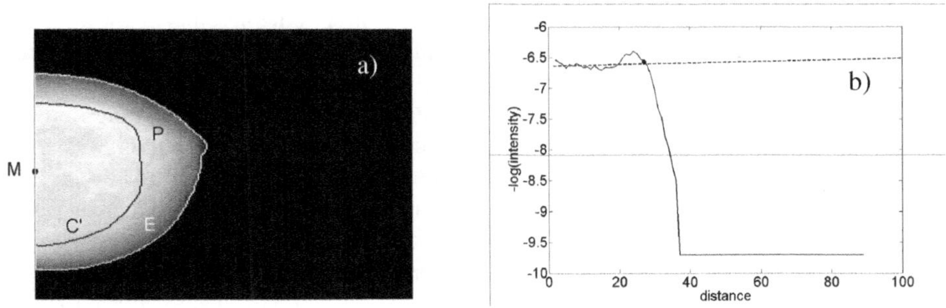

Fig. 1. a) Image of $L(x,y)$ showing the central chest wall point **M**, the inner periphery contour **C′**, the peripheral region P and the outer edge **E**; **b)** plot of $L(r)$ for a given angle φ. The dot indicates the distance r_0 where the periphery C' was located, and the dotted line the linear fit.

2.2 Peripheral Detection on Simulated Mammograms

CT image data sets of 26 volunteers were obtained using a dedicated breast CT scanner [6]. The CT data was segmented into skin, fibroglandular and adipose tissue, and a finite element model was applied to simulate the effects of mammographic compression, as described in Yaffe *et al* [2]. The model squeezes the breast with a rigid compression plate to achieve a target thickness. The deformed volumes were used to simulate a digital mammogram, considering the effects of the polychromatic x-ray spectrum, the primary and scattered energy transmitted through the breast (using the point spread functions of Boone *et al* [7]), the anti-scatter grid and the detector efficiency. A corresponding thickness map, $T(x,y)$, of the breast was obtained by projecting a unity volume of the breast. The volume of dense tissue in the whole breast, V_D, and the peripheral region, V_{PD}, were obtained by using the VBD algorithm of Yaffe *et al* [2] and by determining the true periphery, P, by using a modified version of the peripheral detection algorithm on $T(x,y)$. The total breast volume, V, and the volume, V_P, and area, A_P, in the peripheral region were also determined. The simulated mammograms were then analyzed with the peripheral correction method outlined in section 2.1, and the thickness map $T_P(x,y)$ was obtained from the algorithm. The total volume of dense tissue V'_D, the volume of dense tissue in the periphery, V'_{PD}, the total breast volume, V', and the volume, V'_P, and area, A'_P, of the peripheral region were also determined, respectively. To evaluate the error in the method, the root mean square (rms) of T - T_P was computed over the periphery, P', determined by the algorithm In addition, the "extrema error", the points with density that were outside the range from 0 to 1.0, were identified as a surrogate measure of inaccuracy used to compare the simulation results and the results obtained from the digital mammograms.

2.3 Peripheral Detection on Clinical Mammograms

The peripheral detection algorithm was tested on 209 digital mammograms from a GE Senographe 2000D, and the VBD analysis was performed. The method of Mawsdley and Tyson [3] is used to account for errors in the compressed thickness readout from the mammography unit. In addition, for the uniformly compressed region (within **C**), the compressed thickness is adjusted iteratively in 1 mm increments by minimizing the number of density values that are outside the range from 0 to 1.0 (and thus for which the thickness is over- or underestimated, respectively) to less than 5% of the total number of points in that region. The remaining values on the whole image that where smaller than 0 and larger than 1.0 were set to 0 and 1.0 respectively.

The total dense volume, V_D, total volume, V, the dense volume in the periphery, V_{PD}, and the volume of the periphery, V_P, were computed. The extrema error was also computed on the image in order to evaluate the accuracy of the peripheral correction method. Assuming a sensitivity of 0.046 density per mm (Yaffe *et al* [2]), the estimated thickness error ΔT was computed as $\Delta T = \Delta m/0.046$, where Δm is the calculated erroneous density value minus the upper or lower bound density of 0 or 1.0.

3 Results

3.1 Peripheral Detection on Simulated Mammograms

For the 26 deformed CT cases, on average, the true values were $V = 566$ cm^3, $V_P = 347$ cm^3, $V_D = 108$ cm^3, $V_{PD} = 67$ cm^3. Using the thickness map T_P from the algorithm, we obtain the following averages: $V' = 564$ cm^3, $V'_P = 351$ cm^3, $V'_D = 110$ cm^3, and $V'_{PD} = 71$ cm^3. The respective averages of A_P and A'_P were 76.8 cm^2 and 75.6 cm^2. Fig. 2(a) shows the histogram of the rms of $T(P') - T_P(P')$, with a mean of 3.3 mm. Fig. 2(b) shows the relation between the true VBD and VBD′, the density obtained with the detected periphery. The average VBD and VBD′ were 0.231 and 0.245 respectively. A linear least square fit gave a slope of 1.01, an intercept of 0.011 and a correlation with R^2=0.89. The rms of VBD - VBD′ was 0.043. On average, the fraction of points showing extrema errors was 6% and 9%, when using the true thickness T and when using T_P, respectively.

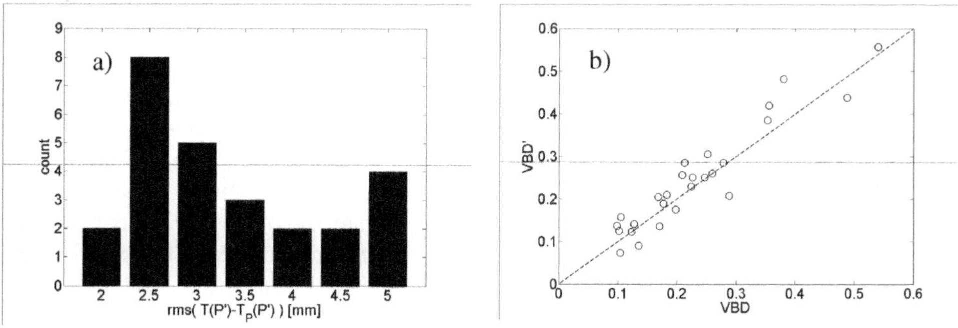

Fig. 2. a) Histogram of the rms of $T(P') - T_P(P')$; **b)** Plot of the true VBD versus VBD′, the density calculated using the detected periphery. The dashed line represents the identity function.

3.2 Peripheral Detection on Clinical Mammograms

For the 209 digital mammograms, the average V_D and V were 114 cm^3 and 811 cm^3 respectively. V_{PD} and V_P were 60 cm^3 and 378 cm^3, respectively. Thus the average

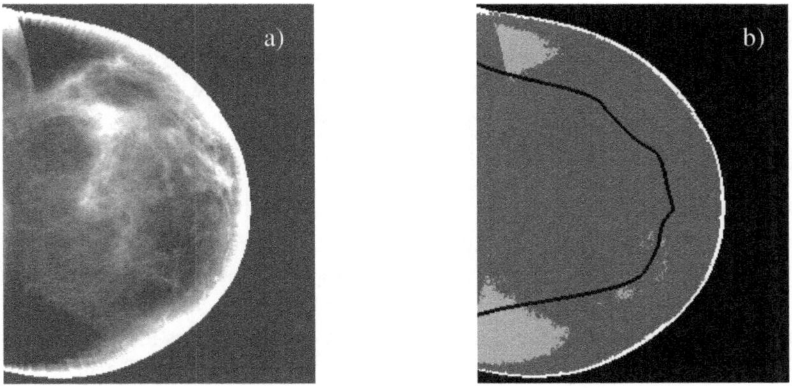

Fig. 3. a) Calculated density map of a digital mammogram. The VBD was 0.130. The image contrast was amplified for clarity; **b)** Corresponding mammogram (dark gray) showing the density values under 0 (light gray), density values above 1.0 (white) and the periphery contour (black line). The outside of the breast is also black.

VBD was of 0.162 for the whole breast and 0.173 in the periphery only. The average fraction of points showing extrema errors was 9%, with 6% and 3% of the errors with density below 0 and above 1.0, respectively and with the majority (8%) of these errors occurring in the peripheral region. Fig. 3a) shows the calculated density map of a mammogram, and Fig. 3b) shows the same mammogram with the extrema errors highlighted, as well as the periphery contour. Fig. 4 shows the distribution (for the 209 images) in estimated minimum thickness error ΔT for the points showing extrema errors. Negative and positive values represent thickness overestimation (density below 0) and underestimation (density above 1.0), respectively, and the average ΔT for those subsets were respectively -2.0 mm and 10.5 mm. The average absolute thickness error was 6.2 mm. The presence of the offset positive ΔT peak is discussed in the next section.

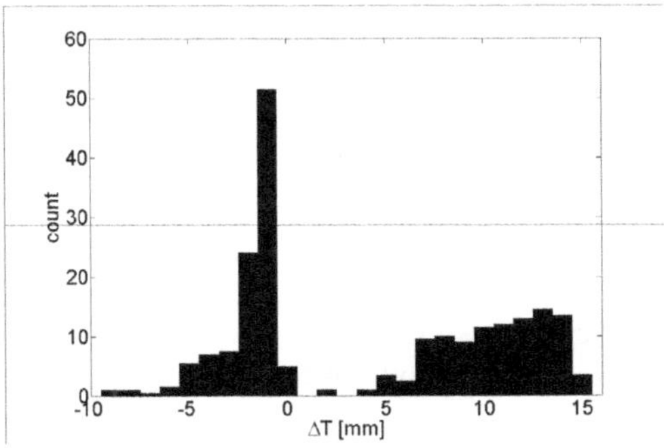

Fig. 4. Histogram of the estimated thickness error for the points with extrema errors, for the 209 digital mammograms

4 Discussion

In comparing the detected peripheral volume and thickness with the truth from the 26 simulated images, the thickness was well predicted, with the rms(T - T_P) = 3.3 mm in the peripheral region. Since the thickness errors were partially compensated by correcting the density to the nearest acceptable value, the total VBD was within 0.014 with rms(VBD - VBD′) = 0.041.

This validation study had limitations. The breast deformation was simulated and thus it is possible that real breasts would be compressed in a different manner, yielding a differently-shaped outer bulge in the periphery for which the thickness model might not apply. Moreover, the simulation of the mammograms and the general VBD algorithm are also susceptible to error. For example, even when the true breast thickness profiles were used from the simulated mammograms, there was an average of 6% of points with extrema errors.

The method was also tested on 209 clinical digital mammograms. The true thickness and VBD information were not available. On average, the fraction of points showing extrema errors was 9%, which is similar to what is observed in the validation study with simulated mammograms. Those erroneous values occurred for the most part in the peripheral region. They can be due to errors in the VBD calculation algorithm or from errors in thickness from the peripheral detection algorithm. Assuming the latter, a lower bound on the error in thickness ΔT can be estimated from the approximate sensitivity of density with respect to thickness, as reported by Yaffe *et al* [2]. We note a distinct peak in the histogram of Fig. 4 for positive ΔT, with a mean of 10.5 mm. Those erroneous points were almost exclusively located near the outer edge of the breast image, as can be seen in Fig. 3b) and occurred on 3% of the image on average. Around the outer edge, the thickness of the breast drops very rapidly to zero, and thus small errors in the position or shape of the thickness profile will induce large errors in the thickness. It would thus be possible to optimize the shape of the thickness profile near the outer edge. However, since the thickness of the breast is small near the outer edge, those large errors in thickness only had a small effect on the calculated volume and dense volume of the breast. Moreover, the presence of skin at the outer edge, which has a larger x-ray attenuation coefficient than that of pure fibroglandular tissue, will cause density values to be above 1.0. The remainder of the errors in thickness occurred mostly in the rest of the periphery, with a mean ΔT of -2.0 mm. We note that the actual thickness errors are likely larger, since we assumed that the true density value for the erroneous points was either 0 or 1.0, while the true density value could be in between that range.

This study indicates that a reasonable estimate of the VBD can be achieved by neglecting the peripheral region altogether. The average total VBD was 16.2%, while the VBD on the central compressed breast region only was 15.1%. However, in our case set, the peripheral region accounted for an average of 50% of the dense volume in the breast and 47% of the total volume. The peripheral volumes detected by the algorithm are large. This is because the periphery is defined to where the thickness starts to drop, and the initial decay of the thickness profile in the periphery is very gradual (the thickness decreases by only 7% roughly half-way between the inner periphery and the breast edge). A better estimate of the peripheral volume could be determined by choosing a larger thickness threshold and we will investigate this.

Another limitation of the peripheral detection algorithm is in the use of radial lines to detect the periphery edge and to apply the thickness profile. This method works well for breast images that have an approximately semi-circular shape, since the radial lines then are approximately orthogonal to the breast edge. For breasts with a more elongated or contracted shape, the radial lines, for small angles with respect to the chest wall, were not orthogonal to the breast edge. For those images and at those locations, the largest errors in the VBD or thickness estimation occurred. It would be possible to improve the algorithm by using conformal lines that intersect the breast edge perpendicularly. Finally, the algorithm didn't perform well in the nipple region and where skin folds occurred, but those regions accounted for a small area of the mammograms.

References

1. Blot, Z., Zwiggelaar, R.: A volumetric approach to glandularity estimation in mammography: a feasibility study. Phys. Med. Biol. 50, 695–708 (2005)
2. Yaffe, M.J., et al.: The myth of the 50-50 breast. Med. Phys. 36, 5437–5443 (2009)
3. Mawdsley, G.E., Tyson, A.H., Peressotti, C.L., Jong, R.A., Yaffe, M.J.: Accurate estimation of compressed breast thickness in mammography. Med. Phys. 36, 569–576 (2009)
4. Snoeren, P.R., Karssemeijer, N.: Thickness correction of mammographic images by means of a global parameter model of the compressed breast. IEEE Trans. Med. Im. 23, 799–806 (2004)
5. Rico, D., Yang, J., Augustine, B., Mawdsley, G.E., Yaffe, M.J.: Peripheral thickness correction for volumetric breast density estimation. In: Proc. 6th Int. Workshop on Digital Mammography (Bremen), pp. 187–91. Springer, Heidelberg (2002)
6. Boone, J.M., Nelson, T.R., Kwan, A.L., Yang, K.: Computed tomography of the breast: design, fabrication, characterization, and initial clinical testing. Med. Phys. 33, 2185 (2006)
7. Boone, J.M., Lindfors, K.K., Cooper III, V.N., Seibert, J.A.: Scatter/primary in mammography: Comprehensive results. Med. Phys. 27, 2408–2416 (2000)

Is CAD Able to Assist in the Detection of Subtle Breast Findings – Lobular Cancers, and T1a/T1b Masses in Dense Breasts?

Isaac Leichter[1,2], Zahava Gallimidi[3], Asnat Heyman-Reiss[3], Nicolas Merlet[1], Eli Ratner[1], Menachem Abramov[1], Inna Stainvas[1], and Richard Lederman[4]

[1] Siemens Computer Aided Diagnosis, Jerusalem, Israel
[2] Dept. of Medical Engineering, Jerusalem College of Technology, Jerusalem, Israel
[3] Dept. of Radiology, Rambam Medical Center, Haifa, Israel
[4] Department of Radiology, Hadassah University Hospital, Jerusalem, Israel

Abstract. 234 pathology-proven FFDM malignant cases and 3872 normal cases were culled retrospectively from 6 screening facilities. For malignant cases, location and size of the biopsied finding and breast density were recorded. All cases were run with a prototype CAD algorithm (Siemens) to evaluate the impact of breast density, lesion size and lesion pathology on CAD performance. The overall CAD sensitivity was 84.2%, with 85.5% sensitivity in "non-dense" breasts and 82.3% in "dense" breasts (p=0.26). No significant difference (p=0.10) was found between CAD sensitivity for ductal lesions (86.4%) and lobular lesions (70.6%). The sensitivity for invasive ductal lesions (86.9%) was slightly higher (p=0.30) than for in-situ lesions (82.6%). The CAD sensitivity for large masses (90.1%) was significantly higher (p<0.001) than for small masses (66.0%). The false mark rate was 1.01. The study indicates that CAD can assist the radiologist in identifying suspicious lesions, independent of breast density, lesion pathology or invasiveness.

Keywords: FFDM, Computer Assisted Detection (CAD), CAD sensitivity, False mark rate, Breast density, Lobular carcinoma, Ductal carcinoma, In-situ lesions, T1a/T1b masses.

1 Introduction

While screening mammography has greatly impacted breast cancer survival, the detection of cancers in dense breasts is problematic and often requires additional imaging modalities (US, MRI). The detection of small cancers is also problematic, especially in dense breasts, while the detection of such cancers is of utmost importance since their timely detection has a direct impact on breast cancer survival [1], [2]. It is also known that lobular cancers are difficult to detect in screening mammography [3], although their detection is of great merit. The importance of detecting non-invasive ductal cancers should also not be underestimated [4]. The purpose of this study was to evaluate the impact of breast density, lesion size, lesion pathology and lesion invasiveness on the performance of a CAD algorithm for screening mammography.

J. Martí et al. (Eds.): IWDM 2010, LNCS 6136, pp. 474–480, 2010.

2 Material and Methods

2.1 Case Acquisition and Review

4106 FFDM cases were culled retrospectively, in a consecutive manner, from 6 screening facilities. 234 of the cases were pathology-proven malignant (159 masses, 43 masses containing calcifications and 32 clustered microcalcifications), and 3872 cases were normal. The normal cases were either negative cases or cases with findings deemed benign by the radiologist, at screening. For the normal cases, no follow-up was available to assure that they indeed represented "true-normal" cases. As shown in figure 1, of the 234 malignant cases, in 199 cases the pathology indicated a ductal process (176 invasive, 23 in-situ), in 17 cases a lobular process, and 18 cases had only cytology results.

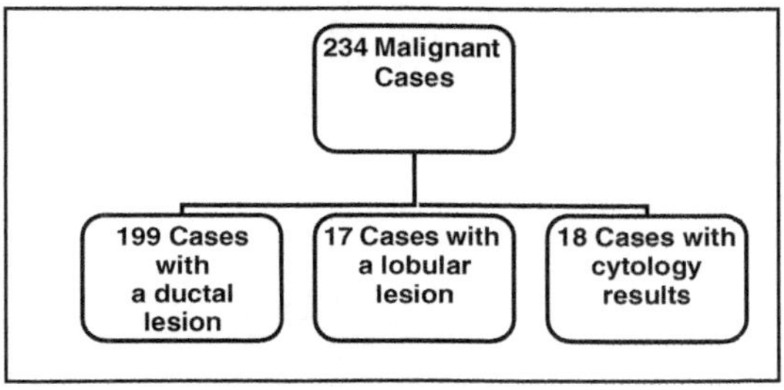

Fig. 1. Histopathology of the malignant lesions

2.2 CAD Methodology

The CAD algorithm is designed to identify and mark on digital mammograms findings suspicious for malignancy. In order to analyze the performance of the CAD algorithm, the prompts generated by CAD have to be compared with the mammographic findings that were proven to be malignant at biopsy. Therefore, for each malignant case, a non-blinded radiologist marked on the digital image the location of the biopsied finding. The same radiologist also recorded the breast density and, in cases of mass lesions, the size of the mass. Cases with breast density categories 1 and 2 were considered "non-dense breasts" (148 cases), while breast density categories 3 and 4 were considered "dense breasts" (86 cases). Masses less than 10 mm in size (T1a and T1b) were considered small (50 cases), while masses greater than or equal to 10 mm in size were considered large (152 cases).

All cases were run on a prototype CAD algorithm (Siemens) that actually consists of two separate algorithms, one for the detection of masses and the other for the detection of clustered microcalcifications. These algorithms have been trained on a large dataset of malignant and normal cases. It should be emphasized that all the cases included in the present study were unknown to the algorithms, and none of the cases were used for training the algorithm.

Detection was assessed by correlating the location of the CAD prompts to the location of the biopsied finding on the digital image. This analysis permitted the calculation of the CAD sensitivity for the detection of malignant findings. Figures 2 and 3 show the CAD prompts marking subtle findings in dense breasts. The CAD prompt for a mass is an ellipse, the size and orientation of which varies according to the size and orientation of the mass. The CAD prompt for a cluster of micro-calcifications is a rectangle, which also varies with size and orientation to conform to the cluster.

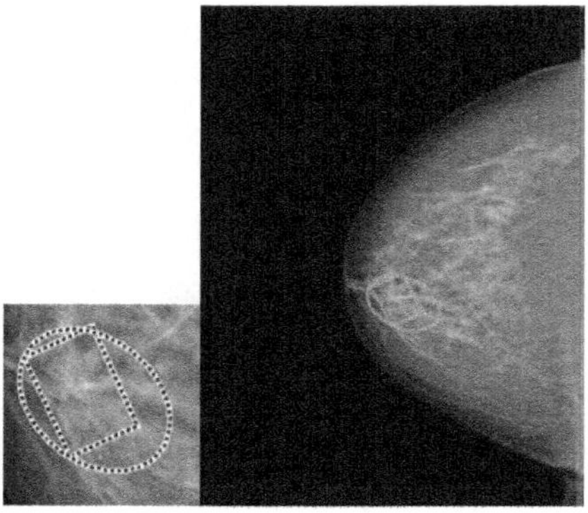

Fig. 2. A small subtle mass with calcifications detected by CAD in a dense breast

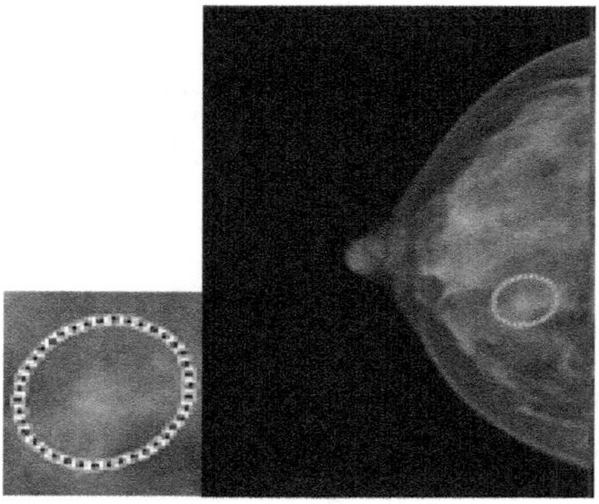

Fig. 3. A barely visible invasive lobular cancer detected by CAD in a very dense breast

2.3 Analysis of the Cases by the CAD Algorithm

The CAD sensitivity was analyzed by breast density, lesion size, lesion pathology and lesion invasiveness in order to evaluate the effect of these factors on the CAD performance. The 95% Confidence Intervals for the CAD sensitivities were also calculated. The sensitivity of the CAD algorithm for ductal lesions was compared with that for lobular lesions and, likewise, the sensitivity for invasive ductal lesions was compared with that of in-situ ductal lesions. To determine the significance of the differences found in the CAD sensitivities, one-tailed p-values were calculated, using the two sample t-test assuming unequal variances.

Although no follow-up was available to assure that the normal cases represented "true-normal" cases, all CAD prompts generated by CAD for the normal cases were considered false marks. Since some CAD prompts for these cases may be "true marks", the false mark rate per case which was calculated, may in fact be an overestimation, and the actual false mark rate may be lower than reported.

3 Results

3.1 CAD Performance by Breast Density

The CAD algorithm correctly detected 197 of the 234 cancers, yielding an overall detection sensitivity of 84.2% [95% CI: (79.5%, 88.9%)]. The CAD sensitivity for cancers in "non-dense" breasts (85.5%) was slightly higher than in "dense" breasts (82.3%), with no significant difference (p=0.26). The CAD sensitivity for the 202 malignant masses was 84.16% [95% CI: (79.08%, 89.24%)] and for the 32 clustered microcalcifications – 84.38% [95% CI: (71.08%, 97.68%)]. As shown in figure 4, breast density did not significantly affect the CAD sensitivity for either masses or clustered microcalcifications. The CAD sensitivity for mass lesions was 85.4% in dense breasts vs. 82.3% in non-dense breasts (p=0.28). Likewise, the CAD sensitivity for clustered microcalcifications was 86.7% in dense breasts vs. 82.4% in non-dense breasts (p=0.37). In dense breasts, the CAD sensitivity for masses and clustered microcalcifications was, in fact, almost identical.

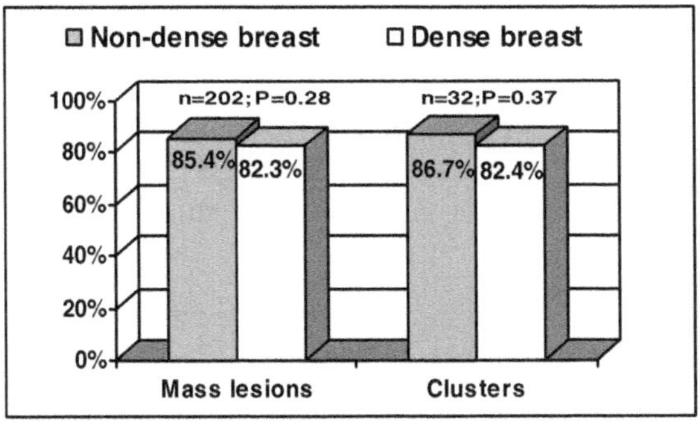

Fig. 4. CAD sensitivity for mass lesions and for clustered microcalcifications by breast density

3.2 CAD Performance by Histopathology

Figures 5 and 6 display the analysis of the CAD algorithm performance by histopathology. As shown in Figure 5, the detection sensitivity for the 199 ductal lesions (86.4%) was higher than for the 17 lobular lesions (70.6%), but the difference did not reach significance (p=0.10).

Figure 6 displays a separate analysis of the ductal carcinomas only, by lesion invasiveness. As shown in this figure, the detection sensitivity of CAD for the 176 ductal lesions with invasive pathology (86.9%) was slightly, but not significantly higher (p=0.30) than for the 23 in-situ ductal lesions (82.6%).

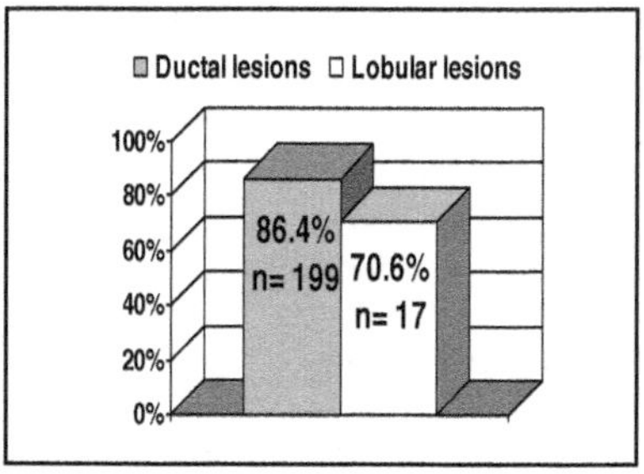

Fig. 5. CAD sensitivity of ductal vs. lobular lesions

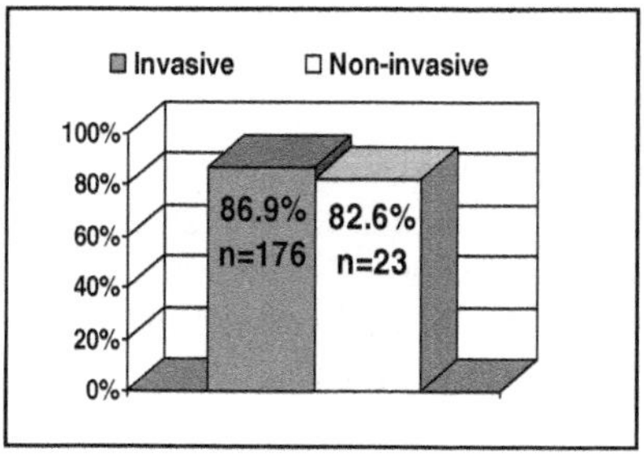

Fig. 6. Ductal lesions - CAD sensitivity by invasiveness

3.3 CAD Performance by Lesion Size

Figure 7 shows the sensitivity of the CAD algorithm by lesion size, in dense and in non-dense breasts.

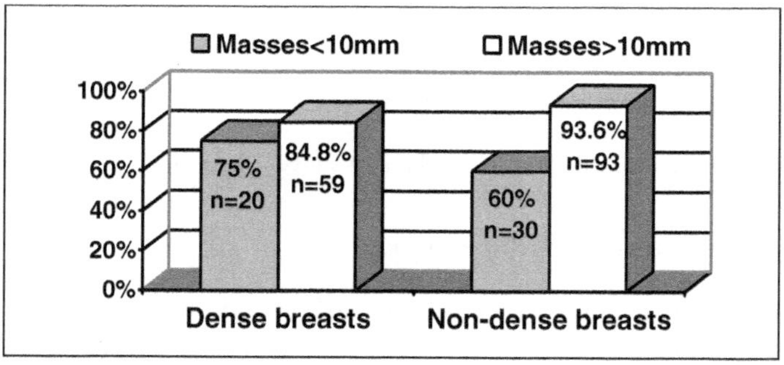

Fig. 7. CAD sensitivity for large and small masses, in dense and in non-dense breasts

It was found that lesion size had a greater impact on the CAD sensitivity than breast composition or lesion pathology. The CAD sensitivity for the 152 large masses (90.1%) was significantly higher ($p<0.001$) than for the 50 small masses (66.0%). As shown in figure 7, this trend is true for both dense breasts and non-dense breasts.

However, in dense breasts the difference in the CAD sensitivity between large and small masses was only 9.8% ($p=0.2$), compared to 33.6% ($p<0.0006$) in non-dense breasts.

For the 3872 normal cases, the false mark rate generated by CAD per case was 1.01. The mass false mark rate was 0.73 while the cluster false mark rate was only 0.28.

4 Discussion

It is especially important for CAD to assist in the detection of cancers in dense breast [5] and in the detection of lobular cancers [3], which are known to have lower mammographic sensitivity when analyzed conventionally. It was found that breast density did not significantly affect the CAD sensitivity for either masses or clustered microcalcifications. Likewise, the CAD sensitivity was not significantly affected by either lesion pathology or lesion invasiveness. It can be, therefore, concluded that CAD has the potential to assist in the detection of subtle breast findings such as lobular cancers, even in dense breasts.

The impact of lesion size on the CAD sensitivity was greater than that of breast composition or lesion pathology. Although the sensitivity for T1a/T1b masses was significantly lower than for larger masses, in the subset of cases with dense breasts the difference was not statistically significant. Surprisingly, it was found that the detection sensitivity of CAD for small masses was higher in dense breasts (75%) than in non-dense breasts (60%). This finding is especially remarkable since small masses in dense breasts are considered one of the great challenges in the conventional interpretation of mammography, and hence CAD may prove to be most beneficial for this type of lesion.

In order to obtain statistical significance, in the present study the ratio of malignant cases to normal cases was enriched to 6%, about ten times higher than expected in screening mammography. Despite the use of the enriched case mix the present study included only a small number of in-situ ductal carcinomas (23) and of lobular carcinomas (17). The small sample size may explain the lack of statistical significance for the rather large difference in CAD sensitivity between ductal carcinoma (86.4%) and lobular carcinoma (70.6%). The lack of statistical significance between the CAD sensitivity for invasive ductal carcinoma (86.9%) and in-situ ductal carcinoma (82.6%) seems to be a result of the quite similar sensitivities, but the small sample size may also contribute to the lack of statistical significance.

It should be noted that the study dataset included 32 malignant clustered microcalcifications which comprised only 13.7% of the malignant cases. However, the detection sensitivity of the CAD algorithm for clustered microcalcifications was similar to that for masses, and nearly identical in dense breasts. Thus, it is unlikely that the ratio of masses to clustered microcalcifications in the study dataset has influenced the results. The higher CAD sensitivity for ductal lesions compared to lobular lesions is apparently also not affected by the higher incidence of micro-calcifications in ductal lesions, since the CAD sensitivity for clustered microcalcifications was similar to that of masses.

The CAD algorithm used in this study is able to assist the radiologist in identifying suspicious lesions, independent of breast density, lesion pathology or invasiveness. In particular, this study indicates that this CAD algorithm can assist in the detection of small cancers in dense breasts, and in the detection of lobular cancers, which are difficult to detect in screening mammography. Such a CAD algorithm with quite a high detection sensitivity and with a relatively low false mark rate should increase the acceptance of the true CAD prompts by the radiologist.

References

1. Tabar, L., Tony Chen, H.H., Amy Yen, M.F., Tot, T., Tung, T.H., Chen, L.S., Chiu, Y.H., Duffy, S.W., Smith, R.A.: Mammographic Tumor Features Can Predict Long-Term Outcomes Reliably in Women with 1-14-mm Invasive Breast Carcinoma. Cancer 101, 1745–1759 (2004)
2. Feig, S.A.: Auditing and Benchmarks in Screening and Diagnostic Mammography. Radiol. Clin. North Am. 45, 791–800 (2007)
3. Evans, W.P., Warren Burhenne, L.J., Laurie, L., O'Shaughnessy, K.F., Castellino, R.A.: Invasive Lobular Carcinoma of the Breast: Mammographic Characteristics and Computer-Aided Detection. Radiology 225, 182–189 (2002)
4. Poplack, S.P., Wells, W.A.: Ductal Carcinoma in Situ of the Breast: mammographic-pathologic correlation. Am. J. Roentgenol. 170, 1543–1549 (1998)
5. Brem, R.F., Hoffmeister, J.W., Rapelyea, J.A., Zisman, G., Mohtashemi, K., Jindal, G., Disimio, M. P., Rogers, S.K.: Impact of Breast Density on Computer-Aided Detection for Breast Cancer. Am. J. Roentgenol. 184, 439–444 (2005)

Towards More Realistic Biomechanical Modelling of Tumours under Mammographic Compressions

Carolina Wessel[1], Julia A. Schnabel[2], and Sir Michael Brady[1]

[1] Department of Engineering Science, University of Oxford
Parks Road, Oxford OX1 3PJ, UK
{cwessel,jmb}@robots.ox.ac.uk
[2] Institute of Biomedical Engineering, Department of Engineering Science, University of Oxford, ORCRB, off Roosevelt Drive, Oxford OX3 7DQ, UK
julia.schnabel@eng.ox.ac.uk

Abstract. In this paper we present a biomechanical model of isolated breast tumours under mammographic compression forces. We apply a range of reported mechanical properties, both linear elastic and hyperelastic. We also introduce a volume of increased density/stiffness around the tumour. These variables have a non-negligible effect on stresses and strains, as shown in this work.

Keywords: breast tumour, finite element analysis, biomechanics.

1 Introduction

Biomechanical modelling of the breast provides constraints for medical image analysis including non-rigid registration. A recent review of the topic can be found in [1]. However, even in the macroscopic domain, realistic breast tumour mechanics pose considerable modelling and computational challenges among which are:

1. the geometrical and structural complexity of the adult female breast, which includes fatty and fibroglandular tissues, as well as muscles and ligaments;
2. published values of mechanical properties vary considerably for each type of tissue [1], and even different constitutive equations are proposed. However, such parameters are needed to develop a realistic model of breast tumour mechanics. Given the significance of these issues in this work, they will be covered in more detail in Section 1.1;
3. the geometrical complexity of tumours has to be considered: it is only necessary to see mammographies or histopathological images of ductal carcinomas (Fig 1 taken from [2]) to appreciate both the need for and the difficulty of this task. In this case, the stellate character of tumours is especially important as spiculated shapes are generally indicative of malignancy [2]. Spicules, as shown in Figure 1, may cover part or the whole surface of the tumour core, show different lengths and base diameters. It should be noted here that other authors [3, 4] have mostly considered smooth uniform tumour surfaces;

J. Martí et al. (Eds.): IWDM 2010, LNCS 6136, pp. 481–489, 2010.

4. it is accepted among clinicians that an increase of breast density is linked to a high probability of a tumour presence. This increase has been experimentally related with total collagen density increase around the tumour in $Col1a1^{tmJae}$ mice in [5]. In turn, an increase in collagen density corresponds to an increase in tissue "stiffness". This suggests that, even normal healthy breast tissues show different mechanical properties in the presence of tumours, even though, to the best of our knowledge, the mechanical properties of stiffened tissues have not been measured.

In this work, the relevant mechanical variables –stresses and strains/displacements- have been determined using Abaqus (*www.simulia.com*) to implement a Finite Element Analysis. Because of item 1 above in this work we consider an idealized tumour in isolation, that is, surrounded by one type of homogeneous tissue, and apply a far field static mammographic compression force. We have used a range of reported mechanical properties, both linear elastic and hyperelastic to analyse the effect mentioned in item 2. Finally, we have introduced a volume of increased "stiffness" around the tumour to study the effect mentioned in item 4.

Following a mechanical description in Section 1.1, in this paper we compare values of stresses and displacements of smooth (case A) and stellate (case B) tumours, for different elastic and hyperelastic mechanical properties. Finally we add the effect of increased-stiffness surrounding materials (case C).

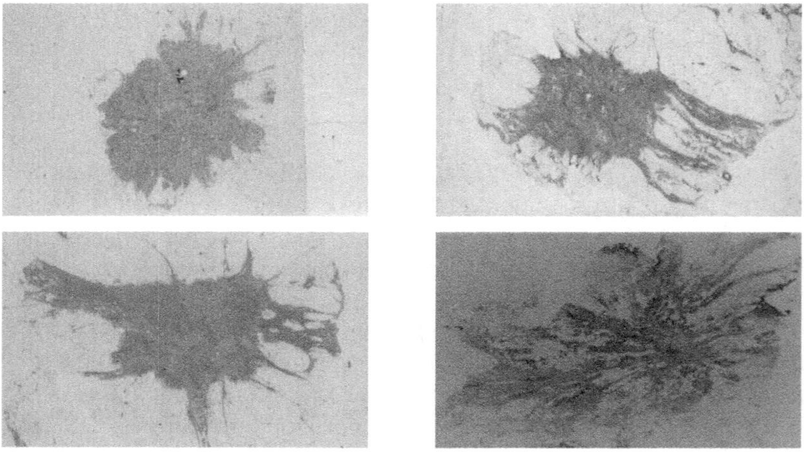

Fig. 1. Histopathological images of breast invasive carcinomas, taken from [2]

1.1 Mechanical Description of the Problem

Our aim is to biomechanically model breast tumours under mammographic compressions, that is, to determine stresses and displacements/strains for the problem shown in Fig 2. The tumour –whether smooth or stellate- is surrounded by one or more materials, all of which are homogenous, isotropic and incompressible and are rigidly attached. This structure lays on the y=0 plane and is subject to a static far field compression load, that is, a uniform force F in the $-y$ direction. These boundary conditions are shown in Fig 2 as white arrows and triangles.

To derive a unique solution to the Mechanics equations subject to the above boundary conditions, constitutive equations must be introduced. Some authors [6, 7] have proposed a linear relationship between stresses and strains or equivalently, between forces and displacements for breast tissues.

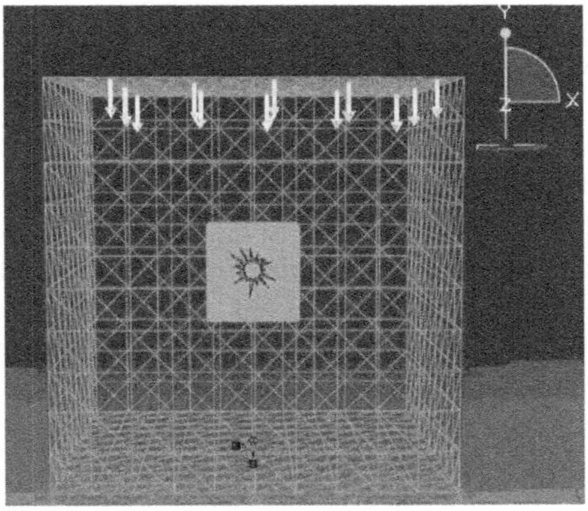

Fig. 2. Idealized breast tumour subject to a far field mammographic compression

On the other hand, Samani and co-workers [8, 9, 10] performed indentation tests on breast tissue samples *ex vivo*. One of the plotted Force vs. Displacement curves is shown in Fig 3. Evidently, there is a non-linear behaviour for these tissues even for forces that are very small compared to those typical of mammography. This strongly suggests that breast tissue behaviour is better described as hyperelastic than linear-elastic specially when subject to mammographic compression forces. In fact, according to Samani, breast tissues can be modelled using various hyperelastic models such a polynomial (n=2) or Arruda-Boyce fits.

In spite of the evidence shown in Fig. 3, in the present work both linear-elastic and hyperelastic constitutive equations have been implemented for the sake of comparison.

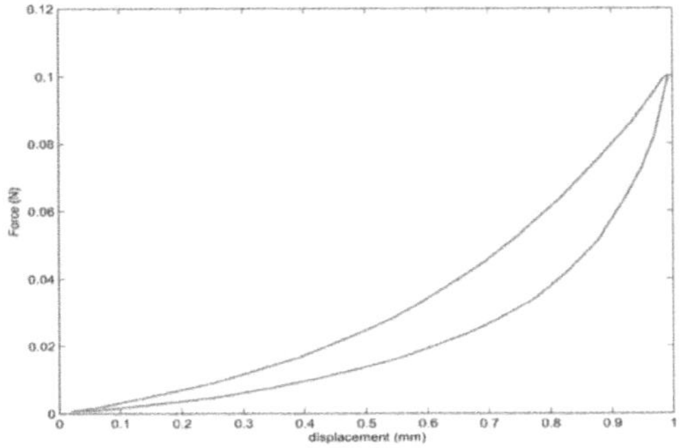

Fig. 3. Force-displacement curve for breast tissue [8, 9]

2 Materials and Methods

We consider three different cases. Each can be represented as part of the structure shown in Fig. 2, with an applied force F of 160N. Case A represents a smooth spherical tumour (Fig 3, a). Case B is a mathematically created stellate mass where all the spicules have random values of base radius and length (Fig. 3, b). Case C is the same stellate tumour used in case B and surrounded by three spherical concentric layers (Fig 3, c). The mechanical properties are constant in each of these layers. They vary linearly from the tumour's to the non-tumour's properties, whether elastic or hyperelastic. Limitations in space hinder us from describing the generation and implementation of the models in Abaqus and the corresponding convergence analysis.

In this work we have used the linear elastic constants reported in [6] for tumours and fatty tissues, the polynomial (n=2) values for fatty and fibroglandular tissues [8] for both Arruda-Boyce [10] and polynomial (n=2) fits for malignant tumours. This gives three sets of material properties that will be denoted by E2, PA and PP, respectively. Given the large uncertainties in these values and given the fact that this is a qualitative study, we have averaged them. Other reported values (such as those shown in [7] and other fits included in [10]) were also explored but they led to convergence issues or rendered physically unrealistic results.

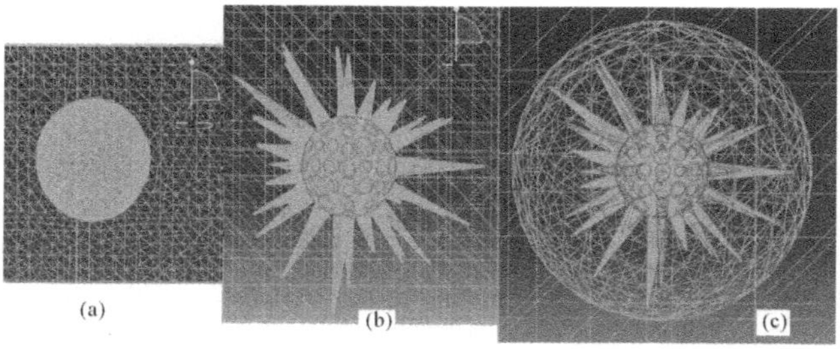

Fig. 4. Smooth spherical tumour (a), randomly mathematical spiculated tumour (b) and spiculated tumour surrounded by three layers of denser/stiffer materials

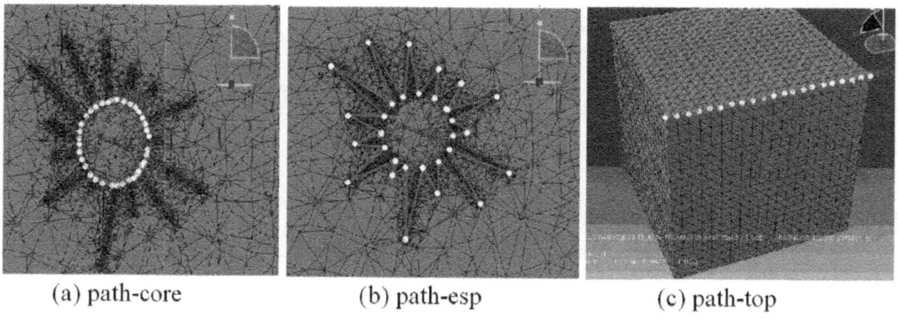

Fig. 5. Three paths where the values of von Mises stresses stresses σ, displacement moduluss u and displacements in the y direction u_2 have been analysed

To analyse the effect of a stellate geometry we compared the results of cases A and B while, to determine the mechanical effect of denser/stiffer materials we have compared cases B and C. Values of von Mises stresses σ [Pa], displacement modulus u [m] and engineering strains in the y direction e_2 have been determined for the three paths located at z=0 shown in Figure 5 as white dots, where the absolute distance between successive points is given in meters.

3 Results

3.1 Comparison of Cases A and B

Figure 6 shows the maps of von Mises stresses σ for the spherical (b) and stellate tumours (a) for the same scale. White represents highest values of σ, that correspond in this case to the spicules oriented nearest the y direction, that is, the direction of the applied force. The comparison of Fig. 6 (a) and (b) clearly shows that a spiculated mass produces an increase in von Mises stresses even within the core of the tumour.

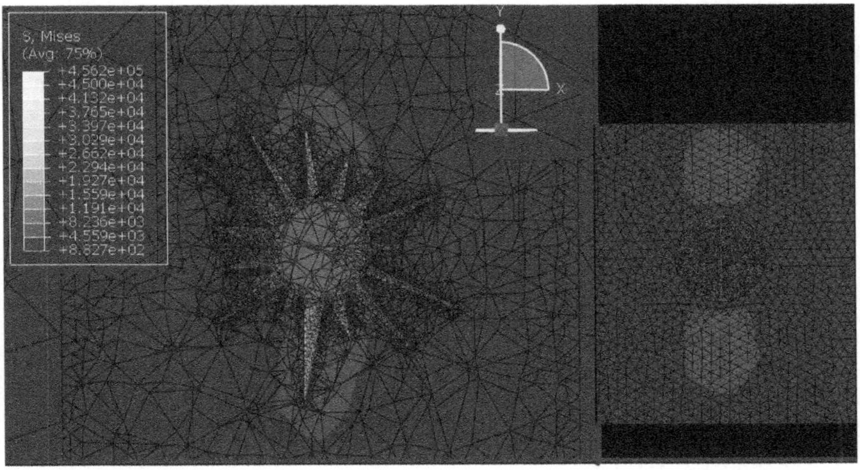

(a) (b)

Fig. 6. Map of the von Mises stresses for spiculated (a) and a spherical tumour (b)

To show this effect the maximum values of σ for the path-esp have been determined for the linear elastic E2, PA and PP sets. These results are shown in Figure 7. The maximum values of σ -expected to appear in the spicules oriented nearest to the y direction- increase dramatically for spiculated masses. It is also interesting to point out that the values for the sets PP and PA are higher than those of the E2 set.

Values of strains in the path-top for the three different sets of material properties have been used to confirm that e_2 does not change significantly from case A and B.

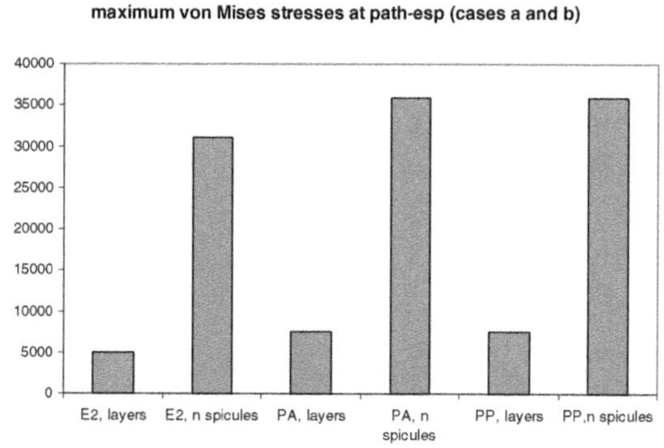

Fig. 7. Maximum von Mises stresses at path-esp for spherical and stellate masses for the three different sets of mechanical propterties

On the other hand, it is worth analysing the effect of spicules in displacements within the tumour. Figure 8 (a) shows the values of u along path-core for the linear elastic E2 values. It is readily observable that the stellate mass introduces only small changes in these values, effect that can also be observed for the other two sets of material properties. Figure 8 (b), on the other hand, displays displacements for the three sets where, in this scale, both PA and PP and both E2 values overlap. Figure 8 (b) shows the dramatic increment observable for the PP and PA sets.

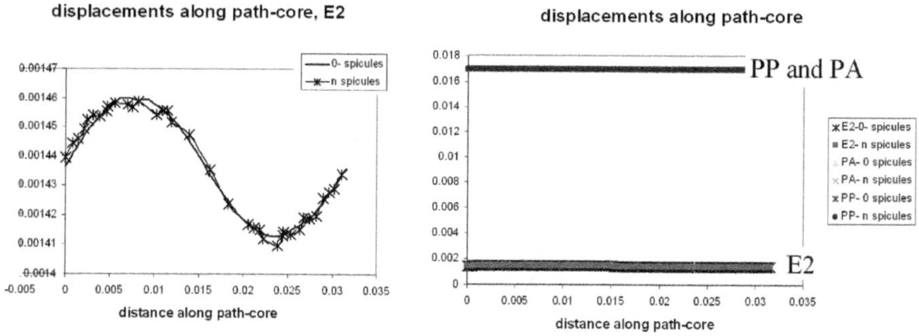

(a) (b)

Fig. 8. Module of the displacements u along path-core for the linear elastic E2 set of values (a) and for all sets of values (b)

3.2 Comparison of Cases B and C

In this section we consider the effect of denser/stiffer layers around tumours. Figure 9 shows the maps of von Mises stresses σ for the stellate tumour (a) and the embedded tumour (b) for the same scale. Here again white represents highest values of σ, that correspond in (a) to the spicules oriented nearest the y direction, that is, the direction of the applied force. It is interesting to point out that the von Mises stresses have decreased in the layers case, generating a stress shield effect.

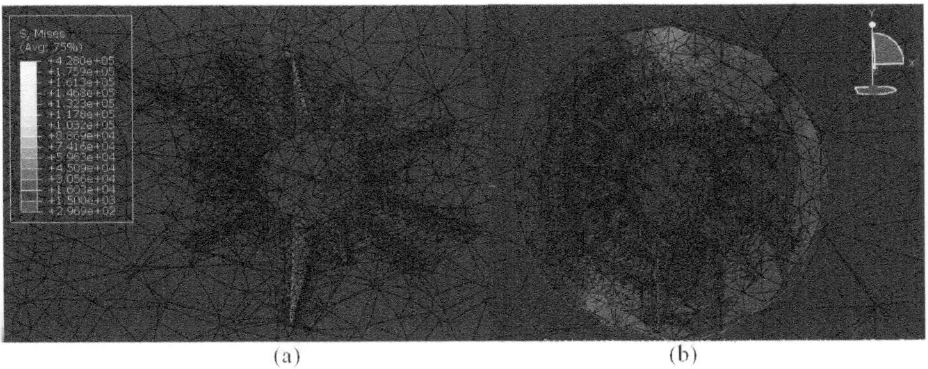

Fig. 9. Map of the von Mises stresses for spiculated (a) and embedded spiculated tumour (b)

This effect can be readily observed in Figure 10 that shows maximum von Mises stresses values along path-esp for all sets of properties. Here again values of PP and PA maximum von Mises stresses are higher the E2 values.

Values of strains in the path-top for E2, PP and PA have been determined to confirm that there is no significant change on e_2 for cases B and C.

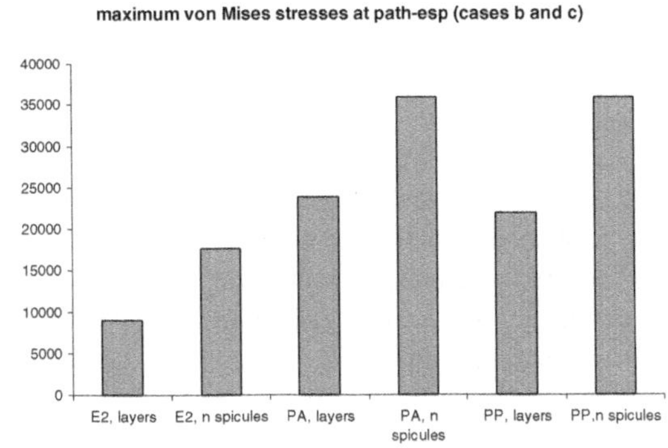

Fig. 10. Maximum von Mises stresses along path-esp for a stellate tumour and an embedded stellate tumour for the three different sets of mechanical properties

Regarding displacements within the tumour, Figure 11 shows values of u for cases B and C. It can be observed that the layers of surrounding material cause a decrease of the values of u. Values in Figure 11 (a) correspond to PA properties but the other sets present a similar behaviour. The difference between sets of properties can be observed in Figure 11 (b) that shows the dramatic increase in displacements for the PA and PP sets.

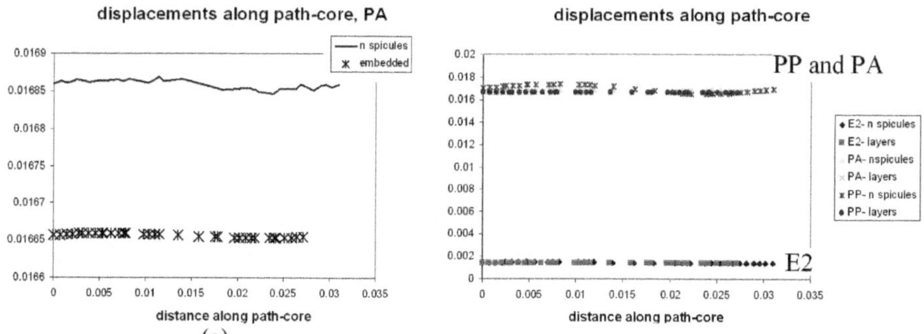

(a) (b)

Fig. 11. Modulus of the displacements u along path-core for the stellate and embedded stellate tumour for the PA set (a) and for all three sets (b)

4 Conclusions

As presented herein, the comparison of smooth and stellate tumours shows that spicules generate a dramatic increase on stresses at a local level. On the other hand the spiculated character of malignant tumours does not, per se, influence displacements within them.

Besides, increasing the stiffness around a fully spiculated tumour generates both a progressive shield of stresses and a decrease of displacements.

All such analyses depend on the choice of linear-elastic and hyperelastic constitutive equations. It has been shown here that von Mises stresses at the spicules' tips and displacements within tumours show higher values for both hyperelastic sets of materials for spherical, stellate and embedded stellate tumours.

The wide range of published values of mechanical properties imposes a considerable limitation for realistic biomechanical modelling. In this work we have shown that not only constitutive equations but also de spiculated character of the tumour and the presence of denser/stiffer material around it render different results even for an isolated tumour. More yet has to be done in order to be able to implement a mechanical registration of cancerous breasts.

We plan nextto investigate the effect of non homogeneous materials a different number of layers.

References

1. Rajagopal, V., Nielsen, P.M.F., Nash, M.P.: Modeling breast biomechanics for multi-modal image analysis, successes and challenges. Wiley Interdisciplinary Reviews: Systems Biology and Medicine (2009)
2. Cherel, P., Becette, V., Hagay, C.: Stellate images: anatomic and radiologic correlations. European Journal of Radiology 54, 37–54 (2005)
3. Pathmanathan, P.: Predicting Tumour Location by Simulating the Deformation of the Breast using Nonlinear Elasticity and the Finite Element Method. University of Oxford (2006)

4. Hipwell, J.H., Tanner, C., Crum, W.R., Schnabel, J.A., Hawkes, D.J.: A New Validation Method for X-ray Mammogram Registration Algorithms Using a Projection Model of Breast X-ray Compression, Medical Imaging. IEEE Trans. 26(9), 1190–1200 (2007)
5. Provenzano, P.P., Inman, D.R., Eliceiri, K.W., Knittel, J.G., Yan, L., Rueden, C.T., White, J.G., Keely, P.J.: Collagen density promotes mammary tumor initiation and Progression. BMC Medicine 6, 11 (2008)
6. Wellman, P.S., Howe, R.D., Dalton, E., Kern, K.A.: Breast tissue stiffness in compression is correlated to histological diagnosis, Tech. Rep., Harvard BioRob Lab (1999)
7. Samani, A., Bishop, J., Luginbuhl, C., Plewes, D.B.: Measuring the elastic modulus of ex vivo small tissue samples. Phys. Med. Biol. 48, 2183–2198 (2003)
8. Samani, A., Plewes, D.B.: A method to measure the hyperelastic parameters of ex vivo breast tissue samples. Physics in Medicine and Biology 49, 4395–4405 (2004)
9. Samani, A., Plewes, D.B.: An inverse problem solution for measuring the elastic modulus of intact ex vivo breast tissue tumours. Phys. Med. Biol. 52, 1247–1260 (2007)
10. O'Hagan, J.J., Samani, A.: Measurement of the hyperelastic properties of 44 pathological ex vivo breast tissue samples. Phys. Med. Biol. 54, 2557–2569 (2009)

Scoring Systems in Computer-Based Training for Digital Mammography

Paul Taylor[1], Mark Hartswood[2], Lilian Blot[3], Rob Procter[4], and Stuart Anderson[2]

[1] Centre for Health Informatics and Multiprofessional Education, University College London
p.taylor@chime.ucl.ac.uk
[2] School of Informatics, University of Edinburgh
[3] Department of Computer Science, Durham University
[4] Manchester eResearch Centre, University of Manchester

Abstract. A computer-based training tool was developed through a collaborative design process. The tool allows trainee radiologists to access a large number of suspicious lesions. The tool employs a certainty-based scoring system in which trainees' responses are scored not just as right or wrong but according to their confidence. Different approaches to providing trainees with feedback were considered: one based on a histogram and one using a line graph of cumulative scores. Following an initial assessment by radiologists, a revised scheme was introduced in which disagreements between trainee and expert are rated according the clinical or pedagogical significance of the error.

1 Introduction

Recent years have seen significant improvements in the technology used to create mammograms. Nevertheless these images are still difficult to interpret and the performance of radiologists is highly variable. [1] It is therefore important to consider whether novel tools can be used to enhance the competence of radiologists. There is considerable scope for computer-based training tools to improve the ability of radiologists to read screening mammograms. The aim of the work reported here is to explore different ways that a computer-based learning environment can add value to conventional training methods.

We have developed a number of tools for different aspects of mammographic image interpretation. One, provisionally termed 'Lesion Zoo', is intended to give trainees access to a large number of abnormalities. The argument is that experience of a wide range of appearances is necessary for the acquisition of visual expertise. [2] In this paper we describe this prototype, paying particular attention to the scoring of trainees' assessments, and report on expert and trainee radiologists' initial experience with the tool.

2 Lesion Zoo Prototype

Lesion Zoo displays a sequence of selected regions of interest, each containing a lesion (either a mass or a microcalcification) and invites the user to classify the lesion

J. Martí et al. (Eds.): IWDM 2010, LNCS 6136, pp. 490–496, 2010.

using the BIRADS descriptors and to assess it, using the BIRADS assessment categories. Users are given case by case feedback on their performance and finally summary statement of their performance over the set of images (see Figure 1).

Lesion Zoo uses a database of 300 annotated images. Each of the cases in the database was selected for inclusion because it had particular value for training.[3] All the selected cases were annotated by an expert radiologist. The radiologist viewed the original films with the associated clinical information and then entered relevant information into the database via a bespoke annotation tool. One element in this involved marking the centre and diameter of any regions of interest. The Lesion Zoo software displays these regions of interest, automatically selected from the database images and invites users to enter a description of the lesion via a menu of descriptors based on the BIRADS terminology.

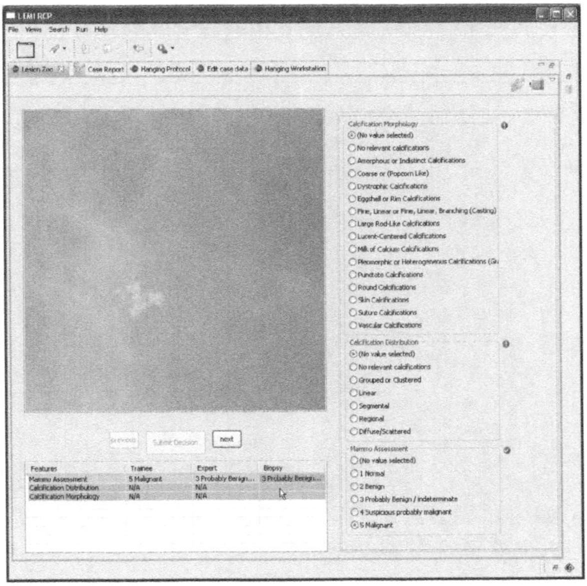

Fig. 1. screen shot of the Lesion Zoo application. The trainee's description of the image is entered via the menu on the right. Feedback is given in the table at the bottom: the trainee disagrees with the biopsy and the expert decision (note the darker red tone for the biopsy).

2.1 Feedback

Feedback provided to trainees uses 'certainty-based marking'.[4] This is a scheme that aims to assess the change in confidence of a learner. Key to the approach is the collection of data not just on the accuracy of a student's responses but also on his or her confidence in that accuracy. The point is not, as in ROC analysis, to gain a threshold-independent measure of accuracy. Instead, data about the confidence a student has in his or her clinical judgment is incorporated into the scoring in order to allow an assessment of the practical value of the judgment. Confident but inaccurate judgments are dangerous. Accurate but unconfident judgments are not useful to the trainee. Useful knowledge leads to responses that are both confident and accurate.

Using this approach, responses are scored according to the scheme in Table 1 and aggregated to create a single score. The weights were chosen to reflect practice using the approach elsewhere and may need to be modified in this setting.

Table 1. Scheme for certainty-based marking

Certainty	1 (Low)	2 (Mid)	3 (High)
Score if correct	1	2	3
Score if incorrect	0	-2	-6

We developed two methods of visualising the results of certainty-based marking to provide feedback in Lesion Zoo: a histogram showing how the responses are distributed across the six categories in Table 1 and a graph showing how a trainee's performance changed as cases were attempted, illustrated in Figure 2.

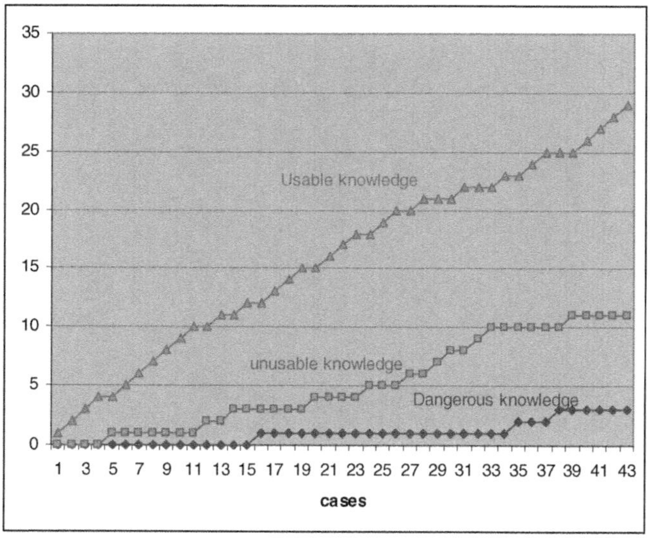

Fig. 2. Number of cases classified as useful knowledge (mid to high confidence, correct response), unusable knowledge (low confidence) and dangerous knowledge (mid to high confidence, wrong response) as a function of number of cases attempted. The x-axis corresponds to the cases done so far by the trainee. Ideally, the green line should rise to a slope of 45° while the red and blue should approach the horizontal.

3 Refining the Prototype

The Lesion Zoo prototype has been developed in close collaboration with staff in two UK NHSBSP screening units. It has been used on an experimental basis by expert and trainee radiologists to collect feedback on the design and the scoring system.

3.1 Methodology

To gather feedback on Lesion Zoo, we used a qualitative, observation-based methodology [5]. We have found in previous studies of the use of computer-based tools in mammography that this approach is very effective for understanding how people use these tools [6].

We gathered qualitative data over four use sessions (two observed and written notes taken, one videoed and one where feedback was elicited from the user after use) and the design team examined the data for emerging themes. Typically the sessions involved encouraging users to work through a set of 10 or so lesions and to verbalize their impressions. These sessions provided the basis for iterative refinement to the lesion zoo tool. Although informal and lightweight we found the approach to be helpful and informative, with users demonstrating consistent and coherent views of the tool's more or less appealing characteristics.

3.2 Results

Some participants responses indicated the sorts of reflection and deliberation that we hoped would be elicited by Lesion Zoo. On other occasions, participants' responses were not so straightforwardly positive. In particular there were a number of occasions where they disagreed with the characterizations of lesions, and were concerned that the scoring mechanism was unfair when penalising and rewarding certain sorts of responses. One participant using the Lesion Zoo pointed out how it can be hard to make a forced choice between BIRADS descriptors when these may not necessarily be mutually exclusive.

> "Lesions can have two appearances – it can be spiculated in one part and ill-defined in another"

That a lesion might not always unambiguously be assigned to a single category, and to penalise disagreement with the expert on the basis of what is seen as a judgement call is seen as unfair. The tool should not be unnecessarily discouraging, for example, by penalising the trainee for trivial classification errors when, otherwise, they are demonstrating competence.

It became increasingly obvious that Lesion Zoo would need to have a notion of which distinctions are clinically important, and which not, and to be able to moderate the marking accordingly. For example, conflating round and oval as descriptors of mass shape is a less critical error than calling a malignant presentation benign. It also became clear that there can be a number of dimensions against which the appropriateness of a response might be judged. For example, while confusing two different benign descriptors would not adversely consequential for the patient, it may indicate problems with the trainee's conceptual grasp of the subject matter.

One trainee radiologist categorically disagreed with the expert, stating that she had recalled a lesion similar to the one presented that had turned out to be a cancer, whereas the expert opinion on the lesion presented was that it was benign.

> "I saw a cancer the other day that looked just like that – I disagree with the expert"

The point here is not whether on this case the classification was appropriate, or the trainee's memory exact, but that there should be some room to allow for disagreement between professionals to be voiced where feeling is strong.

Feedback from the participants suggested that one assumption behind certainty-based marking, that there was a simple right answer in the interpretation of these cases, was inappropriate. We subsequently undertook an exercise to establish what sorts of deviation from expert opinion might be more or less consequential to enable us to moderate the marking appropriately. Three consultant radiologists were asked to complete a confusion matrix for each of the BIRADSs rating questions, scoring the severity of the disagreement for each combination of trainee and expert opinion.

The experts' rankings for BIRADs assessment are shown in Table 2. Similar tables of rankings were obtained for the different descriptors used to characterize masses and microcalcifications. These will be used to weight the scores assigned to disagreements.

Table 2. Experts' (n=3) rankings for the significance of disagreements between expert and trainee over BIRADs assessments

		Trainee				
		M1	**M2**	**M3**	**M4**	**M5**
Expert	**M1**		2	3	4	5
	M2	2		2	4	4
	M3	3	3		2	3
	M4	5	5	2		1
	M5	5	5	4	1	

4 Discussion

Our study provided insights into how the tool would be used in practice and revealed a number of ways in which it could be improved. We have provided a facility for trainees to express their opinion as a free text comment. We also solicited free text comments from the experts assessing the lesions so that they also had room to express how the lesion matched the BIRADS descriptors. Comments could be used to state, for example, that the lesion is typical or atypical member of the chosen category. Expert comments are made available to the trainee after they have made their own assessment of the lesion.

Another issue concerned radiologists' application of the five-point BIRADS scale for suspicion (1. Normal to 5. Benign). One junior film reader suggested that her more experienced colleagues tended to be more confident in calling a lesion M5. A senior radiologist (independently) made the same point when she said that junior readers are more cautious in how they grade lesions. At screening there is no difference in clinical outcomes between grading lesions as M4 or M5, as both will result in a recall for further assessment, thereby providing leeway for junior radiologist to make conservative assessments without impacting on patient care. The less experienced radiologist drew attention to this, partly because she thought she might improve her score if she rated lesions in a way that emulated how she presumed the expert would rate them. In other words, she was attempting to 'second guess' the expert.

This exemplifies a more general problem observed with using game-like environments for education, neatly expressed by Conati and Lehman, who devised a 'microworld' game to teach principles in physics [7]: *"Our observation of student players indicates that it may be possible for a student to become skilled in solving a problem in game terms, i.e. without significantly improving their physics knowledge"*. So, by using a game format, one may introduce elements of fun, competition, ease of playing and focus on a specific task, but risk a worse than might hoped for transfer of skill to the real world domain.

We have learned lessons experimenting with confidence based marking (CBM) in a novel domain. While sympathetic to CBM's goals of reducing trainees' motivation for simply guessing when uncertain, and encouraging reflection on, and awareness of, lacunae in trainees' understanding of a topic, we found its application in a mammography task somewhat tricky in a couple of respects. Part of this stems from the fact that CBM was designed to be applied to domains where it is possible to unambiguously distinguish between correct and incorrect answers. This is evidently not the case in mammography, where variation in lesion appearance in continuous, rather than discrete and always neatly categorisable, as the BIRADs classification scheme might imply. Thus, the evident discomfiture of the participants in the exercise above where reasonable or non-consequential disagreements were penalised, and of one consultant radiologist who completed a session only to have her decision-making described as 'dangerous' by the tool.

5 Conclusions

The evaluation of the Lesion Zoo has enabled us to have a better understanding of how it would be used and to identify criteria for a successful marking scheme for trainees' responses.

First, the scheme must be seen to be fair. BIRADS descriptors are not all mutually exclusive, assigning a lesion to a single category is something of a judgement call. Penalising the trainee for errors in such cases is seen as unfair. It became increasingly obvious that the system had to have a notion of which distinctions are the important ones, and which less so, and to moderate the marking accordingly. For example, conflating 'coarse' and 'eggshell' calcifications is a less critical error than calling a malignant presentation benign. However, while confusing two very different benign descriptors (e.g. 'skin' and 'suture' calcifications) would not have any consequence, it may indicate problems with the trainee's grasp of the subject matter.

Second, we must also allow for disagreement between professionals. We have provided a facility for trainees to express their opinion. This provides room for trainees to express disagreement with the expert classification or a justification for their own. It is hoped that this also will provide a means of softening the impact of the marking scheme by providing an opportunity for readers to voice professional disagreement, as well as creating a valuable resource for understanding how trainees interpret lesions. We also collected a free text comment about each lesion in the zoo from three expert radiologists. The aim of the comment is for the expert to be able to provide a rationale for their classification and rating decisions, for example, if the lesion exemplifies the category, or if its classification is problematic.

It is anticipated that these expert comments (made available to the trainee after completing each lesion) will help reinforce the trainee's grasp of the relation between the lesion's appearance and its classification, as well as alerting them to less clear cut cases where their own opinion might more reasonably deviate from the expert.

A collaborative design process has resulted in a novel tool that includes a sophisticated assessment of the significance of a trainee's disagreement with an expert and assesses this in conjunction with a measure of the trainee's confidence in order to provide a measure of how useful knowledge is developing.

References

1. Goddard, P., Leslie, A., Jones, A., et al.: Error in radiology. Br. J. Radiol. 74(886), 949–951 (2001)
2. Taylor, P.: A Review of Research into the Development of Radiological Expertise: Implications for Computer-based Training. Academic Radiology 14, 1252–1263 (2004)
3. Brady, M., Gilbert, F., Lloyd, S., et al.: eDiaMoND: the UK's Digital Mammography National Database. In: Pisano, E. (ed.) Proceedings of the International Workshop on Digital Mammography (2004)
4. Gardner-Medwin, A.R.: Updating with Confidence: Do your students know what they don't know? Health Informatics 4, 45–46 (1998)
5. Bryman, A.: Social Research Methods. Oxford University Press, Oxford (2008)
6. Alberdi, E., Povyakal, A., Strigini, L., et al.: The use of Computer Aided Detection tools in screening mammography: A multidisciplinary investigation. Br. J. Radiol. 78, 31–40 (2005), special issue on Computer-aided diagnosis
7. Conati, C., Fain Lehman, J., EFH-Soar, J.: Modeling Education in Highly Interactive Microworlds. In: Torasso, P. (ed.) AI*IA 1993. LNCS (LNAI), vol. 728, Springer, Heidelberg (1993)

The Effect of Slab Size on Mass Detection Performance of a Screen-Film CAD System in Reconstructed Tomosynthesis Volumes

Guido van Schie[1], Karin Leifland[2], Matthew Wallis[3], Elin Moa[4], Magnus Hemmendorff[4], and Nico Karssemeijer[1]

[1] Radboud University Nijmegen Medical Centre, Department of Radiology, Geert Grooteplein Zuid 18, 6525 GA Nijmegen, The Netherlands
[2] Unilabs Mammography, Capio S:t Görans Hospital, Department of Mammography, SE-112 81 Stockholm, Sweden
[3] Cambridge Breast Unit and NIHR Cambridge Biomedical Research Centre, Cambridge, United Kingdom, CB2 2QQ
[4] Sectra Mamea AB
Smidesvägen 5, SE-171 41 Solna, Sweden
g.schie@rad.umcn.nl

Abstract. In the development of a computer-aided detection (CAD) system a large database of training samples is of major importance. However digital breast tomosynthesis (DBT) is a relatively new modality and no large database of cases is available yet. To overcome this limitation we are developing a CAD system for mass detection in DBT that can be trained with regular 2D mammograms, for which large datasets are available. We trained our system with a very large database of screen-film mammograms (SFM). Our approach does not use projection images, but only reconstructed volumes, because it is expected that manufacturers of tomosynthesis systems will only store the reconstructed volumes. In this study we developed a method that converts reconstructed volumes into a series of SFM-like slices and combinations of slices, called slabs. By combining slices into slabs, more information of a whole mass, which usually spans several slices, is used and its appearance becomes more similar to a 2D mammogram. In this study we investigate the effect of using slabs of different sizes on the performance of our CAD system. For validation we use a dataset of 63 tomosynthesis cases (245 volumes) consisting of 42 normal cases (163 volumes) and 21 abnormal cases (82 volumes) with a total of 47 malignant masses and architectural distortions. The volumes are acquired with a tomosynthesis system from Sectra and are reconstructed into 0.3 cm thick slices. Results show that performance of our CAD system increases significantly when slices are combined into larger slabs. Best performance is obtained when a slab thickness of 1.5 cm (5 slices) is used, which is significantly higher than using slabs of a single slice, two slices and all slices.

Keywords: Tomosynthesis, mass, computer-aided detection.

J. Martí et al. (Eds.): IWDM 2010, LNCS 6136, pp. 497–504, 2010.

1 Background

A limitation of 2D projection mammography is that superimposition of normal tissue can falsely be seen as a lesion and true lesions can get obscured by overlying breast tissue. Digital breast tomosynthesis (DBT) seems a promising modality to overcome these problems, by reconstructing a 3D volume of the breast from several low dose, limited angle x-ray projections. CAD systems for detecting masses in DBT are being developed to aid radiologists. Singh et al. [1] developed a mass detection system that detects suspicious locations in the projection images and projects them back into the volume to obtain 3D locations of mass candidates. Subsequently a featuresless approach is used on reconstructed slices and slabs of these 3D locations, to reduce false positives (FP). Reiser et al. [2] chose not to use the reconstructed data, and developed a CAD system where initial mass candidate detection and feature analysis for FP reduction, is done directly on the projection images. Another approach is taken by Chan et al. [3] who combine information from both the reconstructed volume and the projection images by merging results from their 2D and 3D CAD systems.

Our approach differs from the systems above in that we aim at making a system that only uses reconstructed volumes and does not use the projection images. The reason herefore is twofold: first it is expected that manufacturers of tomosynthesis systems will not store the individual projection images, but only the reconstructed volumes. Second, for some tomosynthesis systems the projection data is unsuitable for direct interpretation, due to the acquisition method of the system (like for the scanning-multislit system used in this study). Another difference of our approach is that instead of using the limited amount of available tomo data to train the detection algorithm, we use our existing 2D mass detection algorithm that was trained on a large database of screen-film mammography (SFM) images. It is well known that a CAD system improves when it is trained with a larger database. However tomosynthesis is relatively new and no large database of cases is available yet. Kallenberg et al. [4] showed that for full-field digital mammography (FFDM), using a CAD system that was trained on a large SFM database outperformed a CAD system that was trained on a smaller FFDM database. Because our tomosynthesis database is very small (21 abnormal and 42 normal cases) and we have a very large SFM database (636 abnormal and 3262 normal cases), we assume that it is also a good approach to use a CAD system that was trained on this large SFM database, to detect masses in DBT. In this study we developed such a system.

Although in-plane resolution in tomosynthesis is comparable to 2D mammography, resolution in the z direction is low due to the small angle that is used to acquire the projection images. A reconstructed slice usually is about 1 to 3 mm thick (depending on the tomosynthesis system and reconstruction parameters). A reconstructed tomosynthesis volume can therefore be seen as a stack of 2D mammograms and an obvious approach is to apply the 2D CAD system to each individual slice of the reconstructed volume. However most masses are larger than 3 mm in diameter and will therefore span several slices. By combining several slices into a larger slab, more information of the whole mass is used and its appearance might become more similar to a 2D mammogram. In this study we investigate the effect of using slabs of different sizes on the performance of our CAD system.

2 Method

2.1 CAD System

Our CAD system for mass detection in reconstructed tomosynthesis volumes is based on a previously developed system for 2D screen-film mammograms. This system consists of three stages: a preprocessing stage, an initial detection stage and a false positive reduction stage. In the preprocessing stage a mammogram is downsampled to a resolution of 0.2 x 0.2 mm and segmented into three areas: breast tissue, pectoral muscle and background [5]. Candidate mass regions are detected in the second stage as described in [6,7]. In this stage five features are computed that measure the presence of a stellate pattern and central mass in each location of the mammogram. An ensemble of five neural networks (trained on a small separate dataset of 302 images) then assigns a likelihood score to each location, resulting in a likelihood image. This image is smoothed and local maxima are selected as locations of interest. Finally a candidate mass region is segmented for each location of interest by using a segmentation method based on dynamic programming [8]. In the third stage a set of 31 features is computed for each segmented region. This feature set contains the five features from the initial detection stage, the maximum and mean likelihood score of the region, compactness, region size, pectoral overlap, 2 linear texture features, 2 border features and several local context, location and contrast features [9,10]. A second ensemble of five neural networks is used to assign a malignancy score to each region. These networks are trained on a large screen-film database of 636 abnormal cases (2156 images) and 3262 normal cases (9688 images).

In order to make 3D DBT data suitable for a 2D SFM CAD system, we developed a method that converts a reconstructed DBT volume into a series of slabs with an SFM-like representation. A slab can simply be generated by computing the mean voxel value per column of several adjacent slices (Eq. 1). Because voxel values in our reconstructed DBT volumes represent an estimation of the attenuation coefficient (scaled with a constant) at that location, voxel values in a slab now represent the attenuation coefficient of a thicker piece of tissue.

$$\bar{\mu}(x,y,z') = \frac{1}{N} \sum_{z=z'}^{z'+N-1} \mu(x,y,z) \tag{1}$$

Where $\bar{\mu}(x,y,z')$ is the voxel value of a slab at location (x,y,z'), $\mu(x,y,z)$ the voxel value at location (x,y,z) in the DBT volume and N the number of slices per slab. We compute overlapping slabs and construct each slab from the same amount of slices by letting z' run from 0 to the number of slices of the volume $-N$ (see Fig 1). In this study slab size is varied from 1 to 8 slices (0.3 to 2.4 cm)

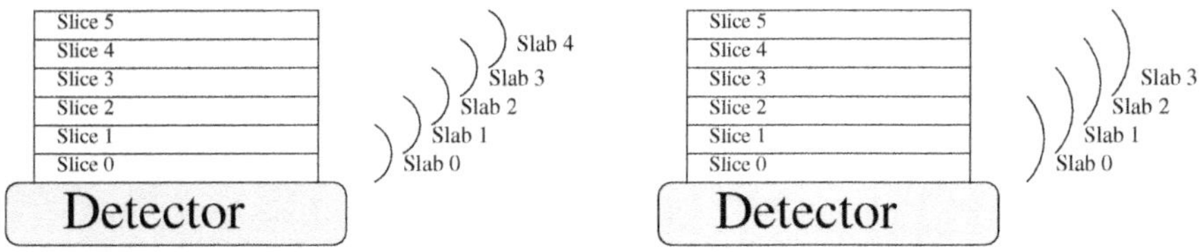

Fig. 1. Schematic drawing of generated slabs of 2 slices (left) and 3 slices (right) of a volume of 6 slices

and also a large slab of all slices of a volume is constructed, which on average is 20 slices (6.0 cm) thick, depending on the thickness of the breast.

Subsequently slabs are converted into an SFM-like representation by using the method described by Kallenberg et al. [4]. This method converts raw FFDM images, where pixel values are linearly related to x-ray exposure (E), into SFM-like images where pixel values represent optical density ($od(E)$), by applying the characteristic curve:

$$od(E) = od_{min} + \frac{od_{max}}{1 + (\frac{E}{s})^{-g}} \tag{2}$$

Parameters g, s, od_{min} and od_{max} respectively represent the gradient and speed of the film, and the minimum and maximum of the curve. The method estimates parameter s from an image at hand by mimicking the automatic exposure control of an SFM system. Voxel values of the tomosynthesis slabs represent attenuation coefficients (μ) and are converted into exposure (E) first, by applying Beer-Lambert law:

$$E = E_0 \cdot e^{-\mu \cdot d} \tag{3}$$

Where E_0 is the incident exposure and d the tissue thickness. Fig 2, 3, 4 and 5 show an example of an original reconstructed tomosynthesis slice (downscaled to a resolution of 0.2 x 0.2 mm) and the converted SFM-like slab of 5 slices.

The SFM-like slabs are used as input to the CAD system described above, resulting in a set of suspicious regions per slab. In order to count multiple findings of the same false positive in adjacent (overlapping) slabs only once and give a single malignancy score to each true positive that was detected in multiple slabs, detection results are merged. For simplicity, depth information (i.e. z-location of a lesion) was not used in this study and overlapping regions of all slabs are merged into a single set of 2D regions per volume. Two regions in different slabs are marked as overlapping when the (2D) center of one region lies in the other. Overlapping regions are merged by keeping the one with the highest malignancy score and discarding the others.

2.2 Validation

For validation we use a dataset of 63 tomosynthesis cases (245 volumes) consisting of 42 normal cases (163 volumes) and 21 abnormal cases (82 volumes) with

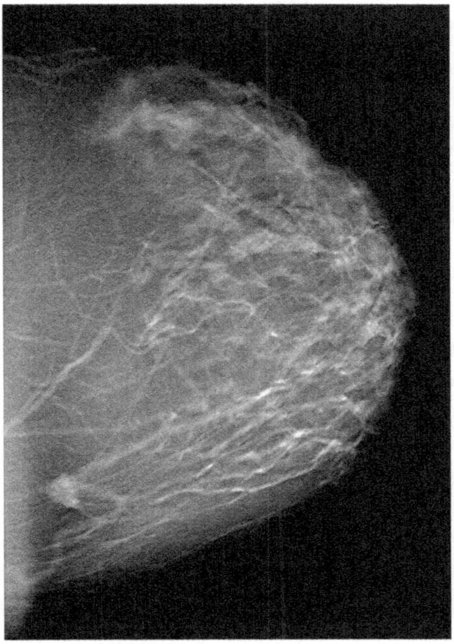

Fig. 2. Original reconstructed tomosynthesis slice, downscaled to a resolution of 0.2 x 0.2 mm. Pixel values represent an estimation of the attenuation coefficients.

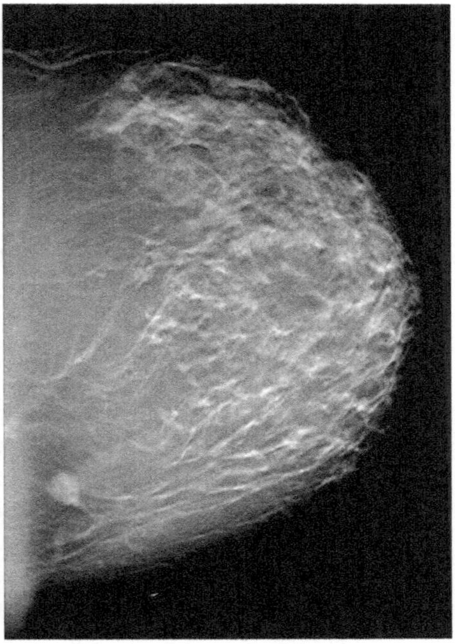

Fig. 3. Transformed SFM-like slab of 5 slices. Pixel values represent an estimation of optical density.

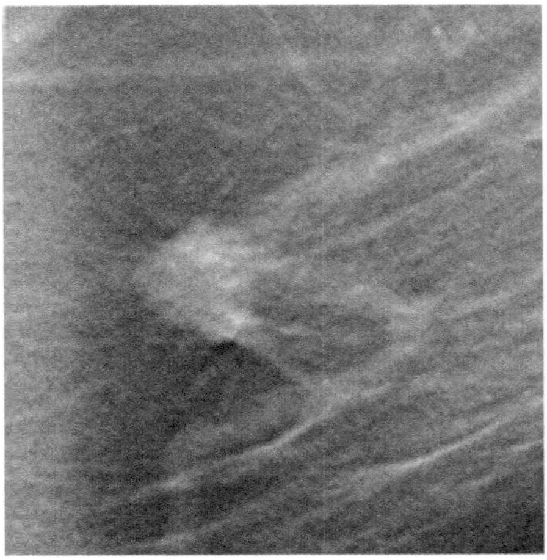

Fig. 4. Magnification of the mass in the tomosynthesis slice of Fig 2

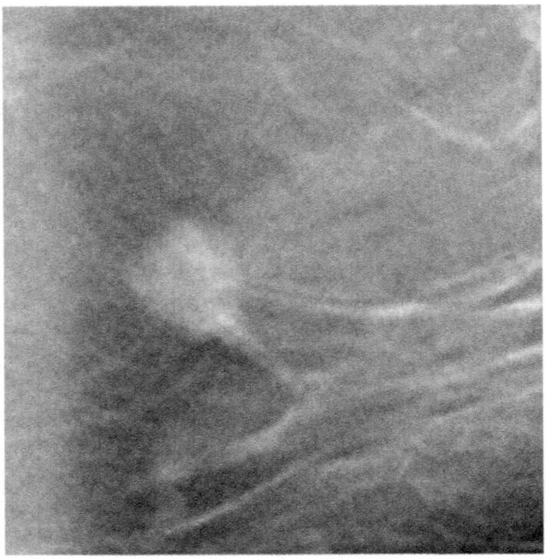

Fig. 5. Magnification of the mass in the SFM-like slab of 5 slices of Fig 3

a total of 47 malignant masses and architectural distortions. The volumes are acquired with a Sectra tomosynthesis system that is equipped with a photon-counting scanning-multislit detector. The system uses a tomo angle of 12 degrees (± 6) and reconstructs 3 mm thick slices at an in-plane resolution of 0.1 x 0.1 mm. All volumes are reconstructed by means of a stochastic model and the convex algorithm for maximum likelihood estimation of the attenuation parameters of the voxels. No postprocessing is used. All lesions are annotated with a contour in the slice where it is most clearly visible.

In this study we carried out a total of 9 experiments, one for each different slab size of 1 to 8 slices and all slices together. In each experiment each volume in the dataset is transformed into a volume of (overlapping) SFM-like slabs of the slab size used in that particular experiment. Subsequently our CAD algorithm is applied to the slabs and the detected regions are merged as described above. An FROC curve is calculated for each experiment to measure performance.

A malignant lesion is counted as a true positive when the CAD system detects a region with its center of mass inside the annotated region of the lesion (in 2D). All other detected CAD regions that do not hit an annotated lesion are counted as false positives. FROC curves are calculated by varying a threshold on the malignancy scores of detected regions. To obtain a single performance measure for each slab size experiment we compute the mean sensitivity in the range of 0.05 to 2.0 false positives per volume, S. To avoid domination of performance at high false positive rates, the mean is calculated on a logarithmic scale.

Significance of the obtained performance differences between the slab sizes is computed by using the bootstrap method. Cases are sampled with replacement to construct a 1000 new sets of cases, where the total number of cases of a set is kept equal to the original set. For each new set, FROC curves and mean sensitivity S is calculated for all different slab size experiments. The difference in mean sensitivity ΔS of two different slab size experiments is calculated for each of the 1000 new sets and p-values are defined as the fraction of ΔS values that are negative or zero. Performance differences are considered significant if $p < 0.05$.

3 Results

Results of slab size on mass detection performance are shown in Fig. 6 where mean sensitivity S is plotted as a function of slab thickness. Results show that performance increases when slabs of more than 1 slice are used upto a slab thickness of 1.5 cm. When more slices are added performance starts to decrease. All slab sizes between 2 and 8 slices (0.6 and 2.4 cm) perform significantly better than a single slice. The best performance is acquired with slabs of 5 slices (1.5 cm) which is significantly better ($p < 0.02$) than performance of slabs of a single slice, 2 slices and all slices. Fig. 7 shows the FROC curves (averaged over all bootstraps) for slabs of 1, 5 and all slices.

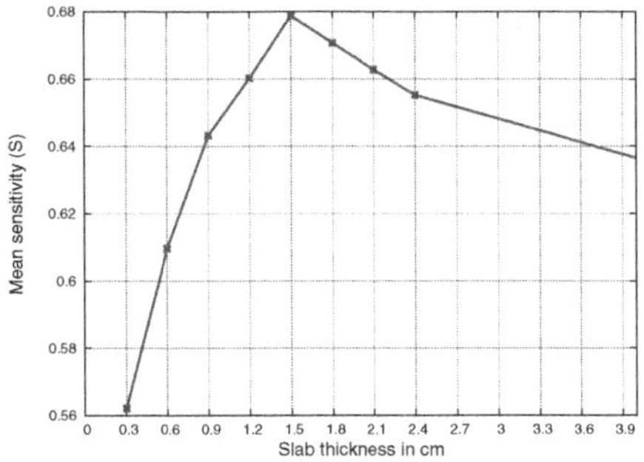

Fig. 6. Mean sensitivity in the range of 0.05 to 2.0 false positives per volume, set of against slab thickness

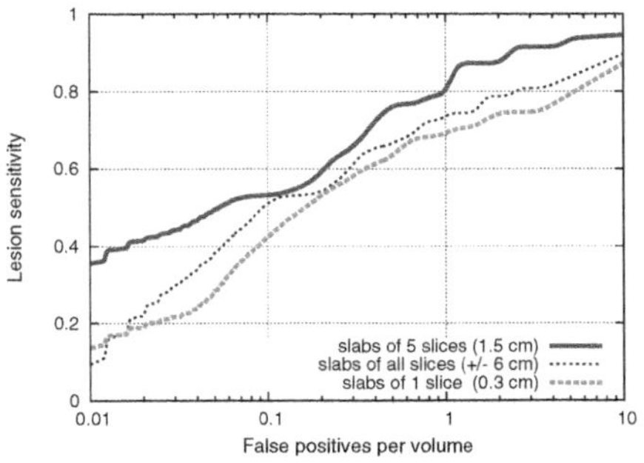

Fig. 7. FROC curves for 3 different slab sizes: 1 slice (0.3 cm), 5 slices (1.5 cm) and all slices (approximately 6 cm, depending on breast thickness)

4 Discussion

Results show that the performance of our CAD system increases when slices are combined into larger slabs. Best performance is obtained when a slab thickness of 1.5 cm (5 slices) is used, which is significantly higher than using slabs of a single slice, two slices and all slices. An explanation for this result can be found in the fact that most masses are larger than 3 mm in diameter and will therefore span several slices. By combining slices, more information of the whole mass is used and its appearance becomes more similar to a 2D mammogram (but without overlapping normal tissue). Features extracted from these slabs resemble the features our SFM CAD system was trained on. In the future other slab constructions will be investigated e.g. one where slab thickness is related to the diameter of the initial detected lesion. Other region merging strategies will also be investigated. Generating thicker slabs by averaging the individual reconstructed slices, will also make the slabs less sensitive to noise and reconstruction artifacts that are present in the individual slices. We therefore assume that our method will also be applicable to volumes generated with other reconstruction algorithms that calculate an estimation of the attenuation coefficient of the voxels. In this study our CAD system only detects masses in the x-y plane, by merging overlapping regions found in the slabs. Therefore, in theory it is possible that a mass is detected by a very suspicious FP region found below or above the true lesion, however we can argue that this is a highly unlikely event, in particular at the low false positive rates considered in this study. 3D localization of suspicious regions is however being developed. Furthermore visual inspection of the individual reconstructed tomosynthesis slices suggest that more improvement can be gained by making more use of spiculation of masses, which is often very clearly visible in individual tomosynthesis slices. This is a topic of interest in future developments.

Acknowledgments. This work has been funded by the EU HighReX project within the Sixth Framework Programme (FP6).

References

1. Singh, S., Tourassi, G.D., Baker, J.A., Samei, E., Lo, J.Y.: Automated breast mass detection in 3D reconstructed tomosynthesis volumes: a featureless approach. Med. Phys. 35(8), 3626–3636 (2008)
2. Reiser, I., Nishikawa, R.M., Giger, M.L., Wu, T., Rafferty, E.A., Moore, R., Kopans, D.B.: Computerized mass detection for digital breast tomosynthesis directly from the projection images. Med. Phys. 33(2), 482–491 (2006)
3. Chan, H.P., Wei, J., Zhang, Y., Helvie, M.A., Moore, R.H., Sahiner, B., Hadjiiski, L., Kopans, D.B.: Computer-aided detection of masses in digital tomosynthesis mammography: comparison of three approaches. Med. Phys. 35(9), 4087–4095 (2008)
4. Kallenberg, M., Karssemeijer, N.: Computer-aided detection of masses in full-field digital mammography using screen-film mammograms for training. Phys. Med. Biol. 53(23), 6879–6891 (2008)
5. Karssemeijer, N.: Automated classification of parenchymal patterns in mammograms. Phys. Med. Biol. 43(2), 365–378 (1998)
6. Karssemeijer, N., Te Brake, G.M.: Detection of stellate distortions in mammograms. IEEE Trans Med. Imaging 15(5), 611–619 (1996)
7. te Brake, G.M., Karssemeijer, N.: Single and multiscale detection of masses in digital mammograms. IEEE Trans Med. Imaging 18(7), 628–639 (1999)
8. Timp, S., Karssemeijer, N.: A new 2D segmentation method based on dynamic programming applied to computer aided detection in mammography. Med. Phys. 31(5), 958–971 (2004)
9. te Brake, G.M., Karssemeijer, N., Hendriks, J.H.: An automatic method to discriminate malignant masses from normal tissue in digital mammograms. Phys. Med. Biol. 45(10), 2843–2857 (2000)
10. Hupse, R., Karssemeijer, N.: Use of normal tissue context in computer-aided detection of masses in mammograms. IEEE Trans Med. Imaging 28(12), 2033–2041 (2009)

Synthesising Malignant Breast Masses in Normal Mammograms

Michael Berks[1], Chris Taylor[1], Rumana Rahim[2], David Barbosa da Silva[3], Caroline Boggis[4], and Susan Astley[1]

[1] Imaging Science and Biomedical Engineeering, University of Manchester, Oxford Road, Manchester, M13 9PT, UK
[2] Department of Clinical Radiology, University of Manchester, Manchester, UK
[3] University of Manchester Medical School, Stopford Building, Oxford Road, Manchester M13 9PT, UK
[4] Nightingale Breast Centre and Genesis Prevention Centre, University Hospital of South Manchester, Southmoor Road, Manchester M23 9LT, UK
michael.berks@manchester.ac.uk

Abstract. Using mammograms in which signs of breast cancer have been synthesised overcomes the problem of obtaining a sufficiently large volume of real data with known ground truth for training and test purposes. This paper describes a fully automated method for generating synthetic spiculated masses. Statistical methods are used to model the appearance and location of a training set of real masses and their effect on surrounding breast tissue. The models are then used to synthesise the appearance of a malignant mass in an otherwise normal mammogram. By virtue of using generative statistical models, the synthesis process can be fully automated. In an observer study in which 10 expert mammogram readers attempted to distinguish between synthetic masses generated by the method and real masses, we report an area $A_z = 0.70 \pm 0.09$ under the receiver operating characteristic.

Keywords: Mammography, breast cancer, breast mass, lesion synthesis, statistical models, DT-CWT.

1 Background

Large sets of digital mammograms displaying a wide range of abnormalities associated with breast cancer are essential for successfully training and assessing radiologists, validating applications such as automatic detection software, and evaluating the performance of visual display units. Synthesising such abnormalities overcomes the problems of obtaining a sufficient volume of real data with known ground truth.

Spiculated masses are one of the most frequent mammographic signs of malignancy, but despite previous attempts, have yet to be synthesised successfully [1-3]. Of the previous work in which spiculated masses have been synthesised, only Saunders et al. [3] and Caulkin et al. [1] describe a quantitative evaluation of masses they have generated. In both cases an observer study was implemented in which expert radiologists assigned

J. Martí et al. (Eds.): IWDM 2010, LNCS 6136, pp. 505–512, 2010.

a realism rating to randomised sets of real and synthetic masses. The ability of radiologists to identify synthetic masses was quantified by fitting an ROC curve to their responses and computing the area-under-curve (A_z). Caulkin et al. reported an A_z of 0.69±0.13 whilst Saunders et al. reported an A_z of 0.65±0.07. This suggests radiologists were able to identify synthetic masses generated by both methods at a rate significantly better than chance.

In this paper, we describe a novel method for generating malignant spiculated masses in mammograms previously displaying no sign of disease, and present the results of an observer study to evaluate the method. Finally, we discuss the advantages of our approach over previous synthesis methods.

2 Data and Methods

2.1 Mammogram Data

Our method for synthesising malignant mammographic masses is based on constructing statistical models that describe the appearance of a set of real training data. It is important the training data contains as wider variation of mass appearance as possible, whilst being a representative sample of the global screening population.

We used a sequential set of mammograms containing a biopsy proven-malignancy, drawn from cases detected during NHS screening at Nightingale Breast Centre, University Hospital of South Manchester and digitized using a Vidar CADPRO scanner at a resolution of 40μm.

Within these data, a set of 101 malignant masses were identified and annotated by a consultant breast radiologist. Where there were two or more mammographic views of a single mass, one view was randomly selected for inclusion in the training data. These masses varied in size from a minimum of ~16mm^2 to a maximum of ~900mm^2 (mean 176mm^2). As would be expected of a set of malignant masses, the majority had irregular borders and displayed signs of spiculations, with the number and length of spicules increasing for larger, more developed masses. Fig. 1 shows three real masses from our training data.

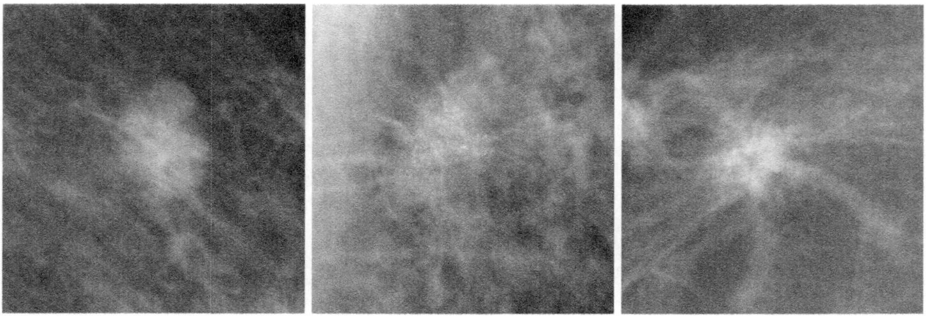

Fig. 1. Three real mammographic masses from our training data

2.2 Overview of Modelling Method

Fig. 2 depicts a flow chart of the modelling steps we have applied to our set of training masses, and subsequently the synthesis steps required to simulate a new malignant mass in an otherwise normal mammogram.

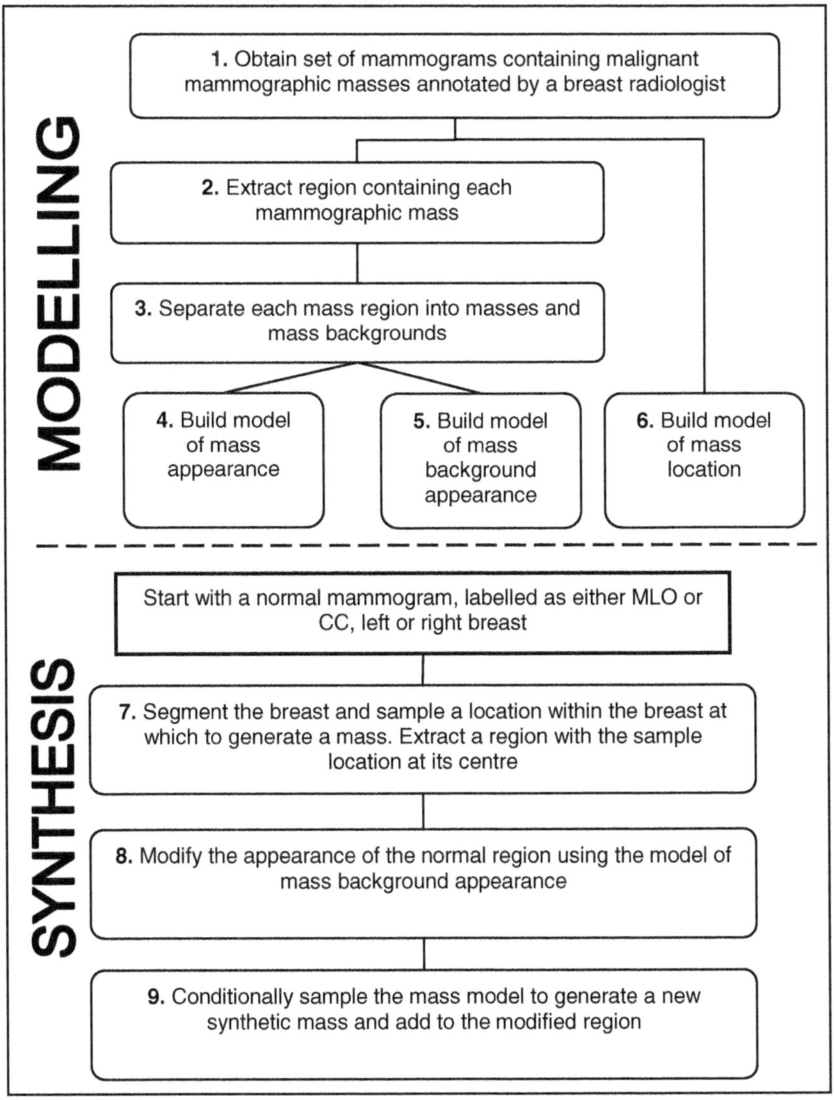

Fig. 2. Flow chart summarising the process for modelling a set of malignant masses and subsequently using the models to generate synthetic masses in normal mammograms

We have previously described how the central density of each mass in the training data can be separated from surrounding breast tissue (boxes 2 and 3 in Fig. 2.) and how a model can be constructed to optimally fit the set of separated central densities (box 4 in Fig. 2.) [4].

In summary, separation is achieved in each region by estimating the underlying background grey-levels in the area occupied by the mass, using a thin-plate spline interpolant. High frequency components are added to the initial estimates, before the background is subtracted from the original region to obtain the separated central density. A principal component based statistical appearance model [5], is fitted to encapsulate the variation in shape, size and texture in the set of separated densities. A unique representation of a central density can be synthesised by randomly sampling from the probability distribution encoded in the model.

In the following sections we describe the remaining steps in our synthesis method.

2.3 Synthesising Mass Background Appearance

Having separated the central density from a real mass region, we are left with an image we label the *mass background*. The background contains all other breast tissue in the region, including any spicules associated with the mass. In many backgrounds the presence of a mass distorts the surrounding breast tissue.

Our aim is to model this distortion so that given a normal region we can modify the region in way that not only generates mass spicules, but also accounts for how structures in the region would be altered by the presence of a mass. To work with the complex textures and structures present in mammographic tissue, we use the dual-tree complex wavelet transform (DT-CWT) [6] to decompose image regions into subbands localised in scale and orientation. We apply the following three stage process to synthesise mass background appearance in a normal region:

1. Compute the DT-CWT of a normal region
2. Modify coefficients in the decomposition to match the properties of DT-CWT coefficients in real mass backgrounds
3. Invert the modified DT-CWT to reconstruct a region in which mass background appearance has been synthesised

For the masses evaluated in this paper, stage 2 of this process was achieved by directly transferring DT-CWT coefficients from a real mass background (randomly selected from the training data) into the transform of a normal region. In the top row of Fig. 3 we depict three normal mammographic regions and in the second row show the result of modifying each region to display mass background appearance.

Further ways of achieving stage 2 are described elsewhere in this volume [7].

2.4 Combining Masses and Mass Backgrounds

Using our earlier work [4] and the method described in the section 2.3, we can synthesise both the central density of a mammographic mass and the distorted appearance of breast tissue displayed in a mass background. The two can be added to complete the appearance malignant mass. However, the size and shape of the central density must match the appearance of the synthesised background. We achieve this by conditionally sampling the central density given the properties of the real mass used as a

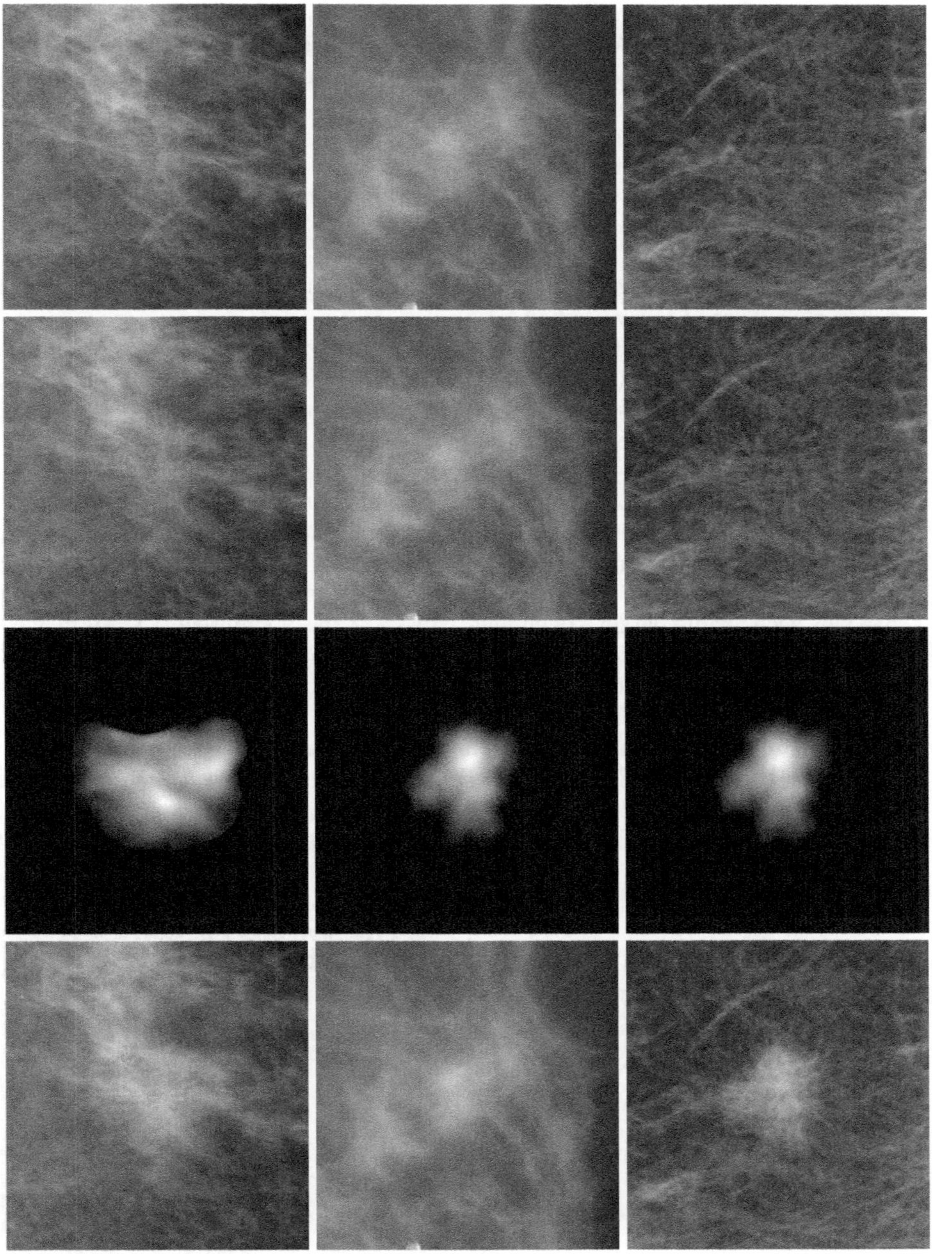

Fig. 3. *Top row*: three normal mammogram regions; 2nd *row*: modified regions displaying mass background appearance; 3rd *row*: central densities conditionally sampled to match the mass backgrounds; *bottom row*: complete synthetic mass regions

template when synthesising the mass background. In the third row of Fig. 3 we show three central densities sampled to match each of the synthetic mass backgrounds displayed in the row above. As a result of adding each central density to its mass background, we generate a complete mass region, as depicted in the bottom row of Fig. 3.

2.5 Modelling Mass Location

Given a method for synthesising the appearance of malignant mammographic mass in a normal region, it remains to show how normal regions can be automatically selected from whole mammograms.

To achieve this, we construct a probabilistic model of mass location that extends an earlier method proposed by Caulkin [1].

First, we apply an automatic segmentation algorithm to locate the skin-air boundary of the breast and for MLO views the pectoral muscle, in each mammogram in our dataset. The segmentations are then used to define a common set of landmarks points to describe the breast shape in each MLO or CC mammogram. We then build point distribution models (PDM) [8] to encapsulate the variation in breast shape and thus define a mean shape to act as reference frame for all mammograms.

Given a mean breast shape for both MLO and CC mammograms, and a training set of mammograms containing (annotated) masses, Caulkin's method [1] can be applied to construct probabilistic maps of mass location. This involves warping the breast shape in each mammogram containing a mass to the appropriate CC/MLO mean shape, and subsequently computing the location of each mass relative to the mean breast shape. This populates the mean shapes with a set of mass centres (see Fig. 4. (a) and (b)).

A probability density can then be computed to describe the spatial distribution of mass centres by placing an adaptively scaled Gaussian kernel on each centre. The probability of finding a mass at a given location in the mean breast shape can then be computed by summing the contribution from each of the Gaussian kernels (see Fig. 4. (c) and (d)).

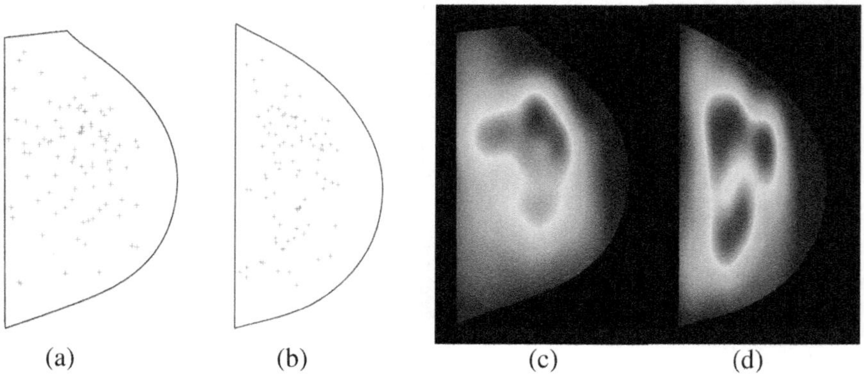

Fig. 4. Mean breast shape with mass centres marked a) MLO b) CC; probability density of mass location c) MLO d) CC

To sample a location at which to synthesise a mass in a normal mammogram, we again warp the breast shape to the appropriate MLO/CC mean shape. A location can then be sampled from the model distribution. Reversing the warp then transforms the sampled location so that its coordinates are relative to the original breast shape.

However, the probability distributions extend to the edge of the skin-air boundary in the mean breast shapes. If a location were sampled here, the synthetic mass would lie over the skin-air boundary and would thus appear unrealistic. Our solution is to mask off (i.e. set to zero) the probability density at some radius inside the skin-air boundary. The size of the masking radius is adapted to the size of the lesion to be sampled (it is assumed this is known as part of the synthesis method) to ensure that no matter how large a lesion is it will not overlap the skin-air boundary.

3 Results

To assess whether the models consistently generated synthetic masses of realistic appearance, we implemented an observer study involving ten expert mammogram readers from the Nightingale Breast Centre. During the study a reader was shown a randomised set of 30 real and 30 synthetic masses, and asked to rate each mass on a scale varying from definitely real to definitely synthetic. The real masses were randomly selected from the training data, whilst the synthetic masses were generated using the method described in section 2. The readers completed the task individually and independently, and were unaware of the exact number of real or synthetic masses in the test set.

We fitted ROC curves to the reader ratings and computed a mean A_z of 0.70±0.09. This shows that the synthetic masses were identified at a rate significantly better than chance and implied that a proportion of synthetic masses generated from our method can be distinguished from real masses. In addition to the ratings assigned to each mass, the readers were asked to provide feedback on masses they had identified as synthetic. This feedback suggested that in the majority of masses correctly identified as synthetic, failures within the mass background synthesis algorithm were responsible. Further details on this study, including a full breakdown of individual A_z scores are given in a recent publication [9].

4 Discussion

Like previous methods, we have yet to generate malignant masses that are indistinguishable from real examples to expert mammogram readers. However we note that of the malignant masses that have been quantitatively evaluated [1, 3], ours are the only ones that have been generated fully automatically. In addition, we have made an attempt to explicitly model the interaction between a mass, spicules associated with the mass and surrounding breast tissue. This crucial aspect of mass appearance was not addressed in previous methods that generate masses and spicules independently of the background they are added to [1-3]. Synthesising the distortion of normal breast tissue due to the presence of a mass is a particularly challenging task, and one our method has not yet solved. However, we have indentified improvements to the method to overcome these limitations. In particular, rather than copying DT-CWT coefficients directly from a real mass background (as described in section 2.3), we propose sampling coefficients from texture models trained on the set of real mass backgrounds. These methods are the subject of ongoing research [7].

References

1. Caulkin, S.: Generating Synthetic Abnormalities in Digital Mammograms Using Statistical Models. Imaging Science and Biological Engineering, PhD. University of Manchester, UK, Manchester, p. 268 (2001)
2. Ruschin, M., Tingberg, A., Bath, M., et al.: Using simple mathematical functions to simulate pathological structures–input for digital mammography clinical trial. Radiat Prot. Dosimetry 114, 424–431 (2005)
3. Saunders, R., Samei, E., Baker, J., et al.: Simulation of mammographic lesions. Academic radiology 13, 860–870 (2006)
4. Berks, M., Caulkin, S., Rahim, R., et al.: Statistical Appearance Models of Mammographic Masses. In: Krupinski, E.A. (ed.) IWDM 2008. LNCS, vol. 5116, pp. 401–408. Springer, Heidelberg (2008)
5. Cootes, T.F., Taylor, C.J.: Statistical models of appearance for medical image analysis and computer vision. In: Proc. SPIE Medical Imaging, vol. 42, pp. 236–248 (2001)
6. Kingsbury, N.: Complex wavelets for shift invariant analysis and filtering of signals. Applied and computational harmonic analysis 10, 234 (2001)
7. Berks, M., Taylor, C.J., Rahim, R., et al.: Modelling Structural Deformations in Mammographic Tissue Using the Dual-Tree Complex Wavelet. In: 10th International Workshop on Digital Mammography. Springer, Girona (2010)
8. Cootes, T.F., Taylor, C.J., Cooper, D.H., et al.: Training Models of Shape from Sets of Examples. In: 3rd British Machine Vision Conference, pp. 9–18 (1992)
9. Berks, M., Barbosa Da Silva, D., Boggis, C.R.M., et al.: Evaluating the realism of synthetically generated mammographic lesions: an observer study. In: SPIE Medical Imaging, San Diego, California, USA (2010)

Comparison of Microcalcification Detection Rates and Recall Rates in Digital and Analogue Mammography

Nicola Barr[1], Caroline Boggis[2], Nicky Barr[2], Mary Wilson[2], Julie Morris[3], Michael Berks[4], and Susan Astley[4,*]

[1] University of Manchester Medical School, Stopford Building, Oxford Road, Manchester M13 9PT, England
[2] Nightingale Breast Centre and Genesis Prevention Centre, University Hospital of South Manchester, Southmoor Road, Manchester M23 9LT, England
[3] Medical Statistics, University Hospital of South Manchester, Wythenshawe Hospital, Southmoor Road, Manchester M23 9LT, England
[4] Cancer and Enabling Science, University of Manchester, Stopford Building, Oxford Road, Manchester M13 9PT, England,
sue.astley@manchester.ac.uk

Abstract. 21158 screening mammograms were obtained, 10024 acquired using full field digital mammography (FFDM) and 11134 acquired using film-screen mammography. For each mammogram, data were collected on recall for further assessment due to detection of microcalcification, use of needle biopsy, and presence of microcalcifications in biopsy specimens. 61.5% of women who had a core biopsy following digital mammography had microcalcifications detected, compared with 65.8% for analogue mammography but this difference was not significant (p=0.71). Rates of detection of microcalcifications in women screened by the two methods were similar. It was also found that the recall rate for assessment for women screened digitally (6.1%) was significantly higher than the recall rate for those screened by analogue mammography (3.3%), 95% confidence interval 2.2% - 3.4%. Screening using digital mammography leads to a higher recall rate for assessment than analogue mammography, although similar rates of detection of microcalcifications occur with both imaging modalities.

Keywords: breast screening, cancer detection, mammographic features, recall rates, digital mammography.

1 Introduction

It has been reported that full field digital mammography (FFDM) results in fewer recalls for assessment than conventional film-screen mammography [1], and the benefits for certain groups of women (such as those with dense breasts) has been demonstrated in the DMIST trial [2]. Clearly, reduced recall is advantageous provided that the detection of significant abnormalities is not compromised. Whilst we would expect that there would be significantly fewer recalls for technical reasons, the effect of the introduction of FFDM on recalls for mammographic abnormality in women from 50-65 is harder to predict.

* Corresponding author.

J. Martí et al. (Eds.): IWDM 2010, LNCS 6136, pp. 513–517, 2010.

Our breast screening centre was in the process of changing from analogue (film) mammography to digital mammography, so it was possible to directly compare the two methods using the same set of film readers who were reading both analogue and digital mammograms during the transition period.

Of particular interest is the detection of microcalcifications as they are often present at an early stage of the disease, and they are seen in the majority of cancers detected by mammography.

The aim of this study was to determine whether digital mammography led to a higher recall rate than analogue mammography, and then to compare the detection of microcalcifications in women screened by the two modalities.

2 Materials and Methods

21,158 screening mammograms were included in this study. They were taken between November 2006 and March 2009. 10,024 of the mammograms were digital mammograms taken at a static breast centre (The Nightingale Breast Centre in South Manchester) and 11,134 of the mammograms were analogue (film-screen) mammograms obtained from a mobile unit which travelled to different areas of the community in Greater Manchester in order to offer easy access to screening. All the mammograms were read in one reading room by the same group of expert readers. Although the women in the study were screened by different methods of mammography and in different locations it was expected that there would be no significant differences between the population that was screened at the static centre and that which was screened by the mobile unit.

Data were collected from the screening database and from patient records when necessary. The data collected included: the number of women screened by digital mammography at the Nightingale Centre between November 2006 and March 2009; the number of women screened by analogue mammography in a mobile unit between November 2006 and March 2009; the number of women recalled for mammographic abnormality in each location; and whether a core biopsy had been performed. 160 patients in the study group had received a stereotactic core biopsy; the results of these biopsies were obtained to determine whether microcalcifications had been found.

The proportion of digital mammograms with detected microcalcifications was compared with the proportion of analogue mammograms with detected microcalcifications. The results of the stereotactic core biopsies taken from women who received digital mammography were also compared with the results of the stereotactic core biopsies taken from women who were screened using analogue mammography. The proportion of patients who were screened by digital mammography and recalled because of a mammographic abnormality was compared with the proportion of patients who were screened by analogue mammography and also recalled for further views. Chi-square tests and 95% confidence intervals were used to test differences between the two groups.

3 Results

Out of the 10,024 women undergoing digital mammography that were included in this study, 608 had been recalled for assessment because of mammographic abnormality and 83 women underwent a stereo core biopsy, with 48 of these biopsies leading to the detection of microcalcifications. Of the remaining 11,134 women who had analogue mammograms 362 had been recalled, with 77 receiving a stereo core biopsy, of which 48 biopsies also showed microcalcifications (Table 1).

Table 1. Recalls and core biopsies in women screened with digital and analogue mammography at a static centre and mobile unit between November 2006 and March 2009

Number of Women	*Digital Mammography*	*Analogue Mammography*
Screened	10,024	11,134
Recalled	608	362
Stereo core biopsy	83	77
Microcalcifications detected	48	48

It is evident from figure 1 that the proportion of women recalled for review in clinic following digital mammography at 6.1% was higher than the proportion recalled for review following analogue mammography at 3.3%,a difference of 2.8% with a 95% confidence interval of 2.2% - 3.4%.. This was a significant difference (p=0.000).

For women recalled by analogue mammography there was a greater chance that they would undergo a stereo core biopsy, with 21.27% of those recalled from the mobile unit having this investigation. In contrast, only 13.65% of those recalled following digital mammography underwent a stereo core biopsy.

61.5% of all women that had received a stereo core biopsy following digital mammography had microcalcifications present, whereas 65.8% of all patients that had received a stereo core biopsy following analogue mammography had microcalcifications present. This was not statistically significant (p=0.712).

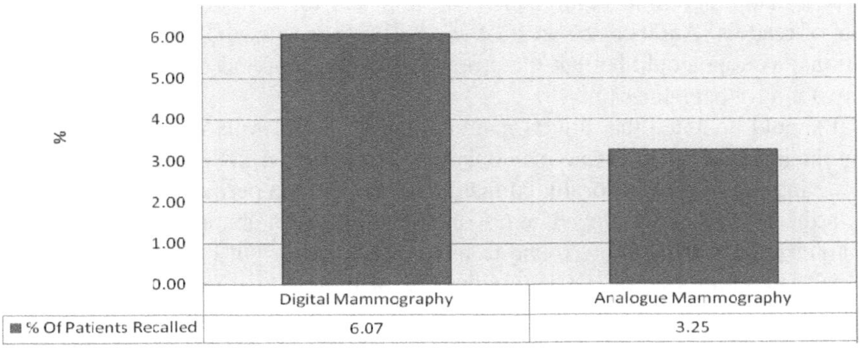

Fig. 1. Percentage of women recalled by digital and analogue mammography

4 Discussion

One of the advantages of digital mammography over analogue mammography should be a reduction in overall recall rate, since the radiographers acquiring the mammograms are able to immediately identify images which do not meet the exacting technical standards required for detection and diagnosis. Any technically inadequate mammograms can thus be repeated without recalling the woman. However, our results show that any reduction in recall rate for technical reasons is offset by an increase in the rate of recall for assessment for mammographic abnormality. We found almost double the number of women being recalled for review with digital mammography than were recalled for review after screen-film mammography. Published data on recall rates with digital mammography shows a wide range of findings. In a study in London in 2006-2007, no significant difference in recall rates was found (3.2% with digital *vs* 3.4% with analogue)[3]. These authors also undertook a meta-analysis of data from eight studies and this confirmed that there was no significant difference either in recall or detection rate.

Many authors have identified a relationship between recall rate and cancer detection [4,5]. Worldwide, mammographic screening recall rates vary widely depending on the way in which the screening programme is operated, including age at which screening commences and screening interval. In the UK, the lower age range is currently being extended to 47, although at the time this study started, women were invited from the age of 50. The screening interval also plays a part; in the UK this is 3 years. Other factors include number of views, number and experience of readers. These factors make comparison between screening programmes difficult, but it is worth noting that the recall rates for both analogue and digital mammography are both well within the normal range for screening programmes. In the UK in 2007-8, the overall recall rate was 4.2%, a decrease of 0.3% from the previous year [6]. During both these years, the vast majority of mammograms acquired were analogue. It is apparent that the recall rate with analogue mammography in our study was slightly lower than the national average. This could be a result of the fact that the Nightingale Centre is one of the busiest screening centres in the UK, and has exceptional breadth and depth of reading expertise. All mammograms are double read with either a consultant radiologist or a breast physician as part of the reading team. If the readers score a mammogram as being equivocal, it is sent for arbitration, usually by another pair of readers. Another reason for the recall rate in our study being lower than the national average could be that the proportion of incident and prevalent round screens may not have been the same.

It should be noted that digital mammography is still in its infancy in the UK, and that the increase in recall rate may only be a temporary effect. This can be verified by observing the recall rate for digital mammography over a period of time.

In the UK, approximately 1 in 6 women who have an abnormality detected during routine breast screening are found to have breast cancer. Of these, about a fifth have ductal carcinoma in situ (DCIS), with 29% of these cases being treated by mastectomy [6].

Our expectation was that the advantages of digital mammography would translate into having a greater rate of microcalcification detection in comparison to analogue mammography. This was not the case in this study. We found no significant difference between the rate of detection of microcalcifications between digital and analogue mammography. The results of this study were inconsistent with previous studies which showed a better detection rate of microcalcifications using digital mammography in comparison with analogue mammography [7,8]. The results of the study appear to be more consistent with the results revealed in a diagnostic setting which also showed digital and analogue mammography to be equivalent in relation to the detection of microcalcifications [9,10]. Our results are also consistent with those of the Digital Mammographic Imaging Screening Trial (DMIST) [2] which showed that digital mammography and analogue mammography both had a similar screening accuracy.

Acknowledgments. The authors would like to acknowledge the contribution of the readers who participated in this study, and of the administrative staff of the Nightingale Breast Centre for their assistance, especially Sue Berry.

References

1. Sala, M., Comas, M., Macia, F., Martinez, J., Casamitjana, M., Castells, X.: Implementation of Digital Mammography in a Population based Breast Cancer Screening Program: Effect of Screening Round On Recall Rate and Cancer Detection. Radiology 252(1) (2009)
2. Pisano, E.D., Gatsonis, C., Hendrick, E., Yaffe, M., Baum, J.K., Acharyya, S., et al.: Diagnostic performance of digital versus film mammography for breast-cancer screening. N. Engl. J. Med. 353, 1773–1783 (2005)
3. Vinnicombe, S., Pereira, S.M.P., Mc Cormack, V., Shiel, S., dos Santos, S.I.: Full-Field Digital versus Screen-Film Mammography: Comparison within the UK Breast Screening Program and Systematic Review of Published Data. Radiology 251, 347–358 (2009)
4. Yankaskas, B.C., Cleveland, R.J., Schell, M.J., Kozar, R.: Association of Recall Rates with Sensitivity and Positive Predictive Values of Screening Mammography Am. J. Roentgenol. 177(3), 543–549 (2001)
5. Otten, J.D., Karssemeijer, N., Hendriks, J.H.C.L., Groenewoud, J.H., Fracheboud, J., Verbeek, A.L.M., de Koning, H.J., Holland, R.: Effect of Recall Rate on Earlier Screen Detection of Breast Cancers Based on the Dutch Performance Indicators. JNCI Journal of the National Cancer Institute 97(10), 748–754 (2005)
6. 2009 Annual Report of the NHS Breast Screening Programme, `http://www.cancerscreening.nhs.uk/breastscreen/publications/nhsbsp-annualreview2009.pdf` (accessed 12.03.10)
7. Fischer, U., Baum, F., Obenauer, S., Luftner-Nagel, S., Von Heyden, D., Vosshenrich, R., et al.: Comparative study in patients with microcalcifications: full-field digital mammography vs screen-film mammography. EurRadiol. 11, 2679–2683 (2002)
8. Del Turco, M.R., Mantellini, P., Ciatto, S., Bonardi, R., Martinelli, F., Lazzari, B., Houssami, N.: Full-Field Digital Versus Screen-Film Mammography: Comparative Accuracy in Concurrent Screening Cohorts. AJR 189, 860–866 (2007)
9. Obenauer, S., Luftner-Nagel, S., von Heyden, D., Munzel, U., Baum, F., Grabbe, E.: Screen film vs full-field digital mammography: image quality, detectability and characterization of lesions. Eur. Radiol. 12, 1697–1702 (2002)
10. Yamada, T., Ishibashi, T., Sato, A., et al.: Comparison of screen-film and full-field digital mammography: image contrast and lesion characterization. Radiat. Med. 21, 166–171 (2003)

Constancy Checking of Digital Breast Tomosynthesis Systems

Jurgen Jacobs[1], Nicholas Marshall[1], Lesley Cockmartin[1], Ehsan Samei[2], and Hilde Bosmans[1]

[1] Department of Radiology, University Hospitals Leuven, Herestraat 49, 3000 Leuven, Belgium
[2] Carl E. Ravin Advanced Imaging Laboratories, Duke University, Durham, NC 27710
jurgen.jacobs@uz.kuleuven.ac.be

Abstract. As the number of installed digital breast tomosynthesis (DBT) systems increases, the need for quality control routines rises. Current work reports on our initial experience with a newly developed method for the analysis of DBT acquisitions of a homogenous phantom. Both the uniformity of the projection as of the reconstruction data is analyzed, together with the in-plane and inter-plane noise variations. The approach was tested in 2 different ways: on DBT projection and reconstructed data of different vendors and via simulations of potential detector artifacts known from 2D mammography into the projection images of the DBT (and followed by reconstruction of the hybrid data). The following potentially disturbing artifacts were observed: localized detector artifacts, modification of reconstruction software settings and synchronization issues. Our results indicate that the proposed method could be an easy and reliable way of performing constancy checking of DBT systems.

Keywords: quality control, quality assurance, mammography, digital breast tomosynthesis, phantom.

1 Background

Digital breast tomosynthesis (DBT) or limited angle tomography for mammography is a new and emerging technology for the detection and diagnosis of breast pathologies. This technique reconstructs images of planar sections in the breast while eliminating superimposition of anatomical background structures. Although still largely experimental, DBT systems are already being installed for study purposes or for initial clinical trials. In parallel to this, the first commercially available systems are entering the market. Today the stability of these systems remains largely untested and the methods on how to test them efficiently are unknown. The need to develop and validate test metrics for this pseudo 3D modality together with the huge size of both projection and reconstructed image series, make this a challenging task. In current work we report on a selection of methods for performing constancy checks of DBT systems. Results of this study will be discussed in the International Breast Phantom Group where a more advanced phantom for constancy check is being developed.

J. Martí et al. (Eds.): IWDM 2010, LNCS 6136, pp. 518–525, 2010.

2 Methods and Materials

2.1 Evaluation Strategy

On a daily basis an homogeneous phantom consisting of a 4-cm-thick polymethylmethacrylate (PMMA) slab with dimensions large enough to cover the complete detector, is imaged twice under clinically relevant exposure settings. This double exposure allows to rule out artifacts in the phantom which could be interpreted falsely as detector problems and helps substantially in the interpretation of the artifacts via this minimal 'reproducibility' test. To check for possible angular deviations based on the used system setup, one image is taken in CC (cranio-caudal) mode and one in MLO (medio-lateral oblique) mode. Often DBT systems can be used for both 2D Full-field Digital Mammography (FFDM) and tomosynthesis acquisitions. Because there could be a difference in setup between 2D mode and tomosynthesis mode (usage of grid, calibration, difference in used spectrum, …), we test a DBT system always in DBT mode. If available, both the projection data as well as the reconstruction data are being used during the evaluation. A possible pre-exposure which is not part of the tomographic series, is not taken into account in present analysis.

2.1.1 Projection Data

To find detector related artifacts, only the projection image closest to 0° is analyzed for homogeneity using the method described in [2-3] (homogeneity of mean pixel value (PV), standard deviation (std. dev.), signal-to-noise-ratio (SNR), variance, skewness, kurtosis, minimum PV, maximum PV and median PV). For all other projections only analysis values of a reference ROI (2x2cm at 6 cm from chest wall side) are calculated. Certain DICOM header values (exposure settings, reconstruction related information, used tomographic angles, ...) are collected and monitored over time.

2.1.2 Reconstruction Data

The homogeneity of all reconstructed planes (mean PV, std. dev., SNR and variance) is calculated using regions-of-interest (ROI) of 2mm x 2mm with a 50% overlap. This results in a volumetric homogeneity analysis that can be used for visualization and quantification of detector artifacts. The evaluation of this volume is done using three different analysis stacks (XY, XZ and YZ). For 6 reference ROIs, *parameter chimneys* are calculated in all reconstructed planes at the same in-plane coordinates to identify and follow up the differences in the homogeneity of reconstructed planes (inter-plane variations) (Fig cc). Although the in-plane noise power spectrum (NPS) would be related directly to the plane-by-plane viewing of DBT reconstructed images [4], the use of a co-variance matrix at a reference location may be a more robust estimate of noise given the shift-variant nature of noise in DBT reconstructed images. The full-width-half-maximum (FWHM) of the co-variance graph with the half-height plane as a reference together with the in-depth NPS in two directions (YZ and XZ), is being used to analyze the inter-plane correlation and thus could indicate modifications in the used reconstruction algorithm.

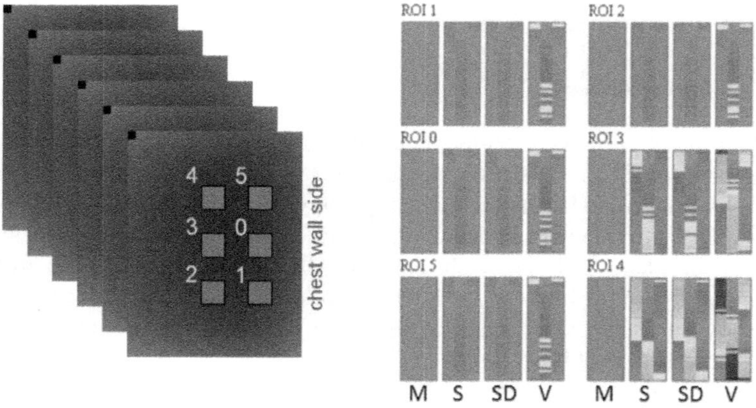

Fig. 1. Example of 4 *parameter chimneys* (mean pixel value, SNR, standard deviation and variance) at 6 different ROIs for system C. Each chimney gives a color coded indication of the deviation of the current plane against the mean of the bottom 5 planes, the mean of the top 5 planes and the mean throughout the complete volume

2.2 DBT Systems Used in Current Study

The proposed method is being tested routinely on three prototype DBT systems (Siemens Novation with BTS option, lab setting – system A; Siemens Inspiration with BTS option, clinical setting – system B; IMS Giotto with BTS option, clinical setting – system C) and one commercially available unit (Hologic Selenia Dimensions, clinical setting – system D). As a proof of concept also a limited number of datasets of two other prototype systems were considered (GE Healthcare Senographe Essential modified for tomosynthesis investigation, lab setting – system E; Sectra photon counting tomosynthesis prototype, clinical setting – system F).

2.3 Simulation of Detector Related Artifacts

Currently it is still not clear if projection data will be available on all DBT systems. Certain quality issues however are directly related to the detector and could therefore probably most easily be detectable in this type of data. In an attempt to check for the influence of detector related artifacts typically seen in 2D mode, on the reconstructed volume of homogeneous phantoms, it was tried to simulate these artifacts for one system (system B). Masks containing simulations of typical detector artifacts were created and multiplied with every projection image of a tomosynthesis series of the homogenous phantom. Vendor specific reconstruction software was used to reconstruct the modified projection images using clinically relevant settings. Subtracting two reference planes of the reconstructed volumes of both the original and the modified projection images, gives an indication of the influence of the simulated artifact on the volume homogeneity (Fig. 2). Artifacts that were simulated, included: dead pixels, dead rows, vibration lines, blooming, lag ghost and failing dead pixel map corrections.

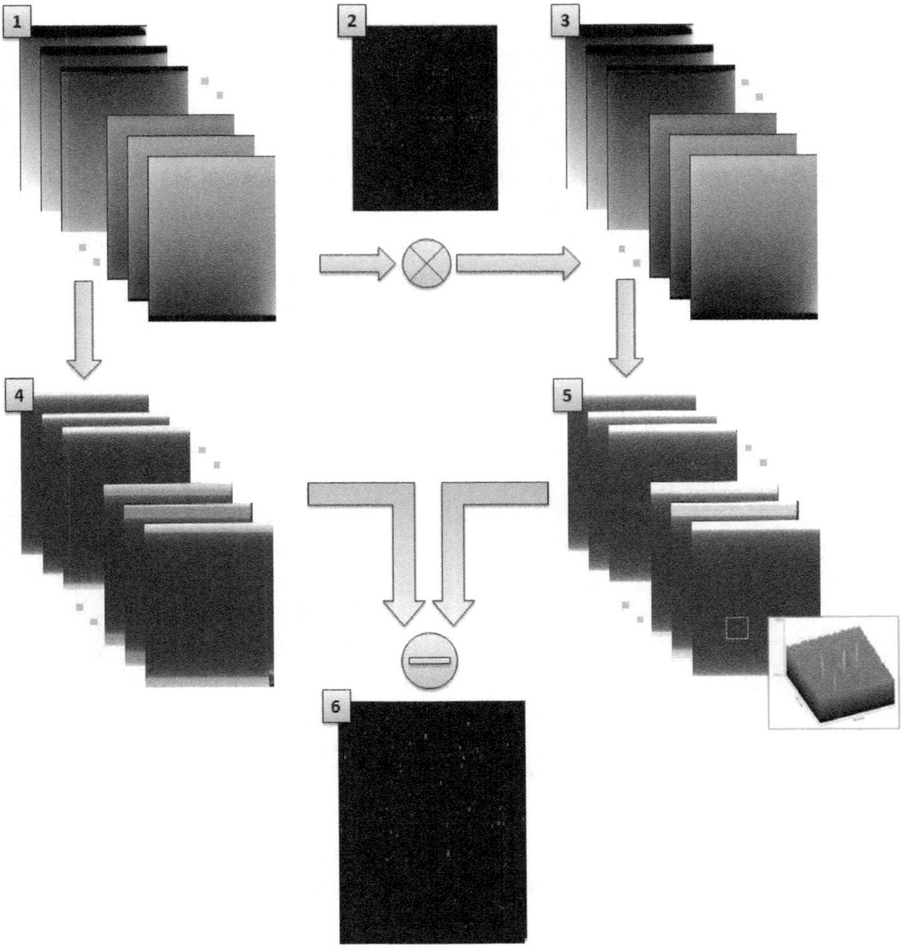

Fig. 2. Schematic overview of the artifact simulation routine. (1) original projection data of homogeneous phantom; (2) mask containing the artifact ; (3) original projection data multiplied with the mask image; (4) and (5) both projection datasets are reconstructed using vendor specific reconstruction software; (6) a reference plane at 20 mm for both datasets is subtracted to investigate the influence of the artifact on the homogeneity.

3 Results

We successfully evaluated the following number of datasets per system: system A: 34 (91 days); system B: 251 (209 days); system C: 12 (6 days); system D: 32 (39 days); system E: 2 (1 day) and system F: 2 (1 day). Due to the prototype character of most tested DBT systems, several image format transformations were needed before datasets of all tested systems could be analyzed and therefore vendors are encouraged to implement DICOM supplement 125 [5] when presenting their final product. For three systems (system D, E and F) it was not possible to retrieve the projection data at all. In our analysis, we couldn't see any difference between the results for datasets made in CC view and datasets made in MLO view for all tested systems. The proposed methods were able to check for the presence of typical detector artifacts (dead pixels, dead lines - Fig. 1) and for their influence on the volumetric homogeneity (Fig. 3) (system A, visible in all evaluations). For system B, a synchronization issue between the exposures and the detector read-outs during a tomographic acquisition was detected (Fig. 4) (2 evaluations) together with an unknown artifact resulting in randomly deviating pixels in the reconstruction dataset (5 evaluations). An example of the influence of simulated lag ghost on the reconstructed dataset can be seen in Fig. 5 (system B). The normalized inter-plane covariance measurements were able to find simulated modifications in reconstruction settings or reconstructed plane thicknesses (Fig. 6) (system A).

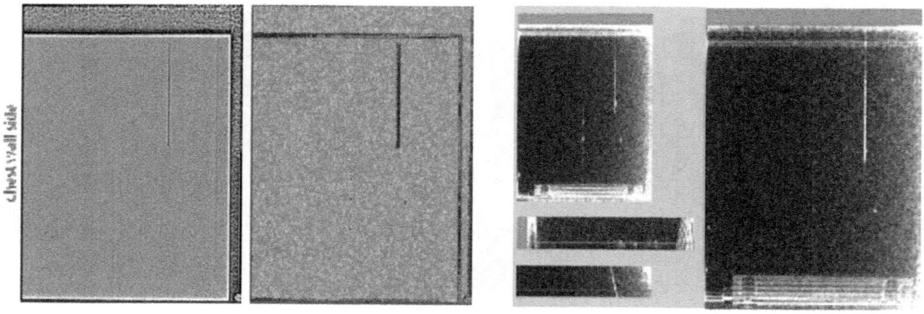

Fig. 3. Dead detector column visible in the projection data (left: variance map) and (middle: color coded variance deviation) and in the three stacks of the reconstructed data (right: variance data) (System A)

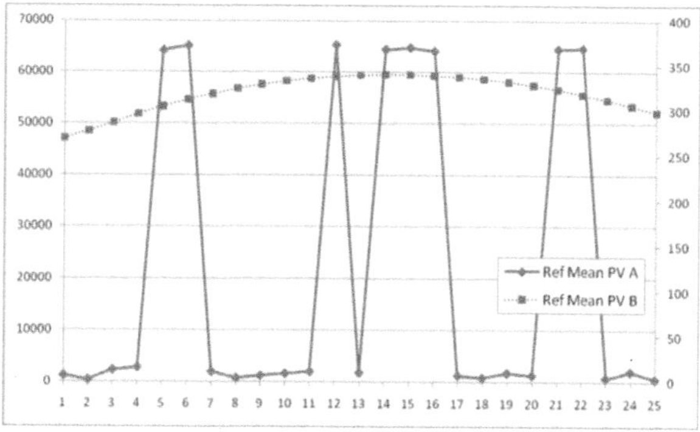

Fig. 4. Reference mean PV of all 25 projections indicating a synchronization problem (A) and a normal projection dataset (B) (System B)

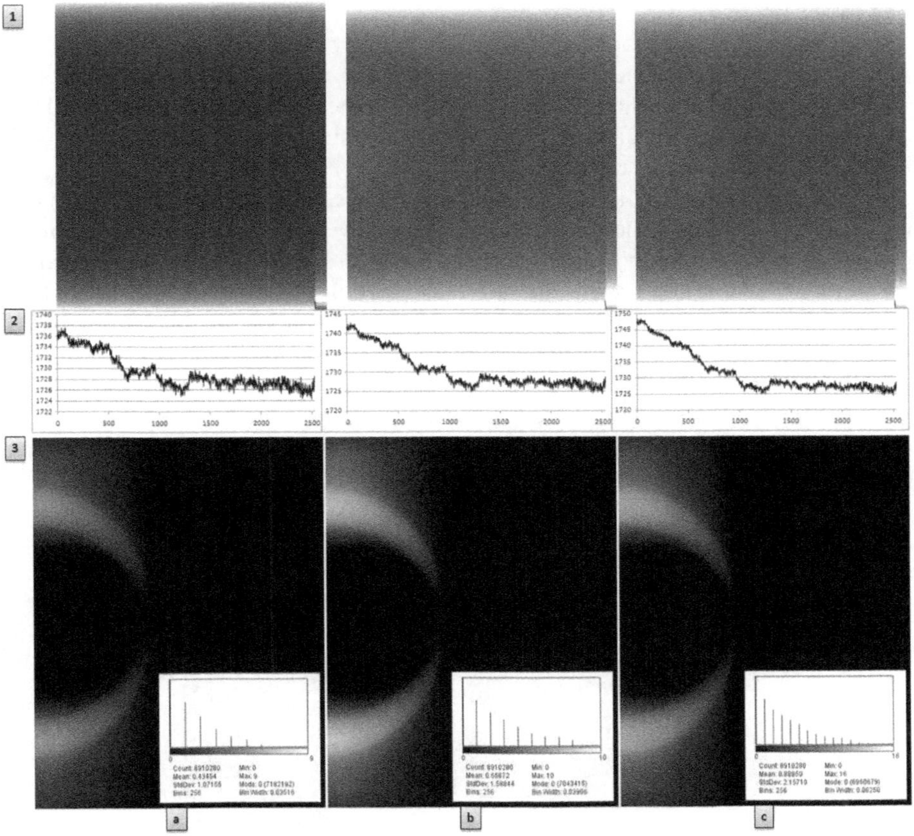

Fig. 5. Example of the influence of simulated ghost on the reconstructed dataset for three different levels of ghost (a: 5%; b: 8% and c: 11% reduction of pixel value). (1) modified projection image at 0° including the ghost; (2) line profile perpendicular to chest wall side; (3) subtraction image of reference plane at 20mm.

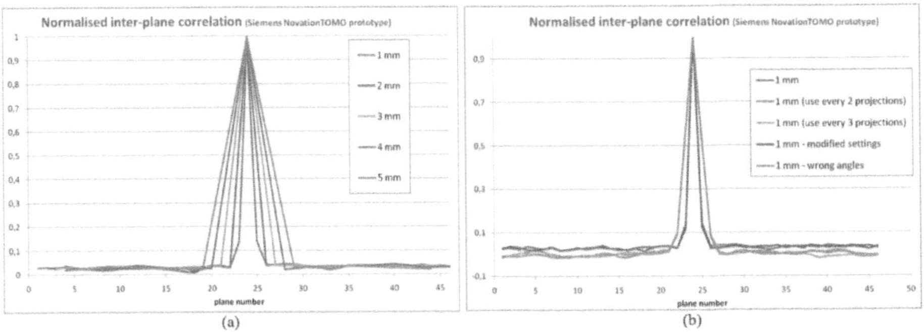

Fig. 6. Influence of changing reconstruction inter-plane distance (a) and modified reconstruction parameters (b) on the normalised inter-plane covariance with half height as the reference plane (system A)

4 Discussion

In current work we presented a fast and easy method to perform constancy checking of DBT systems. This method could be applied to datasets of all tested systems. Typical artifacts known from 2D FFDM (both real and simulated), could be detected in both the projection and the reconstructed datasets.

However, by using only a homogeneous phantom, certain DBT specific aspects are not taken into account. Due to the angular tube movement and the influence of the reconstruction algorithm, potentially missing breast tissue could be an issue. Also the spread of artifacts in the Z-direction or the Z-direction sensitivity profile are not addressed in current method. Therefore recently a more advanced phantom (*Agatha phantom*) has been made in the frame of a new AAPM working group[1], established to develop and harmonize breast imaging phantoms [6]. A picture of this phantom can be seen in Fig. 7. A complete constancy testing protocol could combine the strengths of both approaches: as a quick evaluation the method proposed in current paper could be used, while for a more detailed evaluation the Agatha phantom could be used.

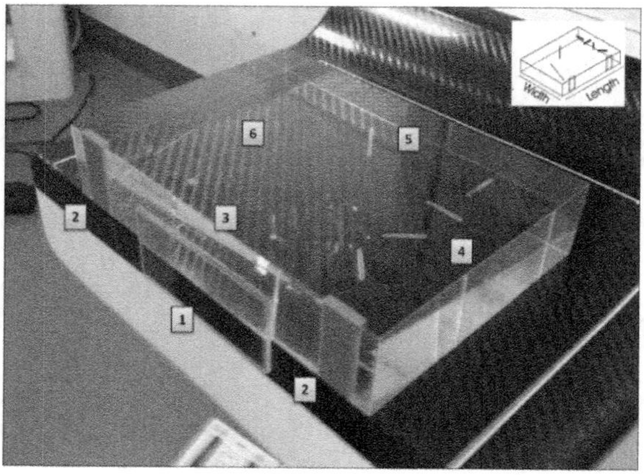

Fig. 7. Example of the *Agatha* phantom. This phantom consists out of: (1) phantom positioning aid; (2) cuboidal inserts to check for missing breast tissue; (3) low and high contrast spheres to check the artifact spread function (ASF) and to measure 3D MTF; (4) in-plane nylon wires to check the line object spread function (LOSF); (5) vertical wire to check for SDNR throughout the volume; (6) tilted Tungsten wire to check Z-direction sensitivity profile.

Acknowledgements

The authors of current work would like to thank the manufacturers and scientists who have helped to get access to the different systems: dr. H. Souchay and dr. R. Klausz (GE Healthcare); dr. A. Smith (Hologic); dr. A. Taibi and dr. S. Vecchio (IMS); prof. M. Danielson and dr. M. Wallis (Sectra) and dr. T. Mertelmeier (Siemens).

The proposed software tools and analysis methods were developed and tested in the frame of a European Commission funded FP7 project, namely BreastCT.

[1] http://www.aapm.org/org/structure/default.asp?committeecode=WGPBI

References

1. Niklason, L.T., et al.: Digital Tomosynthesis in Breast Imaging. Radiology 205, 399–406 (1997)
2. Jacobs, J., et al.: One year of experience with remote quality assurance of digital mammography systems in the Flemish breast cancer screening program. In: Proceedings of the 9th IWDM meeting, Tucson Arizona (2008)
3. Jacobs, J., et al.: Development and First Experience with a Software Framework for Remotely Supervised Constancy Checking of Digital Breast Tomosynthesis Systems. In: RSNA 2009 (2009)
4. Zhao, B., et al.: Experimental validation of a three-dimensional linear system model for breast tomosynthesis. Med. Phys. 36(1), 240–251 (2009)
5. DICOM Standards Committee, Working Groups 02, Projection Imaging –15, Mammography and CAD, Digital Imaging and Communications in Medicine (DICOM) Supplement 125: Breast Tomosynthesis Image Storage SOP Class, Final Text – 2008/08/28
6. Jacobs, J., et al.: Towards an international consensus strategy for periodic quality control of digital breast tomosynthesis systems. In: Proceedings SPIE Medical Imaging 7622-15 (2010)

Mammographic Image Segmentation and Risk Classification Using a Novel Texture Signature Based Methodology

Wenda He[1], Erika R.E. Denton[2], and Reyer Zwiggelaar[1]

[1] Department of Computer Science,
Aberystwyth University, Aberystwyth, SY23 3DB, UK
{wwh06,rrz}@aber.ac.uk
[2] Department of Radiology,
Norfolk & Norwich University Hospital, Norwich NR4 7UY, UK
erika.denton@nnuh.nhs.uk

Abstract. Clinical research has shown that breast cancer risk is strongly related to characteristic mixture of breast tissues as seen on mammographic images. We present an automatic mammographic image segmentation approach, which uses a novel texture signature based methodology, the resultant segmentation can be found useful as a means of aiding radiologists' estimation in mammographic risk assessment. The developed approach consists of four distinct steps: 1) feature extraction use a stack of small detail annotated mammographic patches, 2) Tabár mammographic building blocks are modelled as texture signatures, 3) model selection and reduction is used to remove noise and possible outliers, and 4) mammographic image segmentation. Visual assessment indicates good and consistent segmentation results. The MIAS database was used in a quantitative and qualitative evaluation with respect to mammographic risk assessment based on both Tabár and Birads risk categories. We found substantial agreement ($\kappa = 0.7$ and 0.75 based on Tabár and Birads risk categories, respectively) between classification results and ground truth data. Classification accuracy were 78% and 75% in Tabár and Birads categories, respectively; 86% and 87% in the corresponding low and high categories for Tabár and Birads, respectively.

Keywords: Tabár, breast segmentation, breast classification, mammography, parenchymal patterns.

1 Introduction

Strong evidence has shown that breast cancer risk is closely related to the relative proportion of the characteristic mixture of breast tissues, referred to as mammographic parenchymal patterns. Tabár *et al.* have proposed a mammographic texture modelling scheme based on mixtures of four building blocks composing the normal breast anatomy (i.e.; nodular, linear, homogeneous and radiolucent). Nodular densities mainly corresponds to Terminal Ductal Lobular Units (TDLU); linear densities correspond to either ducts, fibrous or blood

J. Martí et al. (Eds.): IWDM 2010, LNCS 6136, pp. 526–533, 2010.

vessels; homogeneous densities correspond to fibrous tissues which appears as bright areas in mammographic images, and could hide the underlying normal TDLU and ducts as well as their alterations; radiolucent areas are related to adipose fatty tissues, which appears as dark areas in mammographic images [1]. Under influences of Wolfe's work [2], Tabár subdivide mammograms into five risk classes based on the relative proportion of tissue belonging to the proposed four building blocks. The relative proportion of the four building blocks in different patterns and their variations are as follows (using the following order [nodular, linear, homogeneous, radiolucent]): Pattern I [25±10%, 15±5%, 35±17%, 25±14%]; Pattern II/III (Pattern III is similar in composition to Pattern II, with the retroareolar prominent ducts are associated with periductal fibrosis [1]) [2%, 14%, 2%, 82%]; Pattern IV [49±15%, 19±7%, 15±7%, 17±10%]; and Pattern V [2%, 2%, 89%, 7%]. Figure 1 shows example mammographic risk patterns with respect to Tabár's modelling. In mammographic risk assessment, inter and intra observer variability are introduced due to radiologist's subjective appraisal of mammograms. Quantitative defined mammographic risk patterns as described by Tabár's modelling suggest the possibility of using an accurate and repeatable texture based mammographic segmentation technique to automate the mammographic risk assessment, as well as quantification of change of the relative proportion of different breast tissue patterns can be evaluated [1].

Signatures are effectively 2D histograms statistical approaches based on histogram features can be used to encode various image features (*e.g.* coarseness, fineness and orientation). Guliato *et al.* [3] used a signature based on turning angle function of contour of breast masses to encode features, that characterise contours roughness and complexity for breast tumour classification. Zwiggelaar *et al.* [4] investigated scale-orientation signature for labelling of structures in images, and to classify pixels into linear structures, blob-like structures or background texture. A handful of related papers used the same but improved scale-orientation signature as a means of abnormality detection (*e.g.* central mass of spiculated lesions) in digital mammography, and have described signatures as rich descriptors of the neighbourhood around each image pixel [5]. Signatures can be used to describe mammographic tissue structures in terms of their density, size of abnormalities, orientation and thickness of linear structures etc. This is particularly useful as radiologists often categorised mammographic image patterns for risk assessment, based on mixture of characteristic various in tissue

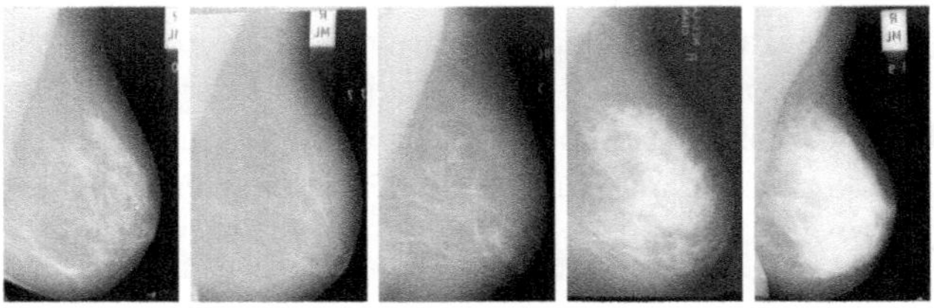

Fig. 1. Example Tabár's mammographic risk patterns. From left to right: Patterns I-V, corresponding from low to high mammographic risk.

type. In this paper we present a novel texture signature approach for mammographic image segmentation, and the resultant segmentation were quantitatively evaluated as a means of mammographic risk classification.

2 Data and Methods

We used the MIAS database [6] which contains 320 available images (file mdb296rl and mdb295ll are excluded for historical reasons). A total of 643 mammographic patches (199 nodular, 253 linear, 70 homogeneous and 121 radiolucent) were subsampled from randomly selected mammograms by an expert mammographic screening radiologist. In addition, 98 mammographic patches were detail annotated. The collection of the patches consists of representative Tabár's mammographic building blocks, with respect to mammographic parenchymal patterns, covering various risk categories.

At the learning stage a stack of small mammographic patches with detailed annotations were used (regardless of the associated risk class for the original mammograms) to model mammographic building blocks as described by Tabár's modelling. The feature extraction involves three 2D histograms encoding distinct texture features (*e.g.* intensity variance, skewness, kurtosis, orientation, elongation). A circular window (63 pixels in diameter) was used to compute local texture features. The diameter for the circular window was determined based on the size of texture features we wanted to capture, and Fourier analysis on the local patches. The first part of the texture signature shows a cumulative histogram based on radial distance to the centre pixel, and grey-level intensity. The signature's y-axis represents measurements for the same grey-level intensity, and the x-axis represents measurements for the same radial distance to the centre pixel. The second part of the texture signature shows a cumulative histogram based on angle between the tangent at segment, and the y-axis of the circular window and grey-level intensity. The signature's y-axis represents measurements for the same grey-level intensity, and the x-axis represents measurements for the same angle between the tangent at segment, and the y-axis of the circular window. The third part of the texture signature is a modified co-occurrence matrix with $d = 1$ and $\theta = 0\,°$. In this case the signature's x-axis and y-axis both represent grey-level information, and the absolute value of intensity difference is accumulated. This part of the texture signature represents not only frequency of some grey-level configuration, but also the magnitude of the variance of two pixels; the configuration varies rapidly in fine textures, and slowly in coarse textures. Finally, the three individual signatures are combined to form the final texture signature. Figure 2 shows example 2D histograms as signatures and final texture signatures. From a visualisation point of view, the texture signature can be considered as a colour image, and the three 2D histograms are the RGB channels of the image. We used signature size of 25×25, this parameter setting was thoroughly studied via algorithm validation, chosen to capture sufficient texture features, and to keep feature dimensionality low.

K-means clustering was used to group similar texture signatures in the feature space as a means of establishing texture models. We generated 120 cluster

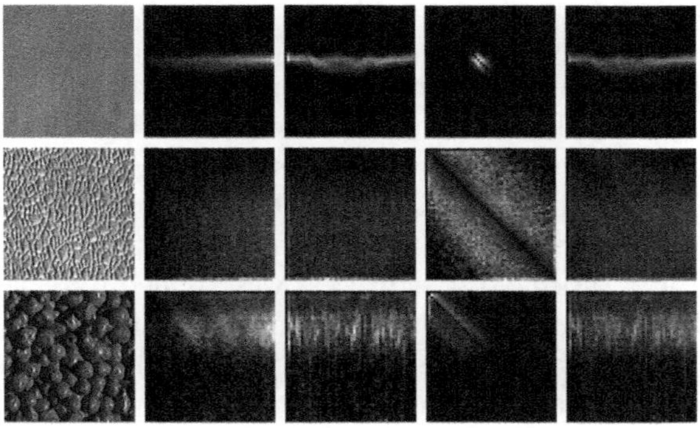

Fig. 2. Example texture signatures for three different texture images, from left to right showing: texture image, texture signature part I - III, and combined texture signature

centres, each tissue class was represented by 30 models. The initial number of cluster centres was determined based on visual appearances of mammographic building blocks, and was over-estimated to facilitate the model selection and reduction process.

In model selection and reduction, we first combined similar models generated in the initial clustering, using a cross voting technique based on the Euclidean distance between texture signatures, and higher votes indicated a group of closely clustered texture signatures. The mean texture signatures were calculated to represent the subgroups of texture signatures, so as to reduce the models. Texture signatures which received no votes were added to the reduced models, because these outliers belong to the same texture class. Next we filtered out some obvious incorrect models based on the standard deviation (normalised from 0 to 1) of the models. For example, high intensity homogeneous tissue and low intensity radiolucent, can be distinguished from the intensity distribution accumulated at two distinct ends of the intensity spectrum in the texture signatures, respectively. Whilst texture signatures for nodular and linear structure contain thicker band and rough peaks, respectively. The threshold values (i.e.; 0.4, 0.5, 0.6 and 0.5 were used for nodular, linear, homogeneous and radiolucent, respectively) were determined empirically, based on visual assessment on segmentation accuracy of the annotated mammographic patches.

The resulting models can be used to classify unseen pixels. We used a distance-weighted K-Nearest Neighbour (KNN) to increase classifier robustness, and reduce misclassification. The modified KNN weights the contribution of each of the K neighbours based on the inverse square of its Euclidean distance, giving greater weight to closer neighbours. Where $K = 3$ was empirically defined and variation indicated robustness.

Once mammographic images are tissue segmented, the relative percentage of different tissues was used in risk classification, as mammographic risk assessment with respect to Tabár's modelling. In Tabár pattern I, the relative proportion of the four building blocks can vary due to involution, which is a process of tissue changes from one to another (*e.g.* total fatty replacement), occurs highly individual, and can be caused by hormone replacement or pregnancy. With involution, pattern I will change to either pattern II or pattern III [1]. This regression also happens in pattern IV. The variations of the pattern I and IV were taken into account during the mammographic risk classification, by adding or subtracting the standard deviation of the relative proportion of a building block accordingly when measuring distance between two tissue composition vectors. To minimise classification error, a threshold post process was applied to reclassify some radiolucent pixels. In particular, if a pixel has very high intensity and was classified as radiolucent, then it was reassigned to either nodular or homogeneous. The threshold values were determined based on the mean value of homogenous and nodular from the collection of detail annotated mammographic patches, respectively. The collection of the mammographic patches is sufficient to reflect the intensity distribution and variation across the whole MIAS database.

3 Results and Discussion

Example segmentation on tissue specific areas can be found in Figure 3 which shows good consistency with the annotated data. Figure 4 shows example segmentation for a mammographic image. The characteristic mixture of breast tissues can be observed. Average relative proportion of the four building blocks and their standard deviation with respect to Tabár's tissue model, based on the segmentation are (in this order [nodular, linear, homogeneous, radiolucent]): Pattern I [19±5.8%, 14±5.3%, 26±15.6%, 41±16.3%]; Pattern II/III [10±9.4%, 12±4.2%, 10±18.8%, 68±21.6%]; Pattern IV [44±11.7%, 11±4.3%, 13±11.5%, 32±12.4%]; and Pattern V [11±3.3%, 7±4.4%, 64±11.0%, 18±10.9%]. The current result has a smaller mean variance of the relative proportion of the four building blocks, in direct comparison with the results presented in [7], which indicates a better match to the Tabár's tissue model. All the available images in the MIAS database were used in the evaluation. Figure 5 (a) shows classification accuracy for discriminating between Tabár's categories. In direct comparison with results presented

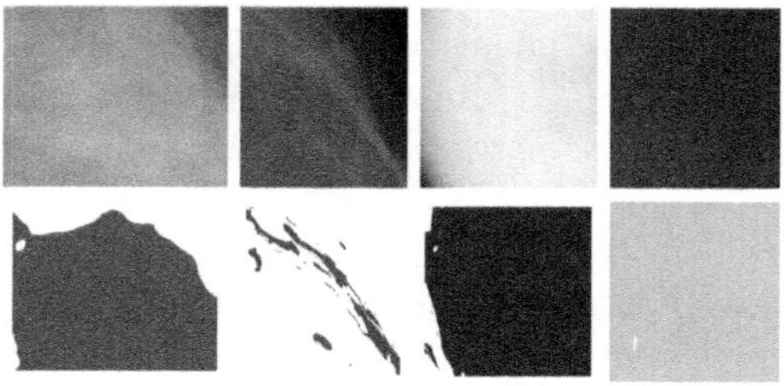

Fig. 3. Example mammographic patch segmentations. Top row from left to right showing example patches for nodular, linear structure, homogeneous and radiolucent. Bottom row showing corresponding segmentations with respect to tissue specific areas (Colour coded areas represent the specific tissue classes while white represents the other tissue classes).

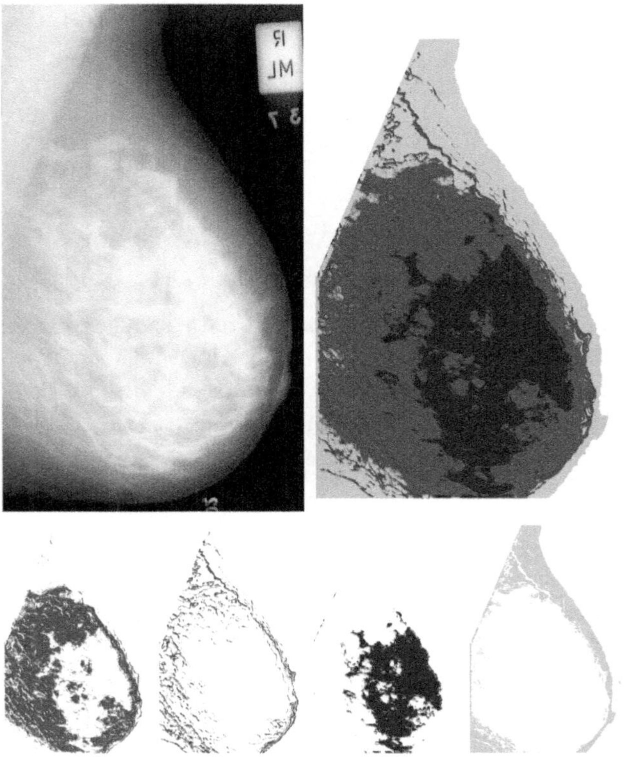

Fig. 4. Example segmentation (mdb108rl, Tabár IV). Tissue composition [42.6%, 8.6%, 25.9%, 22.9%]. Top row showing the mammographic image and its segmentation. Bottom row showing corresponding tissue segmentation from left to right covering: nodular, linear, homogeneous and radiolucent. In each case white regions indicate the other tissue classes (*e.g.* for the nodular case this would constitute the linear, homogenous and radiolucent tissue).

In [7], significant improvements were made in total classification accuracy by 12.8% to 78% ($\kappa = 0.7$). Total classification in the corresponding low (pattern I and II/III) and high (pattern IV and V) categories improved by 15% to 86% ($\kappa = 0.69$). To avoid bias and determine the robustness of the classifier, we also performed classification based on Birads categories. Classification results as seen in Figure 5 (b), achieved in 75% ($\kappa = 0.75$) and 87% ($\kappa = 0.74$) accuracy for Birads four categories, and the corresponding low and high categories, respectively.

All the patches were normalised to zero mean and unit variance during the training stage to reduce intensity distribution variance (*e.g.* contrast and brightness). This could potentially alter the inter and intra class variation, leading to incorrect models from the learning stage.

Misclassification may be caused by the model selection and reduction process which only takes intra class variation into account and fails to remove all the

Tabár Pattern	I	II/III	IV	V	Accuracy
I	102	16	1	0	85.7%
II/III	7	61	11	14	65.6%
IV	9	5	65	1	81.3%
V	1	4	0	23	82.1%

(a)

Birads Pattern	I	II	III	IV	Accuracy
I	41	5	10	3	69.5%
II	1	68	12	5	79.0%
III	0	11	101	29	71.6%
IV	0	0	5	29	85.3%

(b)

Fig. 5. Classification confusion matrix

noise or outliers. The amount of annotated data is relatively small, and may not be adequate to be a strong training dataset. Visual inspection indicates that some of the annotated data are less precise which may be related to hand tremor and other limitations during the manual process, as a consequence the annotation data contain artifacts and noise which is not beneficial for the model selection and reduction.

From a classification point of view, the K-Nearest Neighbour algorithm can be sensitive due to irrelevant features in high dimensional space, misclassification between radiolucent and nodular can be caused by structure and shape features overtaken by intensity features (dominant features). An appropriate dimensionality reduction (i.e.; principal component analysis) may be required to improve the current method. On the other hand, distance weighting allows all training examples to have an influence on the classification of a given pixel. The current classification method is a global method because all training examples were considered, when classifying a new pixel. This may not be the optimal way to deal with inter class variation. A local method cab be used, if only the nearest training examples are considered.

4 Conclusions

The proposed texture signature based mammographic image segmentation has shown good consistency with expert radiologist's annotations, and was able to produce realistic segmentation on tissue specific areas. A quantitative measurement was performed for mammographic image classification based on Tabár and Birads categories over all the available images in the MIAS database. Kappa's coefficient was used as a mean of qualitative evaluation, and indicated strong correlations between classification results and ground truth data. The relative proportion of the four building blocks indicated agreement with Tabár's model. The total classification were 78% and 86% in Tabár's categories and the corresponding low and high categories, respectively; 75% and 87% in Birads categories and the corresponding low and high categories, respectively. The novel aspects in this study are: 1) demonstration for mammographic image segmentation and risk classification using the developed texture signature approach based on the MIAS database; 2) the achieved classification accuracy based on Tabár's categories outperformed previously published results; 3) to our knowledge this is the first attempt to use not only the density information, but also other parenchymal pattern information, as attributes for mammographic risk assessment based on Birads categories; and 4) the produced segmentation results can be used for quantitative measurement of the characteristic mixture of breast tissues in mammographic risk assessment, as well as for quantification of change of relative proportion of different breast tissues. The developed automatic mammographic segmentation method can be found useful as a means of aiding radiologists' estimation in mammographic risk assessment, as well as abnormality detection. Future work will focus on algorithm improvement and possible evaluation in a clinical environment.

References

1. Tabár, L., et al.: Breast Cancer: The Art And Science Of Early Detection With Mamography: Perception, Interpretation, Histopatholigic Correlation, 1st edn. Georg Thieme Verlag (2004)
2. Wolfe, J.N.: Risk for breast cancer development determind by mammographic parenchymal pattern. Cancer 37(5), 2486–2492 (1976)
3. Guliato, D., et al.: Feature extraction from a signature based on the turning angle function for the classification of breast tumors. Journal of Digital Imaging 21, 129–144 (2008)
4. Zwiggelaar, R., Taylor, C.J.: Abnormal masses in mammograms: Detection using scale-orientation signatures. In: Medical Image Computing and Computer-Assisted Intervention, London, UK, pp. 570–577. Springer, Heidelberg (1998)
5. Holmes, A.S., et al.: Transforming pixel signatures into an improved metric space. Image Vision Comput. 20(9-10), 701–707 (2002)
6. Suckling, J., et al.: The mammographic images analysis society digital mammogram database. In: Dance, Gale, Astley, Gairns (eds.) Excerpta medica. International Congress Series, vol. 1069, pp. 375–378. Elsevier, Amsterdam (1994)
7. He, W., et al.: Mammographic segmentation based on mammographic parenchymal patterns and spatial moments. In: 9th International Conference on Information Technology and Applications in Biomedicine, p. 115 (2009)

Stratified Sampling for Case Selection Criteria for Evaluating CAD

Robert M. Nishikawa[1,2] and Lorenzo L. Pesce[1]

[1] Carl J. Vyborny Translational Laboratory for Breast Imaging Research,
Department of Radiology
[2] Committee on Medical Physics
The University of Chicago
5841 S Maryland Ave. MC-2026, Chicago IL 60637, USA
{r-nishikawa,lpesce}@uchicago.edu

Abstract. Ideally, the outcome of any CAD performance assessment should predict how well the system would work if used clinically. In principle, if the selection process draws cases that are "representative" of the general patient population, the study design will be unbiased. In this study we explored the effect of stratified sampling on stand-alone and radiologists' performance using data from an observer study. Although our database was relatively small, 50 cancer cases, no meaningful difference in performance was measured among different stratified sampling schemes or against the whole dataset nor was there any difference in the variance in the measured performance metrics. These results cast doubts on the usefulness of requiring stratified sampling, whose added cost does not seem to be justifiable without empirical evidence. We believe that it is more important to specify how cases should be collected than try to define the range and frequency of the characteristics of patients and cancers to be included the dataset, which we suspect to be prone to actually produce unrealistic samples.

Keywords: computer-aided diagnosis, computer-aided detection, evaluation, observer study, mammography, breast cancer, case selection.

1 Introduction

Case selection is likely to affect strongly the measured performance of CAD schemes and human observers (e.g., radiologists). It is also crucial for interpreting results in the context of clinical utility and ultimately economical viability. When a case sample matches the general population, the stand-alone test performance of CAD can be expected to provide a precise and reasonably unbiased measurement. Creating a database ignoring entirely the composition of the population and how it affected the specific performances is feasible (e.g., a perspective random sample). However, to obtain precise and conclusive measurements for something like breast cancer screening such a dataset is likely to be very large (on the order of thousands or tens of thousands). For an observer study, this is logistically impractical and economically unfeasible, as a very large amount of physician time would be necessary to perform such a study.

J. Martí et al. (Eds.): IWDM 2010, LNCS 6136, pp. 534–539, 2010.

With the exception of one heroic effort (1), the number of cases used in a breast-cancer screening observer study is usually a few hundred or less and typically requires 50-100 cancers to have sufficient statistical power. This sets prevalence around 10-30%, which is substantially higher than the approximately 0.5% observed clinically, but small enough to hope that radiologists will not read too differently from their clinical practice. Given that there might be only 50 cancers in a database, it is impossible for cases to match but a few characteristics of a clinical population and therefore these characteristics have to be chosen carefully. Case selection is critical not only to the outcome of an experiment, but also to its interpretation.

In order to meaningfully reflect the general population, it has been proposed to base the selection of cases on patient characteristics such as breast density and age, and lesion characteristics such as type, histology and size. Then for each characteristic, a distribution of the different subtypes is specified. For example the distribution of cancers could be, by lesion type: 40% masses, 40% clustered microcalcifications, 10% architectural distortions, and 10% asymmetric densities. If the total number of subclasses is large, then to collect 100 cases that fulfill the specified distributions for all the characteristics is likely to be either impossible or require a perhaps unusual combination of factors on a number of cases. This would mean that a lot of reasonable cases might have to be discarded while a number of less common ones might have to be included to reach the desired composition, therefore biasing the results instead of producing a more accurate measurement. Moreover, it is not unreasonable that between 200-1000 cases will need to be collected to obtain the 100 cases needed. This is likely to produce a considerable burden, which should be evaluated in light of ethical, logistical and economical considerations.

2 Method

We used 50 cancer cases from an observer study performed previously. These cases were originally selected because they contained a cancer that was missed clinically, but detectable in retrospect and, therefore, they may not represent the general population (however, this is unlikely to have any relevance with respect to the conclusions of this paper). The characteristics of these cancers are given in Table 1. Five different characteristics were considered in this study:

1. Number of cancers less than 1 cm in size
2. Number of cancers greater than or equal to 1 cm and less than 2 cm in size
3. Number of cancers greater than or equal to 2 cm in size
4. Number of cases with dense breasts
5. Number of cases with either an asymmetric density or an architectural distortion

Using this dataset we created other datasets using a bootstrap-like approach (2) equivalent to considering this dataset as the population. For a given bootstrap sample, a case was randomly drawn from the total 50-case set. The case is returned and another sample is drawn from the 50-case set. This is repeated 50 times creating a new dataset with 50 samples, where some of the cases are represent more than once in the bootstrap dataset. Similarly, for that dataset, readers are randomly selected from total dataset of 8 readers. Therefore, one bootstrap dataset consists of 50 cases and 8 readers, where cases and readers may be represented more than once. In total 100,000 such sets were created.

Table 1. Characteristics of the 50 cancer cases used in this study

Size (cm)	**Number**	**%**
d<1	25	50
1.0<=d<2	14	28
2.0<=d	6	12
Lesion Type		
Mass	28	56
Calcifications	5	10
Other*	17	34
Breast Density		
Dense	27	54
Fatty	23	46

*Other = Architectural distortion or asymmetric density

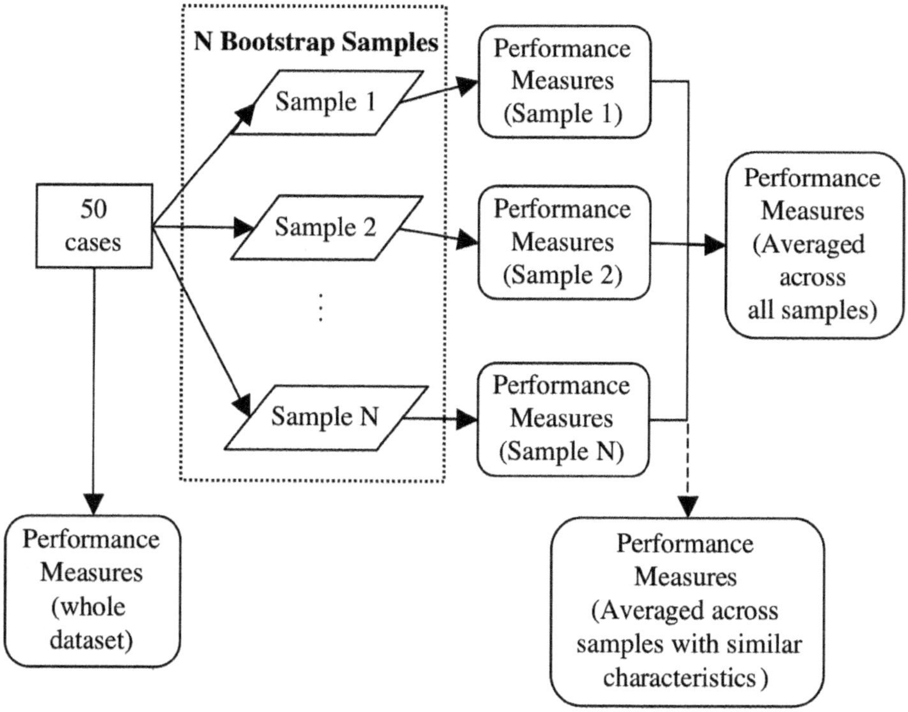

Fig. 1. Flowchart of method used to calculate performance measure for different datasets

Each bootstrap set had a specific distribution of characteristics (such size and breast density). We then stratified these bootstrap datasets to create groups with comparable characteristics (see Fig. 1). Typically, 50-4000 bootstrap samples had comparable characteristics and these were grouped together. For each stratified sample type, composition varied by +/-1 case (+/-2%) within a characteristic. For each stratified sample type, we computed the mean and standard deviation of the stand-alone sensitivity (by case and by image), of the false detection rate and of the radiologists' unaided vs. aided sensitivity.

3 Results

Table 2 shows mean values and standard errors for the standalone sensitivity of the CADe scheme, the false detection rate, the average of the eight radiologists in terms of sensitivity without CADe, with CADe, and the percentage change in sensitivity for the different sample types. The means and standard deviations are nearly the same and we see no reason to expect for any of these slightly different values to be more or less close to the actual population value being the differences among the expected values for each sample types much smaller than the sampling variability.

Table 2. CADe and radiologists' performance for different sets of 50 cases

	Number of Cases per Category						CADe Sensitivity		CADe FP per case	Radiologist Sensitivity		
	Cancer size (cm)			Dense Breasts	ARD + ASD		By Case	By Image		Without CADe	With CADe	increase with CADe
	D<1	1≤d<2	d≥2									
Bootstrap Samples	26	17	8	23	17	Mean	0.52	0.41	0.64	0.60	0.66	8.6%
						σ	0.07	0.05	0.08	0.05	0.05	2.4%
Bootstrap Samples	35	11	5	23	17	Mean	0.52	0.39	0.62	0.61	0.66	8.1%
						σ	0.07	0.05	0.07	0.05	0.04	2.3%
Bootstrap Samples	35	11	5	23	14	Mean	0.52	0.40	0.64	0.61	0.66	7.9%
						σ	0.07	0.05	0.08	0.05	0.05	2.5%
Bootstrap Samples	29	17	5	23	14	Mean	0.53	0.42	0.66	0.61	0.66	8.4%
						σ	0.07	0.05	0.07	0.05	0.05	2.4%
Bootstrap Samples	29	17	5	26	17	Mean	0.53	0.42	0.62	0.60	0.65	9.2%
						σ	0.07	0.05	0.07	0.05	0.05	2.6%
Bootstrap Samples	29	17	5	20	17	Mean	0.52	0.41	0.65	0.61	0.66	8.6%
						σ	0.07	0.06	0.08	0.04	0.04	2.3%
All Bootstrap Samples						Mean	0.52	0.40	0.63	0.61	0.66	8.3%
						σ	0.07	0.06	0.08	0.05	0.05	2.4%
All 50 cases	25	14	6	27	17	Mean	0.52	0.40	0.63	0.61	0.66	8.2%

Table 3 shows the correlation between the average values of a case characteristic and CADe or radiologist performance for the bootstrapped datasets. The correlation appears to be poor between characteristics and performance measures: bootstrap samples with a large fraction of cases with small cancers did not show a major decrease in performance.

Table 3. Correlation coefficients between characteristics of database and CADe and radiologists' performance

Characteristic	CADe Sensitivity		CADe	Radiologist Sensitivity		
	By Case	By Image	FP per case	Without CADe	With CADe	% increase with CADe
average size	-0.06	-0.13	0.02	-0.05	-0.03	0.06
% dense breasts	0.08	0.11	-0.27	-0.19	-0.15	0.19

4 Discussion

While the similarity in the mean values was surprising, it is perhaps more unexpected for samples with the same characteristics to have a standard deviation similar to the whole dataset, which is more heterogeneous. This seems to contradict the very reason why a stratified sampling scheme would be useful in the first place. Further the poor correlation observed between the characteristics of the cases and CADe or radiologist performance seems to indicate that using their values for selecting cases for evaluating CADe may have a much smaller impact than expected, perhaps even nearly negligible.

The database used in this experiment was small, therefore the results may not hold for a larger dataset. Similarly, our observer study had only 8 readers who may not be representative of the general population of MQSA radiologists. However, assembling a larger sample of paired reader-cancer-case samples would require an unusually large reader study.

We believe the size of our study and the composition of our dataset (clinically missed cancers) may explain the lack of effect on stand-alone performance. However, the literature on to the effect of breast density on CADe performance appears to be conflicting at this time, suggesting that further studies are necessary.

However, for observer studies, we believe our results have important ramifications. In an observer study of CADe, only a small fraction of the cases are important in determining the effectiveness of CADe. These are the cases that are missed by the readers in the unaided condition and are found when CADe is used. These cases are most likely to be difficult cases, such as those we used in our experiment. That is, if a larger, more general dataset (e.g., 200 consecutive cancer cases) were used, we argue that the cases that will power the difference between with and without CADe will be the cases represented in our dataset. The other 150 cancers will affect the overall sensitivity and area under the ROC curve (AUC), but not the difference in sensitivity and to a lesser degree the difference in AUC between with and without CADe. Therefore, selecting cases so as to represent the general patient population has merit, at least in principle, but we believe that the burden it generates when designing a CADe observer study is unjustified.

5 Conclusions

Based on our dataset, we conclude the following:

1. The cancer size and breast density does not seem to be a reliable predictor of the performance either of stand-alone CADe or of radiologists reading with or without CADe. Thus, it is not clear how to use these characteristics to select a dataset for clinical evaluations.

2. We could not find evidence that sampling stratified by the characteristics we considered is capable of reducing the standard deviation of performance measures. Given the burden of case collection implied by this type of approach, careful consideration should be applied before using it.

3. Since the performance of radiologists and the CADe scheme showed little dependence upon the exact characteristics of the cases in the database, we believe that it is more important to specify how the cases were selected, in order to prevent selecting a sample whose performance might relate poorly with the actual population for which a clinical question needs to be answered.

For stand-alone testing, a larger dataset with cases more representative should be performed. For observer studies, we believe our dataset is representative of the most relevant cases (i.e., those that will allow a difference to be measured) and therefore we find no evidence supporting the use of stratified sampling in CADe observer studies.

References

1. Efron, B., Tibshirani, R.J.: An Introduction to the Bootstrap. Chapman and Hall, New York (1993)
2. Gilbert, F.J., Astley, S.M., McGee, M.A., Gillan, M.G., Boggis, C.R., Griffiths, P.M., Duffy, S.W.: Single reading with computer-aided detection and double reading of screening mammograms in the United Kingdom National Breast Screening Program. Radiology 241, 47–53 (2006)

Breast Shapes on Real and Simulated Mammograms*

Christine Tanner[1,2], John H. Hipwell[2], David J. Hawkes[2], and Gábor Székely[1]

[1] Computer Vision Laboratory, ETH Zürich, 8092 Zürich, Switzerland
[2] Centre for Medical Image Computing, University College London, UK
tannerch@vision.ee.ethz.ch

Abstract. We investigate the need for anisotropic materials when simulating X-ray mammograms from real 3D MR breast images employing biomechanical models. We previously observed on 3D MRI that the breast in the prone position elongates very little in the anterior-posterior direction even when applying large lateral-to-medial compressions. Improved accuracy was achieved for these 3D deformations when employing transverse-isotropic materials, where the tissue in anterior-posterior direction was stiffer than in the other two directions. We investigate here whether this also holds when simulating cranio-caudal mammograms where the patient is standing. The realism of the simulated breast compressions was judged by comparing the anterior breast shapes of simulated and real mammograms. The anterior breast shape was quantified by the log ratio of the medial-lateral to anterior-posterior diameter of an ellipse fitted to the anterior breast edge. The breast shapes on real and simulated mammograms were on average statistically significantly different (0.48 versus 0.27, P<0.01) when employing isotropic materials. No such difference was observed for transverse-isotropic materials (0.47). The estimated breast thickness, to achieve the breast shape observed on the corresponding real mammogram, was on average very unrealistic for isotropic materials (8.7 mm) while reasonable for anisotropic materials (49.5 mm).

1 Introduction

Relating breast MR images to mammograms is an important but difficult task. Simulation of mammograms from MRI has been proposed as an important step in assessing registration methods for this purpose [4], and to learn more about breast deformations [13]. Several groups have proposed the use of finite element (FE) models for simulating breast compressions [10,9,17,8,3,13]. Evaluation of these methods for large breast compressions (> 38%) have been limited to one recent study [14]. There 3D MR breast images were acquired before

* We would like to thank the Radboud University Nijmegen Medical Centre and the UK MR breast screening study for providing images. We acknowledge funding from EPSRC (EP/E031579/1) and the European 7th Framework Program, HAMAM, ICT-2007.5.3.

J. Martí et al. (Eds.): IWDM 2010, LNCS 6136, pp. 540–547, 2010.

and after lateral-to-medial breast compressions. Analysis of these showed an anisotropic deformation behaviour with the breast expanding less in anterior-posterior (AP) than in the superior-inferior (SI) direction. The study also showed for the first time that FE models employing transverse-isotropic materials can more accurately predict these deformations than models with isotropic material properties.

The study was based on MR images of the patient in the prone position, where the breast is stretched in the anterior-posterior direction due to gravity. The anisotropic behaviour could therefore be an effect of the pre-stretching of Cooper's ligaments. During mammography, the patient is standing and the breast is pulled away from the chest by the radiographer to improve visibility of overlying structures, which might produce a similar effect. This study investigates if the anisotropic deformation behaviour still exists for deformations between prone MR breast images and mammograms.

2 Method

Materials. The first cohort consisted of digital mammograms from a data set of 79 patients from the Radboud University Nijmegen Medical Centre. Breast thickness had been recorded for these image. The second cohort comprised MRI images acquired for the UK MR breast screening study (MARIBS) [6] for 20 women. Corresponding digitized mammograms were available for ten of these cases. Breast thickness had not been recorded for these. The MR images had a voxel dimension of $1.33 \times 1.33 \times 2.5$ mm^3 and a coronal slice orientation. All MRIs were obtained with the women in the prone position.

Biomechanical Model. Breast compressions were simulated for 20 subjects as described in [13,14]. In short we employed high resolution 10-noded tetrahedral meshes, stepwise (0.5%) boundary conditions to model plate contact, rezoning to cope with badly shaped elements, FE package ANSYS and up to 5 tissue types (fat, glandular, tumour, muscle, skin). Isotropic linear and non-linear material models were selected according to published values [12,5,16,11,15], see Table 1. The configuration of the transverse isotropic linear materials was based on the optimal material parameters determined during the MRI compression study [14]. These were mapped onto the new cohort on the basis of similar percentage in glandular tissue, resulting in the material definitions shown in Table 2. The same boundary conditions were employed for the two material models. The breast thickness after compression was selected from a normal random distribution (51 ± 13 mm) [18,1,2]. Breast rolling parameters, where the breast is rotated around the AP-direction before compression, were sampled from a uniform random distribution (range [-11,11]^o) [13]. Simulated mammograms were created by projecting the intensity-inverted masked MRI of the compressed breast onto a 2D image with an isotropic resolution of 0.5 mm.

Table 1. Overview of isotropic material models. Tissue variations were simulated by randomly selecting a breast tissue model (B1-B4), one of the related tumour model (T1-T5), a muscle model (M1,M2) and factors (f_f, f_g, f_t, $f_m \in \{-1, -0.5, ..., 1\}$) which determine the deviation from the mean Young's moduli values [13].

	Young's modulus E in kPa for strain $e \geq 0$. $E(e) = E(-e)$ for $e < 0$	
	B1 [12]	B2 [5]
Fat	$1 + 0.5f_f$	$18.5 + 7f_f$
Glandular	$1.5 + 0.5f_g$	$(27.50 + 9.16f_g)\exp(3.64e)$
Tumour T1	$3.6 + 0.5f_t$	$(82.67 + 24.80f_t)\exp((5.16 + 0.22f_t)e)$
Tumour T2	$10.4 + 1.9f_t$	$(10.91 + 1.62f_t)\exp((16.59 + 0.22f_t)e)$
Tumour T3	$16.5 + 3.5f_t$	$(53.45 + 21.16f_t)\exp((11.08 + 0.92f_t)e)$
	B3 [16]	B4 [11]
Fat	$(4.46 + 2.35f_f)\exp((7.4 + 4.0f_f)e)$	$3.25 + 0.91f_f$
Glandular	$(15.1 + 6.75f_g)\exp((12.3 + 7.4f_g)e)$	$3.24 + 0.61f_g$
Tumour T1	$(17.76 + 4.2f_t)\exp((21.4 + 2.8f_t)e)$	$6.41 + 2.86f_t$
Tumour T2	$(37.57 + 6.05f_t)\exp((20 + 1.4f_t)e)$	$10.4 + 2.6f_t$
Tumour T3	$(33.78 + 6.15f_t)\exp((24.08 + 5.5f_t)e)$	$16.38 + 1.55f_t$
Tumour T4	-	$19.99 + 4.2f_t$
Tumour T5	-	$42.52 + 12.47f_t$
Muscle M1 [15]	$2.25 + 0.25f_m + (10.85 + 0.33f_m)e + (61.88 + 0.5f_m)e^2$	
Muscle M2 [15]	$8.12 + 0.08f_m + (21.05 + 0.08f_m)e + (242.93 + 0.08f_m)e^2$	

Table 2. Transverse isotropic material properties for subjects S1 to S20 from optimal results from [14]. Parameters are expressed as ratios of the Young's modulus (E), namely the ratio to the coronal plane ($r_a = E_{AP}/E_{SI} = 2^{m_a}, E_{ML} = E_{SI}$) and the ratio to fat ($r_k = E_k/E_{\text{fat}} = 2^{m_k}$) for glandular tissue ($r_g$), tumour ($r_t$), muscle ($r_m$) and skin ($r_s$). #Case where the stated value was used instead of the optimal $m_a = -1$ to have stiffer tissue in AP-direction for all cases. *Case without a tumour in [14]. Tumour variation was modelled by $m_t = m_g + m_r$ with m_r sampled from a uniform random distribution (range [-1,11]).

	S1	S2	S3	S4	S5	S6	S7	S8	S9	S10	S11	S12	S13	S14	S15	S16	S17	S18	S19	S20
m_a	1	6	1	#1	2	4	2	3	3	#4	9	3	2	2	3	3	3	2	3	3
m_g	-8	1	0	-6	1	1	0	4	-5	-1	1	0	-1	-1	3	0	0	1	0	3
m_t	8	*1	*11	8	*10	*6	*9	*8	-2	*2	*11	*9	-2	-2	*13	*-1	*1	*11	*9	*5
m_m	-1	-3	0	-2	-4	-5	0	4	0	-1	1	-6	-1	-1	3	-3	-3	-7	-3	3
m_s	2	7	5	5	3	5	5	8	9	8	8	5	7	7	9	5	6	3	6	6

Breast Shape. The digital mammograms were downsampled from 0.1 mm to the same image resolution as the simulated mammograms. The breast was segmented on real and simulated mammograms employing Otsu's method [7]. Visual inspection showed that this worked very well for all cases. The breast shape was measured by first fitting an ellipse to the anterior breast edge up to a constant depth D as illustrated in Fig. 1. After visual inspection, D was set to 19 mm to avoid problems with images where the true breast edge goes off the film and with small breasts. Using the coordinate values of the fitted points, the shape was then quantified by the ratio $e = d_{ML}/(2d_{AP})$, where d_i denotes the maximum range in the ith direction and the medial-lateral (ML) and AP-direction was assumed to be aligned with the image axes. Since e is a ratio, we assess $\log(e)$.

Assessment. The robustness of the shape measure with respect to variations in the set-up (breast rolling, in-plane rotation, etc.) was investigated by comparing the values of the left and right breast for 79 patients. Then we compared the shape distribution of three populations, namely 158 real mammograms and 20 simulated mammograms when using isotropic or anisotropic materials. For 10 cases, where we had mammograms in addition to the simulations, we estimated the breast thickness required to achieve the breast shape observed on the real mammogram. We tested for statistically significant differences in mean values using two-sided (paired) t-test.

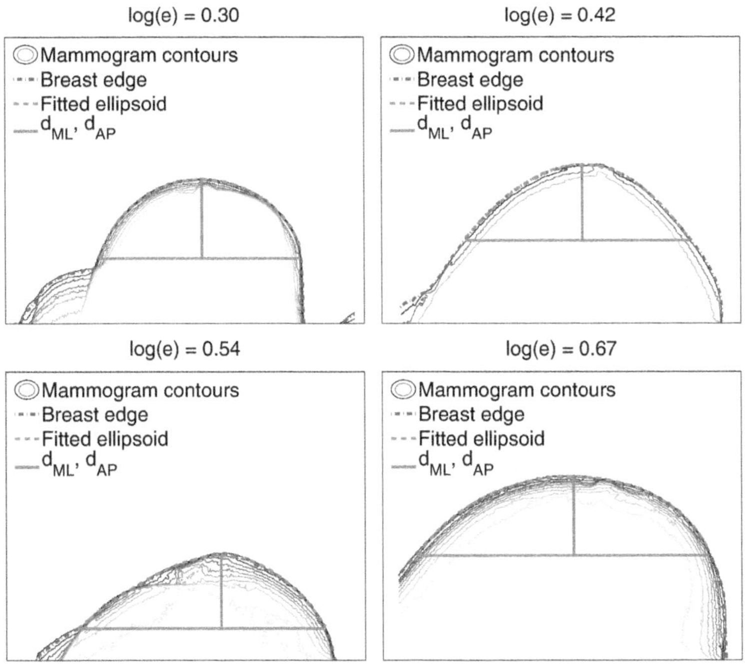

Fig. 1. Examples of anterior breast shapes from real mammograms. An ellipse was fitted to the breast edge up to a depth of 19 mm. The shape was then quantified by log(e), where $e = d_{ML}/(2d_{AP})$.

3 Results

Fig. 1 shows a selection of shapes encountered. The left and right breast were very similar in breast thickness, volume of the anterior breast region (defined by the breast area up to depth D times the breast thickness) and anterior shape ($\log(e)$), see Table 3. The statistics for the three population (real, iso, aniso) are summarized in Table 4. Isotropic materials produced simulated mammograms which were statistically significantly different in mean breast shape (0.48 versus 0.27, $P<0.01$), while transverse isotropic materials did not.

Fig. 2 illustrates the change in breast shape during compression for simulations using isotropic and transverse isotropic materials. Isotropic materials seem unable to recover the breast shape seen on the corresponding real mammogram, while transverse isotropic materials reproduce the real breast shape more accurately. When fitting a quadratic function to the shape values with respect to the breast thickness, see Fig. 3, it can be observed that in most cases the shape of the real mammogram is only achieved for very unrealistic breast thicknesses (mean 8.7 mm, range [-9.7,28.5] mm) including impossible negative values. This is not the case for transverse isotropic materials (mean 49.5 mm, range [34.4,65.7] mm).

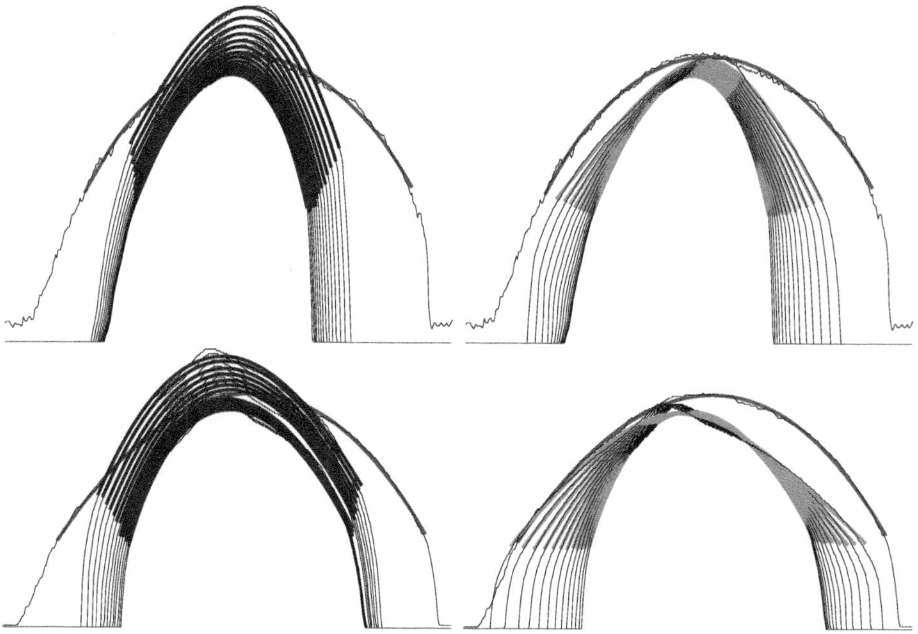

Fig. 2. Illustration of the change in breast shape during simulated compression for isotropic (left) or transverse isotropic materials (right) in comparison to the breast shape of the corresponding real mammogram (red) for two cases (top, bottom). The anisotropic tissue was (top) 2^2 and (bottom) 2^9 times stiffer in anterior-posterior direction.

Table 3. Mean and standard deviation (a,b) of features from X-ray mammograms for (a) left and (b) right breast and (c) of difference in features for contralateral breasts. The distributions had no statistically significant different mean values (paired t-test).

	(a) left	(b) right	(c) left - right
Thickness (mm)	54.67 ± 13.19	55.03 ± 12.77	-0.35 ± 4.55
VolumeRegion (cm^3)	271.10 ± 79.60	271.69 ± 77.88	-0.59 ± 24.50
Shape ($\log(e)$)	0.48 ± 0.07	0.48 ± 0.07	0.00 ± 0.04

Table 4. Mean and standard deviation of features from (a) 132 X-ray mammograms and (b,c) 20 simulated mammograms from MRIs and biomechanical simulations using (b) isotropic or (c) transverse-isotropic material models. †Distributions with statistically significant different mean values to (a) at the $P = 0.01$ level (t-test).

	(a) real	(b) iso	(c) aniso
Thickness (mm)	54.85 ± 12.94	55.51 ± 11.72	55.51 ± 11.72
VolumeRegion (cm^3)	271.40 ± 78.49	226.77 ± 75.90	258.48 ± 78.62
Shape ($\log(e)$)	0.48 ± 0.07	0.27 ± 0.20†	0.47 ± 0.17

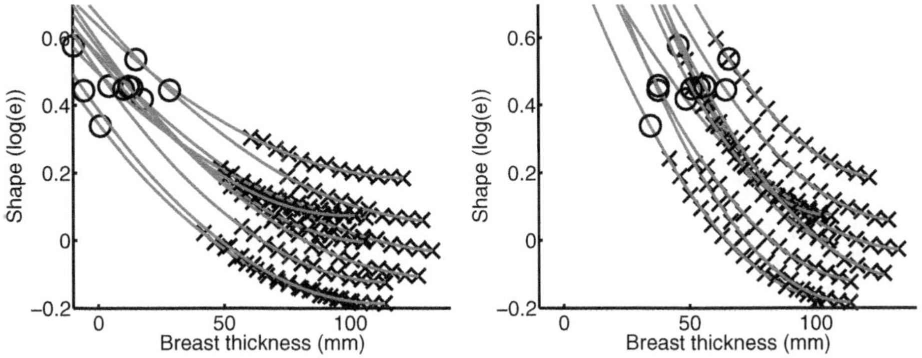

Fig. 3. Relationship between the breast shape and the progression of the simulated compression for (left) isotropic and (right) transverse isotropic materials. A quadratic function (line) was fitted to the thickness-versus-shape values (crosses) measured during a FE simulation at 4% compression intervals. Circles depict the estimated breast thickness for the breast shape of the corresponding real mammogram predicted from the quadratic function. Very unrealistic and even impossible negative breast thicknesses are estimated for models with isotropic materials.

4 Discussion

We compared the breast shape of real and simulated mammograms on a population basis as well as for corresponding images from the same subject. Simulated shapes were significantly different when employing isotropic materials, but not for transverse isotropic materials with an increased stiffness in the AP direction. The simulation progression showed that impossible or unrealistic high compressions would be required for isotropic models to achieve the breast shape of the corresponding mammogram.

These results confirm for the first time the need for material models which are stiffer in AP direction when simulating mammograms from MR images by pure compression. Alternatively, simulations might be based on the more complicated scenario of first removing gravity and then applying the compression for isotropic materials.

For the registration of MRI and mammograms, Ruiter required additional surface displacements to compensate for insufficient elongation [9]. Anisotropic materials are likely to remove this need. Furthermore, the shape measure could prove useful for fast selection of the initial material properties.

References

1. Boone, J.M., Lindfors, K.K., Cooper III, V.N., Seibert, J.A.: Scatter/primary in mammography: Comprehensive results. Medical Physics 27, 2408 (2000)
2. Brnić, Z., Hebrang, A.: Breast Compression and Radiation Dose in Two Different Mammographic Oblique Projections: 45 and 60°. European Journal of Radiology 40, 10–15 (2001)
3. Chung, J., Rajagopal, V., Nielsen, P., Nash, M.P.: Modelling Mammographic Compression of the Breast. In: Proc. Int. Conf. on Medical Image Computing and Computer-Assisted Intervention, p. 758 (2008)
4. Hipwell, J., Tanner, C., Crum, W., Schnabel, J., Hawkes, D.: A New Validation Method for X-ray Mammogram Registration Algorithms using a Projection Model of Breast X-ray Compression. IEEE Transactions on Medical Imaging 26(9), 1190–1200 (2007)
5. Krouskop, T.A., Wheeler, T.M., Kallel, F., Garra, B.S., Hall, T.: Elastic Moduli of Breast and Prostate Tissues Under Compression. Ultrasonic Imaging 20, 260–274 (1998)
6. Leach, M., Boggis, C., Dixon, A., Easton, D., Eeles, R., Evans, D., Gilbert, F., Griebsch, I., Hoff, R., Kessar, P., et al.: Screening with Magnetic Resonance Imaging and Mammography of a UK Population at High Family Risk of Breast Cancer: A Prospective Multicentre Cohort Study (MARIBS). The Lancet 365, 1769–1778 (2005)
7. Otsu, N.: A threshold selection method from gray-level histograms. Automatica 11, 285–296 (1975)
8. Pathmanathan, P., Gavaghan, D., Whiteley, J., Brady, M., Nash, M., Nielsen, P., Rajagopal, V.: Predicting Tumour Location by Simulating Large Deformations of the Breast Using a 3D Finite Element Model and Nonlinear Elasticity. In: Barillot, C., Haynor, D.R., Hellier, P. (eds.) MICCAI 2004. LNCS, vol. 3217, pp. 217–224. Springer, Heidelberg (2004)

9. Ruiter, N.V.: Registration of X-ray Mammograms and MR-Volumes of the Female Breast based on Simulated Mammographic Deformation. Ph.D. thesis, Universität Mannheim (2003)
10. Samani, A., Bishop, J., Yaffe, M.J., Plewes, D.B.: Biomechanical 3D Finite Element Modeling of the Human Breast Using MRI Data. IEEE Transactions on Medical Imaging 20(4), 271–279 (2001)
11. Samani, A., Zubovits, J., Plewes, D.: Elastic moduli of normal and pathological human breast tissues: an inversion-technique-based investigation of 169 samples. Physics in Medicine and Biology 52(6), 1565–1576 (2007)
12. Sarvazyan, A., Goukassian, D., Maevsky, E., Oranskaja, G.: Elastic Imaging as a New Modality of Medical Imaging for Cancer Detection. In: Proc. Int. Workshop on Interaction of Ultrasound with Biological Media, Valenciennes, France, pp. 69–81 (1994)
13. Tanner, C., Hipwell, J.H., Hawkes, D.J.: Statistical Deformation Models of Breast Compressions from Biomechanical Simulations. In: Krupinski, E.A. (ed.) IWDM 2008. LNCS, vol. 5116, pp. 426–432. Springer, Heidelberg (2008)
14. Tanner, C., White, M., Guarino, S., Hall-Craggs, M.A., Douek, M., Hawkes, D.J.: Anisotropic Behaviour of Breast Tissue for Large Compressions. In: Proc. Int. Symposium on Biomedical Imaging: From Nano to Macro, pp. 1223–1226 (2009)
15. Van Loocke, M., Lyons, C., Simms, C.: A validated model of passive muscle in compression. Journal of Biomechanics 39(16), 2999–3009 (2006)
16. Wellman, P.S.: Tactile Imaging. Ph.D. thesis, Harvard University (1999)
17. Yin, H.M., Sun, L.Z., Wang, G., Yamada, T., Wang, J., Vannier, M.W.: ImageParser: A Tool for Finte Element Generation from Three-Dimensional Medical Images. BioMedical Engineering OnLine 3(31) (2004)
18. Young, K.C., Burch, A.: Radiation Doses received in the UK Breast Screening Programme in 1997 and 1998. The British Journal of Radiology 73, 278–287 (2000)

Regularization Parameter Selection in Maximum a Posteriori Iterative Reconstruction for Digital Breast Tomosynthesis

Anna K. Jerebko, Markus Kowarschik, and Thomas Mertelmeier

Siemens AG, Healthcare Sector, Allee am Röthelheimpark 2,
91052 Erlangen, Germany
{Anna.Jerebko,Markus.Kowarschik,Thomas.Mertelmeier}@siemens.com

Abstract. The method presented in this paper addresses the problem of regularization parameter selection in maximum a posteriori iterative reconstruction for digital breast tomosynthesis. The method allows analytically deriving the combination of prior function parameters for noise level expected in the reconstruction without priors and estimated breast density such that it effectively controls the level of noise while preserving the edges of breast structures. Results show reduced noise level and improved contrast to noise ratio compared to filtered back projection and maximum–likelihood iterative reconstruction without penalizing term.

Keywords: breast tomosynthesis, statistical iterative reconstruction, prior functions, maximum a posteriori iterative reconstruction.

1 Introduction

Digital Breast Tomosynthesis (DBT) suffers from incomplete data and poor quantum statistics limited by the total dose absorbed in the breast. Popular filtered backprojection reconstruction methods (FBP) provide high contrast and excellent detail level in reconstructed images [1] but lose the information about relative tissue density. This happens due to removal of low frequency components with some filter kernels (see Fig.1).

Another problem is that out-of-plane artifacts are enhanced by filtering together with the image features. Hence, an appropriate statistical iterative reconstruction approach that maximizes similarity between the calculated and measured projections and enables noise reduction through introduction of priors [2] may have some advantages. One of the disadvantages of this approach is that it requires to empirically determine the optimal parameters of the prior function that usually involves a grid search over a range of values, as it was implemented in [3] and [4]. The latter is not practical for high resolution breast imaging, especially when multiple prior functions need to be evaluated.

In this study we propose a method for analytical statistical criteria based selection of prior function parameters for iterative maximum a posteriori (MAP) statistical reconstruction algorithm for DBT.

J. Martí et al. (Eds.): IWDM 2010, LNCS 6136, pp. 548–555, 2010.

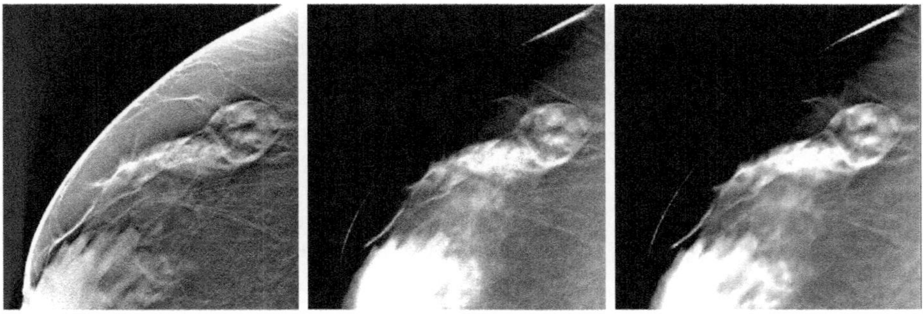

a) b) c)

Fig. 1. Central slice segment of a breast volume a) FBP reconstruction that shows high contrast, high resolution but the brightness of voxels corresponding to peripheral fatty tissue is not different from the denser tissue in the center of the breast. b) Iterative ML-gradient reconstruction without priors and noise reduction gives a more realistic density representation of the breast tissue. c) Iterative MAP reconstruction with Geman prior.

2 Method

The goal of the maximum likelihood (ML) method [2] is to find the expectation value μ (breast volume attenuation coefficients) that maximizes the log-likelihood function $L(\mu)$:

$$\mu_{\max} = \arg\max_{\mu} \{ L(\mu) \}.$$

ML reconstruction generally yields reasonably good results in DBT and converges in 4-5 iterations, but produces overly noisy images without the use of priors. In [2], typical ML iteration schemes are modified to take into account a smoothing prior. The log-likelihood $L(\mu)$ is then changed to the log-posterior $\Phi(\mu) = L(\mu) - \beta U(\mu)$, where $U(\mu)$ is a smoothing prior function penalizing the differences in values of neighboring pixels, and $\beta > 0$ is the regularization parameter. We refer to this approach as maximum a posteriori (MAP) method. For the case of the ML-gradient reconstruction method, for example, Lange et al [2] suggest the following Newton based update method with the simple quadratic smoothing prior $U(\mu)$:

$$\mu_j^{n+1} = \mu_j^n + \frac{\mu_j^n}{\sum_i Y_i l_{ij} + \beta \mu_j^n \frac{\partial^2}{\partial \mu_j^2} U(\mu^n)} \frac{\partial}{\partial \mu_j} \Phi(\mu^n). \quad (1)$$

In this update scheme, j refers to the voxel index, l_{ij} represents the intersection length of the i-th ray with voxel j, and Y_i denotes the measured photon count in projection i.

In the next section we describe a practical method which allows analytically deriving the parameters of the prior function $U(\mu)$. The proposed method can be applied to a variety of prior functions. In transmission tomography and breast imaging particularly, it is important to preserve edges while reducing noise. That is why we choose an edge preserving generalized Geman prior function with convexity condition [5] as an example:

$$U(\mu) = \sum_j \sum_k \frac{\Delta\mu_{jk}^{\ 2}\sigma^m}{2\left(\sqrt{\Delta\mu_{jk}^{\ 2}/2+\sigma^2/2}\right)^m};\tag{2}$$

where $\Delta\mu_{jk} = \mu_j - \mu_k$ is the difference in intensity of the neighboring voxels, $k \in N_j$ - voxels in the neighborhood of voxel *j*, and $m \leq 16/17$ in order to maintain convexity of the prior function and keep its second partial derivative positive [5].

It is important to select an edge-preserving prior function for DBT to be able to suitably visualize clinical features with very important high frequency components, such as calcifications and spiculated masses. The convexity condition, on the other hand, ensures convergence of the algorithm. The selected Geman prior function satisfies these requirements, although we do not claim that it is the only possible choice.

The parameter σ is selected such that the first derivative of the prior function reaches its maximum (for non-convex optimization only) or the highest curvature point (for convex functions, see Fig.2) at values of $\Delta\tilde{\mu}$ that correspond to the expected noise level. Naturally, $\Delta\tilde{\mu}$ must be lower than the weakest gradient of the edges that must be preserved. The penalty for voxel intensity deviations below or equal to the highest expected noise level will be "proportional" to those deviations. At the same time, the edges with differences in voxel intensity higher than that level will only receive a small (same as the highest expected noise) amount of correction or penalty. For non-convex optimization methods it is possible to choose a Geman prior function with *m=2* which allows penalty for edge voxels to be close to 0.

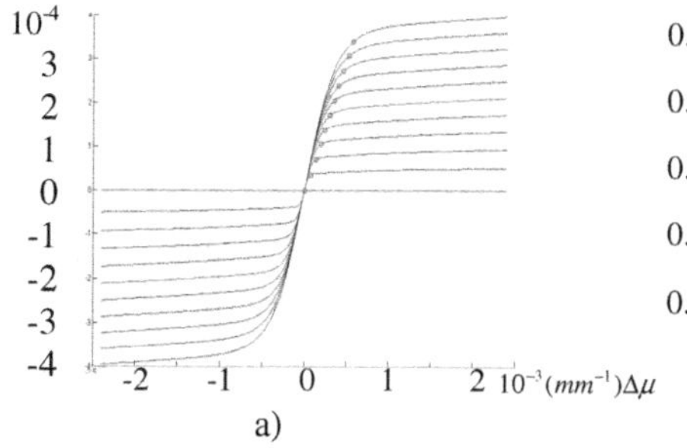

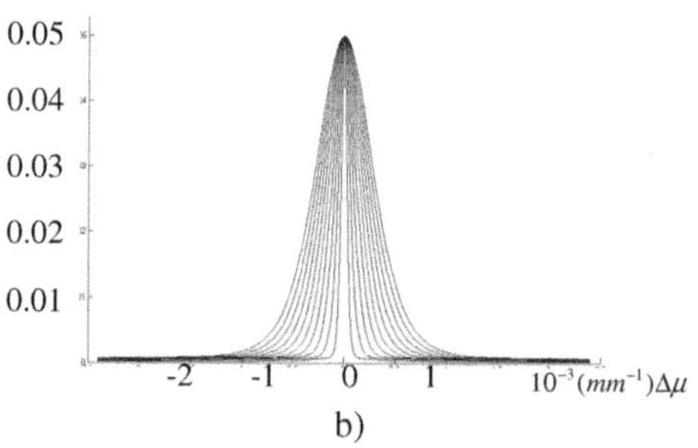

Fig. 2. a) 1st derivative of prior function for $0 \leq \sigma \leq 0.0006$, b) 2nd derivative is positive definite

Furthermore, the parameter β is explicitly computed for the expected noise level and the mean breast volume attenuation estimate $\overline{\mu}$. For the ML-gradient optimization method (1) introduced in [2], β is computed in the following way:

$$\beta = \frac{\Delta\tilde{\mu} \cdot \overline{\sum_i Y_i l_{ij}}}{\bar{\mu} \cdot \frac{\partial}{\partial \mu_j} U(\mu)\Big|_{\Delta\mu_j = \Delta\tilde{\mu}} - \Delta\tilde{\mu} \cdot \bar{\mu} \cdot \frac{\partial^2}{\partial \mu_j^2} U(\mu)\Big|_{\Delta\mu_j = \Delta\tilde{\mu}}}. \tag{3}$$

The average value $\overline{\sum_i Y_i l_{ij}}$ is computed only once before the iterations. The noise level $\Delta\tilde{\mu}$ could be predicted for a given x-ray spectrum as a function of the breast thickness and the tube load (mAs) using regression methods. Generally, the noise level can be kept constant for any breast thickness via selection of the tube load and the photon energy (keV). $\bar{\mu}$ could be predicted using regression methods from photon energy and breast density estimate obtained from segmented original projections using existing methods [6].

The convex optimization update formula [2] also suggested in (1) yields following equation for β:

$$\beta = \frac{\Delta\tilde{\mu} \cdot \sum_i l_{ij} \langle l_i, \bar{\mu} \rangle d_i e^{-\langle l_i, \bar{\mu} \rangle}}{\bar{\mu} \cdot \frac{\partial}{\partial \mu_j} U(\mu)\Big|_{\Delta\mu_j = \Delta\tilde{\mu}} - \Delta\tilde{\mu} \cdot \bar{\mu} \cdot \frac{\partial^2}{\partial \mu_j^2} U(\mu)\Big|_{\Delta\mu_j = \Delta\tilde{\mu}}}. \tag{4}$$

3 Results

The breast tomosynthesis system used in our experiments was a prototype based on the Siemens MAMMOMAT *Novation*^{DR} modified to acquire 25 low-dose projections over an angular range of approximately 45° using a stationary amorphous selenium flat-panel detector. The x-ray tube motion is continuous to avoid mechanical instabilities. During the integration phase of the detector cycle, the x-ray generator is pulsed to acquire the projection data. The system was described in detail in our prior work [1]. The clinical data used for this work was acquired at Duke University Medical Center and Malmö University Hospital.

Our experiments with patient data and phantoms showed that the combination of parameters β and σ effectively controls the level of noise while preserving the boundaries (see Fig. 1c, 3, 6) for both convex and gradient optimization in MAP reconstruction. Whereas ML reconstruction without prior already gives a more realistic density grayscale (see Fig 1.) than FBP, the difference in noise and CNR is not much different. Only when using a penalizing term for noise regularization, SNR and CNR can be improved. We analyzed contrast and CNR in a patient dataset, by measuring it in the smallest visible vessel and a featureless background spot in the central slice.

As it could be seen from the Fig. 4a and b, the relative contrast improves with more iterations, but CNR goes down for ML methods because of the increasing noise level. CNR increases with more iterations for both convex and gradient implementations (see Fig. 4b). Visually, the results of the convex and the gradient implementation look very similar, although the convex method converges more robustly (without

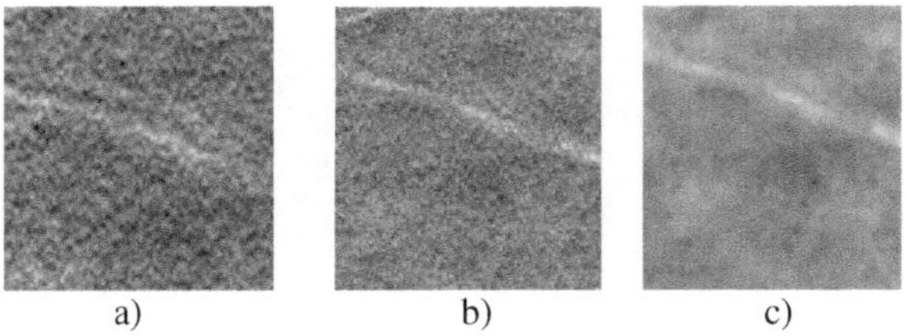

Fig. 3. Faintest smallest visible vessel (90x90 pixels ROI) in the central slice on a featureless background a) FBP results. b) 5th iteration of ML gradient reconstruction without priors. c) 5th iteration of MAP gradient reconstruction with Geman prior.

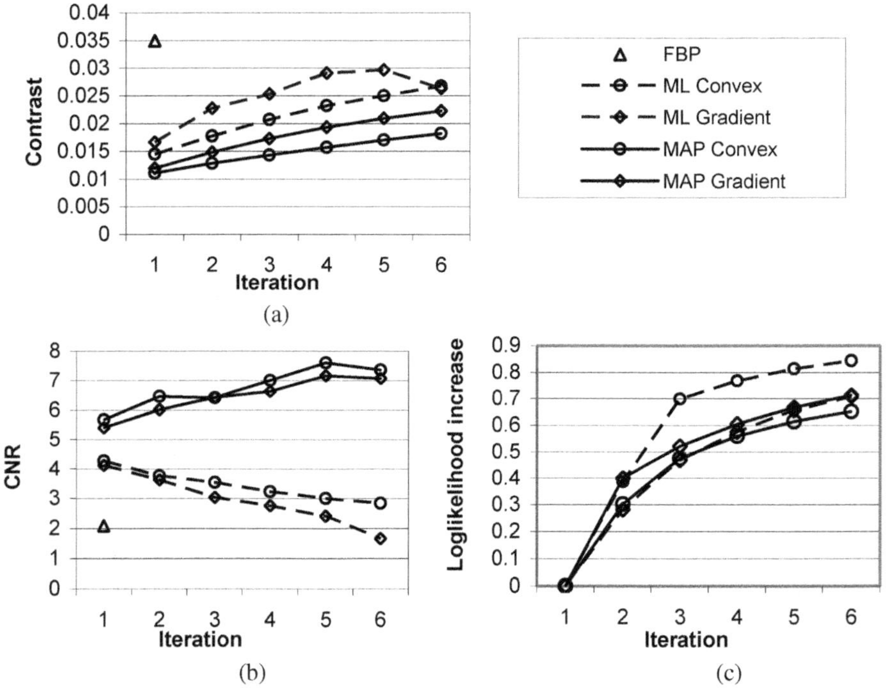

Fig 4. a) Contrast and b) CNR measured in the small vessel and the background featureless spot throughout the iterations. c) Convergence of all investigated algorithms.

visible oscillations) than the gradient algorithm. Although the ML-convex method converges faster than the ML-gradient method (Fig. 4c), this difference disappears as soon as priors are introduced in both implementations.

The spatial resolution was assessed using a German QC phantom [7]. The phantom contains several inserts, one of which is designed to assess spatial resolution of the system using a high-contrast bar pattern (see Fig 5). This experiment was performed with 'extreme' smoothing, when $\Delta\tilde{\mu}$ is set exactly to the noise level measured in the

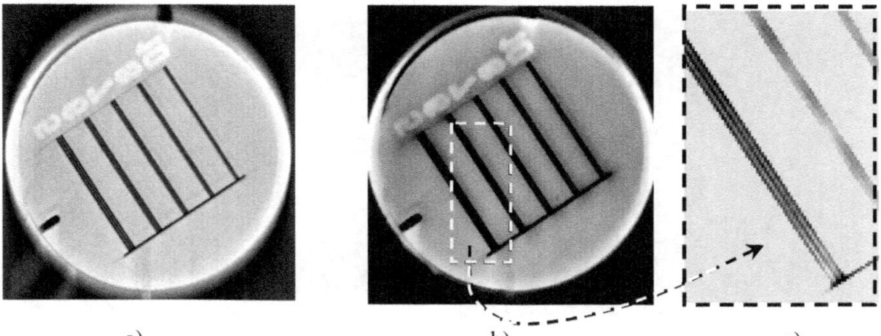

Fig. 5. Magnified image from a lead bar phantom reflecting spatial resolution of the system reconstructed using a) ML-convex method (i.e., without prior). b) MAP-convex (i.e., with prior). c) Enlarged fragment of MAP-convex reconstruction with window-level allowing visualization of the lead bar-slit pattern.

featureless spot of the phantom insert. The visual assessment of the lead-bar phantom reconstructions confirmed that resolution is actually not lost with the introduction of smoothing priors. Boundaries of fine structures in the image are preserved while the image looks considerably smoother.

This strategy, however, should be applied with care when reconstructing real patient images. When $\Delta\tilde{\mu}$ is set exactly to the expected noise level, the reconstructed images look too artificially smooth and 'washed-out', like the examples in Fig.3c and 6c. Some microcalcifications in fact may have gradient level similar to that of noise voxels (see calcification indicated by arrow in Fig. 6b).

With the use of smoothing priors their intensity could be reduced to unacceptable level (Fig. 6c). In practice, in order to preserve important clinical features $\Delta\tilde{\mu}$ should be set to:

$$\Delta\tilde{\mu} = \min(\Delta\tilde{\mu}_{noise}, g_c), \tag{5}$$

where $\Delta\tilde{\mu}_{noise}$ is the expected noise level, and g_c is the minimum gradient level of the clinical features (microcalcifications) that must be preserved. As an example, we chose the smallest visible calcification with the smallest gradient as g_c (where $\Delta\tilde{\mu}_{noise} \rangle\, g_c$) and set $\Delta\tilde{\mu}$ as in eq. 5 and used it for σ and β calculations (see Fig. 2 and equation 4). The resulting reconstruction in Fig. 6d still looks smooth while all the microcalcifications are well visible. Iterative reconstruction in general (Fig. 6 b, c and d) preserves low frequency information well, so that the tissue background of the calcification cluster has a visibly higher density, compared to the FBP reconstruction in Fig. 6a.

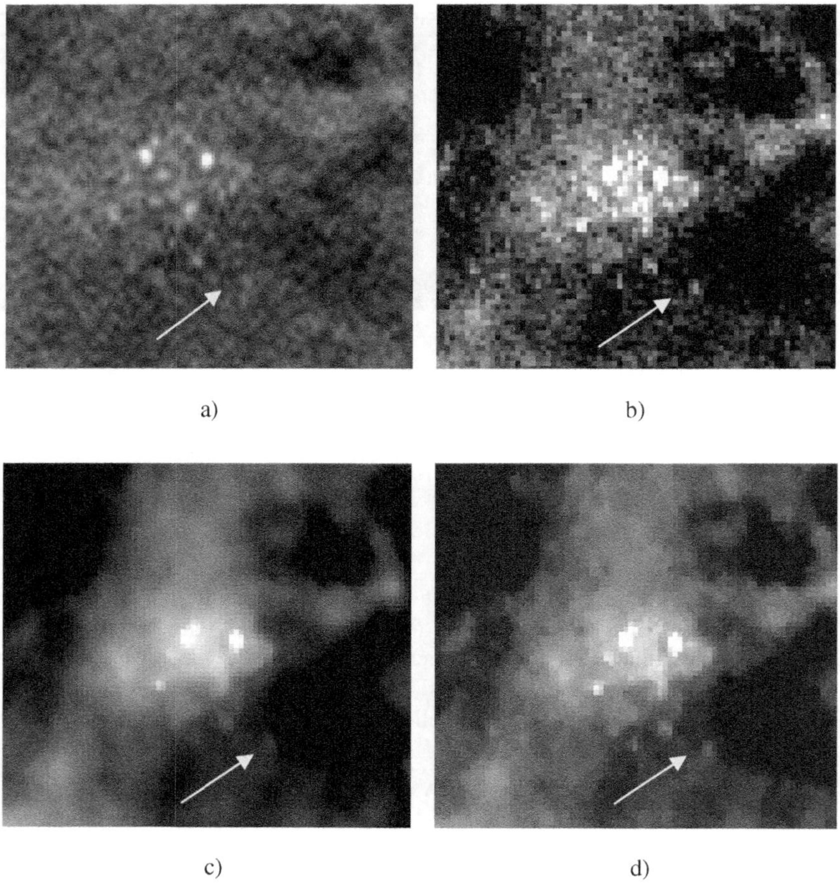

Fig 6. VOI from a patient data set with a microcalcification cluster reconstructed using a) FBP method, b) ML-convex, c) MAP-convex with $\Delta\tilde{\mu}$ set above the noise level measured in image in Fig. 6b, d) MAP-convex with $\Delta\tilde{\mu}$ set to preserve the microcalcification indicated by the arrow.

4 Conclusion

We have presented a method to analytically determine regularization parameters for MAP iteration algorithms for DBT. Our preliminary experiments show that the resulting parameters in fact provide improved CNR, which could lead in turn to the possibility of dose reduction in clinical practice.

However, it is likely that, in practice, the parameters would have to be additionally tuned depending on the feedback of the clinicians in an observer study that would be required because human perception of image diagnostic quality goes well beyond known image quality measures.

References

1. Mertelmeier, T., Orman, J., Haerer, W., Kumar, M.K.: Optimizing filtered backprojection reconstruction for a breast tomosynthesis prototype device. In: Proc. SPIE, vol. 6142 (2006)
2. Lange, K., Fessler, J.A.: Globally convergent algorithms for maximum a posteriori transmission tomography. IEEE Trans. Im. Proc. 4(10), 1430–1438 (1995)
3. Li, Q., Bai, B., Cho, S., Smith, A., Leahy, R.: Count Independent Resolution and Its Calibration. In: 10th International Meeting on Fully 3D Image Reconstr in Rad and Nuclear Med 2009, pp. 232–226 (2009)
4. Fessler, J.A., Rogers, W.L.: Spatial resolution properties of penalized likelihood image reconstruction: space-invariant tomographs. IEEE Trans. Image. Proc. 5, 1346–1358 (1996)
5. De Man, B.: Statistical methods for image reconstruction. In: ICMP (2005)
6. Alonzo-Proulx, O., Tyson, A.H., Mawdsley, G.E., Yaffe, M.J.: Effect of Tissue Thickness Variation in Volumetric Breast Density Estimation. In: Krupinski, E.A. (ed.) IWDM 2008. LNCS, vol. 5116, pp. 659–666. Springer, Heidelberg (2008)
7. PAS 1054: Anforderungen und Prüfverfahren für digitale Mammographie-Einrichtungen. Beuth-Verlag, Berlin (2005), `http://www.artinis.com/PASMAM_description.htm`

Image Quality Phantom in Close Compliance with the ICRU-44 Breast Tissue Substitute Standard and Realistic Background Noise Pattern for Digital Breast Tomosynthesis

Steven Raeymaeckers, Nico Buls, Gert Van Gompel, Cathérine Breucq, and Johan de Mey

Brussels University Hospital, Departement of Radiology, Laarbeeklaan 101, 1090 Jette, Belgium
Steven.Raeymaeckers@vub.ac.be,
{Nico.Buls,Gert.Van.Gompel,Catherine.Breucq, Johan.Demey}@uzbrussel.be

Abstract. Digital breast tomosynthesis (DBT) is an emerging technique that allows for slices parallel with the detector to be constructed through the breast, as opposed to two dimensional mammography. Present phantoms for use on mammography systems are two dimensional in nature and focus mainly on differences in contrast. Since tomosynthesis is a three-dimensional-technique, a suitable phantom for use on a DBT-system should contain different overlying structures, as is the case within a real breast. We present a method to construct such a three dimensional phantom from a liquid polyurethane-basis that contains overlapping masses, microcalcifications and a representative anatomical background. This way a representative quantitative phantom could be constructed for image quality control on a DBT-system.

Keywords: Phantoms, Mammography, Tomosynthesis, Image Acquisition, Image Quality, Systematic Testing/Validation.

1 Background

Digital Breast Tomosynthesis (DBT) is an exciting new technique that is based on a long since understood radiological principle. Multiple projection images of the breast are acquired at different angles of the X-ray tube while the detector remains stationary. These projection images are then processed in such a way to create slices or slabs throughout the breast. These slices have the benefit of an excellent resolution in the plane of view even though the Z-axis remains poor, the result of an incomplete dataset. This way superimposed structures as are commonly seen on two-dimensional mammography can be differentiated from one another, a theoretical advantage in the radiological examination of the dense breast with abundant glandular tissue. Though it is still early days, many research groups have already published interesting data regarding the potential clinical role of DBT in the diagnosis of breast cancer and reduction of recall rates [1, 2].

J. Martí et al. (Eds.): IWDM 2010, LNCS 6136, pp. 556–560, 2010.

However there is an obvious lack of a suitable phantom model that can be used for image quality control and optimization of acquisition settings on DBT-systems. Any existing breast phantom model to this day focuses mainly on differences in contrast and resolution of their different inlaying structures. These models all lack superimposed structures as well as a realistic background noise generated by superimposed glandular tissue. We propose a method to construct a three dimensional phantom in close compliance with the ICRU-44 breast tissue standard.

2 Methods

The main component of the breast phantom we constructed is a polymer with a specific gravity of 1.00 and attenuation factor of 1,23 cm^{-1} The latter was determined experimentally using a typical mammography MoMo 28kV photon spectrum. This specific polymer is created by mixing two liquid components, the mixture hardens within 24 hours.

We constructed a first phantom in the shape of four individual stackable layers by pouring the polymer-mixture into circular molds. Two layers contain masses of varying diameters, one layer contains microcalcifications and a fourth layer acts as a structured noise pattern. These layers can be oriented as such in any way that overlap between different structures ensues. Figure 1 shows the four layers separately, a superimposed mammogram as well as some reconstructed DBT-images.

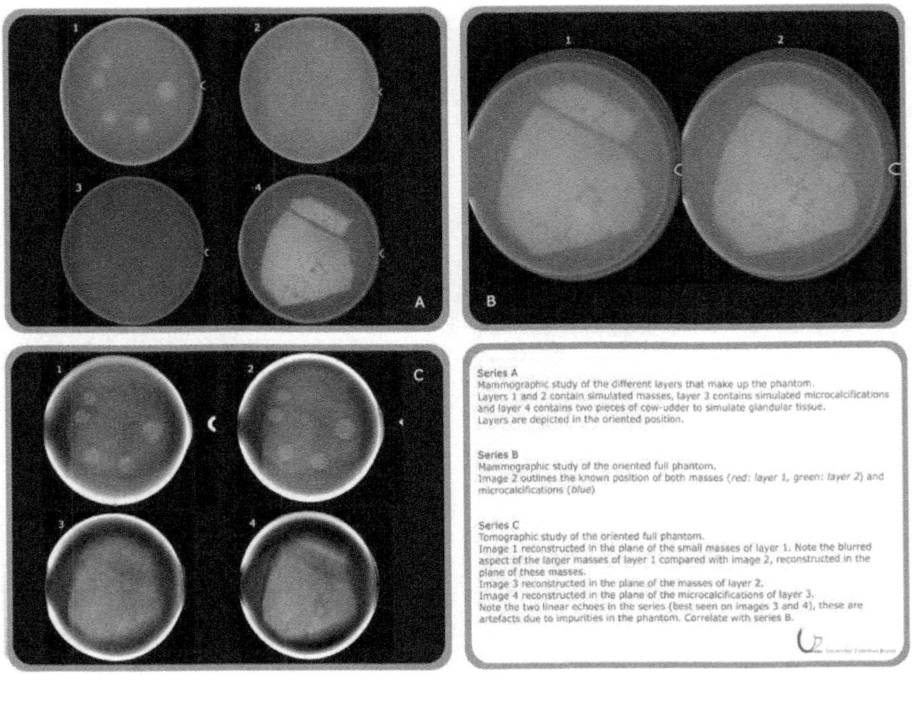

Fig. 1. First phantom model. Mammography of the separate layers (series A) ; Total oriented phantom (series B) ; DBT reconstruction images (series C).

Within the first two layers we simulated masses of different diameters by introducing circular spheres which are removed later on in the process, after the polymer has hardened. The remaining cavities then get filled with a similar type of polymer but one that has a 4% higher specific gravity. The experimental attenuation factor of this polymer was found to be 1,27 cm^{-1} using the same MoMo 28kV photon spectrum. This difference in attenuation is essential in order to ensure that masses and background can be differentiated.

The third layer of this first phantom-model contains pulverized particles of eggshell in order to simulate microcalcifications. These particles are checked on a mammography unit for size (0,1 - 0,3 mm) prior to their integration in the phantom. It is common practice to use particles of eggshell or teeth to mimic microcalcifications [3].

The fourth layer serves as a structured background noise pattern and contains a glandular part of mammal udder. This glandular tissue has first been preserved in a formaldehyde solution. The overlap of this noise pattern ensures a more difficult visualization of the simulated masses and microcalcifications as is the case within a real breast as well.

A second phantom model was constructed later on. We chose to create a single-layer model in the conical shape of a partially compressed breast using the same principles as described above.

Images were acquired in both cases by digital mammography (*GE Healthcare Essential*) and DBT on a prototype system (*GE Healthcare*) using different acquisition parameters.

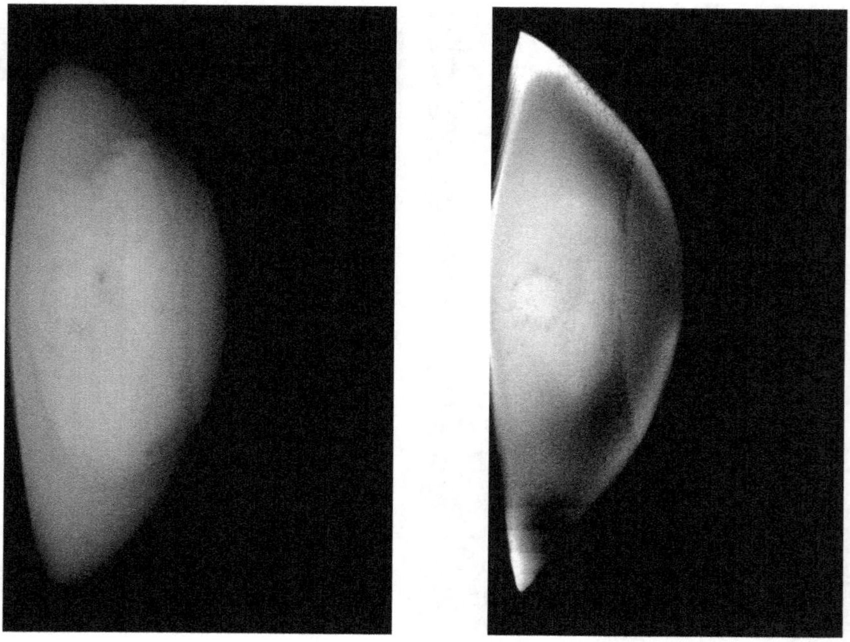

Fig. 2. Second phantom model. Simulated mass seen on mammography (left) and DBT-reconstruction (right).

3 Results

Attenuation experiments on the two rubber compounds we used, showed linear attenuation coefficients of 1,23 cm^{-1} (base material) and 1,27 cm^{-1} (masses) respectively. These values are somewhat higher than tabulated values for ICRU-44 breast tissue (1,11 cm^{-1}), but they are still in adequate agreement.

Standard digital mammography of the multi-layered phantom proved difficult to read. Of all 24 masses the first phantom contained, only five could be differentiated on the mammogram. These masses all had a diameter of more than 1 cm. None of the smaller masses could be detected using mammography. The clusters of microcalcifications were readily detectable on the mammogram though they suffered heavily from masking by the overlying glandular tissue.

The DBT study was performed with various acquisition settings. The angle of motion of the X-ray tube, the number of acquired projection images and the tube settings were altered. Regardless of these changes in acquisition settings we managed to detect virtually all simulated masses, whether overlapping or not. The microcalcifications were readily visualized as well and suffered less loss of contrast due to the overlying glandular tissue. Despite this fact the microcalifications did appear much more blurred and showed extensive signs of loss of anatomical detail compared to the mammography.

We calculated the contrast to noise ratio (CNR) of the reconstructed images of the different DBT-studies. As could be expected the CNR increases with an increase in the number of projection images taken and with an increase in exposure. The CNR also increases as the angle of motion of the X-ray tube increases. The latter would seem to suggest that the larger the angle, the better the quality of the reconstructed images and this at no extra cost in patient dose.

The relatively sharp edges of both the cylindrical model of the first phantom and the parts of glandular tissue caused relatively extensive artefacts on the reconstructed DBT images. Interestingly enough while we found that an increase in the angle of motion of the X-ray tube does attribute to an increase in CNR, this does also worsen the described artifacts as well as furthers blurs the microcalcifications.

A 6-month radiographic follow up of the first phantom model showed no signs of degradation or dehydration. Since the glandular tissue is encased in the polymer, it is kept completely safe from contact with air. This combined with the fact that we sterilized the tissue prior to its integration in the phantom, should ensure long-term preservation.

In order to overcome the artifacts caused by the cylindrical shape of the phantom, we constructed a second model in the form of a partially compressed breast. Because of an algorithm that GE developed for use on their experimental tomosynthesis-unit and that allows for this more natural contour to be taken into account when reconstructing the final images, the described artifacts caused by the edge of the phantom were strongly reduced in the second model.

4 Discussion

The three dimensional phantom model we constructed can be used for image quality assessment and the optimization of acquisition settings on a DBT-system. It is relatively easy and cheap to make and one can add additional components in function of their own personalized needs.

The polymers we used for the construction of our phantom-models are in close compliance with the ICRU-44 breast tissue standard and allow for a certain degree of elasticity. The glandular material in the phantom ensures the addition of realistic background noise. The eggshell particles we used to simulate microcalcifications while visually very realistic, are not easily reproducible. The two preliminary models we constructed are mainly qualitative, however it should not be difficult to add quantitative elements.

Both mammographic and DBT-examinations of our phantoms produced quite realistic images, especially those of the second model. We calculated the contrast to noise ratio for the reconstructed DBT-images of the cylindrical phantom and found that the CNR increases with an increased number of projection images, increased exposure and an increased angle of motion of the X-ray tube. However when the angle of motion increases microcalcifications loose their typical morphology, becoming more blurred. It is then not so easy as to conclude that a larger angle of motion ensures better image quality.

Both phantom models still contain small cavities of air. This could be avoided by mixing the polymers in an airtight environment. Otherwise, if this is not possible, one might vibrate the mixture for a certain amount of time to release as much of the air-bubbles as is possible or let the mold settle in an underpressurized environment.

Further development of this phantom could focus on the testing of different types of polymers, to see if an attenuation factor can be found that matches the ICRU-44 breast tissue standard more closely. Other structures (spheres, iodine contrast agents, geometrical objects) could be introduced that can be used for image quality assessment in other breast imaging applications, such as in contrast enhanced mammography. Also, more reproducible structures could be introduced to add a quantitative dimension to the phantom for QA purposes.

References

1. Poplack, S.P., Tosteson, T.D., Kogel, C.A., Nagy, H.M.: Digital breast tomosynthesis: initial experience in 98 women with abnormal digital screening mammography. AJR Am. J. Roentgenol. 189(3), 616–623 (2007)
2. Andersson, I., Ikeda, D.M., Zackrisson, S., Ruschin, M., Svahn, T., Timberg, P., Tingberg, A.: Breast tomosynthesis and digital mammography: a comparison of breast cancer visibility and BIRADS classification in a population of cancers with subtle mammographic findings. Eur. Radiol. 18(12), 2817–2825 (2008)
3. Obenauer, S., Hermann, K.P., Grabbe, E.: Dose reduction in full-field digital mammography: an anthropomorphic breast phantom study. Br. J. Radiol. 76(907), 478–482 (2003)

The Standard Attenuation Rate for Quantitative Mammography

Christopher E. Tromans and Sir Michael Brady

Wolfson Medical Vision Laboratory, Department of Engineering Science,
University of Oxford, Parks Road, Oxford, UK, OX1 3PJ
cet@robots.ox.ac.uk

Abstract. We introduce the Standard Attenuation Rate (SAR), a quantitative, and normalised measure of radiodensity per unit distance traversed by the primary beam incident on each pixel of an x-ray mammogram is presented. We sketch an algorithm to compute the SAR. The calculation utilises a physics model of image formation, including consideration of photon production in the x-ray tube, photon detection within the image receptor, and photon scattering occurring within the tissues of the breast. Using the model, the difference in the flux incident upon, and exiting from, the breast is quantified relative to a reference material. Experimental validation of the SAR representation is presented, based on a tissue equivalent phantom designed and manufactured specifically for the purpose. The observed performance across the clinical range of acquisition parameters is very promising, supporting the suitability of this approach to form the basis of a next generation of diagnostic techniques based on quantitative tissue measurement.

Keywords: Quantitative mammography, measuring radiodensity, acquisition physics modelling, scatter removal.

1 Introduction

The brightness recorded at any location within a mammogram depends on many factors, including: the specifics of the x-ray source, for example the accelerating potential and the anode material; the exposure time; the filtration of the photon fluence before it exposes the breast; the detector characteristics; any anti-scatter devices present; and, most importantly, the unknown contents of the breast that is being imaged. As a result of this dependency upon a multiplicity of factors, compounded by the unknown tissue structures with the breast, most human and computer analysis is essentially qualitative rather than quantitative. We present a quantitative representation, which we call the Standard Attenuation Rate (SAR), which consists of a normalised measure of the radiodensity per unit distance of tissue traversed between the focal point of an x-ray source and each pixel in an x-ray projection image of a breast: that is the radiodensity per unit distance encountered by the primary beam. This may be thought of as an analogue to the Hounsfield unit used universally in CT, though in this case specifically optimised for imaging the soft tissues of the breast. A measure

J. Martí et al. (Eds.): IWDM 2010, LNCS 6136, pp. 561–568, 2010.

such as SAR has numerous potential applications, for example, to quantify the attenuation (and hence atomic composition and density) of a suspect feature (a lesion or calcification), the microenvironment of a tumour, or the breast as a whole (as is the case when estimating breast density), for which the result may then be compared to that which is known for such a lesion or a cancerous breast. Quantitative tissue composition measurement also has uses in soft-copy display, for example in optimal image display in setting window width and level, as well as in CAD systems which often utilise image normalisation techniques prior to applying detection algorithms. Further applications are possible in the field of digital breast tomosynthesis, where the absolute measure of radiodensity provides a normalised basis between projections upon which reconstruction algorithms may operate.

In 1996 Highnam et al [1-2] published the h_{int} representation which achieves a normalised measure of tissue characteristics through utilising a model of image formation physics to ascertain the thickness of "interesting tissue" (breast tissue that is anything other than fat) above each pixel in a mammogram. The majority of descriptions of other approaches in the literature for the quantification of mammographic images have related to the study of volumetric breast density. Most adopt the technique of comparison with images of a reference object. The choice of reference object material varies: Pawluczyk et al [3] adopts tissue-equivalent plastic, whilst Diffey et al [4] utilise an aluminium step-wedge. Techniques differ in how they measure the image transfer function: some incorporate reference objects into each clinical acquisition [4-5], whilst others image the references alone in order to calibrate the equipment [6], and Pawluczyk et al combines both techniques using different reference objects [3]. Modelling the physics of image acquisition has the advantage that it enables the scattered radiation and effects of beam hardening to be given full consideration, whereas the use of reference objects only includes such effects in as much as they are present in both the calibration and the clinical images. The aluminium step wedge, for example, will result in a significantly different photon energy spectrum incident upon the detector from that exiting the tissue of the breast since significant discrepancies exist between the attenuation coefficients of the two materials across the range of photon energies present in a mammography x-ray beam. Also due to its small size and possible proximity to the breast (depending on breast size) scattering effects will differ. This work follows the approach of Highnam et al [2] in utilising a model of image formation for tissue quantification, however it is a considerably enhanced reworking of their method, giving far greater consideration to the underlying physical phenomena, and resulting in a method that is devoid of many of their simplifying assumptions.

2 Materials and Methods

In order to calculate the SAR, a detailed, though inevitably approximate, model of the physics of mammographic image formation is used to calculate the difference between the primary x-ray photon fluence incident upon the breast and the primary x-ray photon fluence exiting the breast and being subsequently recorded by the image detector, quantified relative to a reference material: in this work a 50/50 mix of adipose and fibroglandular tissue is used, according to the elemental composition and density reported by Hammerstein et al [7]. Other choices may be substituted, according to the application, for example the 70/30 composition reported by Yaffe et al [8] as being a typical breast. The reference material used here, as opposed to other proportions, amounts, in the definition of SAR, to changing a single parameter. The quantification of radiodensity relative to a reference material, rather than the approach of trying to distinguish adipose from fibroglandular in a two tissue model adopted by other authors comes from our previous work [9] in which significant variation was reported between the varying reports in the literature of the elemental composition and density of the two tissue types, suggesting significant variation within the population. Further, the results of the histology study reported by Alowami et al [10] show the dependency of mammographic appearance on stromal composition, in particular the increased collagenous stroma and expression of lumican and decorin in breast tissue associated with high mammographic density, illustrating the need to considerthe variation within the compsition of fibroglandular tissue across the population. Measuring radiodensity supports the inclusion of these effects, which violate the two tissue assumption.

The image formation model consists of three components: a model of the x-ray tube, a model of the image detector, and a model of photon scattering within the tissues of the breast.

The algorithm inputs for a given image are the raw "FOR PROCESSING" DICOM pixel data, the tube voltage (kVp), the anode and filter material, the exposure, and the compressed breast thickness: all values which may be found for modern digital mammography equipment in the DICOM metadata header. The pixels of the output image are the multiplicative scaling factor which must be applied per unit x-ray beam traversal distance to the attenuation of the reference material in order that the signal recorded at the image receptor matches that observed in the acquired image, after the effect of scatter has been subtracted (using the scattering model described herein). Specifically, the image pixel intensity in the standard attenuation rate image is c in:

$$D^{-1}(I) - S = P = \sum_{\varepsilon} e^{-\mu(\varepsilon)x} F = \sum_{\varepsilon} e^{-\mu_{ref}(\varepsilon)cx} F$$

where I is the input image; $D^{-1}(x)$ is the inverse detector transfer function (calculated using the detector model discussed later), specifically the mapping between pixel intensity and incident photon fluence; S is the scattered photon fluence recorded by the detector; P is the primary image; $\mu(\varepsilon)$ is the attenuation of the column of breast tissue between the focal spot and the pixel under consideration at photon energy ε; x is the traversal length through the breast tissue between the focal spot and the pixel in question; F is the incident photon fluence upon the upper surface of the breast (calculated using the tube model discussed herein); and $\mu_{ref}(\varepsilon)$ is the attenuation of the reference material.

The basis of the x-ray tube model is the spectral data reported by Cranley et al[11]. The anode self filtration (heel effect) is considered through the calculation of an effective target angle for each image pixel, that is the target angle in the measurement geometry of the published data[11] for which the self filtration distance within the anode is the same as that for the ray emanating in the direction of the detector pixel under consideration. The beam emanating from the anode surface is attenuated according to the traversal path through the tube window and any additional beam filtration, as well as scaled according to the inverse square law. Two empirical corrections for beam intensity and quality are applied, in order to account for discrepancies between the spectrum calculated by the model, and that observed in reality due to such effects as manufacture tolerances and anode pitting with age.

In order to develop a complete detector model which considers photon transport and signal production, such as that presented by Williams et al [12], a detailed knowledge of the exact construction details is required. Such information is largely unobtainable since it is considered proprietary by manufacturers. Recognising this, we have developed an empirical approach, in which the family of detector transfer functions are sampled so as to capture all the dependant variables, in particular fluence and photon energy spectrum (beam hardening), whilst also being independent of the precise detector technology. Samples are interpolated using a series of regression-fitted analytical functions which are then used to describe the detector response characteristics. A purpose designed breast equivalent attenuator, shown in fig. 1, has been manufactured to streamline the acquisition of the calibration images.

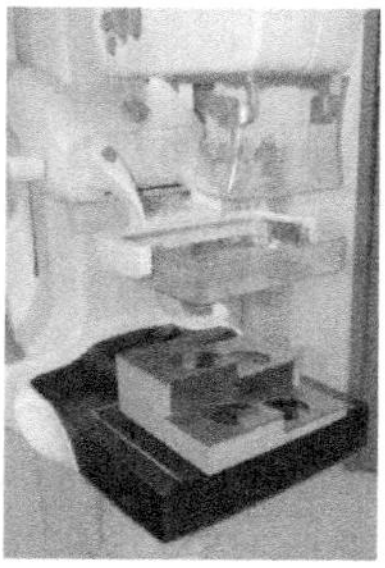

Fig. 1. The phantom used for calibration of the family of detector transfer functions

The detector output is measured as the average pixel intensity for a circle of diameter 13 pixels located at the centre of each aperture shadow, and the incident fluence, including scatter, is calculated using the tube and scatter model to give the input of the transfer function. 70mm lead apertures are used to control the volume of material contributing scatter to the pixels from which the detector output is measured. Four thickness of PMMA (which approximates the attenuation of breast tissue) are employed, 15, 30, 45 and 90mm, so as to capture the variation in detector response with the incident photon energy spectrum, occurring due to hardening of the beam arising from varying imaging thickness. Log-quadratic interpolation is used to interpolate intermediate values between the measured thicknesses. Images of the calibration object are acquired for each anode-filter combination separately, and for at least three tube potentials spanning the clinical range. Interpolation between measured tube potentials is achieved using regression fitting of log-quadratic relations. The cases with and without the scatter grid present are calibrated separately.

The scatter model is based on the molecular form factor and the incoherent scatter function, which together scale the free electron coherent and incoherent scatter relations respectively, to provide the angular scatter characteristics of breast tissues. In order to achieve a clinically realistic computation time, whilst maintaining accuracy, optimal information sampling and interpolation at the limit prescribed by the Nyquist-Shannon sampling theorem is used to calculate the scatter-to-primary ratio at each image pixel. A geometric model of an anti-scatter grid is used to calculate the selective photon transmission probability according to the angle of incidence. The low frequency blurring effect of scatter is included by using a kernel derived from the image acquisition conditions and the underlying fundamental physical photon scattering relations. Subtraction of the calculated scatter signal from the acquired image yields both an approximation of the magnitude of primary signal, and the removal of the image blurring arising as a result of scatter. Further details of the scatter model may be found in our companion conference paper [13] assessing the possibility of replacing physical anti-scatter grids with software correction.

3 Results

In order to validate the performance of the technique, we have designed and constructed a tissue equivalent phantom (manufactured from CIRS supplied plastics), shown in fig. 2, comprising of a pair of interlocking step wedges, one of glandular, the other of adipose, equivalent material, with a further two adipose wedges forming the sides. It should be emphasised that this phantom is used purely for validation, and is not used in the design, calibration, or implementation of the standard attenuation rate algorithm or underlying image formation model.

Validation was performed on a GE Senographe Essential installed in a symptomatic breast clinic which sees everyday use in patient examination and is in no way specially customised for this work. Tube output and half value layer measures were taken from the routine NHS quality assurance surveys. Fig. 3 shows raw images of the validation phantom for a variety of image acquisition parameters spanning the range employed clinically, together with the corresponding SAR image. Fig. 5 illustrates the normalisation quantitatively.

Fig. 2. The validation tissue equivalent validation phantom, white is fibroglandular equivalent, terracotta is adipose

Fig. 3. Validation images of the tissue equivalent test object, raw "FOR PROCESSING" acquired image (window width 764, centre 1452) [top] and SAR image [bottom]; 25kVp Mo-Mo 140mAs [left], 30kVp Mo-Rh 56mAs [centre] and 30kVp Rh-Rh 90mAs [right] (window width 0.4, centre 1.0)

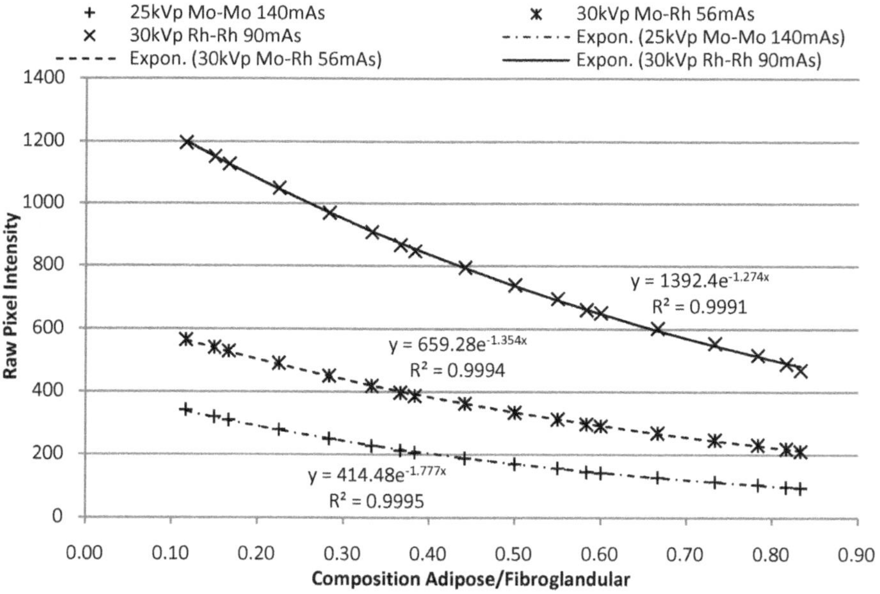

Fig. 4. The median value of a 50 by 50 square of pixels within each step of the experimental validation acquisitions of the tissue equivalent test object in the raw "FOR PROCESSING" image

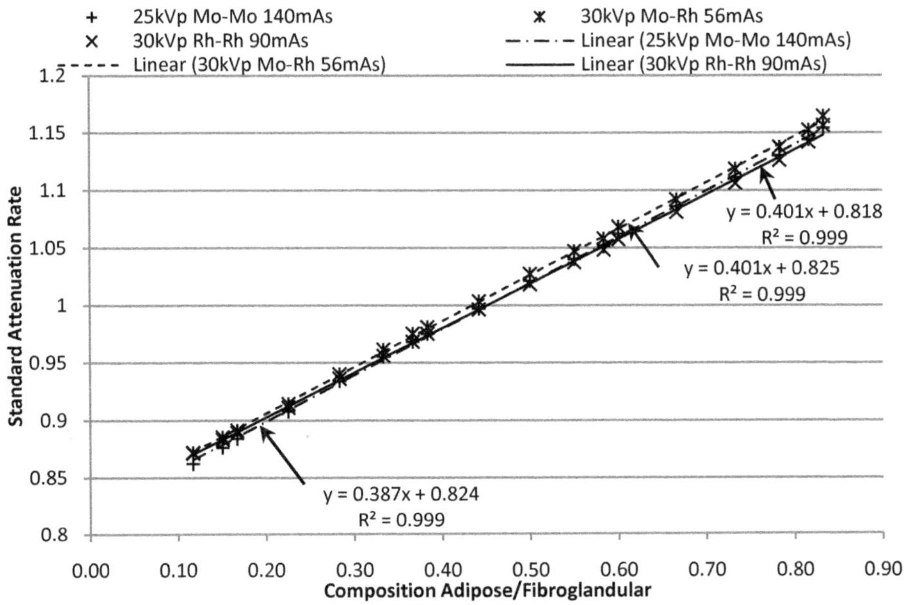

Fig. 5. The median value of a 50 by 50 square of pixels within each step of the experimental validation acquisitions of the tissue equivalent test object in the SAR image

4 Discussion

Qualitative evaluation of fig. 3 reveals a high degree of similarity between the SAR images of the phantom acquired over the range of beam qualities shown, as is to be expected given the constant phantom composition. Quantitative results are shown in fig. 3 and fig. 4, where the coefficients of determination being unity (to 3 significant figures) show the successful transformation from a exponential relation between pixel intensity and composition in the raw image, to a linear relation in the SAR image. The maximum difference between the gradients in the linear relations describing the relation between composition and SAR pixel intensity throughout the tested beam qualities is 3.4%, and the intercepts is 0.86%, showing a high degree of accuracy in the image normalisation and thus in the underlying model of image formation.

5 Conclusion

A method is presented for deriving an image dependant solely on the attenuation characteristics of the underlying breast tissue, from a mammogram. Experimental validation shows the successful decoupling of image contrast, within the bounds of measurement and modelling error, from the x-ray acquisition conditions yielding an image dependant solely on the underlying tissue composition. Thus quantitative measurement is facilitated which could form a basis for novel diagnostic techniques.

Acknowledgments

Thanks to the medical physicists and radiographers at the Breast Imaging Centre at the Churchill Hospital Oxford for their help in facilitating the validation experiments. Christopher Tromans wishes to thank the EPSRC for funding his research fellowship.

References

[1] Highnam, R., Brady, M.: Mammographic image analysis. Kluwer Academic, Dordrecht (1999)
[2] Highnam, R., et al.: A representation for mammographic image processing. Med. Image Anal. 1, 1–18 (1996)
[3] Pawluczyk, O., et al.: A volumetric method for estimation of breast density on digitized screen-film mammograms. Med. Phys. 30, 352–364 (2003)
[4] Diffey, J., et al.: A New Step-Wedge for the Volumetric Measurement of Mammographic Density. In: Astley, S.M., Brady, M., Rose, C., Zwiggelaar, R. (eds.) IWDM 2006. LNCS, vol. 4046, pp. 1–9. Springer, Heidelberg (2006)
[5] Shepherd, J.A., et al.: Novel use of single X-ray absorptiometry for measuring breast density. Technol. Cancer Res. Treat 4, 173–182 (2005)
[6] Kaufhold, J., et al.: A calibration approach to glandular tissue composition estimation in digital mammography. Med. Phys. 29, 1867–1880 (2002)
[7] Hammerstein, G.R., et al.: Absorbed radiation dose in mammography. Radiology 130, 485–491 (1979)
[8] Yaffe, M.J., et al.: The myth of the 50-50 breast. Med. Phys. 36, 5437–5443 (2009)
[9] Tromans, C., Brady, M.: An Alternative Approach to Measuring Volumetric Mammographic Breast Density. In: Astley, S.M., Brady, M., Rose, C., Zwiggelaar, R. (eds.) IWDM 2006. LNCS, vol. 4046, pp. 26–33. Springer, Heidelberg (2006)
[10] Alowami, S., et al.: Mammographic density is related to stroma and stromal proteoglycan expression. Breast Cancer Res. 5, R129-35 (2003)
[11] Cranley, K., et al.: Catalogue of diagnostic X-ray spectra and other data. Institute of Physical Sciences in Medicine, York (1997)
[12] Williams, M.B., et al.: Analysis of the detective quantum efficiency of a developmental detector for digital mammography. Med. Phys. 26, 2273–2285 (1999)
[13] Tromans, C., Brady, M.: Investigating the Replacement of the Physical Anti-Scatter Grid with Digital Image Processing. In: Proceedings of 10th International Workshop on Digital Mammography (IWDM), Girona, Spain (2010)

National Software Supported Quality Assurance Program in Digital Mammography: Experiences and Challenges

Christian Moll[1,*], Andreas Jahnen[1], Alex Meyer[2], Martine Grelot[2], Johannes Hermen[1], Carlo Back[3], Marie-Christine Wagnon[4], and Norbert Rösch[1]

[1] Centre de Recherche Public Henri Tudor, CR SANTEC, 29 avenue John F. Kennedy, 1855 Luxembourg, Luxembourg
[2] Entente des Hôpitaux Luxembourgeois, Physique Medicale, 5 rue des Mérovingiens, 8070 Bertrange, Luxembourg
[3] Direction de la Santé, Division de la Radioprotection, Villa Louvigny, 2120 Luxembourg, Luxembourg
[4] Direction de la Santé, Programme Mammographie, Villa Louvigny, 2120 Luxembourg, Luxembourg

Abstract. Quality assurance in digital mammography, especially in screening environments, needs a lot thoughtfulness. As the images are digital, it provides the possibility of the use of computer assisted tests. This requires adequate computer readable phantoms and powerful software tools, which should support the medical technical assistant (MTA) in carrying out the required constancy tests. An added value of software assisted quality assurance is the ability to generate reports and statistical evaluations of the collected results. The national mammography screening program in Luxembourg has established a sophisticated quality assurance program including automatic reading, ready to use reporting and tailored formation at a national scope during the last years. This paper reports our experiences during the implementation of this common national wide quality control system and discuss the challenges involved.

Keywords: Quality assurance, constancy control, automatic phantom reading.

1 Introduction

In the Luxembourgian mammography screening program 44 000 women between an age of 50 and 69 years are invited every two years to conduct a mammography examination. With an overall acceptance of 64% participating women from all invited the program runs successful. The invited women have the opportunity to go to one of the nine screening centers which are participating in the program. Once an examination is performed, the first reading is done in the screening center. The examination is then transferred for the second reading to the Ministry of Health mammography program. Here a blinded second reading is performed. If the diagnosis match, an appropriate

* Corresponding author.

J. Martí et al. (Eds.): IWDM 2010, LNCS 6136, pp. 569–575, 2010.

report is send. In the case the results do not match, a third reading is scheduled. The communication process, which includes the management and scheduling of screening participants and their registration in the local and remote Radiology Information Systems (RIS) and the secure transfer of the taken images for second reading is done using a custom management system electronically (Figure 1).

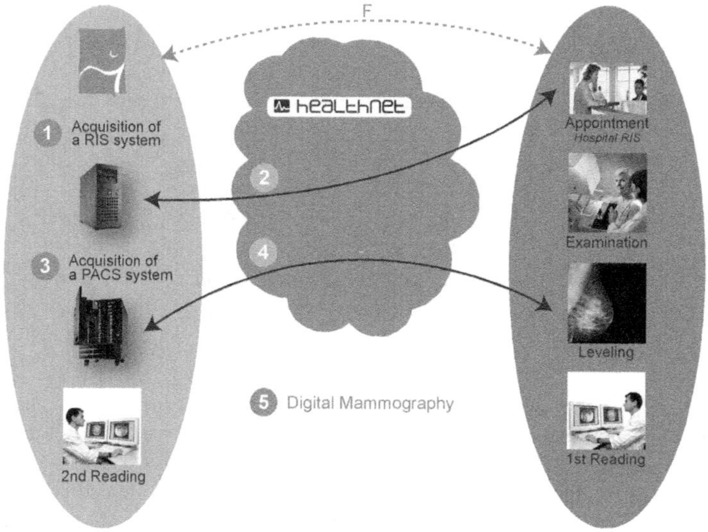

Fig. 1. Workflow of the process of second reading

With the introduction of the Council directive 97/43/EURATOM quality control for all radiographic modalities becomes an obligation in the EU countries [1]. The detection of early cancer in mammography screening programs make demands on cutting edge image quality. As in mammography screening programs potential healthy women are radiated, there is a need for high quality at low dose levels. This high demand on quality resulted in the development of a quality assurance program adapted to the specific needs of the Luxembourgian mammography screening program.

2 Materials and Methods

The developed quality assurance program is based on the recommendations given by the European Reference Organisation for Quality Assured Breast Screening and Diagnostic Services (EUREF) in the European Guidelines for Quality Assurance (QA) in Breast Cancer Screening and Diagnosis [2]. In 2006 the Luxembourgian mammography screening program switched the first mammography unit to a digital mammography system. In parallel a work group developed an adapted quality assurance program to correspond to the new requirements. This QA program needs to take the characteristics of the new digital technology into account: For example in the analog mammography the determination of the noise was only possible with high efforts, it changed in digital completely, because access to the discrete pixel values stored in the image files is possible. The QA not only focuses on the image quality of the mammography system it also has to control the other parts of the image chain into account. This includes the environment of the readers as well as the constancy of the used displays.

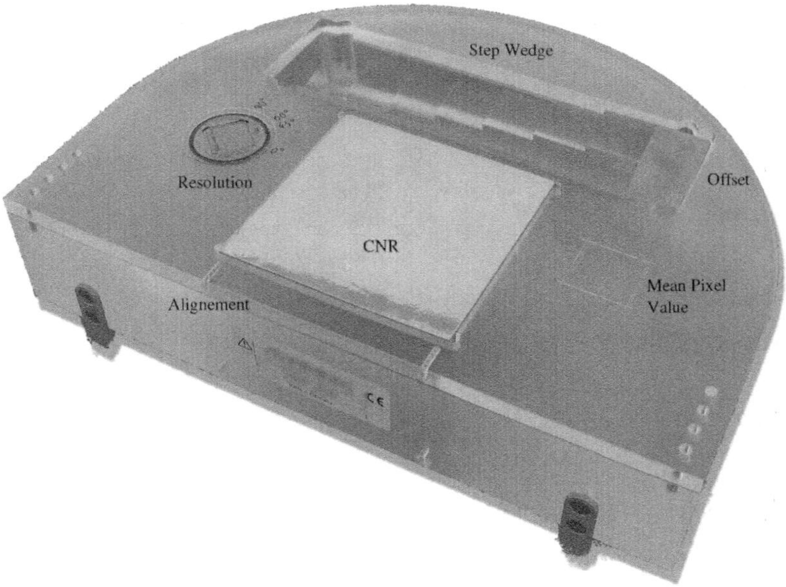

Fig. 2. Phantom accorging to PAS 1054

For the constancy control the method described in the Public Available Specification 1054 (PAS 1054) [3] and the corresponding phantom (Figure 2) was chosen. This phantom is designed in respect to the guidelines established from the EUREF [3+4]. In addition the CDMAM phantom [5], is used to determine the low-contrast visibility of small objects. The PAS 1054 phantom is used for the daily and weekly constancy control tests as it fits well to the requirements and was one of the first commercially available phantom that time. This phantom includes a lot of features by default. It includes, depending on the model, a 14-steps step-wedge made of PMMA or aluminum. The aluminum step-wedge is obligatory for level B tests, but can also be used for the level A tests. The PMMA step-wedge is only applicable for level A tests so we decided to use the aluminum one for all the tests that are performed.

To support the MTA, who are doing the daily test in the sites, and to use the benefits provided by the new technology, the Optimage project was chosen to ease up and reduce the daily work of the MTA. The Optimage software, which is a suite for automated quality control of radiographic modalities, is a module based solution to automatically read phantom measurements and document them probably [6]. That simplifies the procedures in constancy control for the MTA and maximizes the available information for the responsible medical physics expert. Due to the fact that constancy control in

Table 1. Parameters included into the quality assurance program Besides the replacement of the machines and the introduction of the software, also training of the Medical Technical Assistants (MTA) was scheduled. This leaded to the offer of continuing training courses for MTA as well as radiologists [9]. To enable them to understand the changes and different behaviors of the digital technology the offered courses included hands-on workshops.

Parameters	*Frequency*	*Reference levels*
Visual display control	Weekly	
Determine geometrical deformation with AAPM TG18-QC	Weekly	Borders of the grid must be visible and non deformed
Contrast and grey level TG18	Weekly	The 5% and 95% fields must be differentiable, steps must have a linear increase in the luminance
Visibility of artifacts	Weekly	Non disturbing artifacts
Homogeneity of detector	Weekly	SNR >10% change >10%
Artifacts	Weekly	Visible artifacts
Defect elements of detector	Weekly	Pixel >20% mean of ROI
Average Glandular Dose	Daily	>5%
SNR	Daily	>10%
Mean pixel value	Daily	>10%
CNR	Daily	>10%
Contrast	Daily	>10% for every step in the step wedge
Low Contrast visibility (KP-MDP)	Weekly	>= reference
Function of AEC	Monthly	CNR >10% mAs >10%

Luxembourg is performed in collaborative work of the MTA, that are responsible to perform the level A test (high frequency; simple tests) together with the medical physics experts with review the results and report them to the supervising authority. The medical physics experts are also responsible to do the level B tests (infrequent; more advanced)[7].

In addition the phantom enables us to measure the offset of the signal, the high-contrast resolution, mean grey value, signal to noise ratio (SNR) and the accurateness of the thorax side border. This set of measurements can be enhanced by adding inserts to the phantom. In the different set ups the phantom can then also measure contrast to noise ratio (CNR), ghosting and low contrast detail. Therefore different inserts are available, utilizing parts of the CDMAM phantom that are representing the achievable contrast detail curve. Also available are inserts using structures from the ACR or Ackerman phantom, containing small fibrils, masses and micro-calcification simulating objects. For systems that are used in screening contexts it is an obligation to evaluate one of these low contrast detail inserts daily, to monitor this parameter as best as possible [3]. An extract of the performed tests and their frequency for the constancy control of the mammography units in the screening program can be found in table 1.

The Optimage GUI is optimized to do the test and documentation of the measured parameters in just three steps. The user just selects the image to analyze, and the software automatically recommend a suitable profile to be used. Then the user is able to start the verification process. During this step the software uses methods of image recognition to find the features (step wedge, line pairs, ...) embedded in the phantom. If all features can be found the calculation step is enabled.

In the case of a wrong placed or oriented phantom the software stops the measurement with a comment to the user to correct this before he can continue. When the measurement is performed the user gets a short customizable list of the most interesting parameters. More advanced users can also examine the complete list of all measured parameters (>100). From the result a report can be created and the results can be stored in a relational database. These data can later on be used to generate time reports and statistical analysis of the observed machines [8].

3 Results

The work groups reviewed the state of the art in performing constancy control in digital mammography. Thereby the European guidelines proposed from the EUREF and the German PAS 1054 were selected as the bases for the implementation of the quality assurance in the Luxembourgian regulations in general. This implies that all digital mammography systems in Luxembourg have to be tested in the same conditions, whatever it is used for curative or screening purpose. The PAS 1054 was selected because it is the only specification which also defines the phantom (see Figure 2).

Phantom images are evaluated automatically using the Optimage Software package. Due to the grown IT infrastructure in the Luxembourgian hospitals the diversity of used systems is very high. To be able to install the Optimage software in the hospitals the complete flexibility of the software has to be used. It has been installed in every hospital with radiographic department participating in the national mammography screening program. Till march 2010 two centers replaced their mammography units completely to digital full field systems. During this year 3 additional centers will switch to digital mammography systems.

The biggest challenge was to integrate the software in the different environments. The software is installed on Linux servers, Windows Servers, Citrix and dedicated Windows PC.

In all cases the software has to be integrated as a DICOM node to the network infrastructure as well. The success of this is very depending on the manufacturer of the picture archiving and communication systems (PACS). In one case we needed to invest in a license to receive uncompressed images. 2/3 of the DICOM nodes could be set-up by the PACS administrators of the hospital and for the rest of the installations a technician of the manufacture had to asked to set up the node. More complicated was the access to the raw image data. In only 20% of the installations the option to send these kind of data was accessible without the help of a technician.

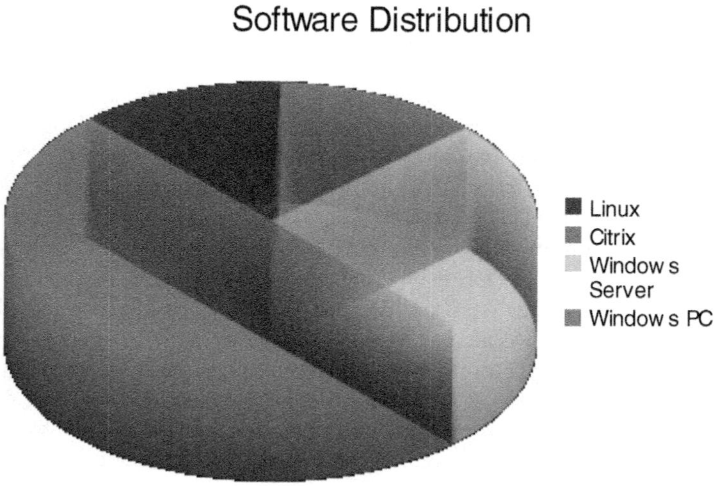

Fig. 2. How the Optimage Software is installed in the screening centers

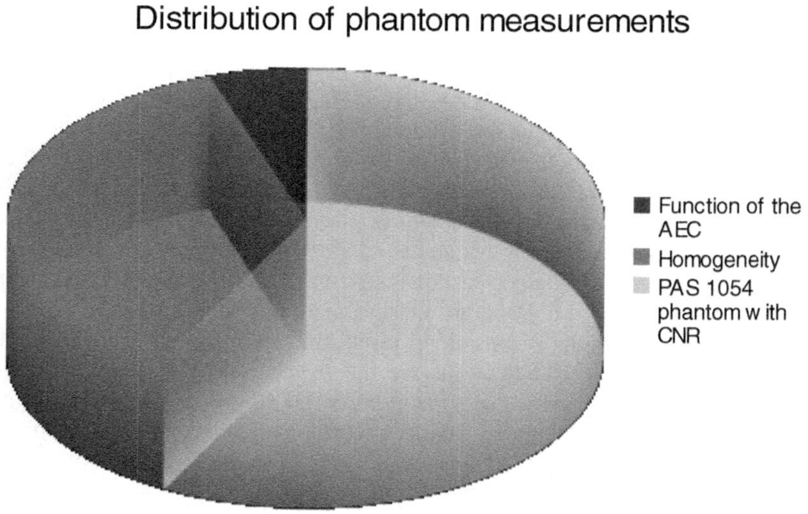

Fig. 3. Measurement distribution of the last 3 years

In 3 years of usage of the Optimage software in the screening centers, they aggregated 943 measurements, for the distribution of the performed phantom measurements see figure 3. The in the software used automatic segmentation algorithms for the feature extraction was improved during this time up to a detection rate 99.5%.

To enable the MTA to do the constancy tests in digital mammography all involved MTA participated in a special training program to be able to do it, with and without the support of the Optimage software. This was done till the end of 2009 in several training sessions. We turned attention on giving high quality courses in collaboration with external experts. Also the hand-on workshops were included to the training. We kept a ratio between tutor and the participants by 1:4 or better.

4 Conclusions

The introduction to the quality assurance program was only possible in close collaboration of all participating parties (IT, PACS administrators, MTA, medical physics experts). Also the manufactures have to be involved, as the access to the raw image data needs additional configuration effort.

This integral approach within our work group enabled us to take the different opinions and requirements into account. This resulted in a high acceptance of the programme. Due to the developed automatic reading software, that was deployed with every digital device from the beginning, daily quality control tests are efficient, objective and centrally documented.

We have been faced with several challenges during our work: The software needs adaptation to the different hospital IT systems. The MTA needed dedicated training to correctly apply the developed QA procedure and use the software without problem. A help desk is useful for the users that have problems performing the tests.

References

1. European Union: Council Directive 97/43/EURATOM (1997)
2. EUREF: European guidelines for quality assurance in breast cancer screening and diagnosis, 4th edn. (2006)
3. Blendl, C., Hermann, K.-P., Mertelmeier, T.: PAS 1054 - Anforderungen und Prüfverfahren für digitale Mammographie Einrichtungen. NAR im DIN. Beuth Verlag, Berlin (2005)
4. Ewen, et al.: European Protocol for the Quality Control of the Physical and Technical Aspects of Mammography Screening - Part B: Digital Mammography und die Abnahme und Konstanzprüfungen nach PAS 1054 im Vergleich. NAR im DIN. Beuth Verlag, Berlin (2005)
5. Bijkerk, K.R., Lindeijer, J.M., Thijssen, M.A.O.: The CDMAM phantom: a contrast-detail phantom for mammography. Radiology 185, 395 (1993)
6. Jahnen, A., Schilz, C., Shannoun, F., Schreiner, A., Hermen, J., Moll, C.: Optimage central organised image quality control including statistics and reporting. Radiation Protrotection Dosimetry 129(1-3), 253–257 (2008)
7. Shannoun, F., Schanck, J.M., Scharpantgen, A., Wagnon, M.C., Ben Daoud, M., Back, C.: Organisational Aspects of Mammography Screening In Digital Settings: First Experiences of Luxembourg. Radiation Protection Dosimetry 129(1-3), 195–198 (2008)
8. Jahnen, A., Hermen, J., Moll, C.: Optimage User Tutorial, Luxembourg (2009), `http://santec.tudor.lu/project/optimage`
9. Jerusalem, N., Moll, C., Jahnen, A.: Training in digital mammography through ImageJ. In: Proceedings of the ImageJ User and Developer Conference 2008, Luxembourg (2008)

Inter and Intra Observer Variability in a Semi-automatic Method for Measuring Volumetric Breast Density

Rosanne Verow[1], Michael Berks[1], Jennifer Diffey[1], Camilla Chung[1], Joanna Morrison[1], Mary Wilson[2], Caroline Boggis[2], Nicky Barr[2], Julie Morris[3], Alan Hufton[1], and Susan Astley[1]

[1] Imaging Science and Biomedical Engineering, School of Cancer and Imaging Sciences, Stopford Building, University of Manchester, Oxford Road, Manchester M13 9PT

[2] The Nightingale Centre and Genesis Breast Cancer Prevention Centre, University Hospital of South Manchester NHS Foundation Trust, Southmoor Road, Wythenshawe, Manchester, United Kingdom, M23 9LT

[3] Department of Medical Statistics, University Hospital of South Manchester NHS Foundation Trust, Southmoor Road, Wythenshawe, Manchester, United Kingdom, M23 9LT

jennifer.diffey@postgrad.manchester.ac.uk

Abstract. If breast density is to be incorporated into breast cancer risk prediction models, the technique used for measurement must be quantitative, accurate, objective and reproducible. We present a semi-automated method that has been used by three independent operators to measure glandular volume from the digitised mammograms of 29 women (116 images). Additionally, one operator used the method on 10 separate occasions on a sample of 24 images. Intra-observer variability was found to be acceptably low, with coefficients of variation ranging from 3.5 – 5.7% depending on mammographic view (intra-class correlation coefficient close to 1 in all cases). However, inter-observer variability was greater with significant differences in glandular volume recorded between observers. This was attributed to the method of breast edge detection. The development of a new automatic breast edge detection algorithm has resolved the issue. The average difference in glandular volume measurement between two independent operators in the cranio-caudal view is -0.89cm^3 (95% confidence interval -2.77 – 0.99 cm^3) using the new method, compared to 5.99cm^3 (95% confidence interval 2.72 – 9.76 cm^3) using the old method.

Keywords: breast density, volumetric technique, observer variability.

1 Introduction

Early detection of breast cancer is essential in providing effective treatment and improving the chance of survival. This is achieved to a significant extent through routine breast screening. However, it is particularly important to identify women at increased risk of the disease and breast cancer risk prediction models have been developed for this purpose, including the Gail [1], Claus [2] and Tyrer-Cuzick [3] models.

Despite evidence of a strong link between breast density and cancer risk, breast density was not included in risk prediction models until relatively recently. Barlow et al [4]

J. Martí et al. (Eds.): IWDM 2010, LNCS 6136, pp. 576–582, 2010.

developed a model incorporating breast density and use of hormone therapy as additional inputs. Breast density was found to be a statistically significant risk factor for breast cancer diagnosis in pre- and post-menopausal women and it is thought that its inclusion in risk prediction models may offer improved accuracy in the identification of women at high risk of developing breast cancer.

The reason for not including breast density in risk prediction models can be attributed to difficulties in measuring this parameter. If breast density is to be measured routinely, it is necessary to use a method which is quantitative, accurate, reproducible and objective. Operator dependency should be minimised in order to reduce intra and inter observer variability. An additional advantage of an automated technique is that it is less time consuming. This is essential if the technique is to be applied to a screening population.

We have developed a semi-automated method for measuring volumetric breast density. Three observers used this method on the mammograms of 29 women who had consented to take part in a study assessing the feasibility of using the method in the UK breast screening programme. Each woman had four mammographic views taken (left and right cranio-caudal, CC, and mediolateral-oblique, MLO) giving a total of 116 images. One operator used the method on 10 separate occasions to analyse a subset of 24 images from this sample. Following improvements to the breast edge detection algorithm, the new method was applied to the same sample of mammograms. We present results of intra and inter observer variability for the original method and inter observer variability for the method with the new automatic breast edge detection algorithm.

2 Materials and Methods

Our method for volumetric breast density measurement been described previously [5 - 7]. A calibrated stepwedge is imaged alongside the breast, with radio-opaque magnification markers on the compression paddle to enable accurate determination of breast thickness, taking into account paddle tilt. Using software written in Matlab, the operator is prompted to select the radio-opaque markers in a digitised mammogram; the location of each marker is determined using a Canny edge detector (Figure 1a). Secondly, they mark and define the edges of two steps on the stepwedge. Finally, they draw an approximate breast outline (Figure 1b).

Initially, the location of the breast edge was determined based on the gradient of a series of greyscale profiles plotted from the user-defined outline in a direction normal to the breast edge. The breast edge algorithm has since been improved to remove any operator dependency. It now uses the Hough transform and is described as follows: an approximate location for the breast edge is determined by i) applying a global threshold based on analysis of the grey-level histogram in each mammogram ii) applying morphological operators to the resulting binary image to isolate the main breast region. The approximate breast edge location is used to initialise an adapted active contour algorithm which then computes a more precise (and locally smooth) demarcation of the breast edge.

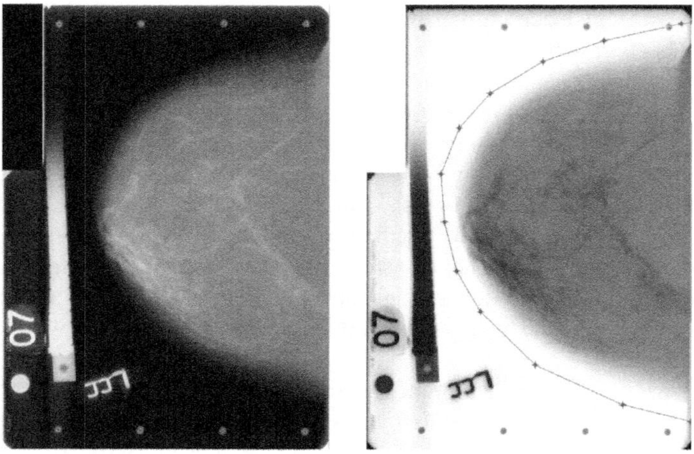

(a) (b)

Fig. 1. (a) original mammogram showing definition of markers; (b) approximate definition of breast edge by operator

Three independent operators were trained in the use of the software and carried out a practice session on 10 images before commencing the study. The same three operators then used the software with the original breast edge detection to calculate glandular volume from the digitised mammograms of 29 women (116 images). The sample was selected to include a range of breast thicknesses and densities (based on radiologist assessment of breast density) and to include both small (18x24cm) and large (24x30cm) format films. Within this sample, a subset of 24 images was analysed on 10 separate occasions by the same operator to enable assessment of intra-observer variability.

The software provides outputs of breast thickness, breast area, breast volume and glandular volume. Our method for breast thickness determination accounts for paddle tilt by measuring the thickness at a number of locations from the chest wall to the nipple edge [8]. The breast thickness measurement considered in this study was that at the chest wall.

The software with the improved breast edge detection was run again on the images marked up by two of the three observers; the only operator input required was the segmentation of the pectoral muscle, as the results of the original marker and step-wedge mark-up had been saved.

3 Results and Discussion

3.1 Intra-observer Variability

There was little intra-observer variability in the determination of glandular volume. Repeatability of the method was considered to be acceptable, with average coefficients of variation (CV) ranging from 3.49% to 5.73% depending on radiographic projection (Table 1). Although the CV for the RCC view exceeds 5%, the intra-class

Table 1. Coefficient of Variation and Intra-class Correlation Coefficient for glandular volume

	View			
	LCC	LMLO	RCC	RMLO
Average co-efficient of variation (%)	3.49	4.01	5.73	4.65
Intra-class correlation coefficient	0.96	0.95	0.97	0.96

correlation coefficient (ICC) is 0.97 which indicates repeatability. In all cases the ICC was close to 1. Results for breast thickness, breast area and breast volume all showed CV less than 5% and ICC close to 1.

3.2 Inter-observer Variability: Original Breast Edge Detection Method

Box and whisker diagrams were plotted for each mammographic view to compare the range of data generated by each operator for breast thickness, breast area, breast volume and glandular volume. Figure 2 shows the range of data for glandular volume in the RCC view for three operators using the original breast edge detection method.

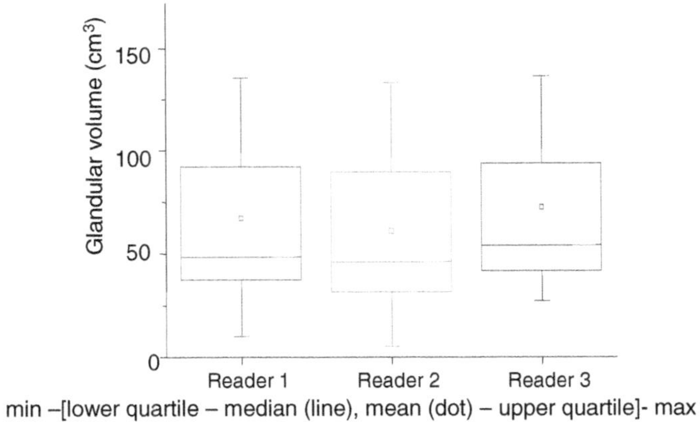

Fig. 2. Box and whisker plot showing the range of glandular volume data generated by each of the three readers for the RCC view

A number of trends were observed. There was greater spread and poorer agreement between operators for the CC views compared to the MLO views, with the greatest discrepancy being found for the RCC view. This was true for breast area, breast volume and glandular volume.

Reader 3 consistently generated the highest glandular volumes and Reader 2 generated the lowest glandular volumes for all views.

The relationship between inter-observer variability with glandular volume was examined further by comparing the individual glandular volume results obtained for each image with the average glandular volume for that image. Figure 3 shows the results for the left MLO view. The spread of data was found to increase significantly ($p<0.05$) as glandular volume increased.

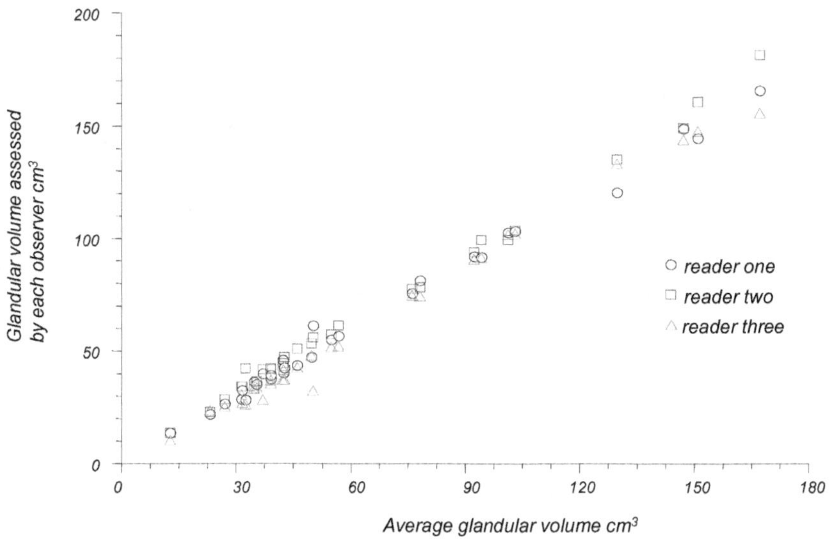

Fig. 3. Glandular volume generated by each observer plotted against the average glandular volume for that image

The comparison of breast thickness between operators was found to exhibit very little variation, with an average difference between readers of just 0.28mm and a maximum difference of 1.93mm. It is therefore thought that the greatest source of discrepancy is the mark-up of the breast edge and further investigation of the images that exhibited the greatest spread in glandular volume revealed that there was a considerable difference in the operator-defined breast edge and the subsequent determination of the breast edge by the software.

3.3 Inter-observer Variability: Automatic Breast Edge Detection Algorithm

The automatic method is highly reproducible; the mean difference in breast area for CC views is 0% when running the algorithm three times on the same set of images. There is still some variation in the breast area of the MLO views as the operator has to mark the pectoral muscle.

A box and whisker plot showing the spread of data for glandular volume between Readers 1 and 2 for the RCC view is shown below in Figure 4. It can be seen that the mean and median values are much closer using the automatic breast edge detection algorithm that those in Figure 1 using the original method. The same was found for breast area and breast volume, as shown in Table 2.

It is interesting to note the effect of the new breast edge detection model in terms of the percentage bias in glandular volume. The mean difference between readers expressed in terms of the discrepancy relative to the median value is 12% (maximum 18%) using the original breast edge detection method, compared to only 1.8% (maximum 5.5%) using the new method.

Table 2. Average and 95% confidence interval for the differences in breast area, breast volume and glandular volume for Reader 1 and Reader 2, RCC view

Parameter	Measurement	Original method	Automatic method
Breast area (cm^2)	Mean difference (Reader 1 – Reader 2)	-19.6	0.0
	95% limits of agreement	-23.7 – (-15.6)	0.0
Breast volume (cm^3)	Mean difference (Reader 1 – Reader 2)	-103.3	-1.31
	95% limits of agreement	-127.4 – (-79.2)	-3.25 – 0.62
Glandular volume (cm^3)	Mean difference (Reader 1 – Reader 2)	5.99	-0.89
	95% limits of agreement	2.72 - 9.26	-2.77 - 0.99

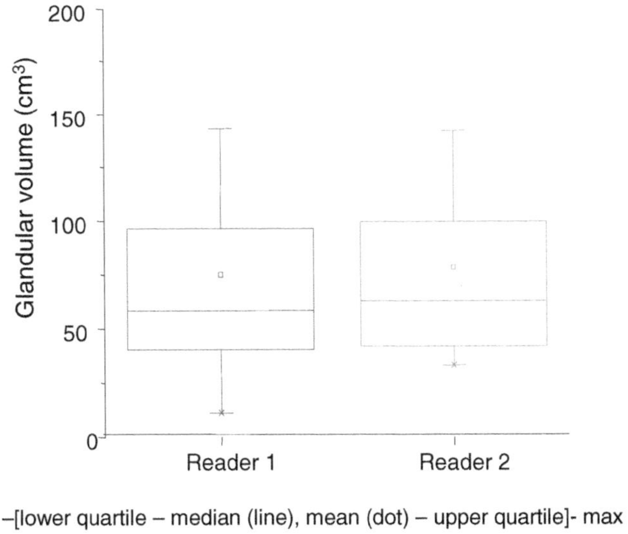

Fig. 4. Box and whisker plot showing the range of glandular volume data generated by Readers 1 and 2 for the RCC view

4 Conclusion

The stepwedge based method gave repeatable results when images were marked up by a single operator. However, there was inter-observer variability in the results for breast area, breast volume and glandular volume, with the largest differences between operators being observed for glandular volume. The disagreement in the measurement of glandular volume generated by each reader was found to increase as breast glandularity increased. The spread in data was consistently higher for the RCC view. It is interesting to note that the greatest intra-observer variability was also found for the RCC view. The reasons for this are not clear but warrant further investigation.

The inter-observer variability was attributed to differences in the way in which each operator marked the breast edge, rather than their definition of other landmarks on the image. This had a pronounced effect on the way in which the software calculated the exact breast edge. Determination of breast thickness exhibited negligible inter-observer variability, with a mean difference of only 0.28mm between observers.

A new automatic breast edge detection algorithm has been developed and has shown promising results in improving inter-observer variability. Work is in progress to improve this algorithm to auto-detect the pectoral muscle as currently this is still defined by the operator on the MLO views.

Acknowledgements

We are very grateful to John Lewis plc and the breast cancer prevention charity Genesis for funding this study. Many thanks go to Tom Hamnett from the mechanical workshop at The Christie Hospital, Manchester for constructing the stepwedges and markers.

References

1. Gail, M.H., Brinton, L.A., Byar, D.P., Corle, D.K., Green, S.B., Schairer, C., Mulvhill, J.J.: Projecting Individualised Probabilities of Developing Breast Cancer for White Females who are being Examined Annually. J. Nat. Can. Inst. 81, 1879–1886 (1989)
2. Claus, E.B., Risch, N., Thompson, W.D.: Autosomal Dominant Inheritance of Early-Onset Breast Cancer: Implications for Risk Prediction. Cancer 73, 643–651 (1994)
3. Tyrer, J., Duffy, S.W., Cuzick, J.: A Breast Cancer Prediction Model Incorporating Familial and Personal Risk Factors. Stats. Med. 23, 1111–1130 (2004)
4. Barlow, W.E., White, E., Ballard-Barbash, R., Vacek, P.M., Titus-Ernstoff, L., Carney, P.A., Tice, J.A., Buist, D.S.M., Geller, B.M., Rosenberg, R., Yankaskas, B.C., Kerlikowske, K.: Prospective Breast Cancer Risk Prediction Model for Women Undergoing Screening Mammography. J. Nat. Can. Inst. 98, 1204–1212 (2006)
5. Hufton, A.P., Astley, S.M., Marchant, T.E., Patel, H.G.: A Method for the Quantification of Dense Breast Tissue from Digitised Mammograms. In: Astley, S.M., Brady, M., Rose, C., Zwiggelaar, R. (eds.) IWDM 2006. LNCS, vol. 4046, Springer, Heidelberg (2006)
6. Patel, H.G., Astley, S.M., Hufton, A.P., Harvie, M., Hagan, K., Marchant, T.E., Hillier, V., Howell, A.: Automated Breast Tissue Measurement of Women at Increased Risk of Breast Cancer. In: Astley, S.M., Brady, M., Rose, C., Zwiggelaar, R. (eds.) IWDM 2006. LNCS, vol. 4046, pp. 131–136. Springer, Heidelberg (2006)
7. Diffey, J., Hufton, A., Astley, S.: A New Stepwedge for the Volumetric Measurement of Mammographic Density. In: Astley, S.M., Brady, M., Rose, C., Zwiggelaar, R. (eds.) IWDM 2006. LNCS, vol. 4046, pp. 1–9. Springer, Heidelberg (2006)
8. Diffey, J., Hufton, A., Beeston, C., Marchant, T., Astley, S.: Quantifying Breast Thickness for Density Measurement. In: Krupinski, E.A. (ed.) IWDM 2008. LNCS, vol. 5116, pp. 651–658. Springer, Heidelberg (2008)

Optimization of the Exposure Parameters with Signal-to-Noise Ratios Considering Human Visual Characteristics in Digital Mammography

Maki Yamada[1], Yuri Kato[1], Naotoshi Fujita[2], and Yoshie Kodera[3]

[1] Graduate School of Medical Sciences, Nagoya University, 1-1-20 Daiko-minami, Higashi-ku, Nagoya 461-8673, Japan
[2] Department of Radiological Technology, Nagoya University Hospital, 65 Tsurumai-Cho, Showa-Ku, Nagoya 466-8550, Japan
[3] School of Health Sciences, Nagoya University, 1-1-20 Daiko-minami, Higashi-ku, Nagoya 461-8673, Japan

Abstract. The use of digital mammography systems has become widespread recently. However, the optimal exposure parameters are uncertain in clinical practice. We need to optimize the exposure parameter in digital mammography while maximizing image quality and minimizing patient dose. The purpose of this study was to evaluate the most beneficial exposure variable—tube voltage for each compressed breast thickness—with these indices: noise power spectrum, noise equivalent quanta, detective quantum efficiency, and signal-to-noise ratios (SNR). In this study, the SNRs were derived from the perceived statistical decision theory model with the internal noise of eye-brain system (SNR*i*), contrived and studied by Loo LN [1], Ishida M *et al.* [2] These image quality indices were obtained under a fixed average glandular dose (AGD) and a fixed image contrast. Our results indicated that when the image contrast and AGD was constant, for phantom thinner than 5 cm, an increase of the tube voltage did not improve the noise property of images very much. The results also showed that image property with the target/filter Mo/Rh was better than that with Mo/Mo for phantom thicker than 4 cm. In general, it is said that high tube voltage delivers improved noise property. Our result indicates that this common theory is not realized with the x-ray energy level for mammography.

Keywords: digital mammography, average glandular dose, noise power spectrum, signal-to-noise ratio, noise equivalent quanta, detective quantum efficiency.

1 Introduction

The number of female breast cancer patients in Japan has been increasing recently. The prevalence of breast cancer has increased especially among women in their middle and early years. The mortality rate of breast cancer is also increasing year by year. Mammography is helpful in the early detection of breast cancer. However, it is said that Japanese women have dense breasts, which makes it difficult to find a tumor on

J. Martí et al. (Eds.): IWDM 2010, LNCS 6136, pp. 583–590, 2010.

mammography. There is very little difference between the x-ray attenuation of the glandular tissue and that of tumor tissue, so the exposure parameter for mammography should be chosen appropriately. It requires maximization of the image quality and minimization of patient dose. The purpose of this study was to evaluate the most beneficial exposure variable—tube voltage for each compressed breast thickness—using these indices: noise power spectrum (NPS), noise equivalent quanta (NEQ), detective quantum efficiency (DQE), and signal-to-noise ratios (SNR). In this study, the SNRs were derived from the perceived statistical decision theory model with the internal noise of eye-brain system (SNRi), contrived and studied by Loo LN [1], Ishida M *et al.* [2] We measured those indices under a fixed average glandular dose (ADG). The NPS, NEQ, and DQE were evaluated with fixed image contrast.

2 Materials and Methods

2.1 Equipment Used in This Study

The mammography equipment used in this study was Mermaid model MGU-100B. The CR reader used was REGIUS V stage, Model 190. The CR plate used was CP1M200 (with columnar crystal phosphors). This equipment were manufactured by Konica Minolta MG. X-ray images of low contrast objects were obtained for the measurement of the SNR*i*. We used a dosimeter, Radiation Monitor Controller model 9015 manufactured by Radcal Corporation, for the evaluation of AGD. To measure the contrasts and to calculate NPS, we obtained images of acrylic steps (thickness: 1–10 mm) and uniformly exposed x-ray images. We used acrylic plates (thickness: 2, 4, 5, and 7 cm) as breast phantoms. Mo/Mo and Mo/Rh were chosen for the target/filter combinations. As the source of low contrast signal, we used a resinous disc (diameter: 4.2 mm, thickness: 4.5 mm). To measure the presampled MTF for NEQ and DQE, we

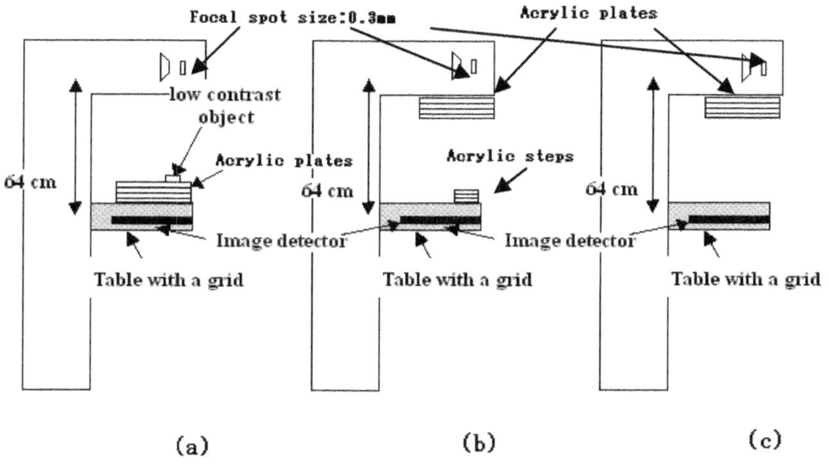

Fig. 1. Geometrical layouts for the measurements of SNRp (a), the image contrasts (b) and uniform exposure to calculate the NPS (c)

obtained images of the tungstenium edge. For the calculation of DQE, we require the number of quanta q, which contributes the image. To survey q, we measured exposed x-ray quanta with the spectrometer RAMTEC 413 manufactured by Toyo Medic Corporation.

2.2 Measurement of the Average Glandular Dose

We measured AGD and chose the exposure level that gave the AGD on EUREF dose acceptable level (Table 1) for every combination of the thicknesses of phantoms and the tube voltages (25, 28, 30, 32, 35 kV). We could not adjust the exposure level for AGD closely because of the restriction of the machine. Therefore we selected the exposure level as near as possible to the EUREF acceptable level dose. Table 2 shows the exposure conditions for 4-cm thick polymethyl methacrylate (PMMA) with Mo/Mo. For another thickness of phantom and target/filter combination, we chose the exposure level in the same way.

Table 1. EUREF dose acceptable level for each thickness of PMMA and its equivalent breast thickness

PMMA thickness [cm]	Equivalent breast thickness [cm]	Dose acceptable level [mGy]
2	2.1	0.6
4	4.5	1.6
5	6.0	2.4
7	9.0	5.1

Table 2. Exposure level for 4-cm-thick PMMA with Mo/Mo

AGD [mGy]	25 kV	28 kV	30 kV	32 kV
	Exposure level [mAs]			
1.3	80	50	40	32
1.7	100	63	50	40

2.3 Measurement of Image Contrasts

NPS depends on the image contrast, and the image contrast changes with the tube voltage. For this reason, we evaluated NPS under fixed image contrast. In order to do this, the image contrast of each tube voltage was measured. We obtained the images of the acrylic plates and the acrylic steps, which were put on the chest wall side of the table for the contrast evaluation (Fig. 2). We read out the raw data of the acrylic step images and found the contrast by the pixel value of each step.

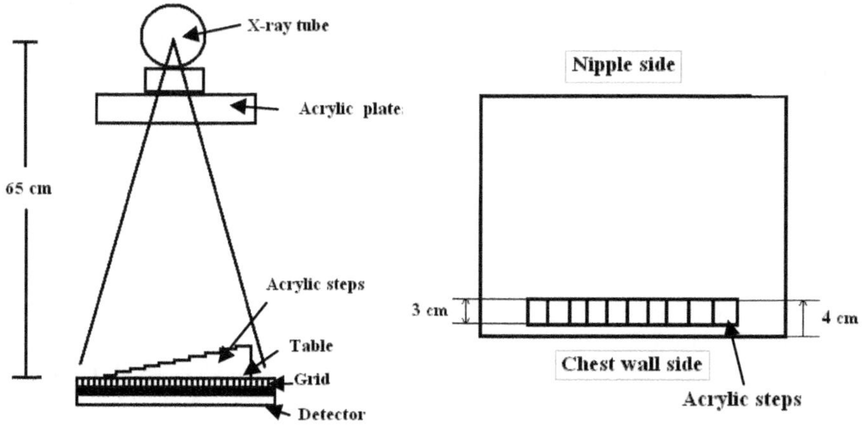

(a) Overview (b) Overhead view

Fig. 2. Layout of measurement of the contrast

2.4 Calculation of NPS

We calculated the NPS by the 2 dimensional fast Fourier transform method. When we evaluated the NPS values at 25, 30, 32, and 35 kV, we adjusted the contrasts under these tube voltages to under 28 kV. To tune the image contrasts, we computed the NPS and multiplied the following correction factor *k*.

$$k = (\text{image contrast at 28 kV / image contrast at another tube voltage})^2 \quad ...(1)$$

2.5 Calculation of NEQ and the DQE

To calculate the NEQ and the DQE for each system, we used the NPS, the presampled MTF and number of x-ray quanta. The MTF was measured with the edge method. The number of x-ray quanta *q*, which contributes the image was quantified with the spectrometer. Using the value of presampled MTF, NPS, and *q*, we computed the NEQ and the DQE as described below.

$$NEQ(u) = \frac{G^2(\log_{10} e)^2 MTF^2(u)}{NPS_{\Delta PV}(u)} \quad ...(2)$$

$$DQE(u) = \frac{NEQ(u)}{q} = \frac{G^2(\log_{10} e)^2 MTF^2(u)}{qNPS_{\Delta PV}(u)} \quad ... (3)$$

Here, u is spatial frequency; G is the gradient of the digital characteristic curve; $NPS_{\Delta PV}$ is NPS calculated with the pixel value; and q is the number of incident x-ray quanta per unit which contribute the image composition. To measure the presampled MTF, an exposure level of 50 mAs, a tube voltage of 28 kV, and a target/filter of Mo/Mo were used for the exposure conditions.

2.6. Calculation of SNRi

SNRi, which takes the spatial frequency response of the human visual system and the internal noise of eye-brain system into consideration, is written as

$$SNR_i = \frac{S_p}{\sqrt{N_p^{\,2} + N_i^{\,2}}} = \frac{SNR_p}{\sqrt{1 + (N_i / N_p)^2}} \quad \ldots(4)$$

In this equation (4), the *SNRp* represents the SNR only with the human visual characteristic, without the internal noise. The SNRp is shown as (5).

$$SNR_p = \frac{S_p}{N_p} \quad \ldots(5)$$

where

$$S_p = \iint [DIS^2(u,v) VRF^2(u,v) du dv]$$

and

$$N_p = \sqrt{\frac{\iint [DIS^2(u,v)VRF^2(u,v)dudv] \cdot [DWS^2(u,v)VRF^2(u,v)dudv]}{\iint [DIS^2(u,v)VRF^2(u,v)dudv]}}$$

DIS(u,v) and *DWS(u,v)* mean the displayed signal spectrum and the NPS, respectively. The visual spatial frequency response of the human observer[3)] is represented by *VRF(u,v).*

The *Ni* shown in equation (4) means the internal noise[4)] that is caused by the noise inherent in an observer, for instance, the noise in relation to neurophysiological instability and to fluctuations of the observer's memory.

3 Results

3.1 NPS and NEQ

The NPS was almost stable when the tube voltage changed between 2-cm and 4-cm-thick phantoms. There were few differences in the NPS values for each tube voltage with 5-cm-thick phantom, and we observed the tendency that high tube voltage improves the noise property of the image. This trend was more remarkable for 7-cm-thick phantom. As the combination of target/filter, Mo/Rh was more beneficial than Mo/Mo for phantoms thicker than 4 cm. This trend was also more conspicuous for thicker phantom.

For 2-cm-thick phantoms, the NEQ did not change with the tube voltage and filtration except for 32 and 35 kV with Rh filter. The NEQ values on these two exposure conditions were lower than that on others. Also, 35 kV with Rh filter gave the highest NEQ for phantom thicker than 4 cm. In these cases, the superiority of Rh filter to Mo became clearer for thicker phantom.

3.2 DQE

The results of the DQE are shown in Figures 3-5. The DQE value was large for the low tube voltage for every thickness of phantom. The difference between the tube voltages was clear for the thick object. The DQE with Mo filter was larger than that with Rh filter. The difference between the filtrations was clear for the thin object.

3.3 SNRi

I show you the result for the SNRi in Figure 6. For 2 and 4cm-thick objects, we recognized the tendency that the SNRi value with Mo filter was larger than that with Rh filter. The relation of the SNRi value and tube voltage was not shown in our result however.

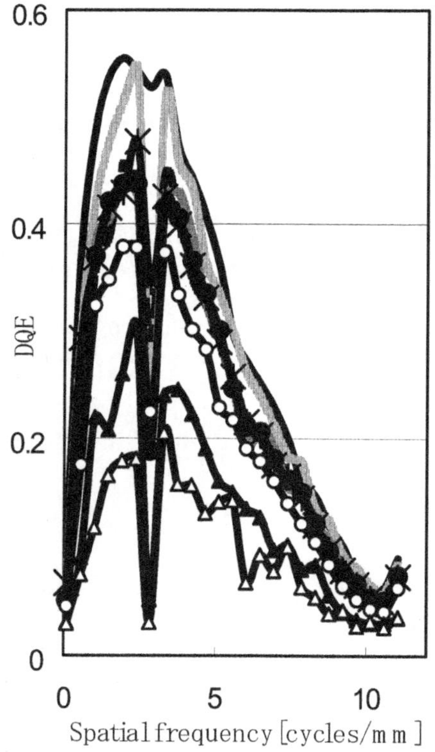

Fig. 3. DQE for 2-cm-thick phantom

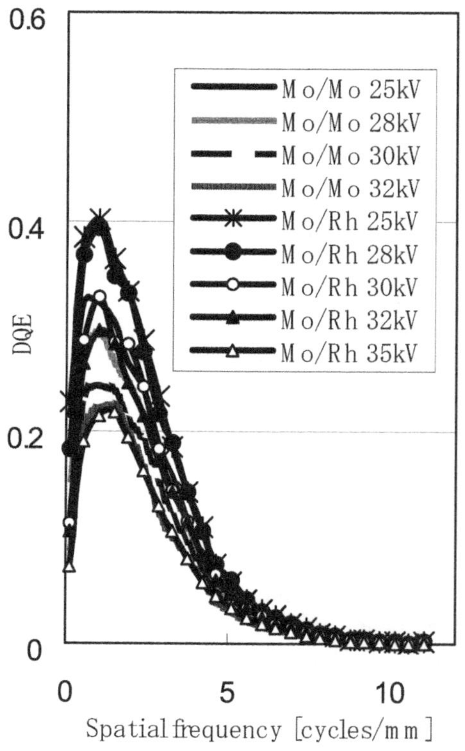

Fig. 4. DQE for 4-cm-thick phantom

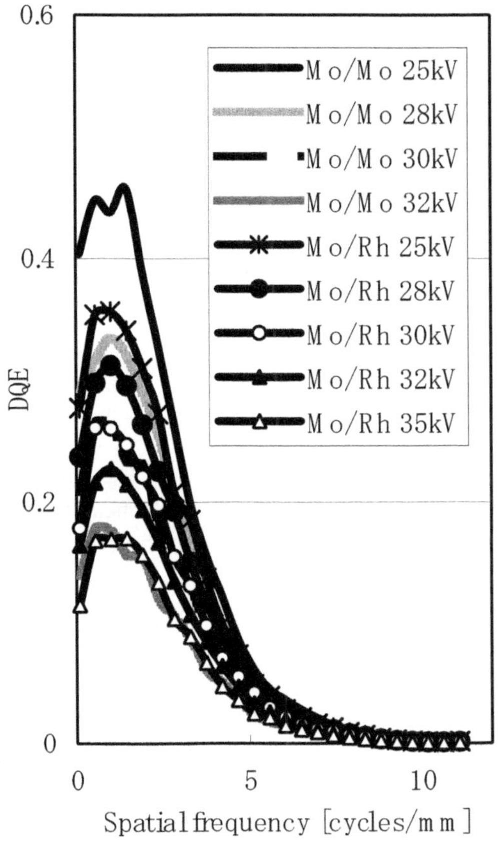

Fig. 5. DQE for 7-cm-thick phantom

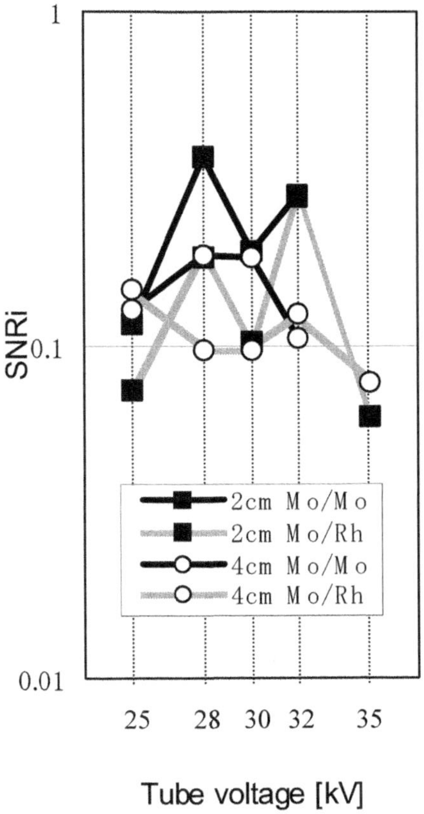

Fig. 6. SNRi for 2 and 4 -cm-thick phantom

4 Discussion

The results of NPS and NEQ indicate that when the image contrast and AGD was constant, for a phantom thinner than 5 cm, an increase of the tube voltage did not improve the noise property of images very much. The results also showed that image property with the target/filter Mo/Rh was better than that with Mo/Mo for the phantom thicker than 4 cm. In general, it is said that high tube voltage delivers fine noise property. The result of our study indicates that this common theory is not realized with the x- ray energy level for mammography.

From the results of DQE, we can say that the contribution of high energy x-ray photons for image is less than that of low energy x-ray photons. The effect of beam hardening makes the difference between tube voltages clear for thick objects.

We could not recognize the relation of the SNRi value to the level of x-ray energy. We should make accuracy high with more many low contrast objects. In this study, the SNR*i* considered in the dimension of pixel value of raw data image but the index needs to be calculated in the dimension of brightness.

5 Conclusion

In conclusion, when the image contrast and AGD was constant, for the phantom thinner than 5 cm, an increase of the tube voltage did not improve the noise property of images very much. The results also showed that image property with the target/filter Mo/Rh was superior to that with Mo/Mo for the phantom thicker than 4 cm. Our result signified that high tube voltages did not improve the noise property of images with the x-ray energy level for mammography.

From the result of DQE, low energy x-ray photons contributed image effectively. This tendency was clear for thick object.

In addition, we need to examine SNRi in brightness with accuracy.

Acknowledgments

We would like to thank Mr. Hiromu Ohara, Dr. Chika Honda, Mr. Hirotaka Hara, Mr. Sumiya Nagatsuka, Mr. Tomonori Gidou, and Ms. Miyuki Mori from Konica Minolta MG for their helpful discussions and advice.

References

1. Loot, L.-N.D., Doi, K., Metz, C.E.: A comparison of physical image quality indices and observer performance in the radiographic detection of nylon beads. Phys. Med. Biol. 29(7), 837–856 (1984)
2. Ishida, M., Doi, K., Loo, L.-N., et al.: Digital image processing: effect on detectability of simulated low contrast radiographic patterns. Radiology 150, 569–575 (1984)
3. Jain, A.K.: Fundamentals of Digital Image Processing, pp. 49–79. Prentice Hall, Englewood Cliffs (1989)
4. Ohara, K., Chan, H.-P., Doi, K., et al.: Investigation of basic imaging properties in digital radiography. 8. Detection of simulated low-contrast objects in subtraction angiographic images. Med. Phys. 13(3), 304–311 (1986)

Ideal Observer Comparison between Tomographic and Projection x-Ray Images of the Breast

John M. Boone[1], Nathan J. Packard[1], and Craig K. Abbey[2]

[1] Department of Radiology, University of California Davis, Sacramento, California, USA
[2] Department of Psychology, University of California Santa Barbara Santa Barbara, California, USA

Abstract. The image data from 348 breast CT studies performed on patient volunteers in a Phase II clinical trial were used in a pre-whitened matched filter (PWMF) model observer study. This computer-based investigation simulated the addition of spherical lesions of various diameters to the breast data sets, and a series of images with different slice thicknesses ranging from 0.3 mm to 44 mm were created by voxel averaging. The PWMF was tailored to each slice thickness, and signal known exactly receiver operating characteristic analysis was performed. A total of 1000 lesions and "non-lesions" were simulated for each diameter on each breast data set. Receiver operating characteristic (ROC) curve analysis was performed, and sensitivity at 95% specificity was computed. Thin slice imaging as with breast CT demonstrated significantly better sensitivity than thick slice images as with mammography, with improvements in sensitivity ranging from 5% to 50%.

Keywords: Breast CT, Ideal Observers, Receiver Operating Characteristic Curves.

1 Introduction

Breast CT has been proposed as an alternative for breast cancer screening and diagnosis. At our institution, two breast CT scanners have been designed, fabricated and are being using to image suspected breast cancer patients in the clinical setting. Over 300 women have been imaged on the two UC Davis scanners to date. The technical performance and dose issues of breast CT has been discussed in previous publications [1-15].

While we are still acquiring patient images for a pending human observer clinical trial, the vast experience that breast imaging physicians have with mammography will inevitably lead to a training bias in any ROC experiments performed with radiologists. We therefore have sought to study the fundamental, "theoretical", performance of breast CT image data sets in comparison to traditional projection imaging. This ideal observer study was conducted to perform such a comparison, using pre-whitened matched filter techniques on the breast CT data sets.

J. Martí et al. (Eds.): IWDM 2010, LNCS 6136, pp. 591–597, 2010.

2 Methods and Materials

The breast CT systems built in our laboratory have been used in both Phase I and Phase II clinical trials, and approximately 300 women have been imaged to date on these two systems. The breast CT systems use a 40 cm x 30 cm flat panel detector system with a readout rate of 30 frames per second, with 2 x 2 detector elements binned together to yield a pixel size of 0.388 mm. Projected to the isocenter, the sampling pitch is about 0.210 mm. A 16.6 s acquisition of 500 images is used to image the breast during patient breath-hold, with the breast hanging in pendant geometry through a hole in the patient couch. No breast compression was used. The x-ray system uses a tungsten anode with 0.2 mm Cu filtration at 80 kVp, and the radiation dose level is adjusted to be the same as for two view mammography for each individual women, based upon her breast size and composition.

The raw data set of 500 projection images is flat field corrected, logged, and Fourier filtered prior to being backprojected to a 512 x 512 x 512 CT image matrix. The CT slice thickness is approximately 0.28 mm, depending on the dimensions of the women's breast. The overall reconstruction process results in a ~512^3 data set of nearly isotropic voxels, with the data converted to Hounsfield Units (HU).

Spherical breast lesions, with HU equal to that of the glandular tissue, were added to each breast CT data set, one at a time. The lesions ranged in diameter from 1.0 to 15 mm, with 6 different diameters used. A number of different methods for placing the lesion into the breast image data set were explored, as shown in Figure 1. Because the breast lesions have approximately the same density as glandular tissue, adding the HU values of the lesion in not possible when there is glandular tissue present. The final procedure used a distance transform and an edge smoothing process, to assure that the inserted lesion is as close to an actual spherical lesion in the breast as is possible. A total of 1000 lesions of each size, and corresponding "non-lesions" (where no gray scale was added – these are "normal" breasts), were added to each breast CT data set.

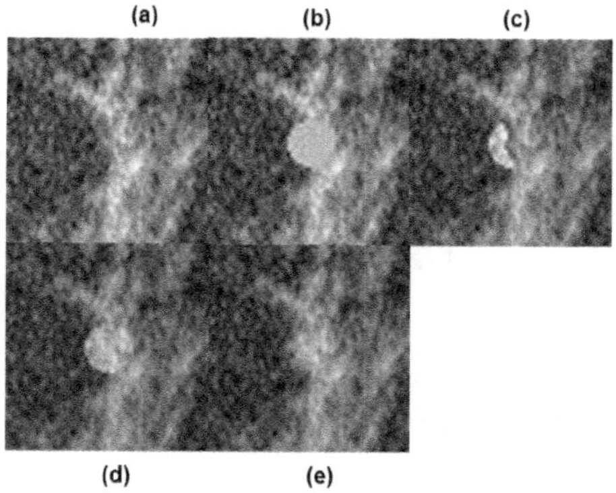

Fig. 1A. Several strategies for lesion insertion, a) background, b) direct intensity replacement, c) intensity addition to adipose tissue only, d) modulation on tissue boundary, e) modulation on tissue and sphere boundary

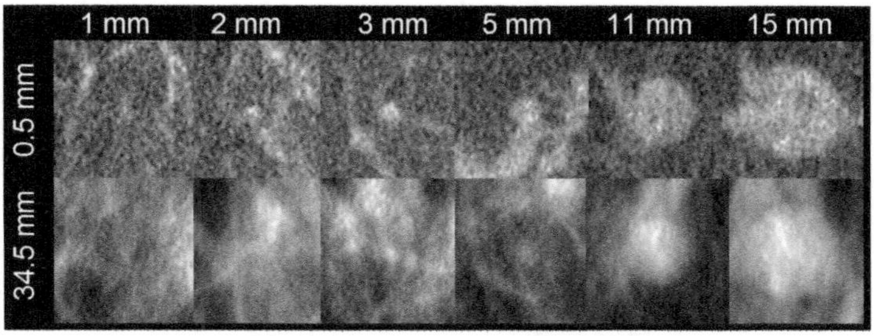

Fig. 1B. The results of the final approach for inserting simulated lesions into breast volumes are shown in this figure, where the 0.5 mm slice thickness corresponds to breast CT, and the 34.5 mm projection thickness is more like mammography in appearance. Lesions of various diameters were inserted, and these increase from left to right in this panel of images.

Once the lesion was synthetically added, the breast image data was projected onto a 256 x 256 region of interest and the signal-known-exactly (SKE) pre-whitened matched filter computation was performed, resulting in a filter response function. The matched filter was computed for each slice thickness studied. For lesions which are larger than the slice thickness, then that image cuts through the lesion and it would appear more disk-like than a spherical projection. Figure 2a illustrates the relative filter response (the sum of the product of the matched filter with the projected image data), and the ROC curve that is generated from this data is shown in Figure 2b.

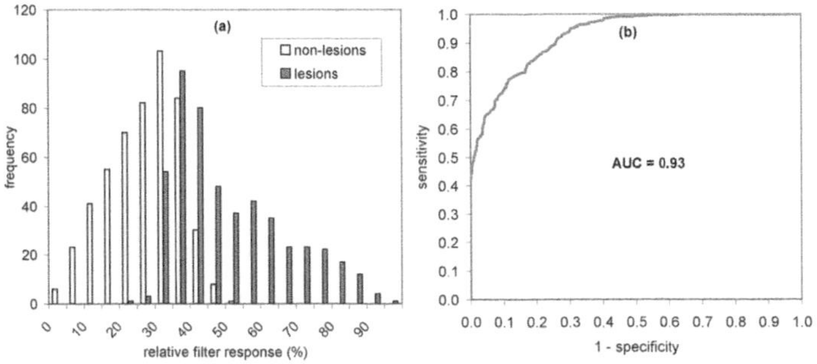

Fig. 2. Distribution of the PWMF response to 11mm lesions and non-lesions (a) with the corresponding ROC curve (b). This data was generated from a scan of a particularly dense breast for thin projections of 0.75 mm (e.g. bCT).

3 Results

Figure 3 illustrates the mean signal to noise ratio (SNR) for lesions of different diameters as a function of projection length. The projection length is simply the slice thickness, and thinner projection lengths (~0.3 mm) correspond to breast CT while thicker projection lengths (~44 mm) are thought to be more representative of mammographic images. The SNR is seen to be near maximum for each lesion diameter

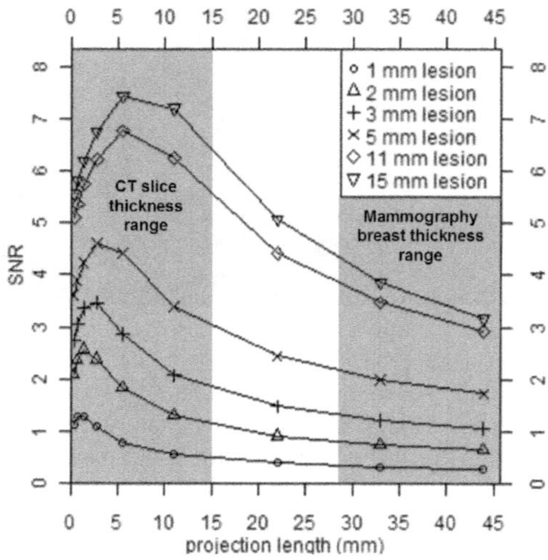

Fig. 3. The signal to noise ratio (SNR) computed for lesions of various diameters (indicated) is plotted as a function of the projection length (slice thickness). Thin slice images such as with breast CT are on the left of this figure, while thick slice images towards the right are typical of mammography. It is seen that the SNR is higher for thinner images, in general.

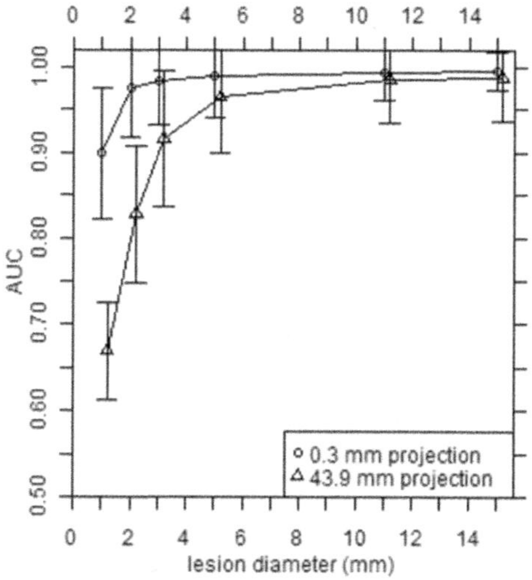

Fig. 4. The average area under the ROC curve, AUC, is shown (with one standard deviation error bars) as a function of lesion diameter for a 0.3 mm projection image (breast CT) and a 44 mm projection ("mammography"). These data were computed in fatty breasts, as determined by radiologist interpretation using the BIRADS scale for fatty breasts.

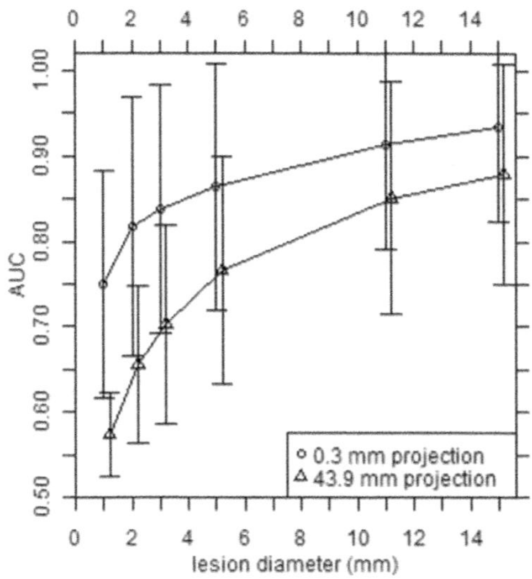

Fig. 5. The area under the ROC curve (AUC) is shown as a function of lesion diameter for 0.3 mm projection (breast CT) and 44 mm projection (mammography) thicknesses. These data are for dense breasts.

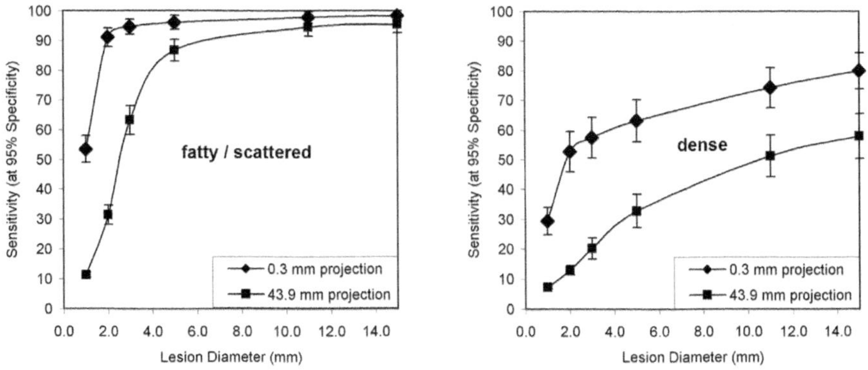

Fig. 6. The sensitivity at a realistic 95% specificity was computed from the ROC curves, and these results are shown for fatty breasts (left) and dense breasts (right).

when the projection length is slightly less than the lesion diameter, which is not surprising. Once reaching a maximum SNR, the SNR continues to decrease as the projection length increases.

An ROC curve using the 2000 lesions and non-lesions was computed for each of the 348 breast volume data sets, and for each lesion diameter. The area under the ROC curve, AUC, was computed and these data are shown in Figure 4 for fatty breasts. The thin (0.30 mm) slice results show significant improvement in AUC compared to the thicker images, and indeed for this cohort of fatty breast data, the AUC for lesions larger than about 5 mm is almost perfect at 1.0, while there is a pronounced drop-off in AUC for the thicker images as the lesion diameter decreases below 10 mm.

Figure 5 illustrates the AUC for dense breasts for both thin and thick slice imaging. While both curves are significantly lower in this dense breast cohort compared to the fatty breast cohort results seen in Figure 4, the thin slice images remain superior in terms of AUC compared to thick slice images. Indeed, the separation between the two curves in Figure 5 (dense breasts) is in general greater than that of Figure 4 (fatty breasts), suggesting that breast CT's performance in dense breasts, while not perfect, will nevertheless be substantially improved over projection imaging techniques.

The sensitivity was computed at 95% specificity for each ROC curve (the data from Figures 4 and 5) and the averages are shown in Figure 6. In Figure 6a, the data are shown for the fatty breasts, and for Figure 6b, the data shown are for dense breasts. It is quite clear from Figure 6 that the thin (0.3 mm) sections corresponding to breast CT out-performed the projected data (44 mm), which simulate mammographic projections. Dramatic increases in sensitivity are observed for all but the largest (and most obvious) lesions, Although the overall performance of breast CT was reduced in denser breasts, the performance remained consistently better than projection imaging across all lesion diameters for the case of dense breasts.

4 Conclusions

Using a pre-whitened matched filter methodology, the results of this computer simulation exercise demonstrate a statistically significant ($p<0.05$) improvement in lesion detection for thin slice (breast CT) images as compared to projection images (mammography). While the improvement of breast CT degrades in dense breasts relative to fatty breasts, the drop in detection performance is less for breast CT compared to projection imaging approaches.

Only mass lesions were studied, and microcalcifications were not. The results of this study, while demonstrated in computer modeling, lend support to the notion that tomographic imaging of the breast may lead to better cancer detection rates for smaller and earlier cancers. While these results are based on computer simulations and not on human observers, the data generated show good consistency and the trends are consistent with clinical experience. Unlike with human observer studies, this investigation used over 7 million simulated lesions for these detection experiments , a number not achievable with human observer studies.

References

1. Boone, J.M., Nelson, T.R., Lindfors, K.K., Seibert, J.A.: Dedicated breast CT: Radiation dose and image quality evaluation. Radiology 221, 657–667 (2001)
2. Boone, J.M., Shah, N., Nelson, T.R.: A comprehensive analysis of DgNCT coefficients for pendant-geometry cone-beam breast computed tomography. Medical Physics 31, 226–235 (2004)
3. Kwan, A.L.C., Boone, J.M., Shah, N.: Evaluation of x-ray scatter properties in a dedicated cone-beam breast CT scanner. Med. Phys. 32, 2967–2975 (2005)
4. Boone, J.M., Kwan, A.L.C., Seibert, J.A., Shah, N., Lindfors, K.K.: Technique factors and their relationship to radiation dose in pendant geometry breast CT. Med. Phys. 32, 3767–3776 (2005)

5. Yang, K., Kwan, A.L.C., Miller, D.F., Boone, J.M.: A general geometric calibration method for cone beam CT systems. Med. Phys. 33, 1695–1706 (2006)
6. Boone, J.M., Lindfors, K.K.: Breast CT: potential for breast cancer screening and diagnosis. Future Oncol. 2, 351–356 (2006)
7. Boone, J.M., Kwan, A.L., Yang, K., Burkett, G.W., Lindfors, K.K., Nelson, T.R.: Computed tomography for imaging the breast. J. Mamary Gland Biol. Neoplasia 11, 103–111 (2006)
8. Yang, K., Kwan, A.L.C., Boone, J.M.: Computer modeling of the spatial resolution properties of a dedicated breast CT scanner. Med. Phys. 34(6), 2059–2069 (2006)
9. Kwan, A.L.C., Boone, J.M., Yang, K., Huang, S.Y.: Evaluation of the spatial resolution characteristics of a cone-beam breast CT scanner. Med. Phys. 34(1), 275–281 (2007)
10. Lindfors, K.K., Boone, J.M., Nelson, T.R., Yang, K., Kwan, A.L.C., Miller, D.W.: Dedication breast CT: Initial clinical experience. Radiology 246, 725–733 (2008)
11. Nelson, T.R., Cervino, L.I., Boone, J.M., Lindfors, K.K.: Classification of breast computed tomography data. Med. Phys. 35, 1078–1086 (2008)
12. Huang, S.-Y., Boone, J.M., Yang, K., Kwan, A.L.C., Packard, N.: Evaluation of breast skin thickness from 46 dedicated breast CT examinations. Med. Phys. 35, 1199–1206 (2008)
13. Xia, J.Q., Lo, J.Y., Yang, K., Floyd Jr., C.E., Boone, J.M.: Dedicated breast computed tomography: volume image denoising via a partial-diffusion equation based technique. Med. Phys. 35, 1950–1958 (2008)
14. Metheany, K.G., Abbey, C.K., Packard, N., Boone, J.M.: Characterizing anatomical variability in breast CT images. Med. Phys. 35, 4685–4694 (2008)
15. Yang, K., Kwan, A.L.C., Huang, S.-Y., Packard, N.J., Boone, J.M.: Noise power properties of a cone-beam CT system for breast cancer detection. Med. Phys. 35, 5317–5327 (2008)

Breast Structural Noise in Digital Breast Tomosynthesis and Its Dependence on Reconstruction Methods

Yue-Houng Hu, Michael Masiar, and Wei Zhao

Department of Radiology, State University of New York at Stony Brook, L-4, 120 Health Sciences Center, Stony Brook, New York, 11793-8460

Abstract. Digital breast tomosynthesis (DBT) is being investigated to overcome the obscuring effect of overlapping breast tissue in projection mammography. To quantify the effectiveness of DBT in reducing overlapping breast structures, it is important to investigate how breast structural noise propagates during the reconstruction process. Others have found that breast structure may be characterized as power law noise of the form κ/f^{β}. We investigate how the power law exponent, β, varies as a function of reconstruction methods. Clinical DBT data sets were used to analyze breast structural noise in both projection and reconstructed domains using different filter schemes of a filtered back projection (FBP) reconstruction algorithm. The dependence on filter settings was compared with cascaded linear system theory. The goal this work is to combine frequency domain analysis of breast structural noise with previous work on quantum noise in DBT and develop a generalized framework to optimize DBT for breast lesion detection.

Keywords: digital breast tomosynthesis, anatomical clutter, NPS, linear system model, ideal observer model.

1 Introduction

Lesion detection in screening mammography is often obscured by overlapping breast tissue structure, which is caused by the projection of three-dimensional (3D) breast anatomy onto a two-dimensional (2D) image. Digital breast tomosynthesis (DBT) is a promising 3D x-ray imaging modality that can reduce the obscuring effect of structural noise. By acquiring a set of x-ray projection images (<50) over a limited angular range (<50 degrees), a 3D volume may be reconstructed with thin image slices oriented parallel to a stationary detector.

Burgess has characterized the structural noise in screening mammography as following a power law in the form of κ/f^{β} [1], where f is the radial spatial frequency and β the exponent, quantifying the correlation of the spectral power. β was found to have a mean of approximately 3.

Metheany et al. studied the power law structural noise of a breast computed tomography (bCT) system [2]. The results showed that β of the reconstructed bCT images is lower than that of mammograms by approximately 1.

Engstrom et al. investigated β of a DBT system with an iterative, maximum likelihood-expectation maximization (MLEM) reconstruction algorithm [3]. They found

J. Martí et al. (Eds.): IWDM 2010, LNCS 6136, pp. 598–605, 2010.

that, on average, the value of β of DBT slices using MLEM was 0.194 lower than the respective projection images. Further, they studied the change in κ as a function of DBT reconstruction, finding a drop of approximately an order of magnitude in value. However, comparison is not straightforward as the values are represented in different units.

Previously our group has developed a cascaded linear system model for analyzing the signal and noise propagation from projection mammography to DBT using linear reconstruction methods (e.g. FBP). This model has been used successfully as a tool to optimize the detector performance, imaging geometry, and reconstruction filters in DBT. Combining the cascaded linear system model with a frequency domain ideal observer detectability index, *d'*, has also provided useful guidance in optimizing angular dose distribution for the detectability of microcalcifications in DBT, which is affected mainly by x-ray quantum noise [4]. The ultimate goal of the present work is to incorporate breast structural noise in the calculation of *d'* so that DBT optimization can be performed for the detection of breast masses. The ideal observer detectability index, *d'*, may be calculated for both 3D volumes and in-plane images using:

$$d'^2_{3D} = \iiint \frac{K_c^2\left[\left|O^2(f_x,f_y,f_z)\right|T^2(f_x,f_y,f_z)\right]}{S_{St}(f_x,f_y,f_z)+S_Q(f_x,f_y,f_z)}df_x df_y df_z \quad , \qquad (1)$$

$$d'^2_{in-plane} = \iint \frac{\int K_c^2\left[\left|O^2(f_x,f_y,f_z)\right|T^2(f_x,f_y,f_z)\right]df_z}{\int\left[S_{St}(f_x,f_y,f_z)+S_Q(f_x,f_y,f_z)\right]df_z}df_x df_y$$

where O is the object signal spectrum, T represents the system modulation transfer function (MTF), and K_c is a contrast constant used to vary the contrast of O. The noise power spectra include both S_Q and S_{St}, which represent the quantum and structural components of the total noise, respectively. Both quantities are expressed in the reconstructed image domain. The *x*- and *z*-directions are defined as the tube travel and slice-thickness directions, respectively, and the *y*-direction orthogonal to the *x*-*z* plane.

In this paper we will analyze the propagation of β through DBT reconstruction using a modified FBP algorithm with a number of different filtering schemes. Understanding the dependence of S_{St} on reconstruction methods will facilitate the implementation of structural noise into a cascaded linear system model, and allow the calculation of *d'* including both structural and quantum noise (Eq. 1).

2 Methods and Materials

2.1 Acquisition and Reconstruction

The clinical images used in the present investigation were obtained using a prototype DBT system (Siemens Novation^{TOMO})[1]. It employs a full field amorphous selenium (a-Se) digital mammography detector with 2816x3584 pixels and 85 μm pixel size. For each cranial-caudal (CC) or medio-lateral oblique (MLO) view, 25 projection

[1] Breast tomosynthesis is an investigational practice and is limited by U.S. law to investigational use. It is not commercially available in the U.S. and its future availability cannot be ensured.

images were acquired over a nominal angular range of ±22 degrees over a stationary detector. The 2x1 detector binning mode was selected for all acquisitions where 2 adjacent pixels in the tube-travel (*x*-) direction were binned. Images were then reconstructed using a modified FBP algorithm [5],[6],[7].

Four different reconstruction filters were implemented as shown in Fig. 1: (1) ramp filter (H_{RA}); (2) spectral apodization filter (H_{SA}); (3) slice-thickness filter (H_{ST}); and (4) polynomial (4th order) filter (H_{POLY}), which replaces H_{RA} and H_{SA} and provides a non-zero response at zero frequency to compensate for signal loss at low frequencies in DBT due to the incomplete angle of sampling [7]. The SA and ST filters are both in the form of Hanning windows applied in the *x*-direction and the *z*-direction, respectively. Their corresponding window widths *A* and *B* are defined in multiples of the Nyquist frequency of the detector (f_{NY} = 5.88 cycles/mm). We used four reconstruction filter combinations in our study: (1) simple backprojection (SBP, i.e. no filters) - designated as filter scheme 1 (FS1); (2) RA filter only (FS2); (3) RA filter and ST filter with *B* = 0.17 (FS3); and (4) POLY filter with response function shown in Fig. 1.

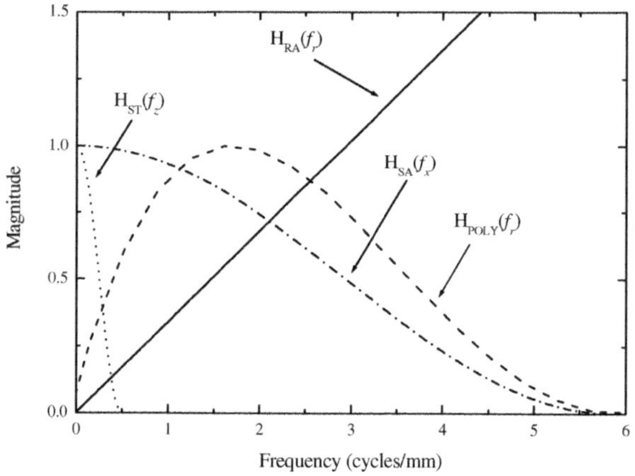

Fig. 1. Four reconstruction filters used in the modified FBP algorithm: H_{RA}, H_{SA}, H_{ST}, and H_{POLY} correspond to ramp, spectral apodization, slice-thickness, and polynomial (4th order) filters, respectively. The direction of filter application is indicated on the figure labels.

2.2 Noise Power Analysis

DBT image datasets from 11 patients were included in the present study, with all except one case consisting of both CC and MLO views. The single exception consisted of an MLO view only. Seven cases were of the right breast and four of the left.

2D noise power spectrum (NPS) analysis was performed on both projection images and reconstructed DBT image slices. The 3D NPS was calculated for the entire reconstructed image volume. The methods used for calculating 2D and 3D NPS were similar to those used previously for x-ray quantum NPS calculations [8].

For 2D NPS calculations of each projection image, the region of the breast with approximately equal thickness was manually selected. This cropped image was then divided into ROIs with 64x64 pixels, overlapping in both *x*- and *y*-directions by 32 pixels. The mean of each ROI was subtracted and a 2D Hanning window was applied to reduce the effects of spectral leakage. The NPS was obtained as the ensemble average of the modulus squared of the 2D Fourier transform (FT) of each ROI.

For 3D NPS calculations of each reconstructed volume, a region as large as 384x384x96 voxels was selected with uniform breast thickness. It was then divided into VOI with 64x64x64 voxels, overlapping in *x*-, *y*-, and *z*-directions by 32 pixels. The mean was subtracted from each VOI before a 3D Hanning window was applied. The 3D FT of each VOI was calculated, from which the 3D NPS was determined.

The 2D NPS was viewed along the tube travel direction f_x with $f_y = 0$ and the 3D NPS was viewed along f_x with $f_y = f_z = 0$. The low frequency region of the NPS was used for the power law fitting with κ/f^{β}.

3 Results and Discussion

Shown in Fig. 2 are examples of the CC and MLO DBT image slices of one patient's right breast, reconstructed using FS4. The NPS results shown in Fig. 3 and Fig. 4 were obtained from the CC view of this DBT image data set.

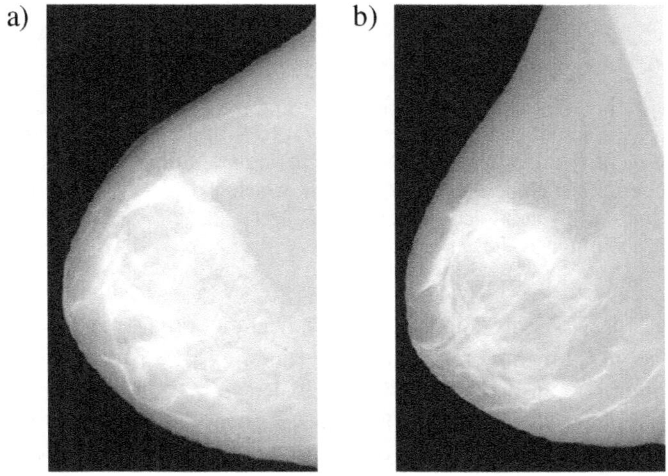

Fig. 2. Example of a single DBT slice from the CC (a) and MLO (b) views of one patient. Subsequent figures (3 and 4) are from the same image data set.

Shown in Fig. 3(a) is the total 3D NPS, which includes both structural and quantum noise, displayed in the *x*-*z* plane with $f_y = 0$. The DBT reconstruction was performed using FS2 (RA filter only). For ease of analysis, the spoke along f_x with $f_y = f_z = 0$, displayed as the solid white line in Fig. 3(a), was chosen for power law fitting and the results are shown in Fig. 3(b). Shown in the same graph for comparison is the NPS of the projection images. As shown in Fig. 3(b), the exponent of power law noise as quantified by β, does not change between the projection NPS and the 3D NPS analyzed along any angle of the sampled 3D spectrum. This finding with limited angle DBT is

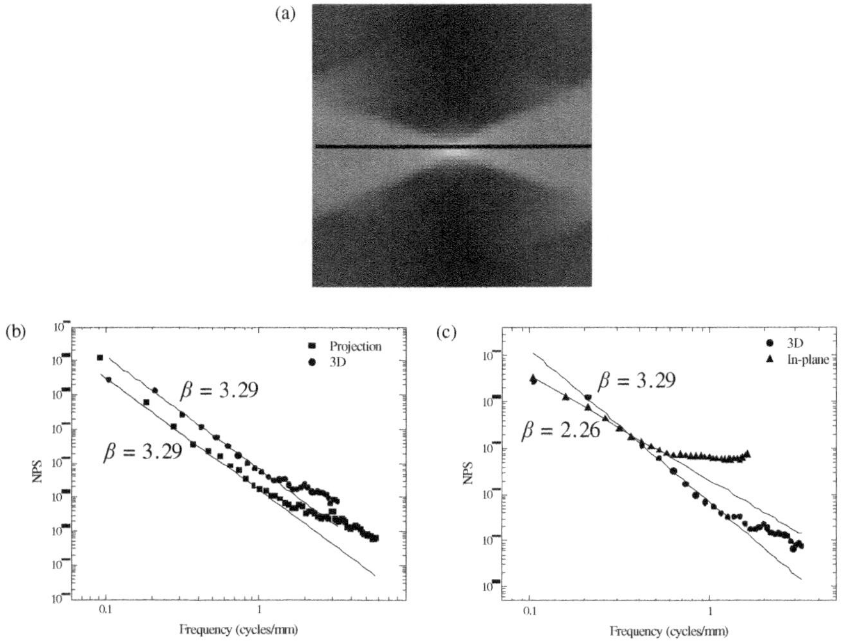

Fig. 3. Results of breast structural noise analysis using the DBT projection and reconstruction data set of the CC view of one patient: (a) the total NPS of the x-z plane of the CC view; (b) log-log plot of the projection and 3D NPS; (c) a log-log plot of the 3D and in-plane NPS. The analyses were performed along f_x, where $f_y = f_z = 0$, which is depicted as the solid, white line in (a). The solid lines in (b) and (c) are power law fits for the data. $\beta = 3.29$, 3.29, and 2.26 for the projection, 3D, and in-plane views, respectively.

consistent with previous investigations by Metheany et al [2] and Tward et al. [9] for structural noise in CT with complete angular sampling, and is expected from the central slice theorem. The in-plane (IP) NPS calculated from each individual reconstructed image slice and plotted along the f_x-axis (with $f_y = 0$) are shown in Fig. 3(c). Plotted for comparison is the 3D NPS along the same axis. Since the IP NPS is equivalent to the summation of the 3D NPS in the f_z-direction, the power law noise exponent, β decreases by approximately 1, despite the lack of complete angular sampling.

It is also important to note that β for the IP NPS in f_y direction should be the same as that in the 3D, i.e. not reduced by 1. This results in asymmetry in the IP NPS. This is shown clearly with the comparison of IP NPS in Fig. 4 (a log-log plot of the NPS along the x-axis, where $f_y = 0$, and the y-axis, where $f_x = 0$).

Shown in Fig. 5 are the total 3D (a) and IP (b) NPS calculated using the same data set as that used in Fig. 3, but with different filter schemes. In the case of SBP (FS1), the value of β in the 3D NPS was 5.21, which is increased by ~2 from that in the projection view (3.29). This is consistent with the blur function in SBP, which is $1/f$. The addition of the RA filter for both FS2 and FS3 result in $\beta = 3.29$, which is equivalent to that of the projection view as expected from the central slice theorem. The addition of the ST filter in FS3 made essentially no difference in β since its effect is to eliminate high frequency (quantum) noise in the z-direction. This has little effect on power law noise, which is heavily weighted in the lower frequencies. The result with POLY filter (FS4) exhibits a β value that is lower than that in SBP, but higher than either of FS2 or FS3. This is due to the retention and relative accentuation of lower frequency information that is characteristic of the POLY filter, which increases the noise at low frequencies. For each reconstruction filter scheme, the IP NPS in Fig. 5(b) exhibits a decrease in β by approximately 1 from the 3D NPS in Fig. 5(a).

The 2D and 3D NPS analyses described above were performed on the DBT acquisitions of 11 patients. The mean and standard deviation (σ) for β over all patient image data sets are shown in Table 1 for projection, 3D, and IP NPS analyses, using

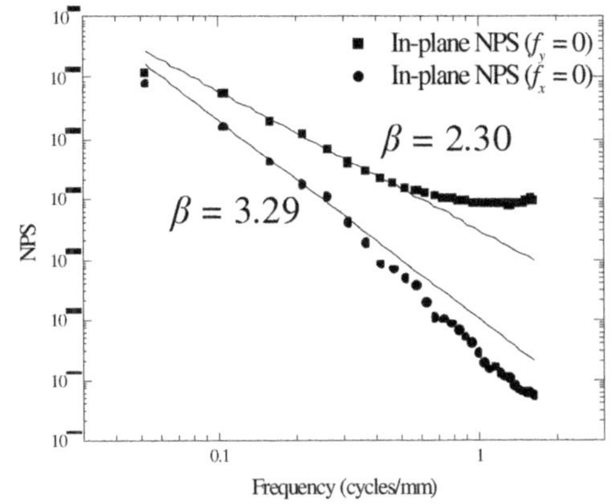

Fig. 4. The comparison of IP NPS (using FS2) along f_x (with $f_y = 0$) and f_y (with $f_x = 0$) with the fitted β values of 2.30 and 3.29, respectively

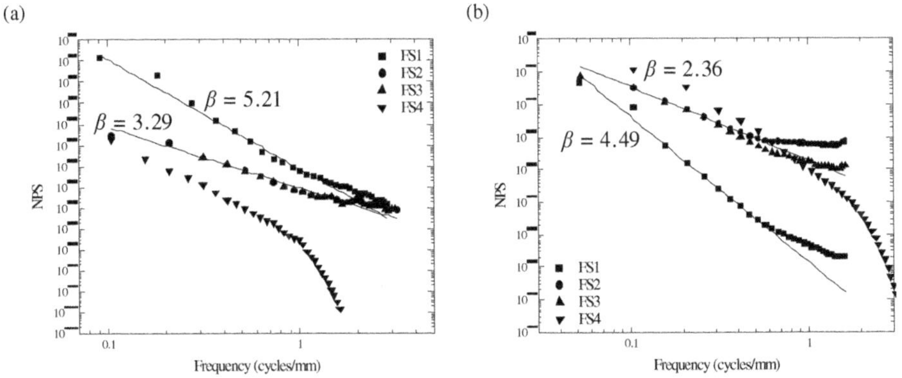

Fig. 5. Log-log plots of the 3D (a) and in-plane (b) NPS of the same example patient data set for each filter scheme. Power law fittings for FS1 and FS3 are shown in solid lines in each graph.

each filter scheme. In general, the β value for IP NPS is decreased by approximately 1 from that of the 3D NPS. However, the absolute value of β changes with the reconstruction filter and the dependence is consistent with the filter function at low spatial frequencies. To achieve the ultimate goal of the present work, the amplitude of power law noise, κ, with respect to the quantum noise needs to be determined. This will allow us to incorporate structural noise in the linear system model and the ideal observer SNR, and optimize DBT for the detection of different type of breast lesions.

Table 1. Mean and standard deviation (σ) of β values obtained from 11 sets of patient data. The results are shown for projection, 3D, and IP NPS with each filter scheme.

		3D				IP			
	Projection	FS1	FS2	FS3	FS4	FS1	FS2	FS3	FS4
β	3.09	5.52	2.83	2.87	4.34	4.45	2.06	2.27	3.10
σ	0.74	0.81	0.53	0.46	0.81	0.68	0.58	0.62	0.66

4 Conclusion

We analyzed the breast structural noise from clinical DBT data sets from 11 patients, using both CC and MLO views. The power law exponent β in the tube travel direction of the 3D NPS is the same as that of the projection NPS, which is to be expected from central slice theorem. For in-plane structural noise, β decreases by approximately 1 from the 3D NPS. The absolute value of β changes with the reconstruction filter, and the dependence is consistent with predictions from linear system theory and the incorporation of filter functions.

Acknowledgements

We gratefully acknowledge the financial support from NIH (1R01EB002655 and 1R01CA148053) and Siemens Health Care.

References

1. Burgess, A.E.: Mammographic Structure: Data Preparation and Spatial Statistics Analysis. Med. Imag. 1999: Image Processing 3661, 642–653 (1999)
2. Metheany, K.G., Abbey, C.K., Packard, N., Boone, J.M.: Characterizing Anatomical Variability in Breast CT Images. Med. Phys. 35(10), 4685–4694 (2008)
3. Engstrom, E., Reiser, I., Nishikawa, R.: Comparison of Power Spectra for Tomosynthesis Projections and Reconstructed Images. Med. Phys. 36(5), 1753–1758 (2009)
4. Hu, Y.-H., Zhao, W.: Nonuniform Angular Dose Distribution in Digital Breast Tomosynthesis for Increased Conspicuity of Small High Contrast Objects. Med. Imag. 2009: Physics of Medical Imaging 7258, 72580Q1–72580Q8 (2009)
5. Mertelmeier, T., Orman, J., Haerer, W., Dudam, M.K.: Optimizing Filtered Backprojection Reconstruction for a Breast Tomosynthesis Prototype Device. Med. Imag. 2006: Physics of Medical Imaging 6142, 61420F1–61420F12 (2006)

6. Lauritsch, G., Harer, W.H.: A Theoretical Framework for Filtered Backprojection in Tomosynthesis. Med. Imag. 1998: Image Processing 3338, 1127–1137 (1998)
7. Ludwig, J., Mertelmeier, T., Kunze, H., Harer, W.: A Novel Approach for Filtered Backprojection in Tomosynthesis Based on Filtered Kernels Determined by Iterative Reconstruction Techniques. In: Krupinski, E.A. (ed.) IWDM 2008. LNCS, vol. 5116, pp. 612–620. Springer, Heidelberg (2008)
8. Dobbins, J.T., Ergun, D.L., Rutz, L., Hinshaw, D.A., Blume, H., Clark, D.C.: DQE (f) of Four Generations of Computed Radiography Acquisition Devices. Med. Phys. 22, 171–181 (1995)
9. Tward, D.J., Siewerdsen, J.H., Fahrig, R.A., Pineda, A.R.: Cascaded Systems Analysis of the 3D NEQ for Cone-Beam CT and Tomosynthesis. Med. Imag. 2008: Physics of Medical Imaging 6913, 69131S1–69141S12 (2008)

Dual-Energy Dividing Mammography

Vjacheslav Gorshkov[1], Nadezhda Rozhkova[2], and Sergey Prokopenko[2]

[1] Space Research Institute (IKI), Profsoyuznaya str. 84/32, 117997 Moscow, Russia
[2] Russian Centre of Rontgenology, Profsoyuznaya str. 86, 117997 Moscow, Russia
vagorshkov@mail.ru

Abstract. The cancers exhibit a higher effective atomic number than the normal tissue. The aim of this study was to improve early detection of microcalcifications, which are the earliest indicators of breast cancer, and to get a better definition of the cancer boundary on the base of visualization of an effective atomic number distribution. Dual-energy subtraction mammography may provide information regarding the product of the atomic number, density and thickness of the breast tissue. The paper presents the dual-energy dividing mammography which is based on the dividing of the logarithms of the low-energy image by the high-energy image. That ratio depends upon only an effective atomic number and does not depend on the density and the thickness of the breast. The study shows that the visualization of that ratio improves the detection of microcalcifications and the sharpness of the breast image.

Keywords: dual-energy mammography, cancer, microcalcifications, atomic number, early detection.

1 Background

The cancerous tissue and the normal tissue are very similar but their atomic number differs. The microcalcifications, as the earliest indicators of breast cancer, have significantly higher atomic number than normal tissue. That is why a cancer with the microcalcifications exhibits a higher effective atomic number than normal breast tissue.

Traditional X-ray mammography visualizes the distribution of the photon number which is registered by the detector film. This visualization presents the non-linear distribution of the product of the atomic number, density and thickness of the breast tissue. It is impossible to determine the reason of the optical density variety at the mammogram, whether it is caused by changes of tissue density or atomic number.

The dual-energy subtraction mammography [1-5] visualizes weighted subtraction of the logarithm of the low-energy image from that of the high-energy image. This subtraction is in proportion to the product of the atomic number, density and thickness of the breast tissue. But this method does not allow to distinguish areas with raised high density from the areas with the high atomic number.

The division of the same logarithms can be more useful in diagnosing malignant growths because it depends only on the effective atomic number.

J. Martí et al. (Eds.): IWDM 2010, LNCS 6136, pp. 606–613, 2010.

2 Method

Despite the fact that the effective atomic number of the breast elements varies from 5.5 (cholesterol) to 12-14 (microcalcifications), the real range of an effective atomic number variation is narrower and varies from approximately 6.5 to 7.5. It is explained by a small concentration of anomalies in the breast [6].

Fig.1. presents the dependence of the division and the subtraction of the mass coefficients for energy 20 and 40 keV on an effective atomic number (Z).

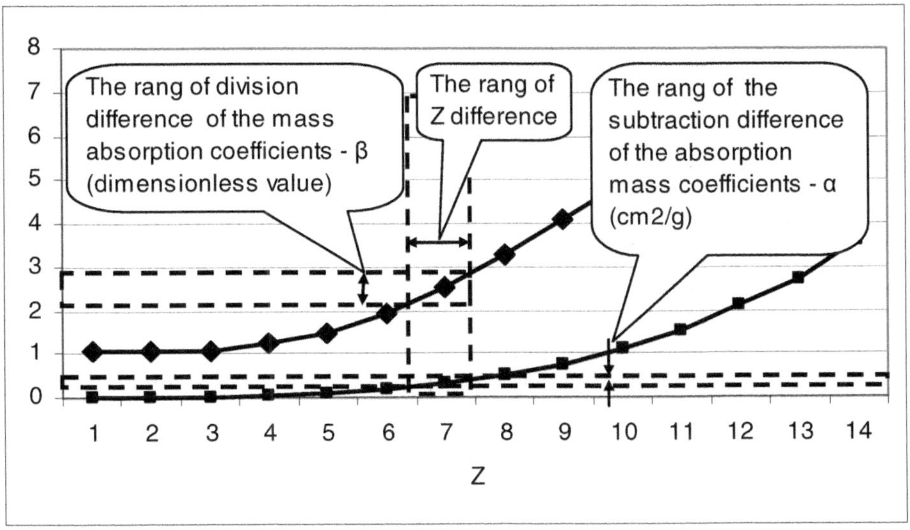

Fig. 1. Dependences of the division and the subtraction of the mass coefficients

In spite of the fact that mass photoelectric absorption coefficient is dependent on the fourth power of the atomic number Z, the subtraction and division of the absorption coefficients for low and high energy are in practically proportion to the effective atomic number in this rang. Hence the subtraction of the logarithms is in proportion to the product of the atomic number, density and thickness of the breast tissue and the division of the same logarithms is in proportion only to the effective atomic number. The next equations illustrate this fact.

$$\alpha = \mu_m^L - \mu_m^H = \frac{1}{\rho d}(\ln\frac{N_0^L}{N^L} - \ln\frac{N_0^H}{N^H}) = k_\alpha Z + a_\alpha, \tag{1}$$

$$\beta = \frac{\mu_m^L}{\mu_m^H} = \ln\frac{N_0^L}{N^L} \Big/ \ln\frac{N_0^H}{N^H} = k_\beta Z + a_\beta, \tag{2}$$

where μ_m is the mass coefficient of absorption, ρ is the density, d is the thickness, Z is the effective atomic number, k and a are the coefficients, N_0 and N are the initial and detected photons numbers, respectively, L and H are the indexes for low and high energy, respectively.

The dual-energy dividing mammography is based on equation (2).

As a rule, the thickness of the breast tissue is constant during mammography checkup. Hence, equations (1) and (2) allow us to calculate the actual density distribution:

$$\rho d = \frac{(\ln \frac{N_0^L}{N^L} - \ln \frac{N_0^H}{N^H})}{k_\alpha Z + a_\alpha} = \frac{(\ln \frac{N_0^L}{N^L} - \ln \frac{N_0^H}{N^H})}{\frac{k_\alpha}{k_\beta}\left(\frac{\ln \frac{N_0^L}{N^L}}{\ln \frac{N_0^H}{N^H}} - a_\beta\right) + a_\alpha} \quad (3)$$

That is why these two methods do not compete but rather supplement each other. Subtraction and division of the logarithms give three additional images: the effective atomic number, the density and their product.

Having two images (at low and high energy) we can apply non-traditional methods of image processing. We present two of them.

The first is connected with artificial decreasing high energy during atomic number calculation. It can be accomplished by artificial increasing of the initial photon number in the equation 2. This corresponds to the increasing of the linear attenuation coefficient that can be interpreted as the increase of the effective atomic number. It gives the opportunity to get a better definition of the cancer boundary.

The second type is concluded in the artificial shift of the low energy image relatively another one. This method is particularly useful for the early detection of the microcalcifications.

3 Results

The reconstruction of the atomic number distribution was carried out for real mammary glands. Eighteen women were enrolled in our study and four of them (patient 1-4) are presented in this paper [7].

For the high-energy beam, an aluminum filter with thicknesses 1mm was placed in the beam. For the low-energy beam, molybdenum filter with thicknesses 0.03 mm was tested.

Two x-ray beam voltages were selected (low-energy beam, 22–33 kVp, high-energy beam, 44–49 kVp). Low and high photon energies were admitted as 22 keV and 30 keV in reconstruction procedure.

Initial photons numbers were measured by means of 2 mm aluminum plate.

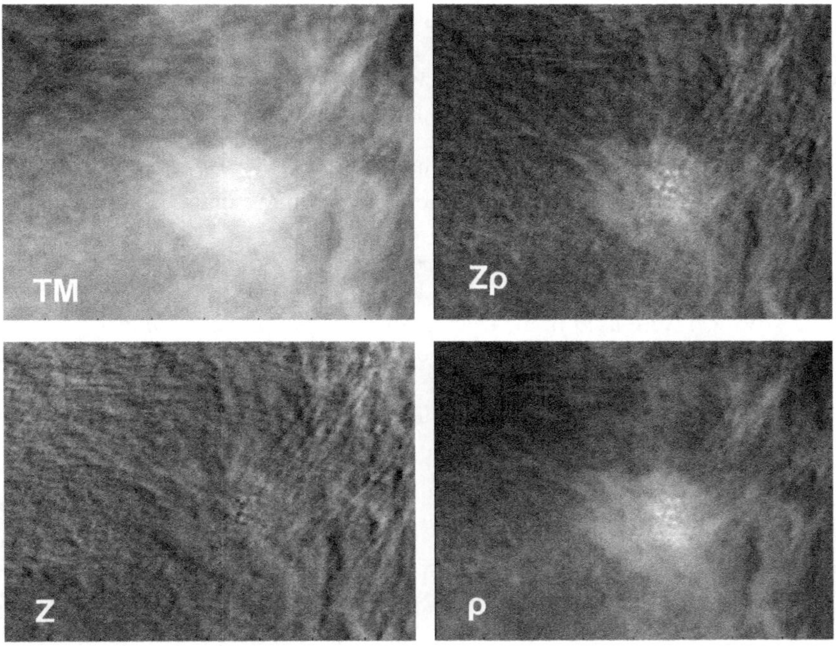

a)

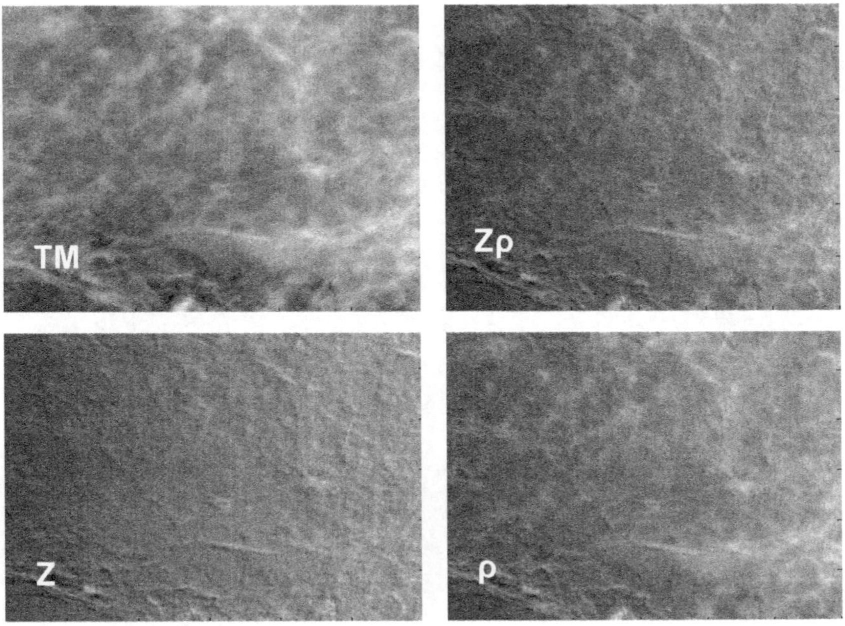

b)

Fig. 2. (a) Cancer with microcalcifications {a},(b) Healthy section of the breast {b}.**TM** - Traditional mammogram, ***Zρ*** - Visualization of the product of the effective atomic number and density, ***Z*** - Visualization of an effective atomic number, ***P*** - Visualization of the density.

Fig.2 (patient 1) presents the traditional mammogram of the cancer with microcalcifications, visualizations of the product of an effective atomic number and density, the effective atomic number. The density was reconstructed by means of formulas (1-3). In spite of the fact that all four images are similar the sharpness and the contrast of the atomic number distributions are better than at the traditional mammograms both for cancer (a) and for healthy section (b) of the breast.

Such effect is seen at the image of the cancer without microcalcifications (fig.3) (patient 2).

Fig.4 presents the first mentioned above method of image processing for this breast. Low and high photon energies 22 keV, 25 keV (z1), 15 keV, 30 keV (z2), 30 keV, 22 keV (z3) were used in the calculation of initial photons numbers. The cancer boundary is just the same but the image contrast is significantly better and there are practically no atomic number variations in the healthy section of the breast. The swelling boundary is almost just the same at these three images but significantly sharper than at fig.3b .

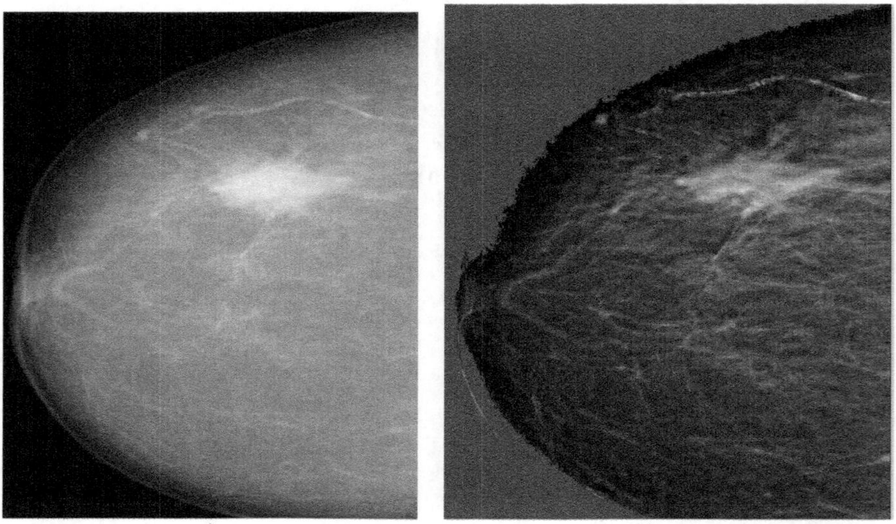

a) b)

Fig. 3. (a) Traditional mammogram {a}, (b) Reconstructed z-distribution {b}

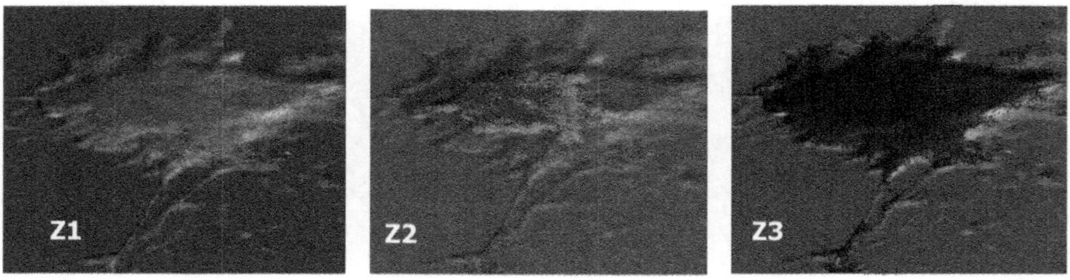

Fig. 4. Z-distribution with various values of high energy, admitted during atomic number calculation

Fig. 5 presents the z-distribution reconstructed without (a) and with (b) initial image shift (along x-axis – 5 pixels and y-axis – 40 pixels) (the second type of image processing) for patient 2.

The darker microcalcification images correspond to the healthy sections of the breast (low energy) against the background of the tissue with the microcalcifications (high energy).

The lighter microcalcification images correspond to the tissue with the microcalcifications (low energy) against the background of the healthy sections of the breast (high energy).

Reconstructed atomic number is determined by division of mass coefficients which obey the following inequalities

$$\frac{\mu_T^L}{\mu_T^H} > \frac{\mu_T^L}{\mu_{Ca}^H} < \frac{\mu_{Ca}^L}{\mu_{Ca}^H}, \quad \frac{\mu_T^L}{\mu_T^H} < \frac{\mu_{Ca}^L}{\mu_T^H} > \frac{\mu_{Ca}^L}{\mu_{Ca}^H} \qquad (4)$$

Hence the double-images of microcalcifications have minimum and maximum of numerical value at the atomic number distribution. This explains the high contrast of microcalcification images at z-distribution reconstructed on the base of shifted mammograms.

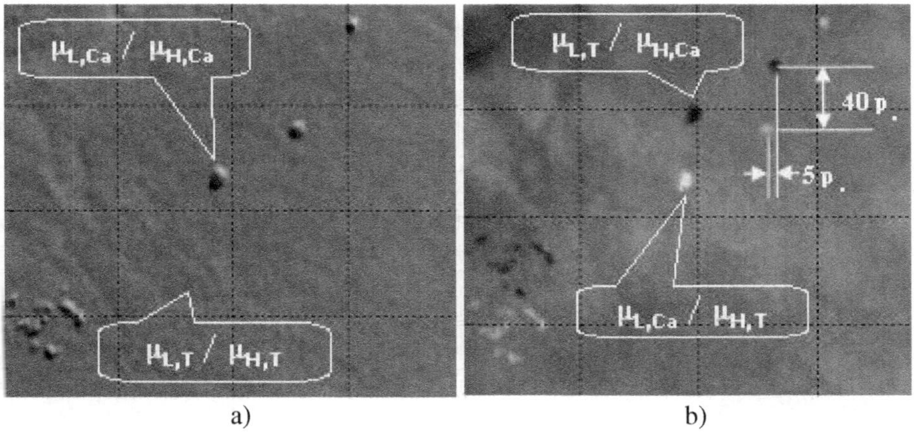

Fig. 5. Z-distribution. (a) without initial image shift {a},(b) with shift {b}

Fig.6 presents results of z-reconstruction of the suspect area with high X-ray absorption marked by the rectangle (patient 3).

Not all light areas in the marked rectangle at the traditional mammogram are shown up at the atomic number distribution. The initial shift along x-axis and y-axis was equal to 5 pixels. The presence of the white points at z-distribution allows supposing that they can correspond to microcalcifications. It is only the supposition because the cancer diagnosis was not set for that patient and the morphological analysis was not done.

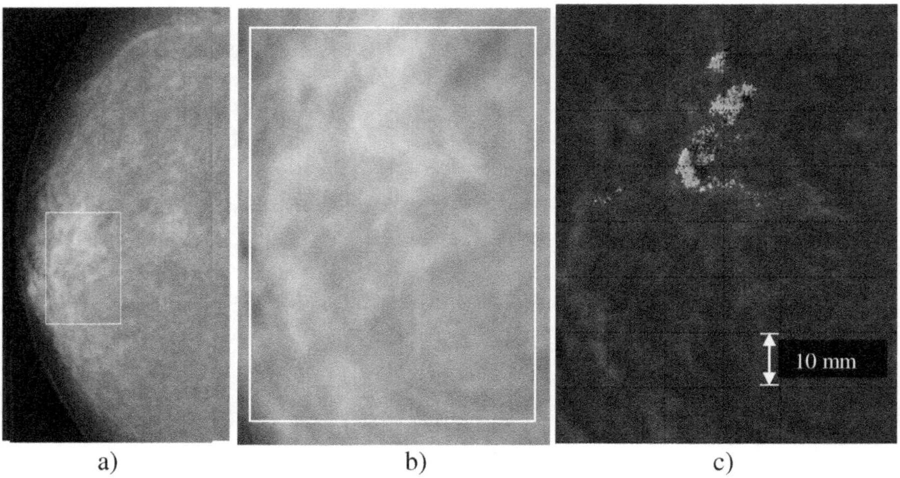

Fig. 6. (a) Traditional mammogram {a}, (b) enlarged area marked by the rectangle {b} , (c) z-distribution in the marked area {c}

The cancer (marked by the rectangle at the fig. 7a) was found in the breast of the fourth patient.

Visualization of the affective atomic number distribution with the first and the second type of image processing allows detecting the microcalcifications (Fig.7b), the most of which are not seen at the traditional mammogram (Fig.7a).

The morphological analysis of the removed organ confirmed the presence of microcalcifications in areas marked by white point at the atomic number distribution.

Microcalcifications are seen not only in the cancer area but in milk ducts of normal part of the breast. It allows considering the possibility of the early detection of the microcalcifications.

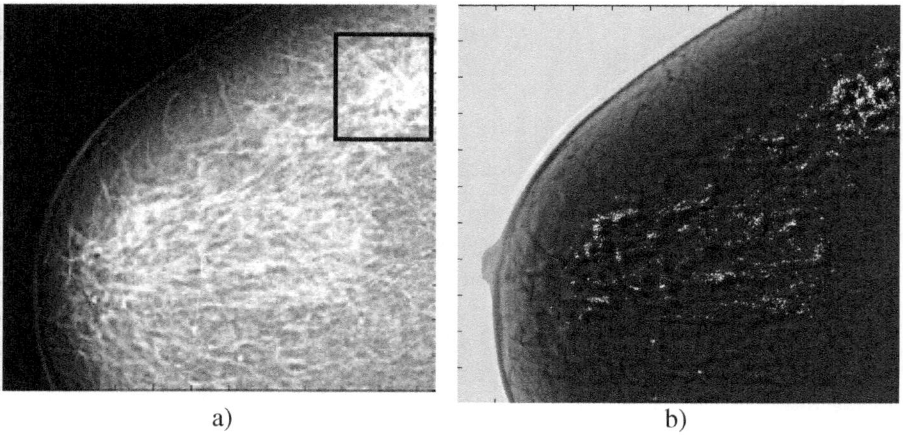

Fig. 7. (a) Traditional mammogram with marked area with the cancer {a}, (b) reconstructed z-distribution {b}

4 Discussion

The traditional mammogram visualized the non-linear product distribution of the atomic number, density and thickness of the breast tissue. The dual-energy subtraction mammography gives the linear distribution of the same product. The dual-energy dividing mammography gives only atomic number distribution. The removal of density variation makes the cancer image sharper.

If two pixels at traditional mammogram are characterized by the same counted photons they will be presented by equal numerical value at the image got by any conventional image processing technique. If the first pixel corresponds to the high density and the second to the high atomic number and have just the same number their numerical values will be different at the atomic number distribution. Conventional image processing technique can be applied to the distribution of the atomic number.

Particularity of microcalcification reconstruction in dividing mammography is concluded in the appearance of tree-dimensional image. This artifact is explained by the impossibility of the ideal synchronization of two mammograms got at low and high energy.

Not all of the known cancers were reconstructed by dividing mammography algorithm. The investigation was not intended to be a definitive test of dividing mammography but rather a demonstration of its feasibility and potential. But presented results allow concluding that the visualization of the effective atomic number on the base of dual-energy dividing mammography enables improving the diagnosis of the mammary gland disease.

References

1. Lewin, J.M., Isaacs, P.K., Vance, V., Larke, F.J.: Dual-energy contrast-enhanced digital subtraction mammography: feasibility. Radiology 229(1), 261–268 (2003)
2. Boone, J.M., Shaber, G.S., Tecotzky, M.: Dual energy mammography: a detector analysis. Med. Phys. 17, 665–675 (1990)
3. Bliznakova, K., Kolitsi, Z., Pallikarakis, N.: Dual-energy mammography: simulation studies. Phys. Med. Biol. 51, 4497–4515 (2006)
4 Brem, R.F., Hoffmeister, J.W., Rapelyea, J.A., Zisman, G., Mohtashemi, K., Jindal, G., DiSimio, M.P., Rogers, S.K.: Impact of Breast Density on Computer-Aided Detection for Breast Cancer. American Journal Roentgenology 184, 439–444 (2005)
5. Bliznakova, K., Kolitsi, Z., Pallikarakis, N.: Dual-energy mammography: simulation studies. Phys. Med. Biol. 51, 4497–4515 (2006)
6. Gorshkov, V.A., Rozhkova, N.I., Prokopenko, S.P.: Visualization of the atomic number in the mammography. Testing Diagnostics 11, 33–35 (2008) (in Russian)
7. Rozhkova, N.I., Gorshkov, V.A., Prokopenko, S.P.: The Features of X-ray Imaging of a Mamma Based on Identification of Effective Atomic Number. Medical Visualization 2, 95–99 (2009) (in Russian)

Effect of BI-RADS Assessment in Improving CAD of Masses

Antonio García-Manso, Carlos J. García-Orellana, Ramón Gallardo-Caballero, Horacio González-Velasco, and Miguel Macías-Macías

Pattern Classification and Image Analysis Group, University of Extremadura, Avd. Elvas s/n. 06006 - Badajoz, Spain
carlos@capi.unex.es

Abstract. In this work we study how the BI-RADS assessment could help to improve the performance of a CAD (Computer Aided Diagnosis) image-based system in the task of masses diagnosis. Our system is based on the use of Independent Component Analysis (ICA) as feature extractor from mammographic images, and Neural Networks as a final classifier. For our tests, the "Digital Database for Screening Mammography" (DDSM) has been used, particularly the subset BCRP_MASS1. The best results were obtained when we used the image data (with feature extraction by means of ICA) together with the BI-RADS assessment provided by DDSM database. Keywords: CAD, breast cancer, mammogram, independent component analysis, neural networks.

1 Background

Different sources [1,2] show that breast cancer treatment in an early stage of development can increment considerably the patient survival chance. Additionally, early breast cancer detection increases the chances for conservative surgery to be carried out instead of mastectomy, the only solution in advanced breast cancers [3].

The screening programs have been very useful in order to reduce the mortality [4]. However, interpreting screening mammograms is not easy. In one screening program, thousands of mammograms have to be processed, from which only a very little quantity will present one cancer in an early stage or an evident cancer. This labour can be tedious and stressful, and can cause radiologists' fatigue leading to a diagnosis error. Both film-screening mammography and digital mammography present challenges when radiologists attempt to distinguish healthy breast tissue from malignancies that are usually tiny.

Double reading of mammograms by two radiologists has been demonstrated to improve the detection rate of breast cancer [5]. However, the resources are limited, either in personal or capital and thus, it is not possible to set up this strategy in a generalized form. CAD (Computer Aided Diagnosis) systems can be seen as an acceptable-cost second "opinion".

J. Martí et al. (Eds.): IWDM 2010, LNCS 6136, pp. 614–621, 2010.

In the pathology of breast cancer there are mainly two types of abnormalities: the clusters of microcalcifications and masses. Many different techniques for breast cancer CAD have been proposed [6] [7][8]. However, in most of the cases, these proposals are difficult to compare because they use different datasets, and the performance of the CAD system is highly dependent on the datasets used [9][8].

Focusing on the masses, as we said above, there are very different approaches to detection, segmentation and diagnosis using CAD techniques. We can find recent reviews of the different techniques in [7][8]. Despite this amount of techniques used in the CAD of masses, new techniques and approaches are constantly proposed. For example, in [10] a statistically based enhancement is performed, followed by a multilevel-thresholding segmentation and region selection. In [11], the authors use Principal Component Analysis (PCA) for detection ("eigendetection") and they perform a false positive reduction using two-dimensional PCA along with the breast density information.

The rest of the paper is organized as follows: in section 2, the method and materials are described in detail. Next, in section 3, we present the results, and finally, in section 4, our conclusions are stated.

2 Method

Our objective in this work is to diagnose previously selected "Regions of Interest" (ROIs) from mammographic images that contain a mass, and also to study the influence of using the BI-RADS assessment besides image features.

To this aim, we have developed a classifier system where the *inputs* are the pixels of ROIs (rescaled to a size of 128x128, 64x64 or 32x32) and/or BI-RADS assessment.

After rescaling the ROIs, the next step in our classifier is a *feature extraction* stage for those rescaled ROIs. This stage has been done using Independent Component Analysis (ICA) [12]. The result of the feature extraction step (besides assessment, when considered) is presented as input to a *Neural Networks classifier*. The *output* of this classifier is related to the probability of malignity of the mass under study.

In Figure 1, we show how the overall process is carried out. In a first stage, we obtain the ICA transformation matrix using a set of images with a mass, resampled to 128x128, 64x64 or 32x32. Before applying the ICA algorithm we make a Principal Component Analysis (PCA) in order to reduce the dimension. The result of this stage is an ICA basis that can be used to represent ROI images with unknown mass. The second step is to train a classifier (in this work, a neural network). As inputs to the classifier, we make a feature extraction with ICA using the previously obtained basis along with non-image features (like age, density, assessment, ...). Finally, in the third step, we diagnose an unknown ROI previously detected using the trained classifier and expressed as a linear combination of the ICA basis extracted in the first step. The result indicates whether the mass is considered as malignant or benign.

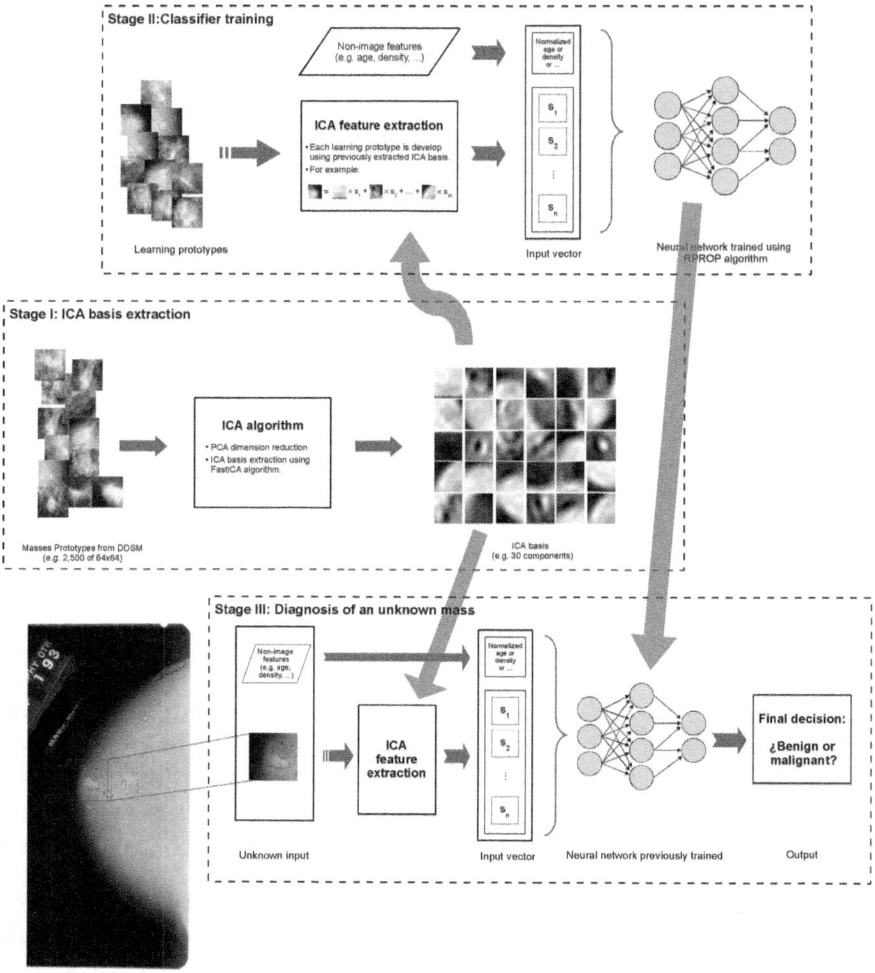

Fig. 1. General process outline. Stage I shows an extracted ICA basis for a dimension reduction to 30 components. In stage II this basis is used to make a feature extraction step from images ROIs with mass, and later a neural network classifier is trained. The stage III shows the process of diagnosing a new detected ROI with a mass.

In the next subsections, we describe the used database (*Digital Database for Screening Mammography* (DDSM)), the method applied for feature extraction (ICA) and the experimental setup.

2.1 Database

Different databases have been used to develop the works of mammographic CAD. Most of them use their own set of cases which is usually collected in near centers, a situation that makes very difficult the comparison of all the results obtained

Table 1. Composition of the two test sets considered in this work with respect to assessment and pathology. Numbers 0, 1, ..., 5 are the assessment. B (benign) and M (malignant) are related to pathology.

Assessment →	0		1		2		3		4		5	
Pathology →	B	M	B	M	B	M	B	M	B	M	B	M
Random set	35	4	0	0	17	0	77	10	160	78	5	92
MASS1+50 set	4	0	0	0	1	0	16	0	29	10	1	66

with different techniques. Within the group of databases of generalized use, the most commonly referenced is probably the MIAS database, at least in oldest works [13]. MIAS group provides a free version with low resolution which may be useful in some applications. Other references make use of the Nijmegen database, very difficult to obtain nowadays, or LLNL/UCSF. Recent developments usually work with the DDSM database [14], which includes 2,620 complete cases with the four typical views (left/right MLO and CC), where each view has truth marks surrounding abnormality extension (if it exists) and assessment using BI-RADS terminology from ACR [15]. Besides, DDSM creators have formed two subsets of mammograms with masses, in order to facilitate the comparison of performance between different CAD systems (they are called BCRP_MASS0 (for training) and BCRP_MASS1 (for test)).

All our work has been done using DDSM database. We have used two different sets: firstly, we have considered ROIs from the whole database (except from BCRP_MASS1) for training (2,197 ROIs) and BCRP_MASS1 subset plus fifty benignant ROIs for test (127 ROIs). We have extended the test set because BCRP_MASS1 only has malignant ROIs (MASS1+50 set). Secondly, we have built the learning and the testing sets by a random selection from whole database, with 2,086 and 244 ROIs respectively (Random set). In Table 1 the composition of test sets with respect to assessment and pathology is shown.

2.2 Independent Components Analysis (ICA)

Independent Component Analysis is a statistical and computational technique for revealing hidden factors that underlie sets of random variables, measurements or signals. ICA defines a generative model for the observed multivariate data, which is typically given as a large database of samples. In the model, it is assumed that the data are linear combinations of unknown latent variables, and the system whereby combined is also unknown. It is also assumed that non-gaussian latent variables are mutually independent, and they are called independent components of the observed data. These independent components, also called sources or factors, can be found by ICA. ICA is slightly related to PCA (Principal Component Analysis). However, ICA is a much more powerful technique, able to find underlying factors or sources when conventional methods fail completely. The data analyzed by ICA may come from very different types of fields, including digital images, documents, databases, economic indicators and psychometric measures. In many cases, the measures are given as sets of parallel signals or time series, the term "blind source separation" is used to characterize this problem. As examples, we can mention simultaneous combination of speeches that have been recorded by several microphones, brain waves stored by multiple sensors, interfering radio signals in a mobile phone, or parallel time series obtained from industrial processes [12]. Let x be a m-dimensional aleatory variable, the method ICA tries to find a transformation W such that, when being applied to x,

$$s = W \cdot x$$

makes the new components of the n-dimensional vector s the most statistically independent.

Supposing that x is the result of combining a group of original signals with diverse sources of noise, this would allow to recover the different components and to discard the noisy components. This technique can be used for the extraction of features, since the components of x can be considered as features that represent the objects [16]. It is usual to make PCA in a first stage, before applying ICA, in order to reduce the original dimension of data.

To apply this technique in feature extraction on mammography, we must suppose that a region of a given size in the mammography (called patch) is a linear combination of a series of independent unknown sources, a priori. These unknown sources can be seen as statistical independent patches, and can be estimated by ICA using samples. The process provides us a base of functions (small squared images in our case) that lets us expand a given patch from the mammography in terms of them. The mentioned procedure can be expressed graphically as follows:

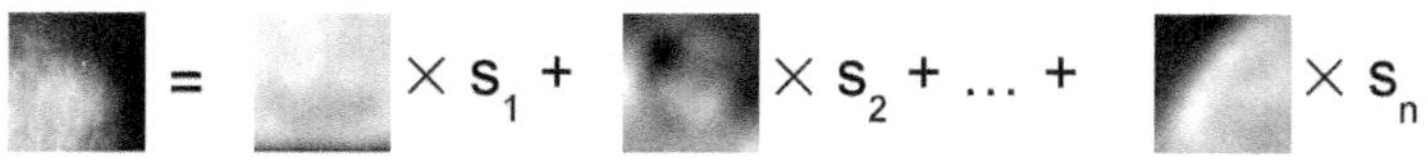

Where s_i coefficients represent the features to extract and which let us characterise a region from its sources.

2.3 Experimental Setup

All the programs needed for this study have been written using R and C++ languages, using Linux as platform.

The training was carried out for the three sizes of ROIs (128x128, 64x64 and 32x32) by varying the number of components retained (from 10 to 60). We have also simulated combinations of these features extracted by ICA with additional input parameters (apart from the assessment, we used the density, age and subtlety). In summary, over 3,000 classifiers were trained. All the simulations needed were performed using our Beowulf cluster with 180 CPU cores, and the results were saved in a SQL database.

3 Results

As we explained above, we have used two different subsets from DDSM to test our system. The first one, includes all the ROIs from BCRP_MASS1 plus fifty benignant ROIs randomly selected (and, obviously, not used in training). We called this set *MASS1+50*, and in total has 127 ROIs. The second set used for testing, with 244 ROIs, was randomly chosen from DDSM ROIs with masses. The results are presented in Table 2, where we show the ROC (Receiver Operating Characteristic) area and the Accuracy, both parameters calculated using the ROCR package for R language [17].

The ROC curves are shown in Figures 2 and 3 for *Random* and *MASS1+50* test sets, respectively. For *Random* test set, the best results are achieved for an ICA configuration of 55 components extracted from 32x32 ROIs. And for the other test set (*MASS1+50*) the best results are obtained with ROIs of 64x64 and 25 ICA components.

Table 2. Comparison of the results obtained over the two sets considered for testing

	MASS1+50 set		**Random set**	
	ROC area	*Accuracy*	*ROC area*	*Accuracy*
ICA	0.727	73.2 %	0.643	64.3 %
ICA+Assessment	0.978	92.9 %	0.846	79.1 %

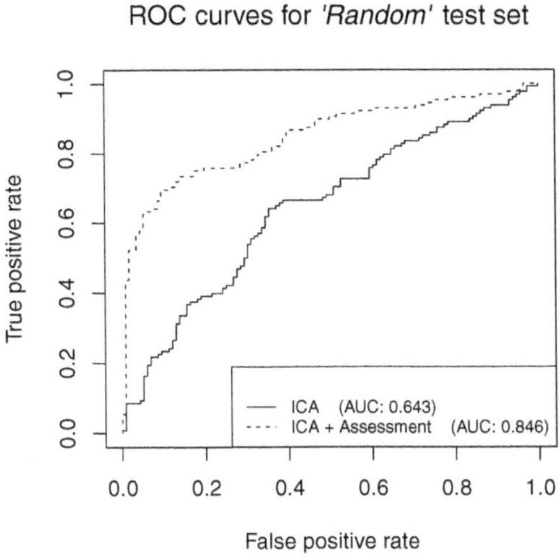

Fig. 2. ROC curves for *Random* test set using only ICA as feature extractor and ICA features plus BI-RADS assessment. The best results for this test set are obtained extracting 55 ICA components from ROIs resized to 32x32.

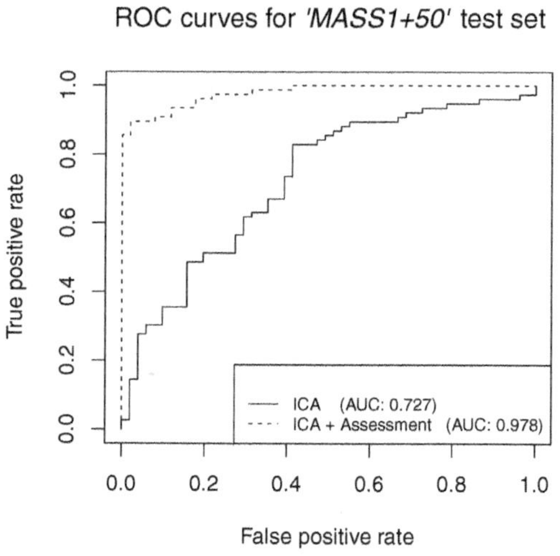

Fig. 3. ROC curves for *MASS1+50* test set using only ICA as feature extractor and ICA features plus BI-RADS assessment. The best results for this test set are obtained extracting 25 ICA components from ROIs resized to 64x64.

4 Discussion

The results show something that could be obvious a priori: the use of BI-RADS assessment as input parameter (along with image features) greatly improves the diagnostic of a ROI with a mass. As usually happens in complex problems of pattern recognition, the results show a significant dependence on the testing set used. In our case the set *MASS1+50* is significantly easier to diagnose than the other, chosen at random, at least for our system. However, it must be noted that similar results are obtained with other ICA configurations different from those referred in the results section. Therefore, to establish more correctly which is the best configuration we should perform a statistical study (for example, by cross-validation).

Acknowledgements

This work has been partly supported by "Junta de Extremadura" and FEDER through project PRI08A092.

References

1. CRUK. Breast cancer factsheet. Technical report, Cancer Research UK, London (December 2009), `http://www.cancerresearchuk.org`
2. `http://www.cancerhelp.org.uk/`

3. Haffty, B.G., Kornguth, P., Fisher, D., Beinfield, M., McKhann, C.: Mammographically detected breast cancer: Results with conservative surgery and radiation therapy. Cancer 67, 2801–2804 (1991)
4. Chung, B., Cady, M.: Mammographic screening: No longer controversial. American Journal of Clinical Oncology: Cancer Clinical Trials 28(1), 1–4 (2005)
5. Wallis, R.G., Moss, M.G., Blanks, S.M.: A comparison of cancer detection rates achieved by breast cancer screening programmes by number of readers, for one and two view mammography. Journal of Medical Screening 5(4), 195–201 (1998)
6. Sampat, M.P., Markey, M.K., Bovik, A.C.: Computer-Aided Detection and Diagnosis in Mammography. In: Handbook of Image and Video Processing, 2nd edn., pp. 1195–1217. Academic Press, London (2005)
7. Rangayyan, R.M., Ayres, F.J., Leo Desautels, J.E.: A review of computer-aided diagnosis of breast cancer: Toward the detection of subtle signs. Journal of the Franklin Institute 344(3-4), 312–348 (2007)
8. Elter, M., Horsch, A.: Cadx of mammographic masses and clustered microcalcifications: A review. Medical Physics 36(6), 2052–2068 (2009)
9. Nishikawa, R.M.: Current status and future directions of computer-aided diagnosis in mammography. Computerized Medical Imaging and Graphics 31(4-5), 224–235 (2005)
10. Domínguez, A.R., Nandi, A.K.: Detection of masses in mammograms via statistically based enhancement, multilevel-thresholding segmentation, and region selection. Computerized Medical Imaging and Graphics 32(4), 304–315 (2008)
11. Freixenet, J., Oliver, A., Martí, R., Lladó, X., Pont, J., Pérez, E., Denton, E.R.E., Zwiggelaar, R.: Eigendetection of masses considering false positive reduction and breast density information. Medical Physics 35(5), 1840–1853 (2008)
12. Hyvärinen, A., Karhunen, J., Oja, E.: Independent Component Analysis. John Wiley & Sons, Chichester (2001)
13. Suckling, J., et al.: The mammographic image analysis society digital mammogram database. In: Exerpta Medica. International Congress Series, vol. 1069, pp. 375–378 (1994)
14. Heath, M., Bowyer, K.W., Kopans, D.B.: Current status of the digital database for screening mammography. In: Digital Mammography. Kluwer Academic, Dordrecht (1998)
15. American Collage of Radiology. BI-RADS – Breast Imaging Atlas – Mammography, 4th edn. (2003)
16. Hyvärinen, A., Oja, E.: Independent components analysis: algorithms and applications. Neural Networks 13, 411–430 (2000)
17. Sing, T., Sander, O., Beerenwinkel, N., Lengauer, T.: Rocr: visualizing classifier performance in r. Bioinformatics 21(20), 3940–3941 (2005)

Analysis of the Benefit/Risk Ratio Obtained in the Valencian Breast Early Detection Program During 2007 and 2008

Ricardo Tortosa[1], Juan Ignacio Villaescusa[2], Miguel Ramos[3], and Gumersindo Verdú[3]

[1] Radiotheraphy Department, Hospital Clinico Universitario, Valladolid, Spain
[2] Radiation Protection Department, Hospital Universitario La Fe, Valencia, Spain
[3] Chemical and Nuclear Engineering Department, Universidad Politécnica de Valencia, Spain
rictortosa@gmail.com, villaescusa_ign@gva.es,
{mirapas,gverdu}@iqn.upv.es

Abstract. The Valencian Breast Cancer Early Detection Program (VBCEDP) started in the Valencian Community (Spain) in 1992. Up to now, 24 mammographic units have been installed all over the region. Mammography is used to aid in the diagnosis of breast cancer diseases in women. There are different types of mammographic equipments involved in the program (analogical, CR and DR). Ionizing radiation implies a health risk in the studied women that has to be estimated and controlled. Software to calculate approximately the cancer induction in the VBCEDP has been developed based on Monte Carlo techniques. This paper attempts to analyze benefit/risk ratios obtained in the program during 2007 and 2008.

Keywords: Benefit/risk, Breast Early Detection Program.

1 Introduction

Screening mammographic programs try to get an early diagnosis of the breast cancer in middle aged women. The European Guidelines of Quality Assurance in Breast Cancer Screening and Diagnosis [1] is the document that regulates this practice, allowing quality on the diagnostic and the comparison between different screening units.

Although screening for the early detection of breast diseases reduces breast cancer mortality, it is well known that the diagnosis by mammography presents risks for women undergoing screening due to the exposition to ionizing radiation. At present, it is considered the mean glandular dose (MGD) in acquiring the mammography as a risk parameter. Then it is possible to use the MGD in order to obtain the risk of induced breast cancers in a screening programme. Risk projection models obtained from data of exposed populations, such as the survivors of the atomic bombs or patients exposed to high doses due to medical reasons have been used. That was the reason of the development of SCREENRISK [2]. This software based on Matlab© [3] and gives an easy way of quantifying risks in mammographic screening programs.

Knowledge of the number of cancers detected and induced by the program in each mammographic unit allows a way of assessing the benefit/risk ratio.

J. Martí et al. (Eds.): IWDM 2010, LNCS 6136, pp. 622–627, 2010.

2 Method

SCREENRISK is software that estimates the risk of exposure-induced cancer and fatal cancer for a specific type of cancer in a given population.

The excess relative risk (ERR) is a parameter used to transport risks between populations which have been exposed to radiation and have different baseline rates. Risk projection models are used in epidemiology in order to estimate incidence and mortality cancer rates in one population under study from data that has been obtained in other populations.

Different studies have been chosen to estimate risks in the Valencian Breast Cancer Early Detection Program. The mortality models are: (1a) Life Span Study cohort (LSS) that includes female bomb survivors between 1950 and 1985; (2a) and the LSS with follow-up until 1990 depending on age at exposure and (3a) depending on attained age. The incidence models for breast cancer are: (1b) the Life Span Study for incidence breast cancer (1958-1993); (2b) the Massachusetts fluoroscopy study, for tuberculosis patients (TBO) and the extension (TBX); (3b) the New York acute postpartum mastitis cohort (APM); and (4b) the benign breast disease treatment in Sweden (BBD).

The excess relative risk is fitted with

$$ERR\left(\vec{z}^{(m)}\right)=\alpha^{(m)}\theta(s)\Phi(\vec{D}_g)\exp[\gamma^{(m)}(t_e-t^{(m)})]\cdot\left(\frac{t_k}{50}\right)^{\beta^{(m)}} \tag{1}$$

where $\Phi(D_g)$ is the dose response with dose d_g to the breast, s is the gender of the individual and $\theta(s)$ is a function that depends on gender and cancer type, equal to unity for breast cancer on female. The covariate vector for the ERR is the same for incidence and mortality, and it includes the variables. The other variables are age of exposition (t_e) and reached age (t_k). Each model has its own parameters based on cases under study.

It has been considered in order to transport the risks an extension of Cox Proportional Hazards to estimate the Excess Absolute Risk (EAR) as

$$EAR^{(m)}(t_k \mid \vec{z}^{(m)})=\lambda(t_k)[ERR(\vec{z}^{(m)})] \quad t_k \geq t_e + L \tag{2}$$

where $ERR^{(m)}$ is the excess relative risk of the model m for breast cancer. In (2), it is observed that $ERR^{(m)}$ is transported to a population with a risk base function $\lambda(t_k)$ for incidence or mortality. L represents the latency period.

The risk of exposure induced cancer (REIC) is defined as the probability of an individual develops a radio-induced cancer, not necessarily mortal, all over his life. The risk of exposure induced death (REID) shows that an individual dies due to a radio-induced cancer. So, deriving a Markov process, REIC and REID can be obtained as [2]:

$$REID\ (t_e \mid \vec{z}_{fbc})=\sum_{j=e+L}^{M}\hat{s}_1(t_j \mid \vec{z}_{fbc})EAR_{fbc}(t_j \mid \vec{z}_{fbc}) \tag{3}$$

$$REIC\ (t_e \mid \vec{z}_{in},\vec{z}_{fbc})=\sum_{j=e+L}^{M}\hat{s}_1(t_j \mid \vec{z}_{fbc})EAR_{in}(t_j \mid \vec{z}_{in}). \tag{4}$$

where $\hat{s}_1(t_j \mid \vec{z}_{fbc})$ is the estimator of the survival function, EAR_{fbc} is the excess absolute risk for mortal breast cancer and $\vec{z}_{in}$ is the vector of covariates for the survival function.

In a breast screening program, women are invited to undergo mammography between an initial age (a) and a final age (b), with a constant screen interval (s) and receiving normally one exposure per breast at each time.

There are different indicators when evaluating the associated cancer risk during breast screening. One of these is the average radiological detriment for breast cancer incidence and mortality in a given instant of the screening and can be estimated as:

$$\Pi_{in}^{(m)} = \sum_{j=a}^{b} v(t_j) REIC^{(m)}(t_j \mid \overrightarrow{d_{gj}}) \cdot \omega(t_j). \quad (5)$$

$$\Pi_{fbc}^{(m)} = \sum_{j=a}^{b} v(t_j) REID^{(m)}(t_j \mid \overrightarrow{d_{gj}}) \cdot \omega(t_j) \quad (6)$$

where $v(t_j)$ is the number of views per breast in each visit, $\omega(t_j)$ is the fraction of population and d_{gj} is the average mean glandular dose per film at an age-at-exposure tj for the cohort m for radioinduced breast cancer.

2.1 The Valencian Breast Early Detection Program (VBCEDP)

The Valencian Breast Cancer Early Detection Program started in 1992 and actually 24 units are working on that. Yearly quality controls are performed in all units with the recommendations of the European Protocol on Dosimetry in Mammography.

The VBCEDP is directed towards asymptomatic women between 45 and 69 years old, with an initial age lower than other screening programs (i.e. UK Screening Program starts at the age of 50 and finishes at 64 years old). The screening examination consists of two exposures per breast; craniocaudal (CC) and mediolateral oblique (OBL). The first time that the woman participates in the program (first round) receives two exposures per breast and a single mammogram OBL per breast in subsequent rounds. The screening rounds are spaced every two years and two independent radiologists read each mammogram.

Each six months population samples are taken from the screening units involved in the program in order to estimate and control the radiological risk.

2.2 Mammographic Systems Used in the VBEDP

The 24 mammographic units of the VBCEDP used different types of obtaining the needed images for the screening. There are analogical and digital systems. The analogical ones use films as both a receptor and a display for the image to produce static and fixed images. Digital mammography uses detectors (similar to those found in digital cameras) that change the x-rays into electrical signals. These signals are then transferred to a digital receptor that converts the x-ray energy to numbers, processes the numbers, and produces an image that can be displayed on a monitor or printed on a high-resolution laser printer.

Digital mammography systems can be classified as [4]:

Direct systems (DR systems); the image is obtained without the use of a laser to digitize it. Those systems are flat-panel phosphor detector, phosphor-CCD and selenium flat panel detector. Direct systems can be divided into two subtypes. If x-rays are transformed directly to electric signal, those are called direct conversion systems (selenium flat panel detector). If a scintillator is needed to obtain light and then transform into electric signal, those are called indirect conversion systems (flat-panel phosphor detector, phosphor-CCD).

Indirect systems; the radiation hits a photo-estimulating phosphor panel that is later digitized with a laser. Those are called CR systems.

In the VBEDP, there are 5 DR systems (3 selenium flat panel, 1 phosphor-CCD and 1 flat panel phosphor), 5 CR systems and the rest are analogical. The equipment has been purchased from different suppliers.

2.3 Estimation of Benefit/Risk Ratio

A possible way of calculating benefit/risk ratio for the VBCEDP can be taken as [5]:

$$\frac{Benefit}{Risk} = \frac{\text{Detection rate}}{\text{Induction rate}} \cdot \frac{(A-B)}{M} \tag{7}$$

where A is the proportion of individuals who survive for any subsequent period of the cancers detected by screening, B is the proportion of individuals who survive for any subsequent period of cancers detected symptomatically by non-screening services and M is the percentage mortality of this group.

In equation (7), (A-B)/M represents conversion factor C. There are different ways of obtaining numerical data for factor C. The easiest to interpret and apply is to observe the change of the Nottingham Prognostic Indicator (NPI) after the introduction of the screening. NPI index is based on size, state and grade of breast tumor when it is detected. It is a numerical value, equal to the sum of (0.2 x size in centimeters) + stage + grade. The lower the NPI, the better the prognosis. It was first derived empirically and later verified in a screened population. Data for 15-year survival of a group of an age-matched population of women for three different value ranges of NPI are given in Table 1.

Factor C has a value of 0.62 for no future screening and 1.62 for future screening of older women[1]. See appendix for calculations.

Table 1. Nottingham Prognostic Indicator (NPI) data [5]

NPI	15-year survival (%)		Proportion of women presenting (%)	
	Actual	Age corrected	Before screening	After screening
< 3.4	80	96	29	76
3.4-5.4	42	51	54	20
>5.4	13	16	17	4

[1] Older women here refers to women over the age of normal routine screening i.e. 64 years (UK Screening Program).

3 Results

Table 2 shows the obtained results of detections/inductions ratio, benefit/risk ratio (C=0.62) and benefit/risk ratio (C=1.62) for different mammographic equipment during 2007:

Table 2. Values benefit/risk obtained for 2007

Mammographic equipment	$\frac{\text{Detection rate}}{\text{Induction rate}}$	$\frac{\text{Benefit}}{\text{Risk}}(C=0.62)$	$\frac{\text{Benefit}}{\text{Risk}}(C=1.62)$
Analogical	54.74	33.94	88.69
CR+DR	68.70	42.60	111.30
CR	79.72	49.32	129.14
DR	52.19	32.36	84.54

Table 3 shows the obtained results of detections/inductions ratio, benefit/risk ratio (C=0.62) and benefit/risk ratio (C=1.62) for different mammographic equipment during 2008:

Table 3. Values benefit/risk obtained for 2008

Mammographic equipment	$\frac{\text{Detection rate}}{\text{Induction rate}}$	$\frac{\text{Benefit}}{\text{Risk}}(C=0.62)$	$\frac{\text{Benefit}}{\text{Risk}}(C=1.62)$
Analogical	49.31	30.57	79.88
CR+DR	46.21	28.65	74.86
CR	52.43	32.50	84.93
DR	42.07	26.08	68.15

4 Discussion

Results obtained in Table 2 indicate that computed radiography systems have a better benefit/risk ratio either it is used a no future screening parameter (C=0.62) or a future screening parameter (C=1.62). Table 3 shows that computed radiography systems have a better benefit/risk ratio in both studied cases. Analogical systems present the second best benefit/ratio and DR systems obtain the poorest results for both studied years.

C-factor depends on detection rate and consequently on the mammographic equipment. This involves that each type of mammographic system would need a different C-factor. Keep in mind that NPI was verified for analogical screening.

It can be appreciated that calculated data for 2008 are lower than those of 2007. This fact could be explained as cancer detection rate was 3.9 detected cancers per 1000 women in 2007 and 3.2 detected cancers per 1000 women in 2008 [6]. But radioinduced cancer rate remained approximately constant.

The difference in detected cancers for two consecutive years is due to the planning of the Program itself. The screening rounds are spaced every two years so a better comparison can be done with two alternative years.

It is important to remark that NPI index is an approximate value for the Valencian program because it was calculated for the National Health System Breast Screening Program where screened women have an interval age of 50-64 years.

As a result of this work, it can be stated that Valencian Breast Cancer Early Detection Program is clearly justified in terms of protection to ionizing radiation and digital indirect systems (CR) show better results than analogical ones.

Acknowledgments. This work is funded with the Project PI06/90502 of the Carlos III Health Institute (Spain). We are also grateful to the Valencian Breast Cancer Early Detection Program Office.

References

1. European Commission 2006. European Guidelines of Quality Assurance in Breast Cancer Screening and Diagnosis (2006), ISBN 92-79-01258-4
2. Ramos, M., Ferrer, S., Villaescusa, J.I., Verdú, G., Salas, M.D., Cuevas, M.D.: Use of risk projection models to estimate mortality and incidence from radiation-induced breast cancer in screening program. Phys. Med. Biol. 50, 505–520 (2005)
3. MATLAB© 6.5. The MathWorks INC 2002 (2002)
4. Pisano, E.D., Yaffe, M.J.: Digital mammography. Radiology 234(2), 353–362 (2005)
5. Law, J., Faulkner, K.: Concerning the relationship between benefit and radiation risk and cancers detected and induced, in a breast screening programme. British Journal of Radiology 75, 678–684 (2002)
6. Pons, C., Miranda, J., Ibañez, J., Salas, M.D.: Informes de Salud. Programas de Prevención del Cáncer de Mama de la Comunidad Valenciana. Informe anual 2007-2008. Dirección General de Salud, ISSN 1139-6873

Appendix: Calculation of C Factor

Calculation of C factor from detection/induction to benefit/risk ratios from Nottingham Prognostic Indicator data in Table 1.

Overall 15-year percentage survival before introduction of screening (from Table 1):

$$(0.29 \text{ x } 96) + (0.54 \text{ x } 51) + (0.17 \text{ x } 16) = 58\% \tag{8}$$

Percentage of mortality in the absence of screening is M = 100-58 = 42%

Overall 15-year percentage survival after introduction of screening (from Table 1):

$$(0.76 \text{ x } 96) + (0.20 \text{ x } 51) + (0.04 \text{ x } 16) = 84\% \tag{9}$$

Percentage of mortality in the presence of screening is M = 100-84 = 16%

Hence, $C = \frac{(84-58)}{42} = 0.62$ in the absence of subsequent screening, or

$C = \frac{(84-58)}{16} = 1.62$ if screening continues for all subsequent years.

Monte Carlo Simulation of Scatter Field for Calculation of Contrast of Discs in Synthetic CDMAM Images

Oliver Díaz[1], Mary Yip[1], Jorge Cabello[1,⋆], David R. Dance[2,3], Kenneth C. Young[2,3], and Kevin Wells[1]

[1] Centre for Vision, Speech and Signal Processing, Faculty of Engineering and Physical Sciences, University of Surrey, Guildford, GU2 7XH, UK
o.diaz@surrey.ac.uk
[2] National Coordinating Centre for the Physics of Mammography, Royal Surrey County Hospital, Guildford, GU2 7XX, UK
[3] Department of Physics, Faculty of Engineering and Physical Sciences, University of Surrey, Guildford, GU2 7XH, UK

Abstract. This paper reports on a further development of an image simulation chain, and in particular, the inclusion of contrast degradation across an image using scatter to primary ratios calculated using Monte Carlo simulation. The Monte Carlo technique, using the Geant4 toolkit, has been implemented to model the scatter conditions when imaging the CDMAM phantom with commercial digital mammography. Observed differences between linear and cellular anti scatter grid are presented and discussed. These results support previous assumptions taken by Yip et al.[1].

Keywords: Digital Mammography, Monte Carlo simulations, Scatter, anti-scatter grids, CDMAM phantom.

1 Background

New digital technologies are being developed for X-ray mammography which have the potential to improve cancer detection rates, but there is a need to investigate how these perform in comparison with conventional 2D digital mammography, and how these should be used in an optimal way. Since clinical trials to assess the performance of any mammography system are expensive, there is great interest in the development of computer-based models of the breast and imaging systems which can be used for evaluation and optimisation.

As a first step towards modelling patient images, and as a validation of the methodology, we are developing models for 2D imaging of the CDMAM phantom. This phantom is widely used for testing the performance of digital mammography systems in Europe. Previously, Yip et al[1] presented contrast detail

⋆ Now at Institute for Corpuscular Physics (IFIC), University of Valencia/CSIC, Valencia, Spain.

J. Martí et al. (Eds.): IWDM 2010, LNCS 6136, pp. 628–635, 2010.

curves calculated using a model incorporating heel effect, geometric and detector blurring and image noise. Loss of contrast due to scatter was assumed to be globally uniform across the image. In the present work, we address this previous assumption by incorporating scatter to primary ratio (SPR) maps calculated from Monte Carlo simulations.

2 Method

Figure 1a shows an adapted diagram from [1] of the simulation chain including the SPR component. The simulation chain has two parts. First, a noise free image is generated. Pixels values are assigned to a binary CDMAM template according to the exposure to be simulated and a heel effect mask is applied. Discs are added to the CDMAM template using the SPR calculated from the Monte Carlo simulations in order to calculate the contrast for each disc as explained below. The image is then blurred with an MTF filter. In the second part, a noise image is simulated using a Gaussian noise field and a noise filter calculated from the relevant Normalised Noise Power Spectrum(NNPS). This image is scaled according to an a-priori mean variance relationship for the detector. The final image is obtained after adding the blurred image and the scaled noise image.

Geometry. For the Monte Carlo simulations, the following components (from top to bottom) have been included in the setup according to their real dimensions and materials taken from manufacturer's specifications as illustrated in Figure 1b: (i) compression paddle, (ii) CDMAM phantom with a block of Perspex on top and bottom, (iii) breast support, (iv) anti-scatter grid with cover on top and bottom and (v) detector with its top cover as well. Selenium, cesium iodide and cesium bromide have been used for the Hologic, GE and Agfa detectors simulated in this work. The CDMAM test object has been simulated using a 3mm thick Perspex layer and 0.5mm thick Aluminium layer. 20mm thick Perspex blocks were placed on top and bottom of CDMAM. The focal spot was considered as a point source 65cm above the centre of chest wall edge of the detector. A top view diagram of the set up used for GE simulation is shown in Figure 2(a). The chest wall (CW) is located at position O. Outer area (1) represents the external dimensions of the system (detector, grid, breast support and compression paddle). (2) illustrates the position of the blocks of Perspex placed on top and bottom of CDMAM and (3) and (4) shows the location of the Perspex and Aluminium layers respectively used to simulate CDMAM.

Spectra. Table 1 describes the spectra used for the three systems simulated: Hologic Selenia, GE Essential and Siemens Mammomat 3000 using an Agfa computerised radiography plate. The energy spectrum has been simulated in Monte Carlo using data calculated from [2] and adjusted mathematically with a specific aluminium filter thickness to match the measured half value layer (HVL). Three simulations of 10^{10} X-ray photons have been simulated in Monte Carlo. These were summated to model each system.

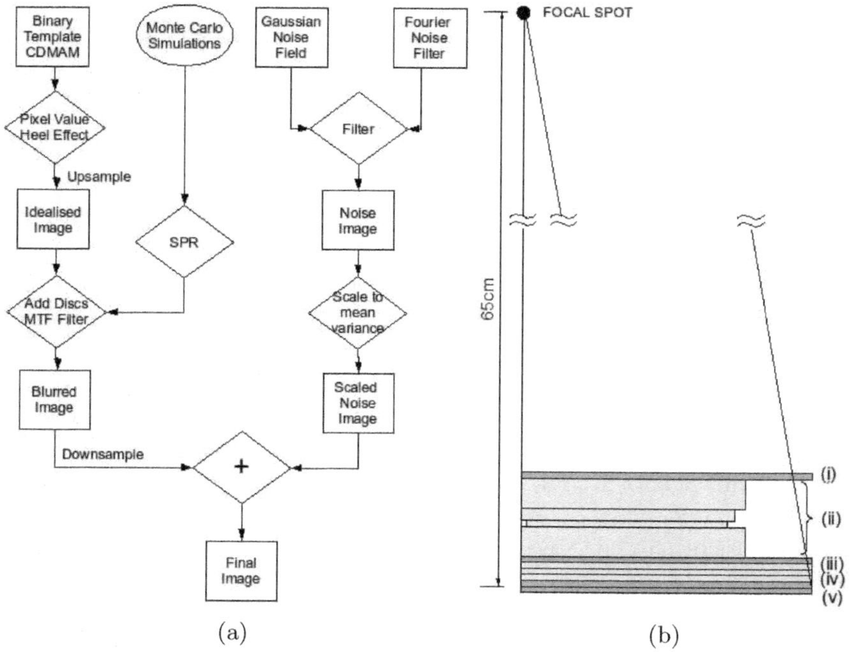

Fig. 1. (a) Diagram of Image Simulation Chain, adapted from [1], where the insertion of SPR map can be observed . (b) illustrates the setup used in Monte Carlo simulations with all the layers commented above. The numbered layers are discussed in the text.

Table 1. Spectra used in the simulations

System	Spectrum(Target/Filter)	HVL
Hologic Selenia	W/Rh @ 30kVp	0.45mm
GE Essential	Rh/Rh @ 29kVp	0.42mm
Siemens/Agfa	Mo/Rh @ 28kVp	0.41mm

Anti-scatter grid. Special attention has been paid to implementing the design of the moving focused anti-scatter grid to match the specifications of particular manufacturers systems. For the manufacturers chosen for this work, GE and Siemens/Agfa use linear grids, and a cellular grid is used by Hologic. In all the Monte Carlo simulations, the anti-scatter grid has been shifted several times along one axis (for the case of linear grids) or 2 axes (in the case of cellular grid) to cover a complete scatter grid unit cell(septa width + interspace width). The simulated anti-scatter grid was sufficiently large to cover the entire detector after every shift. The septa were positioned and angled so as to simulate a focused grid with the focal spot at a distance of 65cm from the edge of the detector. Cover layers on the top and bottom of the grid have been included in the simulations.

Detector. The Monte Carlo simulation used the Geant 4 code [3,4]. Once a photon has left the X-ray source, it is tracked through all the layers above the detector until either it is completely absorbed by one of the layers (including the detector), or it leaves the system. During its passage through the geometry shown in Figure 1b, the photon is tagged as a scatter if it undergoes a Compton and/or Rayleigh interaction in any of the layers above the detector. When a photon reaches the detector, only Compton interactions and photoelectric absorption within the detector are taken into account in order to score the energy deposited. The energy deposited in the detector and its spatial location are stored according to the aforementioned flag (scattering, or primary if it does not interact with any layer above the detector).

SPR Calculation. The SPR was calculated using a similar method to that of [5], where the ratio of the image generated by energy deposited by scattered photons (affected by Compton or Rayleigh) within the detector over the image generated by energy deposited by primary photons gives the scatter to primary map(Equation 1). Here I_{Sca} and I_{Pri} represent the images generated by energy deposited inside the detector affected by scattering and primary photons respectively for one of the position of the grid. $I_{Sca,T}$ corresponds to the sum of all the I_{Sca} images generated for each shifted grid position, while $I_{Pri,T}$ is similarly defined for I_{Pri}.

$$\frac{S}{P} = \frac{\sum I_{Sca}}{\sum I_{Pri}} = \frac{I_{Sca,T}}{I_{Pri,T}} \tag{1}$$

This expression is evaluated separately for each pixel. The SPR is generated using a detector comprising 1mm^2 pixels under the assumption that the scatter distribution does not contain fine structures. Then, using this assumption that the scatter distribution is a slowly varying function, the SPR image is median-filtered by an 11x11 kernel, providing a smoother, less statistically noisy representation of the SPR distribution. As the region of interest in this work is the CDMAM area, the image recorded in the detector is then cropped to the CDMAM's dimensions (inner area (4) shown in Figure 2a) giving a SPR image size of 152x230 pixels using the pixel dimension (1mm^2).

Contrast discs calculation using scatter map. Once the SPR map is generated and cropped to CDMAM's location, this is inserted in the image simulation chain to calculate the new contrast of the discs in the CDMAM as explained above. The contrast of each disc within the CDMAM is calculated using Equation 2, where C_{obs} represents the contrast of a disc observed at some particular location, C_p is the relative contrast of this disc considering only primaries passing the CDMAM and blocks of Perspex on top and bottom with and without the gold disc of specific thickness and $\frac{S}{P}$ is the value of the SPR for the specific coordinates of the disc in the CDMAM image.

$$C_{obs} = \frac{C_p}{1 + \frac{S}{P}} \tag{2}$$

3 Results

Images of energy deposited by primary and scattered photons recorded in the GE detector after Monte Carlo simulations are shown in Figure 2b and 2c respectively. The resultant SPR image is presented also in Figure 2d. The mean value of SPR calculated in the CDMAM area is 0.170 with a standard deviation of 0.008 for this system. Figure 2a shows a top view of the GE system and block of Perspex and CDMAM dimensions are highlighted as mentioned above.

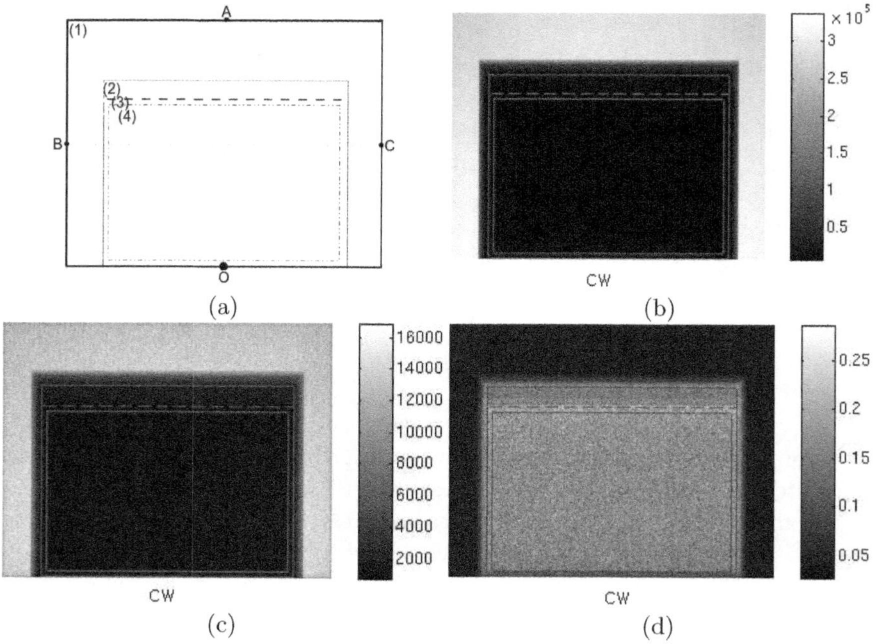

Fig. 2. Images generated from Monte Carlo Simulations: (a) top-view diagram of a GE system where outer area (1) represents system dimension, (2) block of Perspex on top and bottom of CDMAM and (3) and (4) Perspex and Aluminium layers of CDMAM respectively. (b) shows the energy deposited by primary photons in the detector $I_{Pri,T}$, (c) represents energy deposited in the detector due to scattered photons $I_{Sca,T}$ and (d) represents the SPR calculated using (b) and (c). Note that all layers shown in (a) are highlighted in (b),(c) and (d).

A 3D plot of the median-filtered SPR map shown in Figure 2d is plotted in Figure 3a, where the the various edges of the components of the phantom are also shown. Figure 3b and 3c illustrate SPR values calculated by the mean of 10 profiles along planes parallel and perpendicular to the anode-cathode axis (profiles OA and BC of Figure 2a).

For the Siemens/Agfa system, the mean SPR value across the CDMAM location was 0.172 (standard deviation of 0.008) and for the Hologic system a mean SPR value of 0.06 (standard deviation of 8.22x10^{-4}) was found.

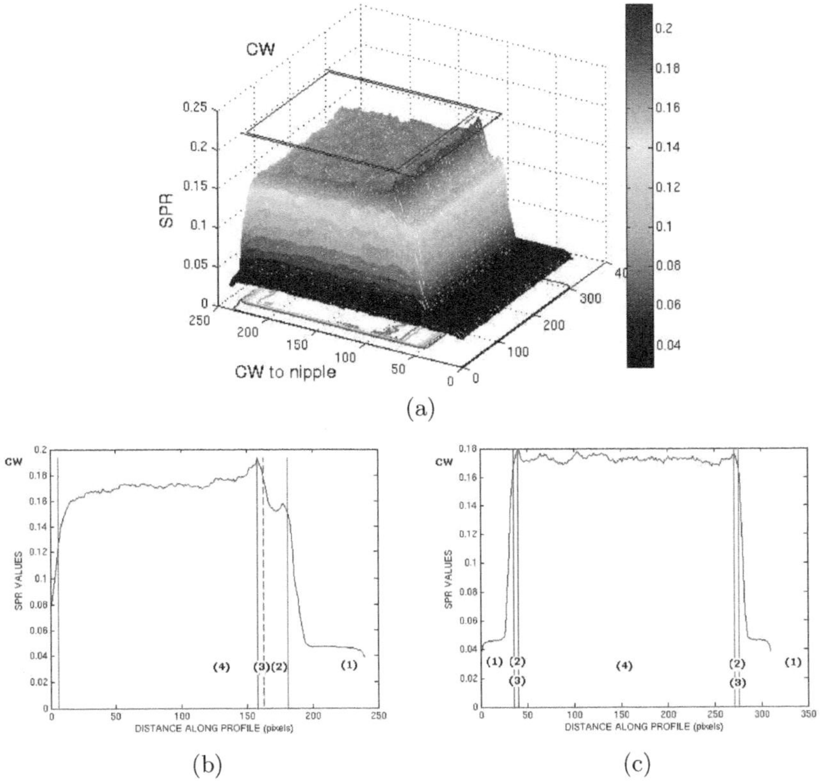

Fig. 3. (a) shows a 3D plot of the SPR map in the detector where CDMAM areas are highlighted. (b) and (c) represents the mean SPR values of 10 profiles along profile OA and BC illustrated in Figure 2a respectively. Vertical lines represents changes in dimension along the profiles OA and BC shown in Figure 2a. Numerical labels (1-4) corresponds to regions shown in Fig. 2a.

A study of the anti scatter grids used by GE (linear) and Hologic (cellular) was also undertaken. Figure 4 illustrates the equivalent profiles as seen above but in this case, SPR is plotted for both GE and Hologic system with and without anti scatter grid. Contrast improvement factors (CIF) of 1.50 and 1.60 was measured for GE and Hologic anti scatter grids respectively. Ratios of CIF between linear and cellular grids found by [6] varies from 0.91 to 0.95 for different thickness of breast phantoms for 30kVp. In this work, the ratio of the CIF values of the GE and Hologic grids is 0.937.

Automated scoring software for CDMAM using the CDCOM framework [7][8] has been used in this work for evaluation. A comparison of experimental acquired data, previous simulation data using a globally uniform SPR [1] and results from this work are shown in Figure 5 for GE systems and three different mean glandular doses(MGD).

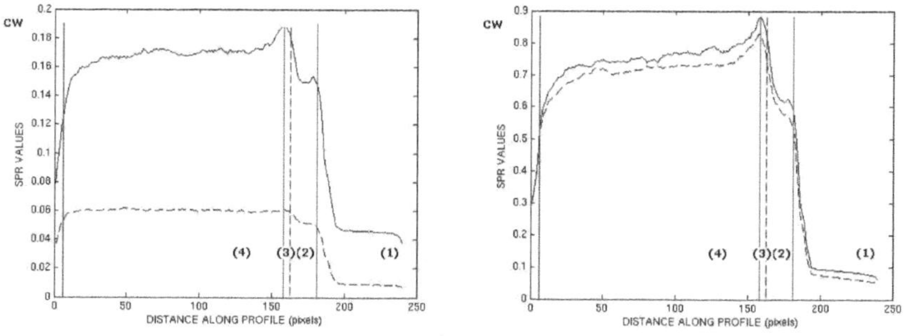

(a) (b)

Fig. 4. Grid Study:In both figures, solid lines represent the SPR from the GE system and dashed lines from the Hologic system. (a) and (b) show the profiles OA with and without anti scatter grid. Vertical lines represents changes in dimension along the profiles OA and BC shown in Figure 2a. Numerical labels (1-4) corresponds to regions shown in Fig. 2a.

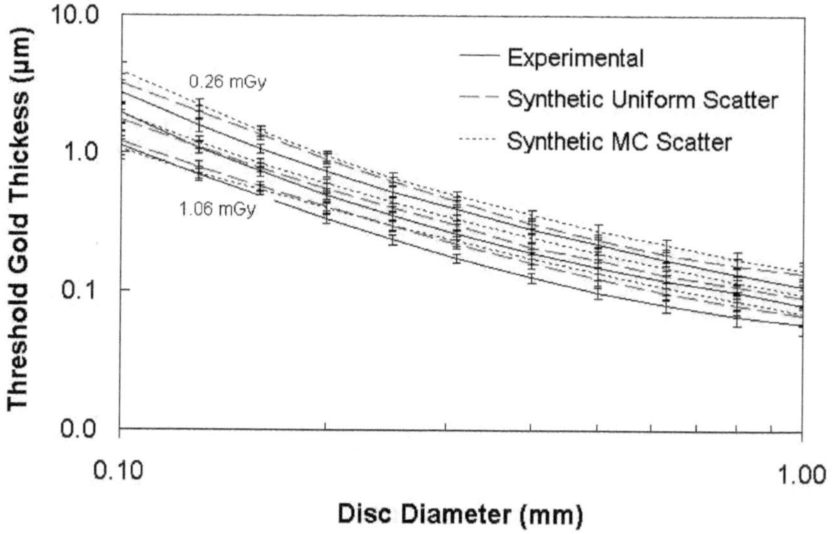

Fig. 5. Contrast detail curves for GE system. From top to bottom: 0.26, 0.52 and 1.06 mGy MGD. Solid lines illustrate contrast detail measured from experimental data. Dashed and dotted lines represent contrast detail considering uniform scatter (from [1]) and calculating SPR from Monte Carlo simulations respectively. Errors bars indicate 2 standard errors of mean.

4 Discussion

Considering Figure 3b, an increase of the SPR values is observed near the edge from area (1) to (2) and from (2) to (3)(4). Previously, [9] introduced this effect due to the compression paddle. A similar effect of increasing SPR at the edges is also described in [5]. Thus, components of the system such as the compression paddle or breast support may increase somewhat the scattering on the edge of phantoms and should be taking into account for scatter correction.

The cellular grid implemented here has better performance than the linear grid as expected. The very low scatter found with this set up when comparing with other cellular grid [6] can be attributed to wider septa and interspace. This also eliminates the increased scatter at the edges of CDMAM discussed previously.

It is observed in Figure 5 that images simulation using both uniform scatter and scatter calculated from Monte Carlo are within the error bars, so the results shown in this work support previous assumption used by [1]. Moreover, the simulation chain still have to be improved.

Acknowledgments. This work is part of the OPTIMAM project and is supported by CR-UK & EPSRC Cancer Imaging Programme in Surrey, in association with the MRC and Department of Health (England). The author would like to thank GE, Hologic, Siemens and Agfa for providing technical information used in the simulations.

References

1. Yip, M., Alsager, A., Lewis, E., Wells, K., Young, K.C.: Validation of a digital mammography image simulation chain with automated scoring of CDMAM images. In: Krupinski, E.A. (ed.) IWDM 2008. LNCS, vol. 5116, pp. 409–416. Springer, Heidelberg (2008)
2. Boone, J., Fewell, T., Jennings, J.: Molybdenum, rhodium and tungsten anode spectral models using interpolating polynomials with applications to mammography. Med. Phys. 24(2), 1863–1874 (1997)
3. Agostinelli, S., et al.: Geant4 – a simulation toolkit. Nuclear Instruments and Methods in Physics Research 506(3), 250–303 (2003)
4. Allison, J., et al.: Geant4 development and applications. IEEE Transactions on Nuclear Science 53(1), 270–278 (2006)
5. Sechopoulos, I., Suryanarayanan, S., Vedantham, S., D'Orsi, C., Karellas, A.: Scatter radiation in digital tomosynthesis of the breast. Med. Phys. 34(2), 564–576 (2007)
6. Rezentes, P.S., de Almeida, A., Barnes, G.T.: Mammography Grid Performance. Radiology 210, 227–232 (1999)
7. Visser, M., Karssemeijer, N.: Manual CDCOM version 1.5: software for automated readout of CDMAM 3.4 images, `http://www.euref.org`
8. Young, K.C., Cook, J.J.H., Oduko, J.M.: Automated and human determination of threshold contrast for digital mammography systems. In: Astley, S.M., Brady, M., Rose, C., Zwiggelaar, R. (eds.) IWDM 2006. LNCS, vol. 4046, pp. 266–272. Springer, Heidelberg (2006)
9. Dance, D.R., Day, G.J.: The computation of scatter in mammography by Monte Carlo methods. Physics in Medicine and Biology 29(3), 237–247 (1983)

Classifying Breast Masses in Volumetric Whole Breast Ultrasound Data: A 2.5-Dimensional Approach

Gobert N. Lee[1,*], Toshiaki Okada[2], Daisuke Fukuoka[3], Chisako Muramatsu[2], Takeshi Hara[2], Takako Morita[4], Etsuo Takada[5], Tokiko Endo[6], and Hiroshi Fujita[2]

[1] School of Computer Science, Engineering and Mathematics, Flinders University, Sturt Road, Bedford Park, Adelaide SA 5042, Australia
gobert.lee@flinders.edu.au

[2] Department of Intelligent Image Information, Division of Regeneration and Advanced Medical Sciences, Graduate School of Medicine, Gifu University, 1-1 Yanagido, Gifu 501-1194, Japan

[3] Technology Education, Faculty of Education, Gifu University, 1-1 Yanagido, Gifu 501-1193, Japan

[4] Department of Mammary Gland, Chunichi Hospital, 3-6-38 Marunouchi, Naka-ku, Nagoya, Aichi 460-0002, Japan

[5] Division of Medical Ultrasonics, Center of Optical Medicine, Dokkyo Medical University, 880 Kitakobayashi, Mibu, Tochigi 321-0293, Japan

[6] Department of Radiology, National Hospital Organization Nagoya Medical Center, 4-1-1 Sannomaru,Naka-ku, Nagoya, Aichi 460-0001, Japan

Abstract. The aim of this paper is to investigate a 2.5-dimensional approach in classifying masses as benign or malignant in volumetric anisotropic voxel whole breast ultrasound data. In this paper, the term 2.5-dimensional refers to the use of a series of 2-dimensional images. While mammography is very effective in breast cancer screening in general, it is less sensitivity in detecting breast cancer in younger women or women with dense breasts. Breast ultrasonography does not have the same limitation and is a valuable adjunct in breast cancer detection. We have previously reported on the clinical value of volumetric data collected from a prototype whole breast ultrasound scanner. The current study focuses on a new 2.5-dimensional approach in analyzing the volumetric whole breast ultrasound data for mass classification. Sixty-three mass lesions were studied. Of them 33 were malignant and 30 benign. Features based on compactness, orientation, shape, depth-to-width ratio, homogeneity and posterior echo were measured. Linear discriminant analysis and receiver operating characteristic (ROC) analysis were employed for classification and performance evaluation. The area under the ROC curve (AUC) was 0.91 using all breast masses for training and testing and 0.87 using the leave-one-mass-out cross-validation method. Clinically significance of the results will be evaluated using a larger dataset from multi-clinics.

Keywords: ultrasound breast mass, classification, geometric feature, echo feature.

* Corresponding author.

J. Martí et al. (Eds.): IWDM 2010, LNCS 6136, pp. 636–642, 2010.

1 Introduction

Mammography is very effective in breast cancer detection. It is the routine technique used in breast cancer screening in women who have no symptom of breast cancer. However, mammography is less sensitivity in detecting breast cancer in younger women or women with dense breasts. This is due to the inherited limitations of x-ray employed in the image acquisition in mammography. Breast ultrasonography is another long-standing technique in breast imaging and is a valuable adjunct in breast cancer detection. Distinguished from mammography, the technique employs acoustic waves and does not have the same limitation as mammography. However, it is not without its shortcomings.

Currently, ultrasound breast examination is routinely performed by an ultrasonographer or ultrasonologist. A small hand-held probe of size about 4 cm is used and the ultraonographer/ ultrasonologist runs the probe over the entire breast or pre-identified regions during an examination. The technique can provide very valuable information in the hands of experienced examiners but is in general time consuming. Results are operator independent and reproducibility is poor. A novel breast scanning system that can acquire the data of the entire breast quickly, systematically and repeatedly with precision will be of great advantage.

We have previously introduced a prototype whole breast ultrasound scanner for auto-acquisition of volumetric breast ultrasound data [1]. Diagnostic value of the data was investigated [2]. The volumetric ultrasound data of a whole breast consist of a stack of two-dimensional images, each depicting an axial slice image of the breast. In exploiting the benefit of volumetric data, three-dimensional analysis was used in our previous study in classifying malignant and benign breast masses [3].

One issue noted in our previous three-dimensional analysis was that the data was anisotropic. Anisotropic data are generally computationally cumbersome. One of the common practices would be to resample the data to create isotropic voxel. However, this would not be a good practice for our volumetric whole breast data as the resolution in one direction (z-direction, normal to the axial plane) is about 8 to 10 times lower than that in the other two directions. The discrepancy is large and a reliable model for interpolation cannot be guaranteed. Another option is to increase the number of data points in the z-direction in the raw data. This could be achieved by reducing the interval between adjacent slice images. Options for slice intervals are 2 mm, 1 mm and 0.5 mm. Corresponding unilateral breast study contains 84, 168 and 336 (axial) images, respectively, with acquisition time increases from 20, to 40 and 80 seconds, respectively. The increase in number of axial unnecessarily burdens the interpreters while longer acquisition time leads to problems such as image blurring due to patient movement. Neither of the above options is desirable in this situation as the first one relied on interpolated slice images of which accuracy of the image details to be employed in the computer-aided image analysis cannot be guaranteed. The second one imposes on a clinical practice to collect extra data which is a burden to the practice at no clear clinical benefits. After taken the above into consideration, this paper investigates the efficacy of a 2.5-dimensional analysis, a step between 2-dimensional and 3-dimensional analyses.

2 Method

2.1 Ultrasound Data

Volumetric full-breast ultrasound data were used in this study. The data included 63 breast masses. Of them 33 were malignant and 30 (16 cysts; 14 fibroadenomas) were benign. The malignant and benign masses were related to 29 and 24 breasts, respectively. All the masses were annotated by a radiologist experienced in breast ultrasound and the malignant masses were proven by biopsy. With the patient in prone position, a diagnostic ultrasound system Prosound-II SSD-5500 (Aloka Co., Ltd, Japan) and a prototype full-breast scanner ASU-1004 (Aloka Co., Ltd, Japan) (Figure 1) were used to acquire the full-breast images. The scanner ASU-1004 was equipped with a 5-10 MHz 6 cm linear probe. Operating in a fixed pattern, the probe scanned an area of $16 \times 16\ cm^2$ in 3 sweeps, covering the full-field of a breast. The original scan images were B-mode breast section images in DICOM format with an overlap margin of 1 cm on each of the 'stitching' side. Volumetric full-breast data were generated by 'stitching' corresponding images in the 3 sweeps together (Figure 2). Details of the scanner can be found in [4-6].

The full-breast ultrasound scans were performed in the period 2003-2004 at the Center of Optical Medicine, Dokkyo University School of Medicine, Tochigi, Japan where a prototype full-breast scanner ASU-1004 was located. The size of each (stitched) B-mode image in the constructed volumetric full-breast data was 694×400 pixels with a spatial resolution of 0.23 mm/pixel and a slice-to-slice interval of 2 mm.

The images had a gray scale resolution of 8 bits. For each mass, a series of axial slice images containing that mass is employed in the 2.5-dimensional analysis (Figure 3). Features are measured individually on each slice image. The same feature measured on difference slice images are combined at a later stage.

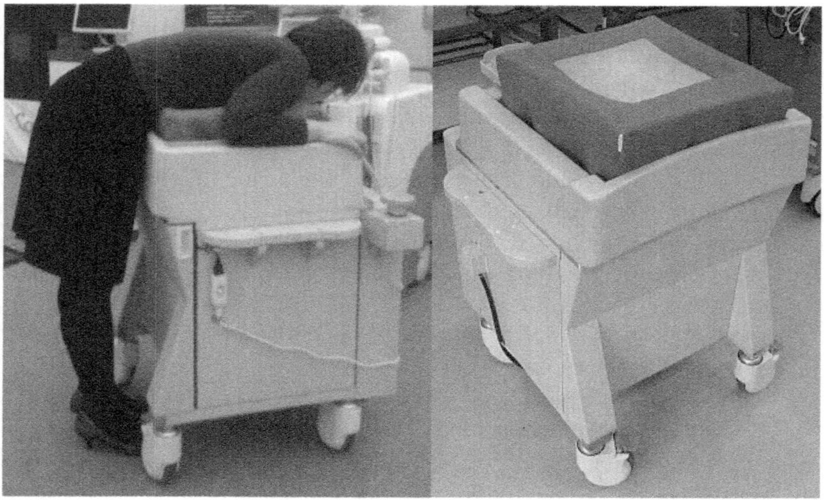

Fig. 1. The prototype full-breast scanner ASU-1004 (right) with a patient in prone position (left)

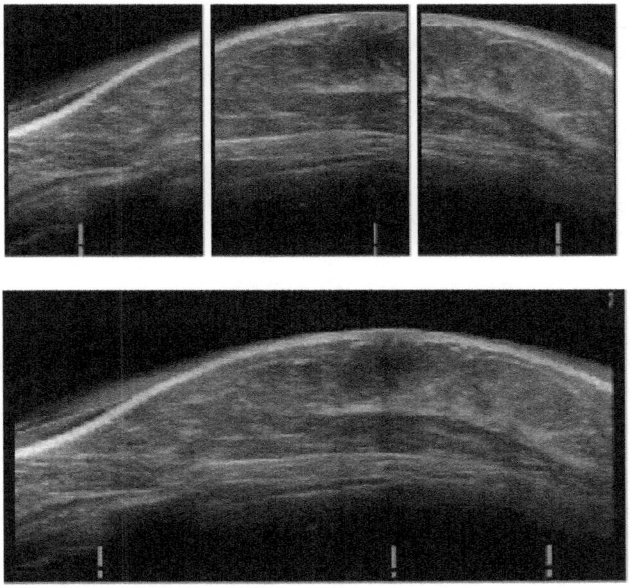

Fig. 2. Corresponding breast section slice images in the 3 sweeps (above) are 'stitched' together to form a slice image in the volumetric full-breast data (below)

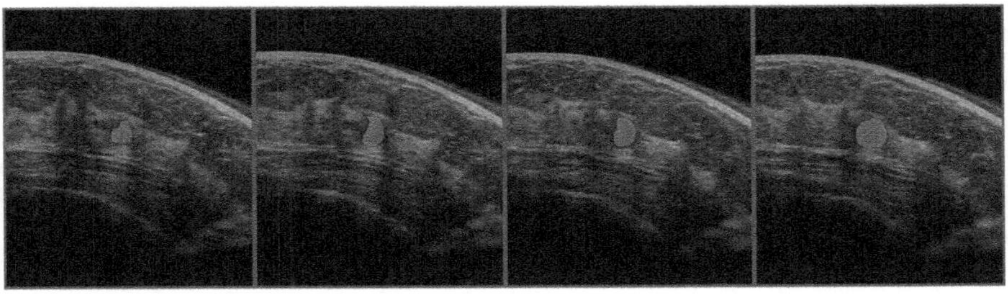

Fig. 3. 2.5-dimensional analysis. In this example, a series of 4 images containing the mass were used in the analysis. The same set of features is measured on each images and is combined at a later stage according to a set rule.

2.2 2.5-Dimensional Analysis

For each mass lesion, a number of axial images containing the mass lesion were identified. The number of associated slice images depends on the size of the mass and the slice-to-slice interval. With the use of a 2 mm slice-to-slice interval and the lesion size ranges from 5 mm to over 3 cm in this study, the number of associated slice image for a mass varies from a minimum of 1 to 2 images to a maximum of over 10 images, with the majority being 4 to 6 images. With the series of slice images containing the mass lesion identified, lesion boundaries were delineated manually and lesions in each of the slice images were segmented.

After the segmentation process, features were measured on the lesions depicted in each of the slice images. Six features were defined. They were compactness, orientation, shape, depth-to-width ratio, homogeneity and posterior analysis. These six features were similar to the features selected in our previous study [3] and were based on the image features that radiologists' found useful and routinely consulted in breast ultrasound images interpretations. A summary of the six features is given in the next paragraph.

In general, compactness (*C*) measures the degree of roundness of an object and is given by

$$C = 4 \times \pi \times A / S^2 \tag{1}$$

where *A* and *S* are the area and circumference of the object, respectively. Benign masses are usually round in shape while malignant masses are more likely to be irregular or oval in shape. Orientation measures the angle (in degrees) between the horizontal axis and the major axis of the ellipse that has the same second-moments as the object and is given by

$$\tan^2 \theta + \frac{M(2,0) - M(0,2)}{M(1,1)} = 0 \tag{2}$$

where $M(p, q)$ is the pq^{th}-order moment and is given by

$$M(p,q) = \sum_{i,j} i^p j^q .$$

Depth-to-width (DW) ratio is another feature that can provide information of the orientation of an (elongated) object. This feature can be simply defined as the ratio of the height to the width of the smallest bounding box containing the mass. Homogeneity of the mass is computed using the variance of the intensity inside a mass. Benign masses such as cysts generally display homogeneity (small variance) inside the mass. Posterior echo is also another feature to distinguish benign and malignant lesions. The absence of posterior echo is an indicator of malignant lesion.

The above six features were measured on the lesions in each individual slice images. In other words, the six features were repeatedly measured on a series of mass cross-sectional images separated at a fixed interval of 2 mm.

The 2.5-dimensional analysis is based on features measured in a series of 2-dimensional images. For each breast mass, measurements of the same feature measured on a series of images are combined according to a rule which is feature-specific. For example, the depth-to-width (D/W) ratio measures the depth (vertical extent) of a mass to the width (horizontal extent) of a mass in a 2-dimensional image. Malignant lesions are more rigid and less compressible when subject to external force, hence the D/W ratio of malignant lesions is generally high. On the other hand, benign lesions such as cysts, which are usually filled with fluid or lipids, are more compressible and deformable. Hence, their D/W ratios are generally low. In other words, higher the D/W ratio, more likely is the lesion malignant. So in a 2.5-dimensional analysis, the maximum of the D/W ratios measured on a series of 2-dimensional images of a lesion is the strongest evidence for malignancy. Table 1 listed the rules in combining the multi-slice measurements of the same feature towards 2.5-dimensional analysis assuming strongest evidence for malignancy.

Table 1. Rules for combining multi-slice feature measurements in 2.5-dimensional analysis

FEATURES	*2.5-DIMENSIONAL ANALYSIS*
Compactness	minimum
Orientation	maximum
Depth-to-width ratio	maximum
Posterior echo	minimum
homogeneity	maximum
Shape	maximum

3 Results

Linear discriminant analysis and receiver operating characteristic (ROC) analysis were employed for classification and performance evaluation. Discriminative powers of the six 2.5-dimensional features (combined over slice images) were analyzed in Table 2. The discriminative power of individual feature was indicated by the area under the ROC curve (AUC) obtained when using that feature alone in classifying the mass as benign or malignant. Both the resubstitution AUC using all breast masses for training and testing and the leave-one-mass-out cross-validation AUC are depicted. Table 2 shows that among the six features, three of them have strong discriminative power, namely, orientation, depth-to-width ratio and posterior echo.

When using all the six features for classification, the area under the ROC curve (AUC) was found to be 0.91 using all breast masses for training and testing (resubstitution) and 0.87 using the leave-one-mass-out cross-validation method.

Among a number of classifiers, linear discriminant analysis was chosen for its robustness. Its hyperplane decision surface makes it less susceptible for over-training which is preferable for studies with small samples.

Table 2. Discriminative powers of the six features indicated by the area under the ROC curve (AUC)

FEATURES	*AUC (resubstitution)*	*AUC (leave-one-mass-out)*
Compactness	0.64	0.64
Orientation	0.82	0.79
Depth-to-width ratio	0.83	0.84
Posterior echo	0.84	0.84
homogeneity	0.66	0.50
Shape	0.60	0.58

4 Discussion and Conclusion

The classification based on 2.5-dimensional analysis in this study resulted in high accuracy in discriminating malignant and benign lesions in volumetric breast ultrasound data with anisotropic voxel. AUC indices in this study are in general high and similar to that based on 3-dimensional analysis in our previous study [2]. However, direct comparisons cannot be made. This is because the sample sizes in the two studies were different (63 masses in this study and 36 in the previous 3-d study) and shape feature was introduced in the current 2.5-dimensional analysis but not in the previous 3-dimensional study. In addition, though features definitions are very similar in the two studies, different algorithms were used to compute the features in the two studies. Slight variations in the interpretation of individual features may exist.

Plan for further work in this project is two-folded. (1) a larger database is required to confirm the results in this study. (2) Classification categories will also be extended to include normal breast tissue lumps and other artifacts in the breast which are the false positives found in the detection stage.

Acknowledgments. This research was supported in part by a grant for the Knowledge Cluster Gifu-Ogaki (KCGO), referred to as the "Robotics Advanced Medical Cluster," from the Ministry of Education, Culture, Sports, Science and Technology, Japan.

References

1. Takada, E., Ikedo, Y., Fukuoka, D., Hara, T., Fujita, H., Endo, T., Morita, T.: Semi-Automatic Ultrasonic Full-Breast Scanner and Computer-Assisted Detection System for Breast Cancer Mass Screening. In: Proc. of SPIE Medical Imaging 2007: Ultrasonic Imaging and Signal Processing, vol. 6513, pp. 651310-1–651310-8. SPIE, Bellingham (2007)
2. Lee, G.N., Morita, T., Fukuoka, D., Ikedo, Y., Hara, T., Fujita, H., Takada, E., Endo, T.: Differentiation of Mass Lesions in Whole Breast Ultrasound Images: Volumetric Analysis. In: Radiological Society of North America, 94th Scientific Assembly and Annual Meeting Program, p. 456 (2008)
3. Lee, G.N., Fukuoka, D., Ikedo, Y., Hara, T., Fujita, H., Takada, E., Endo, T., Morita, T.: Classification of Benign and Malignant Masses in Ultrasound Breast Image Based on Geometric and Echo Features. In: Krupinski, E.A. (ed.) IWDM 2008. LNCS, vol. 5116, pp. 433–440. Springer, Heidelberg (2008)
4. Fukuoka, D., Hara, T., Fujita, H., Endo, T., Kato, Y.: Automated Detection and Classification of Masses on Breast Ultrsonograms and its 3D Imaging Technique. In: Yaffe, M.J. (ed.) IWDM 2000: 5th International Workshop on Digital Mammography, pp. 182–188. Medical Physics, Madison (2001)
5. Fukuoka, D., Hara, T., Fujita, H.: Detection, Characterization, and Visualization of Breast Cancer Using 3D Ultrasound Images. In: Suri, J.S., Rangayyan, R.M. (eds.) Recent Advances in Breast Imaging, Mammography, and Computer-Aided Diagnosis of Breast Cancer, pp. 557–567. SPIE, Bellingham (2006)
6. Ikedo, Y., Fukuoka, D., Hara, T., Fujita, H., Takada, E., Endo, T., Morita, T.: Development of a Fully Automatic Scheme for Detection of Masses in Whole Breast Ultrasound Images. Med. Phys. 34(11), 4378–4388 (2007)

A supplement to the European Guidelines for Quality Assurance in Breast Cancer Screening and Diagnosis

Ruben E. van Engen, Kenneth C. Young, Hilde Bosmans, Barbara Lazzari, Stephan Schopphoven, Patrice Heid, and Martin A. Thijssen

European Reference Organisation for Quality Assured Breast Screening and Diagnostic Services (EUREF), P.O. box 6873, 6503 GJ Nijmegen, The Netherlands
r.vanengen@euref.org

Abstract. In 2006 the fourth edition of the European Guidelines for Breast Cancer Screening and Diagnosis was published by the European Commission. Due to the fast developments in the field of digital mammography and the experience with digital mammography systems over the past years a supplement to the technical quality control procedures proved necessary. This paper describes important changes compared to the Guidelines and their rationale. Testing methods which are new or have changed include: the size of the standard region of interest (ROI), the thickness compensation measurement, noise evaluation, threshold contrast visibility, an AEC measurement which simulates local dense area. A paragraph on evaluation of image processing has been added. With these changes European quality control procedures are again up-to-date with current knowledge.

Keywords: mammography, quality control.

1 Introduction

In 1996 the third edition of the European Guidelines for quality assurance in mammography screening was published, fully based on screen-film mammography. Part of these guidelines is the European protocol for the quality control of the physical and technical aspects of mammography screening. Due to the introduction of digital mammography an addendum to this edition was published in 2003. In this addendum technical quality control procedures for digital mammography systems were given. This addendum is incorporated as chapter 2b in the fourth edition of the European Guidelines for Quality Assurance for Breast Cancer Screening and diagnosis, which has been published by the European Commission in 2006 (Perry 2006).

Since 2006 developments in the field of digital mammography and the current knowledge of failures in mammography equipment necessitated a supplement to this protocol. In this paper the changes in the forthcoming supplement compared to the fourth edition of the Guidelines are listed and their rationale is given.

J. Martí et al. (Eds.): IWDM 2010, LNCS 6136, pp. 643–650, 2010.

2 Adaptations Compared to the Fourth Edition of the European Guidelines

A number of quality control tests have been updated: the size of the standard region of interest (ROI), the thickness compensation measurement, the noise evaluation, computer readout of threshold contrast visibility. Furthermore an AEC measurement which simulates local dense area has been introduced and a paragraph on evaluation of image processing has been added. Additionally the dosimetry section has been adapted to accommodate the recently introduced X-ray spectra. This will not be part of this paper but is described elsewhere (Dance 2009).

2.1 Region-of-Interest

In the fourth edition of the Guidelines the standard ROI is a square with dimensions of 2 cm x 2 cm. The centre of this square is positioned 60 mm from chest wall side and centered laterally. In this ROI pixel value and standard deviation are measured to calculate Signal-to-Noise-Ratio (SNR) and Contrast-to-Noise-Ratio (CNR). In practice, for systems with specific pixel value trends over the imaging field, like the heel effect, this means that the measured standard deviation includes this trend. To reduce the influence of these trends a number of options are available. The first option is that the physicist performs a trend correction on all images. However this is quite complicated and probably time-consuming. Therefore a more realistic option is to decrease the size of the standard ROI, thus minimizing the influence of non uniformity, see table 1. (Al Sagar 2008).

Table 1. Error in CNR for different sizes of ROI on a CR system

ROI Size (mm)	CNR	Error compared to non uniformity corrected image (%)
20 x 20	12.3	10
10 x 10	13.1	4
5 x 5	13.3	2.5
2.5 x 2.5	13.6	0.2

Reduction of non uniformity is obtained with very small ROIs, but in practice, the ROI should include a sufficient number of pixels for reliable measurement of SD. In Table 2 the coefficient of variation (COV) of SD in the reference ROI is given for 10 images made under identical exposure conditions (Bouwman 2009). It is clear that the choice of ROI size is a compromise between the reduction of trends in the X-ray field end the accuracy of SD measurements. Therefore a standard ROI with dimensions 5 mm x 5 mm has been chosen.

Table 2. COV of SD for two mammography systems for different sizes of ROI

ROI Size (mm)	COV in SD of DR system (pixel size 100 micron) (%)	COV in SD of CR system (pixel size 50 micron) (%)
20 x 20	0.4	5.6
10 x 10	0.8	2.0
5 x 5	0.9	1.0
2.5 x 2.5	3.4	1.5

2.2 Thickness Compensation Measurement

It is practice to simulate the exposure to different breast thicknesses by PMMA slabs. Dance (2000) calculated the PMMA thicknesses that correspond in terms of X-ray attenuation to specific compressed breast thicknesses. However the slabs are thinner than the corresponding typical breasts. On most mammography systems, the X-ray spectrum selection is based on the height of the compression paddle. Simulating the attenuation of a breast with PMMA will therefore introduce a difference in X-ray spectrum compared to a typical breast with the same attenuation. We propose to use spacers to equal the height of the compression paddle to the height of the typical breast with the same attenuation, see table 3. The spacers should be positioned outside the area, which determines the exposure factors.

Table 3. Typical breasts thickness equivalence to PMMA thickness (Dance 2000)

PMMA thickness (mm)	Equivalent typical breast thickness (mm)	Thickness of spacer (mm)
20	21	None
30	32	2
40	45	5
45	53	8
50	60	10
60	75	15
70	90	20

In addition it is proposed to use a smaller piece of aluminum (10 mm x 10 mm, 0.2 mm thick) for the CNR evaluation. Next to a change in ROI size (see paragraph 2.1), it is proposed to calculate the signal in a 5 mm x 5 mm ROI of the aluminium insert and the background in four 5 mm x 5 mm background ROIs, see figure 1. The four background ROIs are used to minimize the influence of trends in the X-ray field. The background pixel value (PV) and SD are the average of the four background ROIs.

$$PV(background) = \frac{\sum_1^4 PV(ROI_n)}{4}; \quad SD(background) = \frac{\sum_1^4 SD(ROI_n)}{4} \tag{1}$$

The CNR is calculated using the formula from the fourth edition of the Guidelines and the limiting values from the fourth edition remain unchanged.

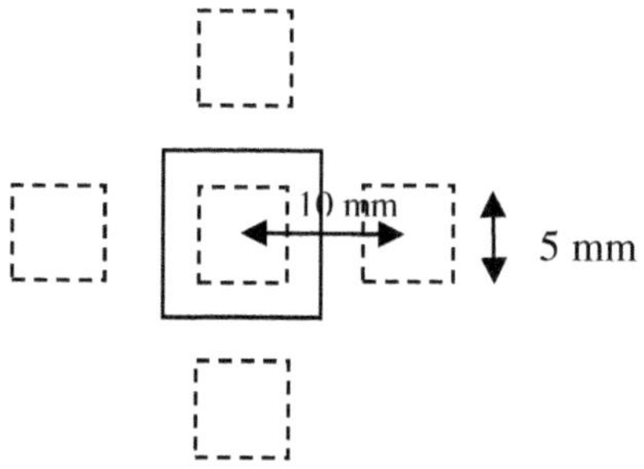

Fig. 1. schematic drawing of the ROI in the aluminium sheet and the background ROIs in the thickness compensation measurement

2.3 Noise Evaluation

The noise evaluation of the fourth edition of the European Guidelines proved to be insufficiently sensitive to additional noise, therefore a completely different approach is suggested. Noise in images can be subdivided in components: electronic noise, quantum noise and structure noise:

$$SD^2 = k_e^{\,2} + k_q^{\,2} \cdot PV + k_s^{\,2} \cdot PV^2 \tag{2}$$

SD = standard deviation in reference ROI
k_e = electronic noise coefficient
k_q = quantum noise coefficient
k_s = structure noise coefficient
PV = average pixel value in reference ROI

Electronic noise is assumed to be independent of the exposure level, structure noise is assumed to be proportional with exposure and quantum noise is assumed to be related with the square root of the exposure. Due to this difference in behavior against dose the components can be separated. All removable parts (e.g. compression paddle, covers and anti-scatter grid) are removed from the X-ray beam and a 2 mm thick aluminium attenuator is positioned as close as possible to the X-ray tube. In manual mode the X-ray spectrum, which is chosen for a standard image, is set and images are made at different mAs-values (10 values) over the whole range of available values.

The dose on the detector surface is calculated from the tube output and mAs values. For systems with a non-linear response, the pixel data is linearized before further analysis. In the reference ROI, pixel value and SD are measured. SD^2 against detector dose is plotted. The measured data is fitted using equation 2 and the noise coefficients are determined. The detector dose range for which quantum noise is the largest noise component is determined. For the clinically used detector dose levels quantum noise should be the largest component, see figure 2.

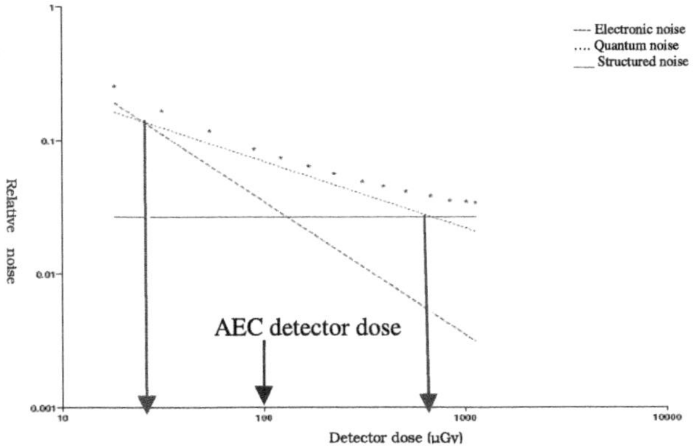

Fig. 2. Example of the proposed noise analysis

2.4 Threshold Contrast Visibility

The method described in the fourth edition shows intra and inter reader variability. In addition, scoring threshold visibility images by human observers is time consuming. Therefore computer readout is introduced.

Threshold contrast visibility is determined for cylindrical details with diameters in the range from 0.1 to 1 mm. As with human reading, the details have to be imaged on a background object with a thickness equivalent (in terms of attenuation) to 50 mm of PMMA. The details must be positioned at a height of 20 to 25 mm above the breast support table. Use the exposure factors that would be selected for a 60 mm average breast. Make sixteen images of the details and move the phantom slightly between the successive exposures to obtain images with different relative position of the details and the detector elements.

Score the images using CDCOM (latest version downloadable from www.euref.org) and calculate the detection matrix. For each diameter a psychometric curve is fitted using equation (3) to determine the threshold contrast for each diameter (Veldkamp 2003):

$$p(d) = \frac{0.75}{1+e^{-f(C-Ct)}} + 0.25 \qquad (3)$$

C = logarithm of signal contrast $C = \log(1 - e^{-\mu d})$.
C_t = signal contrast at the threshold of 62.5%
f = fitting parameter
p(d) = probability of detection p(d) of an object with size d

A threshold at 62.5% correct response is used to determine the threshold contrast. Results for which the psychometric curve is fitted with only a few data points are disregarded. In order to use the limiting values from the fourth edition, the resulting thresholds for each diameter have to be converted to human readout. This can be done using the relationship in Young (2008), that predicts the human readout from the computer readout.

$$TC_{predicted} = a[TC_{auto}]n \qquad (4)$$

$TC_{predicted}$ = Predicted human readout threshold contrast
TC_{auto} = Computer readout (using CDCOM) threshold contrast
a and n = fitting parameters

Pooling the data (series of images from 113 systems) from the Guildford, Nijmegen and Leuven centres (Young 2008), provided the following factors:

n = 0.888
a = 1.17

For images with high noise the conversion factors are not validated yet, system dependence of the factors n and a was not observed in this dataset. The resulting predicted human readout threshold contrasts are fitted with a third order polynomial function to obtain a contrast-detail curve.

$$Tc = a + b\,x^{-1} + c\,x^{-2} + d\,x^{-3} \qquad (5)$$

T_c = nominal threshold contrast (%) calculated at 28 kV Mo/Mo
x = detail diameter (mm)
a, b, c and d = coefficients adjusted to achieve a least squares fit, and are ≥ 0

The fitted curve is checked against the limiting values for human readout as published in the European Guidelines, 4th edition.

2.5 Simulating a Local Dense Area

In digital mammography most X-ray units incorporate an automatic exposure controller (AEC) system using a pre-exposure technique. With the information from this pre-exposure, the area (or areas) with highest attenuation is determined and the exposure factors are tuned to this area. However, it was observed that some AEC systems tune the exposure to a very large area of the breast, in some cases even the whole breast. This potentially leads to underexposure of high attenuation areas i.e. glandular tissue. This potential underexposure is not detected with the current AEC quality control tests, which use plates of PMMA covering the total area of the detector. Therefore a method has been set up in which an area with higher attenuation is simulated, see figure 3.

Three PMMA plates (1 cm thick) and two spacers are positioned on the bucky. The plates are positioned such that the area in which the exposure factors are determined is fully covered, the spacers are positioned outside this region. By adding 1 to 10 small PMMA plates of dimensions 20 x 40 mm (2 mm thick), a small area with increasing glandularity is simulated. The AEC system should detect the extra attenuation and exposure factors should be adapted accordingly, so pixel value or SNR (depending on the workings of the AEC) should be approximately equal for all images in the area of extra attenuation.

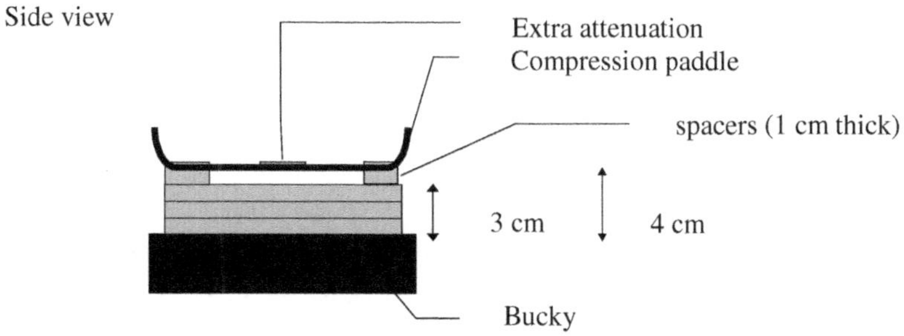

Fig. 3. Setup for the measurement simulating a local dense area

2.6 Image Processing

The lack of methods to evaluate the effect of image processing algorithms is a gap in the quality evaluation in digital mammography. One of the problems in evaluating image processing is that image characteristics, like pixel value distribution (histogram), shape etc. are used in processing algorithms. Therefore phantoms whose characteristics differ from those of a breast cannot be used to evaluate image processing. The difference in characteristics causes the processing of phantom images to be different from images of breasts and artefacts may also be introduced because image processing algorithms presume a breast edge (skin line) that may not be present in technical test objects.

Therefore evaluation of image processing can only be performed subjectively by scoring mammograms by radiologists. After installation of a system and the acceptance test by a physicist, it is advised to carefully evaluate a series of clinical images of ca. 50 patients. If possible, images from the installed system should be compared to previous images from an established modality of the same women. The following list of image characteristics might be taken into account when comparing images (van Ongeval 2008):

1. The visualization of the skin line
2. The visibility of vascular structures through dense parenchyma
3. The visualization of vascular and fibrous structures and pectoral muscle
4. The visualization of structures along the pectoral muscle
5. The visualization of Coopers ligaments and vascular structures in the low and high pixel value areas of the image
6. The edges of microcalcifications
7. The noise in the low and high pixel value areas of the image
8. The contrast in the low and high pixel value areas of the image
9. The appearance of glandular tissue
10. The appearance of background area
11. The confidence of the radiologist with the representation of the image
12. The presence of artefacts.

It must be realized that in individual cases visibility of structures might differ due to e.g. differences in positioning. Conclusions should not be drawn on small number of cases.

3 Discussion and Conclusions

With the adaptations of quality control tests described in this paper the measurements have become more powerful to detect problems (noise evaluation, simulating a local dense area), less reader dependent and time consuming (threshold contrast visibility) and the effect of the Heel effect is minimized (changes in size of standard ROI and the setup of the thickness compensation measurement). It is still not possible to evaluate the quality of image processing algorithms objectively, although not perfect the paragraph on image processing evaluation does provide some tools for this evaluation. Additionally the dosimetry section has been adapted to accommodate the recently introduced X-ray spectra (Dance 2009). With this supplement, it is expected that the European Guidelines are up-to-date with regard to the physical and technical aspects of mammography screening.

References

1. Perry, N., et al. (eds.): European Guidelines for quality assurance in breast cancer screening and diagnosis, 4th edn. European Communities (2006)
2. Alsager, A., Young, K.C., Oduko, J.M.: Impact of heel effect and ROI size on the determination of contrast-to-noise ratio for digital mammography systems. In: Samei, E., Hsieh, J. (eds.) Medical Imaging 2008: Physics of Medical Imaging, vol. 69134i (2008)
3. Bouwman, R., Young, K., Lazzari, B., Ravaglia, V., Broeders, M., van Engen, R.: An alternative method for noise analysis using pixel variance as part of quality control procedures on digital mammography systems. Physics in Medicine and Biology 54, 6809–6823 (2009)
4. Veldkamp, W.J., Thijssen, M.A., Karssemeijer, N.: The value of scatter removal by a grid in full field digital mammography. Med. Phys. 30, 1712–1718 (2003)
5. Dance, D.R., Skinner, C.L., Young, K.C., Beckett, J.R., Kotre, C.J.: Additional factors for the estimation of mean glandular breast dose using the UK mammography dosimetry protocol. Phys. Med. Biol. 45, 3225–3240 (2000)
6. Dance, D.R., Young, K.C., van Engen, R.E.: Further factors for the estimation of mean glandular dose using the United Kingdom, European and IAEA breast dosimetry protocols. Phys. Med. Biol. 54, 4361–4372 (2009)
7. Van Ongeval, C., Van Steen, A., Geniets, C., Dekeyzer, F., Bosmans, H., Marchal, G.: Clinical image quality criteria for full field digital mammography: A first practical application. Radiation Protection Dosimetry 129(1-3), 265–270 (2008)

Investigation of Practical Scoring Methods for Breast Density

Baorui Ren, Andrew P. Smith, and Julian Marshall

Hologic Inc., 35 Crosby Drive, Bedford, MA 01730 USA
Hologic / R2, 2585 Augustine Drive, Santa Clara, CA 94054 USA
bren@hologic.com

Abstract. Breast density is known as a strong risk factor for breast cancer. Clinically, physicians often use the BI-RADS or the Boyd categories to describe the density of breast, measured by observing mammograms. More accurately, breast density is measured by the percentage of glandular tissue in a breast. For all these methods, there might be more easily interpretable clinical value if the breast density was reported with scoring methods which are correlated to the patient distribution. In this paper two practical scoring methods will be discussed. The first one is population-based, with each segment of the continuous scores matching the patient BI-RADS distribution found in large scale clinical study. The second one is statistics-based, with the breast density result compared with the mean and the standard deviation from a reference population. Both methods will be described in details, together with preliminary results from an evaluation study with a total of 942 patients.

Keywords: volumetric breast density, BI-RADS, Z-score, V_{bd}-score, ACR-score.

1 Background

Mammography density is a reflection of the amount of glandular tissue as opposed to other fatty tissue in the breast. It is measured using either area based or volume based methods, with the breast density (BD) result expressed in percentage. Studies [1] have shown that the percentage breast density is a better predictor of breast cancer risk than the breast density classified into discrete categories. However, physicians are more familiar with the conventional BI-RADS or Boyd type descriptions of breast density. There might be greater clinical value if the breast density is also reported with scoring methods related to the existing categorical classifications and patient distribution. The new quality scores generated from breast density algorithm could provide physicians with more information and help them to better interpret the breast density result. We have been investigating two new scoring methods that will be discussed in this paper. The first method is developed based on patient BI-RADS population distribution and the second one is based on statistical distribution of breast density. Our goal is to generate additional information from the existing breast density result, to assist application of breast density in clinical practice.

J. Martí et al. (Eds.): IWDM 2010, LNCS 6136, pp. 651–658, 2010.

2 Methods

We carried out the study with a database of 942 patients acquired from several clinical sites and with BI-RADS score available for each patient. The breast density of each mammogram was calculated with an FDA-approved commercial software package – Quantra [2]. Quantra^{TM} was developed based on physical modeling of mammography system, and performs volumetric assessment of breast tissue compositions. In this paper, breast density refers to the volumetric breast density (VBD or V_{bd}) only, different from the more common radiological area-based percentage breast density. The V_{bd} results of all mammograms of one patient were averaged first, and the mean value was used as her breast density. The BI-RADS distributions of our database were compared with the population distributions in the ACRIN DMIST [3] study and good agreement was established between the two, suggesting that our database represented a good screening population. With both the breast density and the BI-RADS score results available, we developed two practical scoring methods to generate new quality scores based on breast density results.

The first method is a look-up-table based mapping method, which translates the percentage breast density into another continuous score that can better follow the BI-RADS classification of a mammogram. We refer the new score as the "ACR-score" in the paper for the ease of discussions. While for each individual patient, the ACR-score may not always match her BI-RADS number, we constrain and regulate the results such that over the entire group of study population, the ACR-score distribution matches the BI-RADS patient population distributions. In this way we expect that overall performance of the algorithm will be satisfactory. The look-up-table derived from the study population can be applied to other individual or patients population for general applications. In this algorithm, each unit segment of the continuous ACR-score will be corresponding to each BI-RADS category; and the number of patient in each unit segment of ACR-score will match the number of patient in each BI-RADS category. Using this new ACR-score, physician could conveniently link the breast density result to the BI-RADS category. Since BI-RADS score was given to each patient, and not to each breast or each mammogram, we used the averaged breast density of a patient to develop the mapping function to get her ACR-score.

The second method is a statistical method, which the breast density of a patient is compared with the mean and the standard deviation (SD, or σ) of breast density from a reference population. The offset value of a patient from the mean breast density of the reference population in unit of SD is reported as a new breast score, which is referred as the "V_{bd}-score" in the paper. The V_{bd}-score can help to reveal how the breast density of a particular patient stands against the reference population. For example, a V_{bd} -score of +1.0 means her breast density is at 1σ greater than the mean, or alternatively speaking, there is about 84% of the population having a breast density lower than hers. This method is proposed based on the same concept of the T-score or Z-score methods for bone mineral density (BMD) in bone densitometry, and may potentially become similarly useful in the field of breast density.

3 Results

3.1 Characteristics of the Database

The database in this study consists of 3307 mammograms of 1669 breasts from 942 patients (not all patients have all four mammograms). The mean patient age is 53.1 year old (σ=12.2) with a range from 21 to 96 year old. The mean compressed breast thickness is 5.4 cm (σ=1.6 cm). The mean breast density is 17.5% (σ=7.1%). In Table 1, the BI-RADS breast density distributions and mean age are compared with the ACRIN DMIST study. The patient populations in the database are considered similar to the screening populations in the DMIST study.

Table 1. Patient distributions in the Hologic database and in DMIST study

	Hologic (pts)	Hologic (%)	DMIST (pts)	DMIST (%)
BI-RADS 1	112	11.9%	5184	10.5%
BI-RADS 2	434	46.1%	21171	42.9%
BI-RADS 3	340	36.1%	19089	38.7%
BI-RADS 4	56	5.9%	3690	7.5%
Mean age	53.1		54.6	

3.2 The "ACR-Score" Method for BI-RADS Type Output

As shown in Table 1, the BI-RADS distributions in our database are: 11.9%, 46.1%, 36.1% and 5.9% for BI-RADS category 1, 2, 3 and 4, respectively. The normalized probability distribution function (PDF) of breast density is shown in Fig. 1. Theoretically values of BD are between 0% and 100% but in our database few patients have BD over 50% so the plot is showed only between 0% and 50% for clarity. The shape of the distribution curve looks noisy, which is probably due to the relatively small number of patients in the study. The mean and σ of BD are 17.5% and 7.1%, respectively. We divide the study population into four segments according to their breast density values from small to large, and make the percentage of patient population in each segment to match the distribution of each BI-RADS category. We find the limiting breast density values are about 9.3%, 16.4%, and 28.8% between adjacent segments. In Fig.1 we show the four segments under the PDF curve.

Since BI-RADS density by definition has only 4 discrete categories but the value of breast density is continuous from 0% to 100%, we would like to generate a continuous ACR-score, with each unit segment of the ACR-score corresponding to the each BI-RADS category. We assume that the breast density from 0% to 100% will be mapped to a numerical value range of 0 to 4 monotonically, and each unit segment between 0 and 4 will be corresponding to the BI-RADS category from 1 to 4. Under these assumptions, an empirical mapping curve is developed and is given in Fig. 2. The mapping curve can convert a patient's breast density to her ACR-score. It should be noted that the BD limits in Fig. 2 are adjusted, and are slightly different from those limits in Fig. 1. We find it is necessary as the PDF curve is noisy and the adjustment can help us to get better matching of ACR-scores to BI-RADS later.

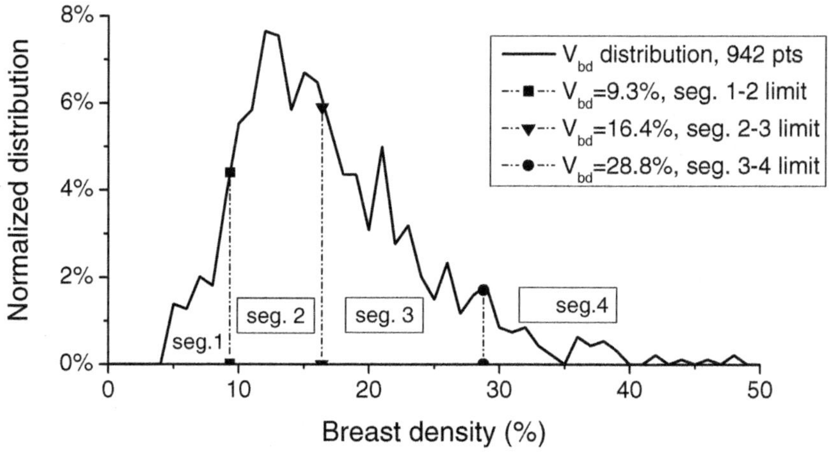

Fig. 1. Normalized BD distribution function and its four segments. The patient number in each segment from left to right matches the patient number in each BI-RADS category from 1 to 4.

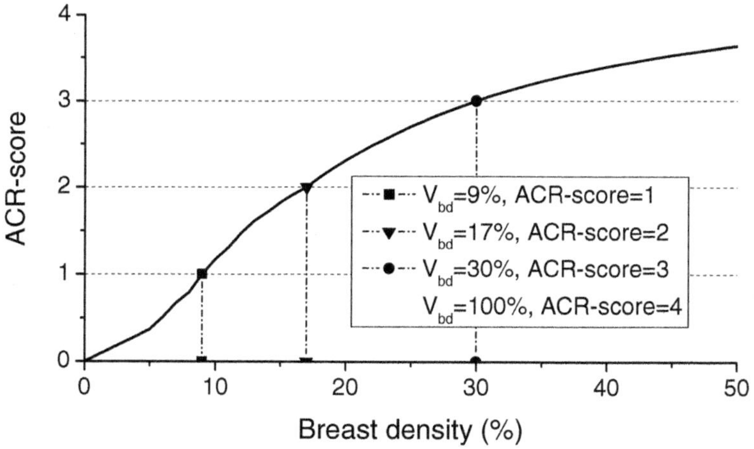

Fig. 2. A mapping curve that converts V_{bd} values to the ACR-scores, with each segment of 0-1, 1-2, 2-3, and 3-4 in ACR-score corresponds to BI-RADS category 1, 2, 3 and 4, respectively

3.3 The Statistical V_{bd}-Score (Z-Score) Method

The definition of the V_{bd}-score is given by equation 1. The mean *M*, standard deviation σ and the power term *L* come from a reference patient population through LMS fitting method [4], which could be either age and race matched group or a general population. In the study through age-independent LMS fitting, we find that $L = -0.596$, $M = 18\%$, and $\sigma = 6.6\%$. The distribution of V_{bd} is skewed so value of the power term *L* is different from *1.0*. It should be noted that values of the mean *M* and standard deviation σ obtained through LMS fitting method are slightly different from other direct (non-LMS) method. (Unless specifically mentioned, the mean and σ in this paper refer to these derived from non-LMS method.) If we use the database as a reference to itself, then V_{bd} for typical V_{bd}-score of Z=0, +1, and +2 is 18%, 27%, and 47%, respectively. In Fig.3, these three characteristics lines are co-plotted with PDF curve to show their relationships to the BI-RADS segmentations.

$$V_{bd} - score = Z = \frac{M((V_{bd}/M)^L - 1)}{L\sigma} \qquad \text{Eq. 1}$$

With the new V_{bd}-score method, a person with a V_{bd}-score of Z=0, or +1, or +2 will have about 50%, or 84%, or 97.7% of population with breast density lower than her, respectively. On the other hand, given a person's B-score, together with the known values of the power term, mean and standard deviation of the reference population, the breast density of the person can be calculated. For example, for V_{bd}-score equals to 1.5, her V_{bd} value is 35%.

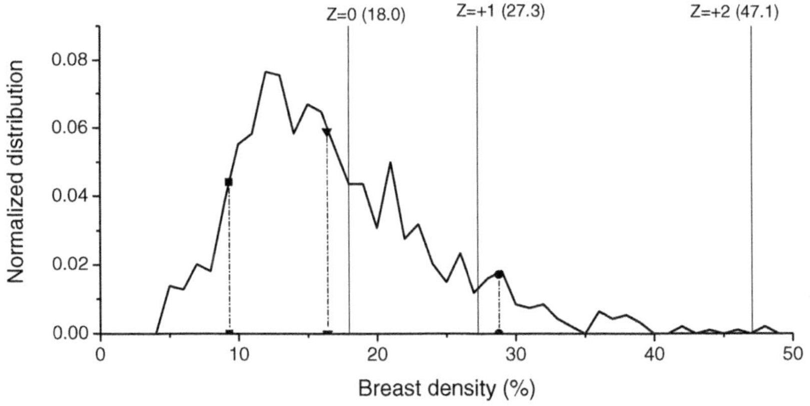

Fig. 3. The same PDF plot from Fig. 1 but with V_{bd} locations for several different V_{bd}-scores labeled. The V_{bd} are at 18%, 27.3%, and 47.1% for Z= 0, +1, +2, respectively.

4 Discussions

4.1 The Variation of Breast Density

The database in this study consists of 3307 mammograms, 1669 breasts from 942 patients. Our previous studies in the past have suggested the database consists of fairly representative screening population, which support the purpose of this study. In Table 2, we also compare our patient population with a paper from Yaffe [5]. We find the mean breast densities between the two study populations are close (19.3% vs. 17.5%) but the standard deviations differ (12.1% vs. 7.1%).

4.2 Correspondence of ACR-Score and BI-RADS

We want to study the correspondence of the ACR-score to radiologist's BI-RADS scores. As a point of comparison, we started by studying the inter-radiologist variability of BI-RADS scores. BI-RADS is a subjective score method using only 4 discrete steps. When doctors score clinical images, there is a large variation in the reported density scores. To characterize the inter-reader variation among radiologists, we performed an analysis with some data obtained as part of a clinical trial. The results are included in the appendix, which show that the inter-reader agreement is about 61%. What this means is that two radiologists on average will rate a patient with the same BI-RADS density score 61% of the time.

Table 2. Comparison of mean and σ of patient population of this study versus that of Yaffe [5]

	Yaffe mean (σ)	Hologic, mean (σ)
number	2831 patients	942 patients
VBD	19.3% (12.1%)	17.5% (7.1%)
age	59.3 (??)	53.1 (12.2)
cm	5.9 (1.6)	5.4 (1.6)

In Fig. 4, we show the histograms of patient distribution versus the ACR-score in each of the four BI-RADS categories. The distributions of ACR-score are also summarized in Table 3. The result shows that the ACR-score has an overall agreement rate of 62% to BI-RADS, and the agreements for the BI-RADS 2 and 3 patients are higher than the other two categories. The agreement of ACR-score to BI-RADS of 62% is remarkably the same as the agreement between two radiologists rating BI-RADS. It is known that BI-RADS number is assigned clinically based on both the density and the texture pattern of breast tissues. Since our ACR-score method does not take breast texture as an input, the agreement of the ACR-score and BI-RADS is compromised due to lack of information.

Table 3. Patient distributions vs. ACR-score in each BI-RADS breast density category

	ACR-score 0 - 1	ACR-score 1 - 2	ACR-score 2 - 3	ACR-score 3 - 4	Total pts num.	agreement ratio
BIRADS 1	33	74	5	0	112	29%
BIRADS 2	24	313	92	5	434	72%
BIRADS 3	4	73	226	37	340	66%
BIRADS 4	0	5	40	11	56	20%
Total					942	62%

The BD-score method is developed with a constraint to match large scale distribution of patient population, and overall agreement has been shown to be 62%. So our ACR-score method is comparable to the clinical performance of radiologist in predicting BI-RADS scores. However, our BI-RADS method uses decimal number to score a breast, which would allow better differentiation of breast than the discrete BI-RADS scores. For example breasts with ACR-score of 2.9 and 3.1 do suggest great similarity

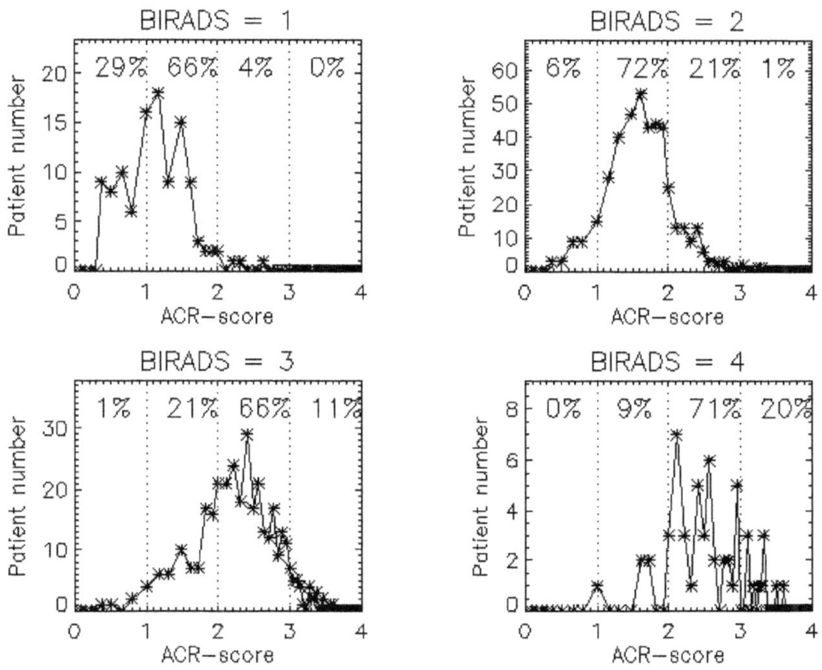

Fig. 4. Patient distribution versus the ACR-score in each BI-RADS category

in breast density, but two such breasts could be classified into adjacent BI-RADS categories, and fail to reveal their proximity in breast density.

4.3 The V_{bd}-Score Method

Regarding the V_{bd}-score method, a quick literature search suggests that it might be a new concept to the breast density field since no similar discussion could be found in existing literatures. Further studies on this method are needed to determine how to best select appropriate reference population for the V_{bd}-score calculation, and to correlate the V_{bd}-score to the risk of breast cancer. As we know, the T-score or Z-score methods have been well established in the field of bone densitometry. In particular, the World Health Organization (WHO) has a formal definition for osteoporosis based on the T-score value of a patient (when T<-2.5). Therefore we expect that a similar V_{bd}-score defined for breast density would inherit many advantages of this statistical method established in the BMD field.

5 Conclusion

In this paper, we describe two new scoring methods for breast density, and evaluate them with a database of 942 patients. Our first method aims to generate a BI-RADS type score to assist doctors to correlate to breast density to BI-RADS category number.

The current algorithm has an agreement rate of 62% matching the BI-RADS score from doctors. This performance is similar to how radiologists agree among themselves with BI-RADS scores. Our second method aims to provide statistical information on how a patient's breast density result is compared with a reference population, thus adding insight into the breast density result. Both methods need to be further evaluated to justify them as new methods for breast density.

References

[1] McCormack, V.A., dos Santos Silva, I.: Breast density and parenchymal patterns as markers of breast cancer risk: a meta-analysis. Cancer Epidemiol Biomarkers Prev. 15, 1159–1169 (2006)
[2] Hartman, K., et al.: Volumetric Assessment of Breast Tissue Composition from FFDM images. In: Krupinski, E.A. (ed.) IWDM 2008. LNCS, vol. 5116, pp. 33–39. Springer, Heidelberg (2008)
[3] Pisano, E.D., et al.: Diagnostic Performance of Digital versus Film Mammography for Breast-Cancer Screening. NEJM 353, 1773–1783 (2005)
[4] Cole, T.J.: The LMS method for constructing normalized growth standards. Eur. J. Clin. Nutr. 44, 45–60 (1990)
[5] Yaffe, M.J., et al.: The Myth of the 50-50 breast. Med. Phys. 36, 5437–5443 (2009)

Appendix: Study on Inter-reader BI-RADS Score Agreement

In this study, 309 images were scored by 15 radiologists using BI-RADS breast density. We evaluated the inter-reader variation between any pairs of two readers. There were a total of 15x14/2=105 pairs, and about 105x309=32445 individual data points. We made a 4x4 matrix to record all these data point through histogram method and the results are shown in Table 4. We found radiologists tend to agree with each other better for BI-RADS 2 and 3 cases, and the overall agreement rate between two radiologists was 61% in BI-RADS results.

Table 4. Inter-reader BI-RADS agreements, tested with 309 images rated by 15 radiologists

R1 \ R2	BIRADS 1	BIRADS 2	BIRADS 3	BIRADS 4	Total #	agree. (%)
BIRADS 1	698	1122	70	0	1890	37%
BIRADS 2	1122	8246	3418	115	12901	64%
BIRADS 3	70	3418	9262	1621	14371	64%
BIRADS 4	0	115	1622	1546	3283	47%
Total					32445	61%

Performance of the New Double Layer Amorphous Selenium Detector for Digital Mammography Compared to the FDA Approved CR System for Digital Mammography

Rafael Salvador, Rafael Salvador-Izquierdo, Leonardo Romero, Xavier Salvador-Izquierdo, and Manuel Salvador

Department of Radiology, CDI Imagine, Hospital Universitari Vall d'Hebrón, C/Valencia 223, 08007-Barcelona-Spain

Abstract. The performance of a new double layer amorphous Selenium detector for digital mammography is compared to the FDA approved CR system for digital mammography in a two step study. A study comparing radiation between both systems is first done to obtain the best settings for a clinical study. In a second step the results in terms of quality are evaluated by three readers comparing the final images. A minimal reduction of dose (20%) is obtained and a better definition of glandular structures is demonstrated with the new device.

Keywords: Breast, Breast Cancer, Digital Mammography.

1 Purpose

The purpose of the study is to establish if there is any improvement in mammography with the use of a DR double layer amorphous selenium detector (A) compared to the FDA approved CR (P) for digital breast imaging.

2 Method and Material

The systems compared are a novel CE approved DR System for mammography. The Amulet (FUJIFILM, Japan), mainly based on a double Amorphous Selenium layer detector, and an optical switch reading of the electronic virtual imaging, against a known CE and FDA approved CR System for mammography, the Profect (FUJI-FILM, Japan), based on a removable imaging plate, and a laser beam reading of the electronic virtual imaging.

A two step study has been designed. In the first step, a study analysis by three observers, comparing the films obtained in both systems under different conditions (mAs and kvp), as well as anode filter combinations) with two different standard phantoms RMI 156 and CDMAM. We used the CDMAM to quantify the image

J. Martí et al. (Eds.): IWDM 2010, LNCS 6136, pp. 659–663, 2010.

quality, where the Image Quality Figure (IQF) is calculated according to the formula:

$$\mathrm{IQF} = \sum_{i-1}^{15} \mathrm{Ci{\cdot}Di,th}$$

Where Di, denotes the threshold diameter in contrast-column i. Summation over all contrast-columns yields the IQF [1].

In the other phantom RMI156 result evaluation we used the recommended criteria described in the Mammography quality control manual published by the ACR [2].

This part has been designed to obtain the best conditions for the second part of the study.

The second step has been the clinical study. A total of 104 patients have been included in the study. To avoid any bias, patients have been collected consecutively without any selection, but only if they signed an informed consent agreement. Patients were examined for breast cancer screening (60), previous check out to breast augmentation surgery(7), follow up of previously detected nodules, asymmetries or calcifications (27), actual clinical problem (7), and other (prosthesis, lumpectomy, axillary tumor) (3).

The patient data included were age and breast density following the four ACR BI-RADS types. Type 1, <25%, type 2, 25-50%, type 3, 50-75%, and type 4, >75%.

All the patients had a CC and a MLO view both in the right with Profect (P), and left breast with Amulet (A). Where the Kvp were fixed at 28 and the mAs were automatically selected by the Automatic Exposure Control (AEC) device in the mammographic equipment (P). Then the same recorded mAs and kvp were fixed manually in each patient in the DR system (A). Just to compare with exactly the same figures. Three independent readers were (RST) a radiologist with more than 20 years experience years in breast imaging; (RSI) a young radiologist with less than 5 experience years in breast imaging; and (XSI) a radiology resident in training in radiology. The results from the comparison of the different structures, quality, artifacts, etc, were classified using a five point rating scale where: (1) P is much better than A; (2) only better; (3) equal; (4) worse; (5) much worse; and of no value if it were impossible to see.

Vascular structures, pectoral muscle contour, microcalcifications, the skin line definition, sub-areolar structures, fibroglandular tissue, pectoral muscle penetration, artifacts, and electronic or digital noise were assessed and compared in every patient between both systems.

3 Results

The first part of the study produced the results given below.

1. RMI 156 Results

28kV Mo/Mo	Dose	Fiber	Calc	Mass
PROFECT	100%	5,0	4,0	4,5
AMULET	100%	5,8	4,0	4,5
	80%	5,0	4,0	4,3
	60%	5,0	3,8	4,0

a) With the exposure conditions of 28kV and Mo/Mo,

Fibre and Calcium in image A with 80% dose shows the same quality as image P with 100% dose.

Mass in image A with 100% dose shows the same quality as image P with 100% dose.

29kV Mo/Rh	Dose	Fiber	Calc	Mass
PROFECT	100%	4,3	4,0	4,5
AMULET	100%	5,0	4,0	4,5
	80%	5,0	4,0	4,5
	60%	5,0	4,0	4,0

b) With the exposure conditions of 29kV and Mo/Rh,

Fibre, Calcium and Mass in image A with 80% dose shows the same quality as image P with 100% dose.

c) These results led to the suggestion that A performs 20% dose reduction compared to P with 29kV Mo/Rh.

2. CD-MAM Results

a) The CD curve of A shows better performance than P at a smaller diameter.

b) IQF of A 80% dose image is better than or equal to IQF of P 100% dose image with either 28kV Mo/Mo or 29kV Mo/Rh.

It was confirmed that the dose was almost the same with the same mAs, kVp and anode/filter between Amulet and Profect. In this image evaluation, the breast thickness was almost the same between Amulet and Profect. Therefore the AGD was approximately the same.

[2]Clinical test

In the clinical test, of the 155 patients consecutively attending from March 13th to March 23rd; only 104 were included, having excluded some because of mastectomy, gynecomastia, or simply patient refusal.

The results are summarized in chart 3.

a) A shows better performance for “VASC”, “PM”, “MICRO”, “RETROAR”, “TISSUE”, “PMPENET”, and “NOISE” by means of the 5 point scale

ITEM	TOTAL	VASC	PM	MICRO	SKIN	RETROAR	TISSUE	PMPENET	ARTIF	NOISE
sample size	4049	613	351	242	624	624	624	340	7	624
average	3,69	3,71	3,62	3,62	3,14	3,63	3,99	3,85	3,43	3,96
sigma	0,71	0,57	0,66	0,65	0,77	0,70	0,51	0,60	1,05	0,68
t-ratio	0,98	1,25	0,93	0,96	0,18	0,90	1,93	1,42	0,41	1,41
P-value	0,16	0,11	0,18	0,17	0,43	0,18	0,03	0,08	0,35	0,08

b) Regarding "MICRO", A shows higher performance when breast density becomes lower.

c) Regarding "TISSUE", A shows higher performance when breast density becomes higher.

The sharpness of fibro-glandular tissue was better seen with the new system (p value below 0.05, statistically significant). The other items studied (Pectoral muscle contour, Microcalcifications, sub-areolar structures, pectoral muscle penetration, and electronic or digital noise) were better on DR but did not reach statistical significance (p value in a range between 0.08 and 0.18). Artifacts and skin line definition were similar in both systems.

4 Conclusion

There is a lot of literature on different types of detectors, based or not on Selenium, applied to Full Field Digital Mammography [3-8]. Here we compare a new Selenium based detector to a CR system for breast imaging.

From the first part of the study we conclude:

1. RMI156

a)- Because A has a higher MTF and DQE than P especially at high frequency, image A shows Fibre and Calcium more finely and clearly. Regarding Mass, A has the same performance as P.

b)- There is a different X-ray energy dependence between A and P. A shows better X-ray conversion efficiency with higher energy than P. Therefore an amount of dose reduction using A is higher with 29kV, Mo/Rh than with 28kV, Mo/Mo because 29kV, Mo/Rh beam has higher energy than 28kV, Mo/Mo beam.

2. CD-MAM

These results led to the suggestion that A performs with a 20% dose reduction compared to P, either with 28kV Mo/Mo or 29kV Mo/Rh.

Amulet has potential for about 20% dose reduction, but further study is needed to show if there is significant difference. From this phantom study, it is confirmed that Amulet image quality is at least the same as Profect. Therefore the clinical study was performed under the same conditions between Amulet and Profect.

In the clinical study in our series, under any breast density condition, only the sharpness of fibroglandular tissue was statistically better seen with the new DR system. Microcalcifications were better seen in the Amulet equipment when breast density was lower than 50% (types 1 and 2 BIRADS). Other aspects, such as skin line, digital noise, pectoral muscle definition, and penetration, etc., were to be found better but of no statistical value. Further studies are required to show this trend. This new detector improves the resolution applied to breast imaging. The size of the sample in our series was unable to demonstrate big differences in many aspects, but the quality of the new system based on a double layer selenium detector is at least equal if not better than the FDA approved CR system for breast imaging diagnosis.

References

1. Thijssen, M.A.O., Bijkerk, K.R., Van der Burght, R.J.M.: Project Quality Assurance in Radiology. Manual CDRAD-phantom type 2.0. Artinis Medical Systems B.V. Andelst. The Netherlands (2003)
2. ACR Mammography Quality Control Manual, Reston, Va. American College of Radiology (1999)
3. Mahadevappa, M.: AAPM/RSNA Physics Tutorial for Residents, Digital mammography: An Overview. Radiographics 24, 1747–1760 (2004)
4. Bick, U., Diekmann, F.: Digital mammography: what do we and what don't we know? Eur. Radiol. 17, 1931–1942 (2007)
5. Veip, Y.A., Mazurov, A.I.: Analysis of various technologies for digital X-ray image detector design. Biomedical Engineering 42(5), 261–264 (2008)
6. Wolbarst, A.B., Hendee, W.R.: Evolving and experimental technologies in medical imaging. Radiology 236(1), 16–39 (2006)
7. Körner, M., Weber, C., Wirth, S., Pfeifer., K.-J., Reiser, M.F., Treitl, M.: Advances in Digital Radiography: Physical Principles and System Overview. Radiographics 27, 675–686 (2007)
8. Reznik, N., Komljenovic, P.T., German, S., Rowlands, J.A.: Digital radiography using amorphous selenium: Photoconductively activated switch (PAS) readout system. Med. Phys. 35, 1039–1050 (2008)

Machine Learning Techniques and Mammographic Risk Assessment

Neil Mac Parthaláin and Reyer Zwiggelaar

Aberystwyth University, Aberystwyth, Ceredigion, SY23 3DB, Wales, UK
{ncm,rrz}@aber.ac.uk

Abstract. Breast tissue characteristics are widely accepted as important indicators of the likelihood of the developing breast cancer. Methods which have the ability to automatically classify breast tissue distribution therefore provide important tools in assessing the risk to which patients are exposed. This paper examines the machine learning techniques employed for knowledge discovery in a recent approach to mammographic risk assessment. A number of weaknesses for selected classification techniques are identified and examined. Additionally, important trends in the data such as decision class confusion and how this affects the ability to perform accurate knowledge discovery on the extracted image data are also explored. The paper is concluded with some ideas as to how the identified trends in the data and weaknesses in the classification approaches could be addressed.

1 Introduction

The approach described in [14], employs a number of methods for the segmentation and the extraction of features from mammographic images for the task of risk assessment. A number of machine learning methods are used in this approach and the authors achieve good empirical results. Further investigation however has revealed that there are a number of areas where the approach could be improved. More specifically, a number of important trends within the data are ignored, also the leave-one-out cross-validation strategy as well as the chosen classifier learners have a number of weaknesses which are not addressed or justified. The areas which are examined in this paper relate to: 1) The classification methods employed, 2) The classification validation approach and 3) Class confusion within the extracted data

The remainder of the paper is structured as follows. Section 2 summarises the background of the approach under consideration and discusses the points outlined above. Section 3 describes some methods for dealing with the problems and weaknesses identified in section 2. Section 4 concludes the paper with a short discussion of future work.

2 Background and Analysis

The problem considered in [14] is that of mammographic risk analysis, where mammographic breast tissue density information extracted from images is used to assess how likely a woman is to develop breast cancer. The steps involved are described in detail in

J. Martí et al. (Eds.): IWDM 2010, LNCS 6136, pp. 664–672, 2010.

[14], but are summarised here. The initial stages involve the segmentation and filtering of the mammographic images: all mammograms are pre-processed to identify the breast region and remove image background, labels, and pectoral muscle areas.

In the second phase, a feature extraction step is performed, where the fuzzy c-means (FCM) algorithm [2] is employed which results in the division of the breast into two clusters. A co-occurrence matrix is then used to derive a feature set which results in 10 features which describe morphological characteristics and 216 for the texture information (226 total). This feature set is then labelled using the consensus opinion of 3 experts to assign a label to each object mammogram using the BIRADS [1] classification. The dataset employed in the empirical evaluation is the Mammographic Image Analysis Society (MIAS) database [18].

2.1 Detailed Analysis

In this section, a more detailed examination of the results obtained using the approach employed in [14] is made. This highlights a number of areas which require further investigation both in terms of the methods employed and more importantly the data generated from the mammographic images. These aspects are examined in the following sections.

Classification. In [14] the authors employ a number of different methods in order to classify the image data: k Nearest Neighbours (kNN), C4.5, and a Bayesian combination of the two previously described methods.

Whilst the results obtained in [14] offer interesting findings there are some problems with the classification methods employed. The first of these relates to the use of leave-one-out cross-validation (LOOCV). At first, the LOOCV method may appear to be attractive solution as the greatest possible amount of data is used for training in each case. However, because the test data cannot be stratified, LOOCV guarantees a non-stratified object. Stratification involves ensuring adequate representation of objects of each decision class in the test set data. However this is impossible when the test data consists of only a single object.

The second downside to LOOCV relates to the classifiers which are learned. As LOOCV uses all of the data to train and only one instance to test, this can often result in models which are not robust [11] and classifiers which have a tendency to overfit. This occurs as LOOCV tends to include unnecessary components in the generated models [17], which in turn translates into poor generalisation [11]. Furthermore, the method does not work well for data with strong clusterisation, [6] and also tends to underestimate the true predictive error [13].

Further problems which relate to the classification methods adopted in [14] relate to parameter tuning in the case of kNN, and the inability of both kNN and C4.5 to deal with the problem of class imbalance (which is discussed the next section in detail). The specification of any subjective parameter will involve period of 'tuning' in order to discover an optimal value. The effort required in order to 'learn' such optimal values can entail extensive testing and or computation which may not be feasible for large datasets. Furthermore, the specification of a parameter means that a subjective value is imposed upon the data rather than learning using only information contained in the data itself. Both the kNN and combined Bayesian approach of [14] employ such subjective tunable parameters.

Although it is not clear from [14], it would also appear that a wrapper approach is employed for feature selection. Wrappers for feature selection employ the induction algorithm in order to determine feature 'goodness' and can result in good classification accuracies, but they also have a strong tendency to 'overfit' especially in the presence of class imbalance, thus leading to classifiers which are non-robust [12].

Data. The datasets employed in [14] are generated from the raw image data using the method described in section 2. This data includes extracted features which are generated in two different steps. In the first, the image is segmented using the fuzzy c-means (FCM) [2] clustering method. This results in the image being divided into two clusters, one containing the bright pixels and the other dark pixels. It is assumed that the bright pixels are representative of dense regions, while the dark pixels represent fatty regions. From each cluster a set of morphological and texture features which relate to area, mean intensity, standard deviation of intensity, etc. are extracted. For the texture related features, statistics derived from the co-occurrence matrices and laws energy filters are employed. These concatenated features are then used to describe the complete mammogram.

The results generated in [14] provide a comprehensive and detailed examination of the the data through the use of confusion matrices in order to examine classifier performance with respect to consensus expert opinion. Indeed the misclassification of objects is explored on a class-by-class basis for each of the decisions and an overall statistical evaluation metric (kappa coefficient) is employed which attempts to provide a summary of the performance for each classification method. There are however some important characteristics and trends within the data which are ignored in the analysis. Perhaps the most important of these is the class confusion observed for BIRADS classes II and III for all classifiers except for the SFS+kNN method (where there also seems to be significant confusion between classes I and III - this could be attributed to the incorrect removal of important features which are necessary for the classifier to discriminate between these classes however). It would seem that there are two likely explanations for this confusion: 1) There is significant divergence of expert opinion relating to the labelling of objects for decision classes II and III, and 2) There is insufficient information present in the extracted data to enable a classifier learner to to discriminate between objects of these two different decision classes.

Although it is true that human expert opinion will always differ to varying degrees, this does not account for the decision class confusion demonstrated in [14] with respect to consensus opinion and classifier performance. Indeed, if the results are examined closely, it can be seen that the most frequent and significant divergence in consensus opinion amongst experts occurs in relation to classes I and II (50.5% for expert A and 48% for expert B). However, if the results for classifier performance in terms of consensus opinion are examined, it can be seen that the most likely misclassification error or confusion for objects of decision classes II and class III are those from the adjacent decision class. It could be argued that since these decision classes are conceptually 'close' that experts tend to disagree to a greater extent as to their true classification. The comparison provided in [14] of consensus opinion with respect to individual opinion does

Table 1. Confusion matrices and classification accuracies for the MIAS dataset classification using six different classifier learners

Consensus opinion

JRip (*Classification accy* = 66.78%)

	I	II	III	IV
I	72	10	4	1
II	15	65	21	2
III	2	30	56	7
IV	0	4	11	22

PART (*Classification accy* = 63.98%)

	I	II	III	IV
I	64	17	4	2
II	19	64	17	3
III	5	25	56	9
IV	3	2	10	22

FNN (*Classification accy* = 62.42%)

	I	II	III	IV
I	58	20	9	0
II	16	67	20	0
III	2	29	60	4
IV	1	2	18	16

Consensus opinion

FRNN (*Classification accy* = 69.90%)

	I	II	III	IV
I	74	12	1	0
II	12	67	22	2
III	0	26	64	5
IV	2	3	12	20

FRNN-O (*Classification accy* = 66.14%)

	I	II	III	IV
I	66	18	3	0
II	16	62	25	0
III	2	21	66	6
IV	0	2	16	19

VQNN (*Classification accy* = 71.75%)

	I	II	III	IV
I	74	11	1	1
II	13	68	20	2
III	0	22	70	3
IV	2	2	14	19

not support this statement however. This leads to the the second point noted above; there is insufficient information contained in the data in order to be able to discriminate between objects of these classes and hence perform accurate and reliable knowledge discovery. This would seem to be the case for the MIAS dataset where both classes II and III are easily misclassified using C4.5 and SFS+kNN. In order to investigate whether this trend was present in other classifier learners an experimental evaluation has been carried out. A number of different classifiers have been utilised - JRip, PART, Fuzzy NN [19], FRNN [9], FRNN-O, and VQNN [5]. The leave-one-out cross validation method has also been used here such that the results can be compared with those of [14]. Tables 1 shows that in spite the classifier learner employed, the class confusion for classes II and III remains. This combined with the fact that consensus expert opinion does not contain the same level of class confusion for classes II ad III strongly suggests that there is insufficient information available in the extracted data to be able to discriminate between these decision classes.

A feature selection step is often utilised as an effective way of both reducing data dimensionality, and removing irrelevant and redundant information. Indeed a good feature selection method should select only those features which lead to the decision class and retain or even improve classification accuracy [12]. In [14] only one such approach is employed which is based around the kNN algorithm. Here, a number of different feature selection and classification approaches are employed to investigate the impact of various feature selectors on classification accuracy. All of the methods employed in Table 2 (consistency-based feature selection [12], Fuzzy-Rough Feature Selection, CFS [7] and DMTRS)are filter type methods which avoid the negative tendency of wrapper approaches to overfit.

Table 2. Results of various FS and classification techniques

FS Method	Classifier Learner	Classification Accy.	Subset size	FS Method	Classifier Learner	Classification Accy.	Subset size
CS	FRNN	66.46	11	FRFS	FRNN	61.49	7
CS	FNN	55.59	11	FRFS	FNN	51.24	7
CS	FRNN-O	66.15	11	FRFS	FRNN-O	65.84	7
CS	VQNN	70.19	11	FRFS	VQNN	65.22	7
CFS	FRNN	72.37	32	DMTRS	FRNN	55.28	6
CFS	FNN	60.25	32	DMTRS	FNN	62.73	6
CFS	FRNN-O	69.88	32	DMTRS	FRNN-O	62.11	6
CFS	VQNN	74.22	32	DMTRS	VQNN	59.00	6

One other aspect of the data which affects prediction capability and can frustrate classifier learning capability considerably is class imbalance. The problem of class imbalance occurs in classification domains where one or more classes are represented by a small number of objects when compared to the numbers of objects representing other class/classes. The negative impact that class imbalance has on classifier performance has been the subject of much research in the machine learning community [15]. The problem is of great importance since it is very common in real-world domains. Traditional machine learning approaches such as C4.5, MPL, and support vector machines are often biased towards the majority class or classes and thus, may predict minority class objects [8] incorrectly.

This is because many such approaches are designed with the naive assumption that datasets are well-balanced. Examining the dataset employed in [14], it can be seen that classes I and IV are under-represented (87 and 37 objects respectively as opposed to 103 and 95 objects for classes II and III) or imbalanced. This poor representation or imbalance not only means poor performance for minority classes but may also affect the discriminatory ability of better represented classes. Additionally, it is highly likely that this is a contributory factor for the class confusion mentioned previously. Indeed the classifiers employed for the experimental evaluation in [14] are of the type which are known to suffer from poor performance in the presence of class imbalance.

3 Proposed Solutions

One of the first disadvantages of the methods adopted in [14] identified in the previous section was the use of LOOCV as an approach to model selection. The adoption of validation methods such as stratified 10-fold cross validation or other stratifiable validation methods would ensure more robust models but also help to address the underestimation of classification error [11], [3], [19]. A number of tentative experiments have been carried out using this method and a summary is presented below. The RMSE is a combined measure of bias/variance and is useful for gauging the generalisation of a learned classifier. Low RMSE values are an indicative of a robust classifier. In the case of (almost) unbiased methods such as LOOCV, RMSE is a measure of variance. This choice of validation schemes stems from the findings in [11], and [3] where it is shown that stratification and n-fold cross-validation are useful methods for addressing the problem of bias and variance reduction in classifier learning. The validation schemes employed for the experimentation described here are: 1)Stratified 10 fold cross-validation (10FCV), 2) Leave-one-out-cross-validation (LOOCV), 3) Stratified two fold cross-validation (2CV), 4) Stratified three fold cross-validation (3CV), 5) Stratified four fold cross-validation (4CV), 6) Stratified five fold cross-validation (5CV) and 7) A random 66/33% training/test split of the data.

It can be seen from the results that the RMSE for LOOCV is higher for both classifiers indicating very high variance, while stratified 10 fold cross-validation (10FCV) manages lower values and in the case of PART also has better classifier accuracy. For C4.5, LOOCV has a higher RMSE value than 10FCV, yet LOOCV is unbiased, therefore this value relates only to high variance thus indicating a tendency to overfit. This would account for the higher classification accuracy figures achieved by LOOCV.

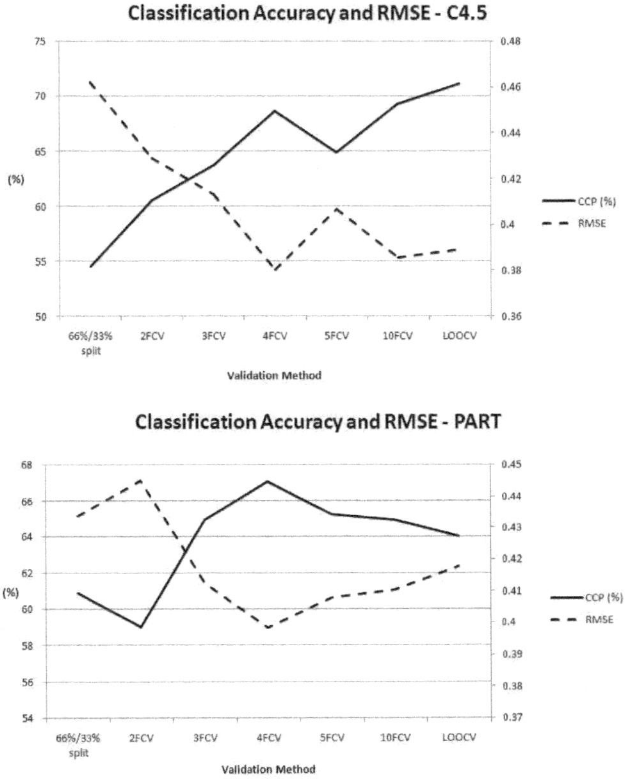

Fig. 1. Classification accuracy and RMSE using 2 classifier learners

Stratified four and five fold cross-validation show low values of RMSE and high classification accuracies thus confirming the observations of [11] that these methods are best for model selection, while 10FCV is best for classification accuracy. A more comprehensive investigation involving methods such as .632 bootstrap, bagging, etc. is beyond the scope of this paper but would form the basis for a complete study of the classifiers learned for this dataset.

Parameter tuning, as mentioned previously can be both time-consuming and have a negative impact on the data under consideration. The adoption of methods which rely only on the information contained in the data will avoid both of these undesirable aspects. Fuzzy-rough set-based classification methods such as [9] have the ability to consider real-valued data and do not require any subjective thresholding or parameter tuning.

3.1 Proposed Methods for Dealing with Extracted Data

The most challenging problem identified in section 2.1 is that of addressing the lack of information contained in the data generated using the feature extraction method of

[14]. As discussed previously, it is apparent that the class confusion noted for decision classes II and III arises from a lack of information within the data to perform accurate classification. This is apparent as the feature selection step employed serves only to introduce further confusion amongst other classes. This would indicate that there is a loss in discriminatory ability when feature selection is employed. However the full dataset does not perform much better indicating perhaps that there is insufficient information contained within the extracted image features

Classical methods for addressing the class imbalance problem have primarily focused on a number of solutions both at data level and algorithm level. However, many new lines of research have been proposed for the general problem of class imbalance. Broadly speaking these can be divided into four groups which: 1)Resample objects in order to rebalance the class distribution of the dataset, 2) Modify existing learning algorithms, 3) Use classifier performance metrics to in an attempt to perform knowledge discovery in imbalanced domains, 4) Examine the relationship between class imbalance and other data complexity characteristics.

In mammographic risk assessment where it is desirable to employ methods which result in robust models, a more suitable technique for dealing with class imbalance is described in [4]. This is an approach for over-sampling the minority class/es which rather than simply replicating objects which belong to the minority class/es, generates new synthetic minority data objects by interpolating between several objects that are similar in some respect. It allows the classifier to build larger decision regions which contain similar objects from the minority decision class/es. There are other undersampling methods which rely on noise removal or the removal of objects from majority classes in an attempt to redress the lack of data objects for minority classes. Other approaches which may be useful in addressing the class imbalance problem for mammographic risk assessment include one-class classification techniques [16], or a combination of one-class classifiers and resampling [10]. These methods use classification techniques in an attempt to determine the distribution of the minority classes and use this information to guide the induction algorithm.

4 Conclusion

This paper has examined a recent approach to mammographic risk assessment [14] and identified some important areas which require further investigation. Some of these areas relate to the extracted data while others to the machine learning methods employed at the knowledge discovery stage. In terms of the data itself, the initial experimentation carried out here suggests that there is insufficient information contained within the data to predict classes II and III accurately. In an attempt to overcome this problem the authors of [14] have combined the classifier learners using a Bayesian method which alleviates this problem somewhat. However, the fact that these classifier learners employ LOOCV which results in non-robust and overfitting models is ignored. This may result in poor performance for unseen data and ultimately 'break' the classifier model.

Other problems such as class imbalance, subjective threshold specification have also been discussed but no empirical evaluation has been made as such a level of investigation is beyond the scope of this paper. An extensive and thorough examination of the machine learning methods and a more complete look at the data in [14] and other such work would therefore form the basis for further investigation.

References

1. American College of Radiology. Illustrated Breast Imaging Reporting and Data System BIRADS, 3rd edn. American College of Radiology (1998)
2. Bezdek, J.C.: Pattern Recognition with Fuzzy Objective Function Algorithms. Plenum Press, New York (1981)
3. Braga-Neto, U., Dougherty, E.R.: Is cross-validation valid for small-sample microarray classification? Bioinformatics 20(3), 374–380 (2004)
4. Chawla, N.V., Bowyer, K.W., Hall, L.O., Kegelmeyer, K.P.: SMOTE: synthetic minority over-sampling technique. Journal of Artificial Intelligence Research 16, 321–357 (2002)
5. Cornelis, C., De Cock, M., Radzikowska, A.: Vaguely Quantified Rough Sets. In: An, A., Stefanowski, J., Ramanna, S., Butz, C.J., Pedrycz, W., Wang, G. (eds.) RSFDGrC 2007. LNCS (LNAI), vol. 4482, pp. 87–94. Springer, Heidelberg (2007)
6. Eriksson, L., Johansson, E., Muller, M., Wold, S.: On the selection of the training set in environmental QSAR analysis when compounds are clustered. Journal of Chemometrics 14, 599–616 (2000)
7. Hall, M.A.: Correlation-based feature selection machine learning. Ph.D. Thesis, Department of Computer Science, University of Waikato, Hamilton, New Zealand (1998)
8. Japkowicz, N., Stephen, S.: The Class Imbalance Problem: A Systematic Study. Intelligent Data Analysis 6(5), 429–450 (2002)
9. Jensen, R., Cornelis, C.: A New Approach to Fuzzy-Rough Nearest Neighbour Classification. In: Chan, C.-C., Grzymala-Busse, J.W., Ziarko, W.P. (eds.) RSCTC 2008. LNCS (LNAI), vol. 5306, pp. 310–319. Springer, Heidelberg (2008)
10. Juszczak, P., Duin, R.P.W.: Selective Sampling Methods in One-Class Classification Problems. In: Kaynak, O., Alpaydin, E., Oja, E., Xu, L. (eds.) ICANN 2003 and ICONIP 2003. LNCS, vol. 2714, pp. 140–148. Springer, Heidelberg (2003)
11. Kohavi, R.: A Study of Cross-Validation and Bootstrap for Accuracy Estimation and Model Selection. In: Proceedings of the International Joint Conference on Artficial Intelligence (IJCAI 1995), pp. 1137–1143 (1995)
12. Liu, H., Motoda, H. (eds.): Computational Methods of Feature Selection. Data Mining and Knowledge Discovery Series. Chapman & Hall/CRC, Boca Raton (2008)
13. Martens, H.A., Dardenne, P.: Validation and verification of regression in small data sets. Chemometrics and Intelligent Laboratory Systems 44(1), 99–121 (1998)
14. Oliver, A., Freixenet, J., Marti, R., Pont, J., Perez, E., Denton, E.R.E., Zwiggelaar, R.: A Novel Breast Tissue Density Classification Methodology. IEEE Transactions on Information Technology in Biomedicine 12(1), 55–65 (2008)
15. Orriols, A., Bernad-Mansilla, E.: The class imbalance problem in learning classifier systems: a preliminary study. In: Proceedings of the 2005 Workshops on Genetic and Evolutionary Computation (2005)
16. Raskutti, B., Kowalczyk, A.: Extreme rebalancing for svms: a case study. SIGKDD Explorations 6, 60–69 (2004)

17. Stone, M.: An Asymptotic Equivalence of Choice of Model by Cross-Validation and Akaike's Criterion. Journal of the Royal Statistical Society, Part B 38, 44–47 (1977)
18. Suckling, J., Partner, J., Dance, D.R., Astley, S.M., Hutt, I., Boggis, C.R.M., Ricketts, I., Stamatakis, E., Cerneaz, N., Kok, S.L., Taylor, P., Betal, D., Savage, J.: The Mammographic Image Analysis Society digital mammogram database. In: International Workshop on Digital Mammography, pp. 211–221 (1994)
19. Witten, I.H., Frank, E.: Data Mining: Practical machine learning tools with Java implementations. Morgan Kaufmann, San Francisco (2000)

Automatic BI-RADS Description of Mammographic Masses

Fabián Narváez, Gloria Díaz, and Eduardo Romero

Bioingenium Research Group, Departament of Medicine,
National University of Colombia, Bogotá, Colombia
{frnarvaeze,gmdiazc,edromero}@unal.edu.co
http://www.bioingenium.unal.edu.co

Abstract. This paper presents a CBIR (Content Based Information Retrieval) framework for automatic description of mammographic masses according to the well known BI-RADS lexicon. Unlike other approaches, we do not attempt to segment masses but instead, we describe the regions an expert selects, after the series of rules defined in the BI-RADS lexicon. The content based retrieval strategy searches similar regions by automatically computing the Mahalanobis distance of feature vectors that describe main shape and texture characteristics of the selected regions. A description of a test region is based on the BI-RADS description associated to the retrieved regions. The strategy was assessed in a set of 444 masses with different shapes and margins. Suggested descriptions were compared with a ground truth already provided by the data base, showing a precision rate of 82.6% for the retrieval task and a sensitivity rate of 80% for the annotation task.

Keywords: Automatic Annotation, BI-RADS, Computer Aided Diagnosis, Content-based Image Retrieval.

1 Introduction

Breast cancer is the most frequent type of cancer in women and is considered as the largest public health problem in women population [1,2]. This disease is fully curable if diagnosis is achieved early and mammography is the more efficient method for visualizing these first abnormalities [3,4]. However, mammographic interpretation is really hard, studies have shown that between 10% and 25% of breast cancers are not detected [5]. An agreement to reduce variability between radiologists resulted in the *Breast and Imaging Report and Database System* (BI-RADS), designed by the the American College of Radiology , as a standard description to report breast lesions for allowing to categorize different pathologies as well as their severity level [6]. This standard established that basic descriptors for masses are shape, margin and density. Thus, automatic mammography categorization based on BI-RADS descriptors is becoming important. Most of related works have classified the tissue density of mammograms based on the four BI-RADS categories [7,8,9]. But, shape and margin descriptors there are

J. Martí et al. (Eds.): IWDM 2010, LNCS 6136, pp. 673–681, 2010.

not considered. Recently, an approach to classify the shape and margin of mammographic mass into different categories was presented [10]. In this case masses were previously segmented in order to extract the boundaries that were then characterized and classified, results reported a precision lower than 70%.

On the other hand, development of Computer Assisted Diagnosis Systems (CAD) for mammography has decreased the variability effects, becoming a well accepted clinical practice to assist radiologists interpreting mammograms when they search and identify micro-calcification clusters [11]. However, the relatively low performance of CAD schemes in mass detection [12], make them less accepted as mass diagnosis tools. As an alternative, the interactive CAD systems, based on content-based information retrieval schemes [13,14,15], identify visually similar mass lesions that eventually are clinically relevant to the actual lesion [16]. Actually, CBIR-based CAD schemes have a potential to provide radiologist with visual aid and increase their confidence in accepting CAD-cued results in the decision making process. However, their main drawback is that they are also based on the segmentation of mass regions, which is a very difficult task especially for masses with blurred boundaries.

This paper proposes a new approach to support and assist the task of describing mass lesions from a set of Regions of Interests (RoIs). Given a particular query or region under examination, the method finds the most similar regions from a database, according to the BI-RADS lexicon, using the two most important diagnostic features for describing masses: morphology and texture. Morphology is described using the Zernike moments [17], and texture is captured via the Neighborhood Gray Tone Difference Matrix (NGTDM) [18]. Once these basic features are computed, a further reduction of dimensionality is achieved using a standard Principal Component Analysis (PCA), assembling a descriptor of 15 dimensions. Finally, a multiclass retrieval algorithm based on a k-NN scheme is constructed for the shape, margin descriptions and pathology classification.

The rest of this article is organized as follows: after this introduction, next section presents the methodology, then results are shown and last section discusses future work and conclusions.

2 Methods

An overview of the proposed framework is shown in Fig. 1. Firstly, a radiologist manually selects a region of interest as the queried RoI, which is preprocessed to improve the mass visual details. Afterward, morphology and texture features are extracted, compared with the information stored in the database (reference database) so that the most similar regions are retrieved. Finally, the BI-RADS is used to assign the most probable description to these regions.

2.1 RoI Pre-processing

Mammography analysis generally must deal with regions difficult to interpret [19] since they are associated to hard acquisition conditions. Every image was enhanced and two resulting images were prepared for analysis, the former aimed to

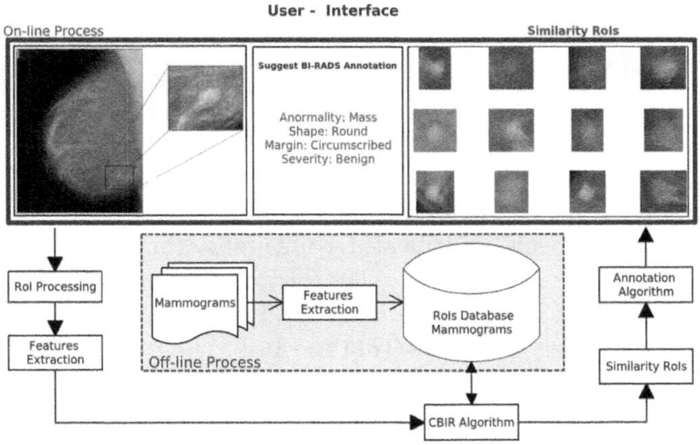

Fig. 1. Method overview

highlight mass edges by stretching the maximum and minimum gray level values to the interval [0, 255], followed by a bin reduction from 256 to 12 bins. The latter captured differences between patterns associated with mass and parenchyma tissue by adaptively equalizing the histogram so that structural details were preserved. In both cases, resultant images were smoothed out by a median filter in order to remove remaining noise [20]. Mass descriptors were drawn from the former image while texture features were extracted from the latter.

2.2 Feature Extraction

According to the BI-RADS lexicon, morphology and texture are the most important criteria for mass diagnosis [5,21]. Tradionally, Zernike moments, a class of statistical moments [22], have shown to be very effective for morphological representation i.e. rotation-invariant and robust to the noise [23]. Furthermore, representation with Zernike moments allows reconstruction with minimal losses and constitute a classical multiresolution representation for shapes [24,23]. The Zernike polynomials are a set of complex polynomials that form an orthogonal complete set $V_{pq}(x, y)$ within the unitary circle, which are defined by the equation 1.

$$V_{pq}(x, y) = R_{pq}(r)e^{jq\theta}, r \in [-1, 1] \tag{1}$$

where $r = \sqrt{x^2 + y^2}$ is the vector magnitude and $\theta = tan^{-1}\left(\frac{y}{x}\right)$ its angle.

The complex Zernike moments are derived from the real-valued radial polynomials, given by [2].

$$R_{pq}(r) = \sum_{s=0}^{(p-|q|)/2} (-1)^s \frac{(p-s)!}{s!(\frac{p+|q|}{2} - s)!(\frac{p-|q|}{2} - s)!} r^{p-2s} \tag{2}$$

where p and q are subject to $p - |q|$ is even, $0 \leqslant |q| \leqslant p$, and $p \geqslant 0$. Then, the complex Zernike moments of order p, with q repetitions for an image intensity function f(x,y) are given by [3].

$$Z_{pq} = \frac{p+1}{\pi} \sum_x \sum_y V^*_{pq}(x,y) f(x,y) \tag{3}$$

where $*$ stands for the conjugated complex of $V_{pq}(x,y)$.

In this work the Hosny implementation [25] of Zernike moments was used for describing shapes. For achieving so, the RoI was mapped onto the unitary circle so that the image center coincides with the unitary circle center, subjected to the condition that every pixel is within the RoI. The number of order moments used for generating the shape descriptor was selected by minimization of the reconstruction error ε given by the equation 4. The first 60 order moments were heuristically selected for generating a descriptor of 961 features.

$$\varepsilon_n = \sum_{i=0}^{N-1} \sum_{j=0}^{N-1} \{ \frac{[f(i,j) - F(i,j)_n]^2}{[f(i,j)]^2} \} \tag{4}$$

where $f(i,j)$ is the original image and $F(i,j)_n$ is the image reconstructed using n first order moments, as presented by Chong et al. [17].

On the other hand, the essential texture features were captured via the Neighborhood Difference Gray Tone Matrix [18] as follows: a neighborhood is firstly set and the absolute difference of the central pixel with its neighborhood average is computed. This average difference constitutes an occurrence that is stored in a histogram whose bins 1, 2, 3, 4 and 5 corresponded to the neighborhood sizes. So, this descriptor allows to capture pattern differences around to mass boundary in an area of 25 pixels, larger neighborhood sizes were evaluated but poor performance was obtained. So, five histograms of 256 positions were generated and five features were calculated from each, as described in [18]:

1. Contrast $= \left(\frac{1}{N_g(Ng-1)} \sum_{i=0}^{G_h} \sum_{j=0}^{G_h} p_i p_j (i-j)^2 \right) \left(\frac{1}{n^2} \sum_{i=0}^{G_h} s(i) \right)$
2. Busyness $= \left(\sum_{i=0}^{G_h} p_i s(i) \right) / \left(\sum_{i=0}^{G_h} \sum_{j=0}^{G_h} i p_i - j p_j \right)$
3. Complexity $= \sum_{i=0}^{G_h} \sum_{j=0}^{G_h} [|i-j|/(n^2(p_i + p_j))] \, [p_i s(i) + p_j s(j)]$
4. Texture strength $= \left[\sum_{i=0}^{G_h} \sum_{j=0}^{G_h} (p_i + p_j)(i-j)^2 \right] / \left[\epsilon \sum_{i=0}^{G_h} s(i) \right]$
5. Coarseness $= \epsilon + \sum_{i=0}^{G_h} p_i s(i)$

where N_g is the total number of different gray levels with $n = N - 2d$, for an $N \times N$ image, G_h is the highest gray tone in the image, p_i is the probability of occurrence of the ith gray tone, $s(i)$ is the histogram value at i^{th} gray tone and ϵ is a small number that prevents these values become infinite.

2.3 Automatic Mass Description

A content based information retrieval strategy that uses the K-NN rule (K-Nearest Neighbor [26]), was implemented. Such algorithm uses the information associated a certain number of images retrieved from the reference data base to assign a label to a particular RoI. Given the knowledge of N prototype RoIs (each marked by an expert radiologist) and their correct classification into several classes, it assigns an unclassified RoI to the class that is most heavily represented among its k nearest neighbors. The algorithm used a weighted Mahalanobis distance (w_d) to measure the similarity among the feature vectors describing both the database and queried RoIs. Once a set of K RoIs are retrieved, each shape, margin and pathology BI-RADS description is set using the decision rule in equation 5, where $Label(S)$ assigns the value description with the largest weight.

$$Label(S) = \arg\max_{S_i} |S_1, ..., S_n|, \quad S_i = \sum_{i=1}^{K} w_d^{s_i} \tag{5}$$

where $w_d = 1/d(\boldsymbol{x}, \boldsymbol{y})$ is the relative weight of each possible value description and S_n corresponds to number of possible labels for each shape, margin and pathology BI-RADS description.

3 Experimental Results

A total of 444 regions extracted from the *Digital Database for Screening Mammography (DDSM)* [27], which were previously BI-RADS annotated by a group of breast radiologists, were used to evaluate the performance of the proposed approach. These regions were split into training (344 RoIs) and testing (100 RoIs) sets. Both the content-based image retrieval scheme and the automatic annotation algorithm were independently assessed.

Before applying the retrieval process, a PCA analysis was used to reduce the feature vector dimensionality. From 986 extracted features (961 from shape descriptor and 25 from texture descriptor), 15 were selected, the ones which reported the larger eigen-value variability in the training subset.

3.1 Content-Based Image Retrieval Evaluation

Retrieval performance was assessed by computing the relevance of the recovered images, according to the ground truth of DDSM mammogram databases, identified by an experienced radiologists. Furthermore, shape, margin and pathology characteristics are taken into consideration in this evaluation. Therefore, We used four levels to describe the degree of relevance , namely $Score = 1$ for three correct targets, $Score = 0,66$ for two, $Score = 0,33$ for one and $Score = 0$ for zero. The Precision-Recall (P-R) graph was used for evaluating the performance of CBIR scheme. The precision (P) was defined as the relevance of images that the system was able to find among all images retrieved by the system, while recall (R) was the relevance of images that the system was able to find among all the relevant images stored in the database.

$$P = \frac{\sum_{i=1}^{k} S_i}{K}, \qquad R = \frac{\sum_{i=1}^{k} S_i}{\sum_{i=1}^{n} S_i} \tag{6}$$

where S_i is the score assigned to the ith RoI, K is the number of retrieved images and n is the total number of images in the reference database.

Retrieval performance for a set of 100 image queries was evaluated, using the 15 most similar images for annotation. An average precision and recall of respectively 82% and 42%, were obtained.

Figure 2 shows the Precision-Recall curve obtained when the average precision and recall measures were computed for a number of retrieved regions which varied from 1 to 15, with incremental steps of 1.The first point, the leftest curve point represents the average precision and recall rates for the first returned image. The second point corresponds to the precision and recall for the first two retrieved images, and so forth. It is observed from the graph that the first retrieved image report a relevance up to 80%. As expected, as long as the number of retrieved images increases, this precision decreases. However and interestingly, precision was higher than 65% in general. On the other hand, high recall values were found basically because the number of relevant images in the database was lower than the number of queried images.

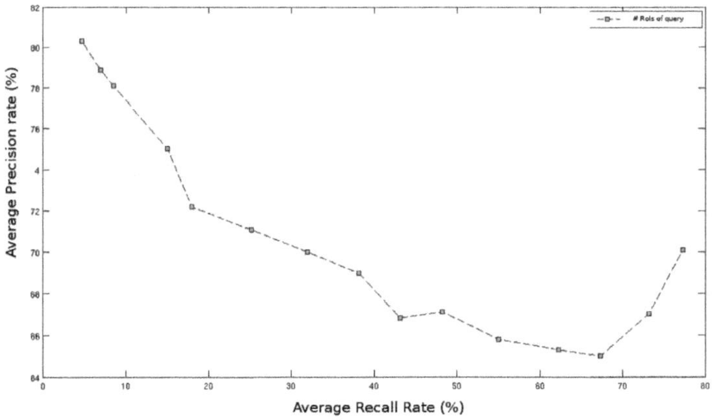

Fig. 2. Precision-Recall Average Curve

3.2 Automatic Annotation Assessment

Each shape, margin and severity BI-RADS description was assigned to each ROI query, based on equation 5. The optimal number of images used to assign each BI-RADS mass description was estimated by a 10-fold cross validation assessment. Results showed that a minimal of 7 images are needed for annotating shape

Table 1. Performance of Automatic BI-RADS description

	Accuracy	*PPV*	*TPR*
Shape	80.3	0.791	0.80
Margin	75.1	0.752	0.76
Pathology	85.3	0.861	0.87
Average	80.23	0.801	0.81

and margin descriptions while 9 are required for establishing a severity level. Table 1 shows the accuracy (ACC), positive predictive value (PPV) and true positive rate (TPR) computed from the automatic annotation for each BI-RAD description. The most difficult annotation was the margin, probably because of the blurred boundaries presented in many mass classes. However, average performance annotation is higher than 80%, a figure very acceptable for a system that tries to assist a diagnosis task.

4 Conclusions

In this paper a new strategy for assisting the diagnosis of mammography, based on a content based image retrieval scheme was proposed, implemented and evaluated. This strategy provided a BI-RADS mass description of a region of interest, which was supported by a set of diagnosed images that were shown to the expert. Instead of attempting to segment masses, we proposed a mass feature description, based on its internal structure with no explicit mass boundary detection.

The proposed approach was evaluated on a public image database (DDSM). Retrieval results have shown that this approach is successfully able to find the most similar images of a RoI query in a reference database. Likewise, the annotation results have also shown the capability of the method for generating the shape, margin and severity descriptions associated to the RoI, according to the BI-RADS lexicon, especially for discriminating the severity label i.e to decide whether a mass is benign or not.

These preliminary results have opened up new strategies for the computer assisting tools based on CBIR schemes in mammographic diagnosis, although a validation of the impact on diagnostic improvement of inexperienced radiologists is required.

References

1. Scottish Intercollegiate Guidelines Network : Management of breast cancer in women. A national clinical guideline 84, 1–3 (2005)
2. American Cancer Society : American cancer statistics (2007); Revisado el 2 Septiembre (2008)
3. Verma, K., Zakos, J.: A computer-aided diagnosis system for digital mammograms based on fuzzy-neural and feature extraction techniques. IEEE Transactions on Information Technology in Biomedicine 16, 219–223 (2002)

4. Buseman, S., Mouchawar, J., Calonge, N., Byers, T.: Mammography screening matters for young women with breast carcinoma. Cancer 97, 352–358 (2003)
5. Bird, R., Wallace, T., Yankaskas, B.: Analysis of cancers missed at screening mammography. Radiology 178, 234–247 (1992)
6. American College of Radiology (ACR): Illustrated Breast Imaging Reporting and Data System (BI-RADS). ACR (1998)
7. Marias, K., Linguraru, M.G., Ballester, M.G., Petroudi, S., Tsiknakis, M., Sir, M.: Automatic labelling and bi-rads characterisation of mammogram densities. In: Proceedings of the IEEE Engineering in Medicine and Biology (2005)
8. Petroudi, S., Kadir, T., Brady, M.: Automatic classification of mammographic parenchymal patterns: A statistical approach. In: IEEE International Conference on Engineering in Medicine and Biology Society (2003)
9. Bovis, K., Singh, S.: Classification of mammographic breast density using a combined classifier paradigm. In: Medical Image Understanding and Analysis (2002)
10. Tao, Y., Lo Shih-Chung, B., Freedman Matthew, T., Erini, M., Xuan, J.: Automatic categorization of mammographic masses using bi-rads as a guidance. In: Proceedings of SPIE, the International Society for Optical Engineering (2008)
11. Nishikawa, R.M.: Current status and future directions of computer-aided diagnosis in mammography. Computerized Medical Imaging and Graphics 31, 224–235 (2007)
12. Gur, D., Stalder, J.S., Hardesty, L.A., Zheng, B., Sumkin, J.H., Chough, D.M., Shindel, B.E., Rockette, H.E.: Computer-aided detection performance in mammographic examination of masses: assessment. Radiology 233, 418–423 (2004)
13. Tao, Y., Lo, S.B., Freedman, M.T., Xuan, J.: A preliminary study of content-based mammographic masses retrieval. In: Proc. SPIE, vol. 6514, p. 65141Z (2007)
14. Zheng, B., Mello-Thoms, C., Wang, X.H., Abrams, G.S., Sumkin, J.H., Chough, D.M., Marie, G.A., Lu, A., Gur, D.: Interactive computer aided diagnosis of breast masses: Computerized selection of visually similar image sets from a reference library. Academical Radiology 14, 917–927 (2007)
15. Rosa, N.A., Felipe, J.C., Traina, A.J., Rangayyan, R.M., Azevedo-Marques, P.M.: Using relevance feedback to reduce the semantic gap in content-based image retrieval of mammographic masses. In: Conf. Proc. IEEE Med. Biol. Soc., pp. 406–409 (2008)
16. Rangayyana, R.M., Ayres, F.J., Desautelsa, J.E.L.: A review of computer-aided diagnosis of breast cancer: Toward the detection of subtle signs. Journal of the Franklin Institute 344, 312–348 (2007)
17. Wee, C.Y., Paramesran, R.: On the computational aspects of zernike moments. Image and Vision Computing 25, 967–980 (2007)
18. Amadasun, M., King, R.: Textural features corresponding to textural properties. IEEE Transactions on systems, man and Cybernetics 19, 1264–1274 (1989)
19. Cheng, H.D., Shi, X.J., Min, R., Hu, L.M., Cai, X.P., Du, H.N.: Approaches for automated detection and classification of masses in mammograms. Pattern Recognition 39, 646–668 (2006)
20. Petrick, N., Chan, H.P., Sahiner, B., Wei, D.: An adaptive density weighted contrast enhancement filter for mammographic breast mass detection. IEEE Trans. Med. Imaging 15(1), 59–67 (1996)
21. Maggio, C.D.: State of the art of current modalities for the diagnosis of breast lesions. Eur. J. Nucl. Med. Mol. Imaging 31(suppl.1), S59–S69 (2004)
22. Teague, M.R.: Image analysis via the general theory of moments. J. Optical Soc. Am. 70, 920–930 (1980)
23. Belkasim, S., Hassan, E., Obeidi, T.: Radial zernike moment invariants. In: The Fourth Int. Conf. on Computer and Information Tech., vol. 1, pp. 790–795 (2004)

24. Kim, H., Kim, J.: Region-based shape descriptor invariant to rotation, scale and translation. Signal Proc.: Image Communication 16, 87–93 (2000)
25. Hosny, K.M.: Fast computation of accurate zernike moments. J. Real-Time Image Proc. 3, 97–107 (2008)
26. Cover, T.M., Hart, P.E.: Nearest neighbor pattern classification. IEEE Trans. on Inf. Theo. IT-13, 21–27 (1967)
27. Heath, M., Bowyer, K., Kopans, D., Moore, R., Kegelmeyer, W.P.: The digital database for screening mammography. In: Yaffe, M.J. (ed.) Proceedings of the Fifth International Workshop on Digital Mammography, pp. 212–218. Medical Physics Publishing (2001)

A Clustering Method for the Extraction of Microcalcifications Using Epipolar Curves in Digital Breast Tomosynthesis

Candy P.S. Ho, Christopher E. Tromans, Julia A. Schnabel, and Sir Michael Brady

Wolfson Medical Vision Laboratory,
Robotics Research Group, Department of Engineering Science,
University of Oxford, Parks Road, Oxford OX1 3PJ, UK
{psho,cet,jmb}@robots.ox.ac.uk
{julia.schnabel}@eng.ox.ac.uk

Abstract. DBT provides significantly more information than mammography. This offers new opportunities to improve existing microcalcification detection methods. In a companion work in this volume, we showed that the use of epipolar curves can improve both the sensitivity and specificity of microcalcification detection. In this paper, we develop a clustering algorithm to form epipolar curves from candidate microcalcifications (which may be noise points), obtained after applying a detection algorithm to each individual projection. This enables the subsequent 3D analysis for the classification of microcalcification clusters.

Keywords: digital breast tomosynthesis, microcalcifications, clustering, epipolar curves, DBT.

1 Introduction

The detection of microcalcifications, microcalcification clusters, and their subsequent classification are important tasks in the early detection of breast cancer. In [1], we demonstrated the use of epipolar curves for the detection of microcalcifications, showing how they can help improve the sensitivity and specificity detection visually. Following this, we are also working on the estimation of the 3D positions of microcalcifications from epipolar curves with sufficient accuracy to enable 3D analysis of the shape of a microcalcification cluster, without first reconstructing the data into an explicit 3D representation, since reconstruction methods continue to be developed. It is expected that the outcome of this work can help automate the classification of breast abnormalities as malignancy and benign, as opposed to simply detecting such abnormalities.

As a key component of the automation of the process of detecting microclacifications using epipolar curves, this paper develops a clustering algorithm to group points found in the individual projection images into the epipolar curves, alternatively rejects them as noise.

J. Martí et al. (Eds.): IWDM 2010, LNCS 6136, pp. 682–688, 2010.

2 Methods

The first step in our detection process is the application of a detection algorithm to each of the individual projection images in order to identify candidate microcalcifications, similar to the existing methodology applied in mammography, except now that we apply the detection algorithms to multiple projections. Of course, the poor signal-to-noise ratio of the individual projection images means that some of the candidates will turn out to be noise points, and, equally, some microcalcifications may be missed in some views. Either way, the input to the clustering algorithm is a list of the 2D positions of all the microcalcification candidates. The output of the clustering algorithm is the set of clusters of 2D points, each of which corresponds either to an epipolar curve (hence a microcalcification tracked over most of the projections) or is identified as a noise point. The algorithm also indicates those microcalcifications that have been missed in which (small number of) views. It also indicates the likely positions in the projection images, potentially enabling a subsequent application of the detection algorithm with adapted parameters, such as local thresholds.

In the study reported here, in order to evaluate our method against ground truth, we have used the X-ray simulation software developed by Tromans et al [2] to generate simulated DBT views of a cluster containing 15 spherically shaped microcalcifications with radius ranging from 0.0355 to 0.075mm in a curvilinear arrangement. Statistical noise of the amount expected with current detectors in each acquisition image is added to better simulate reality. As a demonstration, we selected the middle 7 DBT views out of the total 15 views, taken at angles -8.65°, -5.81°, -2.99°, -0.16°, 2.67°, 5.5°, 8.33° (we refer to these as tomo4, tomo5, tomo6, tomo7, tomo8, tomo9, tomo10 respectively). Note that, as in [1], the angles used follow a plausible DBT acquisition geometry. We have deliberately chosen one of the simplest detection algorithms, namely a top hat transformation, which effectively equates microcalcifications with small, locally bright points, and is expected to miss fainter microcalcifications [3].

Recall from [1] that an epipolar curve can be defined as:

$$\mathbf{x}^s_{+i} = \begin{bmatrix} a\left(\dfrac{L+D\cos\theta_i}{c+D\cos\theta_i}\right)+u \\ b\left(\dfrac{L+D\cos\theta_i}{c+D\cos\theta_i}\right)+v \end{bmatrix} + \frac{D\sin\theta_i}{c+D\cos\theta_i}\begin{bmatrix} c-L \\ 0 \end{bmatrix} \quad \mathbf{x}^s_{-i} = \begin{bmatrix} a\left(\dfrac{L+D\cos\theta_i}{c+D\cos\theta_i}\right)+u \\ b\left(\dfrac{L+D\cos\theta_i}{c+D\cos\theta_i}\right)+v \end{bmatrix} - \frac{D\sin\theta_i}{c+D\cos\theta_i}\begin{bmatrix} c-L \\ 0 \end{bmatrix} \tag{1}$$

In our plausible DBT acquisition system, D = 660mm, L = 40mm, u = 143.36mm, v = 232.96mm and detector resolution = 0.07mm. We have also used these settings in the demonstration. Also, in our demonstration, the *x*-, *y*- coordinates are directly obtained from the projections and the coordinate frame used is the one used by the simulated software. A coordinate frame transformation is applied to transform it to our sensor coordinate frame used in the derivation of the epipolar curve. Here, for *x*-coordinate, we mean the transformation of the 2nd element in Equation (1); for *y*-coordinate, we mean the transformation of the 1st element in Equation (1).

Our clustering technique is based on the following observations, which would have analogues in any similar acquisition geometry:

1. Equation (1) implies that the DBT views generated from equal but opposite directions (θ_i and - θ_i) will have the same value in one of the coordinates. By assigning typical values used in DBT into the variables in the equation, this implies that the x values from the 7 projections are nearly equal.
2. The values in the other axes will follow a specific order with respect to the angles taken to generate the DBT views. In our case, the y values are in descending order with respect to the angles taken to generate the DBT views and the differences in y values between neighbouring projections are very similar for the same microcalcification.
3. Due to the size of microcalcifications and the viewing angles, the number of projection points for each microcalcifications captured in each DBT view are different.
4. Since noise is assumed to be random and is statically generated randomly, it is expected that a noise point in a projection appear as a single point.
5. The superimposition problem in mammography implies that one projection point may be generated from 2 or more microcalcifications in the breast. However, this situation is unlikely in all but a minority of DBT views, due to different acquisition angles and differences in the 3D positions of the microcalcifications. In Observation 2, we mentioned that the differences in y values between neighbouring projections are very similar for the same microcalcification. For different microcalcifications at different 3D positions, this "differences" varies.

Taking account of the above, our clustering method comprises 6 steps:

1. Sort all the points in ascending order of x values and descending order of y values. (Observations 1, 2)
2. Classify the points into noise points and into groups of same x values.
   (a) If only one point has a particular x value, label this point as a noise point. (Observation 4)
   (b) Assign points having same x values into same group. (Observation 1)
3. Determine noise points within a group. The noise points are those points whose y values do not follow the specific order with respect to the angles. (Observations 2)
4. Combine groups with x and y values within a preset range into a cluster. This is because a microcalcification can be larger than one pixel in a projection. (Observation 3)
5. Check each cluster and see whether they should be split due to projection points being too close together but actually they are coming from different microcalcifications. (c.f. superimposition problem in mammography or microcalcifications in the breast are being too close to each other.) (Observation 2, 5)
6. Recheck the noise points. If they are within a preset distance from a cluster, include it in the cluster. One cluster corresponds to one microcalcification. (Observation 3)

3 Results

Fig. 1 shows the ground truth of all the projection points of 15 microcalcifications from 7 DBT views obtained visually and manually. We have assigned a no. 1 to 15 to each microcalcification for identification. The inputs of our clustering technique are points detected by top hat in each DBT view. The detection results by top hat in each individual DBT view are recorded in Table 1:

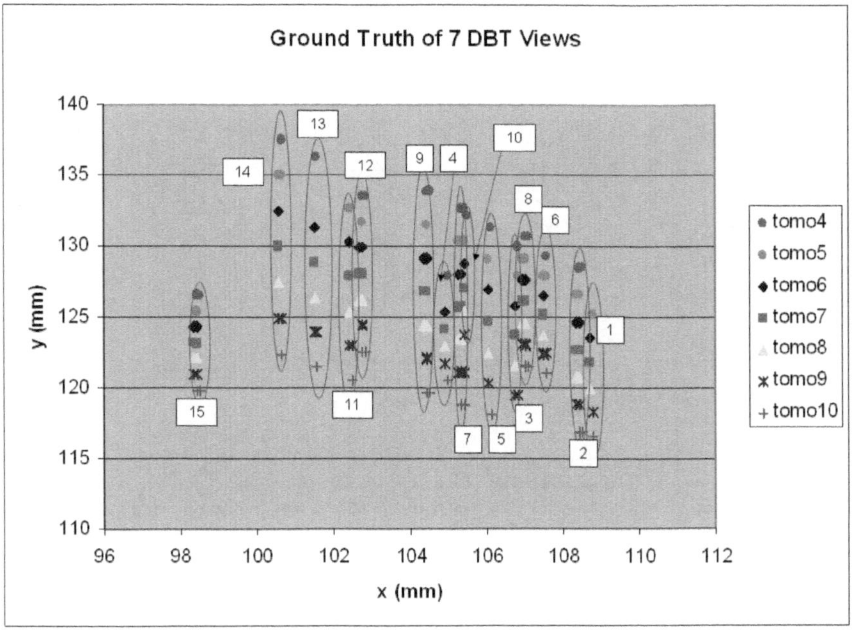

Fig. 1. Ground truth of all the projection points of 15 microcalcifications from 7 DBT views obtained visually and manually. (Note: Some microcalcifications cannot even be visualized by human eyes in tomo4 (microcalcification 1), tomo5 (microcalcifications 4, 13) and tomo10 (microcalcification 10) due to limitations in acquisition.)

Table 1. Detection results of individual projections (tomo4 to tomo10) using top hat transformation

DBT view	Microcalcification detected	No. of microcalcifications detected in each view	No. of noise points detected
tomo4	2,6,7,8,9,12,14,15	8	2
tomo5	1,2,5,6,7,8,9,11,12,14,15	11	4
tomo6	2,5,6,7,8,9,10,12,14,15	10	1
tomo7	2,7,8,9,11,12,13,14,15	9	5
tomo8	2,3,5,6,7,8,9,11,12,15	10	7
tomo9	2,3,7,8,9,10,11,15	8	1
tomo10	2,7,8,9,11,12,14,15	8	0
		Max no. detected: 11	Total (3 views: 13 7 views: 20)

To assess the effects of using different number of DBT views, we have applied our clustering algorithm using 3 and 7 DBT views. The detection results are shown graphically in Fig. 2 and are summarized in Table 2:

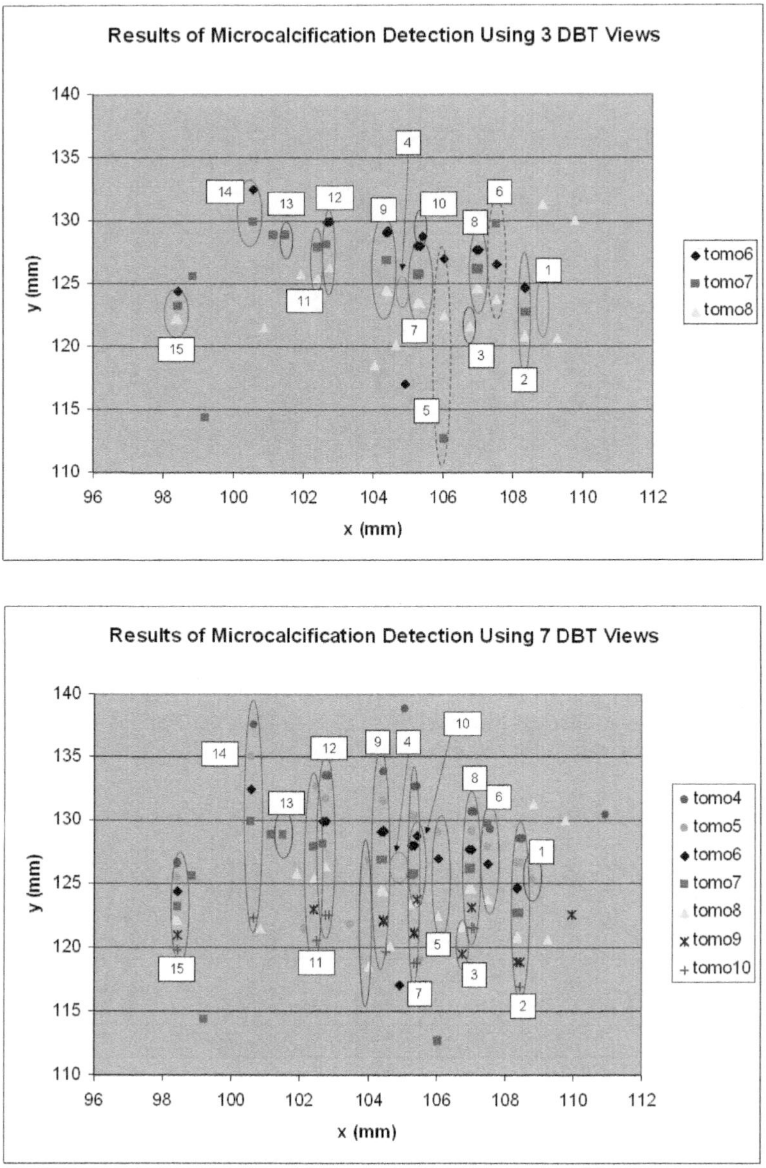

Fig. 2. Detection results using epipolar curves. Top: 3 views; Bottom: 7 views. (Red Circles: True Positives; No Circle: True Negatives; Black Circles: False Positives; Green/Blue Circles: False Negatives; Dotted Red Circles: True Positives/False Negatives (depends on implementation.)

Table 2. Detection results of using epipolar curve approach. (The effects of using different no. of views are also shown.)

No. of Views used	True Positives (out of 15 microcalc.) (counted as no. of clusters)	True Negatives (counted as isolated points)	False Positives (counted as no. of clusters)	False Negatives (out of 15 microcalc.) (counted as no. of clusters)	Uncertain (counted as no. of clusters)
3	8 (2, 7, 8, 9, 11, 12, 14, 15)*	11	0	5 (1, 3, 4, 10, 13)*	2 (5, 6)*
7	12 (2, 3, 5, 6, 7, 8, 9, 10, 11, 12, 14, 15)*	18	1	3 (1, 4, 13)*	0

* refers to the microcalcifications.

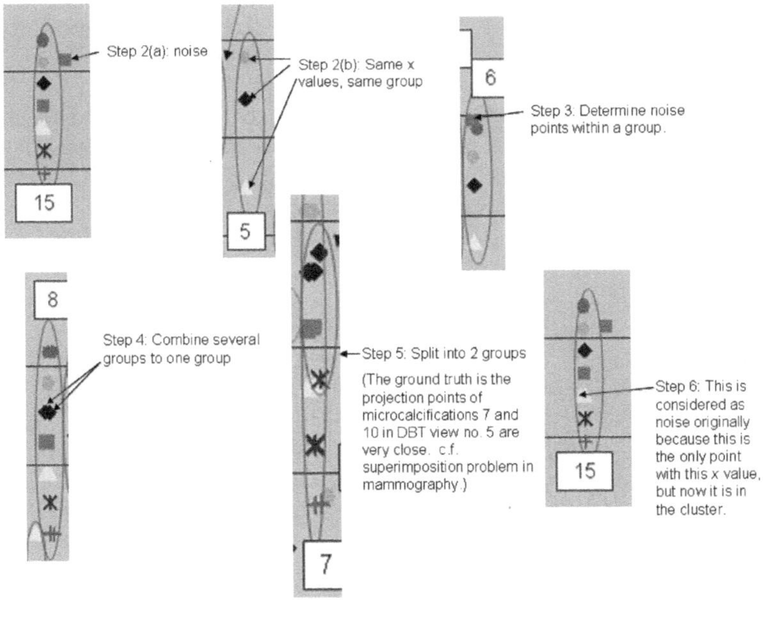

Fig. 3. Explanation of the clustering algorithm step-by-step pictorially (Extracted from Fig. 2 Bottom)

The algorithm is explained pictorially by magnifying the detection results using 7 DBT views (Fig. 2 Bottom) in Fig. 3.

It can be seen that the epipolar curve approach can greatly improve the detection process. Most of the noise points (false positives) can be distinguished. If we use more views (7 in this example), the results are better than those in any single projection. We can detect 12 microcalcifications using 7 views while at most 11 microcalcifications are detected in any single projection. In addition, it tells us which points are missing in the view by looking at the clusters. This allows us to adapt our detection algorithms for better results.

4 Discussion

DBT provides significantly more information than mammography. This offers new opportunities to improve existing microcalcification detection methods. We demonstrated here that one simple clustering technique can be implemented to extract microcalcifications using the epipolar curve approach automatically. Also, we want to point out that the discussion here highlights the cases and conditions which may be encountered during epipolar clustering. It demonstrates the feasibility in computer-aided detection of microcalcifications. It is noted, however, that this is only an initial study and further study and fine-tuning are required in order to accommodate various peculiar cases and extreme cases.

By using the epipolar curve approach, we can now obtain the groups belonging to the same microcalcifications, we can then estimate the 3D positions of the microcalcifications and perform 3D analysis. This will facilitate the classification of breast into malignant and benign.

Acknowledgments. This work forms part of a project funded by the Technology Strategy Board (TSB) and by the Engineering and Physical Sciences Research Council (EPSRC). CET is funded by an EPSRC Fellowship. Special thanks to all the involved parties in this project and people in the Wolfson Medical Vision Lab for their support and valuable inputs.

References

1. Ho, C.P.S., Tromans, C., Schnabel, J.A., Brady, M.: A clustering method for the extraction of microcalcifications using epipolar curves in digital breast tomosynthesis. In: Proceedings of 10th International Workshop on Digital Mammography (IWDM), Girona, Spain (2010)
2. Tromans, C., Brady, M.: The Standard Attenuation Rate for Quantitative Mammography. In: Proceedings of 10th International Workshop on Digital Mammography (IWDM), Girona, Spain (2010)
3. Rick, A., Muller, S., Bothorel, S., Grimaud, M.: Quantitative Modelling of Microcalcification Detection in Digital Mammography. In: Taylor, C., Colchester, A. (eds.) MICCAI 1999. LNCS, vol. 1679, pp. 32–41. Springer, Heidelberg (1999)

Software Framework for Simulating Clusters of Microcalcifications in Digital Mammography

Federica Zanca, Guozhi Zhang, Nicholas Marshall, Eman Shaheen, Elena Salvagnini, Guy Marchal, and Hilde Bosmans

Department of Radiology, University Hospitals Gasthuisberg, Herestraat 49, B-3000 Leuven, Belgium
{federica.zanca,guozhi.zhang,nicholas.marshall,eman.shaheen, elena.salvagnini,guy.marchal,hilde.bosmans}@uzleuven.be

Abstract. Observer performance experiments for lesion detection are an accepted means of assessing the imaging performance of radiological imaging systems. Simulation methods for clusters of microcalcifications have been proposed for creating images with abnormal pathology for its use in such experiments. We report on a software tool that can generate simulated clusters of microcalcifications for different exposure parameters and different digital mammography systems. The effect of the simulation steps on microcalcification templates, (namely exposure settings, breast thickness, modulation transfer function (MTF) and pixel size) is demonstrated and validated. Results were evaluated in terms of the cluster's peak contrast (PC) for three cases: for different exposure conditions within a given system, for different systems and for different system MTF calculation methods. As expected, with higher tube voltage and for insertion into thicker breast simulating material, the lesion contrast decreases while the position of the peak remains unchanged. When different systems are considered with the same exposure settings, the observed difference in the PCs is related to the blurring due to the different MTF and the pixel size of the systems; a shift in the peak position is also observed, due to resampling. This functional and user-friendly system could be used by other researchers for performing comparative studies of mammographic imaging systems.

Keywords: simulation, microcalcifications, observer performance experiments, digital mammography.

1 Introduction

The effectiveness of a mammography system lies in its ability to improve the detection of relevant findings, like microcalcifications. Observer performance experiments for lesion detection are therefore often used to assess the performance of radiological systems. In screening mammography, abnormal cases are relatively rare (approximately 3–6 cancers per 1000 women), and hence it is difficult to assemble an adequate number of images to conduct such experiments. Furthermore, establishing ground truth via pathology examination is time consuming. To address this problem, simulation models for single microcalcifications or clusters have been proposed [1-3];

J. Martí et al. (Eds.): IWDM 2010, LNCS 6136, pp. 689–696, 2010.

they allow the creation of abnormal images containing realistic but artificial lesions, for which position, morphology and characteristics are known. Such hybrid images can be used in detection experiments if the simulations are appropriate [1,3-4]. This paper presents a software tool that can generate simulated clusters of microcalcifications for different exposure parameters and different digital mammography systems. The scientific validation of the different steps in the simulation is described.

2 Materials and Methods

2.1 System Description

The software tool, *mctam*, Micro-Calcification Template Adapter, has been developed in MATLAB (MathWorks, Natick, MA) and allows to simulate clusters of microcalcifications under different exposure conditions and/or for different detector systems. Such simulated lesions can be subsequently multiplied with linear raw mammographic data to create abnormal images, to be used in detection observer performance experiments. *Mctam* is an automated and generalized approach of the method reported by Zanca et al. [3]. It uses a database of magnified high resolution images of clusters of microcalcifications and creates templates for other systems and exposure conditions, as described in detail in the original methodology. An example of a digital image of a large core needle biopsy specimen containing a real microcalcification cluster is shown in Figure 1. Such clusters were used to build up the database of templates. To simulate microcalcifications for a variety of target system parameters and breast thicknesses, specific adjustments (beam quality, detector modulation transfer function (MTF) and detector pixel size) have to be applied to cope with the differences between the acquisition (input) system (used to image the real cluster) and the target system.

2.2 Software Modules

The method has three steps: (a) the creation of an ideal template (for an imaging system with MTF = 1); (b) the expression of the template gray values in terms of Aluminum equivalent thickness and (c) the correction for the difference in the system MTF and detector pixel size between the acquisition and the target system.

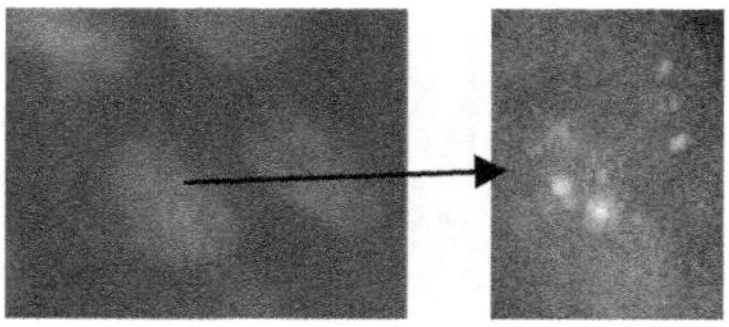

Fig. 1. Example biopsy specimen containing a real cluster of microcalcifications (left). Magnified image of the same cluster (right), extracted following the methodology proposed by Zanca et al. [3].

The steps can be summarized as follows: an input template T_1 (a real cluster of microcalcifications) has been acquired with an input system having pixel size α_1, MTF_1 and in magnification m_1. This template is then corrected (in the Fourier domain) for the MTF_1 of the system to get an ideal template (IT_1). Then the gray values of the ideal template IT_1 are expressed in terms of Aluminum equivalence (AE) [5], to retrieve AE_1. The template is subsequently corrected for any differences in beam quality compared to the input conditions, to obtain the ideal template IT_2 for the target conditions. Next, IT_2 is adjusted towards the MTF_2 of the target system (which includes all pre-sampling components, e.g., focal spot size and magnification), to produce the template T_{2MTF2}. Finally, the template is re-sampled for pixel size β_2 of the target system and for magnification factor m_2 (typically m_2=1), to produce the final template T_2.

2.3 Software Testing

The realistic appearance of the simulated microcalcifications and of the simulated clusters [3] has been proven in a clinical setting where five experienced radiologists compared 59 pairs of simulated and real microcalcifications in a two alternative forced choice (2-AFC) task, designed to test their ability to distinguish the real from the simulated lesions. They also classified the shapes of the microcalcifications according to a standardized clinical lexicon [6]. The radiologists were unable to distinguish the lesions and the shape classification revealed substantial agreement with the truth, showing that we were able to accurately simulate the lesion morphology. While this experiment confirmed our ability to accurately simulate realistic lesion while preserving their morphology, it does not ensure that a given physical microcalcification cluster would be correctly reproduced for different target system parameters and breast thicknesses. This is due to the fact that the study was involved with one system and was not extended to different systems. Indeed, it is important to illustrate the correct behavior of the methodology and of the software tool beyond the system that had been used in our institution before, if the simulation method has to be used in comparative system tests. The different steps in the simulation procedure are validated here by studying major aspects of the imaging chain separately: X-ray attenuation, MTF, and binning for a detector with a different pixel size.

3 Results and Discussion

Table 1 illustrates all the steps in the simulation process and their effects on the input template. This is done by tracing the peak contrast (PC) of the template, where PC is the maximally attenuated pixel of the cluster image divided by the background signal (for normalization to 1). The example of the input template (second row) that is used to track the changes applied through the different steps had been acquired from a specimen with m_1=2, for a specific exposure setting (27 kV MoMo 4 cm Lucite (PMMA)) and system parameters (MTF_A, pixel size 70 μm) and has a PC value of 0.79 at a spatial position (x,y) = (8,24) in the template image. In the second column of Table 1, the input template is recalculated for target conditions identical to the input conditions but for magnification 1, so the MTFs of input and target system are the same, except for the magnification factor. The ideal template IT_1 has a higher contrast with the background (lower PC, 0.66, 34%), due to the removal of the blurring effect

Table 1. Different simulation steps, for different exposure settings, PMMA and detector pixel size. T_1= input template; IT_1= ideal template; AE_1= Aluminum equivalence template; IT_2= ideal template corrected for different exposure settings; T_{2MTF2} = template corrected for the target system MTF_2; T_2 = re-sampled template; PC= Peak Contrast; MoMo =Molybdenum/Molybdenum; WRh= Tungsten/Rhodium; PMMA= Lucite; m_1, m_2 = magnification.

Input system	27 kV MoMo, 4 cm PMMA, System A 70µm, MTF_A m_1=2			
Input Template T_1	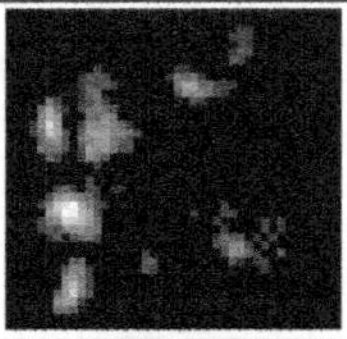 PC= 0.79, (x=6, y=24)			
Ideal template IT_1 (200%)	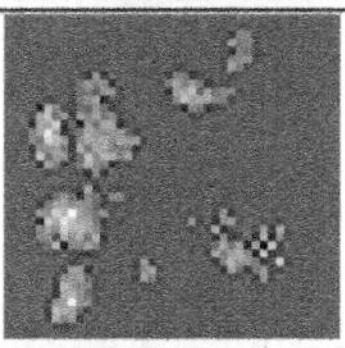 PC= 0.34			
Al equivalence AE_1 (200%)	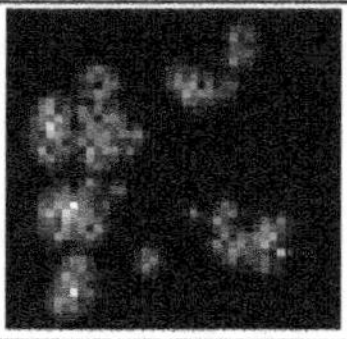 PC=0.47			
Target systems	27 kV MoMo, 4 cm PMMA, System A 70µm, MTF_A m_2=1	30 kV MoMo, 4 cm PMMA, System A 70µm, MTF_A m_2=1	32 kV WRh, 7 cm PMMA, System A 70µm, MTF_A m_2=1	27 kV MoMo, 4 cm PMMA, System B 70µm, MTF_B m_2=1
Correction exposure settings: IT_2 (200%)	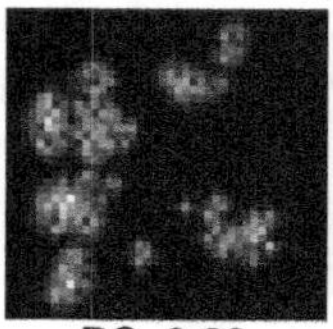 PC=0.66	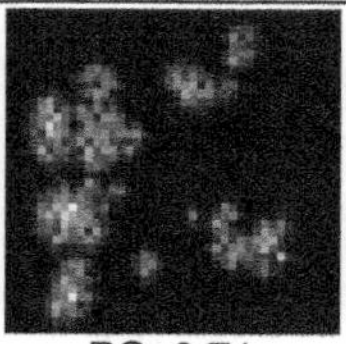 PC=0.71	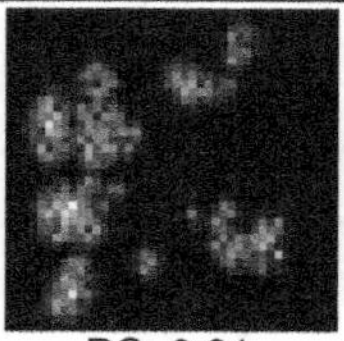 PC=0.81	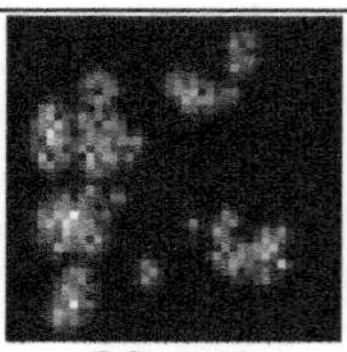 PC=0.66
Correction for target MTF: T_{2MTF2} (200%)	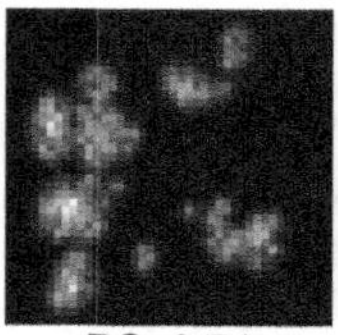 PC=0.74	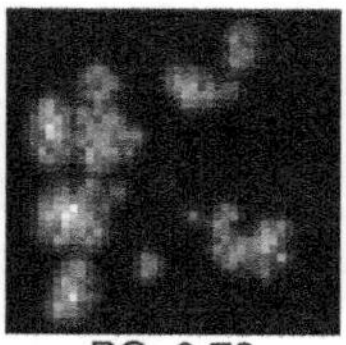 PC=0.78	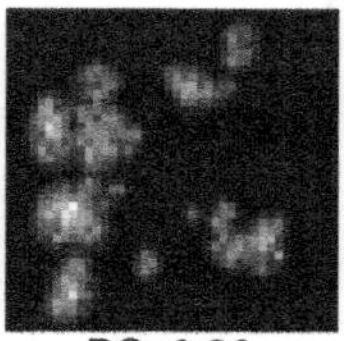 PC=0.86	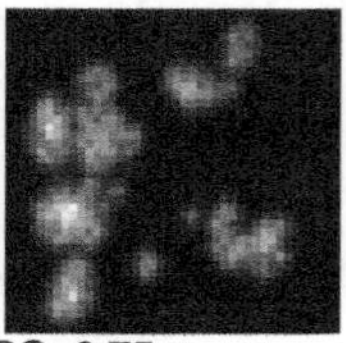 PC=0.75
Re-sampled template: T_2 (400%)	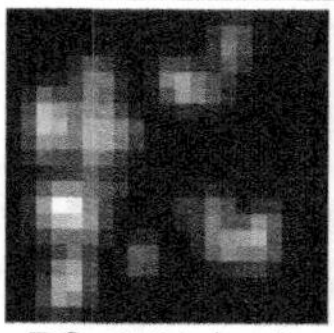 PC=0.81 (x=3, y=12)	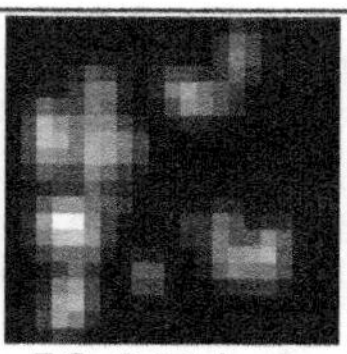 PC=0.84 (x=3, y=12)	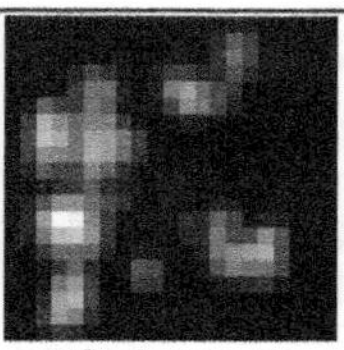 PC=0.9 (x=3, y=12)	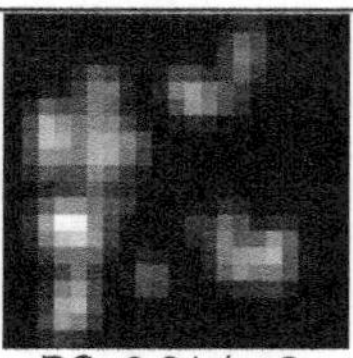 PC=0.81 (x=3, y=12)

caused by the MTF of the input system. The peak Al equivalence (AE_I) of this template is 0.47 mm. The template corrected for the target exposure settings IT_2 gives a PC (0.66, 34%) identical to the PC value of IT_1, as expected. Subsequently, by correcting for the target system MTF (MTF_A m_2=1), the output T_{2MTF2} has a PC of 0.74, which reflects a decreased contrast, from 34% to 26%, of the microcalcification with respect to template IT_2 for a system with an ideal MTF. Finally, when the template is re-sampled (T_2) the PC is very close to the one of T_1 but slightly higher (0.81 instead of 0.79, difference of 2%). The position of the PC in terms of pixel coordinates remains unchanged when going from m_1=2 to m_2=1, reflecting the fact that the relative position of the microcalcifications within the clusters is preserved. Similarly, an increase of only the kV from 27 kV to 30 kV (third column in Table 1), decreases the contrast (from 21% to 16%), and an increase in breast thickness and kV also decreases contrast (from 21% to 10%). The position of the PC remained unchanged. The last column shows the effect of a different target MTF (two different systems), but with the same pixel size. Compared to T_2 in the first column, the output T_{2MTF2} has a slightly changed PC (a difference of 1%) due to the fact that the MTF of system A is higher than the MTF of system B, (0. 75 @ 3 lp/mm vs 0.67 for system B). This difference is washed out when the template is binned, as the PC of the final template is 0.81, as for system A. Again, the position of the peak contrast remained unchanged. Table 2 reports on the same cluster recalculated for different systems at the same exposure settings of the input template, to show the combined effect of the detector pixel size and MTF. As the exposure conditions are not changed with respect to the input template, IT_2 is identical for all systems investigated, with a PC=0.66. The second column of table 2 reports again the reference example: the input template recalculated for target conditions identical to the input conditions, as in Table 1. In the third column, the input template is calculated for a system (system C) having a larger pixel size than the input system A, namely 100 μm and a lower MTF (0.4 @ 3 lp/mm compared to 0.75 of system A). The effect of a lower MTF_C is reflected in T_{2MTF2}, with a decrease in contrast with respect to IT_2 (from 34% to 21%). After re-sampling, the final template T_2 has again a decreased contrast with respect to T_1 (from 21% to 15%) and to system A (from 19% to 15%). The position of the PC is shifted due to the binning (from x=3, y=12 to x=2, y=8).

For system D (column 4), having a pixel size smaller than system A, namely a change from 70 μm to 50 μm, and a lower MTF (about 0.33@ 3 lp/mm compared to 0.75 of system A), T_{2MTF2} shows a strong decrease in contrast with respect to IT_2 (20%). The final template T_2 has the same PC as T_{2MTF2} showing that in this condition the binning doesn't have a large effect on the contrast, probably due to the smaller pixel size than system A, but the position of PC is shifted (from x=3, y=12 to x=5, y=17). Finally, in the last column of Table 2 a system with pixel size 85 μm is considered. The decrease in contrast with respect to IT_2 (from 34% to 25%) is comparable to that of system A, because of the similar pixel size (85 μm vs. 70μm) and a similar MTF (about 0.7 @ 3 lp/mm compared to 0.75 for system A). After resampling, the final template T_2 has again a decreased contrast with respect to T_1 (from 21% to 18%) but as expected very similar to system A. The position of the PC is shifted, due to the binning (from x=3, y=12 to x=3, y=10), but the shift for system E is less important compared to the example of system D, because of the smaller pixel size of system D. Table 3 reports an example of two templates simulated for

Table 2. Presentation of the different simulation steps, for different systems (different MTF and pixel size) but same exposure conditions. All abbreviations are like in Table 1.

Input system	27 kV MoMo, 4 cm PMMA, System A 70μm, MTF_A m_1=2			
Input Template T_1	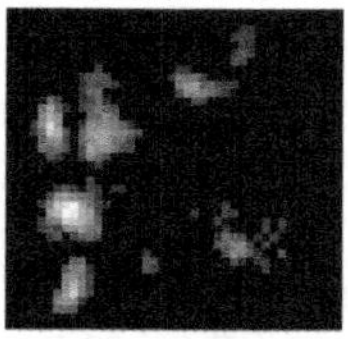 **PC= 0.79, (x=6, y=24)**			
Ideal template IT_1 (200%)	 PC= 0.34			
Al equivalence AE_1 (200%)	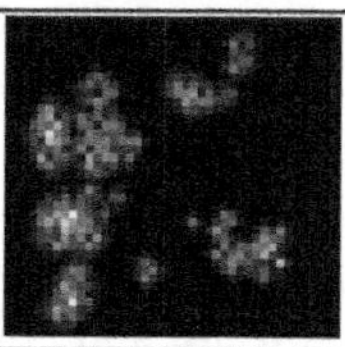 PC=0.47			
Target systems	27 kV MoMo, 4 cm PMMA, System A 70μm, MTF_A	27 kV MoMo, 4 cm PMMA, System C 100μm, MTF_C m_2=1	27 kV MoMo, 4 cm PMMA, System D 50μm, MTF_D m_2=1	27 kV MoMo, 4 cm PMMA, System E 85μm, MTF_E m_2=1
Correction exposure settings: IT_2 (200%)	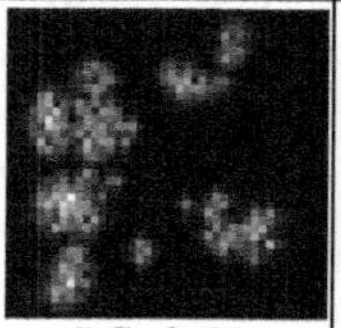 PC=0.66	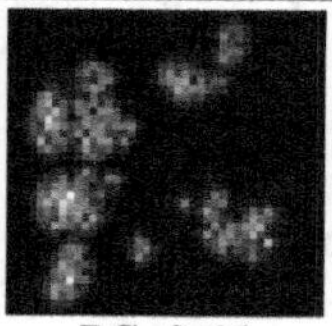 PC=0.66	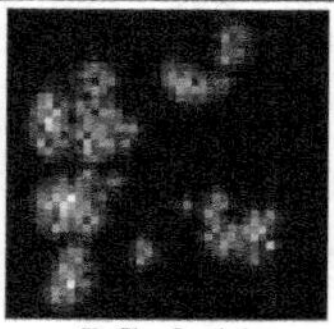 PC=0.66	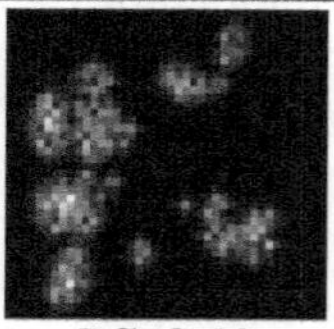 PC=0.66
Correction for target MTF: T_{2MTF2} (200%)	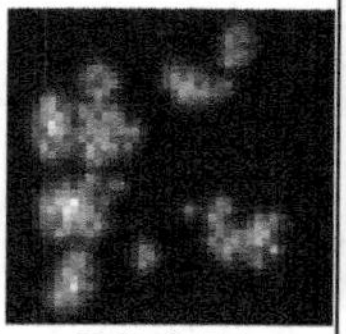 PC=0.74	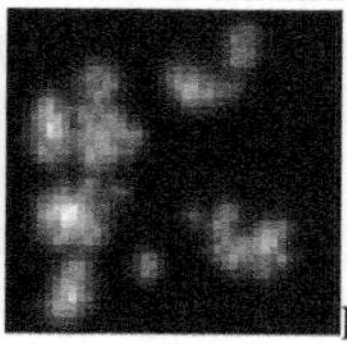 P C=0.79	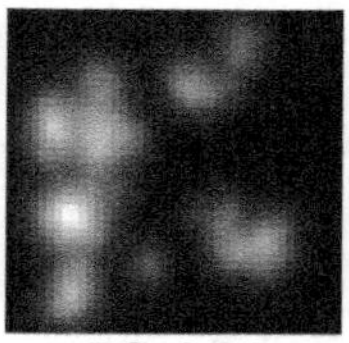 PC=0.86	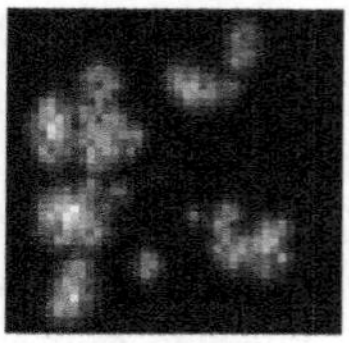 PC=0.75
Re-sampled template: T_2 (400%)	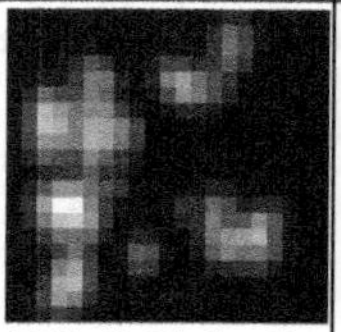 **PC=0.81 (x=3, y=12)**	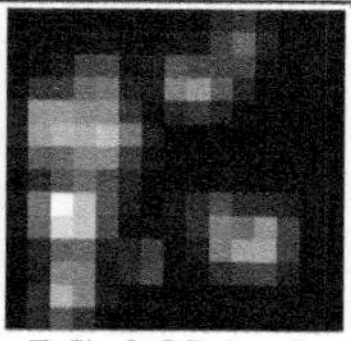 **PC=0.85 (x=2, y=8)**	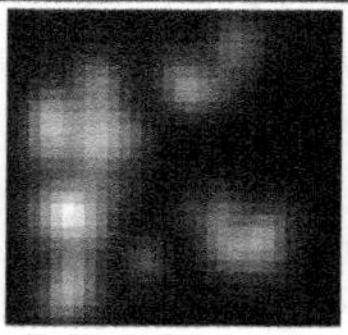 **PC=0.86 (x=5, y=17)**	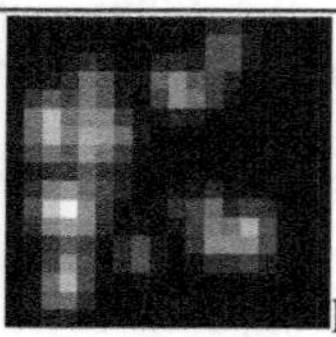 **P C=0.82 (x=3, y=10)**

Table 3. Effect of a measured system MTF_{SYSm} and of a calculated system MTFsysc (from measured detector MTF and calculated scatter MTF, MTFsysc=MTFdetec*MTFscatter) on the contrast of a simulated clusters of microcalcifications. The first column reports the results for MTFsysm while the third column reports the results obtained from MTFsysc. All abbreviations are like in Table 1.

Input system	27 kV MoMo, 4 cm PMMA, System A 70μm, MTF_A m_1=2	
Input Template T_1	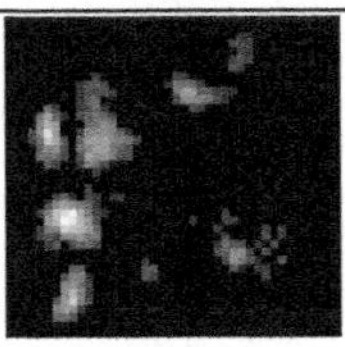 **PC= 0.79, (x=6,** 24)	
1-Ideal template IT_1 (200%)	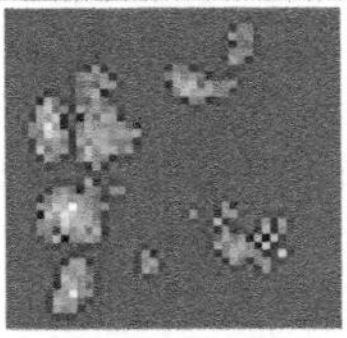 PC= 0.34	
Al equivalence AE_1 (200%)	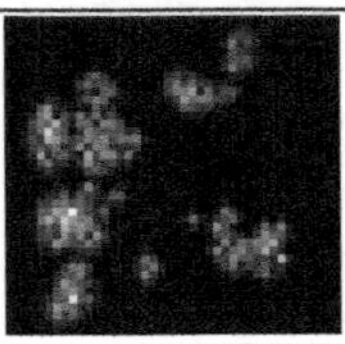 PC=0.47	
Target systems	28 kV WRh, 4 cm PMMA, System A 70μm, MTF_{SYSm} m_2=1	28 kV WRh, 4 cm PMMA, System A 70μm, MTF_{sysc} m_2=1
Correction exposure settings: IT_2 (200%)	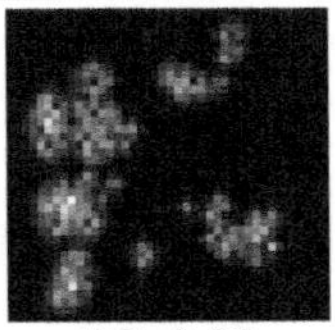 PC=0.66	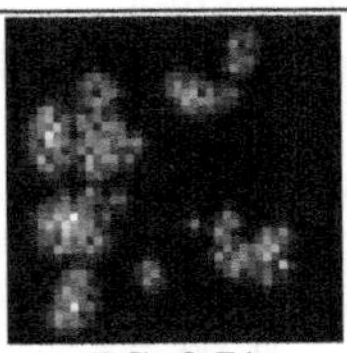 PC=0.71
Correction for target MTF: T_{2MTF2} (200%)	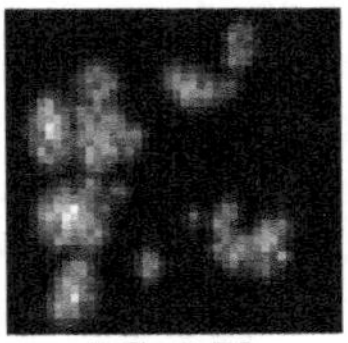 PC=0.79	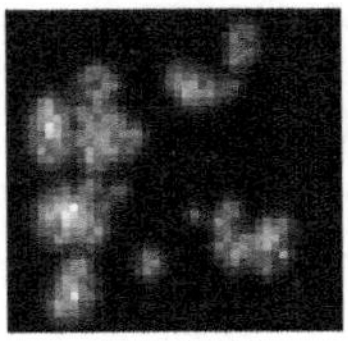 PC=0.79
Re-sampled template: T_2 (400%)	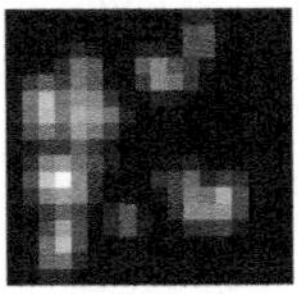 **PC=0.84 (x=3, y=10)**	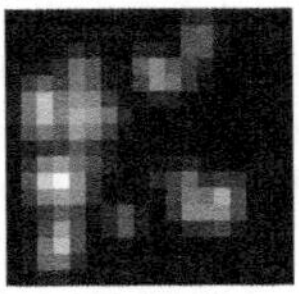 **PC=0.84 (x=3, y=10)**

identical target condition, but using in one case (second column) the system MTF calculated following the edge method as proposed in [7] (and as currently implemented in *mctam*) and in the other case a fitted system MTF, calculated as cascade of measured detector MTF and calculated scatter MTF [8]. Results show that there is no change in contrast nor shift in the PC. The approximation of using a system MTF calculated as in [7] seems therefore not to introduce a large error, although further investigation may be necessary.

4 Conclusion

A software tool is developed to allow for the generation of templates that represent a specific cluster of microcalcifications as originally imaged and adapted to any target system or exposure setting. Evaluating the simulation procedure for different target systems showed that the detector MTF and pixel size have the most effect on the final template in terms of PC and its spatial position within the template image. Which of these templates is best visible after their insertion in real mammogram is an open question, especially when sophisticated image post-processing is applied on the raw images.

Acknowledgment. Part of the project was supported by MEVIC, an IBBT-project.

References

1. Carton, A.K., Bosmans, H., Van ongeval, C., Souverijns, G., Rogge, F., Van Steen, A., Marchal, G.: Development and validation of a simulation procedure to study the visibility of micro calcifications in digital mammograms. Med. Phys. 30, 2234–2240 (2003)
2. Saunders, R., Samei, E., Baker, J., Delong, D.: Simulation of mammographic lesions. Acad. Radiol. 13, 860–870 (2006)
3. Zanca, F., Chakraborty, D.P., van Ongeval, C., Jacobs, F., Claus, G., Marchal, G., Bosmans, H.: An improved method for simulating microcalcifications in digital mammograms. Med. Phys. 35, 4012–4018 (2008)
4. Zanca, F., Van Ongeval, C., Jacobs, J., Marchal, G., Bosmans, H.: A quantitative method for evaluating the detectability of lesions in digital mammography. Radiat. Prot. Dosimetry. 129, 214–218 (2008)
5. Zanca, F., Van Ongeval, C., Marshall, N., Meylaers, T., Michielson, K., Marchal, G., Bosmans, H.: The relationship between the attenuation properties of breast microcalcifications and aluminium. Phys. Med. Biol. 55(4), 1057–1068 (2010)
6. Le Gal, M., Chavanne, G., Pellier, D.: Diagnostic value of clustered microcalcifications discovered by mammography. Bull. Cancer (Paris) 71, 57–64 (1984)
7. Carton, A.K., Vandenbroucke, D., Struye, L., Maidment, A.D., Kao, Y.H., Albert, M., Bosmans, H., Marchal, G.: Validation of MTF measurement for digital mammography quality control. Med. Phys. 32, 1684–1695 (2005)
8. Kyprianou, I.S., Rudin, S., Bednarek, D.R., Hoffmann, K.R.: Generalizing the MTF and DQE to include x-ray scatter and focal spot unsharpness: application to a new microangiographic system. Med. Phys. 32(2), 613–626 (2005)

Improved Microcalcification Detection for Breast Tomosynthesis Using a Penalized-Maximum-Likelihood Reconstruction Method

Mini Das, Howard C. Gifford, J. Michael O'Connor, and Stephen J. Glick

Department of Radiology, University of Massachusetts Medical School, Worcester, MA 01655
mini.das@umassmed.edu

Abstract. In this paper we explore the use of a penalized maximum likelihood (PML) based reconstruction method to improve the image quality and microcalcification detectability in digital breast tomosythesis (DBT). To evaluate performance, a human observer psychophysical study was performed with computer simulated images. The simulation used realistic structured breast models derived from CT scans of surgical mastectomy specimens giving sufficient statistical variability in terms of breast background structural noise. Sensitivity and specificity of microcalcification detectability measured with PML reconstruction was compared to that obtained with the filtered back projection (FBP) method for simulated breast tomosynthesis images. An observer study conducted using localized receiver operating characteristic (LROC) analysis showed significantly better sensitivity and specificity using the PML reconstruction method for simulated mean glandular dose levels of 1.0 mGy for a 5 cm compressed breast. This study suggests that MC detection accuracy is improved using PML reconstruction technique and that it might be feasible to reduce the imaging dose of DBT using this technique.

Keywords: Breast Tomosynthesis, Reconstruction techniques, Dose.

1 Introduction

DBT is a 3D imaging modality for breast which reduces tissue overlap and is hence expected to enhance the visibility of malignancies in comparison to digital mammography (DM). In DBT, the projections are acquired over a limited angular range leading to an ill conditioned and underestimated inverse problem. One of the main issues that may stand against DBT being accepted as a screening modality would be its possibly low performance in detecting small microcalcifications (MC) [1]. Although there are several factors like patient motion, misalignment and reconstruction artifacts which could contribute to MCs being blurred or hard to detect in the 3D image, it is important to make sure that the noise in the reconstruction itself is low enough to improve MC detectability if the above mentioned issues were absent or could be corrected for. FBP has been the most commonly used method for such tomographic reconstructions which yield noisy images when the acquisition dose is low. Heavy post reconstruction filtering operations can reduce noise but will then make it hard to see

J. Martí et al. (Eds.): IWDM 2010, LNCS 6136, pp. 697–703, 2010.

small microcalcifications. Recently, some research groups have proposed and implemented iterative reconstruction methods for DBT that seem to improve the visibility of MCs [2, 3]. In this paper we use a penalized maximum likelihood method and compare it with FBP. In this paper, the merit of each of these techniques is assessed by a human observer study performed using an ensemble of simulated images for various dose levels of acquisition.

2 Methods

To setup a clinically realistic psychophysical human observer study, we used a rigorous computer simulation that incorporated realistic models of the x-ray spectra, breast anatomy, and the signal and noise transport through the breast and an indirect conversion (CsI based) flat-panel detector. Focal spot and detector blurring were also modeled. In this section, each of these components of our DBT system simulation is described as are our 3D models of MC clusters (MCC).

2a. Breast and MC Model

We have generated an ensemble of realistic looking breast models from the CT scans of freshly obtained surgical mastectomy specimens[4]. For this particular study we have used 24 compressed breast models each one of 5 cm thickness. Fig. 1 shows one example of a compressed breast model (coronal, transverse, and saggital slices), and the resulting simulated mammogram generated from this phantom. Simulated MC clusters were distributed randomly in these compressed breast phantoms. At least 2 MCC's were located in each breast phantom. Some of the phantoms had MCC's located in 3 independent slices. X-ray attenuation properties of hydroxyapatite were used to model attenuation of malignant MCC's. There were 7 MCs in each cluster where each microcalcification was modeled as a sphere with diameter of size 150 microns.

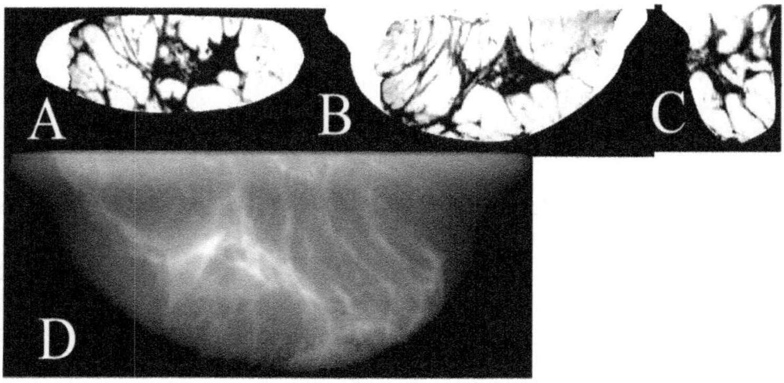

Fig. 1. A,B,C) Three orthogonal slices of a compressed breast phantom displayed as a 'negative' (adipose tissue values light and glandular values dark). D is a simulated mammogram using this compressed phantom. The simulation technique used a Mo/Mo 30kVp spectra at 1.5mGy mean glandular dose with simulated detector characteristics of 100 micron^2 pixel and CsI thickness of 100 microns.

2b. X-Ray Spectrum and Detector

The X-ray spectrum was modeled as generated from a molybdenum anode source, and was scaled to provide the desired total mean glandular dose (MGD) to a 5 cm thick compressed breast at 30 kVp. Focal spot blurring was modeled using a Gaussian function and a 300 micron focal spot. Breast tomosynthesis using a rotating source-and-detector geometry was modeled, with 21 projection views obtained over a 60 degree angular range. Propagation of the signal and noise through an indirect 100 micron thick CsI based flat-panel detector with 100 micron pixel size was simulated using a serial cascade mode [5]. The scintillator blurring was modeled using an empirically measured pre-sampling MTF.

2c. Projections

Projections were obtained using Siddon's ray tracing algorithm with attenuation coefficients for adipose tissue and fibroglandular tissue obtained from the empirical measurements [6]. A projection set with MGD of 1.0 mGy was generated. The total dose was distributed equally between the 21 projections.

2d. Reconstruction

Each projection set was reconstructed using: 1) a PML objective function which was maximized using a separable paraboloidal surrogate (SPS) coordinate ascent algorithm and 2) a filtered-backprojection (FBP) algorithm using 3D post-reconstruction Butterworth filtering applied for regularization. Reconstructions were made with rectangular voxels of 1 mm thickness and 100 micron in plane resolution.

Penalized Maximum Likelihood Algorithm (PML): The penalized likelihood method was adapted from Fessler et.al [7] and uses a **separable paraboloidal surrogate (SPS)** optimization transfer principle to maximize the objective function. In this paper we use a penalized maximum likelihood objective function where the roughness penalty and the edge preserving threshold can be controlled. An estimate μ of the true attenuation coefficient map μ_{true} is obtained using a method that maximises a penalized likelihood objective function of the form,

$$\mu \cong \arg\max(\Phi(\mu))_{\mu \geq 0}; \Phi(\mu) \cong L(\mu) - \beta R(\mu), \ldots\ldots\ldots \quad .(1)$$

where the objective function Φ includes a roughness penalty $R(\mu)$. The parameter β controls the trade off between spatial resolution and noise. Since maximizing Φ is not an easy task due to its non-quadratic nature we have chosen to replace Φ with a paraboloidal surrogate function that is easier to maximize. Several authors have proposed the use of various forms of surrogate functions. Here we have chosen a Huber function that is of the form $\phi(\mu) = \beta\sqrt{1+\frac{x^2}{u^2}}$ where β is the smoothness prior and u controls the threshold in edge preserving. The simultaneous update algorithm updates all the pixels simultaneously with feedback from the backprojection method. The final form of the paraboloidal surrogate algorithm used can be seen in [7] (Chapter 6, sec 6.2 Eq. 99). We noticed that the smoothness prior and the roughness penalty do not impact tomosynthesis images drastically for a wide range of values tested, unlike in the case of a CT image with more complete angular sampling. The smoothness effect observed for the range of β and u values are very small as opposed to that observed in CT images. We have chosen the 69th iteration from each reconstruction.

Filtered back projection (FBP): Another reconstruction method tested was the widely used FBP. A Feldkamp FBP algorithm was used with a ramp filter. For each FBP reconstructed volume a 3D Butterworth filter was applied as a post reconstruction step. The cut off frequency of Butterworth filter was subjectively selected such that MC visibility was the best while maintaining the visibility of the background structures. Very heavy smoothing could reduce the overall noise in the image but seemed to reduce MC visibility. A cut off frequency of 0.4 /pixel was used for these simulations. This was determined using a pilot observer study and was shown to yield the best MC detectability. A lower cut off frequency (like 0.25 /pixel) seemed to yield a smoother or visually appealing image which might be better for mass detectability. A cut off frequency higher than 0.4 /pixel yielded very noisy image and again decreased the detectability of MCs.

2e. LROC Observer Study

An LROC observer study performed using 20 training images and 98 study images in each session presented single slices to the observer with 4 confidence ratings to choose from. The observer had to locate the MCC and rate their confidence level of its presence. Half of the study images had no MC cluster present. During the training session, feedback was provided to the reader after each selection. Immediately after the training session was completed, the comparison study was performed. The observer confidence rating and the suspected MCC location were recorded for all the images displayed. Swensson's method [8] was used to fit the data and obtain the area under the LROC curve. Four observers participated in this study.

A two-way Analysis of Variance (ANOVA) study was performed to estimate the statistical significance of the differences in techniques and also to estimate the statistical significance in the performance difference between the 3 observers.

3 Results and Discussion

Images obtained from PML reconstruction appeared to have lower noise and better visibility of MCC's in comparison to those obtained using FBP. We performed pilot studies for dose levels higher than 1.5 mGy and found that both techniques performed equally well in MC detection. We decided to do a larger observer study for acquisition dose of 1.0 mGy which is lower than the current dose used by some clinical systems [9, 10] to image a 5 cm compressed breast. Fig. 2 shows a comparison of PML vs FBP when the imaging dose was 2.5 mGy. Fig. 3 shows a comparison of images

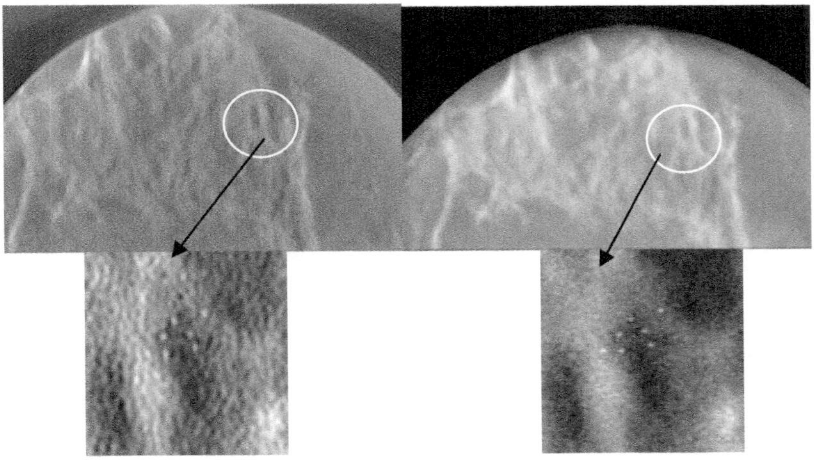

Fig. 2. Comparison of FBP (left) vs PML (right) reconstruction for simulated MGD of 2.5 mGy

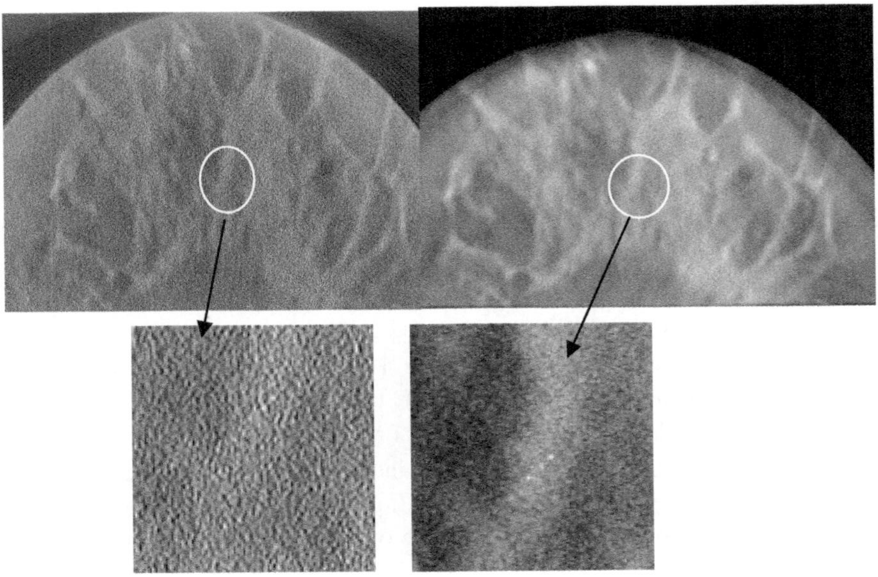

Fig. 3. Comparison of FBP (left) vs PML (right) reconstruction for MGD of 1.0 mGy

when the imaging dose is 1.0 mGy. It can be seen that with a decrease in the total imaging dose, the noise is higher in the reconstructed image making the visibility of MCs more challenging. Visibility of MCs also depends on the breast density and the location in breast where the MCs are present.

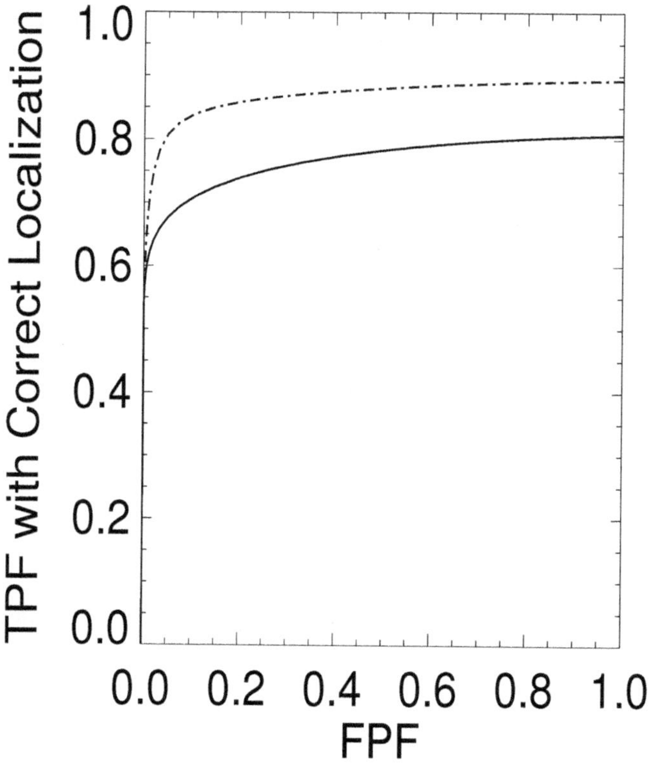

Fig. 4. Comparing the average LROC curves for all 4 observers for PML (dashed) vs FBP(solid) for 1 mGy acquisition

Figure 4 shows the average LROC curves for all four observers plotted for both PML and FBP reconstruction methods for the 1 mGy acquisition dose. The sensitivity and specificity obtained using PML technique is higher than that obtained using FBP. The average area under the LROC curves for PML technique was 0.93 ± 0.05 and for FBP technique was 0.7 ± 0.07. The uncertainties represent the standard deviation in observer performance. A two way ANOVA found statistically significant difference (at a significance level of 0.05) between the two reconstruction strategies with a p= 0.014 for the dose level of 1.0 mGy. There was no statistically significant difference (at a significance level of 0.05) in observer performance for each study with a p value of 0.48

4 Conclusions

Our study shows that images generated using PML reconstruction technique resulted in improved MC detectability in comparison to those generated using FBP. An LROC study conducted using four observers for an acquisition dose level of 1.0 mGy yielded an average area under the LROC curve of 0.93 and 0.7 for PML and FBP techniques respectively. We have chosen to test acquisitions with 1 mGy dose since it this is a bit lower than the standard breast imaging dose level and presents a more challenging scenario. A 2 way ANOVA study suggested statistically significant difference between the perfomance of the two strategies. Studies with multiple dose levels of acquisition will be conducted in future. In conclusion, our results are very encouraging and point towards the possibility of improving MC detectability using PML techniques and also towards the possibility of reducing the imaging dose. A more detailed study which also incorporates low contrast mass detection along with MC detection will be conducted in future.

Acknowledgements

This research was supported by the National Cancer Institute under grants NIH-K25CA140858-01 and NIH-CA102758. The contents of this paper are solely the responsibility of the authors and do not represent the official views of the NIH.

References

1. Paplack, S.P., Tosteson, T.D., Kogel, C.A., Nagy, H.M.: Digital Breast Tomosynthesis: Initial experience in 98 women with abnormal digital screening mammography. AJR 189, 616–623 (2007)
2. Sidky, E.Y., Pan, X., Reiser, I.S., Nishikawa, R.M., Moore, R.H., Kopans, D.: Enhanced imaging of microcalcifications in digital breast tomosythesis through improved image-reconstruction algorithms. Medical Physics 36(11), 4920–4932 (2009)
3. Wu, T., Moore, R.H., Rafferty, E.A., Kopans, D.B.: A comparison of reconstruction algorithms for breast tomosynthesis. Med. Phys. 31(9), 2636–2647 (2004)
4. O'Connor, J.M., Das, M., Didier, C., Mah'D, M., Glick, S.J.: Using mastectomy specimens to develop breast models for breast tomosynthesis and CT breast imaging. In: Medical Imaging 2008: Physics of Medical Imaging. SPIE, San Diego (2008)
5. Siewerdsen, J.H., Antonuk, L.E., et al.: Empirical and theoretical investigation of the noise performance of indirect detection, active matrix flat-panel imagers (AMFPIs) for diagnostic radiology. Med. Phys. 24(1), 71–89 (1997)
6. Johns, P.C., Yaffe, M.J.: X-ray characterisation of normal and neoplastic breast tissues. Phys. Med. Biol. 32(6), 675–695 (1987)
7. Fessler, J.A.: Statistical image reconstruction methods for transmission tomography. In: Sonka, M., Fitzpatrick, J.M. (eds.) Handbook of Medical Imaging, Medical Imaging. Processing and Analysis, vol. 2. SPIE, Bellingham (2000)
8. Swensson, R.G.: Unified measurement of observer performance in detecting and localizing target objects on images. Med. Phys. 23(10), 1709–1725 (1996)
9. Gentry, J.R., DeWard, L.A.: TLD measurements of in vivo mammographic exposures and the calculated mean glandular dose across the united states. Med. Phys. 23(6), 899–903 (1996)
10. Chevalier, M., Moran, P., Ten, J.I., Soto, J.M.F., Capeda, T., Vano, E.: Patient dose in digital mammography. Med. Phys. 31(9), 2471–2479 (2004)

Microcalcification Detection in Digital Breast Tomosynthesis Using an Epipolar Curve Approach

Candy P.S. Ho[1], Christopher E. Tromans[1], Julia A. Schnabel[2], and Sir Michael Brady[1]

[1] Wolfson Medical Vision Laboratory,
Robotics Research Group, Department of Engineering Science,
University of Oxford, Parks Road, Oxford OX1 3PJ, UK
[2] Institute of Biomedical Engineering, Department of Engineering Science, Old Road Campus Research Building, University of Oxford, Headington, Oxford OX3 7DQ, UK
{psho,cet,jmb}@robots.ox.ac.uk
{julia.schnabel}@eng.ox.ac.uk

Abstract. The detection of microcalcifications is a key task in the early detection of breast cancer. Digital breast tomosynthesis (DBT) offers new opportunities to improve existing microcalcification detection methods. By utilizing the multiple projections in DBT, and a model of the DBT acquisition system, we propose the use of epipolar curves to constrain the position of a microcalcification in the multiple DBT views. We show how this can improve both the sensitivity and specification of microcalcification detection.

Keywords: digital breast tomosynthesis, microcalcifications, detection, epipolar curves, standard attenuation rate, DBT.

1 Introduction

The detection of microcalcifications is a key task in the early detection of breast cancer. Existing techniques e.g. local adaptive thresholding and feature extraction [1], machine learning approaches [2], and the physics-based approach based on breast composition [3], all assume that there are at most two X-ray projections in order to detect microcalcifications. However, a single 2D mammogram cannot represent a complete picture of the 3D breast, tissue superimposition being one of the major limitations of mammography. All microcalcification detection algorithms (published or commercially confidential) have limited sensitivity and specificity. This paper shows how digital breast tomosynthesis (DBT) provides a novel opportunity to improve microcalcification detection, yielding improved sensitivity and specificity.

DBT generates multiple projections, for example by rotating the X-ray arm over a limited angular range. Microcalcifications that are hidden in conventional mammography may be visible in some DBT projections. There are two obvious approaches to microcalcification detection in DBT: using reconstructed slices [4]; or analyzing the individual projections, followed by reconstruction [5]. In this paper, we adopt the latter approach; but stress that it can be combined with the former in an expectation-maximisation framework. A substantial advantage of the approach taken here is that we can detect microcalcifications and the clusters they form without reconstruction, since the latter requires considerable further development.

J. Martí et al. (Eds.): IWDM 2010, LNCS 6136, pp. 704–711, 2010.

2 Methods

We sub-divide the detection process into two steps: microcalcification detection in the individual projections; and combination of the individual projection detection results using the concept of epipolar curves, that are familiar from computer vision in stereo or structure from motion. The two steps are iterative in the sense that the latter step may help identify a number of noise points or missing microcalcifications during detection in some projections in the first step and this subsequently can further improve whatever detection algorithm is employed.

Our emphasis in this paper is the derivation of epipolar curves and their application in the detection of mirocalcifications. In the first part, for the purpose of completeness, we outline one of the detection algorithms using breast composition approach we are currently working on. It is noted that further study is required for optimal detection results in real clinical use. We then focus on our epipolar curve approach in the second part. We first present its derivation, based on a plausible DBT geometry, though again we emphasise that this is merely for illustration; the method can straightforwardly be adapted to any geometry. Finally, we demonstrate how the epipolar curves can help locate the true microcalcification points (true positives), noise points (false positives), missing points (false negatives) in the detection process.

2.1 Microcalcification Detection in Individual Projections

Notwithstanding the decreased signal-to-noise ratio of DBT images, existing detection algorithms in mammography can be extended and used straightforwardly in DBT, though of course with modification. Our current work is based on our previous work aimed at determining breast composition. Tromans et al. [6] have introduced the standard attenuation rate, which measures the ratio of any suitable reference material (which may be, for example, a chosen proportion of adipose to fibroglandular tissues) to the thickness of the breast along the X-ray path to the pixel location. In simple terms, the standard attenuation rate provides the sum of the breast composition along the path to the pixel location including the sums of all forms of adipose tissue, fibroglandular tissues and calcifications. Its value is smaller if the path contains more adipose tissue, larger for dense tissues, much larger for calcifications.

Our detection algorithm, based on that of Yam et. al. [3], starts by identifying suspected microcalcification regions from each projection, by extracting those pixels with locally high contrast values. Morphological operations close the holes within such regions and remove isolated pixels, which are unlikely to be microcalcifications. The standard attenuation rate is used to calculate a number of volume measures: v_{ref}, the volume of the reference material of the region; v_{surr}, the volume of the surrounding tissues ; $v_{diff(ref-surr)}$, the difference in these volume measures, which gives the additional volume in the region when compared with the surrounding. Those regions for which $v_{diff(ref-surr)}$ exceeds a preset threshold are the candidate microcalcification regions that are further analysed using epipolar curves.

2.2 Epipolar Curves

DBT generates multiple 2D projections of a 3D breast. By analyzing the geometry of the DBT acquisition system, we can derive the "epipolar curve" formed from the 2D projection in the multiple projections in DBT generated from the same 3D point within the breast. Fig. 1 shows the geometry of a simplified version of a plausible DBT acquisition system, though, as we noted earlier, this is presented simply as an illustration and our approach can readily be extended to any real system, in the same way that any published stereovision algorithm can be realised in conjunction with a suitable calibration and image distortion reduction algorithm. The hypothetical DBT system consists of a sensor plane *S*, a pivot point *P* about which the X-ray arm rotates and the focal spot f_i. *L* is the height of *P* above the sensor plane, and *D* is the length of the X-ray arm. Two coordinate frames are used during the derivation of the epipolar curve: the world coordinate frame *w* with origin at *P* and the sensor coordinate frame *s* with origin at O_s.

Next, we derive the equations of the projection points given a 3D position of microcalcification **m** at $(a, b, c)^w$ and θ_i. Of course, subsequently, we will use this analysis in an inverse sense to detect microcalcifications and their positions in 3D.

The line that contains $f_0, \mathbf{m}$ is given (in world coordinates - superscript *w*) by:

$$\mathbf{x}_0^w = f_0 + \lambda(\mathbf{m} - f_0) = \begin{bmatrix} 0 \\ 0 \\ -D \end{bmatrix} + \lambda \begin{bmatrix} a \\ b \\ c+D \end{bmatrix} \tag{1}$$

This line intersects the sensor plane $z^w = L$ when $\lambda = \dfrac{L+D}{c+D}$. Inserting this into the above equation:

$$\mathbf{x}_0^w = \begin{bmatrix} a\left(\dfrac{L+D}{c+D}\right) \\ b\left(\dfrac{L+D}{c+D}\right) \\ L \end{bmatrix}$$

$$\mathbf{x}_0^s = \begin{bmatrix} a\left(\dfrac{L+D}{c+D}\right)+u \\ b\left(\dfrac{L+D}{c+D}\right)+v \end{bmatrix} \tag{2}$$

Now consider projection points from equal and opposite focal points f_{+i}, f_{-i}. It is straightforward to show that:

$$\mathbf{x}^{s}_{+i} = \begin{bmatrix} D\sin\theta_i + \left(\dfrac{L + D\cos\theta_i}{c + D\cos\theta_i}\right)(a - D\sin\theta_i) + u \\ b\left(\dfrac{L + D\cos\theta_i}{c + D\cos\theta_i}\right) + v \end{bmatrix}$$
$$\mathbf{x}^{s}_{-i} = \begin{bmatrix} -D\sin\theta_i + \left(\dfrac{L + D\cos\theta_i}{c + D\cos\theta_i}\right)(a + D\sin\theta_i) + u \\ b\left(\dfrac{L + D\cos\theta_i}{c + D\cos\theta_i}\right) + v \end{bmatrix} \quad (3)$$

Finally, for the focal spots at θ_i and - θ_i, the 2D coordinates in the sensor coordinate frame are represented by the vectors x^{s}_{+i} and x^{s}_{-i} given the microcalcification point positioned at *(a, b, c)* as follows:

$$\mathbf{x}^{s}_{+i} = \begin{bmatrix} a\left(\dfrac{L + D\cos\theta_i}{c + D\cos\theta_i}\right) + u \\ b\left(\dfrac{L + D\cos\theta_i}{c + D\cos\theta_i}\right) + v \end{bmatrix} + \frac{D\sin\theta_i}{c + D\cos\theta_i}\begin{bmatrix} c - L \\ 0 \end{bmatrix} = \mathbf{w}^{s}_{i} + \mathbf{u}^{s}_{i}$$
$$\mathbf{x}^{s}_{-i} = \begin{bmatrix} a\left(\dfrac{L + D\cos\theta_i}{c + D\cos\theta_i}\right) + u \\ b\left(\dfrac{L + D\cos\theta_i}{c + D\cos\theta_i}\right) + v \end{bmatrix} - \frac{D\sin\theta_i}{c + D\cos\theta_i}\begin{bmatrix} c - L \\ 0 \end{bmatrix} = \mathbf{w}^{s}_{i} - \mathbf{u}^{s}_{i} \quad (4)$$

This is the epipolar curve at *(a, b, c)*, named by analogy with stereovision and structure from motion. An example of an epipolar curve representing a microcalcification is shown in Fig. 2.

Knowing that a microcalcification in the breast can be represented as an epipolar curve, it is straightforward to use epipolar curves to represent a cluster of microcalcifications (Fig. 3 (Top)). In reality, where there is substantial noise in each acquisition image, and where the detection algorithms are in consequence less than perfect, it may be expected that some DBT views have noise points and some microcalcifications are missed in some DBT views. In such cases, our epipolar curves enable identification of noise points and missing microcalcification points by putting all the microcalcification points from different projections found in the detection step into the same 2D coordinate frame (See Fig. 3 (Middle and Bottom).).

Currently we choose the detector plane as our 2D coordinate frame; but, of course, other choices are possible, and are related to the one we propose by a simple homography. The steps of detection and epipolar clustering can be iterated.

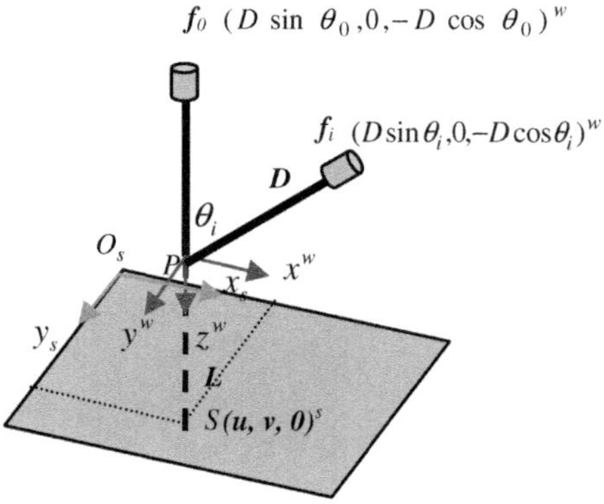

Fig. 1. Geometry of DBT

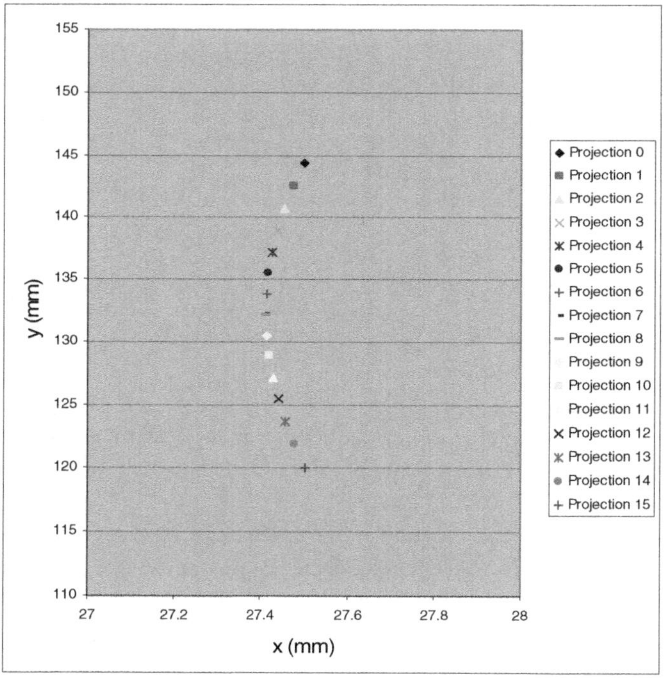

Fig. 2. An example of an epipolar curve representing a microcalcification

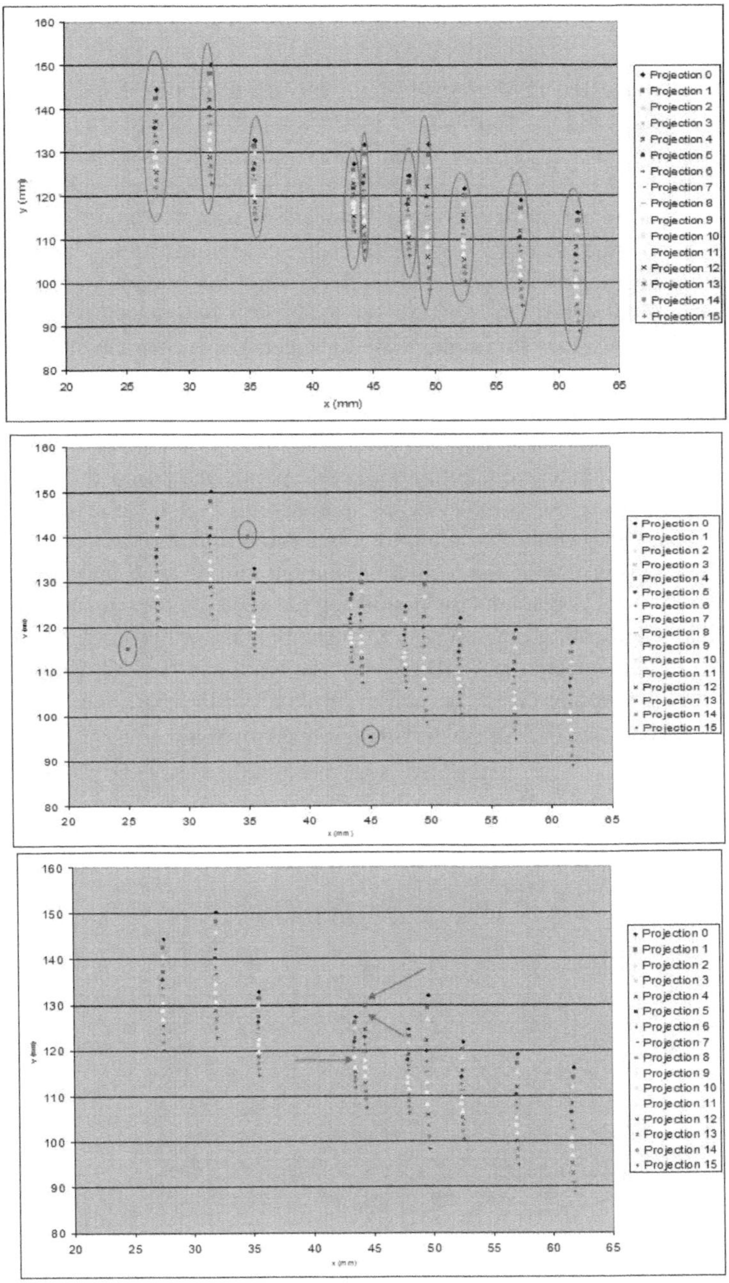

Fig. 3. Top: An example of 10 epipolar curves representing a cluster of 10 microcalcifications in the ideal case (One red circle represents one microcalcification) (True Positives). Middle: The same example with noise points detected in some projections (indicated by red circles) (False Positives). Bottom: The same example but some points cannot be detected in some projections (indicated by red arrows) (False Negatives).

3 Results

In order to evaluate the method against ground truth, testing has been carried out both on simulated DBT projections, as generated by the X-ray simulation software developed by Tromans et al [6], and real X-ray projections of a phantom comprised of perspex with *12* carefully drilled holes containing Calcium substances. Initial results show that detection in individual projections using standard attenuation rate approach is promising, in that *11* of the *12* Calcium regions can be distinguished out of *289* regions segmented as suspected regions, while the remaining one can also be extracted with *7* false positive regions. As mentioned before, we only demonstrate the feasibility of detection in the individual projections in this paper. Further development in the detection algorithms is required.

To demonstrate the use of epipolar curves, we generated *15* simulated DBT views containing *2* clusters of microcalcifications. By putting all the detection results in the same 2D coordinate plane, we identify both noise points and points that are missing in some projections using our epipolar curve approach. In Fig. 4, we show the detection result of *7* DBT views (tomo4, tomo5, tomo6, tomo7, tomo8, tomo9, tomo10) showing one of the clusters of *15* microcalcifications. It can be seen that epipolar curves can improve the detection results by identifying the noise points (isolated points in the figure.) and supplementing the failure in the detection algorithms (e.g. Microcalcification *14* can only be detected in *5* DBT views and is missed in DBT views tomo8 and tomo9. With the epipolar curve, we can even point out the expected locations of the missing points in those DBT views in which they are missed.)

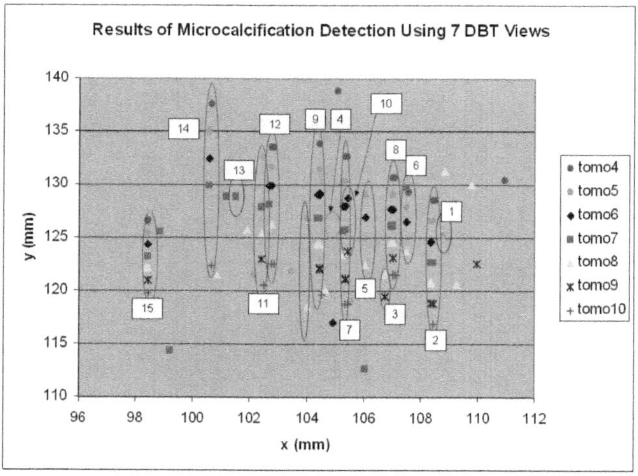

Fig. 4. Detection results of a cluster of 15 microcalcifications using 7 DBT views. (tomo4-tomo10 represents the chosen middle 7 DBT views out of the 15 views; No. in the rectangles are the microcalcification identifiers.) (Red Circles: True Positives; No Circle: True Negatives; Black Circles: False Positives; Green/Blue Circles: False Negatives).

4 Discussion

DBT provides significantly more information than mammography. This offers new opportunities to improve existing microcalcification detection methods. In this paper, we presented evidence that epipolar curves enable the determination of noise points detected (false positives) and the identification of missing microcalcification points (false negatives) in some projections. By combining detection in individual projections and epipolar curves approach iteratively, the detection algorithms can be further fine-tuned and improved detection results can be achieved.

In order to automate the detection processes using epipolar curves, we have also investigated a simple clustering algorithm, an epipolar clustering algorithm, to group microcalcifications, to identify noise points and missing microcalcification points. Readers who are interested can refer to [7].

Acknowledgments. This work forms part of a project funded by the Technology Strategy Board (TSB) and by the Engineering and Physical Sciences Research Council (EPSRC). CET is funded by an EPSRC Fellowship. Special thanks to all the involved parties in this project and people in the Wolfson Medical Vision Lab for their support and valuable inputs.

References

1. Davies, D.H., Dance, D.R.: Automatic computer detection of clustered calcifications in digital mammograms. Phys. Med. Biol. 35(8), 1111–1118 (1990)
2. El-Naqa, I., Yang, Y., Wernick, M.N., et al.: Support vector machine learning for detection of microcalcifications in mammograms. In: Proc. IEEE International Symposium on Biomedical Imaging, pp. 201–204 (2002)
3. Yam, M., Highnam, R., Brady, J.M.: Detecting calcifications using the hint representation. In: Proc. 13th International Conference on Computer-Assisted Radiology and Surgery, pp. 373–377 (1999)
4. Schie, G.v., Karssemeijer, N.: Noise model for microcalcification detection in reconstructed tomosynthesis slices. In: Proc. SPIE Medical Imaging, vol. 10(1) (2009)
5. Wheeler, F.W., Perera, A.G.A., Claus, B.E., et al.: Micro-calcification detection in digital tomosynthesis mammography. In: Proc. SPIE, vol. 6144, pp. 671–682 (2006)
6. Tromans, C., Brady, M.: The Standard Attenuation Rate for Quantitative Mammography. In: Proceedings of 10th International Workshop on Digital Mammography (IWDM), Girona, Spain (2010)
7. Ho, C.P.S., Tromans, C., Schnabel, J.A., Brady, M.: A clustering method for the extraction of microcalcifications using epipolar curves in digital breast tomosynthesis. In: Proceedings of 10th International Workshop on Digital Mammography (IWDM), Girona, Spain (2010)

Spiculated Lesions and Architectural Distortions Detection in Digital Breast Tomosynthesis Datasets

Giovanni Palma[1,2], Isabelle Bloch[2], and Serge Muller[1]

[1] GE Healthcare, 283 rue de la minière 78530 Buc, France
[2] Télécom ParisTech (ENST) CNRS LTCI, 46 rue Barrault, 75013 Paris, France
giovanni.palma@ge.com,
isabelle.bloch@enst.fr,
serge.muller@ge.com

Abstract. Digital Breast Tomosynthesis (DBT) is a new 3D imaging technique aiming at overcoming some limitations of mammography. A computer aided detection system may help the radiologist to process the increased amount of data of this new modality. In this paper we propose to address the detection of masses and architectural distortions in DBT datasets. To achieve this task, we propose a detection scheme composed of two separate channels, each of them being dedicated to the detection of one of the target radiological signs.

We propose a description of these channels as well as a validation on clinical data. We also compare the performance with existing approaches.

1 Introduction

Mammography is a widely used technique to diagnose breast cancer. Nonetheless, due to the nature of these images, superimposition of tissues may lead to obscured lesions or false alarms. Digital Breast Tomosynthesis (DBT) is a new 3D imaging technique that potentially overcomes this limitation, but that also increases the amount of data to be reviewed by the radiologist. A computer aided detection (CAD) system dedicated to this new kind of data may help the radiologist to achieve his detection task, and increase his sensitivity.

In this paper we focus on the detection of masses and architectural distortions, which are suggestive of malignancy. This detection is performed with two independent channels, whose results are aggregated afterward. We also validate our algorithms on a database composed of 101 breast volumes (53 containing one or more biopsy proven lesions, and 48 containing no lesion).

First, we describe the method we implemented to perform the detection task, then we expose and discuss the performance the proposed approach achieved.

2 Methods

As said previously, our approach is composed of two channels dedicated to the detection of dense kernels and convergence areas, respectively. These two radiological signs are likely to represent masses and architectural distortions.

J. Martí et al. (Eds.): IWDM 2010, LNCS 6136, pp. 712–719, 2010.

2.1 Masses Detection

This channel is composed of several steps. First we process the volume slice by slice in order to detect focal-densities. This detection is performed using a fuzzy connected filter as introduced in [1], whose discriminant attributes are size, compacity and contrast-based measures. This kind of filters is suitable to mark potential masses because of their great variability. Actually, fuzzy connected filters allow to define non crisp criterion thresholds, which can help in dealing with border line structures. In the next step, the most suspicious regions are selected and grouped in 3D through a pseudo-connected component labeling. The main difference with a regular connected component labeling is that we introduce a maximum shape variability criterion to be met by the produced 3D connected components. This allows disconnecting some components and thus enables to ensure that each resulting one corresponds to only one potential lesion and does not aggregate two distinct structures together. Then for each suspicious region, the most representative slice is selected and segmented using a dynamic programming segmentation approach [2,3]. Finally, some attributes are extracted in order to feed a SVM classifier [4], which will provide the final decision: the current region is or is not suspicious. Attributes that were used mainly rely on morphological properties of the findings (compactness, size, etc.) and on statistical analysis of its neighborhood (probability of suspicious convergence, statistical measures on orientations, etc.).

2.2 Architectural Distortions Detection

The second channel aims at detecting suspicious convergence areas. In order to achieve this task, we used an a contrario modeling of the problem [5,6]. Briefly the idea is to define a convergence measure similar to the one proposed in [7] and to select realizations in real images that are unlikely to appear in healthy breasts. This last detection step is performed again slice by slice. After applying an aggregation step, a feature extraction procedure associated with a classification stage using a SVM classifier is performed. This last step is done for each slice of each finding, thus each finding is considered suspicious if at least one slice is classified as suspicious by the classifier. The features used during this classification step mainly correspond to the analysis of orientations within the neighborhood of the finding.

2.3 Final Output

The two channels previously presented aim at detecting different types of radiological signs. For this reason, a disjunctive-like aggregation step is implemented to merge their results. Nonetheless, the border between these two types of population may not be well defined. For instance, a highly spiculated mass may in certain cases be detected by the convergence detection channel. For this reason, such an aggregation strategy may improve the performance of the overall chain in comparison with a simple combination of the performance of the two channels.

3 Results

We validated the previously described CAD system on clinical DBT volumes containing biopsy proven lesions. We discuss here the performance of each channel as well as the performance of the overall system.

3.1 Database

The database we used is composed of 101 breasts, 53 with a biopsy proven cancer and 48 with no pathology. Since DBT is a relatively recent technique, this is already a quite large database. DBT volumes corresponding to these breasts were reconstructed using iterative techniques [8] from low dose projections acquired over an angular range of 40 degrees. Slice interspacing was set to 1mm. The 53 cancerous breasts contained a total of 56 lesions, 7 irregulars masses, 4 lobulated masses, 39 spiculated masses, and 3 architectural distortions. Three more lesions were discarded because they were not representative. These lesions require a dedicated detection channels to be designed when more similar data will be available. The other lesions were used to assess the performance of our detection scheme. The density detection channel was evaluated using all irregular and lobutated, as well as the less spiculated masses (29 lesions). The remaining lesions were used to compute the performance of the second channel. The choice to split the spiculated lesions into two pools has been motivated by the idea that the main characteristic of such lesions is sometimes their stellate pattern rather than their density. More practically such a choice also enables to have a more accurate validation of the convergence detection channel, even if a 13 lesions database is not enough to be actually conclusive on its real performance. But still, its allows evaluating the validity of our processing.

3.2 Masses

The performance of the channel dedicated to the detection of masses is presented trough the FROC curve of Figure 1. This curve was obtained using cross-validation techniques on the first part of the database, and corresponds to the assessment of the whole chain (marker selection and false positives reduction). More precisely, the performance of the chain was obtained using a leave-one-out strategy. During the assessment of each finding (the one left out), the classifier was trained using a n-fold procedure in order to provide a good generalization of the learning database (all the cases except the one discarded). Now from a performance stand point, depending on the user needs, several operating points can be used. For instance, we can mention that for a sensitivity of 90%, this detection scheme achieves a specificity of 1.23 false positives per volume.

An example of detection is illustrated in Figure 2. This example proposes the detection map obtained as a result of the fuzzy connected filter and the final decision obtained after false positive reduction. The intermediate result can be interpreted as follows: each pixel is associated with a gray level, which codes a membership degree (black means 0 while white means 1). This membership

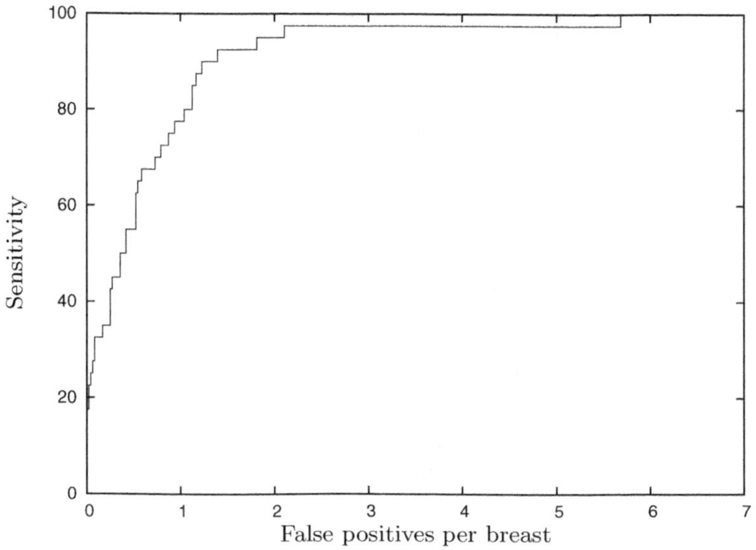

Fig. 1. FROC curve of the suspicious density detection channel

value is the degree to which the pixel is suspicious according to the filter. Now, on this particular example, we can note that the cancer is associated with a degree close but not equal to 1. This actually means that the values of the discriminant criterion were border line. Nonetheless, while the lesion does not perfectly fit the suspiciousness criterion, it is considered for further processing. The last image allows seeing the gain brought by the false positive removal stage. Here, we keep the lesion and one false positive. Let us finally state that while the example shows only one slice, 3D information is taken into account as mentioned earlier.

3.3 Convergence Patterns

The second channel was evaluated on the remaining lesions. The performance is illustrated in Figure 3. This FROC curve has also been obtained using the same cross-validation scheme as the one used for mass detection. Let us mention that this detection scheme achieved a sensitivity of 92% at 0.48 false positives per volume. However, since the amount of data is rather small, the confidence interval associated with the actual performance of this channel is probably pretty large. Nonetheless, this allows us to draw some intermediate conclusions on the validity of the approach.

An example of detection is illustrated in Figure 4. This example presents the result of the marker stage based on a contrario modeling as well as the final decision of the channel. Here the DBT slice contains an architectural distortion, which is retrieved by the marker stage and kept by the false positive removal stage. Example of false positives that are kept or suppressed during the whole processing are also provided.

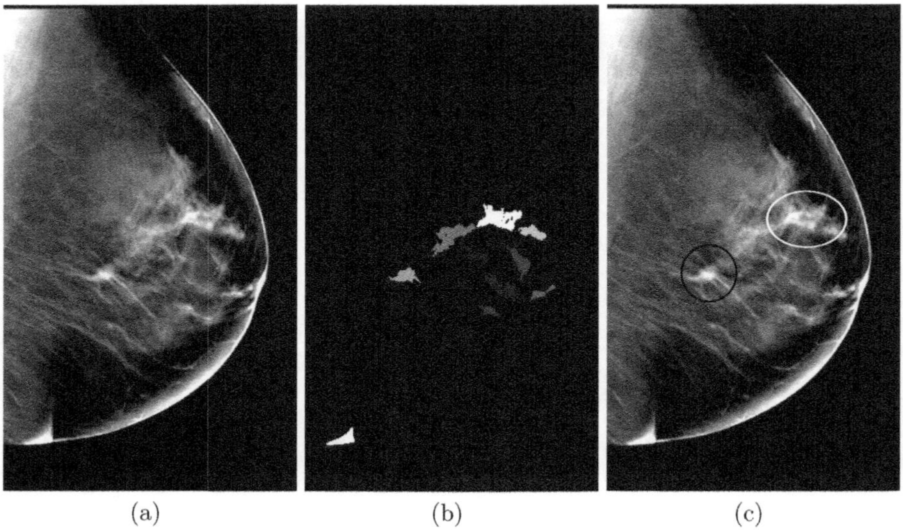

Fig. 2. Example of detection using the mass detection channel. (a) DBT slice containing a spiculated lesion. (b) Output (markers) of the fuzzy connected filter. (c) Final result: the lesion (in black) is detected as well as a false positive (in white).

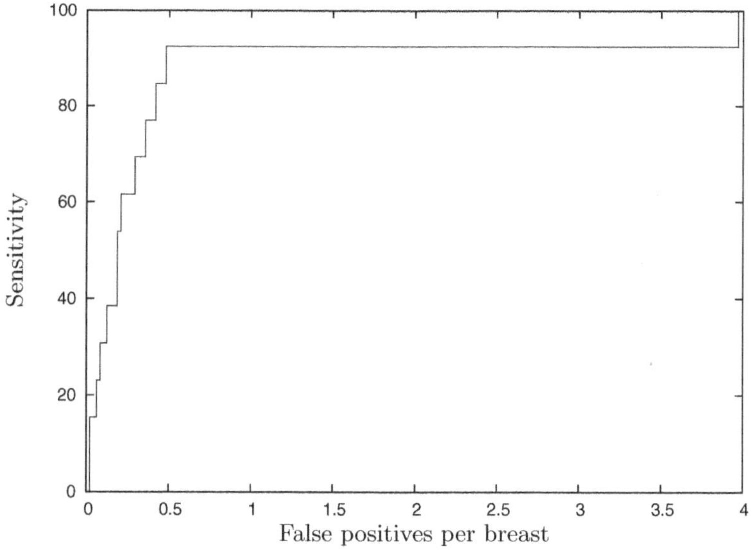

Fig. 3. FROC curve of the suspicious convergence detection channel

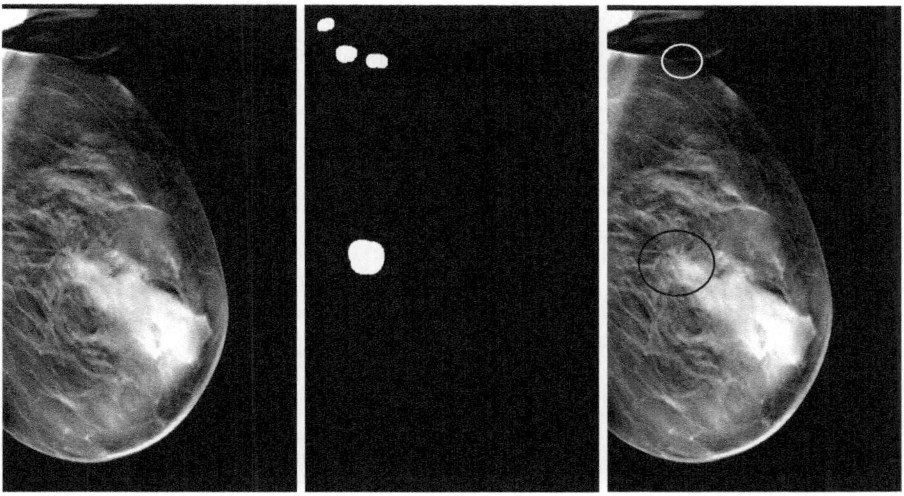

(a) (b) (c)

Fig. 4. Example of detection using the suspicious convergence detection channel. (a) DBT slice containing an architectural distortion. (b) A contrario detection result. (c) Final result: the lesion (in black) is detected as well as a false positive (in white).

3.4 Overall performances

As mentioned earlier, the target populations of the two previous channels may overlap. Table 1 presents operating points for common sensitivities, and illustrates the gain in specificity in comparison to a direct combination of the two channels performance.

The performance we obtain is comparable to other results of CAD systems for DBT datasets [9,10], while we address specifically architectural distortions in addition to masses. In [9], the authors propose a combination of detections within the projections and within the volume in order to obtain a new detection scheme. They validated it on a database containing a little bit more cancers than ours. As we can see in Table 1, our approach achieves a false positive rate of 1.31 for a 81% specificity. For similar specificity (80%), the method proposed by [9] reaches a specificity of 0.84 false positive per volume, which is better. Nonetheless, if we move to another operating point, we can get similar performances: for a sensitivity about 90%, both methods produce 1.6 false positive per breast. Now, if we increase the target sensitivity to 96%, the specificity of our approach becomes 1.81, while the performance of the other approach go beyond 2 false positives per breast. In [10], the authors propose a detection scheme whose amount of false positives is reduced using information theoretic principles. Unfortunately they only reach a sensitivity of 85% for 2.5 false positives per case.

Table 1. Performance of the combination of the two channels

Sensitivity (%)	Specificity (false positives per breast)
81,1	1,31
90,6	1,60
96,2	1,81

Nonetheless, because of the databases sizes/discrepancies, a comprehensive comparison of these approaches is hazardous. However, we can conclude that they are comparable, with a slight advantage (resp. disadvantage) of our approach for higher (resp. lower) sensitivity values, while stating validation on a larger database should be performed in order to compare them in a reliable manner.

4 Conclusion

We have introduced a new scheme for detection of some suspicious radiological signs in DBT volumes. It is composed of two distinct channels dedicated to the detection of masses and highly convergent signs, respectively. Both of them are relying on a two steps detection: a marker extraction stage followed by a false positive reduction stage. While the second step is similar for both channels, the first one is using completely different tools. Thus markers corresponding to masses are retrieved using fuzzy connected filters tools, and suspicious convergence areas are extracted using a contrario modeling based methods. We proposed a validation of the whole system and showed that its performance is comparable to other systems in the literature.

The further steps of the presented work will be to enlarge the database in order to obtain a more accurate validation. Assessing other existing approaches on such a database would enable us to actually compare them with our method. In addition to that, we would like to address other suspicious radiological signs using additional detection channels.

Acknowledgment. This work was realized under CIFRE convention 20061165 (GE Healthcare/Télécom ParisTech).

References

1. Palma, G., Bloch, I., Muller, S.: Fuzzy connected filters for fuzzy gray scale images. In: International Conference on Information Processing and Management of Uncertainty in Knowledge-Based Systems (IPMU), Malaga, Spain, June 2008, pp. 667–674 (2008)
2. Timp, S., Karssemeijer, N.: A new 2D segmentation method based on dynamic programming applied to computer aided detection in mammography. Medical Physics 31(5), 958–971 (2004)

3. Apffel, L., Palma, G., Muller, S., Bloch, I.: Fuzzy segmentation of masses in digital breast tomosynthesis images based on dynamic programming. In: International Conference on Imaging Theory and Applications (IMAGAPP), Angers, France (June 2010)
4. Vapnik, V.N.: The nature of statistical learning theory. Springer, New York (1995)
5. Palma, G., Muller, S., Bloch, I., Iordache, R.: Convergence areas detection in digital breast tomosynthesis volumes using a contrario modeling. In: SPIE Symposium on Medical Imaging, Lake Buena Vista, FL, USA (February 2009)
6. Palma, G., Muller, S., Bloch, I., Iordache, R.: Fast detection of convergence areas in digital breast tomosynthesis. In: IEEE International Symposium on Biomedical Imaging (ISBI), Boston, MA, USA, June 2009, pp. 847–850 (2009)
7. Karssemeijer, N., te Brake, G.M.: Detection of stellate distortions in mammograms. IEEE Transactions on Medical Imaging 5(5), 611–619 (1996)
8. Andersen, A.H., Kak, A.C.: Simultaneous algebraic reconstruction technique (SART): A superior implementation of the ART algorithm. Ultrasonic Imaging 6(1), 81–94 (1984)
9. Chan, H.P., Wei, J., Zhang, Y., Helvie, M.A., Moore, R.H., Sahiner, B., Hadjiiski, L., Kopans, D.B.: Computer-aided detection of masses in digital tomosynthesis mammography: Comparison of three approaches. Medical Physics 35(9), 4087–4095 (2008)
10. Singh, S., Tourassi, G.D., Baker, J.A., Samei, E., Lo, J.Y.: Automated breast mass detection in 3D reconstructed tomosynthesis volumes: A featureless approach. Medical Physics 35(8), 3626–3636 (2008)

An Observer Model for Lesion Detectability in Contrast-Enhanced Digital Mammography

Melissa L. Hill[1], James G. Mainprize[2], and Martin J. Yaffe[1,2]

[1] Dept. of Medical Biophysics, University of Toronto, Sunnybrook Health Sciences Centre, 2075 Bayview Ave., Toronto, Ontario, Canada, M4N 3M5
[2] Imaging Research, Sunnybrook Health Sciences Centre, 2075 Bayview Ave., Toronto, Ontario, Canada, M4N 3M5
melissa.hill@sri.utoronto.ca

Abstract. Lesion detectability in contrast-enhanced digital mammography (CEDM) was studied using a model observer for task-based system evaluation. Digital mammography (DM) and CEDM images were simulated to include a realistic mammographic appearance and contrast enhancement based on simple physiological assumptions of iodinated contrast agent uptake. A Laguerre-Gauss Channelized Hotelling observer (LG-CHO) was implemented to compute the lesion detectability, d', according to the signal-to-noise ratio of the CHO test statistic. Iodine concentrations from 0.3 to 3 mg/ml in simulated lesions resulted in greater d' for log-subtracted CEDM images (acquisition technique of Jong *et al.* Radiology 2003) compared to full dose DM (standard screening DM acquisition technique). Doubling mean glandular radiation dose yielded only minimal relative gains in d' for all but the lowest lesion contrast agent uptake levels considered, suggesting that quantum noise was not the limiting factor for lesion detectability.

Keywords: mammography, contrast enhancement, image perception, image quality, observer model.

1 Introduction

Contrast-enhanced digital mammography (CEDM) can be used to identify the presence of tumour angiogenesis with an intravenous injection of an iodinated contrast agent [1]. In clinical pilot studies this modality has shown promise as an adjunct to mammography to improve lesion specificity, to determine the extent of disease and to detect mammographically occult lesions, particularly in the dense breast [2-4]. Optimization of CEDM has been reported where the signal difference-to-noise ratio (SDNR) is maximized while minimizing mean glandular dose [5-7]. The Rose criterion was then used to predict iodine concentration detection limits [5-7]. However, the SDNR is a simplistic image quality metric that does not account for several factors that influence radiologists' detection performance in mammographic images, including anatomical clutter, lesion shape and margin appearance.

Several investigators are currently developing task-based model observers that can predict human performance at detecting lesions in medical images [8,9]. These models

J. Martí et al. (Eds.): IWDM 2010, LNCS 6136, pp. 720–727, 2010.

could potentially be used for efficient evaluation and optimization of CEDM image quality. In this study we apply the Laguerre-Gauss (LG) channelized Hotelling observer (CHO) [10] to CEDM. The CHO has performed well for approximating the ideal observer in mammography and it has been shown to correlate well with human reader performance [11].

In this work we have used a clustered lumpy background (CLB) [12], which represents mammographic texture closely and allows for the efficient creation of a large number of realizations. Simple assumptions regarding the uptake of contrast agent in normal tissues and lesions were applied in the simulation of single-energy subtraction CEDM. Recently Castella *et al.* evaluated human observer performance of malignant mass detection in both CLB images and real mammograms and found that performance was significantly better with real backgrounds [13]. Hence, we believe that using the CLB will yield conservative estimates of lesion detectability. To demonstrate the utility of this task-based framework for CEDM optimization we characterize lesion detection performance for the imaging conditions used in the study by Jong *et al.* [3], and over a clinically relevant contrast agent uptake range in each of the lesion and the surrounding normal tissues.

2 Methods

2.1 Simulation of Mammographic Backgrounds

We implemented the isotropic double-layered CLB approach of Castella *et al.* [12] to emulate mammographic backgrounds. Raw CLB image intensity was scaled to generate linear attenuation maps consistent with breast tissue x-ray attenuation characteristics [14]. The intensity of each lump was scaled such that the mean linear attenuation in a CLB realization would be equivalent to a 20% fibroglandular (FG) breast [15]. For comparison, uniform 20% FG backgrounds were also generated. A 5 mm diameter Gaussian shape with the attenuation of infiltrating ductal carcinoma [14] was inserted into the centre (signal known exactly) of half the images to represent a mass lesion.

2.2 Contrast Agent Uptake

The attenuation corresponding to 0.3, 0.5, and 3.0 mg/ml of iodine was added to the lesions, a range observed clinically by Jong *et al.* [3]. Iodine contrast was also added to the background tissue with the iodine concentration proportional to the reported distribution volumes (V), the total of the vascular volume and extravascular extracellular space in FG and adipose tissues ($V_{adipose}=V_{FG}/3$ [16, 17]). We will describe the relative proportion of uptake between the lesion and the background in terms of a parameter called the lesion-to-fibroglandular iodine concentration ratio (LFR). For each lesion iodine concentration, LFRs of 1.67, 2.5, 5, 10, and 20 were considered, a range estimated from the results of Jong *et al.* [3].

Table 1. One example CLB realization with *(left)* and without *(right)* a lesion inserted and the corresponding images under DM and CEDM exposure conditions

	Lesion present	No lesion
Noise free attenuation map		
DM exposure conditions		
CEDM exposure conditions, no iodine		

2.3 Image Simulation

X-ray transmission was modeled with a mono-energetic (DM = 18 keV, CEDM = 35 keV) parallel beam geometry through a 5 cm thick breast. The image size was 256×256 pixels with 100 μm detector elements. The entrance surface air kerma (ESAK) used for the CEDM images was 0.34 mGy, equal to that used by Jong *et al.* (Mo anode, 0.3 mm Cu/0.3 mm Al filters, 45 kV, 140 mAs) [3]. The ESAK for the DM images was 4.8 mGy, that selected by the automatic exposure control on a GE Senographe 2000D for a 5 cm, 20% FG breast (Mo anode, Rh filter, 27 kV, 67 mAs). Appropriate scatter fractions and an anti-scatter grid were incorporated [18]. Quantum noise was based on the number of photons that interact with a 100 μm thick layer of CsI.

2.4 Lesion Detectability

Following the approach of Gallas and Barrett [10] a LG-CHO was implemented to compute lesion detectability. A test statistic, *t*, is calculated for two cases: 1) signal

Table 2. CEDM images with 3 mg/ml iodine uptake in the lesion for the same example CLB realization shown in Table 1 (*anatomical*) and for a uniform background. CEDM images are shown for two example LFRs, 1.67 (*left*) and 20 (*right*).

	Lesion uptake 1.67 × fibroglandular (FG)	Lesion uptake 20 × FG
Post-contrast anatomical image		
Log-subtracted anatomical image		
Log-subtracted uniform image		

absent (*ns*) and 2) signal present (*s*). The detectability, d', is the SNR of the test statistic and is calculated according to:

$$d' = \frac{\left|\bar{t}_s - \bar{t}_{ns}\right|}{0.5\left[\mathrm{var}(t_s) + \mathrm{var}(t_{ns})\right]} \quad . \tag{1}$$

A set of 2000 independent CLB realizations was generated. From this image set the required attenuation maps and corresponding images were generated for each case under examination. A group of 500 signal absent and 500 signal present images were used to estimate the CHO weights, and then these weights were used to calculate d' using the remaining 500 image pairs. Lesion detectability was calculated for DM images (no iodine) and logarithmically subtracted CEDM images. Likewise, using a set of 2000 uniform background realizations d' was calculated for uniform log-subtracted CEDM images.

3 Results

One randomly selected CLB realization and its associated images are shown in Table 1 with and without a lesion inserted for each of the DM and CEDM exposure conditions. In this illustration, the backgrounds are identical so that the lesion is apparent to the reader, but in simulation, a unique CLB realization was generated for each signal-present and signal-absent image. Note that all images presented here are windowed from the minimum to the maximum intensity in each image.

The images in Table 2 illustrate the effect of background tissue iodine contrast uptake on the visual appearance of CEDM images. The images on the left-hand side of Table 2 have an LFR of 1.67, with an LFR of 20 on the right.

Fig. 1 plots d' as a function of the LFR for each of the contrast uptake conditions studied. Results are shown for log-subtracted CEDM images with anatomical (CLB) and uniform backgrounds. For comparison, d' was also calculated for a standard mammographic screening case where DM images were acquired of a CLB phantom with no contrast agent present. The DM detectability level is indicated in Fig. 1 by the dashed line.

Fig. 2 demonstrates the fractional change in d' observed in log-subtracted anatomical CEDM images acquired with increased radiation exposure, as a function of LFR for each lesion contrast uptake level studied. The mean glandular dose was doubled from 0.64 mGy to 1.28 mGy.

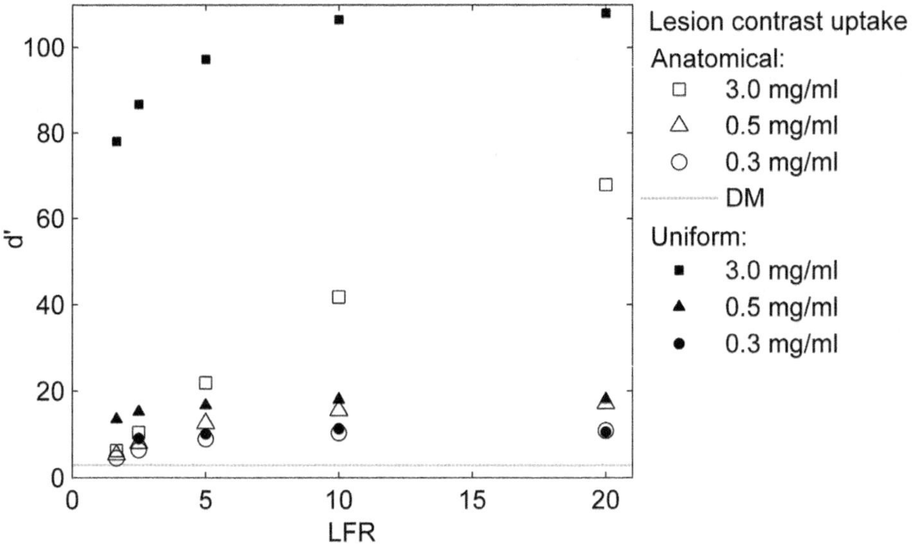

Fig. 1. Lesion detectability in log-subtracted CEDM images for the given lesion iodine uptake as a function of LFR. *Open* (□) markers denote anatomical backgrounds, *filled* (■) markers denote uniform backgrounds. The dashed line (--) is the DM lesion detectability.

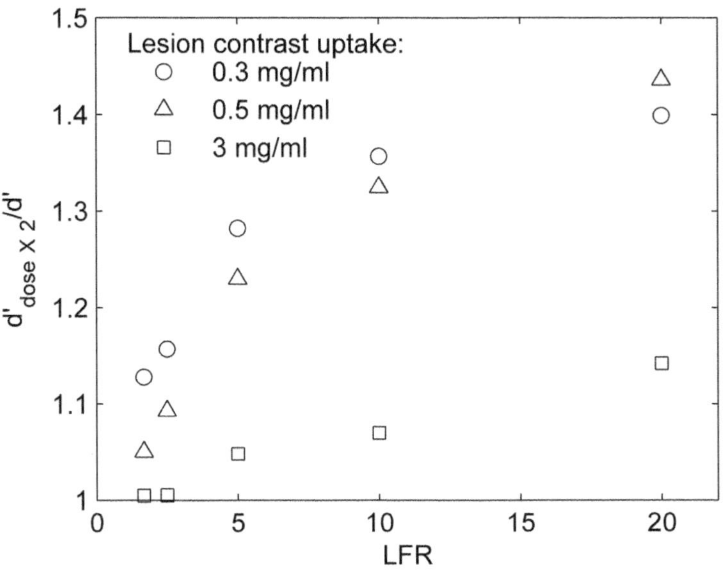

Fig. 2. Fractional increase in lesion detectability for log-subtracted anatomical CEDM images acquired with twice the mean glandular dose at the given lesion iodine uptake level and LFR

4 Discussion

The images in Table 1 illustrate qualitatively that the lesion is not readily apparent in the CLB background for the DM images or the pre-contrast CEDM images. This presents a challenging detection task, such as in a dense breast where FG tissue may mask or mimic a lesion. The potential of subtraction CEDM to improve lesion detectability compared to DM in an anatomical background is shown quantitatively in Fig. 1, where DM is outperformed by CEDM in terms of d'.

In Fig. 1 the detectability in a uniform background is consistently greater than in an anatomical background. This demonstrates that lesion detectability is sensitive to anatomical clutter and suggests that previous estimates of minimum iodine detectability in CEDM may have been overestimated when a uniform background was assumed. For each background type, d' begins to converge as the LFR increases, suggesting that the influence of anatomical clutter decreases as LFR increases. This result is also seen qualitatively by a comparison between the log-subtracted CEDM images with anatomical and uniform backgrounds in Table 2.

Jong *et al.* kept the x-ray dose very low for each image acquisition so that a five time point CEDM exam (six images) could be completed for a total dose approximately equal to that of one screen-film mammographic image [3]. Since single-energy CEDM is restricted to a single view [15], it may be appropriate to image with a total dose equal to that for two mammographic views to get the best possible image quality. The benefits of imaging with an exposure that increases the mean glandular dose by a factor of two are presented in the results in Fig. 2. Although we might expect detectability to improve by a factor of $\sqrt{2}$ due to quantum noise reduction, this magnitude of detectability increase is only attained at the highest LFR for the lowest contrast agent uptake concentrations considered. This result suggests that for most clinically relevant levels of contrast agent uptake, quantum noise is not the limiting factor for detectability, and that the background contrast uptake can be a limiting factor if the LFR is small.

Future work will focus on the validation of the use of the LG-CHO for lesion detectability estimation by comparison with human observers and refinement of the contrast agent uptake model.

Acknowledgments. This work was conducted with the support of the Ontario Institute for Cancer Research through funding provided by the Ontario Ministry of Research and Innovation. MLH is supported by the Natural Sciences and Engineering Research Council of Canada Postgraduate Scholarship.

References

1. Dromain, C., Balleyguier, C., Muller, S., Mathieu, M.-C., Rochard, F., Opolon, P., Sigal, R.: Evaluation of Tumor Angiogenesis of Breast Carcinoma Using Contrast-Enhanced Digital Mammography. A.J.R. 187, W528–W537 (2006)
2. Lewin, J.M., Isaacs, P.K., Vance, V., Larke, F.J.: Dual-energy contrast-enhanced digital subtraction mammography: feasibility. Radiology 229, 261–268 (2003)
3. Jong, R.A., Yaffe, M.J., Skarpathiotakis, M., Shumak, R.S., Danjoux, N.M., Gunesekara, A., Plewes, D.B.: Contrast-enhanced digital mammography: Initial clinical experience. Radiology 228, 842–850 (2003)
4. Dromain, C., Balleyguier, C., Adler, G., Garbay, J.R., Delaloge, S.: Contrast-enhanced digital mammography. Eur. J. Radiol. 69, 34–42 (2009)
5. Ullman, G., Sandborg, M., Dance, D.R., Yaffe, M.J., Skarpathiotakis, M., Alm Carlsson, G.: A search for optimal X-ray spectra in iodine contrast media mammography. Phys. Med. Biol. 50, 3143–3152 (2005)
6. Baldelli, P., Bravin, A., Di Maggio, C., Gennaro, G., Sarnelli, A., Taibi, A., Gambaccini, M.: Evaluation of the minimum iodine concentration for contrast-enhanced subtraction mammography. Phys. Med. Biol. 51, 4233–4251 (2006)
7. Skarpathiotakis, M., Yaffe, M., Bloomquist, A., Rico, D., Muller, S., Rick, A., Jeunehomme, F.: Development of contrast digital mammography. Med. Phys. 29, 2419–2426 (2002)
8. Gifford, H.C., King, M.A., de Vries, D.J., Soares, E.J.: Comparison of the channelized Hotelling observer and human observers for lesion detection in hepatic SPECT imaging. J. Nucl. Med. 41, 514–521 (2000)
9. Richard, S., Siewerdsen, J.H.: Comparison of model and human observer performance for detection and discrimination tasks. Med. Phys. 35, 5043–5053 (2008)
10. Gallas, B.D., Barrett, H.H.: Validating the use of channels to estimate the ideal linear observer. J. Opt. Soc. Am. A. 20, 1725–1738 (2003)
11. Chen, L., Barrett, H.H.: Task-based lens design with application to digital mammography. J. Opt. Soc. Am. A. 22, 148–167 (2005)
12. Castella, C., Kinkel, K., Descombes, F., Eckstein, M.P., Sottas, P.-E., Verdun, F.R., Bochud, F.O.: Mammographic texture synthesis: second generation clustered lumpy backgrounds using a genetic algorithm. Optics Express 16, 7595–7607 (2008)

13. Castella, C., Eckstein, M.P., Abbey, C.K., Kinkel, K., Verdun, F.R., Saunders, R.S., Samei, E., Bochud, F.O.: Mass detection on mammograms: influence of signal shape uncertainty on human and model observers. J. Opt. Soc. Am. A. 26, 425–436 (2009)
14. Johns, P.C., Yaffe, M.J.: X-ray characterisation of normal and neoplastic breast tissues. Phys. Med. Biol. 32, 675–695 (1987)
15. Yaffe, M.J., Boone, J.M., Packard, N., Alonzo-Proulx, O., Huang, S.Y., Peressotti, C.L., Al-Mayah, A., Brock, K.: The myth of the 50-50 breast. Med. Phys. 36, 5437–5443 (2009)
16. Brooksby, B., Pogue, B.W., Jiang, S., Dehghani, H., Srinivasan, S., Kogel, C., Tosteson, T.D., Weaver, J., Poplack, S.P., Paulsen, K.D.: Imaging breast adipose and fibroglandular tissue molecular signatures by using hybrid MRI-guided near-infrared spectral tomography. P.N.A.S. 23, 8828–8833 (2006)
17. Yankeelov, T.E., Lepage, M., Chakravarthy, A., Broome, E.E., Niermann, K.J., Kelley, M.C., Meszoely, I., Mayer, I.A., Herman, C.R., McManus, K., Price, R.R., Gore, J.C.: Integration of quantitative DCE-MRI and ADC mapping to monitor treatment response in human breast cancer: initial results. Mag. Res. Im. 25, 1–13 (2007)
18. Boone, J.M., Lindfors, K.K., Cooper III, V.N., Seibert, J.A.: Scatter/primary in mammography: Comprehensive results. Med. Phys. 27, 2408–2416 (2000)

Fast Deformation Simulation of Breasts Using GPU-Based Dynamic Explicit Finite Element Method

Lianghao Han[1], John H. Hipwell[1], Zeike A. Taylor[2], Christine Tanner[1,3], Sebastien Ourselin[1], and David J. Hawkes[1]

[1] CMIC, University College London, Gower Street, London, UK
[2] MedTeQ centre, School of Information Technologhy &Electrical Engineering, The University of Queensland, ALD, 4072, Australia
[3] Computer Vision Laboratory, ETH Zürich, 8092 Zürich, Switzerland
{l.han,j.hipwell,c.tanner,s.ourselin,d.hawkes}@cs.ucl.ac.uk, z.taylor@itee.uq.edu.au

Abstract. In this study, we investigated the applicability of a Graphics Processing Unit (GPU)-based dynamic explicit finite element (FE) program for fast quasi-static deformation simulations of breasts, and proposed an optimisation-based method to estimate material parameters of *in vivo* breast tissues in the context of nonlinear hyperelastic models. Due to its high-speed execution, the GPU-based FE program was used as a forward solver in the optimisation process. A hybrid simulated annealing algorithm for global optimisation was employed to find the optimised material parameters by mininising the Euclidean distance between FE predicted displacements and estimated displacement from image registration at the selected landmark positions. The proposed method can be used for fast FE analyses of soft tissue deformations in medical image analyses and surgical simulations.

Keywords: Soft Tissue Deformation, Finite Element Method, Image Registration.

1 Introduction

Biomechanical models using finite element methods (FEMs) have been used to predict breast deformations in surgical simulations and in medical image analyses for assisting breast cancer diagnosis [1], [2], [3]. Typically the breast deformation during mammography [1], [3] is considered as a quasi-static problem and has been analysed using static implicit finite element methods (e.g. ANSYS [4]). In static implicit FEMs, the implicit integration scheme is employed to solve finite element equations using iterative methods (e.g. the Newton-Raphson method). Although the static implicit integration method is unconditionally stable and a bigger increment time step can be taken in the solution, it can encounter numerical difficulties converging to a correct solution during an analysis involving large deformations, highly non-linear material behaviour or contact, requiring a large number of iterations. Dynamic explicit FEMs, extensively applied to solve dynamic problems, have proved valuable in solving quasi-static problems when inertial effects can be neglected [5]. In dynamic

J. Martí et al. (Eds.): IWDM 2010, LNCS 6136, pp. 728–735, 2010.

explicit FEMs, the explicit time integration method is used, and the FE equations are solved by explicitly advancing the kinematic state from the previous increment, without iteration. Therefore convergence problems are not an issue. It is also suitable for parallel execution because the computations can be conducted at the element level. Recently, a GPU-based dynamic explicit FEM algorithm with total Lagrangian formulations [6] was implemented via highly parallel graphics hardware for nonlinear deformation analyses of soft tissues. One aim of the current study is to investigate the applicability of this method for fast quasi-static breast deformation simulations.

Nonlinear hyperelastic models have been widely used for describing breast tissues [2], [3], [6]. However, the material parameters used for breast deformation simulations are generally obtained from *in vitro* tests, which are commonly different from *in vivo* data. Therefore, the second aim of this study is to develop a material parameter identification method to estimate *in vivo* material properties of breast tissues in the context of nonlinear hyperelastic models.

2 Method

Like most biological soft tissues, breast tissues exhibit nonlinear, anisotropic and time-dependent response under large deformation [7]. When breast tissues are subjected to small deformations (less than 2-5%), conventional anisotropic linear elastic models are adequate to model their mechanical behavior. However, large deformations are often involved in clinical practice, such as surgery, mammographic examinations or during ultrasound scanning etc, and linear elastic models are no longer valid for these materials. Soft tissues under large deformation often experience large recoverable elastic deformation, therefore hyperelastic models have been widely used to model the nonlinear and anisotropic behavior of these materials. The constitutive behaviour of hyperelastic materials is defined in terms of strain energy potential. By using different forms of strain energy function, several hyperelastic models including the incompressible/nearly incompressible, viscoelastic, hyperelastic and anisotropic behavior of soft tissues have been implemented in the GPU-based dynamical FE program [6].

2.1 Constitutive Model

In this study a transversely isotropic hyperelastic model was chosen to model the anisotropic behavior of breast tissues, and the strain energy potential had the following form [6]:

$$\psi = \frac{\mu}{2}(\overline{I}_1 - 3) + \frac{k}{2}(J-1)^2 + \frac{\eta}{2}(\overline{I}_4 - 1)^2 \tag{1}$$

where μ denotes the initial shear modulus; $\overline{I}_1$ represents the first deviatoric strain invariant; k stands for the bulk modulus; J denotes the total volume change; η represents a material parameter with units of Pa; $\overline{I}_4 = \mathbf{a}_0 \cdot \overline{\mathbf{C}} \cdot \mathbf{a}_0$ stands for the pseudo-invariant of $\overline{\mathbf{C}} = J^{-2/3}\mathbf{C}$, $\mathbf{C}$ denotes the right Cauchy-Green strain tensor, and $\mathbf{a}_0$ is the preferred direction to present the transversely isotropic response. In the model, μ

and k can be determined from another two elastic parameters, Young's modulus E and Poisson's ratio v, through the relationships of $\mu = E/(2(1+v))$ and $k = E/(3(1-2v))$. If the preferred direction of a tissue can be pre-determined, e.g. it is much stronger in the z direction, so $\mathbf{a}_0 = [0,0,1]$, only three parameters, (E, v, η), are required to completely define this anisotropic model. If $\eta = 0$, Equation (1) becomes the well-known Neo-Hookean isotropic model.

2.2 Finite Element Model of Breasts

In breast deformation analyses, FEMs were used to calculate the displacement field that satisfied the applied boundary conditions and a given set of material parameters. Both static implicit FEM program (ANSYS) and dynamic explicit FEM programs (ABAQUS/Explicit and GPU-based dynamic explicit FE program) were employed for breast deformation simulations. The geometrical model for FE analyses was constructed by segmenting MR image volumes into different tissues and then meshed into tetrahedral elements with ANSYS in order to account for the irregular breast shape. The material type of each element was assigned according to the segmentation. The displacement boundary conditions were applied by constraining the surface nodes of the breast, which were extracted by registering the MR images of the compressed breast to the MR images of the uncompressed breast using a 3D non-rigid image registration method. The detailed description on how to produce a FE model of the breast can be found in Ref [1].

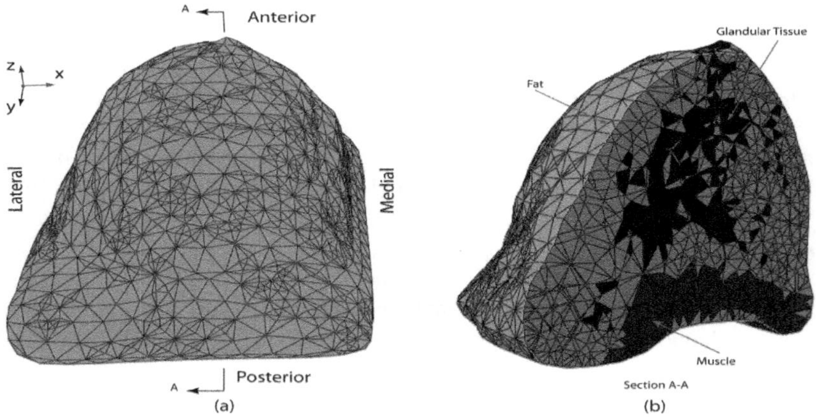

Fig. 1. A typical FE model: (a) Geometric model and FE mesh of undeformed breast (b) Distribution of breast tissues

Fig. 1(a) shows a typical geometric model and FE mesh of an undeformed breast, which will be compressed with a fixed medial plate and a moveable lateral plate. The breast compression by the two plates can be simulated by two approaches: (1) employ a contact model with/without friction effect to simulate the contact between the plates and the breast, and apply a displacement to the compression plates to compress breasts; or (2) directly apply displacement boundary conditions to the breast surface on both medial and lateral sides to simulate frictionless contact between the plates and tissues, without using actual compression plates and contact definition. The second approach is simple and fast but may lead to unrealistic deformation, e.g. swelling on the surface [8]. At present, the contact model has not been implemented in the GPU-based dynamical FE program. Unless otherwise stated, all the breast compression simulations in this paper were performed by using the second approach to model the contact between the plates and the breast. As shown in Fig. 1(b), the breast is segmented into fat, fibroglandular tissue and pectoral muscle. This model consisted of 54025 4-node tetrahedron elements.

2.3 FE-Based Material Parameter Identification Method

The material parameters used for breast deformation simulations are generally obtained from *in vitro* tests, which are different from *in vivo* data. Therefore, an inverse algorithm was proposed to estimate the material parameters of *in vivo* breast tissues. Displacements of landmarks were estimated by image registration between the compressed and uncompressed MR images [1]. The material parameters that best fit the FEM predicted displacements to the estimated displacements of landmarks were found by solving a constrained optimization problem. The objective function was defined as:

$$\arg\min \|\mathbf{u}(\mathbf{p}) - \mathbf{u}_0\|^2 \qquad \text{subject to } \mathbf{lb} < \mathbf{p} < \mathbf{ub} \tag{2}$$

where $\mathbf{p} = [E, v, \eta]^T$ was the material parameter vector with the lower bound constraint $\mathbf{lb}$ and upper bound constraint $\mathbf{ub}$; $\mathbf{u}(\mathbf{p})$ and $\mathbf{u}_0$ were the FE predicted displacements and the estimated displacements, respectively. The inverse problem was solved by using a hybrid simulated annealing algorithm. The global search was performed by use of the MATLAB function, *simulannelbnd,* to find parameter values near the optimum; then with these parameter values as initial values, the local search was performed by calling the MATLAB function, *fmincon*, at the end of iteration of the simulated annealing solver to find the optimization parameters. Since a number of iterations were involved in the inverse reconstruction of material parameters, it was not realistic to use a commercial FE program such as ABAQUS or ANSYS as an FE solver, particularly in case of 3D FE simulations with a typical CPU execution time of several hours. Therefore, due to its high speed execution, the GPU-based dynamic explicit FE program [6] was used as the FE solver in this study.

3 Results

The performance of the GPU-based explicit FE program on quasi-static deformation simulations of the breast was investigated by analysing the breast deformation under compression, as shown in Fig. 1. For comparison, the same FE model was also analysed by two commercial FE packages, ANSYS (static implicit FEM) and ABAQUS/Explicit (dynamic explicit FEM). To make the quasi-static deformation simulation valid when using explicit FEMs [5], the kinetic energy was monitored to ensure that the ratio of kinetic energy to internal energy was less than 5%; that is, the dynamic effect could be neglected. Fig. 2 shows displacement distributions of the breast shown in Fig. 1(a) using three FE programs. The frictionless contact between the compression plates and the breast was simulated by applying a displacement boundary to the breast surface.

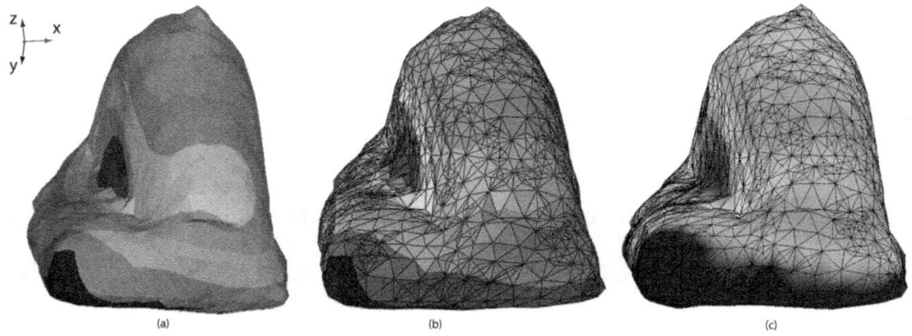

Fig. 2. Total displacement distributions calculated from (a) Static implicit FE (ANSYS 11.0) (b) Dynamic explicit FE (ABAQUS 6.8) and (c) Dynamic GPU-based FE program

As shown in Fig. 2, all three FE programs give a consistent displacement distribution. Although a large deformation (30% compression) was involved, the two explicit FE programs had no difficult converging, and unlike the implicit FE program, ANSYS, no intervention was required during simulation. However, an execution time of 15.6s with the GPU-based dynamic explicit FE program (run on a 2.4GHZ Intel Core 2 CPU PC with a NVIDIA GeoForce GTX280 1GB graphics card) was much less than an execution time of 6.0h with ABAQUS/Explicit (run on a 2.4GHZ Intel Core 2 CPU PC with an integrated graphics card). Much longer computation time and additional processing to handle convergence difficulties (through re-mesh and solution mapping procedures) were required to complete the same simulation using ANSYS.

The estimation of the property parameters of breast tissues was performed with the method described in Section 2.3. Seven FE models were created from the MR data of seven subjects (denoted as S1 to S7) before and after compressing the breast with two plates, following the method described in Section 2.2. For the sake of simplicity and avoiding unrealistic deformation due to partial constraints on the surface [8], all the surface nodes of the breast models were constrained by applying a displacement boundary condition directly extracted through registering deformed MR volume images to undeformed MR volume images of the breast. The GPU-based FE program was employed as a FE solver for breast compression simulations. In the FE-based reconstruction procedure, the material property parameters of fat, glandular tissue and muscle were considered as variables. Since only displacement boundary conditions were applied, the displacement field was determined by the relative values of the material property parameters of tissues. Here we chose fat as the reference material with an initial Young's modulus of 1kPa [1]. Both transversely isotropic and isotropic hyperelastic models were considered. For the anisotropic hyperelastic model, the variables included (v_f, η_f) for fat, (E_g, v_g, η_g) for fibroglandular tissue and (E_m, v_m, η_m) for muscle, and we assumed that the z direction was the preferred direction, $a_0 = [0,0,1]$. For the isotropic Neo-Hookean model, $\eta_f = \eta_g = \eta_m = 0$. The material parameters were estimated by minimising the difference between FE predicted displacements and estimated displacements computed via image registration at the landmark positions. That is, the estimated displacements at the landmarks were considered as ground truth. 80% of all nodes within the regions of glandular tissue and muscle were randomly chosen as landmarks. Fig. 3 shows typical outputs during the material parameter identification procedure for subject S1. Fig. 4 presents a comparison of deformed tissue structures from the image registration and the FE prediction by using optimized material parameters, as shown in Fig. 3(b).

The material parameter identification process was a process to find the best match of deformed internal tissues between the FE prediction and the estimation with image registration through optimising material property parameters. Table 1 lists the mean

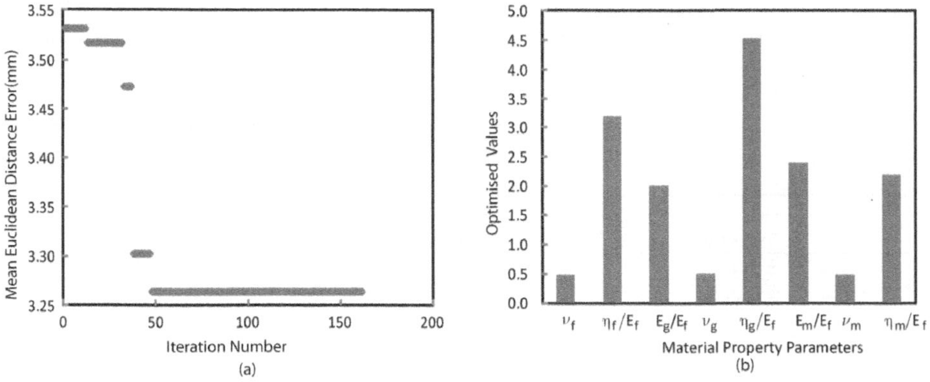

Fig. 3. Typical outputs in the material parameter estimation process for subject S1: (a) Error change with increasing iterations (b) optimized material property parameters

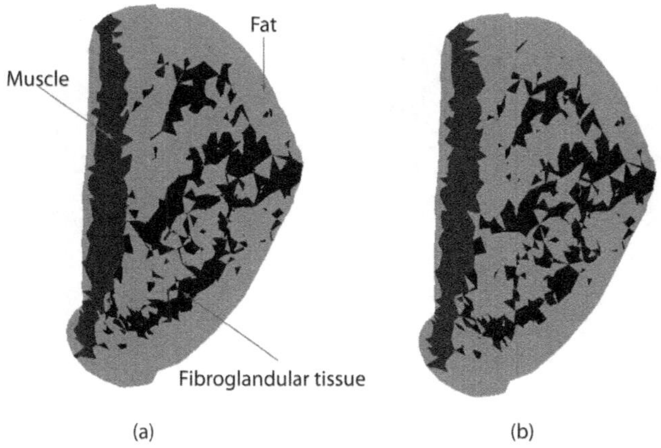

Fig. 4. Deformed tissue structures on the slice through the nipple: (a) Estimation from image registration and (b) FE prediction

Euclidean distance errors (or matching errors) after optimisation. Considering that the displacements of landmarks were estimated from image registration and registration errors can not be avoided, the GPU-based explicit FE program gave a reasonable prediction of internal tissue deformations. The results listed in Table 1 also show that the FE prediction accuracy could be improved by considering the anisotropic effect of breast tissues, although the performance improvement was limited (up to 10%). The limited performance improvement may lie in the fact that the tissues have less freedom to deform due to constraining all the surface nodes of the breast in this study. It is expected that the anisotropic effect of tissues will be more obvious when the contact model was directly used to simulate the breast deformation by compression plates, as shown in Fig. 5. ABAQUS/Explicit was used for these simulations. The contact between the breast and the compression plate was modelled by defining contact pairs on the surface. The optimised material parameters from the material parameter estimation process were employed for both the isotropic model and the anisotropic model. A clearly greater elongation in the z direction was observed from the isotropic model.

Table 1. Mean Euclidean distance error between FE predicted and estimated displacements at the landmark positions using different material models

Subject	Error using isotropic hyperelastic model (mm)	Error using anisotropic hyperelastic model (mm)	Performance Improvement
S1	3.518	3.263	7.2%
S2	2.945	2.935	0.45%
S3	2.086	2.038	2.29%
S4	3.015	2.79	7.36%
S5	2.122	2.076	2.12%
S6	4.583	4.238	7.51%
S7	3.635	3.271	10.0%

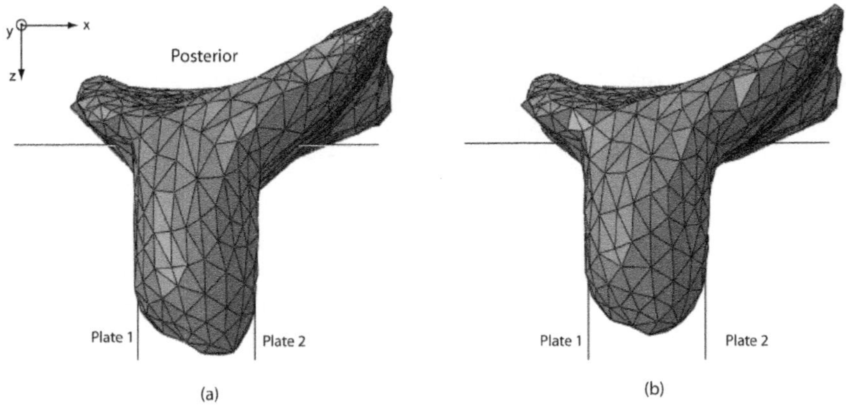

Fig. 5. Breast compression simulations of subject 3 using different material models: (a) Isotropic hyperelastic model (2) Anisotropic hyperelastic model

4 Discussion and Conclusion

The GPU-based dynamic explicit FEM algorithm developed was capable of simulating large nonlinear quasi-static deformations of breast tissues at high speed and there was no convergence difficulty.

By taking advantage of the high speed execution of the algorithm, a general material parameter identification method was developed to estimate the material property parameters of *in vivo* breast tissues. Since only displacement boundary conditions were considered, the material parameters obtained were relative values. However, the same method could be used to estimate real material property parameters by minimising the difference between predicted force-displacement response and experimentally measured force-displacement response. When anisotropic effects of soft tissues were included, the matching errors of internal tissue deformations between the FE prediction and the estimation by image registration decreased for all cases, showing that the anisotropic hyperelastic models provided more accurate simulations of breast deformations. This confirms a similar observation made for linear transverse isotropic materials [1].

Although the frictionless contact between compression plates and the breast can be simulated by applying surface displacement boundary conditions, it may lead to unrealistic deformation simulations on the surface [8]. To avoid this problem and facilitate wider applications such as surgery simulation, the contact model will be implemented into the GPU-based FE program to simulate the frictional/frictionless contact between soft tissues and medical equipment/devices in the future.

Acknowledgments. The study was supported by the HAMAM project funded under the 7th Framework Program for Research, ICT-2007.5.3.

References

1. Tanner, C., White, M., Guarino, S., Hall-Craggs, M.A., Douek, M., Hawkes, D.J.: Anistropic Behaviour of Breast Tissue for Large Compressions. In: 6th IEE International Symposium on Biomedical Imaging, pp. 1223–1226. IEEE Press, Boston (2009)
2. Carter, T.J., Sermesant, M., Cash, D.M., Barratt, D.C., Tanner, C., Hawkes, D.J.: Application of Soft Tissue Modelling to Image-guided Surgery. Med. Eng. Phys. 27, 893–909 (2005)
3. Pathmanathan, P., Gavaghan, D.J., Whiteley, J.P., Chapman, S.J., Brady, J.M.: Predicting Tumor Location by Modeling the Deformation of the Breast. IEEE Trans. Biomed. Eng. 55, 2471–2480 (2008)
4. ANSYS v.11, ANSYS Inc. (2007)
5. ABAQUS, ABAQUS Online Documentation: Version 6.8. HKS, Inc. (2008)
6. Taylor, Z.A., Comas, O., Cheng, M., Passenger, J., Hawkes, D.J., Atkinson, D., Ourselin, S.: On Modelling of Anisotropic Viscoelasticity for Soft Tissue Simulation: Numerical Solution and GPU Execution. Med. Image Anal. 13, 234–244 (2009)
7. Han, L.H., Noble, J.A., Burcher, M.: A Novel Ultrasound Indentation System for Measuring Biomechanical Properties of in vivo Soft Tissue. 29, 813–823 (2003)
8. Ruiter, N.V.: Registration of X-ray Mammograms and MR-Volumes of the Female Breast based on Simulated Mammographic Deformation. University of Mannheim, PhD Thesis (2003)

Breast Image Registration by Combining Finite Elements and Free-Form Deformations

Angela W.C. Lee[1], Julia A. Schnabel[2], Vijay Rajagopal[1], Poul M.F. Nielsen[1,3], and Martyn P. Nash[1,3]

[1] Auckland Bioengineering Institute, The University of Auckland, New Zealand
angela.lee@auckland.ac.nz
[2] Institute of Biomedical Engineering, Department of Engineering Science, University of Oxford, UK
[3] Department of Engineering Science, The University of Auckland, New Zealand

Abstract. During breast cancer diagnosis, the breasts undergo large deformations due to gravity or compression loads. It is therefore non-trivial to recover the deformation and register medical images of the breast in different orientations (e.g. prone versus supine). Free-form deformations and biomechanical finite element models have been used to non-rigidly register breast images from prone to supine, but with limited success. In this paper, we demonstrate that the use of a finite element model to predict the deformation of the breast from prone to supine provides a significantly more accurate registration compared to free-form deformation methods. We also show that the use of this biomechanical model prediction as a prior to free-form deformation provides a significantly more accurate match than does the use of either method independently.

1 Background

There has been considerable interest in applying image registration techniques to warp one medical image of the breast to match another [1]. Non-rigid registration techniques can give rise to unrealistic deformations, such as implausible changes in the volume of the tissues, since they do not account for the physical characteristics of the breast [2]. The finite element method (FEM) can be used to apply physically realistic constraints on breast deformations [3, 4]. Though simulations using the FEM may be physically plausible, reasonable assumptions regarding the tissue properties and the loading conditions are required in order to predict realistic breast deformations. We examined the use of a biomechanical finite element (FE) model as a prior to an intensity-based non-rigid registration algorithm. We assessed localisation accuracy using sets of internal features identified from MRI data of a breast phantom and of the breasts of five volunteers.

2 Methods

Magnetic resonance imaging

Five volunteers were recruited and scanned on a 1.5T Siemens MR scanning system. T_2-weighted images of the breasts were acquired with the volunteers positioned in

J. Martí et al. (Eds.): IWDM 2010, LNCS 6136, pp. 736–743, 2010.

prone and supine orientations. Although the standard sequence for breast MR imaging is T_1, the T_2 imaging protocol used in this study is more appropriate for future studies regarding tissue heterogeneity in the breasts, as it allows greater distinction between the fat and fibro-glandular tissue.

The image dimensions were 512 x 512 pixels spanning a 350mm x 350mm field of view with approximately 60 slices of 2.5mm thickness. Similar imaging was performed on a CIRS triple modality breast phantom containing twelve inclusions. The phantom was imaged before and after a 45% compression, using a T_1-weighted FL3D pulse sequence with a slice thickness of 0.75mm.

Image registration

MRI data for the uncompressed volunteers' breasts and the breast phantom were segmented and used to create personalised FE models (Figs 1A and 2A, respectively) for FEM based registration. Further details on the procedure to create the FE models may be found in [5]. The FE models were based on isotropic, homogeneous, and incompressible mechanical response, as defined by the neo-Hookean constitutive relation $W=V_n(I_1-3)$, where I_1 is the first principle invariant of the right Cauchy-Green deformation tensor and V_n is the stiffness parameter for the nth volunteer's breast: V_1=80Pa; V_2=100Pa, V_3=200Pa, V_4=160Pa, V_5=120Pa.

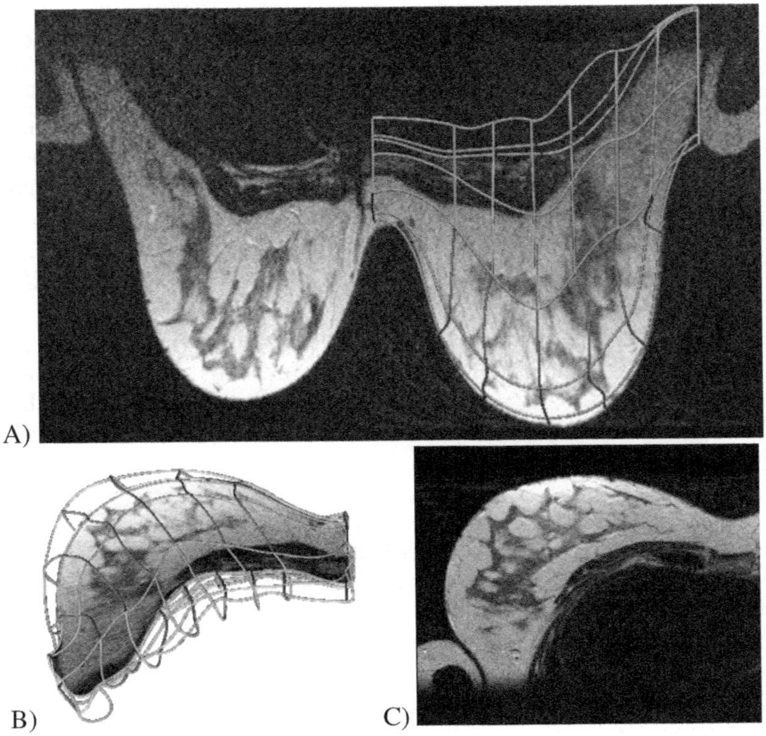

Fig. 1. A) Clinical MR images of the prone breast were used to create a personalised FE model for each volunteer. B) This model was used to predict the supine orientation, and thus warp the prone MRI data. This model-warped image was then compared against the clinical supine image shown in (C).

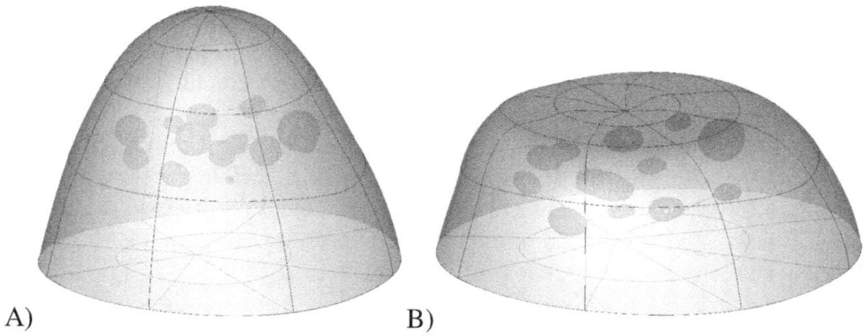

Fig. 2. Finite element model of the breast phantom with inclusions indicated A) before and B) after simulated compression

The breast tissues were assumed to be fixed at the ribs surface for simulating the prone to supine reorientation. FE implementation[1] of finite deformation elasticity was used to simulate the large deformations that the breast tissues and phantom underwent between imaging. Breast deformations were modelled using gravitational body forces to simulate the reorientation from prone to supine (Fig 1B). To mimic the biophysical conditions, forces were not applied to the skin surface as has been the case in previous studies [4]. Frictionless contact constraints were used to represent the interactions between the phantom and the loading plates during simulated compression (Fig 2B).

Non-rigid registration was performed using a free-form deformation (FFD) method based on B-spline warping of the images [6]. We used normalised mutual information as the voxel-based objective function for the FFD registration. Initially, the rigid registration of the images was determined through alignment of the rib structures (which did not deform substantially between scans). A global affine deformation was then used to capture the gross deformation between the images, followed by local registration using B-spline FFDs. We used the multi-resolution capability of the IRTK code[2] to recover the local deformations. The initial coarse resolution (40mm×40mm×40mm) FFD mesh was iteratively refined to a 10mm isotropic mesh.

In this study, we compared the FFD and FE methods with a combined method (FEM+FFD), where the deformation is initially predicted using a FE model. The FEM warped image and the clinical image of the deformed object were then registered using the B-spline FFDs.

Image comparison measures

After applying the three methods described above, we quantitatively evaluated the image registration on a regional basis over the entire breast or phantom. In the phantom, there were 12 embedded inclusions (diameters of 2mm-10mm) that were distinct under MRI. To compare the three registration algorithms, we calculated the target registration error [7] of the centroids of the inclusions between the model-warped images and the images of the compressed phantom. The volume overlap (Dice coefficient, *D*) and surface distance (symmetric mean absolute surface distance, *SMAD*) were also evaluated for each of the inclusions.

---

[1] www.cmiss.org

[2] www.doc.ic.ac.uk/~dr/software/

$$D_j = \frac{2|A_j \cap B_j|}{|A_j| + |B_j|} \times 100, \; j = 1..\text{number of inclusions}$$

$$SMAD = \frac{1}{n_a + n_b}\left(\sum_{n_a} |d_i^{ab}| + \sum_{n_b} |d_i^{ba}|\right)$$

where A_j is the jth inclusion segmented from the target image, B_j is the same inclusion segmented from the warped image, d_i^{ab} is the minimum distance between the ith surface voxel on A and the surface voxels on B, and n_a, n_b are the number of surface voxels in A and B, respectively.

In contrast to the phantom, landmark features inside the breasts are typically not well defined between clinical MR images. It was therefore advantageous to consider similarity measures on a regional basis across the entire breast. We used a block matching algorithm [8] to compare model-warped images with the target images of the supine breasts. This algorithm computes the optimal rigid translation for iteratively smaller 3D blocks (to 5mmx5mmx5mm volume) between the two images by maximising the normalised cross correlation between the voxel intensities in these blocks. The error between the two images was obtained using displacement measures. In this study, the displacement errors across the breast tissues for the different registration schemes were analysed by fitting a linear model and analysis of variance (ANOVA), followed by *post-hoc* comparisons (Tukey's test) using the statistical package *R*[3] to determine whether there were significant differences between the accuracy of the three registration methods.

3 Results

A one-way ANOVA test on the target registration error of the centroids of the 12 inclusions in the phantom indicated that there were significant differences between the three methods (p<0.001). *Post-hoc* comparisons revealed that the FEM+FFD method performed significantly better than the FEM (mean±SD: 1.8mm±0.7mm vs 2.7mm±1.0mm, respectively, p<0.001), which in turn outperformed FFD (mean±SD: 5.5mm±2.1mm, p<0.001 compared to FEM). The other accuracy measures (surface overlap, volume overlap and block matching; Table 1) indicated consistent outcomes. The relatively low Dice index values (FFD: 36%; FEM: 49% and FEM+FFD: 58%) showed that the lesions were not accurately mapped for all three registration methods (Fig 3).

The distributions of the block matching error vector magnitudes were skewed, thus statistics were performed using a square-root transform of the error vector magnitudes. For the individual breast image comparisons, the block matching results led to similar conclusions to the phantom case, whereby FEM outperformed FFD (p<0.001), and FEM+FFD was significantly (p<0.001) more accurate than FEM or FFD alone for each and every individual (Table 1). The volunteer data was

---

[3] www.r-project.org

grouped by registration method and pooled across all volunteers, then analysed by fitting a linear model. One-way ANOVA on the linear models revealed significant differences between the three registration methods (p<0.001). *Post-hoc* comparisons indicated that the FEM performed significantly better than FFD (squared linear model intercepts: 3.45mm vs 4.23mm, respectively, p<0.001), and that the FEM+FFD method outperformed the FEM (squared intercepts: 1.87mm vs 3.45mm, p<0.001) (Table 2).

Table 1. The measures of error for the 3 different registration methods used to deform the clinical images of the breast/ breast phantom to match images of the breast/ phantom under gravity and compression loads, respectively.

	FFD	FEM	FEM+FFD
Breast Phantom (internal landmarks)			
Centroid location (mm)	5.5 ± 2.1	2.7±1.0	1.8 ± 0.7
SMAD (mm)	2.0 ± 0.6	1.4 ± 0.4	1.0 ± 0.3
Dice Index (%)	36.3 ± 18.9	49.1 ± 14.0	58.0 ± 14.3
Breast Phantom (block matching) (mm)	1.0 ± 1.4	0.9 ± 1.2	0.5 ± 1.0
Volunteer 1 (mm)	12.4 ± 8.9	4.3 ± 3.2	3.5 ± 2.0
Volunteer 2 (mm)	8.9 ± 9.8	8.7 ± 8.8	5.2 ± 4.7
Volunteer 3 (mm)	2.4 ± 4.0	2.0 ± 2.2	1.4 ± 2.1
Volunteer 4 (mm)	6.2 ± 8.4	6.1 ± 5.4	2.2 ± 3.3
Volunteer 5 (mm)	5.9 ± 5.4	3.8 ± 3.6	1.8 ± 2.5

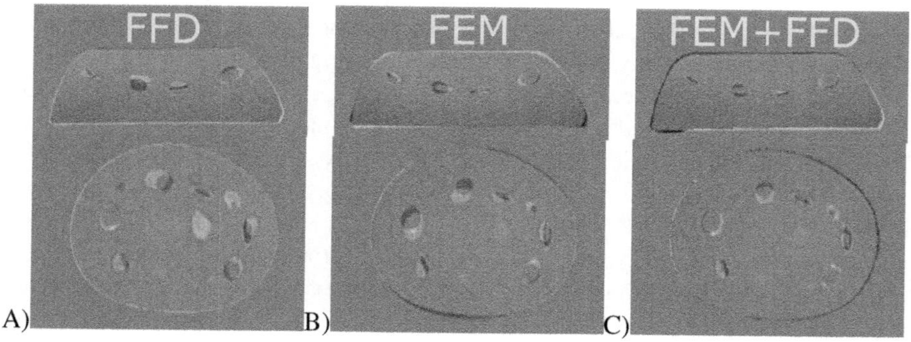

Fig. 3. The difference images for the simulated compression of the breast phantom using A) FFD, B) FEM and C) FEM+FFD minus the experimental compressed image

Table 2. Statistical comparison between the 3 methods for the registration of prone to supine images of the breast. A linear model was fitted to the square root of the magnitude of the block matching error vectors. The linear model intercepts, and standard errors of the mean (SE) the squares of the intercepts are listed. All pair-wise comparisons were significantly different (*p<0.001*).

Method	Squared intercept (mm)	Squared confidence interval (mm)
FFD	4.23	(4.21 ; 4.26)
FEM	3.45	(3.43 ; 3.46)
FEM+FFD	1.87	(1.86 ; 1.89)

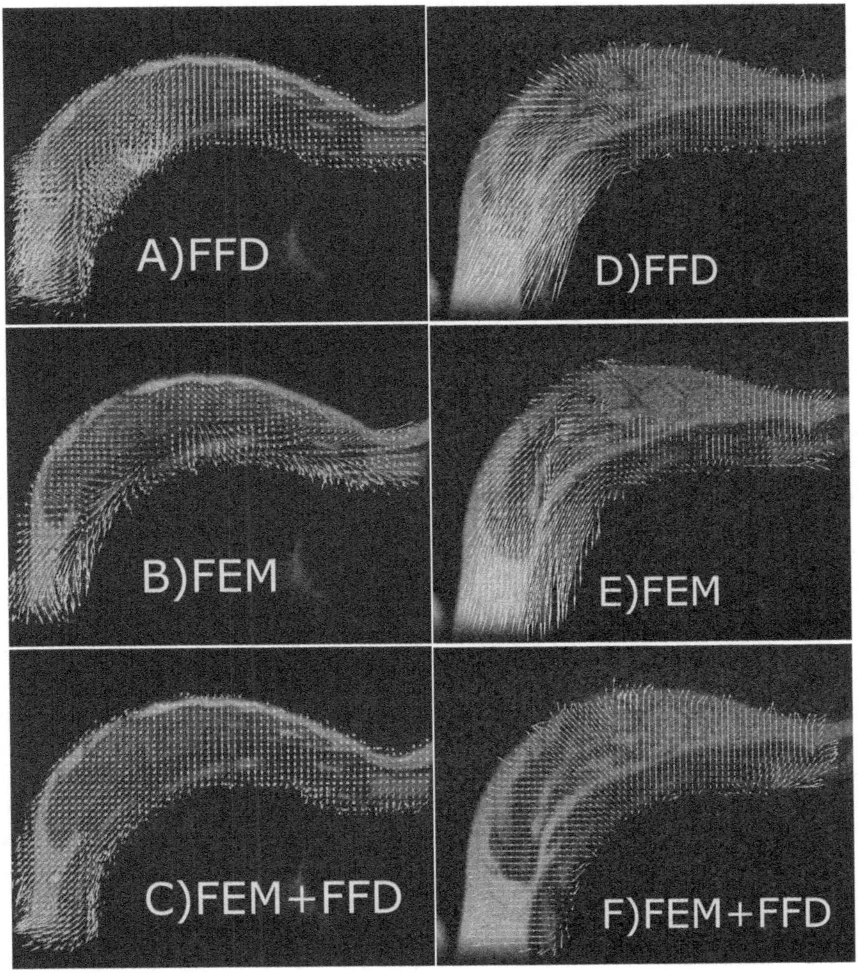

Fig. 4. Registration error vectors on a supine clinical image for volunteer 1 A) FFD; B) FEM; C) FEM+FFD and volunteer 2 D) FFD; E) FEM; F) FEM+FFD

4 Discussion

We have analysed the ability of three non-rigid registration algorithms to predict large deformations of a breast phantom, and the breasts of five volunteers. We registered images of the uncompressed and compressed phantom, whilst deformations were from prone to supine for the breast studies. Tables 1 and 2 demonstrate that using a biomechanical FE model as a prior to localised FFD warping significantly improves the image registration accuracy in all cases. Moreover, the FE model consistently outperformed FFD.

In all of our experiments, the FEM positioned the skin surface above the supine image data. This is reflected in the error distribution for the FEM model and was not improved by decreasing the breast tissue stiffness parameter. We will most likely be able to improve on the results based on the FE predictions by refining modelling assumptions, such as the boundary constraints. Examination of the resulting error vectors for all volunteer cases indicated that using fixed boundary contraints was too restrictive, with the maximum errors for the FEM model occurring along the axilla region of the breast model (Fig 4). The breast tissues are attached via Cooper's ligaments to the chest wall, which allows for some degree of sliding (as opposed to being rigidly fixed). To address this, we plan to investigate the use of sliding boundary constraints at the rib surface in future studies.

From Table 1, we observed that the Dice index for the breast phantom inclusions was rather low for all three methods (FFD: 36%; FEM: 49% and FEM+FFD: 58%), even though the block-matching results indicated a good match. This was also evident in the higher values for the inclusion centroid localisation errors compared to the block matching results. Thus, even though the FFD, FEM and FEM+FFD methods all reliably reproduced the overall deformation of the compressed breast phantom, they less accurately mapped the local deformations of the individual inclusions within the phantom (Fig 3). These inclusions are either cystic or dense masses, which have lower or higher stiffness values, respectively, compared to the bulk material of the phantom. Not accounting for these material heterogeneities may have led to the poor deformation match for the inclusions. In particular, the FE models used in this study were assumed to consist of homogeneously stiff materials, which most likely led to errors in modelling the localised deformations of the inclusions. The FE model of the phantom may thus be improved by representing the spatial differences in the material properties within the phantom. Similarly, breasts are not homogeneous, but they are made up of several tissue types, including fat and fibroglandular tissues surrounded by skin. We are presently developing techniques to reliably represent these tissue heterogeneities inherent in the breast and breast phantom in order to improve model deformation predictions.

In this paper, we have shown that the use of a biomechanical model coupled with an image-based non-rigid registration technique can improve the image registration of large breast deformations, such as from prone to supine. We have also identified ways to improve the FE model predictions and thus registration accuracy (particularly if combined with FFD). Such improvements are expected to increase the clinical applicability of these methods for reliably registering medical images of the breast.

Acknowledgements

This work was supported by the Foundation for Research Science & Technology. A.W.C.L. is supported by a University of Auckland Doctoral Scholarship and M.P.N. is supported by a James Cook Fellowship administrated by the Royal Society of New Zealand on behalf of the New Zealand Government. Dr Chris Bradley provided valuable discussion.

References

1. Guo, Y., Sivaramakrishna, R., Lu, C.C., Suri, J.S., Laxminarayan, S.: Breast image registration techniques: a survey. Med. Biol. Eng. Comput. 44, 15–26 (2006)
2. Tanner, C., Schnabel, J.A., Chung, D., et al.: Volume and shape preservation of enhancing lesions when applying non-rigid registration to a time series of contrast enhancing MR breast images. In: Delp, S.L., et al. (eds.) MICCAI 2000. LNCS, vol. 1935, pp. 327–337. Springer, Heidelberg (2000)
3. Rajagopal, V., Nielsen, P.M.F., Nash, M.P.: Modeling breast biomechanics for multi-modal image analysis - successes and challenges. Wiley Interdisciplinary Reviews in Systems Biology (published online) (2009)
4. Carter, T., Tanner, C., Beechey-Newman, N., Barratt, D., et al.: MR navigated breast surgery: method and initial clinical experience. In: Metaxas, D., et al. (eds.) MICCAI 2008, Part II. LNCS, vol. 5242, pp. 356–363. Springer, Heidelberg (2008)
5. Rajagopal, V., Lee, A., Chung, J.H., Warren, R., Highnam, R.P., Nash, M.P., Nielsen, P.M.F.: Creating individual-specific biomechanical models of the breast for medical image analysis. Academic Radiology 15(11), 1425–1436 (2008)
6. Rueckert, D., Sonoda, L.I., Hayes, C., Hill, D.L., Leach, M.O., Hawkes, D.J.: Non-rigid registration using free-form deformations: application to breast MR images. IEEE Trans. Med. Imaging 18, 712–721 (1999)
7. Fitzpatrick, J.M., West, J.B.: The distribution of target registration error in rigid-body point-based registration. IEEE Trans. Med. Imaging 20, 917–927 (2001)
8. Lee, A.W.C., Rajagopal, V., Nielsen, P.M.F., Nash, M.P.: Method for validating breast compression models using normalised cross correlation. In: Miller, K., Nielsen, M.F. (eds.) Computational Biomechanics for Medicine, pp. 63–72. Springer, Heidelberg (2010)

The Effect of Motion Correction on Pharmacokinetic Parameter Estimation

Andrew Melbourne, John H. Hipwell, and David J. Hawkes

Centre for Medical Image Computing, University College London

Abstract. A Dynamic Contrast Enhanced MRI dataset consists of many imaging frames, often both before and after contrast injection. Due to the length of image acquisition, patient motion is likely and image re-alignment or registration is required before further analysis such as pharmacokinetic model-fitting. Non-rigid image registration procedures may be use to correct motion artefacts, however careful choice of registration strategy is required to reduce mis-registration artefacts associated with enhancing features. This work compares the results of two model-fitting algorithms with two registration methods. Results show changes to the fitted parameters after motion correction within enhancing regions. This preliminary work indicates the importance of careful registration algorithm selection.

1 Introduction

Patient movement between the acquisition of contrast enhanced MRI images can cause difficulties for subsequent analysis and problems for the model-fitting procedures used for pharmacokinetic analysis [1]. Although conceivable that motion artefacts will result in biased parameter estimates from the model-fitting routine, it remains a challenge to demonstrate any improvement due to automatic image alignment algorithms. Some effort has been made, for instance in [2,3,4,5] but there remains no accepted standard of image registration accuracy in DCE-MRI.

In general, automatic image registration algorithms find the registration of local contrast change difficult. This is due to the image-similarity measure being minimised for identical images and therefore not between images with contrast change. As a result it is difficult to analyse changes due to image registration by inspection of an increase (or otherwise) of image-similarity. Due to inappropriate image-similarity measures, a common mis-registration artefact is volume change of enhancing regions. This is often counteracted by a penalty term reflecting the local Jacobian determinant of the deformation field [6], thus this also presents some proxy for registration performance. However since it is normally incorporated as a constraint, this precludes its use as a measure of assessment.

Improvements in the accuracy of parameter estimation and a quantitative analysis of factors affecting registration performance have the potential to enhance the targeting and assessment of therapeutics. The accurate determination of summary parameters in contrast enhanced MR mammography is vital for

J. Martí et al. (Eds.): IWDM 2010, LNCS 6136, pp. 744–751, 2010.

diagnostic and prognostic purposes. Image registration may be able to remove visible motion but the importance is the impact of registration algorithms on the extraction of accurate pharmacokinetic data. This preliminary work highlights some of the differences that can be generated using different model-fitting and registration algorithms emphasising some of the differences between motion artefacts and mis-registration by the registration algorithm.

2 Methods

The chosen registration algorithms are built on the b-spline image registration algorithm developed in [7]. This highly optimised algorithm is suitable for the high resolution 3D datasets used in this study. The four post-contrast volumes are registered to the pre-contrast volume (Reg 1) (See (1) where each timepoint t, in a dataset A, is registered, $\mapsto$, to the first image $A(1)$). The results in this abstract do not penalise the registration algorithm for local volume change, leaving the final deformation field result to be assessed, but do include the conventional bending energy penalisation. A second registration method (Reg 2) incorporates this algorithm into the DCE-MRI registration method in [4] incorporating a data-driven iterative routine for minimising sensitivity to predictable contrast change. The algorithm is shown in (2), in which the result (the new best registered data $\mathbf{A}^{n+1}$) at a given iteration n, is given by the registration of the best registered data from the previous step $\mathbf{A}^n$, registered to artificial images generated from a temporal principal components analysis of the best registered data from the previous step, rebuilt using n principal components $\mathbf{U}_m$ generated at that iteration. All registrations are 3D. Registrations incorporate normalised mutual information (NMI) as an image similarity measure: although due to its generality it is considered suitable for 'multi-modality' images it is not necessarily suitable for images with time-varying tissue contrast.

$$\mathbf{A}(t) \mapsto A(1) \tag{1}$$

$$\mathbf{A}^{n+1}(t) = \mathbf{A}^n(t) \mapsto \sum_{m=1}^{n<T} (\mathbf{U}_m^n \cdot \mathbf{A}^n)\mathbf{U}_m^n(t) \tag{2}$$

Model-fitting is carried out using two heuristic models. This is in contrast to more physiological models that require accurate T1 estimation, which are themselves susceptible to additional motion artefact errors. Here, a modulated sigmoid function is used to fit the signal directly (3). In general the parameters of the modulated sigmoid function reflect the properties we are interested in: namely the enhancement rate represented by μ_{in} and the fade-out rate modulated by μ_{out}. The maximum enhancement is encoded in the S_0 parameter.

A second model is also compared, although the interpretation of the units is slightly different in (4) compared to (3). This form has a close relationship to the dual-exponential input model used in [1] although the derivation and interpretation are different. Model 2 will have poor behaviour at the bolus arrival time and requires the extra condition that $S(t < t_{onset}) = 0$.

In contrast, model 1 maintains a continuous gradient throughout bolus arrival. An important difference of these two models is the interpretation of the bolus onset: bolus arrival at $t = 0$ coincides with the rise from zero for model 2 but corresponds to the half-maximum of model 1, hence care is required to avoid the possibility of systematic bias. Models 1 and 2 could be used to reduce the interpretation needed in a registration and fitting method such as [3]. The models are also less susceptible to errors due to mis-registration between the images used to estimate T1 values. Example curves are plotted in Figure 1 for parameters: model 1; $[S_0, \mu_{in}, \mu_{out}] = [1, 3min^{-1}, 0.01min^{-1}]$ and model 2; $[S_0, \mu_{in}, \mu_{out}] = [4.64, 4.71, 0.009]min^{-1}$.

$$S(t) = \frac{S_0 e^{-\mu_{out} t}}{1 + e^{-\mu_{in} t}} \quad (3)$$

$$S(t) = \frac{S_0}{\mu_{in} - \mu_{out}} (e^{-\mu_{out} t} - e^{-\mu_{in} t}) \quad (4)$$

Direct image registration assessment is carried out by inspection of the average intra-dataset, inter-volume similarity. This provides an indicator of registration performance. The deformation field Jacobian (in the absence of a volume preserving modification to the registration algorithm) may also be analysed, which may show discrepancies between registration algorithms. We also inspect changes to the parameter maps and attempt to correlate these with motion artefacts by generating a time series of synthetic images from the parameter maps.

Five high resolution 3D volumes from four patients are analysed, three at 3T and two at 1.5T similar to the protocol in [8]. A pre-contrast volume is acquired followed by four subsequent volumes after contrast bolus injection over a time of approximately 18 minutes. The datasets thus allow comparison of registration performance alongside model-fitting evaluation. Pixel sizes are $0.92x0.92x1mm$

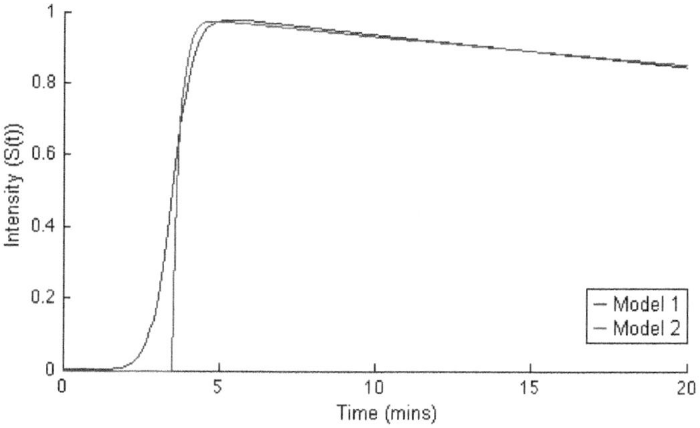

Fig. 1. Example time intensity curves for blue) model 1 (Equation3) and green) model 2 (Equation 4)

for 3T and $0.66x0.66x1.3mm$ for 1.5T. Additional low spatial resolution data is acquired immediately after bolus injection, thus an estimate of bolus arrival time can be used to remove a parameter from the model-fitting.

3 Results

Summary registration and model-fitting parameters are shown in Table 1 for a single slice in each dataset. The total model-fit residual for both model 1 and 2 is lower for registration method 2 (Reg 2) than for no registration and method 1 (Reg 1). The fitting of model 1 often has a lower final model-fit residual than for fitting model 2. The large residual for dataset 5 appears to be the result of mis-registration by method 1 in an area of pectoral muscle. For the registration statistics, the average inter-volume NMI is improved by image registration, more so for registration method 2 than 1, perhaps due to the iterative update of the target images used in Reg 2. Assessment of the standard deviation of the per-pixel percentage volume change - found by taking the log of the local Jacobian determinant for each pixel - shows only a small volume change in addition to smaller volume change statistics seen for Reg 2 than Reg 1. Specific results of the model-fitting for dataset 2 are shown in the histograms of Figure 2. Figures 2a and b show histograms of the distribution of all fitted values of $-\mu_{in}$ and $-\mu_{out}$ found by model 1 and their change under image registration (shown as a line for improved visibility). Correction of motion artefacts visibly changes the parameter distribution across the whole breast, corresponding to improved visual alignment in some areas of the images requiring further investigation.

Inspection of Dataset 2 (associated with Figure 2) reveals differences due to image registration. The first column of Figure 3 shows timepoint 2 registered by each method. Whilst superficially similar in all cases, there are motion artefacts in the original dataset and mis-registration artefacts between timepoints in the central enhancing region for Reg 1 when compared with other images in the dataset. this is true when inspecting neighbouring slices for through-plane effects and for the remaining timepoints. Column 2 contains corresponding images to

Table 1. Registration performance comparison. Average model fit residuals for models 1 (Fit1) and 2 (Fit2) (units of intensity, for a single slice). Also shown are the average inter-volume NMI for each dataset ($\bar{NMI}$) and the per-pixel standard deviation log of the Jacobian determinant (Jac).

	NoReg		Reg1		Reg2		NoReg		Reg1		Reg2	
Dataset	Fit1	Fit2	Fit1	Fit2	Fit1	Fit2	$\bar{NMI}$	Jac	$\bar{NMI}$	Jac	$\bar{NMI}$	Jac
1	0.043	0.046	0.028	0.030	0.026	0.027	1.56	0	1.59	0.017	1.59	0.012
2	0.053	0.056	0.041	0.042	0.028	0.030	1.56	0	1.58	0.016	1.58	0.013
3	0.028	0.30	0.023	0.025	0.020	0.021	1.55	0	1.55	0.018	1.57	0.016
4	0.024	0.027	0.022	0.024	0.022	0.024	1.55	0	1.54	0.022	1.56	0.022
5	0.018	0.018	0.031	0.032	0.013	0.013	1.57	0	1.57	0.021	1.58	0.028

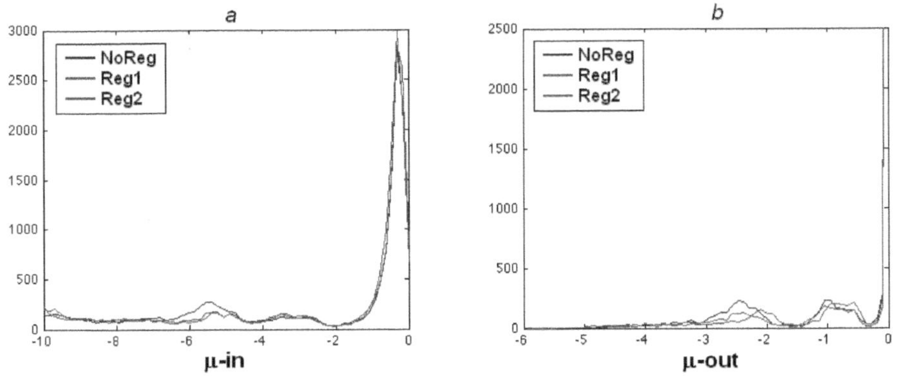

Fig. 2. Histogram counts of all values of μ_{in} and μ_{out} over the manually segmented right breast for dataset 2 using model 1 and varying registration method. *a*: histogram of fitted parameter values for μ_{in} and *b*: for μ_{out}

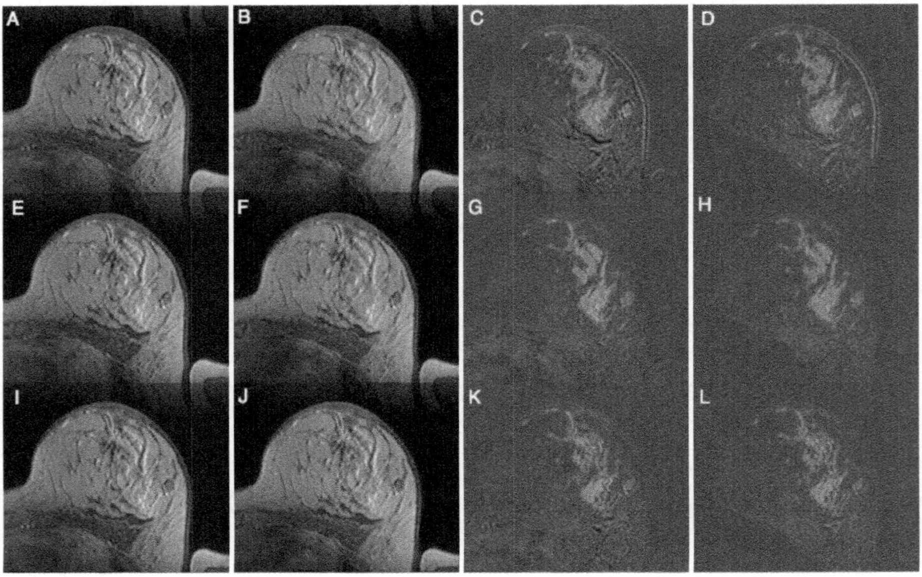

Fig. 3. Differences in model-fitting after each image registration method (rows) (Dataset 2, timepoint 2). First column [a,e,i]: registered timepoint 2, second column: synthetic images at timepoint 2 generated from model-fitting, third column: difference of first column with first (pre-enhancement) image in dataset and fourth column: difference of second column with pre-enhancement synthetic images. Rows: 1) No registration 2) Reg 1, 3) Reg 2.

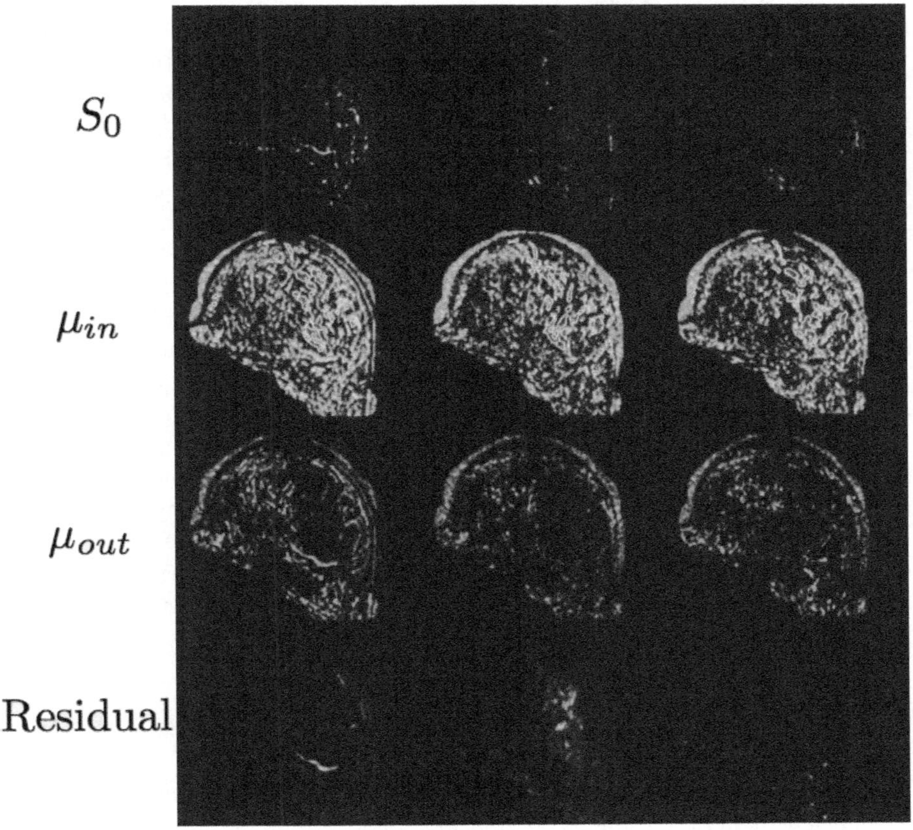

Fig. 4. Analysis of parameter maps for dataset 2. Rows: parameters for model 1, $S_0, \mu_{in}, \mu_{out}, R]^T$. Columns: Unregistered data, Reg1, Reg2. Images smoothed by Gaussian of width $\sigma = \sqrt{2}$. Figure best viewed electronically. All rows scaled to maximum value = maximum intensity (red).

column one, but these images have been generated from the parameters found by the model-fitting routines (here model 1). We begin to see discrepancies in the resolution of small enhancements (compare Figure 3F and 3J). The third and fourth columns of Figure 3 are difference images with the corresponding version of timepoint 1 (original, registered, synthetic as required); differences between registration methods are now enhanced. Motion artefacts associated with the lateral breast are present in the unregistered case: these are reduced by both registration methods. However, in the case of Reg 1, there is some loss of resolution and detail in the model-fitting of the enhancing regions, this is likely to be due to the mis-registration artefacts in the remainder of the dataset. These mis-registration artefacts are not present under Reg 2 resulting in finer model-fitting in the enhancing breast. Thus, care must be taken with the registration algorithm. Possible solutions are the registration of all images within a dataset to one another, in a scheme such as Reg 2, or more careful choice of the fixed image, for instance registering all images to the second (enhancing) image, $A(2)$.

Parameter maps for dataset 2 are shown in Figure 4. The motion artefacts in Figure 3 column 3 around the lateral breast boundary are associated with larger values of the μ_{in} parameter. This reflects the breast motion in the enhancing images relative to the first image in the dataset; thus the intensity change due to motion is fitted by non-zero model-parameter values. Mis-registration artefacts using Reg 1 manifest as the area of large residual in the final row. This feature is not present in the corresponding images for no registration and for Reg 2 and is likely to be the result of registration of enhancing features to an inconsistent position in the pre-enhancement image. In general, after image registration, existing features are enhanced relative to their surroundings and there are examples of higher regions of μ_{in} in small areas after motion correction that are not the result of through plane artefacts.

A summary analysis of the deformation magnitude (using an *l2-norm*) for each timepoint by each registration method reveals: for Reg 1, increasingly substantial deformations are needed with increasing time from timepoint 1 and for Reg 2, each timepoint has a largely equivalent deformation magnitude. This is commensurate with the registration scheme used, since Reg 1 registers each image to the space of the first image, whilst the Reg 2 method registers each image towards an average position.

4 Discussion

This work presents an initial investigation of the effect of image registration on pharmacokinetic parameter estimation. Early results suggest that a reduction of motion artefacts can be linked to changes in parameter estimation over registered regions, confirming that motion correction is an important step in contrast-enhanced MR post-processing. In addition, results suggest that performance differences can be achieved when varying the registration method (hence it is important to choose an optimal registration strategy given the data). In particular, if the chosen registration strategy is sub-optimal, mis-registration artefacts may occur, as shown in Figure 3. The emphasis must be on a registration strategy that reflects the time-series nature of the data or carefully chooses a suitable fixed image to which to register, however validation of differences in registration algorithms is challenging without a standardised framework. Further work will assess the contribution of registration algorithms incorporating a volume change penalty in the registration and algorithms involving iterative model-fitting as part of the registration procedure. Performance differences will be investigated alongside monitoring of the changes to parameters during the registration procedure.

Acknowledgements. This work was funded by EPSRC grant EP/E031579/1. The authors would like to thank the Radboud University Nijmegen Medical Centre.

References

1. Tofts, P.S.: Modeling tracer kinetics in dynamic Gd-DTPA MR imaging. J. Magn. Reson. Imaging 7(1), 91–101 (1997)
2. Hayton, P., Brady, M., Tarassenko, L., Moore, N.: Analysis of dynamic MR breast images using a model of contrast enhancement. Med. Image Anal. 1(3), 207–224 (1997)
3. Buonaccorsi, G.A., O'Connor, J.P.B., Caunce, A., Roberts, C., Cheung, S., Watson, Y., Davies, K., Hope, L., Jackson, A., Jayson, G.C., Parker, G.J.M.: Tracer kinetic model-driven registration for dynamic contrast-enhanced mri time-series data. Magn. Reson. Med. 58(5), 1010–1019 (2007)
4. Melbourne, A., Atkinson, D., White, M.J., Collins, D., Leach, M., Hawkes, D.: Registration of dynamic contrast-enhanced MRI using a progressive principal component registration (ppcr). Phys. Med. Biol. 52(17), 5147–5156 (2007)
5. Milles, J., van der Geest, R.J., Jerosch-Herold, M., Reiber, J.H.C., Lelieveldt, B.P.F.: Fully automated motion correction in first-pass myocardial perfusion mr image sequences. IEEE Trans. Med. Imaging 27(11), 1611–1621 (2008)
6. Rohlfing, T., Maurer, C.R., Bluemke, D.A., Jacobs, M.A.: Volume-preserving non-rigid registration of MR breast images using free-form deformation with an incompressibility constraint. IEEE Trans. Med. Imaging 22(6), 730–741 (2003)
7. Modat, M., Ridgway, G.R., Taylor, Z.A., Lehmann, M., Barnes, J., Fox, N.C., Hawkes, D.J., Ourselin., S.: Fast free-form deformation using graphics processing units. In: Comput. Methods Programs Biomed. (2009)
8. Veltman, J., Stoutjesdijk, M., Mann, R., Huisman, H.J., Barentsz, J.O., Blickman, J.G., Boetes, C.: Contrast-enhanced magnetic resonance imaging of the breast: the value of pharmacokinetic parameters derived from fast dynamic imaging during initial enhancement in classifying lesions. Eur. Radiol. 18(6), 1123–1133 (2008)

Improved Detection of Cancer in Screening Mammograms by Temporal Comparison

Fei Ma, Mariusz Bajger, Simon Williams, and Murk J. Bottema

Flinders University, Bedford Park 5042, SA, Australia
{feim,Mariusz.Bajger,swilliam,murkb}@csem.flinders.edu.au

Abstract. A method is presented for including information from the preceeding mammogram in a scheme for automatically detecting malignant masses in screening mammograms. The method circumvents the inherent difficulty of registering temporal mammograms by replacing image registration by graph matching. The scheme incorporates a single image mass detection algorithm and so the contribution of the temporal analysis can be measured. At a true detection rate of 80 percent, the single image scheme results in 1.02 false positive detections per image while the temporal scheme results in 0.96 false positives. At 90 percent true detection, the false positive rates per image are 1.84 and 1.63 respectively.

Keywords: computer-aided mammography, breast cancer, mass detection, graph matching, temporal analysis.

1 Introduction

In reading screening mammograms, the availability of a previous mammogram provides important information regarding the likelihood of breast cancer. If a suspicious region in the current mammogram matches closely in appearance and location to a suspicious region in the previous mammogram, then the anomaly is not likely to be associated with cancer. If there is no corresponding anomaly in the previous mammogram or if there is significant change, then this argues for cancer. Accordingly, inspecting the previous mammogram helps to reduce the number of false positive detections and allows the identification of cancer that might be missed in reading the current mammogram alone.

Automating the process of incorporating previous mammograms could play an important role in computer-aided mammography but has not been adopted widely. This is due, in part, to the difficulty of the image registration problem, in particular, the task of automatically identifying anomalies in one image with corresponding regions in the previous image.

One approach is to find a function that maps the breast in one image to match the breast in the other image. Hasegawa et al. started with a rigid body alignment followed by registration of the dense regions of the mammogram using deformation of a B-spline control point grid [3]. Richard and Cohen used a variational formulation to find the smooth function ϕ that best explains the deformation needed to map one mammogram to another and applied the method

J. Martí et al. (Eds.): IWDM 2010, LNCS 6136, pp. 752–759, 2010.

to bilateral pairs [11]. While their method is very satisfying from a mathematical point of view, the difficulty with methods based on warping functions is that, in practice, the true map between two images of the same breast is usally not smooth and need not even be well defined.

Another approach is to search locally in one breast for a matching anomaly found in the other breast. Sanay-Gopal et al. used a search region based on location of the mass in the current mammogram relative to the centroid of the breast and the nipple [12]. Sahiner et al. investigated the classification of masses as benign or malignant based on comparing attributes of the current and previous mammogram, but relied on radiologist defined regions of interest to search for associated anomalies in the mammograms [2].

A natural extension is to combine global and local registration. Timp et al. performed initial rigid body global registration by aligning the centers of mass (after removing the pectoral muscle) of the two breasts and then used local searches to align suspicious regions in the current breast [14]. Marias et al. also used two stages. In the first stage, the boundaries of the breast were matched. In the second stage, thin plate slpines were used to match the multiresolution representations of the internal structure of the breasts [9].

Comparing these approaches is difficult because results were reported on different data sets. A comparison of four methods: nipple alignment, center of mass alignment, maximizing mutual information after allowing deformation by shifting, rotating, scaling and vertical shear and warping suggested that the method using mutual information was best [15]. However, the problem of temporal comparison cannot be considered as solved.

This paper extends the strategy introduced in [7], namely that of replacing image registration by graph matching. The potential advantage of this method is that registration is restricted only to image features relevant to breast cancer. This means there is no opportunity for errors in registration of features not related to cancer to contribute to the overall failure of the registration process. The main steps of the method are; (1) segmentation of the current and previous image, (2) assigning mass-like scores to components to reflect the likelihood that the component corresponds to a malignant mass, (3) applying graph matching algorithm to associate candidate masses in the current and previous image and (4) adjusting the range of the mass-like score to reflect the information from both images.

Although the study in [7] demonstrated a slight improvement in detection performance when temporal analysis was included, this has to be interpreted with care for two reasons. First, the performance was reported on a component by component basis and not on the true and false positive detections per image or per case. Second, the performance of the initial segmentation scheme was not competitive with state of the art schemes, leaving open the possibility that the improvement seen in applying temporal analysis only made up for deficiencies in the original segmentation scheme and would not actually improve screening performance in practice.

The present study follows the same steps as in [7] but upgrades part of the segmentation process by using sublevel set analysis [13], [1], [4]. This results in improved segmentation performance for detection of cancer in single images and hence provides a more reliable base for judging the contribution of the temporal analysis.

2 Methods

Initial segmentation was accomplished using an adaptive pyramid (AP) method [5]. This graph theory based method reliably identifies components associated with masses but does not provide accurate boundaries of masses needed to compute characteristics such as size, shape and contrast parameters. The components found by the AP method were refined using sublevel set analysis [13], [1], [4]. This involves forward extraction of sublevel sets, excluding those not relevant to mass detection, backward extraction and contour identification [8]. Sublevel set analysis does produce accurate boundaries but only once locations of interest are provided. Hence, initial segmentation by AP is viewed as seeding segmentation by sublevel sets. Next, a mass-like score was computed based on 17 features (Appendix) measured on the components found in the segmentation step. The mass-like score was used to restrict further attention to candidate malignant masses only. The cut-off was set sufficiently low that a very large number of true malignant masses are likely to be included (in principle all, but this cannot be guaranteed). Many non-malignant anomalies were included too.

A full graph was constructed with the surviving candidate masses as vertices and with fuzzy spatial relationships used as edge weights [10]. More explicitly, the four relationships *to the right of*, *to the left of*, *below*, and *above*, were computed for two points p and q by

$$\mu_{\text{right}} = \begin{cases} \cos^2\theta \text{ if } \frac{-\pi}{2} \leq \theta \leq \frac{\pi}{2}, \\ 0 \qquad \text{otherwise,} \end{cases}$$

$$\mu_{\text{below}} = \begin{cases} \sin^2\theta \text{ if } 0 \leq \theta \leq \pi, \\ 0 \qquad \text{otherwise,} \end{cases}$$

$$\mu_{\text{above}} = \begin{cases} \sin^2\theta \text{ if } -\pi \leq \theta \leq 0, \\ 0 \qquad \text{otherwise,} \end{cases}$$

$$\mu_{\text{left}} = \begin{cases} \cos^2\theta \text{ if } -\pi \leq \theta \leq \frac{-\pi}{2}, \frac{\pi}{2} \leq \theta \leq \pi, \\ 0 \qquad \text{otherwise,} \end{cases}$$

where θ is angle between the positive x-axis and the line segment joining p and q. For two components $A = \{a_1, a_2, \ldots, a_n\}, B = \{b_1, b_2, \ldots, b_m\}$, the multiset $\Theta = \{\theta_{ij} = \angle(a_i b_j), i = 1, 2, \ldots, n, j = 1, 2, \ldots, m\}$ was constructed and the histogram associated with Θ was defined as

$$H_\Theta(A, B) = \{(\theta, f_\theta)\},$$

where $f(\theta)$ is the number of pairs (a_i, b_j) with $\angle(a_i, b_j) = \theta$ and f is normalized by $\max_\theta f(\theta)$.

To evaluate the degree to which a spatial relation between two regions holds, the histogram H is treated as an unlabeled set, which represents the spatial relation between A and B. The problem of to what extent H is described by the four spatial relations is then treated as a problem of compatibility of fuzzy sets [10]. If G denotes the fuzzy set of one of the spatial relations and μ_H and μ_G denote the membership functions of H and G, the compatibility of H to the fuzzy set G is the fuzzy set $CP(H;G)$ whose membership function is defined as

$$\mu_{CP(H;G)}(v) = \begin{cases} \sup_{s,v=\mu_G(s)} \mu_H(s) & \text{if } \mu_G^{-1}(v) \neq \emptyset, \\ 0 & \text{if } \mu_G^{-1}(v) = \emptyset. \end{cases}$$

The final degree to which a spatial relation holds is defined as the center of gravity of the compatibility fuzzy set, $CP(H;G)$.

In addition, the boundary of the breast was included as special vertex in the graph as this allowed a connection between the relative positions of masses and the global shape of the breast. A graph matching scheme similar to [6] was used to match the resulting graphs of the current and previous images. The algorithm was adjusted to compensate for the fact that matching solutions of different lengths must be compared in this application.

3 Data

Cases were obtained from the archives of BreastScreen SA (Adelaide, South Australia) subject to the criteria that (1) one breast contained a malignant mass found at screening and verified by histopathology, (2) a previous mammogram from a visit 2 - 3 years earlier and judged to be normal was available. This resulted in 95 images with cancer and 90 images without cancer. Images were digitized at 48μm resolution and 12 bit depth using a Vidar Diagnostic Pro Advantage digitizer. For the segmentation step, images were reduced by replacing 8×8 pixel patches by a single pixel having intensity equal to the mean intensity of the patch.

4 Results and Discussion

Every image was assigned a score equal to the maximum mass-like score of any of its components. For a given threshold on the mass-like score, the percentage of images containing true malignant masses and correctly identified as such, was computed, as well as the number of false positive components within each current image (Table 1). This resulted in a measure of the performance of the process based on the current screening visit only. To incorporate the previous image, a candidate mass in the current image for which no match was found in the previous image, retained its current image mass-like score. Candidate masses for which a similar anomaly was found in the previous image were assigned the

Table 1. The entries show the average number of false positive detections per image or per case for the given true detection rate

	per-case		per-image	
Detection	Single	Temporal	Single	Temporal
80%	1.04	1.00	1.02	0.96
90%	1.62	1.43	1.84	1.63

minimum of the two scores. This step incorporates the prior knowledge that the appearance of the candidate mass in the previous image decreases the likelihood that it is associated with cancer. Also, if there is a match, but the mass has changed significantly (lesion growth, for example) then this score reflects such a change and the lesion is more likely to be assigned as cancer. These adjusted mass-like scores were used to compute the number of false positive detections per image for fixed true detection rates as described above (Table 1). Similarly, a case was considered a false positive case if at least one image contained at least one false positive.

The performance of the detection scheme applied to current images only (Single) is comparable to results published for the best mass detection algorithms. Hence the reduction of false positive detections per image seen when the previous image is compared to the current image (Temporal) represents a true, albeit small, gain in performance.

4.1 Examples

Visual inspection of images indicates that plausible matches were found in most cases including cases where difference in positioning resulted in different apparent sizes of the breast (Figure 1) or substantial difference appearance (Figure 2).

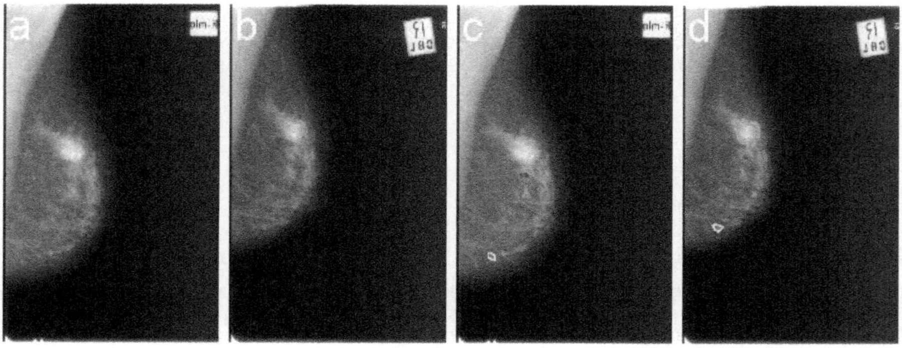

Fig. 1. Examples of match results. (a) The current mammogram. (b) The same breast from the previous visit. (c) The same image as (a) but with outlines showing matched regions. (d) The same image as (b) but showing the matched regions color coded to indicate the correspondence between the regions in (c). The matches seem to be correct despite substantial difference in the size and position of the breast.

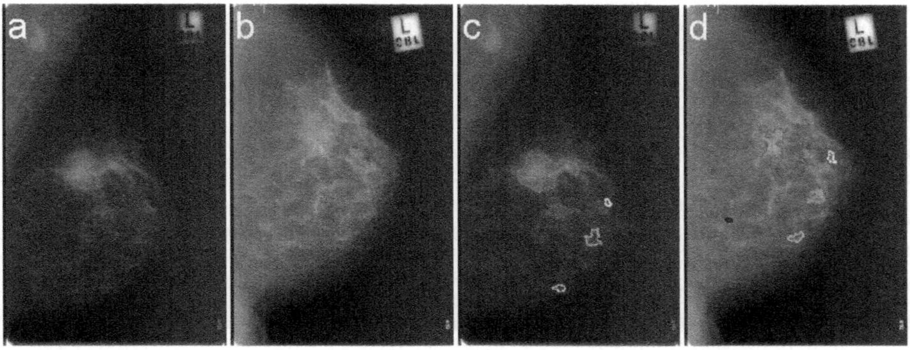

Fig. 2. (a) The current mammogram. (b) The same breast from the previous visit. (c) Same as (a) but with matched regions annotated. (d) Same as (b) but showing matched regions color coded to indicate associations with regions in (c). The matching process correctly identified the regions shown despite substantial differences in relative location, appearance of the anomalies and overall positioning of the breast.

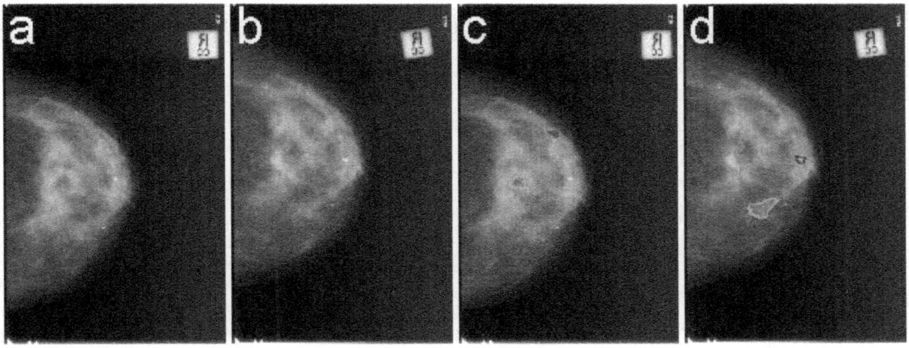

Fig. 3. (a) The current mammogram. (b) The same breast from the previous visit. (c) Same as (a) but with regions annotated. (d) Same as (b) but showing matched regions color coded to indicate associations with regions in (c). The matches are not correct in this case. The global distribution of the dense tissue was very different in these images and lead to incorrect association of regions.

In both these examples, regions which may appear suspicious in the current mammogram were successfully identified in the previous mammogram thus improving the chance that the regions are correctly identified as not being associated with cancer.

The matching process failed conspicuously in only one case (Figure 3). This failure was likely due to very difference of appearance of the dense tissue in the current and previous mammogram. In a clinical setting, the result of such an error would be to alert the radiologist that there were many mass-like regions in the current mammogram with no obvious corresponding anomaly in the previous mammogram and so that these regions should be considered carefully.

Acknowledgments. This work is supported by National Breast Cancer Foundation (Australia). The authors would like to thank BreastScreen SA for providing screening film mammograms.

References

1. Eltonsy, N.H., Tourassi, G.D., Elmaghraby, A.S.: A concentric morphology model for the detection of masses in mammography. IEEE Transactions on Medical Imaging 26(06), 880–889 (2007)
2. Hadjiiski, L., Sahiner, B., Chan, H.-P., Petrick, N., Helvie, M.A., Gurcan, M.: Analysis of temporal changes of mammographic features: Computer-aided classification of malignant and benign breast masses. Medical Physics 28(11), 2309–2317 (2001)
3. Hasegawa, A., Neemuchwala, H., Tsunoda-Shimizu, H., Honda, S., Shimura, K., Sato, M., Koyama, T., Kikuchi, M., Hiramatsu, S.: A tool for temporal comparison of mammograms: Image toggling and dense-tissue-preserving registration. In: Krupinski, E.A. (ed.) IWDM 2008. LNCS, vol. 5116, pp. 447–454. Springer, Heidelberg (2008)
4. Hong, B.-W., Brady, M.: A topographic representation for mammogram segmentation. In: Ellis, R.E., Peters, T.M. (eds.) MICCAI 2003. LNCS, vol. 2879, pp. 730–737. Springer, Heidelberg (2003)
5. Jolion, J.M., Montanvert, A.: The adaptive pyramid: A framework for 2d image analysis. Computer Vision, Graphics, and Image Processing 55(3), 339–348 (1992)
6. Krissinel, E.B., Henrick, K.: Common subgraph isomorphism detection by backtracking search. Softw. Pract. Exper. 34, 591–607 (2004)
7. Ma, F., Bajger, M., Bottema, M.J.: Temporal analysis of mammograms based on graph matching. In: Krupinski, E.A. (ed.) IWDM 2008. LNCS, vol. 5116, pp. 158–165. Springer, Heidelberg (2008)
8. Ma, F., Bajger, M., Bottema, M.J.: Automatic mass segmentation based on adaptive pyramid and sublevel set analysis. In: DICTA 2009, Melbourne, Australia. IEEE, Los Alamitos (2009)
9. Marias, K., Behrenbruch, C., Parbhoo, S., Seifalian, A., Brady, M.: A registration framework for the comparison of mammogram sequences. IEEE Transactions on Medical Image Analysis 24(6), 782–790 (2005)
10. Miyajima, K., Ralescu, A.: Spatial organization in 2d segmented images: Representation and recognition of primitive spatial relations. Fuzzy Sets and Systems 65 (2-3), 225–236 (1994)
11. Richard, F.J.P., Cohen, L.D.: A new image registration technique with free boundary constraints: application to mammography. Computer Vision and Image Understanding 89, 166–196 (2003)
12. Sanay-Gopal, S., Chan, H.-P., Wilson, T., Helvie, M., Petrick, N., Sahiner, B.: A regional registration technique for automated interval change analysis of breast lesions on mammograms. Medical Physics 26(12), 2669–2679 (1999)
13. Shiffman, S., Rubin, G.D., Napel, S.: Medical image segmentation using analysis of isolable-contour maps. IEEE Transactions on Medical Imaging 19(11), 1064–1074 (2000)
14. Timp, S., Karssemeijer, N.: Interval change analysis to improve computer aided detection in mammography. Medical Image Analysis 10, 82–95 (2006)
15. van Engeland, S., Snoerten, P., Hendriks, J., Karssemeijer, N.: A comparison of methods for mammogram registration. IEEE Transactions on Medical Imaging 22(11), 1436–1444 (2003)

Appendix: Features for Mass-Like Scores

R denotes the region to which features are assigned, $R(i,j)$ is the intensity of pixel (i,j) and $|\cdot|$ denotes size.

- Solidity: $\frac{|R|}{|C(R)|}$, where $C(R)$ denotes the convex hull of R.
- Ratio of lengths of the major and minor axis of the ellipse that has the same normalized second central moments as R.
- Standard deviation of the radial distance of the boundary of R.
- Intensity variance along the boundary: $\frac{|E(R)-E(O_2)|}{E(R)+E(O_2)}$, where O_t represents the set of pixels outside component R but within t pixels distance. $E(\cdot)$ is the mean intensity.
- Contrast measure 1: $\frac{(E(R)-E(O_d))^2}{\sigma(R)+\sigma(O_d)}$, where $d = \sqrt{\frac{|R|}{\pi}}$ and $\sigma(R)$ is the std of the intensity in R.
- Contrast measure 2: $\sum_i |H(R,i) - H(O_d,i)|$, where $H(R,i)$ represents the proportion of pixels in component R with intensity i.
- Entropy of the intensity distribution: $\sum_{k=1}^{256} H(R,k)\log(H(R,k))$.
- Energy: $\frac{\sum_{i,j} R(i,j)^2}{|R|}$.
- Luminosity inertial momentum: $\sum_{i,j} R(i,j)D(i,j)^2$ where $D(i,j)$ is the Euclidean distance between (i,j) and center of luminosity. The center of luminosity is defined by $x_{lc} = \frac{\sum_{i,j} R(i,j)i}{\sum_{i,j} R(i,j)}$ and $y_{lc} = \frac{\sum_{i,j} R(i,j)j}{\sum_{i,j} R(i,j)}$.
- Anisotropy: the distance between the geometric center and the center of luminosity.
- Seven second and third order central invariant moments of R.

Combined Reconstruction and Registration of Digital Breast Tomosynthesis

Guang Yang[1,⋆], John H. Hipwell[1], Matthew J. Clarkson[1,2], Christine Tanner[1,3], Thomy Mertzanidou[1], Spencer Gunn[4], Sebastien Ourselin[1], David J. Hawkes[1], and Simon R. Arridge[1]

[1] Centre for Medical Image Computing, Department of Computer Science and Medical Physics, University College London (UCL), London, WC1E 6BT, UK
G.Yang@cs.ucl.ac.uk
[2] Dementia Research Centre, UCL Institute Of Neurology, London, WC1N 3BG, UK
[3] Computer Vision Laboratory, Eidgenössische Technische Hochschule Zürich (ETH), 8092, Zürich, CH
[4] Dexela Ltd, London N1 7EU, UK

Abstract. Digital breast tomosynthesis (DBT) has the potential to enhance breast cancer detection by reducing the confounding effect of superimposed tissue associated with conventional mammography. In addition the increased volumetric information should enable temporal datasets to be more accurately compared, a task that radiologists routinely apply to conventional mammograms to detect the changes associated with malignancy. In this paper we address the problem of comparing DBT data by combining reconstruction of a pair of temporal volumes with their registration. Using a simple test object, and DBT simulations from in vivo breast compressions imaged using MRI, we demonstrate that this combined reconstruction and registration approach produces improvements in both the reconstructed volumes and the estimated transformation parameters when compared to performing the tasks sequentially.

1 Introduction

Digital breast tomosynthesis (DBT) is an X-ray modality in which a small number of low dose X-ray images (typically between 10 and 50) are acquired over a limited angle and reconstructed into a 3D volume [1]. A key issue in the creation of DBT images is the algorithm used to perform the reconstruction. This has been a topic of substantial research with many algorithms being proposed including traditional shift-and-add (SAA) [2], filtered back-projection (FBP) [3], algebraic reconstruction technique (ART) [4], maximum-likelihood expectation maximization (MLEM) [1], and matrix inversion tomosynthesis (MITS) [5]. In addition surveys have been published comparing and contrasting the relative merits of each approach [2] [4].

⋆ This work has been funded by DTI Project *Digital Breast Tomosynthesis* TP/7/SEN/6/1/M1577G. The authors would like to thank the UK MR Breast Screening Study (MARIBS) [6] for providing the data for this study.

J. Martí et al. (Eds.): IWDM 2010, LNCS 6136, pp. 760–768, 2010.

Reconstructed 3D DBT images have high in-plane resolution but low out-of-plane resolution and exhibit reduced superposition of overlying tissue structures as compared to conventional X-ray mammography. Whilst the added depth information offered by DBT has the potential to enhance detection and diagnosis of breast cancer [7]; the greater volume of data, relative to X-ray mammography, increases the need for automated tools to aid the reading process. This is of particular importance if DBT is to be adopted in the high workload screening context.

In this paper we address the problem of comparing temporal DBT volumes via registration. This is a challenging task due to the significant artefacts associated with DBT reconstructions. These are generated by the limited field of view of the acquired images and the correspondingly large null-space in the frequency domain. Rather than registering the images after reconstruction therefore, we investigate the benefits of combining both reconstruction and registration, and test the hypothesis that the performance of each task will be enhanced as a result. We propose an iterative method of least squares optimisation for our combined reconstruction and registration scheme. This avoids the implicit assumption of missing data being equal to zero in algorithms such as in FBP.

In recent relevant research on SPECT imaging [8] Schumacher et al. present a method to combine reconstruction with motion correction using a rigid transformation. We have developed an iterative algorithm which alternates between optimising the reconstructed intensities at each time point and the affine transformation parameters between time points.

2 Method

Two sets of limited angle X-ray acquisitions, $\boldsymbol{y_1} \in \Re^{N_2}$ and $\boldsymbol{y_2} \in \Re^{N_2}$, obtained at different times, can be expressed in terms of a 3D volume, $\boldsymbol{x} \in \Re^{N_3}$, in two positions related by the transformation, R, with parameters, $\boldsymbol{\zeta_p} \in \Re$, and the system matrix $A : \Re^{N_3} \mapsto \Re^{N_2}$ via

$$\boldsymbol{y_1} = A\boldsymbol{x}, \tag{1}$$

and

$$\boldsymbol{y_2} = A\boldsymbol{x}^\dagger = AR_{\zeta_p}\boldsymbol{x}. \tag{2}$$

We solve equations 1 and 2 with respect to estimates $\boldsymbol{x_1}$ and $\boldsymbol{x_2}$ of $\boldsymbol{x}$ and the registration parameters $\boldsymbol{\zeta_p}$, by alternating an incomplete optimisation (*i.e.* n iterations) of the reconstructed volumes $\boldsymbol{x_1}$ and $\boldsymbol{x_2}$:

$$\boldsymbol{x_1^*} = \arg\min_{\boldsymbol{x_1}} \left(\Phi_{Rec1} = \frac{1}{2}\left\|A\boldsymbol{x_1} - \boldsymbol{y_1}\right\|_2^2 \right) \tag{3}$$

$$\boldsymbol{x_2^*} = \arg\min_{\boldsymbol{x_2}} \left(\Phi_{Rec2} = \frac{1}{2}\left\|A\boldsymbol{x_2} - \boldsymbol{y_2}\right\|_2^2 \right) \tag{4}$$

with the registration of the current estimates $\boldsymbol{x_1^*}$ and $\boldsymbol{x_2^*}$ with respect to the registration parameters $\boldsymbol{\zeta_p}$:

$$\zeta_p{}^* = \arg\min_{\zeta_p} \left(\Phi_{Reg} = \frac{1}{2}\left\|R_{\zeta_p}\boldsymbol{x_2^*} - \boldsymbol{x_1^*}\right\|_2^2 \right). \tag{5}$$

After each registration iteration (Eq. 5), and prior to the next iteration of the reconstructions (Eqs. 3 and 4), the reconstruction estimates are updated as follows (Eqs. 6 and 7).

$$\boldsymbol{x}_1 = R_{\zeta_p}\boldsymbol{x}_2^* \tag{6}$$

$$\boldsymbol{x}_2 = \boldsymbol{x}_2^*. \tag{7}$$

This "outer loop" of reconstruction followed by registration is repeated m times. The last iteration outputs $\boldsymbol{x}_1 = \boldsymbol{x}_1^*$, $\boldsymbol{x}_2 = \boldsymbol{x}_2^*$ and $R_{\zeta_p}\boldsymbol{x}_2^*$.

The reconstruction is performed via a nonlinear conjugate gradient search engine and the registration currently via a simple hill-climbing optimisation method. The following analytical gradients are used for $\boldsymbol{x}_1$ and $\boldsymbol{x}_2$

$$\Psi_{\boldsymbol{x}_1} = A^T(A\boldsymbol{x}_1 - \boldsymbol{y}_1) \tag{8}$$

$$\Psi_{\boldsymbol{x}_2} = A^T(A\boldsymbol{x}_2 - \boldsymbol{y}_2). \tag{9}$$

The preceding combined reconstruction and registration method is summarised by

Algorithm 1. Iteratively Combined Reconstruction and Registration

Input: $\boldsymbol{y}_1$, $\boldsymbol{y}_2$.
Output: $\boldsymbol{x}_1$, $\boldsymbol{x}_2$, $R_{\zeta_p}\boldsymbol{x}_2$.

```
begin
  % Initialization of x1 and x2
  x1^{0,0} := 0; x2^{0,0} := 0; ζp^0 := 0;

  % Outer loop for the registration
  for (i = 0; i < m; i++) do
    % Inner loop for the reconstruction
    for (j = 0; j < n; j++) do
      % Ψx is the analytical gradients of the x and CG solver
      Ψ_{x1^{i,j}} := A^T(A x1^{i,j} − y1);
      Ψ_{x2^{i,j}} := A^T(A x2^{i,j} − y2);
      x1^{i,j+1} := x1^{i,j} + (A^T A)^{-1} Ψ_{x1^{i,j}};
      x2^{i,j+1} := x2^{i,j} + (A^T A)^{-1} Ψ_{x2^{i,j}};
    % Run a simple hill-climbing optimisation
    ζp^{i+1} := arg min_{ζp^i} (1/2) || R_{ζp^i} x2^{i,j+1} − x1^{i,j+1} ||_2^2;
    x1^{i+1,j+1} := R_{ζp^{i+1}} x2^{i,j+1};
    x2^{i+1,j+1} := x2^{i,j+1};

  % Output x1, x2, and R_{ζp} x2
  x1 := x1^{i,j+1};
  x2 := x2^{i+1,j+1};
  R_{ζp} x2 := x1^{i+1,j+1} := R_{ζp^{i+1}} x2^{i,j+1}.
end
```

3 Results

In the following three experiments we compare the performance of (a) *sequential* reconstruction and registration, in which $n = 100$ iterations of the reconstruction of projection images, $\boldsymbol{y_1}$ and $\boldsymbol{y_2}$, are followed by a single registration of the reconstructed volumes $\boldsymbol{x_1}$ and $\boldsymbol{x_2}$ ($m = 1$), and (b) our *iterative* method in which $n = 10$ iterations of the reconstruction are followed by a registration, and this is repeated $m = 10$ times. In both cases the total reconstruction iterations are the same ($m \times n = 100$); however, there are 10 registrations in our iterative approach rather than the single registration used in the sequential method. For each pair of test volumes, $\boldsymbol{x}$ and $\boldsymbol{x}^\dagger$, 11 projections covering ± 25 degrees are created to simulate the pair of temporal DBT acquisitions $\boldsymbol{y_1}$ and $\boldsymbol{y_2}$.

In the first experiment a 3D toroidal phantom image was created and rigidly transformed via parameters R_{ζ_p} using a translation of $T_{x,y,z} = [10, 0, -20]$ mm and a rotation about the y axis of -30 degree (Fig. 1). As seen in Fig. 1. (f) and (h), the *iterative* results are more compact and accurate than the *sequential results* Fig. 1. (b) and (d), and the out of plane blurring is reduced (coloured squares). The sum of squared differences (SSD) $\|x_1 - x\|_2^2$ is decreased by an order of magnitude (10^{11} to 10^9); however, for the *iterative* method this value of 4.32×10^9 is superior to the *sequential* result of 6.89×10^9. In the second experiment the same transformation was applied to a 3D breast MR image that obtained similar behaviour (*iterative* 1.25×10^8 vs *sequential* 1.42×10^8 decreased from 1.71×10^{11}) illustrated in Fig. 2. There is a black region with sharp edge at the bottom of both Fig. 2 (h) and (d) due to the transformed image Fig. 2 (e) falling outside of the field of view. However, a better reconstruction for the

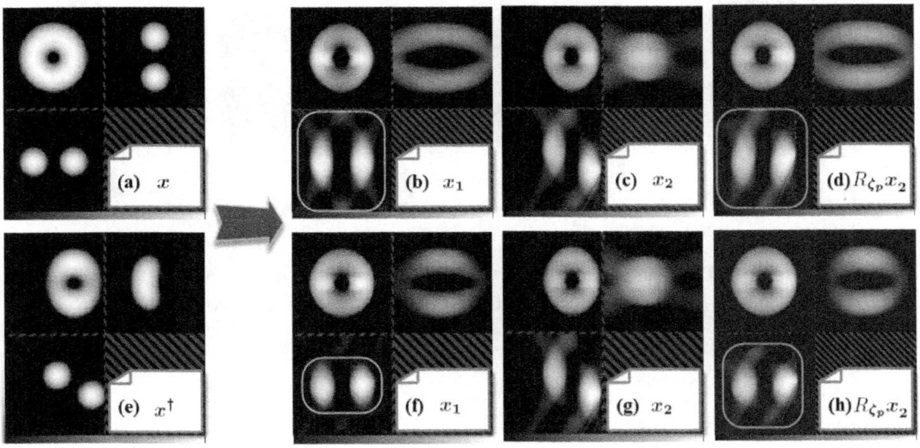

Fig. 1. (a) Original test volume $\boldsymbol{x}$; (e) Transformed test volume $\boldsymbol{x}^\dagger$; Sequential results (b)-(d): (b) reconstruction $\boldsymbol{x_1}$, (c) reconstruction $\boldsymbol{x_2}$, and (d) transformed reconstruction $R_{\zeta_p}\boldsymbol{x_2}$; Iterative results (f)-(h): (f) reconstruction $\boldsymbol{x_1}$, (g) reconstruction $\boldsymbol{x_2}$, and (h) transformed reconstruction $R_{\zeta_p}\boldsymbol{x_2}$

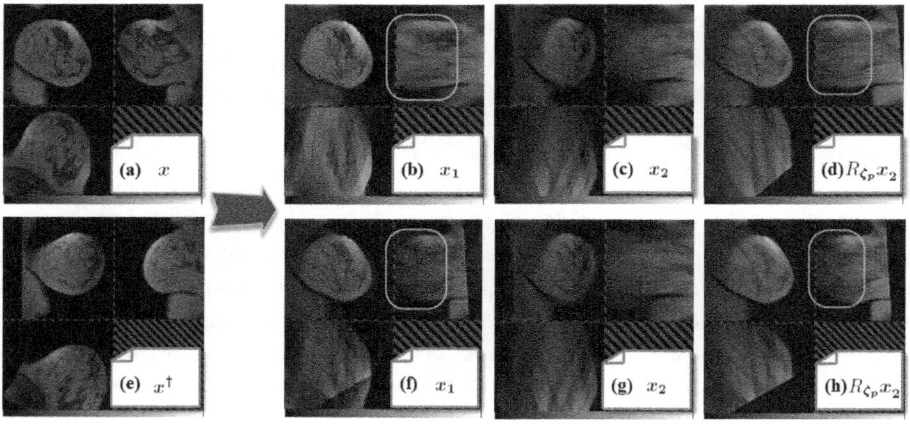

Fig. 2. As Fig. 1 but for a 3D uncompressed breast MR image

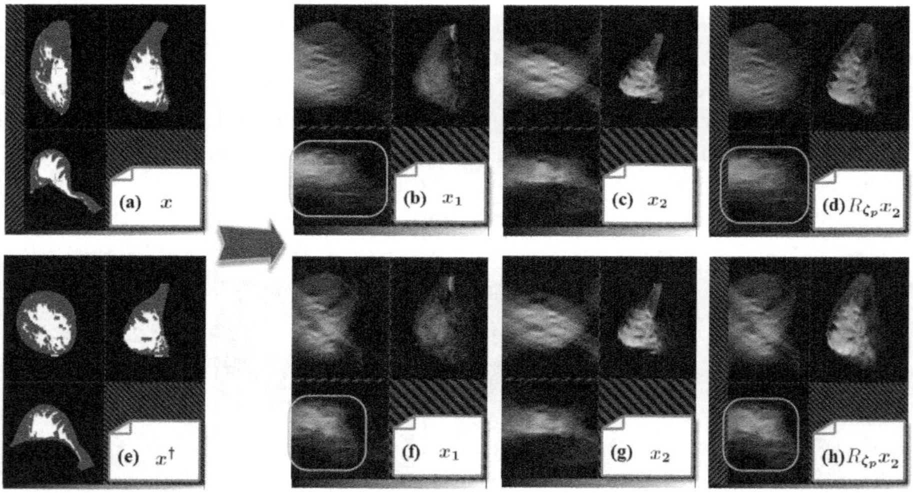

Fig. 3. As Fig. 1 but applied to in vivo MRI acquisition of a breast before and after plate compression (Images have been segmented and mapped to effective X-ray attenuation)

missing data in Fig. 2 (f) is obtained due to our incorporation of all the X-ray acquisitions into the reconstruction of x_1.

In a third experiment we tested the methods using two MRI acquisitions obtained before and after application of a lateral-to-medial plate compression of the breast (Fig. 3). The SSD between reconstruction, $\boldsymbol{x}_1$, and the original volume, $\boldsymbol{x}$, indicates that the *iterative* method produces a more accurate reconstruction of the data (*iterative* 5.9×10^9 vs *sequential* 7.6×10^9 decreased from 6.91×10^{11}). In addition, the affine transformation model is insufficient for the compression deformation which may degrade the reconstructed results; however, measurement of the target registration error for a set of 12 user defined landmarks, indicates that the *iterative* method also produces a more accurate registration result (4.6mm vs 8.6mm, given an initial misregistration of 23.6mm). All the numerical results of the three experiments above are shown in the Table 1 below,

Table 1. Numerical results of the three experiments. ($SSD = \|x_1 - x\|_2^2$)

	Initial	**Combined Method**	**Sequential Method**
Toroid SSD	4.51×10^{11}	4.32×10^{9}	6.89×10^{9}
Uncompressed MRI SSD	1.71×10^{11}	1.25×10^{8}	1.42×10^{8}
Compressed MRI SSD	6.91×10^{11}	5.90×10^{9}	7.60×10^{9}
Misregistration (mm)	23.6	4.6	8.6

Plots of the cost function $\Phi_{Rec1} = \|A\boldsymbol{x}_1 - \boldsymbol{y}_1\|_2^2$ represented in equation 3 for both sequential and combined methods are shown in Figures 4, 5, and 6.

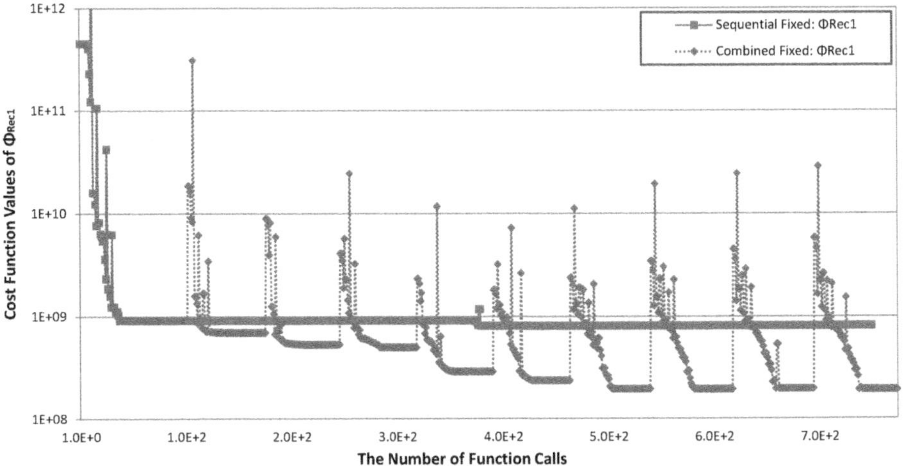

Fig. 4. Plot of the cost function $\Phi_{Rec1} = \|A\boldsymbol{x}_1 - \boldsymbol{y}_1\|_2^2$ for the 3D toroid experiment

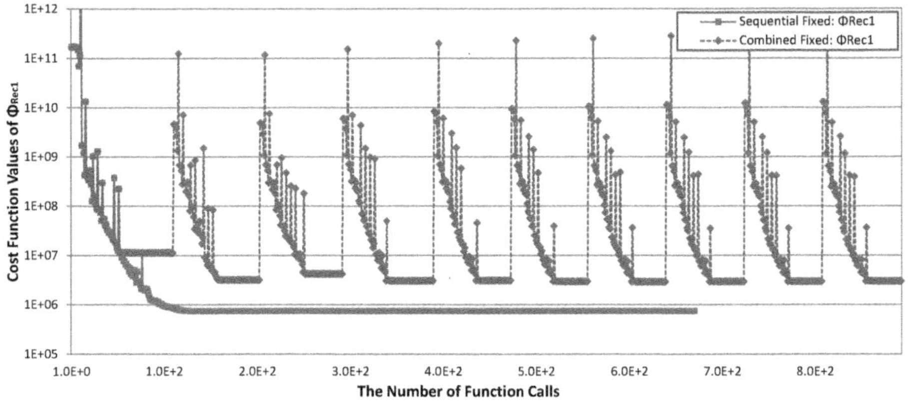

Fig. 5. As Fig. 4 but for the 3D uncompressed breast MR image

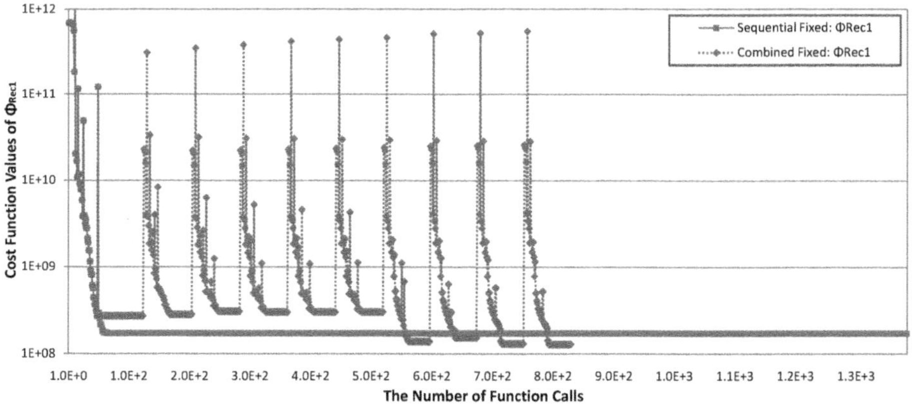

Fig. 6. As Fig. 4 but for the in vivo compressed MR experiment

4 Discussion

Our iterative method was found to produce superior results in optimised cost function value, registration accuracy and reconstructed image appearance. This is illustrated in Fig. 7. We attribute this to the fact that the iterative approach uses all the X-ray acquisition data (both y_1 and y_2) to reconstruct volume x_1. This leads to an improvement in the reconstruction of x_1 which in turn enables a more accurate registration to reconstructed volume x_2 to be achieved.

An implicit assumption in this approach is that there is no change in the breast (such as the growth of a tumour or due to the differences in image acquisition parameters) between the two time-points being reconstructed and registered, justifying the use of SSD as the registration similarity metric. Given this approach, we could envisage a subsequent step where we compare reconstruction volume x_1 with the original acquisitions, y_1 and y_2, to detect change.

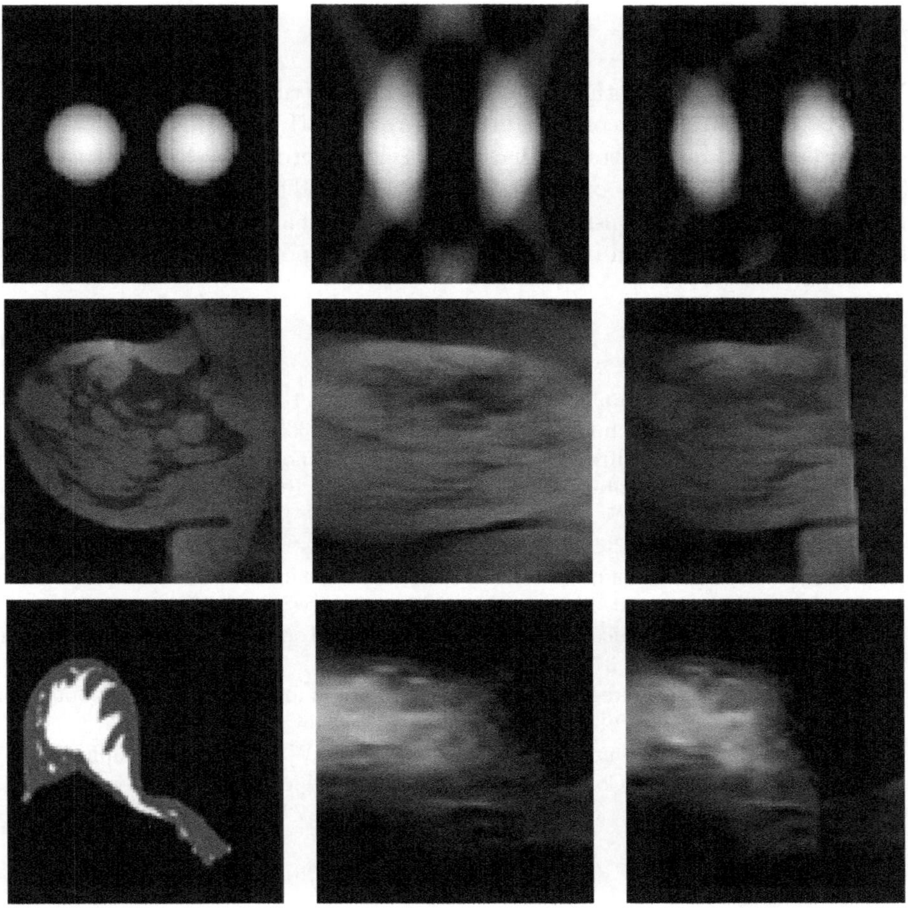

Fig. 7. Magnified results of the three tests above. (a), (b) and (f) of figures 1, 2 and 3. Left to right: Original fixed image x; Results of the sequential method x_1; Results of the iterative method x_1. Only one of the out-of-plane slices has been shown accordingly

The iterative method updates x_1 with the transformation of x_2, $R_{\zeta_p}\boldsymbol{x}_2^*$, after 10 iterations of the reconstruction and a single registration. This results in the 10 peaks in the cost function plot for the iterative method when compared to the smooth plot for the sequential method, Figs. 4, 5 and 6. In Fig. 5, the final cost function value of the sequential method is less than our iterative method because the MR volume has been transformed beyond the field of view ($\boldsymbol{x}^\dagger$ in Fig. 2). This region is visible in the simulated projection images, y_1, however, because the 3D transformation is applied in the world coordinate frame. The result is that the sequential method produces a lower value of the cost function, $\Phi_{Rec1} = \|A\boldsymbol{x}_1 - \boldsymbol{y}_1\|_2^2$, due to greater image overlap despite the reconstruction (and registration) being less successful.

5 Conclusion

We have presented a method to iteratively reconstruct and register temporal DBT data sets. We have compared this approach with performing the two tasks sequentially and demonstrated that the former improves both the registration accuracy and the quality of the reconstructed datasets. In future work we will investigate alternative non-rigid transformations and address the issue of change in the breast tissue which may occur between time points.

References

[1] Wu, T., et al.: Tomographic Mammography Using a Limited Number of Low-dose Cone-beam Projection Images. Medical Physics 30, 365–380 (2003)
[2] Dobbins III, J.T., Godfrey, D.J.: Digital X-ray Tomosynthesis: Current State of the Art and Clinical Potential. Phys. Med. Biol. 48, R65–R106 (2003)
[3] Lauritsch, G., Haerer, W.: A Theoretical Framework for Filtered Back-Projection in Tomosynthesis. In: Proceedings of SPIE, vol. 3338, pp. 1127–1137 (1998)
[4] Zhang, Y., et al.: A Comparative Study of Limited-angle Cone-beam Reconstruction Methods for Breast Tomosynthesis. Medical physics 33, 3781–3795 (2006)
[5] Warp, R.J., et al.: Applications of Matrix Inversion Tomosynthesis. In: Proceedings of SPIE, vol. 3977, pp. 376–383 (2000)
[6] Leach, M.O., et al.: Screening with MRI and mammography of a UK population at high familial risk of breast cancer. The Lancet 365, 1769–1778 (2005)
[7] Kopans, D.B.: Breast Imaging, 3rd edn. Lippincott W. & W., Philadelphia (2007)
[8] Schumacher, H., et al.: Combined Reconstruction and Motion Correction in SPECT Imaging. IEEE Transactions on Nuclear Science 56, 73–80 (2009)

Author Index

GPSR Compliance

The European Union's (EU) General Product Safety Regulation (GPSR) is a set of rules that requires consumer products to be safe and our obligations to ensure this.

If you have any concerns about our products, you can contact us on ProductSafety@springernature.com

In case Publisher is established outside the EU, the EU authorized representative is:

Springer Nature Customer Service Center GmbH
Europaplatz 3
69115 Heidelberg, Germany

Batch number: 10712242

Printed by Printforce, the Netherlands